Lecture Notes in Mechanical Engineering

Lecture Notes in Mechanical Engineering (LNME) publishes the latest developments in Mechanical Engineering—quickly, informally and with high quality. Original research reported in proceedings and post-proceedings represents the core of LNME. Volumes published in LNME embrace all aspects, subfields and new challenges of mechanical engineering.

To submit a proposal or request further information, please contact the Springer Editor of your location:

Europe, USA, Africa: Leontina Di Cecco at Leontina.dicecco@springer.com
China: Ella Zhang at ella.zhang@springer.com
India: Priya Vyas at priya.vyas@springer.com
Rest of Asia, Australia, New Zealand: Swati Meherishi at swati.meherishi@springer.com

Topics in the series include:

- Engineering Design
- Machinery and Machine Elements
- Mechanical Structures and Stress Analysis
- Automotive Engineering
- Engine Technology
- Aerospace Technology and Astronautics
- Nanotechnology and Microengineering
- Control, Robotics, Mechatronics
- MEMS
- Theoretical and Applied Mechanics
- Dynamical Systems, Control
- Fluid Mechanics
- Engineering Thermodynamics, Heat and Mass Transfer
- Manufacturing Engineering and Smart Manufacturing
- Precision Engineering, Instrumentation, Measurement
- Materials Engineering
- Tribology and Surface Technology

Indexed by SCOPUS, EI Compendex, and INSPEC.

All books published in the series are evaluated by Web of Science for the Conference Proceedings Citation Index (CPCI).

To submit a proposal for a monograph, please check our Springer Tracts in Mechanical Engineering at https://link.springer.com/bookseries/11693.

Chinese Society of Aeronautics and Astronautics
Editor

Proceedings of the 6th China Aeronautical Science and Technology Conference

Volume I

Springer

Editor
Chinese Society of Aeronautics and Astronautics
Chinese Society of Aeronautics and Astronautics
Beijing, China

ISSN 2195-4356 ISSN 2195-4364 (electronic)
Lecture Notes in Mechanical Engineering
ISBN 978-981-99-8860-0 ISBN 978-981-99-8861-7 (eBook)
https://doi.org/10.1007/978-981-99-8861-7

This Springer imprint is published by the registered company Springer Nature Singapore Pte Ltd.
The registered company address is: 152 Beach Road, #21-01/04 Gateway East, Singapore 189721, Singapore

Paper in this product is recyclable.

Contents

Research on Carbon Abatement Strategies of Airport Based on LEAP-Airport Model

Wenjing Ye, Lili Wan(✉), Zhanpeng Shan, and Yangyang Lv

Nanjing University of Aeronautics and Astronautics, Nanjing 211106, China
wanlili@nuaa.edu.cn

Abstract. Achieving airport carbon emission peak is crucial for the civil aviation industry to reach carbon emission peak on schedule for airports are important infrastructure of air transportation. The LEAP-Airport model is developed based on the Low Emissions Analysis Platform (LEAP) model to forecast long-term carbon emissions of airport. Taking Nanjing Lukou International Airport as an example, 11 scenarios in 6 categories are set up according to abatement strategies, which are grouped into operational improvements, structural optimization, technological progress, and alternative fuels strategy. By simulating carbon emission trends from 2023 to 2035 under different scenarios based on the LEAP-Airport model, this paper analyses the effectiveness of various abatement strategies, providing a basis for formulating carbon abatement strategy for the airport. The results show that the Lukou airport will not be able to reach the carbon emission peak on schedule if no emission reduction measures were taken; the airport will still not be able to achieve the carbon peak target if the emission reduction efforts were made according to the existing policies; the airport is expected to achieve the carbon peak in 2030 with 574 thousand tons if the abatement measures were strengthened on the basis of the existing policies. Among the four types of airport emission reduction strategies, technological progress and alternative fuels strategies should be implemented as a priority, and the strength of existing policies should be increased to reduce emissions to achieve the peak target on time.

Keywords: Airport abatement strategy · Carbon emission peak · LEAP-Airport Model

1 Introduction

1.1 A Subsection Sample

Climate change is a common global concern. China has formally proposed the medium and long-term strategic goals of reaching carbon peak by 2030 and carbon neutrality by 2060. With the rapid development of China's civil aviation industry and the increasing volume of aviation business, 2% of the total global man-made CO_2 emissions come from aviation industry [1], carbon emission reduction in the civil aviation industry is urgent. As the important infrastructure of aviation industry, achieving carbon emission peak at airports is crucial for the civil aviation industry to achieve the carbon emission peak on schedule.

Chinese Society of Aeronautics and Astronautics (Ed.): CASTC 2023, LNME, pp. 1–9, 2024.
https://doi.org/10.1007/978-981-99-8861-7_1

The current research methods of carbon emission forecasting in civil aviation industry can be divided into three categories. The first category is energy consumption and carbon emission forecasting models based on the causal relationship with several major influencing factors, such as econometric models [2] and kaya identity [3], etc. Such models are comprehensive in considering the influencing factors of carbon emission, but the future development of each influencing factor is difficult to be accurately determined. The second type is the energy consumption and carbon emission prediction models based on the impact of macroeconomic system changes on the energy system, such as system dynamics models [4], etc. These models do not take emission reduction as the starting point of research, and are prone to underestimate the impact of technological development, which is a weak guide for emission reduction work; the third category is the energy consumption and carbon emission prediction models based on the detailed description and analysis of various aspects of the energy system, such as LEAP model [5], which can analyze the level of technological emission reduction in detail but cannot reflect the interaction between macroeconomic policies and carbon emissions.

In order to accurately implement technology and policy to carbon emissions, this paper argues that the third category is more compatible with the objectives of airport carbon emission forecasting studies, and no scholars have applied such models to the field of airport carboF emission prediction yet. Among them, the LEAP model has been widely used in carbon emission forecasting research in the fields of electricity, building and transportation due to its low initial data requirement, flexibility and simplicity of use, and ability to simulate policy impacts. Therefore, this paper selects the LEAP model to carry out carbon emission prediction research for the first time. Taking Nanjing Lukou International Airport as the research object, the LEAP-Airport model is constructed according to the traffic operation characteristics of the airport. Combined with scenario analysis method, 11 scenarios in 6 categories under different types and strengths of abatement strategies are set up, so as to find the best abatement strategy and appropriate reduction efforts, providing decision support for airports to carry out carbon abatement work.

2 Introduction

2.1 LEAP-Airport Model

The LEAP model is a long-term energy alternatives planning model jointly developed by Boston University and Stockholm Institute, and is widely used for energy policy analysis in the fields of electricity, buildings, and transportation. The structure of the LEAP model is usually built on a tree consisting of four levels, divided into sectors, end-use, energy and emission.

The LEAP-Airport model framework constructed for airports is shown in Fig. 1. The civil aviation industry generally only focuses on aircraft carbon emissions, the airport scope not only has aircraft, but also includes many carbon emission sources. This paper divides the airport carbon emission sources into four sectors: aircraft, auxiliary power unit (APU), ground support equipment (GSE), and ground approach vehicle (GAV). Where, Aircraft is divided into widebody, single aisle and regional aircrafts; APU is an auxiliary power unit which works when the aircraft engine is not started to provide power for the aircraft, and is divided into APU for widebody or single aisle; GSE are required to provide corresponding services for aircraft during approach and departure, including cargo loader, fuel truck and water service truck, etc.; GAV is an important way for passengers to arrive at the airport, which is divided into airport bus, taxi and car. Among them, the main fuel types used by aircraft and APU are aviation gasoline and jet fuel, and the future alternative fuels are sustainable aviation fuels (SAFs); the main fuel types used by GSE are gasoline and diesel, and the main alternative fuels are electricity, compressed natural gas (CNG), and liquefied petroleum gas (LPG); the main fuel types used by GAV are gasoline and diesel, and the alternative fuels are electricity, hydrogen, and CNG. The above-mentioned types of emission sources generate carbon emissions from fuel consumption.

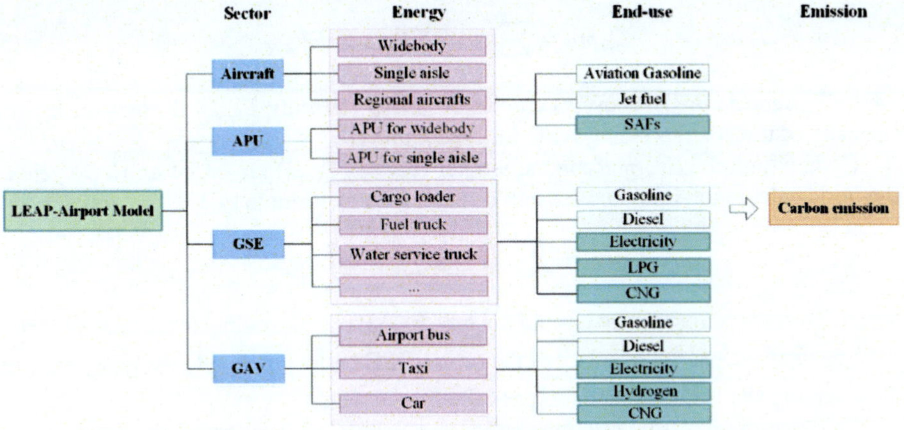

Fig. 1. LEAP-Airport model framework structure.

2.2 Airport Carbon Emission Calculation Method

Airport carbon emissions are the sum of emissions from the four types of sources and are expressed as Eq. (1):

$$C = C_{A/C} + C_{APU} + C_{GSE} + C_{GAV} \tag{1}$$

where C denotes airport carbon emissions; $C_{A/C}$ denotes carbon emissions from aircraft; C_{APU} denotes carbon emissions from auxiliary power units; C_{GSE} denotes carbon emissions from airport ground support equipment; C_{GAV} denotes carbon emission from ground approach vehicle.

Aircraft carbon emissions ($C_{A/C}$) are based on the landing and take-off (LTO) cycle as the basic unit, and can be expressed as formula (2):

$$C_{A/C} = \sum_{j}^{i} AC \times \theta_j^i \times \frac{1}{2} \times \frac{1}{S_j} \times \partial_j^i \times EI_j^i \times EF_{A/C} \tag{2}$$

where i represents the aircraft class, divided into passenger and cargo aircraft; j represents the aircraft type, divided into widebody, single aisle and regional aircrafts; AC denotes airport capacity, divided into passenger throughput and cargo throughput; θ_j^i denotes the percentage of class i type j aircraft; ∂_j^i denotes load factor; S_j is the type j aircraft passenger or cargo capacity; EI_j^i denotes the energy intensity of class i type j aircraft; $EF_{A/C}$ denotes aircraft fuel emission factor.

In the formula, the trend of airport capacity (AC) is consistent with the forecast results of China Aviation Industry Group for China's civil aviation market [6], and the growth rate of passenger throughput is 18.4% from 2023 to 2024 and 5.93% after 2025, the growth rate of cargo throughput is 9.2% from 2023 to 2024 and 5.5% after 2025.

Carbon emissions from auxiliary power units (C_{APU}) are based on the LTO cycle as the basic calculation unit, and can be expressed as formula (3):

$$C_{APU} = \sum\nolimits_j T^j \times EI_{APU}^j \times EF_{A/C} \tag{3}$$

where T^j denotes the LTO cycles amount for class j aircraft; EI_{APU}^j denotes the energy intensity of the corresponding APU for type j aircraft.

Carbon emission from ground approach vehicle (C_{GSE}) are based on LTO cycle as the basic calculation unit too, and can be expressed as formula (4):

$$C_{GSE} = \sum\nolimits_j^k T_j^k \times OT_j^k \times EI_j^k \times EF^k \tag{4}$$

where T_j^k denotes the activity level of the type k GSE corresponding to the type j aircraft; OT_j^k denotes the energy intensity of the type k GSE corresponding to the type j aircraft; EF^k denotes the fuel emission factor of the type k GSE.

Carbon emission from ground approach vehicle (C_{GAV}) are based on passenger-km as the basic calculation unit, and can be expressed as formula (5):

$$C_{GAV} = \sum\nolimits_l PT \times (1 - \delta) \times (1 - \phi) \times \theta^l \times D \times EI^l \times EF^l \tag{5}$$

where l represents vehicle type; PT denotes passenger throughput; δ represents the share of transit passengers; ϕ represents the share of rail transportation trips; θ^l denotes the share of class l vehicles; D represents average distance travelled; EI^l denotes the energy intensity of class l vehicle; EF^l denotes the fuel emission factor of class l vehicle.

2.3 Scenario Setting

In order to better guide the management of airport carbon emission reduction work, this paper sets up scenarios from carbon abatement strategies. Firstly, the emission reduction measures related to airports are categorized into four strategies: operational improvement, structural optimization, technological progress, and alternative fuels. Secondly, the abatement measures with good emission reduction effect and high feasibility are sorted out. Third, two rates of change are set for abatement measures. The low change rate is set according to the existing policy strength, and the high change rate represents greater abatement efforts relative to existing policies, which is mainly based on the historical data development trends.

Operational Improvement Strategy. Operational improvement strategy focus on the process of emission source operations to reduce emissions, and the parameter settings are shown in Table 1. Continuous climb and descent operations (CCO\CDO) are aircraft operating techniques for executing a flight profile optimized to the operating capability of the aircraft, which are an effective means of reducing carbon emissions [7]. Only a few airports currently operate CCO\CDO procedures, assuming that Lukou Airport will introduce the procedure in 2025, and drawing on the industry's current implementation rate of about 20%-40% [8], set the low and high operating rates of 20% and 40% respectively in 2025; In order to save money during the epidemic, the current single-engine taxiing rate implemented by China Eastern is 61.82% [9], and it is expected that the this rate will return to the pre-epidemic level in the future in order to improve operational efficiency; through airport surface operation optimization, the aircraft ground taxiing time could be reduced by a minimum of 27% and a maximum of 35% by 2035.

Table 1. Rate of change settings for operational improvement measures.

Object	Abatement measure	Low			High		
		2025	2030	2035	2025	2030	2035
Aircraft	CCO\CDO	20%	100%	100%	40%	100%	100%
	Reduced engine taxiing	50%	10%	20%	60%	20%	30%
	Reduced taxi time through surface operation optimization	58%	68%	73%	63%	73%	78%

Structure Optimization Strategy. Structural optimization strategy refers to the optimization of the transport structure of various carbon emission sources, and the parameter settings are shown in Table 2. Existing policy points out that for airports with an annual passenger throughput of more than 5 million passengers, the utilization rate of APU replacement needs to be stabilized at more than 95% in 2030, so the low percentage is set at 95% and the high percentage at 100%; the proportion of electric vehicles in the airport will be more than 25% by 2025, so the low proportion in 2025 will be 25% and the high proportion will be 28%; the rail transportation of Nanjing is developed, so the low proportion of public transportation in 2025, 2030 and 2035 will be 58%, 68% and 73%, and the high proportion will be increased by 5% on this basis; the current proportion of electric vehicles in Nanjing is 8%, under the encouragement of policy guidance, more residents will choose electric vehicles.

Technological Progress Strategy. Technological progress strategy mainly improves the energy efficiency of carbon emission sources, and the parameter settings are shown in Table 3. The International Civil Aviation Organization (ICAO) proposed that the global aviation industry should strive to improve fuel efficiency by 2% per year from 2021 to 2050 [10], and since it is difficult to achieve this goal, this paper assumes that the low growth rate is 1.5% and the high growth rate is 2%; based on historical data, combined

Table 2. Rate of change settings for structural optimization measures.

Object	Abatement measure	Low			High		
		2025	2030	2035	2025	2030	2035
APU	APU replacement	90%	95%	100%	95%	100%	100%
GSE	Electrification	25%	30%	35%	28%	35%	40%
GAV	Increase the proportion of public transportation	58%	68%	73%	63%	73%	78%
	Electrification	12%	20%	30%	12%	26%	40%

with the development trend of energy conservation and environmental protection in transportation, it is set that the low energy efficiency improvement rate of GSE and GAV is 2%, and the high energy efficiency improvement rate is 4%.

Table 3. Rate of change setting of measures in the category of technological progress.

Object	Abatement measure	Low	High
		2023–2035	
Aircraft	Improving energy efficiency	1%	2%
GSE		2%	4%
GAV		2%	4%

Alternative Fuels Strategy. Alternative fuels strategy focuses on the use of sustain able fuels for emission sources, and the parameter settings are shown in Table 4. ICAO proposed that the share of SAFs should reach 2% by 2025 and 32% by 2040 [11], taking into account the actual application of SAFs development in China, the low and high shares are set at 12.8% and 16.8% respectively in 2035; at present, a few airports use alternative fuels such as LPG in GSE equipment, and Lukou Airport is expected to introduce GSE alternative fuels in 2025. In the field of ground vehicles, hydrogen fuel is expected to be promoted after 2025 and gradually popularized after 2030, with a low and high percentage of 10% in 2035.

Scenario Description. In order to determine the impact of different emission reduction measures and different degrees of emission reduction on airport carbon emissions, this paper sets up a total of 11 scenarios in 6 categories from the perspective of emission reduction strategy. As shown in Table 5, the Baseline Scenario (BAS) assumes the airport does not take any abatement measures in this scenario; the Operational Improvement Scenario (OIS) assumes that only emission reduction measures are taken for the airport operation process; the Structure Optimization Scenario (SOS) only optimizes the transportation structure of emission sources; the Technological Progress Scenario (TPS)

Table 4. Rate of change settings for alternative fuel measures.

Object	Abatement measure	Low			High		
		2025	2030	2035	2025	2030	2035
Aircraft	Use of sustainable fuels	1.8%	6.4%	12.8%	2%	8.4%	16.8%
GSE		0%	2%	5%	0%	5%	10%
GAV		0%	5%	10%	0%	10%	15%

assumes that only the technology of each sources has been improved; the Alternative Fuels Scenario (AFS) assumes that only fuels have been replaced; the Comprehensive Development Scenario (CDS) is a combination of the above sub-scenario based on the strength of emission reduction measures. Each main scenario is divided into two sub-scenarios, corresponding to two rates of change, high and low.

Table 5. Parameter settings for different scenarios.

Main Scenario	Sub scenario	Operational improvement	Structure optimization	Technological progress	Alternative fuels
BAS	-	-	-	-	-
OIS	OIS1	Low	-	-	-
	OIS2	High	-	-	-
SOS	SOS1	-	Low	-	-
	SOS2	-	High	-	-
TPS	TPS1	-	-	Low	-
	TPS2	-	-	High	-
AFS	AFS1	-	-	-	Low
	AFS2	-	-	-	High
CDS	CDS1	Low	Low	Low	Low
	CDS2	High	High	High	High

3 Results

The results of carbon emission projections under different abatement efforts for Lukou Airport from 2023 to 2035 are shown in Fig. 2. Under the BAS, the airport carbon emissions continue to grow and will reach 1074 thousand tons in 2035. In addition to the CDS, carbon emissions have been significantly reduced under TPS2 and AFS2, showing that technological progress and alternative fuels are effective strategies. Under the CDS1, the carbon emission will reach 654 thousand tons in 2035, reduced by 420 tons compared to the baseline scenario. Nevertheless, airport still could not achieve carbon emission peak, so the abatement efforts of existing policies are not enough. Only under the CDS2 can airport achieve the carbon peak target in 11 scenarios, and the carbon emission will peak in 2030 with 543 thousand tons. Finally, as shown in the gray area in Fig. 2, the airport has a reduction potential of approximately 400 to 500 thousand tons by comparing emissions under BAS with CDS1 and CDS2.

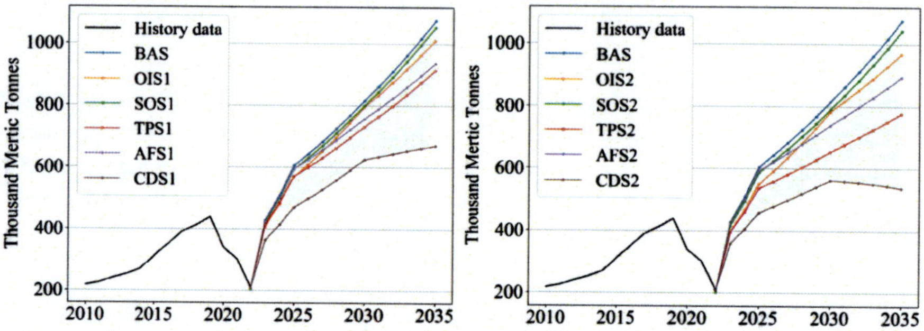

(a) Current policy abatement efforts (b) Enhanced abatement from current policy

Fig. 2. Projected results of airport carbon emissions under different abatement efforts

4 Conclusion

This paper constructs an airport carbon emission prediction (LEAP-Airport) model based on the LEAP model, and sets up 11 scenarios in 6 categories from carbon abatement strategies to predict the development trend of carbon emission in Lukou Airport. The main conclusions are as follows:

(1) If no emission reduction measures are taken, the carbon emissions of Lukou Airport under the BAS will reach 1074 thousand tons in 2035, and the carbon peak target cannot be accomplished. If the airport implements emission reduction measures according to the existing policy, it will still not be able to achieve the carbon peak target, and the carbon emission in 2035 will be 632 thousand tons.
(2) In order to accelerate the carbon peak process, the airport should continue to increase emission reduction efforts based on the existing policy, and the airport has the potential to reduce emissions by about 500 thousand tons under the CDS2.
(3) In the airport carbon emission reduction strategy, priority should be given to the implementation of technological progress and alternative fuels strategies, with particular emphasis on aircraft emission reduction.

References

1. Sridhar, B., Chen, N.Y., Ng, H.: Aircraft Trajectory Design Based on Reducing the Combined Effects of Carbon-Di-Oxide, Oxides of Nitrogen and Contrails. In: AIAA Modeling and Simulation Technologies Conference, p. 0807 (2014)

2. Zhao, J.N., Li, X.M.: Grey prediction of civil aviation carbon emission intensity based on fractional order and Markov optimization. Adv. Transp. Stud. **58**, 121–134 (2022)
3. Yu, J., Shao, C., Xue, C.: China's aircraft-related co2 emissions: decomposition analysis, decoupling status, and future trends. Energy Policy **138**, 111215 (2020)
4. Mayor, K., Tol, R.: Scenarios of carbon dioxide emissions from aviation. Glob. Environ. Chang. **20**(1), 65–73 (2010)
5. Xu, J.H., Wang, K.: Medium- and long-term carbon emission projections and technical emission reduction potential analysis of China's civil aviation industry. China Environ. Sci. **42**(7), 13 (2022)
6. COMAC. www.comac.cc/fujian/2022-2041nianbao_en.pdf. Accessed 13 Nov 2022
7. Vasco, R., Khammash, L.: How can airports influence airline behavior to reduce carbon footprints?. The Air Transportation Industry. Elsevier, New York (2022)
8. Yang, Y.C. https://mp.weixin.qq.com/s/W2sPk9N6GUbVwBRsK1TMvA. Accessed 4 Feb 2021
9. China Civil Aviation News. https://www.163.com/dy/article/HGRICIJP0553RYPC.html. Accessed 9 Sept 2022
10. Kozuba, J., Ojciec, M.: Overview of historical and future trends of commercial aircraft fuel efficiency. Acta Avionica J. 12–17 (2019)
11. Shahriar, M.F., Khanal, A.: The current techno-economic, environmental, and policy status of sustainable aviation fuel (SAF). Soc. Sci. Electron. Publishing **325**, 124905 (2022)

General Design and Aerodynamic Analysis for an Original Flying-Wing and Moving-V-Tail UCAV

Yuke Dai(✉), Tianyou Luo, and Jiaojiao Chen

Section of Air Vehicle General Design, Department of General Design, AVIC Chengdu Aircraft Design and Research Institute, Chengdu 610041, China
245719279@qq.com

Abstract. The general layout design of aircraft has been significantly developed in the aviation history of the past five score years. However, it is apparent to witness the bottleneck and hard to break through traditional creation in the last decade. In this paper, aiming at future's war mode of systematic battle and considering the engineering application potential of cutting-edge technology, the general design of an original flying-wing and V-tail UCAV, which is mainly used for electromagnetic attack is described and modeled; Based on computational flight mechanics and principle of stealth, the performance parameters are provided and structure design is carried out; According to the division criteria of the fighter general arrangement, bay definition of the whole UCAV is launch. Besides, applying CFD means, the main aerodynamic characteristics especially the efficiency of two degree of freedom moving V-tail layout are analyzed. The results show the excellent cruise performance and acceptable maneuver characteristics of this platform which could be realized for a variety of task demands such as distributed detection, systematic countermeasures and focused killing. Moreover, this paper could be an effective reference for innovative aircraft design and exploration.

Keywords: Flying-wing · moving-V-tail · general design · aerodynamic analysis

1 Introduction

As is widely recognized, modern war has developed into a multi-dimensional pattern with an integrated system of land, sea, air, space and electricity [1]. In addition to the traditional military equipment, the demand for multi-platform and multi-function equipment complementary cooperation has increased sharply [2]. Particularly, as the main way of situation awareness and electromagnetic attack, electronic warfare has become an indispensable part of modern war [3]. The practice of several local wars since Gulf War has proved that relying on electronic warfare aircraft to carry out effective airborne electronic attacks and obtain electromagnetic control is an important guarantee for achieving accurate identification and accurate strike [4], and is also a necessary condition for the final victory.

Chinese Society of Aeronautics and Astronautics (Ed.): CASTC 2023, LNME, pp. 10–23, 2024.
https://doi.org/10.1007/978-981-99-8861-7_2

With the increasing highlight of our external combat demands, it is gradually urgent to break through the electromagnetic defense of the one-island chain and the long-distance two-island chain, and to continue the systematic construction of the combat equipment to control the sea and air in our wide ocean and western border. At present, we have vigorously developed aviation weapons and equipment while gradually forming a multi-mission aircraft platform with complete functions and capabilities.

Facing the development of intelligent air combat system, in this paper, a high-performance unmanned electronic warfare aircraft (hereinafter collectively referred to as sky-bat) which is capable of carrying multiple electronic attack payloads to jam and suppress enemy electromagnetic radiation equipment, and destroying enemy radar is generally designed and the platform performance of which is analyzed. Meanwhile, the following typical combat scenarios are proposed: (1) Distributed electronic investigation, which aims at carrying out distributed, all-weather, broadband, active and passive electronic reconnaissance missions to grasp the real-time situation; (2) Systematic electronic countermeasure, which aims to cooperate with the fighter, the bomber and the AWACS in service to provide efficient electromagnetic shielding and electronic support for our high-value equipment. (3) Focused electron killing, which aims to apply airborne full altitude and full range anti-radiation missiles to launch accurately attack, destroy the enemy's electronic warfare and communication network, seize the advantage of BVR air combat, and eventually control the battlefield electromagnetic power.

According to the demands above, several detailed designs of sky-bat is narrated in the following content. Section 2 introduces the concept and general design of sky-bat, including layout and stealth, technic performance and structure design, in which the innovative platform layout of flying-wing with two degree-of-freedom moving V-tail [5] is described. Section 3 specifically focuses on aerodynamic analysis, which consists cruise characteristics and maneuver characteristics. Section 4 is an overall conclusion of the whole paper (Fig. 1).

Fig. 1. Conceptual scheme of sky-bat

2 Concept and Design

In order to satisfy the demands proposed in Introduction, the platform of sky-bat is supposed to possess following characteristics:

(1) High stealth

Sky-bat usually performs electronic reconnaissance, electronic jamming and other tasks, and are always at the forefront of the enemy's investigation range. The shape design of UAV needs to be highly stealth so that sky-bat could find enemy rapidly and attack first.

(2) Efficient cost ratio

Sky-bat needs to achieve a higher efficient cost ratio to cope with the war trend from a single strong platform to multi-platform distributed and clustered operations. On the one hand, the platform uses structure and functional integration technology [] to reduce cost, on the other hand, intelligent energy management and other technologies are utilized to comprehensively improve the engine extraction power and support the long-term operation of the electronic warfare system.

(3) Long Cruise

Sky-bat needs to be in standby for a long time to adapt to the 24h battlefield environment. In the absence of air refueling conditions, the platform needs higher fuel capacity and lift-drag ratio to improve cruise time.

Under the above design input, the layout of sky-bat is designed as follows.

2.1 Layout and Stealth

Following the design idea of light weight, strong stealth, less control surfaces, and high lift-drag ratio, the whole aircraft uses composite materials and new intelligent materials, adopts flying wing layout, and has a two-degree-of-freedom non-decoupling V-tail to ensure outstanding aerodynamic characteristics and stealth performance. The front three-point retractable landing gear and multifunctional buried payload bays are arranged on the belly of the aircraft symmetrically (Fig. 2).

In order to increase the trim capability, the empennage of sky-bat adopts trapezoidal empennage with large wing surface. The empennage could rotate around the longitudinal axis and fold up, with an angle of 60°. At the same time, it takes the ability to rotate along the transverse axis, also with an angle of 60°, which can be used as a speed brake as well. The empennage remains horizontal during cruise, and the leading edge of the empennage would coincide with the trailing edge of the wing, further improving the stealth performance during cruise and reducing interference resistance. When taking-off, landing and maneuvering, the empennage would deflect upward in a V-shape to prevent the aerodynamic center from moving backward and reduce trim resistance under supersonic conditions (Fig. 3).

Since the main mission scenario of sky-bat does not involve maneuver flight at super high angle of attack, the air intake mode adopts back intake, effectively reducing the forward RCS value. In addition, a DSI inlet is arranged at the front section of the back

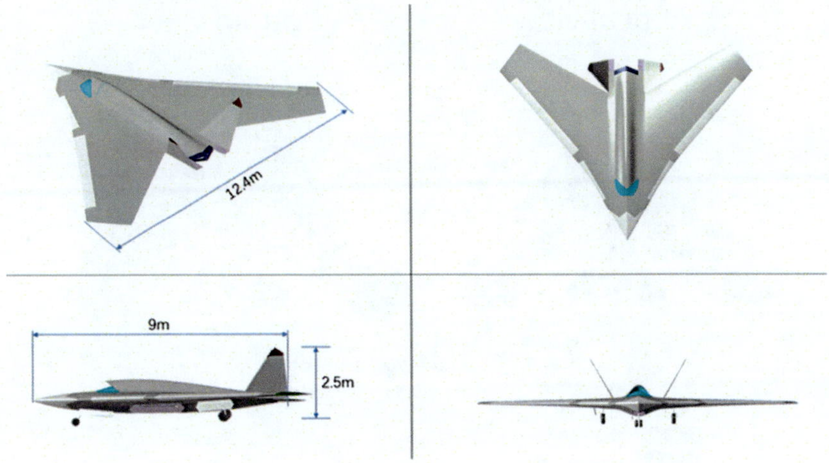

Fig. 2. Three-view drawing of sky-bat

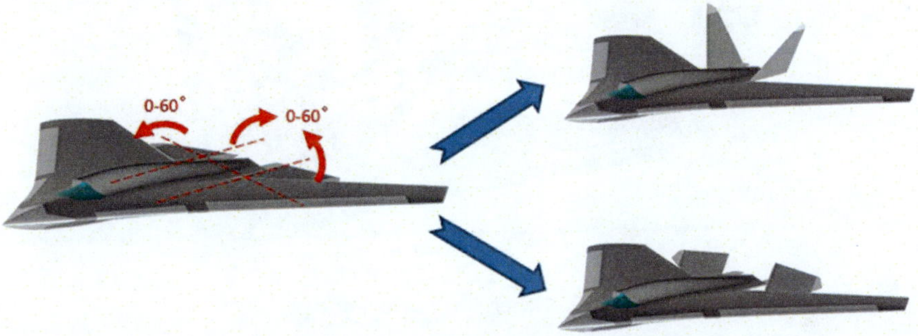

Fig. 3. Schematic diagram of tail morphing

of sky-bat, so that it has excellent total pressure recovery coefficient. The inlet lip and barrel are made of absorbing materials to reduce weight and reduce the lateral RCS of the whole aircraft (Fig. 4).

An adaptive cycle engine is equipped, and the rear of the fuselage is equipped with a binary adjustable vector nozzle to improve the pitching trim ability and maneuverability. The rear edge of the nozzle is hidden before the afterbody edges, and is equipped with advanced infrared absorbing devices, which forms effective shielding and improves infrared stealth characteristics. Fuel tanks are arranged on both sides of the engine and the remaining space inside the wing, which meets the requirements of UAV long-distance cruise (Fig. 5).

The leading-edge sweep angle of the front wing is 60°, the trailing edge sweep angle is 50°, and the trailing edge sweep angle is 30°. According to the aerodynamic evaluation, sky-bat has excellent supersonic/subsonic cruise capability. The principle of parallel design is adopted. The trailing edge of the engine nozzle, the leading edge of the tail wing, and the wing tip are parallel to each other. All hatch doors and joints are

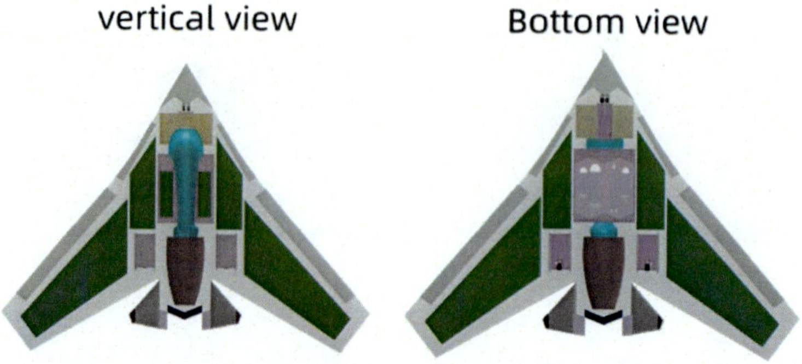

Fig. 4. General layout perspective

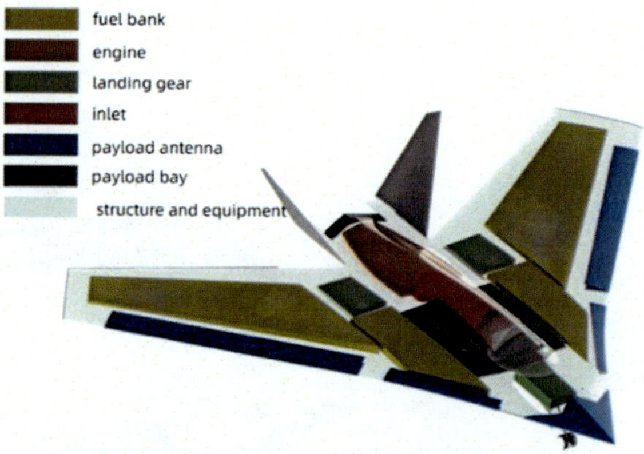

Fig. 5. General bay layout design

designed in zigzag shape, parallel to the wing edge, and the process quality is strictly controlled to ensure the excellent stealth performance of the whole aircraft, so that the forward RCS is less than 0.001m 2, Lateral RCS < 0.012.

2.2 Technic Performance

According to the top-level requirements, comparing the main indicators of the existing UCAVs and look forward to the overall plan of the UCAV in the near future and based on the performance estimation of the sky-bat mission profile, the following traction indicators are obtained (Table 1).

According to the operational requirements and aerodynamic characteristics, using knowledge from flight mechanics as Bregate formula.

$$Rang(N.Mi) = (V/C)(L/D)ln\left(W_{initial}/W_{final}\right) \tag{1}$$

where Rang(N.Mi) is air-range, V is cruise velocity, C is fuel consumption rate, L/D is left-drag ratio, and W is weight of the aircraft. The performance characteristics of sky-bat is calculated (Table 2).

Table 1. Main parameters and indicators

Item	Value
Long (m)	9
Wingspan (m)	12.4
Hight without LG (m)	1.6 (V-tail off), 2.5 (V-tail on)
Area of left surface (m^2)	60
Area of two V-tail (m^2)	8
Empty Weight (kg)	5.5
Max. Take-off Weight (kg)	12
Max. Landing Weight (kg)	10
Max. Inner fuel (kg)	4.5
Max. Payload weight (kg)	2

Table 2. Performance characteristics of sky-bat

Item	Value
Max. Mach	2.2
Subsonic cruise Mach	0.85
Supersonic cruise Mach	1.6
Max. Indicated airspeed (km/h)	1800
Min. Indicated airspeed (km/h)	200
Designed Overload (g)	−4~10
Practical ceiling (km)	20 km
Combat radius (km)	1200
Take-off Distance (m)	≯300 (ground), ≯80 (on warship)
Landing Distance (m)	≯500 (ground), ≯100 (on warship)

2.3 Structure Design

According to its configuration characteristics, the whole aircraft structure is divided into four areas: the front fuselage, the middle fuselage, the left tail and the right tail. Furthermore, as the attribute of flying-wing aircraft, there is no separation surface between the wing and fuselage (Fig. 6).

The structure of the fuselage section cancels the wing-body butt joint, which has the structural advantages of the UAV without rudder surface. Based on the principle of multiple use of one frame, the main frame of the fuselage is arranged at the two ends of different bay as well as where the large concentrated load acts. The extension of the fuselage main frame is used as the main beam of the wing to transfer and balance the

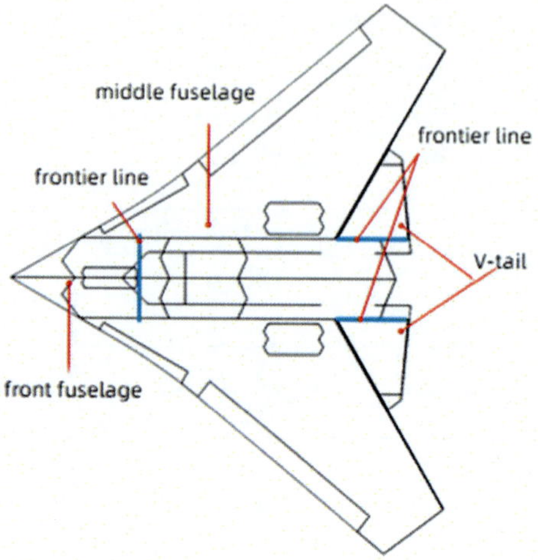

Fig. 6. Schematic diagram of structural separation surface

bending moment and shear force, and the full-height longitudinal main beam at the wing-body fusion is used as the root rib to maintain continuity in the longitudinal direction, while maintaining the section rigidity of the wing (Fig. 7).

Fig. 7. Structural design of the front fuselage

The biaxial variant tail adopts structure-function integration design. High-strength materials are applied, and the actuators of the V-tails are integrated with the load-bearing structure to reduce the space cost and weight while meeting the strength requirements

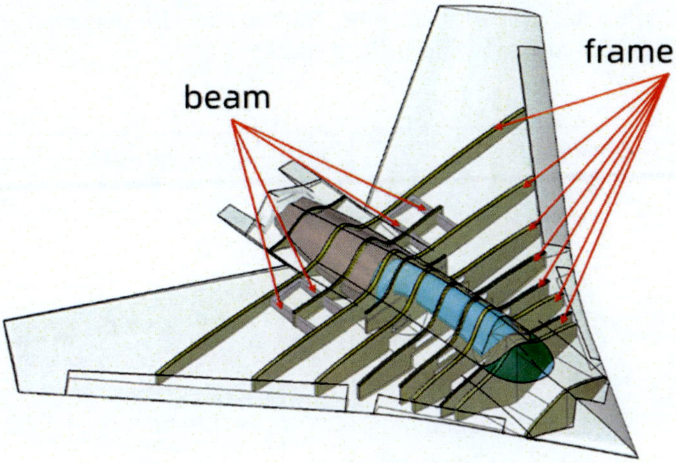

Fig. 8. Structural design of the middle fuselage

and ensuring that structural failure will not occur due to overload (Fig. 8), aeroelasticity and other problems under high maneuvering conditions (Fig. 9).

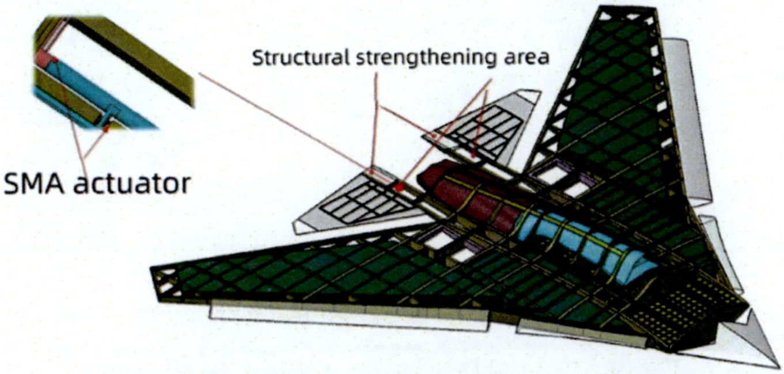

Fig. 9. Structural design of the rear edges

3 Aerodynamic Analysis

3.1 Cruise Characteristics

During subsonic and supersonic cruise, the tail remains horizontal.

In the range of angle of attack from - 4° to 10°, the lift line slope increases with the increase of Mach number. The subsonic lift line slope is 0.045, and the supersonic lift line slope is more than 0.05.

The drag coefficient at low angle of attack at subsonic speed is only 0.007, and that at supersonic speed is less than 0.02. The maximum lift-drag ratio at subsonic speed reaches 11.5 when the angle of attack is 4°. The maximum lift-drag ratio at supersonic speed is 5.5, and the corresponding angle of attack is 6°.

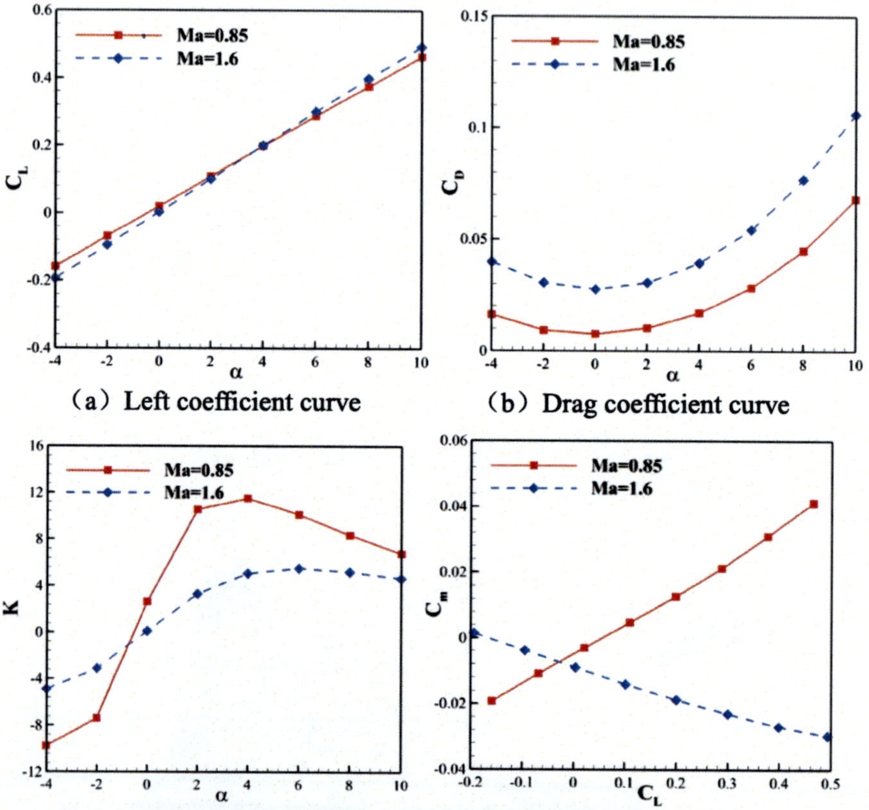

(a) Left coefficient curve (b) Drag coefficient curve

(c) Left/drag ratio curve (d) Pitching moment coefficient curve

Fig. 10. Left and Drag characteristics while cruising

In addition, the linearity of pitching moment coefficient under subsonic and supersonic cruise conditions is good (Fig. 10), which is conducive to longitudinal control. The subsonic longitudinal instability is 9% at small angle of attack while that is 5% for the supersonic longitudinal stability (Fig. 11).

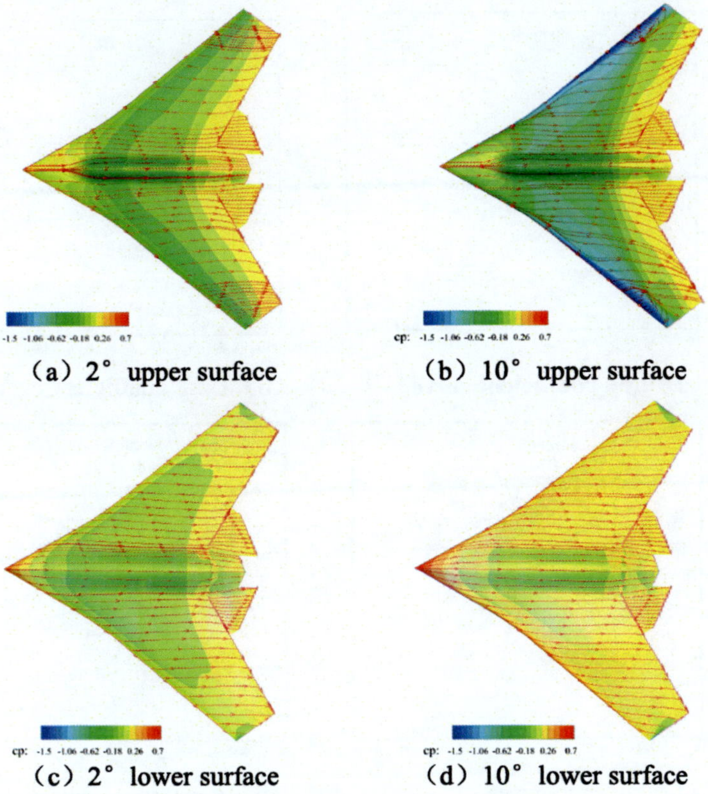

Fig. 11. Diagram of Surface streamline (M0.85)

At a small angle of attack, there is no separation on the wing surfaces. At a high angle of attack of 10°, there is separation on the leading edge of the wing when subsonic and separation on the trailing edge when supersonic.

3.2 Maneuver Characteristics

When sky-bat is flying with high maneuverability (under M1.2), the horizontal empennages rotate to V-tails (Fig. 12).

The stall starts at an angle of attack of 15°nevertheless the process is mild with the maximum lift coefficient of 1.43. The maximum lift-drag ratio is 6.2, and the corresponding angle of attack is 5°.

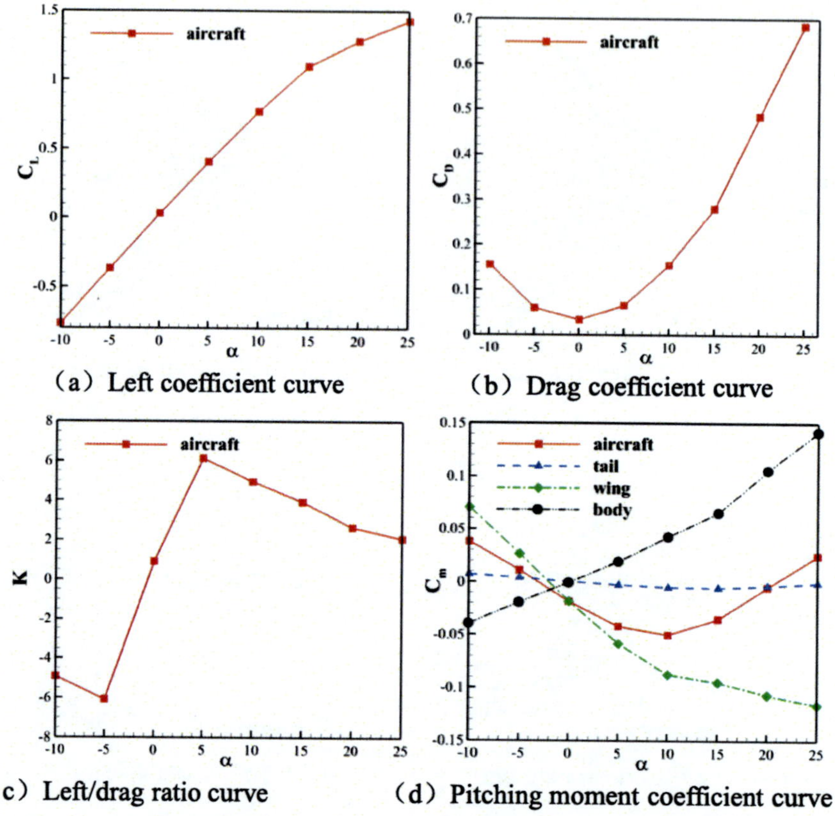

(a) Left coefficient curve (b) Drag coefficient curve

(c) Left/drag ratio curve (d) Pitching moment coefficient curve

Fig. 12. Left and Drag characteristics while maneuvering

Within the range of high angle of attack, sky-bat is longitudinally stable at an angle of attack of 10° with a static stability of 7%. The direction of pitching moment at a high angle of attack is positive, and the streamline is separated at an angle of attack of 10°. In this circumstance (Fig. 13), the wing cannot provide enough bow moment (Fig. 14). When the angle of attack increases to 20°, the vortex above the wing is rolled up, and the lower surface of the V-tail is separated (Fig. 15).

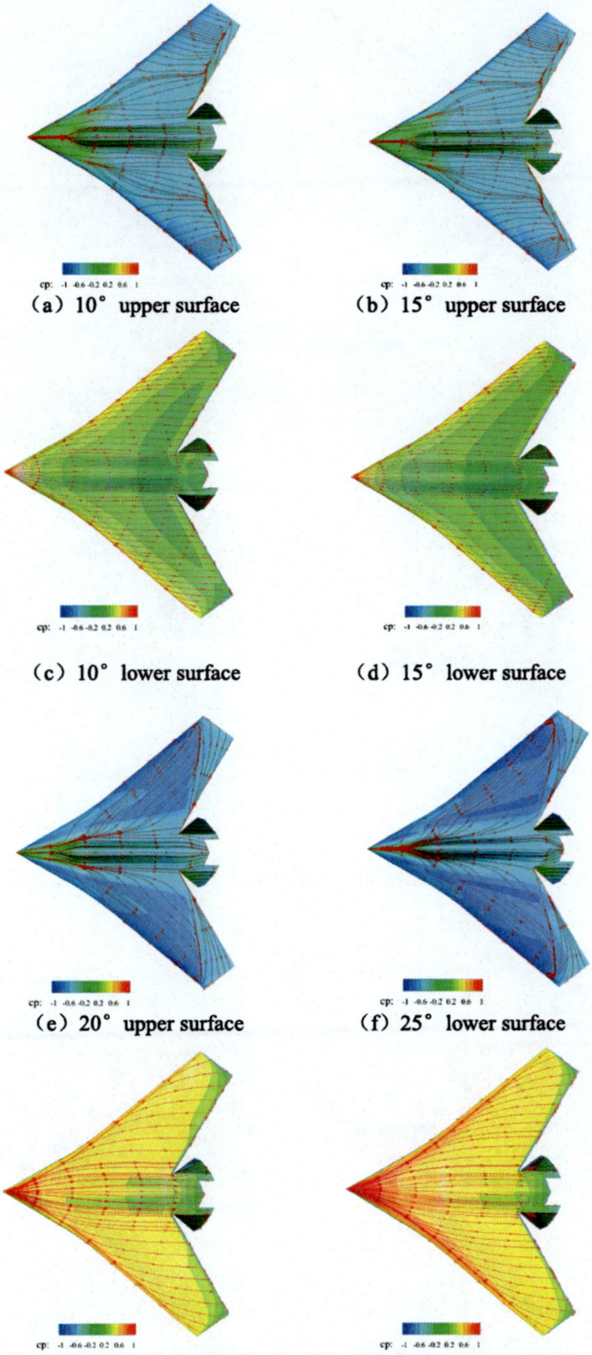

(a) 10° upper surface (b) 15° upper surface

(c) 10° lower surface (d) 15° lower surface

(e) 20° upper surface (f) 25° lower surface

(g) 20° lower surfce (h) 25° lower surface

Fig. 13. Diagram of Surface streamline (M1.2)

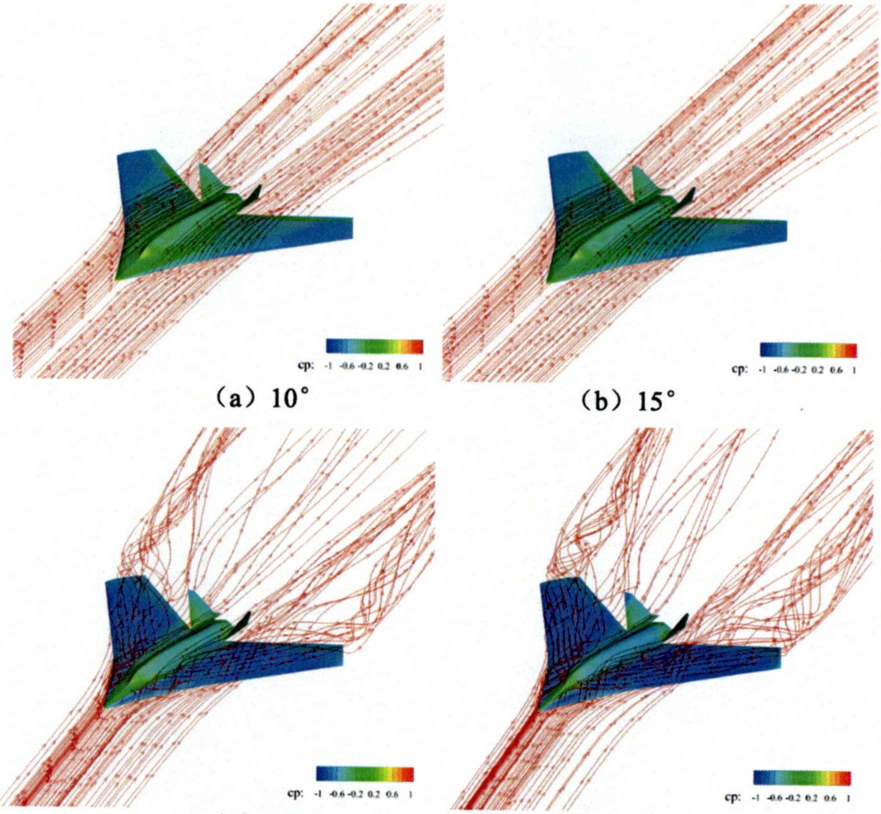

(a) 10° (b) 15°

(c) 20° (d) 25°

Fig. 14. Diagram of Spatial streamline (M1.2)

(a) 10° (b) 20°

Fig. 15. Diagram of Q criterion (M1.2)

4 Conclusions

This paper completes the general design and aerodynamic analysis of a highly stealth, cost-effective and intelligent UCAV called sky-bat, which is basically original and feasible. According to the results of aerodynamic analysis, sky-bat has high lift-drag ratio, small drag coefficient, decent longitudinal aerodynamic characteristics and strong cruise ability under subsonic and supersonic cruise conditions, meeting the requirements of long-distance and long-endurance. Besides, the maneuver ability is acceptable under its specified task envelope. The study of combination of flying-wing and moving V-tail provides valuable data for relevant research and lays foundation for similar application in engineering. With the rapid progress of technology, the future battlefield environment will become increasingly complex and changeable, which would put forward higher requirements for UCAV general design. In this aspect, the investigation of new layout such as sky-bat is of great significance to promote the appearing of potential aircraft general design scheme which could break through and flourish in the future.

References

1. Zhang, K.D., Zhou, Q.S., Wang, J., et al.: A method for jamming waveform design in precision electronic warfare scenarios. IET Signal Proc. **5**, 16 (2022)
2. Gross, M.L.: Medicalized WEAPONS & Modern WAR. Hastings Center Report, **40**(1), 34–43 (2010)
3. Yue, K., Liu, W., Li, G., et al.: Numerical simulation of RCS for carrier electronic warfare airplanes. J. China Aviat. **2**, 11 (2015)
4. Quagliotti, F., Pralio, B., Guglieri, G.: Preliminary design of a UAV configuration. AIAA J. (2001)
5. Abzug, M.J.: V-tail stalling at combined angles of attack and sideslip. J. Aircr. **36**(4), 729–731 (1999)

Transient Temperature of Shape Memory Alloy Valve for Aero Engines

Xiaodong Pang(✉), Yongxiang Ren, Yiqiang Liang, Xiaozhe Zhang, and Xiao Yu

AECC Shenyang Engine Research Institute, Shenyang 110015, People's Republic of China
pangxiaodong2014@foxmail.com

Abstract. Aiming at the higher heat dissipation demand of future high-performance aero engine, this paper proposes a high temperature shape memory alloy thermal control valve, which expands the application scenario of thermal control valve without changing the shape memory material. Through the unsteady solid thermal analysis calculation, the temperature field distribution of the internal structure of the valve core at different times was obtained, and the excess temperature curve of the shape memory alloy spring in the insulated sandwich of the shape memory alloy valve was obtained. By analyzing the curve, the time constant of the selected shape memory alloy thermal regulating valve was calculated to be 81 s. The obtained time constant can be used to guide the material selection of heat insulation structure and memory alloy spring in the process of engineering development.

Keywords: Shape memory alloy · Valve · Temperature response · Heat exchanger

1 Introduction

With the improvement of the comprehensive performance of advanced aircraft and engines, the aero-engine intelligent heat exchanger has become an important requirement and technical feature of the power of future advanced aircraft due to the increase of aerodynamic thermal load and the substantial increase of system cooling requirements [1]. Intelligent heat exchanger based on memory metal, the shape memory alloy (SMA) material has the characteristics of adjustable structure, variable flow form, can be adjusted according to the state of the engine, in order to meet the different design requirements of the engine, the heat transfer performance in different states, to achieve different heat transfer requirements. Shape memory effect means that SMA will undergo large plastic deformation under external force and return to the state before external force after temperature rise [2–4].

SMA alloy has a mature application in the air inlet structure and wing structure [5–8]. SAMPSON program is proposed by the United States, one of the purposes of which is to constantly adjust the geometrical shape optimization of the inlet according to the actual flight conditions so as to improve the flight benefits. Research on the application of SMA in engines [9–12] mainly focuses on noise suppression and nozzle deformation

Chinese Society of Aeronautics and Astronautics (Ed.): CASTC 2023, LNME, pp. 24–32, 2024.
https://doi.org/10.1007/978-981-99-8861-7_3

of turbofan engines. The deformable nozzle enables the engine to obtain greater edge effect during takeoff and landing and higher fuel efficiency during high-speed cruise [13].

Existing studies on shape memory alloys at home and abroad are mostly based on structure and materials, while few studies on transient temperature response of shape memory aggregate materials wrapped in heat insulation layer. In this paper, a steady-state fluid-structure coupling model of shape memory alloy thermal control valve is established, and the calculated results are taken as initial values to carry out transient solid thermal analysis and calculation, and the excess temperature curve and time constant are obtained. Grasp the transient temperature response rule of shape memory alloy thermal control valve.

2 Thermal Control Valve Design

2.1 Performance Calculation of Thermal Regulating Valve

Shape memory alloy thermal control valve can be used in aero-engine entrainment system. Due to the limitation of the existing technical conditions, the temperature range of the shape memory alloy can be reliably deformed is normal temperature and low temperature, so it is necessary to install the shape memory alloy thermal control valve in the appropriate temperature zone. According to the existing material temperature resistance level, choose to install the shape memory alloy thermal control valve in the fuel flow path.

The shape memory alloy valve and the air-oil heat exchanger are installed in parallel on the fuel flow path. In high temperature days, the high temperature of the fuel leads to the poor cooling of the fuel, so the large flow of fuel is needed for air cooling, so the whole system is also designed in accordance with the most severe conditions of the heat exchanger. However, in low temperature days, the fuel temperature is low, the fuel cooling capacity is strong, only need less fuel can achieve the purpose of cooling, however, the heat exchanger structure is fixed, there is no way to adjust the internal fuel flow, resulting in fuel flow path actual flow greater than the minimum cooling needs, low temperature fuel viscosity nonlinear increase, resulting in greater pressure drop loss more pump work, Reduce engine performance.

If the shape memory alloy valve is applied to the fuel flow path of air-oil heat exchanger, close the valve at high inlet fuel temperature, most of the fuel all through the air-oil heat exchanger, open the valve at low inlet fuel temperature, and properly reduce the fuel flow through the air-oil heat exchanger. After the installation of shape memory alloy valve, it can not only meet the requirements of heat transfer, but also achieve the purpose of reducing drag.

Firstly, the shape memory alloy valve is designed under the design point of air - oil heat exchanger. According to the temperature restriction conditions of bleed air, the minimum fuel flow required at different inlet fuel temperatures at the design point was calculated, and the calculation results were shown in Fig. 1. According to the results, at low fuel temperature, only 1/5 of the fuel at high fuel temperature can meet the heat dissipation design requirements of bleed air.

At the same time, the pressure difference under different fuel flows is calculated, and the calculation results are shown in Fig. 2. As can be seen from the results, the fuel side pressure drop is 50kPa at low fuel flow and more than 500kPa at high fuel flow, which is bound to consume more pump work, resulting in the decline of engine performance.

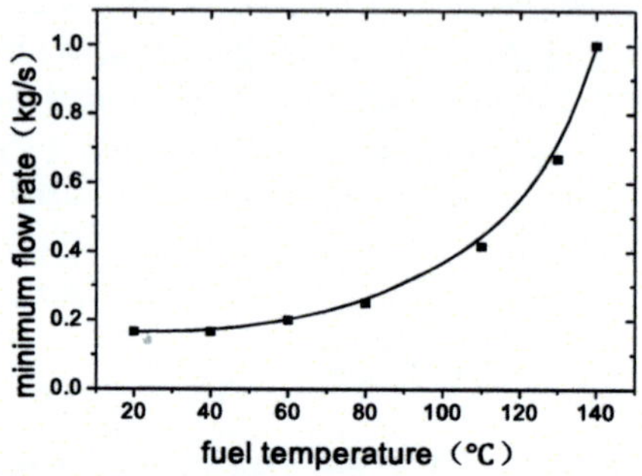

Fig. 1. Minimum fuel flow at design point state of heat exchanger

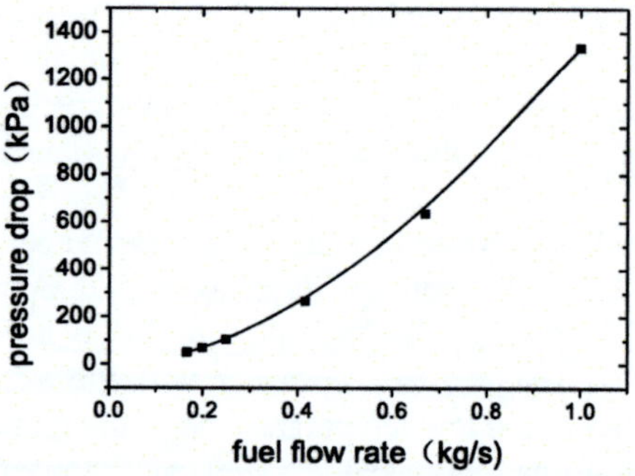

Fig. 2. Flow differential pressure characteristics of heat exchanger

According to the theoretical formula of valve, the optimal valve opening is designed. When the valve opening is in the optimal design result, the internal fuel flow of heat exchanger under different temperature conditions is just in the heat exchanger design point state of the minimum fuel flow curve, not only to meet the heat dissipation requirements, but also to maximize the purpose of reducing drag. The optimal valve opening design results are shown in Fig. 3. The actual valve opening curve as a function of temperature is expected to be as close to the optimal valve opening curve as possible below the curve.

2.2 Structure Design of Heat Regulating Valve

In the process of designing the memory alloy temperature sensing valve, the valve structure of U-shaped heat insulation layer is innovatively proposed. The geometric structure mainly includes: valve shell, SMA spring, U-shaped insulation layer, outer fin, valve

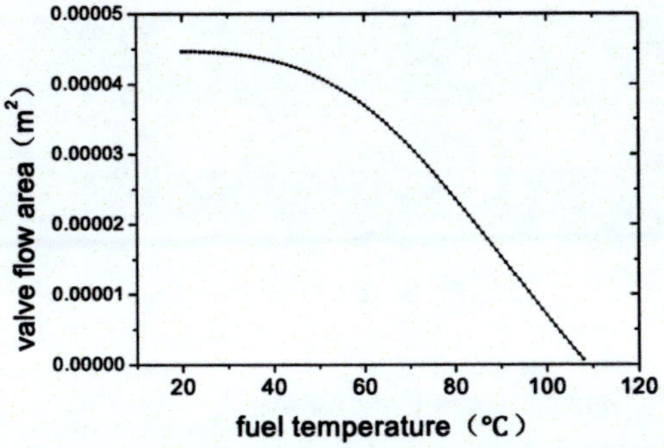

Fig. 3. The optimal valve area varies with fuel temperature

body sandwich, sealing slider, ordinary spring, etc. Its shape memory alloy spring is located in the middle of U-shaped heat insulation layer (see Fig. 4), and the felt temperature is much lower than the fluid temperature in the main flow area, which expands the application of thermal regulating valve, and grants the patent "A thermal regulating valve for heat exchanger in high temperature environment" (ZL201910963910.2).

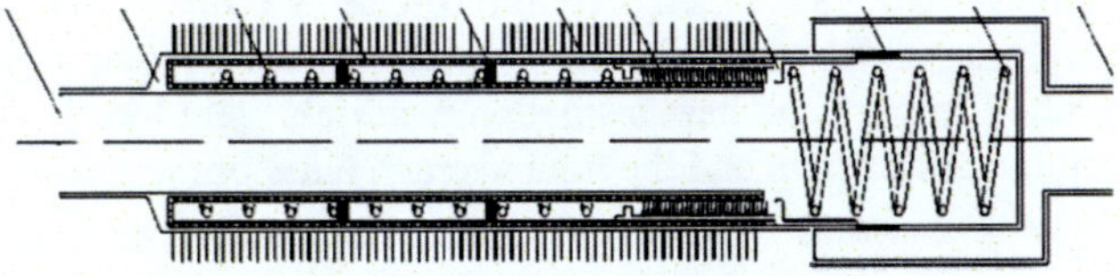

Fig. 4. Valve construction with U-shaped insulation

3 Thermal Response Calculation

3.1 Steady State Fluid-Heat Coupling Calculation

According to the structure of the thermal regulating valve selected in the previous chapter, the simulation analysis is carried out. The fluid type inside the valve is oil, and the fluid type outside the valve is air. In this calculation, the boundary conditions of total pressure and temperature are set at the inlet, and the boundary conditions of mass flow are set at the outlet.

Figure 5 shows the solid temperature field results of the valve shell and SMA spring. According to the results, the inner surface of the valve shell is close to the inlet oil temperature of 180°C, and the outermost surface of the fin is close to the air temperature of 20°C. The simulation results show that the U-shaped heat insulation structure can significantly reduce the temperature of SMA spring.

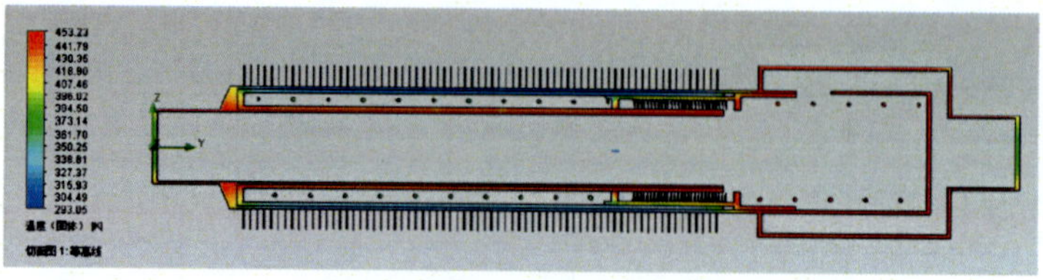

Fig. 5. Solid temperature field

3.2 Unsteady Solid Temperature Calculation

The steady-state temperature calculation results were assigned to the non-steady-state calculation model as initial values, as shown in Fig. 6. The material properties of valve material, U-shaped sandwich material and fluid medium in U-shaped sandwich are defined respectively in the engineering data, which mainly include the material thermal conductivity, density, specific heat capacity and other parameters. The material and boundary conditions are assigned separately in the unsteady model. The structural materials such as valve shell, fin, slide block, ordinary spring and shape memory alloy spring are set as the physical property of stainless steel, the material property of U-shaped insulation material is set as the material property of glass wool feltt, and the material property of fluid medium in U-shaped sandwich is set as the material property of oil. The unsteady solid thermal analysis model was divided into grids, and the smaller structures such as shape memory alloy springs and heat insulation materials were locally encrypted. The grid was divided by workbench, and the final average mass of the grid was 0.76.

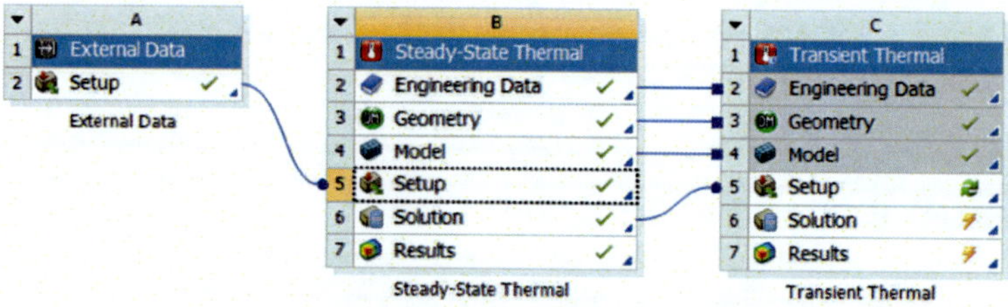

Fig. 6. Temperature assignment process

The temperature change of the fluid passing through the valve is set as the excitation source in response. The change rule of fluid temperature boundary conditions over time is shown in Fig. 7. In steady-state calculation, the fluid boundary condition was set at 180°C, and in unsteady calculation, the fluid temperature was set at 100°C at 0s and maintained at 600s. Set the fluid temperature to 180°C at 600s and keep it for another 600s, same as the initial condition.

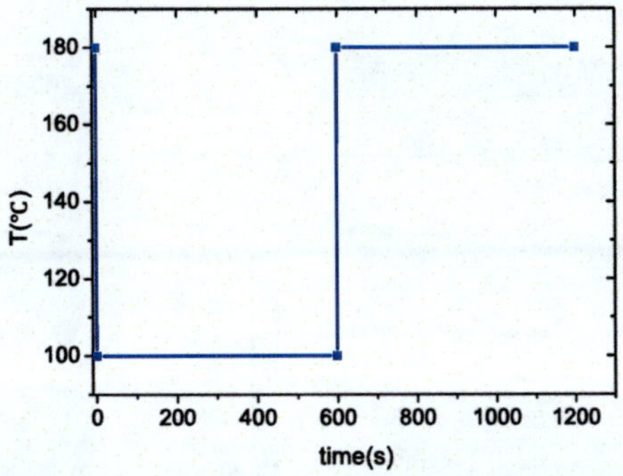

Fig. 7. Fluid temperature boundary changes with time

4 Temperature Response Analysis

4.1 Temperature Cloud Map Results

See Fig. 8 for the internal temperature cloud diagram of the valve under the conditions of 0s, 50s, 100s and 300s respectively. As can be seen from the figure, with the passage of time, the internal surface of the valve in contact with the oil is the first to be affected. The thermal disturbance gradually diffuses into the insulation material, and the overall temperature level inside the valve gradually decreases.

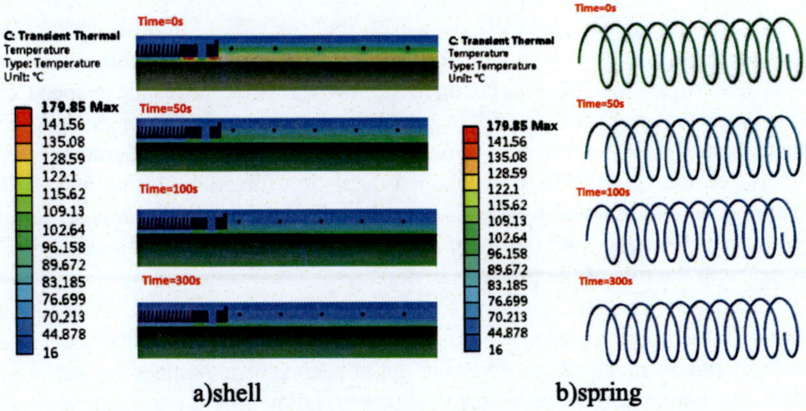

Fig. 8. Internal temperature field results of the valve at different times

The temperature cloud diagram of insulating material, fluid sandwich under the conditions of 0s, 50s, 100s and 300s is shown in Fig. 9. As can be seen from the figure, with the passage of time, the overall temperature level of shape memory alloy spring gradually decreases.

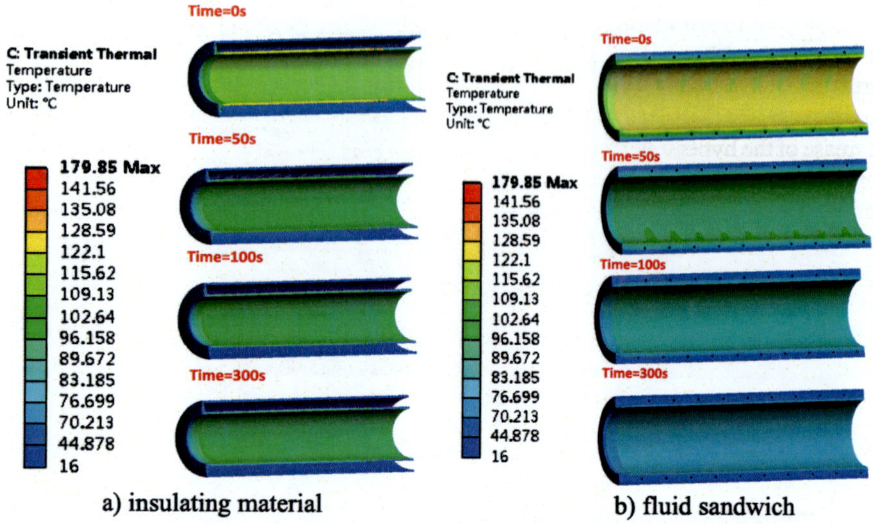

Fig. 9. Temperature field results of insulating material and fluid sandwich

4.2 Temperature Over Time

The average temperature changes over time of shape memory alloy spring, U-shaped heat insulation material and valve shell are shown in Fig. 9. As can be seen from the figure, within 0 ~ 600s, the temperature is affected by the change of boundary conditions, and the temperature gradually decreases, and the change trend becomes slower. The valve housing has the fastest temperature response because it is a thin-walled metal material. Due to the low thermal conductivity, the temperature response speed of the heat insulation material is in the middle, which is much slower than the shell. Although the shape memory alloy spring has good thermal conductivity, the temperature response is slowest because it is inside the U-shaped insulation material. During 600 ~ 1200s, the temperature was affected by the change of boundary conditions, and the temperature gradually increased, and the trend of change gradually slowed down, which was consistent with the rule of the previous period.

The concept of the excess temperature θ (the difference between the temperature at a given moment and the temperature after reaching stability) is introduced, and the excess temperature of the object found in the above curve is reduced to 36.8% of the initial excess temperature value, where the corresponding time is a time constant. Thus, the time constant of the shape memory alloy spring in the valve is 81s.

The solution of unsteady heat conduction problem essentially boils down to solving heat conduction differential equations under specified initial conditions and boundary conditions. In one-dimensional calculation, the Bi number is used as the criterion. Generally speaking, the smaller the Bi number is, the smaller the internal thermal resistance or the greater the external thermal resistance, and the temperature results analyzed by the centralized parameter method are more close to the actual results.

However, in the shape memory alloy valve, in order to improve the shape memory alloy working ability in high temperature medium, the use of U-shaped heat insulation structure. In order to improve the thermal insulation effect of U-shaped insulation layer structure, materials with low thermal conductivity are often used, but this will lead to the increase of the hypersonic number, at which time the centralized parameter analysis is no longer applicable. Therefore, it is necessary to process the temperature change results of the internal structure of the valve over time and convert them into dimensionless curves of excess temperature to evaluate the transient temperature response of the shape memory alloy valve (Fig. 10).

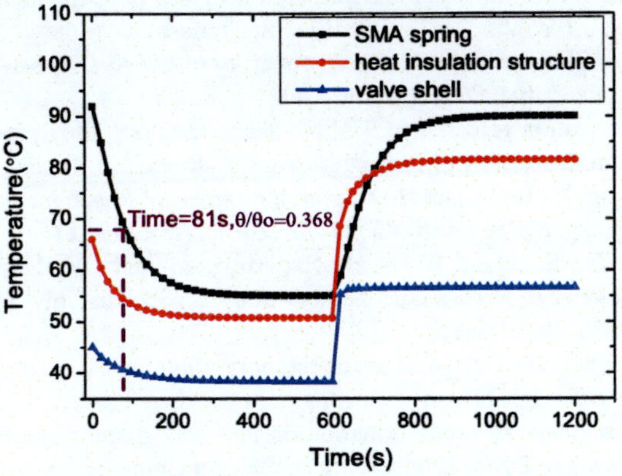

Fig. 10. Temperature changes of different structures with time

5 Conclusion

In this paper, an innovative high-temperature shape memory alloy thermal control valve is proposed. By combining the memory alloy with the heat exchanger, the excess temperature curves of the insulated sandwich and shape memory alloy spring in the shape memory alloy valve ah are obtained through the unsteady solid thermal analysis calculation. The results show that:

1) The calculation results show that the temperature of SMA spring is between fuel temperature and air temperature. The U-shaped heat insulation structure can significantly reduce the temperature of SMA spring, and the cooling range is about half of the temperature difference between high temperature medium and environment.
2) The fluid temperature change through the valve is set as a response excitation source. With the passage of time, the internal surface of the valve in contact with the oil is the first to be affected, and the thermal disturbance gradually diffuses to the interior of the thermal insulation material, and the temperature change rate gradually slows down.

3) Dimensionless processing was carried out on the change rule of temperature of the shape memory alloy spring wrapped in heat insulation material over time, and the result of the time constant of the shape memory alloy spring was obtained, which was about 81s.

References

1. Hao, X., Li, C., Jingquan, Z.: Steady state simulation of integrated thermal energy management system for fighter aircraft. Aircr. Desi. **30**(3), 5 (2010). (in Chinese)
2. Zhirong, H., Chao, Z., Lin, L., et al.: Research progress of shape memory alloys and their applications. Cast. Technol., **38**(2), 5 (2017). (in Chinese)
3. Zhu, X.: Study on preparation and related properties of Cu-Al-(Ni)-Mn shape memory alloy. Central South University (2007). (in Chinese)
4. Mingze, J.: Application research of TiNiNb shape memory alloy pipe joint in aircraft oil circuit connection. University of Aeronautics and Astronautics, Beijing (2004). (in Chinese)
5. Dong, Y., Boming, Z., Jun, L., et al.: Research progress of Shape memory alloys in aviation industry. Met. Funct. Mater. **014**(006), 27–32 (2007). (in Chinese)
6. Han, J., Xiang, X., Baochen, W., et al.: Experimental study on thermomechanical cyclic deformation behavior of Ni-Ti shape memory alloy. J. Sichuan Univ. Sci. Technol. Nat. Sci. Ed. **30**(1), 5 (2017). (in Chinese)
7. Anming, C., Xuejun. Q.: Design of shape memory alloy helical spring. Mach. Des. Manuf. Eng. **28**(1), 3 (1999). (in Chinese)
8. Wei, J., et al.: Influence of predeformation on two-way shape memory Strain of TiNi and TiNiCu alloys Nuclear Power Engineering (2005). (in Chinese)
9. Chau, E.T.F., Friend, C.M., et al.: Technical and economic appraisal of shape memory alloys for aerospace applications. Mater. Sci. Eng. A, 438–440, (SPEC. ISS), 589 (2006)
10. Turner, T.L., Cabell, R.H., Cano, R.J., et al.: Design, fabrication, and testing of a SMA hybrid composite jet engine chevron. Int. Soc. Opt. Photonics (2006)
11. Turner, T.L., Buehrle, R.D., Cano, R.J., et al.: Design, fabrication, and testing of SMA-enabled adaptive chevrons for jet noise reduction. In: Proceedings of SPIE - The International Society for Optical Engineering, p. 5390 (2004)
12. Calkins, F.T., Mabe, J.H., Butler, G.W.: Boeing's variable geometry chevron, morphing aerostructure for jet noise reduction. In: Proceedings of SPIE - The International Society for Optical Engineering, p. 2142 (2006)
13. Barooah, P., Rey, N.: Closed Loop Control of a Shape Memory Alloy Actuation System for Variable Area Fan Nozzle. In: Smart Structures and Materials 2002: Modeling, Signal Processing, and Control. United Technologies Research Center MS, pp. 129–65, 411 Silver lane, East Hartford, CT 06108 (2002)

Study on Heat Transfer Model Modification of Air Oil Radiator for Aircraft Generator

Xiaodong Pang(✉), Yiqiang Liang, Meng Zhao, Jianjun Zhao, and Xiaofei Ding

AECC Shenyang Engine Research Institute, Shenyang 110015, People's Republic of China
pangxiaodong2014@foxmail.com

Abstract. With the continuous development of aero engine technology, the heat dissipation requirements of lubricating oil system and high-power components of engine increase. It is necessary to cool a stream of high-temperature lubricating oil by using low-temperature external culvert air through air lubricating radiator. In order to verify the heat dissipation effect of the air lubricating radiator, it is necessary to carry out a series of component level heat transfer characteristic tests under typical working conditions, and modify the calculation model of the radiator based on the test results. In order to improve the calculation accuracy of the heat transfer model under all working conditions, the evaluation objective of the selection of the correction coefficient was proposed to be the average value of the absolute deviation between the test results of the heat transfer characteristics of the oil outlet temperature index of the air lubricating radiator and the calculated results of the model under typical state points. The evaluation objectives under a series of correction coefficients were calculated respectively, and the correction coefficients under the minimum condition of the evaluation objectives were finally selected. The screened correction coefficients were put into the original calculation model of air lubrication radiator heat transfer, so as to improve the calculation accuracy of the model under all working conditions.

Keywords: Aircraft Generator · Air Oil Radiator · Heat Transfer Model

1 Introduction

With the improvement of the comprehensive performance of advanced aircraft and engines, aero-engine fuel thermal management has become an important requirement and technical feature of advanced aircraft power due to the increase of aerodynamic thermal load and system cooling demand. Thermal management system analysis has gradually become an important content of aircraft and engine design research. With the comprehensive design of aircraft heating management, the thermal management of aircraft motors should be considered in the design of engine fuel system thermal management. With the continuous improvement of aircraft performance indicators, the power of Integrated Drive Generator (IDG) keeps increasing [1, 2], which is followed by the problem of heat dissipation of the generator [3]. The cooling oil of the aircraft generator is cooled by both cold air from the engine fan and engine fuel. In CFM56-7B, PW4000, V2500 and other engines, different cooling schemes are adopted. The hot lubricating

Chinese Society of Aeronautics and Astronautics (Ed.): CASTC 2023, LNME, pp. 33–39, 2024.
https://doi.org/10.1007/978-981-99-8861-7_4

oil at the outlet of the generator is first cooled by the air lubricating oil radiator, then flows through the IDG lubricating oil radiator for further cooling, and then returns to the generator [4].

Liao Yidai [5] designed an air lubricant radiator to effectively improve the cooling capacity of the aircraft engine lubricant system and solve the problem of huge heat generated by the engine when the aircraft is taxiing near the ground at a low speed. Numerical simulation method was used to analyze the change law of the pressure drop of the radiator and the incoming air velocity under the condition with or without air side flow. Zhang Qin [6] established a mathematical model of an air oil radiator, designed a three-flow heat exchanger, used fuel oil, oil and air for heat exchange, analyzed the dynamic response of the heat exchanger, and carried out an in-depth study on the influence of the efficiency of multi-flow heat exchanger. Sun Huguo [7] analyzed the explosion phenomenon of the air-oil radiator used for helicopters, conducted blasting tests and pressure tests to verify the structural strength of the radiator, and confirmed that the joint weld at the cover plate was the origin of the failure.

CAI Huikun [8] designed a face-type air-oil radiator using the engine inlet as the heat transfer surface, and carried out research on heat transfer characteristics of the radiator and equivalence analysis of block simulation. Zhao Peiguo [9] took the cross-flow fin type air/oil radiator as the research object, carried out numerical calculations on the 3D models of the serrated fins on the oil side and the corrugated fins on the air side respectively, and revealed the local Nussel-number variation law of the serrated fins on the oil side and the corrugated fins on the air side. Zhao Ning [10] applied genetic algorithm to optimize the structure of oil radiator. Four objectives of radiator mass, heat transfer coefficient, pipe pressure drop and shell pressure drop were set, and structural parameters of radiator were taken as design variables. The genetic algorithm optimization program is developed.

At present, most correction methods for calculation models of heat transfer of air lubricating oil radiators make use of test results of design points, and do not revise the calculation models of heat transfer in the full range of working conditions based on effective evaluation indexes based on the application scenarios of radiators. When the performance parameters of the medium are far away from the design points, the calculation errors of heat transfer models are large, and high calculation accuracy cannot be achieved. In this paper, the evaluation objectives of the selection of correction coefficients are proposed, and a series of evaluation objectives under the correction coefficients are calculated respectively, and the optimization of the correction coefficients is carried out to modify the calculation model of the heat transfer of the air oil radiator under the whole working condition.

2 Air Oil Radiator Model

The type of air oil radiator on aero engines is usually plate-fin or flat-tube finned radiator (see Fig. 1). Taking the flat tube finned radiator as an example, the fluid in the flat tube is lubricating oil and the section of the channel is square, which needs to be converted into equivalent diameter for heat transfer calculation. See Eqs. (1–2) for the correlation equation of turbulent heat transfer in the tube [11–13], and Eq. (3) for the calculation formula of lubricating oil viscosity.

The fluid on the side of the outer fin of the flat tube is air, and the heat transfer correlation of the fluid across the flat tube is shown in Eq. (4), and the efficiency formula of the fin is shown in Eqs. (5–6).

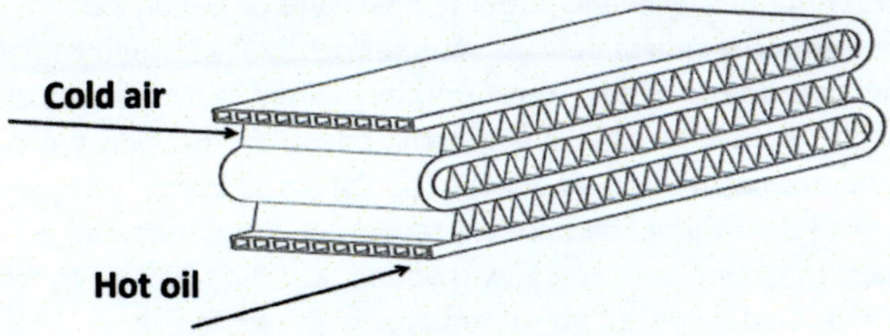

Fig. 1. Core body of air oil radiator

$$Nu = \frac{(f/2)(Re - 1000)Pr}{1 + 12.7(f/2)^{0.5}(Pr^{2/3} - 1)} \quad (1)$$

$$f = (1.58\ln(Re) - 3.28)^{(-2)} \quad (2)$$

$$\lg\lg(\upsilon_t \times 10^{-6} + 0.6) = 0.5679 - 0.3126\lg t \quad (3)$$

$$Nu = 0.236Re^{0.62}Pr^{1/3} \quad (4)$$

$$\eta_f = \frac{\tanh(mh)}{mh} \quad (5)$$

$$m = \sqrt{\frac{2\alpha}{\lambda\delta}} \quad (6)$$

where: Re is Reynolds number, Nu is Nussel number, f is friction factor, ν_t is oil motion viscosity, t is temperature, η_f is fin efficiency, m is fin parameter, h is fin height, α is surface heat transfer coefficient, λ is fin thermal conductivity, δ is fin thickness.

3 Model Modification Method

According to the actual application scenario of the air oil radiator on the engine, the design point and the typical inflection point of the task profile are selected as the typical state point of the air oil radiator test. The inlet parameters of the air side and oil side of the air lubricating oil radiator at typical state points were calculated. Air side inlet parameters include: temperature, pressure, flow rate. Oil side inlet parameters include: temperature, pressure, flow rate.

The inlet parameters of the air side and the oil side of the air lubricating oil radiator calculated at typical state points were taken as the boundary conditions of the test, and the test bench carried out the heat transfer characteristic test of the air lubricating oil radiator according to the given inlet parameters. The platform data acquisition system is used to measure the inlet parameters of the air side at a series of typical status points, including temperature, pressure and flow rate. The inlet parameters of the oil side include temperature, pressure, flow rate and outlet temperature of the oil side.

The calculation model of heat transfer of air lubrication radiator is established according to the efficiency-number of heat transfer units method. According to the inlet parameters measured in the heat transfer characteristic test, the boundary conditions of the component model for calculation were set to be the same as the inlet parameters of the test. On the basis of the original component model for simulation, the correction coefficient C of the heat transfer coefficient K was added, and the new heat transfer coefficient was K*C (The modified model of air oil radiator is shown in Fig. 2).

The evaluation objective of the model deviation of the heat transfer calculation of the air lubricating radiator is determined. The evaluation objective is the average value of the absolute deviation between the test results and the model calculation results of the heat transfer characteristics of the oil outlet temperature index of the air lubricating radiator under typical state points.

The selection principle of correction coefficient is to minimize the evaluation target. The specific implementation process is to select the minimum Cmin, maximum Cmax and step value data of the first round traversal of the correction coefficient. From the minimum Cmin of the correction coefficient, the step value data traversed to the maximum Cmax according to the step value data, a series of evaluation objectives under the correction coefficient were calculated in turn, and the evaluation objective was selected as the minimum correction coefficient C. Generally, Cmin ranges from 0.25 to 0.9, Cmax ranges from 1.1 to 4, and data ranges from 0.001 to 0.1.

By substituting the correction coefficient C into the original calculation model of air oil radiator, the optimal calculation model of air oil radiator heat transfer is obtained, and the calculation model of air oil radiator heat transfer is modified based on all typical state test data.

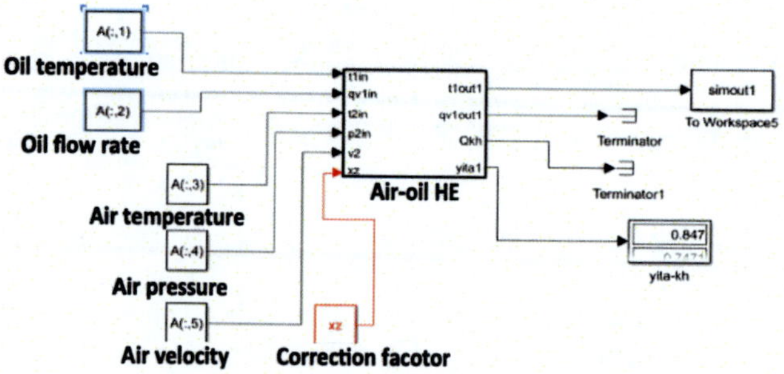

Fig. 2. The modified model of the air oil radiator

4 Model Modification Result

The heat transfer model is modified based on the measured data of a certain type of air oil radiator. The correction factor varies from 1.0 to 1.5 in step size 0.001. Figure 3 shows the deviation results of the air lubricating radiator model under different correction coefficients, and Fig. 4 shows the comparison of results before and after the correction of the air lubricating radiator model. The deviation between the calculated results of the model and the experimental value decreases first and then increases. When the correction coefficient is 1.232, it is the optimal correction coefficient, and the deviation decreases from 2.53 °C to 0.29 °C. The temperature calculation result before correction is larger than the test result on the whole, and the data distribution of the temperature calculation

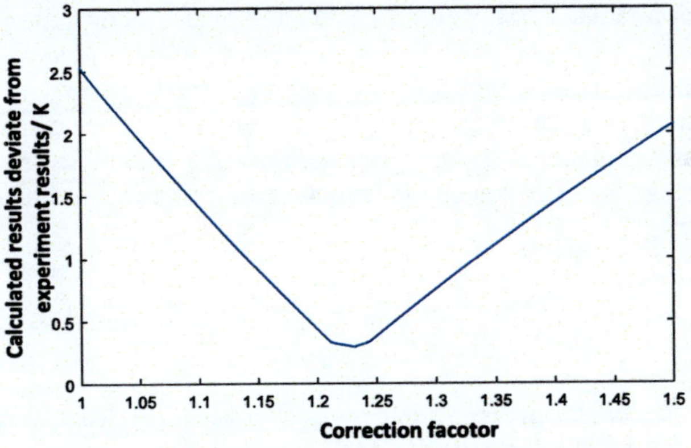

Fig. 3. Deviation results of the air oil radiator model

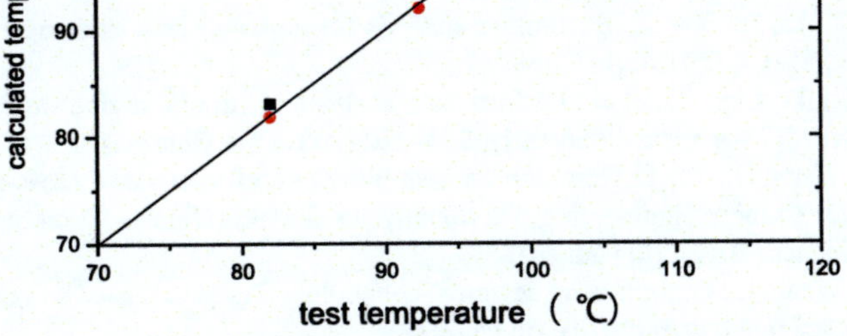

Fig. 4. Comparison of results before and after modification of the air oil radiator model

result after correction is located on both sides of the measured result, which is closer to the measured result on the whole.

5 Conclusion

1) The method for selecting test status points proposed in this paper is to select typical status points to carry out heat transfer characteristic tests according to the actual application scenario of the air lubricating radiator. Each selected test status point has representative significance to avoid carrying out repeated heat transfer characteristic tests.
2) The radiator heat transfer model modification method proposed in this paper makes full use of all test data and includes all test data into the evaluation objectives of the air lubrication radiator heat transfer calculation model modification to avoid waste of test data.
3) The revised heat transfer model of the radiator not only guarantees a better calculation accuracy near the design point, but also improves the calculation accuracy of the heat transfer model of the air lubrication radiator under all working conditions. By optimizing the correction coefficient, the temperature deviation is reduced from 2.53°C to 0.29°C.

References

1. Milosovski, G., Bil, C., Simon, P.: Improvement of aircraft accident investigation through expert systems. J. Aircr. **46**(1), 10–24 (2009)
2. Dai, C., Dai, C.: Load management technology research of aircraft integrated drive generator based on the upper and lower-bit machine architecture. Aircr. Design **29**(4), 36–40 (2009). (in Chinese)
3. Lv, Q., Liu, H., Zhong, J.: Design of propulsion system for testing platform of large bypass turbofan engine. Aeronaut. Sci. Technol. **26**(12), 20–24 (2015). (in Chinese)
4. Linke-Diesinger, A.: Systems of Commercial Turbofan Engines,Springer, Heidelberg, pp. 70–72 (2008). https://doi.org/10.1007/978-3-540-73619-6
5. Liao, Y.: Design and performance study of surface air-oil heat exchanger. Xiamen University (2018) (in Chinese)
6. Zhang, Q.,Liu, Y-F.: Dynamic characteristic analysis of lubricating oil radiator. Mech. Electr. Equipment (004), 37–42 (2015) (in Chinese)
7. Sun, H.-G., Li, B., Xie, Z.-B.: Failure analysis for air-to-oil heat exchanger cracking. Fail. Anal. Prev. **8**(4), 4 (2013). (in Chinese)
8. Cai, H., Su, L., Liao, Y., et al.: Performance analysis of aircraft engine surface oil Radiator. J. Southwest Jiaotong Univ. **056**(001), 214–220 (2021). (in Chinese)
9. Zhao, P., Wang, S., Ai, J.: Research on heat transfer performance of cross-flow fin Air/Oil radiator. In: Chinese Engineering Thermophysics Society. Chinese Society of Engineering Thermophysics (2012). (in Chinese)
10. Zhao, N., Wang, L.: Optimization design of helicopter oil radiator based on genetic algorithm. J. Mach. Design **31**(1), 6 (2014). (in Chinese)
11. Yu, J.: Principle and Design of Heat Exchanger, Beiing University of Aeronautics and Astronautics Press, Beijing, pp. 49–51 (2014). (in Chinese)

12. Shi, M.: Principle and Design of Heat Exchanger, Southeast University Press, Nanjing, pp. 57–60 (2009). (in Chinese)
13. Wu, Q.: Numerical simulation of coiled heat exchanger with convergent-divergent tubes. South China University of Technology (2011). (in Chinese)

Multi UAV Trajectory Tracking Optimization Method Based on Fuzzy Graph Convolution Neural Network

Ziyuan Ma(✉), Huajun Gong, and Xinhua Wang

College of Automation Engineering, Nanjing University of Aeronautics and Astronautics, Nanjing, China
maziyuan@nuaa.edu, {ghj301,xhwang}@nuaa.edu.cn

Abstract. Multi-UAV coordinated trajectory tracking is the most popular research problem at present. It is necessary to optimize and adjust the combined course of multiple UAVs and the motion trajectory of UAVs in real-time, and constantly pursue the dynamic optimal configuration between multiple UAVs and targets to improve and stabilize the positioning accuracy. Given the shortcomings of traditional centralized optimization methods, this paper proposes an advanced fuzzy graph convolution neural network model (FGTT). Because of the multi unmanned multi-source target and the irregular geometric structure, the graph convolution neural network can better optimize and process the trajectory tracking optimization method in non-European space. Several groups of verification show that this method can achieve higher tracking accuracy faster than traditional methods, The stability and accuracy of location and tracking are improved.

Keywords: Fuzzy Graph Neural Network · Multiple UAVs · Trajectory Tracking · Trajectory Optimization

1 Introduction

With the UAV technology entering the civil market. UAVs are widely used in military and civil fields. It is increasingly important in landform exploration, search and rescue, emergency monitoring, etc. Environmental awareness is the key to the autonomous navigation of UAVs, which is used to execute flight missions safely and stably. With the rapid development of UAV cluster technology, modern electronic warfare technology and aviation communication technology, multiple UAVs have great application potential in military reconnaissance, environmental monitoring, anti-terrorism and stability maintenance, and emergency rescue in passive collaborative positioning of mobile signal emitter targets. In the moving state, the dynamic configuration between multiple UAVs and targets, that is, the real-time distance between each UAV and the target, and the real-time angle between any two UAVs and the target, directly affect the passive location accuracy and stable tracking effect of multiple drones [1, 2].

The UAV must react to the surrounding obstacles in real-time during its movement. In recent years, experts and scholars at home and abroad have carried out a lot of research

Chinese Society of Aeronautics and Astronautics (Ed.): CASTC 2023, LNME, pp. 40–48, 2024.
https://doi.org/10.1007/978-981-99-8861-7_5

on autonomous navigation flights of UAVs. This research can be divided into two aspects: on the one hand, UAV can detect the surrounding environment in real-time during the movement process, and detect the surrounding obstacles to achieve obstacle avoidance; On the other hand, the trajectory planning algorithm is used to plan the trajectory when the environmental obstacles are known. The planning algorithm based on trajectory map, artificial potential field, and element decomposition is three flight path planning methods when the environmental obstacles are known. The method based on a trajectory graph is to transform the trajectories that can pass without obstacles into a collection of directed line segments according to some given methods, and then find more consistent trajectories in these collections. For the artificial potential field method, the attractive force and repulsive force should be assumed. The attractive force guides the UAV to approach the target, and the repulsive force prevents the UAV from approaching the obstacle [3]. For the planning algorithm of cell decomposition, the flight space should be divided into cells of the same size, and then the appropriate algorithm should be used to search for cells. Finally, select the cell sequence that matches the flight. However, these three route planning methods must be carried out when the environmental obstacles are known [4].

In the current UAV trajectory tracking and fuzzy neural network, nonlinear control strategies or intelligent control methods are often used to design flight controllers, such as sliding mode control [7, 8], backstepping [9, 10], fuzzy control [11], model predictive control [12], automatic disturbance rejection control [13, 14], reinforcement learning control, neural network, etc. However, the above methods rely more on accurate mathematical models or a large number of flight data and have poor adaptability to unmodeled dynamic characteristics and unknown disturbances. In addition, most of the common control algorithms cannot guarantee that the system state variables can converge in a finite time, resulting in a slow response speed.

In this regard, in order to improve the efficiency of trajectory similarity calculation, this research scheme proposes a trajectory calculation optimization strategy based on an advanced fuzzy graph neural network. Fuzzy graph neural network is a trajectory representation learning framework for spatial network similarity calculation, which combines the preliminary analysis method of hierarchical optimization with the graph neural method of fuzzy hierarchy, tests and detects in the experimental data set, and compares other different optimization methods, Our approach is the most advanced [5, 6].

2 A Stepwise Analysis Method for UAV Trajectory Optimization

The idea of the forward method is to introduce the variables one by one until no variables can be introduced. It is required that the variable introduced each time is a subset of the remaining independent variables that has the greatest impact on y, and can pass the regression coefficient test. The specific implementation methods are as follows:

● Step 1: Establish n unary regression equations for n independent variables $x_1, x_2, x_3 \ldots, x_n$ and y respectively, and calculate the F test statistic value of the regression coefficient corresponding to each independent variable, record it as

$F_1^1, F_2^1, F_3^1 \ldots, F_n^1$, set $F_{i_1}^1 = max\{F_1^1, F_2^1, F_3^1 \ldots, F_n^1\}$. For the predetermined significance level a, we know the critical value $F_a(1, n-2)$. If $F_{i_1}^1 \geq F_a(1, n-2)$, then introduce its corresponding variable x_{i_1} into the regression equation; otherwise, end the variable selection [8].

● Step 2: Establish n-1 binary linear regression equations for the selected independent variable subsevariablesdent vriable $x_j(1 \leq j \leq n, j \neq i_1)$ with the dependent variable y, and calculate the F statistic value:

$$F_j^1 = (j = 1, 2, \ldots, n, j \neq i_1) \tag{1}$$

Step 3: Repeat the selected independent variable subset according to step 2 to obtain the independent variable subset selected by the forward method.

● Step 4: Establish equations for n independent variables $x_1, x_2, x_3 \ldots, x_n$ and y respectively, calculate the F test statistic value of the regression coefficients corresponding to the n independent variables for $x_1, x_2, x_3 \ldots, x_n$, and record it as $F_1^1, F_2^1, F_3^1 \ldots, F_n^1$. Let $F_{i_1}^1 = min\{F_1^1, F_2^1, F_3^1 \ldots, F_n^1\}$. For a predetermined significance level a, check the critical value $F_a(1, m-n-1)$. If $F_{i_1}^2 \leq F_a(1, m-n-1)$, remove the corresponding variable x_{i1} from the regression equation; otherwise, the algorithm ends.

● Step 5: Establish a regression equation for the eliminated independent variable subsets $\{x_{i_1}\}$ and y, and perform an F test on the regression coefficients of these m-1 independent variables, and mark F as $F_j^1 = (j = 1, 2, \ldots, n, j \neq i_1)$. Remember $F_{i_1}^1 = min\{F_1^1, F_2^1, F_3^1 \ldots, F_n^1\}$, if $F_{i_1}^2 \leq F_a(1, m-n)$, remove the corresponding variable x_{i2} from the regression equation; otherwise, the algorithm ends. By analogy, until there are no variables to be eliminated, a subset of the independent variables selected by the backward method is obtained [9].

3 Neural Network Analysis of Fuzzy Hierarchical UAV Diagram

When introducing fuzzy theory, Professor Zaden, the founder of the fuzzy theory, once gave an example of parking. Even a novice can easily park a car between two cars after practicing several times while using a computer that is good at finding accurate values requires a laborious model. In actual production and life, there are often a large number of fuzzy problems. Although high-speed computers can solve some problems, a large number of imprecise controls can be easily solved through fuzzy judgment, experience, and reasoning. Traditional solutions based on accurate mathematical models have natural defects.

Fuzzy theory and artificial neural networks are developed to solve these problems. Fuzzy Neural Network (FNN) combines fuzzy theory and artificial neural network to give full play to their advantages and make up for each other's shortcomings. It shows excellent ability in dealing with large-scale fuzzy application problems. A fuzzy neural network combines a fuzzy system and a neural network. Its essence is to change the input of the neural network into a fuzzy input signal and fuzzy weight value after being processed by the fuzzy system and to anti-blur the output of the neural network, which is called an intuitive effective value. Specifically, in the fuzzy neural network, the input and output of the neural network represent the input and output of the fuzzy system. The membership function and fuzzy rules of the fuzzy system are added to the hidden nodes of the neural network, giving full play to the parallel processing ability of the neural network and the reasoning ability of the fuzzy system.

The neural network is a bridge to build UAV trajectory tracking and positioning. Because UAV involves many optimization parameters during operation, but many parameters cannot be obtained through conventional calculation. The neural network is a high-dimensional calculation method. Although its interpretability is a "black box", it can optimize the data through a neural network to obtain parameters with weight values, based on parameter optimization, the best combination of parameters is further refined, and then the trajectory tracking of UAV is optimized. Secondly, the neural network also has its limitations. In the process of processing, the neural network is deficient due to gradient explosion and other structural deficiencies. Therefore, the structure of the neural network is optimized and improved by the fuzzy graph neural method, The application of neural networks in UAV trajectory tracking is further improved.

Fuzzy hierarchical UAV diagram comprehensive evaluation is a comprehensive evaluation method that uses fuzzy theory to comprehensively evaluate multi-factor and multi-level fuzzy concepts. The method that combines the weights determined by the analytic hierarchy process with the fuzzy comprehensive evaluation for evaluation is the fuzzy analytic hierarchy process (FAHP). The evaluation process of fuzzy analytic hierarchy process:

First, determine the factor set M of the evaluation object. That is, the indicators of each level in the evaluation indicator system are expressed in the form of a set M. The factor set is $M = \{m_1, m_2, ..., m_n\}$;

Second, determine the comment set P. That is, the set of evaluation criteria for each factor is generally evaluated by five levels: {very strong, strong, general, weak, very weak}, the corresponding value is {5 points, 4 points, 3 points, 2 points, 1 min}. The comment set is $P = \{P_1, P_2, ..., P_L\}$;

Third, determine the judgment matrix R. The determination of the evaluation matrix R is completed in two steps. The first step is to evaluate a single factor to obtain a single factor evaluation set $\{r_{i1}, r_{i2}, ..., r_{im}\}$; the second step is to compose a matrix of N single factor sets, that is, the evaluation matrix R.

$$R = \begin{bmatrix} r_{11} & r_{12} & \cdots & r_{1l} \\ r_{21} & r_{22} & \cdots & r_{2l} \\ \vdots & \vdots & \ddots & \vdots \\ r_{n1} & r_{n2} & \cdots & r_{nl} \end{bmatrix} \quad (2)$$

Fourth, determine the weight set A of each factor.

Fifth, carry out the fuzzy comprehensive evaluation. Combining the weight A and the evaluation matrix R of the evaluated object into A ∘ R, the fuzzy comprehensive evaluation result vector B ∘ B = A $\circ B = \{b_1, b_2, \cdots, b_m\}$ of the evaluated object can be obtained [13].

Among them, "∘" represents the fuzzy composition operator. There are mainly four commonly used fuzzy composition operators: $M(\wedge, \vee)$, $M(g, \vee)$, $M(\wedge, \oplus)$, $M(g, \oplus)$, operator a is called the main factor determinant operator, and the result of using this operator to judge only depends on the total The factor that plays a major role in the evaluation, and other factors do not affect the evaluation result. The mathematical symbols $M(\wedge, \vee)$, $M(g, \vee)$, $M(\wedge, \oplus)$, $M(g, \oplus)$ only represent the calculation definitions of different model structures and have no more significance.

Operators $M(g, \vee)$ and $M(\wedge, \oplus)$ are the main factors. The prominent operators are relatively close to the operator $M(\wedge, \vee)$. They highlight the main factors while taking into account other factors to a certain extent;

Operator $M(g, \oplus)$ is called a weighted average operator. This operator balances all factors according to their weights. The evaluation result includes the joint effect of all factors, which truly embodies tcomprehensiveness andis suitable for situations requiring overall indicators. Choose $M(g, \oplus)$ to use the weighted average operator to operate the fuzzy set.

Sixth, multi-level fuzzy comprehensive evaluation. The evaluation index system often has more than one layer, but a multi-layer system requires multi-level fuzzy comprehensive evaluation based on the single-layer evaluation. After completing the previous five steps, a single-factor fuzzy evaluation vector can be obtained. We combine the obtained single-factor evaluation vectors to form a new evaluation matrix K.

$$K = \begin{bmatrix} k_{11} & k_{12} & \dots & k_{1l} \\ k_{21} & k_{22} & \dots & k_{2l} \\ \vdots & \vdots & \ddots & \vdots \\ k_{n1} & k_{n2} & \dots & k_{nl} \end{bmatrix} \tag{3}$$

Then assign weights A1 to these major categories of factors. The combination of A1 and the evaluation matrix K is the second level of comprehensive evaluation J. If there are several levels of indicators, perform several fuzzy comprehensive evaluations according to the above method, and the final fuzzy vector is the total Fuzzy comprehensive evaluation results [14].

$$J = A_1^{\circ} K = A_1^{\circ}(A^{\circ} B) = A_1 \begin{bmatrix} AB_1 \\ AB_2 \\ \vdots \\ ABn \end{bmatrix} \tag{4}$$

Seven, single value processing of evaluation results. The result of fuzzy hierarchy evaluation is a multi-dimensional fuzzy vector. To express the fuzzy evaluation result with a specific numerical value more intuitively, the fuzzy vector needs to be single-valued (Fig. 1).

In the second step, we determined the comment P. By assigning a value to the comment set and then combining the final fuzzy evaluation vector with the comment, a single value can be obtained, which makes the evaluation result more intuitive. The single value $F = K * P^T$ is the synthesis method, that is, the fuzzy evaluation vector is multiplied by the transposed vector of the comment set, and the obtained value is the final evaluation result of something.

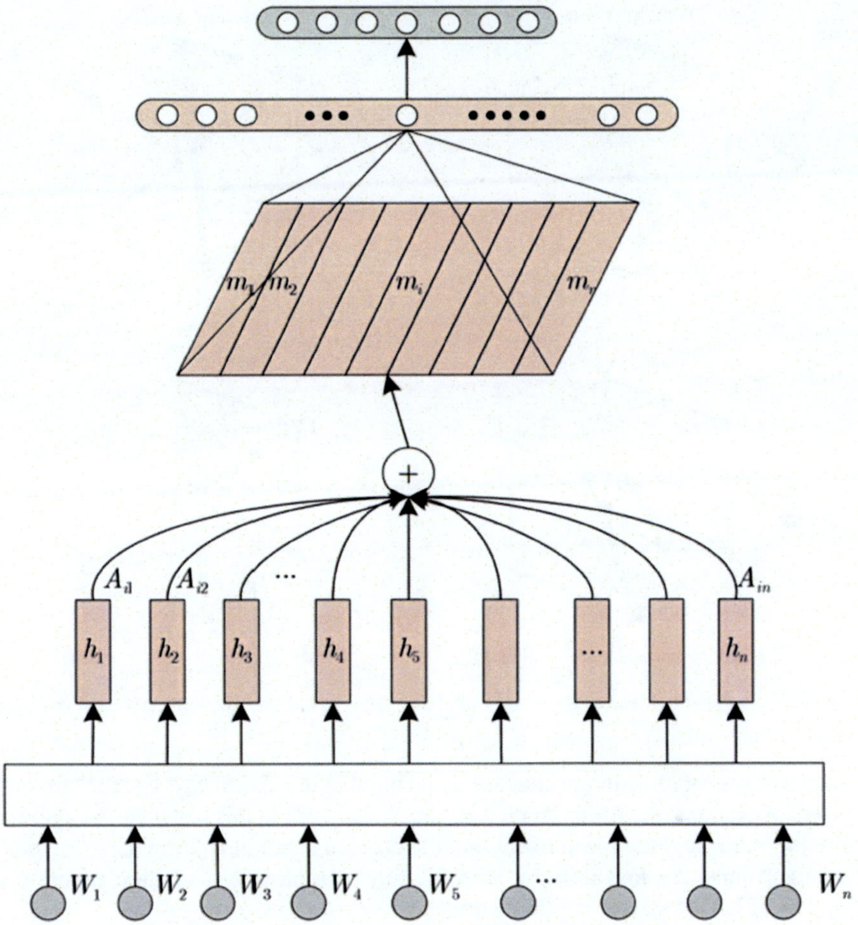

Fig. 1. Advanced Neural Network with Fuzzy Decision

4 Experiments and Analysis

In order to facilitate the establishment of the dynamic model of the aircraft, the following reasonable assumptions are proposed: Assumption 1: the aircraft is regarded as a rigid body. Assume that the center of mass of the aircraft coincides with the origin of the airframe coordinate system. Assume that the 3 aircraft are of symmetric structure, which can ensure that the inertia matrix is a diagonal matrix.

In the path-tracking optimization of multiple UAVs based on a graph convolution neural network, we compared and observed the reliability test rate of decision-making UAVs at different levels and in different numbers. Through the fusion of fuzzy hierarchy analysis in the improved graph neural network, we set the path optimization scheme of different UAVs, calculate the weight calculation of UAVs in different modes, and optimize the UAV path reasonably, efficiently, and quickly (Fig. 2).

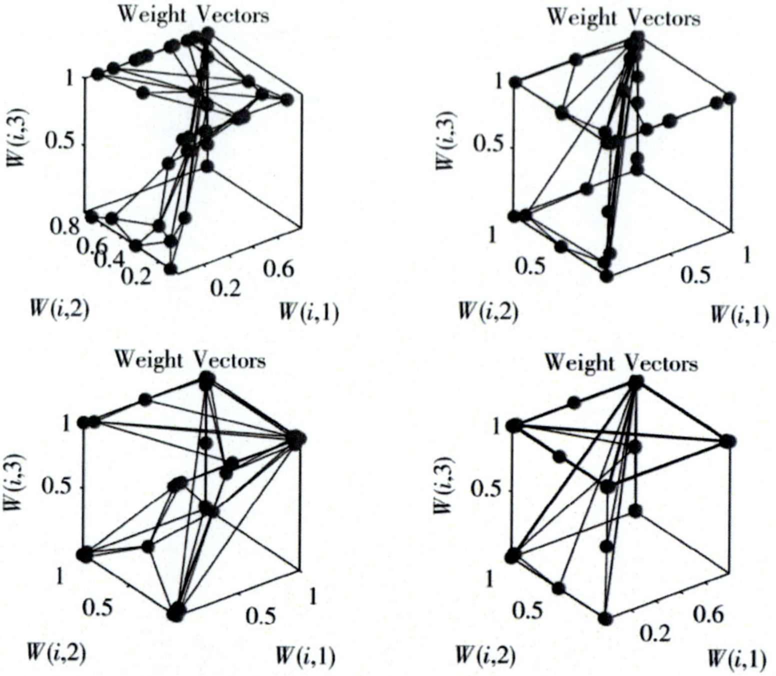

Fig. 2. Calculation of trajectory weight of space UAV based on fuzzy graph neural network

It can be seen that, with the increase in the number of experiments and the number of UAV collaborations, the fuzzy graph neural network used in this paper can be stabilized at a stable value, and the speed of solving the optimal solution is faster. Compared with other algorithms, the loss is faster, and the time complexity is smaller, ensuring the reliability of UAV trajectory optimization.

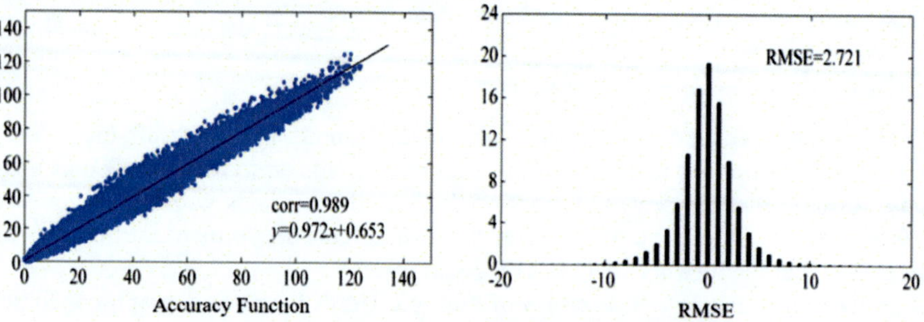

Fig. 3. Accuracy and root mean square error analysis

Through the fitting calculation of our FGTT algorithm model, we can see from Fig. 3 that our algorithm has high accuracy, the root mean square error effect is also very good, and the accuracy function equation is better fitted within the confidence interval due to its perfect characteristics in non-European space geometry (Table 1).

Table 1. Accuracy of trajectory optimization compared with different neural network algorithms

	10 Fold	20 Fold	30 Fold	40 Fold	50 Fold	60 Fold
S-LSTM [5]	90.3	91.6	90.5	89.0	87.1	91.5
S-Atten [6]	89.5	92.5	89.7	91.4	90.0	89.1
SoPhie [11]	91.6	90.7	87.9	92.7	92.0	92.9
Next [12]	92.5	89.5	91.2	91.9	90.7	93.0
GABP [13]	87.6	90.6	87.5	90.0	89.1	92.6
Paper Work	94.1	93.5	95.9	94.5	95.1	96.0

Through the comparison of the above different algorithms, it can be seen that the FGTT algorithm (Paperwork) proposed in this paper has significantly improved accuracy in different iterative experiments. It is most obvious in the fortieth iteration. Compared with the more advanced algorithms such as S-Atten, the algorithm proposed in this paper pays more attention to the geometric arrangement of UAVs in non-European space, greatly simplifies the time complexity from the perspective of non-European space, and greatly improves the accuracy.

This paper further verifies that under different trajectory optimization conditions for UAVs and targets, compared with the difference between this paper and other methods in trajectory optimization time and positioning accuracy, this paper has shorter global trajectory tracking optimization time and higher positioning accuracy than other methods under any kind of conditions. Through the optimization of the fuzzy graph-based neural network algorithm proposed in this paper and the comparison of six different algorithms, we can see the accuracy of the algorithm in data. The algorithm proposed in this paper has high accuracy and reliability and has profound value and significance.

5 Conclusions

With the increasing speed of UAV and target, the global trajectory optimization time of this method and other methods is increasing, but the time of other methods is increasing more; At the same time, the positioning error of other methods is also increasing, and the error increase is also greater.

The algorithm proposed in this paper can further reduce the delay caused by communication distribution; It increases the survivability and robustness of the UAV network under actual combat conditions. In the next step, it is also necessary to further consider the distributed cooperative autonomous heading generation method under aircraft instantaneous maneuver constraints, three-dimensional terrain environment, complex electromagnetic environment conditions, multi-aircraft cooperative passive location, and multi-mobile emitter targets. Combined with the corresponding data, it can be seen that the method proposed in this paper has a good improvement through comparison.

References

1. Xu, S., Dogancay, K., Hmam, H.: Distributed pseudolinear estimation and UAV path optimization for 3D AOA target tracking. Signal Process. **133**, 64–78 (2017). https://doi.org/10.1016/j.sigpro.2016.10.012
2. Wang, W., Bai, P., Liang, X., Wang, Y.: Optimal deployment of sensor-emitter geometries for hybrid localization using TDOA and AOA measurements. IET Sci. Meas. Technol. **13**(5), 622–631 (2020)
3. Gupta, A., Johnson, J., Li, F., et al.: Social GAN: socially-acceptable trajectories with generative adversarial networks. In: 2018 IEEE/CVF Conference on Computer Vision and Pattern Recognition, p. 2255. IEEE, Salt Lake City (2018). https://doi.org/10.1109/CVPR.2018.00240
4. Hasan, I., Setti, F., Tsesmelis, T., et al.: MX-LSTM: mixingtracklets and violets to jointly forecast trajectories and head pose. In: 2018 IEEE/CVF Conference on Computer Vision and Pattern Recognition, p. 6067. IEEE, Salt Lake City (2018). https://doi.org/10.1109/CVPR.2018.00635
5. Lanctot, M., Zambaldi, V., Gruslys, A., et al.: A Unified Game - Approach to Multiagent Reinforcement Learning (2017). https://arxiv.org/abs/1711.00832
6. Feng, Y., You, H., Zhang, Z., Ji, R., Gao, Y.: Hypergraph neural networks. Proc. AAAI Conf. Artif. Intell. **33**(01), 3558–3565 (2019). https://doi.org/10.1609/aaai.v33i01.33013558
7. Xiong, J.J., Zhang, G.: Discrete-time sliding mode control for a quadrotor UAV. Optik **127**(8), 37183722 (2016)
8. Mofid, O., Mobayen, S., Wong, W.K.: Adaptive terminal sliding mode control for attitude and position tracking control of quadrotor UAVs in the existence of external disturbance. IEEE Access **9**, 3428–3440 (2020)
9. Yu, Y., Guo, Y., Pan, X., et al.: Robust backstepping tracking control of uncertain MIMO nonlinear systems with application to quadrotor UAVs. In: 2015 IEEE International Conference on Information and Automation, pp. 2868−2873, Lijiang, China (2015)
10. Glida, H.E., Abdou, L., Chelihi, A., et al.: Optimalmodel-free backstepping control for a quadrotor helicopter. Nonlinear Dyn. **100**(4), 3449–3468 (2020)
11. Bhatkhande, P., Havens, T.C.: Real-time fuzzy controller for quadrotor stability control. In: 2014 IEEE International Conference on Fuzzy Systems, Beijing, China, pp. 913−919 (2014)
12. Alexis, K., Papachristos, C., Siegwart, R., et al.: Robust model predictive flight control of unmanned rotor-crafts. J. Intell. Rob. Syst. **81**(3–4), 443–469 (2016)
13. Xiong, J.J., Zhang, G.B.: Global fast dynamic terminal sliding mode control for a quadrotor UAV. ISA Trans. **66**, 233–240 (2017)
14. Hua, C.C., Wang, K., Chen, J.N., et al.: Tracking differentiator and extended state observer-based nonsingular fast terminal sliding mode attitude control for a quadrotor. Nonlinear Dyn. **94**(1), 343–354 (2018)

Performance of Hearing Protectors in Ejection Seat Parachute Gun Firing

Wujun Xie, Liwu Min, Jun Li(✉), and Hui Li

Key Laboratory of Science and Technology on Life-Support Technology, AVIC Aerospace Life-Support Industries Ltd., Xiangyang, Hubei, China
csz081314LJ@hotmail.com

Abstract. During ejection rescue, the parachute gun attached to a rocket-assisted ejection seat generates jet noise that may damage a pilot's sensitive auditory organs; hence the effectiveness of hearing protectors, primarily a helmet and earmuffs, is evaluated. A parachute gun firing experiment is designed, and overpressure data with and without protection are measured simultaneously during firing using piezoelectric pressure sensors inside and outside dummy ears. A comparative study of the measured data is conducted using time domain analysis and wavelet transformation methods, and the results show that the jet noise generated by a parachute gun firing is a broadband signal, consisting of shock waves and impulse noise; different from the attenuation characteristics in the continuous noise environment, in the jet noise environment, the effect of hearing protectors is significant for extenuating the high-frequency components that are prone to injury, but not significant for extenuating the low-frequency components.

Keywords: Ejection Rescue · Parachute Gun · Jet Noise · Hearing Protectors · Wavelet Transformation

1 Introduction

There are two different sources of gunfire noise: one is the mechanical noise of collision and friction of various motion mechanisms, and the other is the gaseous dynamic noise of the muzzle in the firing process, the latter of which is much higher than the former [1]. China's revised auditory organ safety standards define the noise consisting of one or more bursts of less than 1 s duration with a maximum overpressure peak of less than 6.9 kPa as the impulse noise wave; the maximum overpressure peak of not less than 6.9 kPa is defined as the shock wave [2]. Although the impulse noise has a low overpressure peak and does not pose a threat to human internal organs in general, it is still essentially a weak shock wave or an attenuated strong shock wave caused by long-range propagation [3]. Considering the differences between muzzle shock waves and impulse noise in terms of propagation laws and hazards, shock waves and impulse noise are distinguished in this work.

The primary function of a rocket-assisted ejection seat fitted in fast jet combat aircraft is to push the seat and pilot away from an uncontrolled aircraft; after reaching a safe height

Chinese Society of Aeronautics and Astronautics (Ed.): CASTC 2023, LNME, pp. 49–57, 2024.
https://doi.org/10.1007/978-981-99-8861-7_6

for parachute opening, the human restraint system is released, and the parachute kept in the mounting box is shot out by the gun, defined as the parachute gun, which ensures a safe landing for the pilot [4, 5]. At the instant of parachute gun firing, high-temperature and high-pressure powder gas is ejected from the muzzle, resulting in a shock wave and impulsive noise that exceeds the air pressure. Shock waves and impulsive noises with a high peak and a short duration may damage the pilot's auditory organs permanently, long-term, or temporarily [6]. Therefore, to ensure the health of pilots, the impulse noise and shock wave overpressure measurement and analysis during the parachute gun firing is one of the necessary test items for the rocket-assisted ejection seat.

A large number of data analysis methods and experimental studies have been completed in order to study the aerodynamic characteristics of impulse noise and shock waves generated by guns firing or explosions and to assess the effectiveness of protectors. Wavelet analysis, with the capability of multi-scale analysis in both time and frequency domains, has become an important tool for dealing with non-stationary noise signals. For example, the complex components and propagation characteristics of jet noises generated from a large-caliber light weapon and a rocket motor have been analyzed and studied by this method [3, 7]. A comparison of blast shock wave protection for World War I and current military helmets demonstrated that almost all helmets tested offered invaluable protection against basic blast brain injury, with modern variants providing no more pressure attenuation than historical helmets [8]. Hearing protectors made for industrial noise were applied in high-peak impulse noise and shock wave environments (guns firing or explosions) in [9], and the results suggested that some protector design modules that are extremely effective against low-peak and continuous noise may prevent protection from high-peak impulse noise. However, the aerodynamic characteristics of guns firing noise during ejection lifesaving and the effect of hearing protectors on the hearing organs of pilots have not been reported.

Hence, an experimental method and data analysis model are conducted on the parachute gun firing noise protection problem of a rocket-assisted ejection seat. The wavelet transform is used to decompose and reconstruct the inside and outside test dummy ear aerodynamic noises, which obtain their time-frequency characteristics. On this basis, the overpressure values of the inside and outside test dummy ears are compared and analyzed to assess the effectiveness of the pilot helmet and earmuffs for protection against shooting noise generated from the parachute gun. The results have certain theoretical significance and engineering value, and can provide data support for the design improvement of hearing protective equipment.

2 Parachute Gun Firing Experiment

2.1 Experiment Equipment

The parachute is kept in the storage box at the back of the ejection seat headrest, and the parachute gun is fixed on the outside of the storage box, as shown in Fig. 1 (a). A 98th male dummy, similar to a Hybrid III dummy [10], is used as a pilot surrogate, which is designed and manufactured based on anthropometric measurements to represent the 98th percentile adult male pilot human dimensions, primarily for ejection rescue, parachute and crash safety limit assessments.

High-intensity acoustic ICP pressure transducers (PCB, Model 113B27; Buffalo, New York, USA) with built-in amplification and signal conditioning circuits, featuring ease of use and a high signal-to-noise ratio, are used. The measurement range is 0~690 kPa, the resonance frequency exceeds 500 kHz, the rise time does not exceed 1 μs, and the operating temperature range is −73~135 °C. As shown in Fig. 1, four pressure transducers are installed in the left ear of a dummy, right ear of a dummy, left side of a helmet, and right side of a helmet, to measure the overpressure value with and without hearing protectors, consisting of a helmet and earmuffs, during the parachute gun firing of an ejection seat.

The overpressure value is sampled at 130 kHz using a LAN-XI (BK, Model 3050, Denmark) data acquisition hardware module. The BK-connect software is used to complete hardware control, data acquisition, playback, and processing operations.

(a) (b)

Fig. 1. (a) Parachute gun, storage box and pressure transducer outside-helmet, (b) Dummy head and pressure transducer in-ear.

2.2 Experiment Procedure

In order to simulate a real parachute gun firing process and to ensure the safety of the experiment, the experiment procedure is stipulated as follows:

(1) Dress the test dummy in appropriate protective clothing, helmets, masks, and other pilot protective equipment, and adjust the position of the earmuffs to ensure that the dummy's ears are completely covered.
(2) Secure the dummy with the seat restraint system, connect the pressure sensor test cable and the parachute gun ignition cable.
(3) Use a crane and special spreader to suspend the human-chair system, not less than 2.5 m from the ground, and the surrounding area of the test site should be unobstructed, adjust the seat tilt so that the parachute gun is parallel to the ground, as shown in Fig. 2.
(4) After completing the test equipment commissioning and confirming that the equipment is error-free, the testers evacuate the site and conduct a parachute gun firing.

Fig. 2. Field experiment

3 Results

3.1 Time-Domain Analysis

In the interest of saving space, the same pattern presented by experiments is analyzed using the overpressure-time data collected by the left ear test transducers in one of the experiments with and without hearing protectors as an example.

The Safety Limit for Hearing Organ Damage. After a large number of light and heavy weapon noise comparison tests, it has been found that the impulse noise and shock wave of the two types of weapons have different A-durations (defined as T_A, the time that the overpressure rises from the ambient pressure to its maximum positive overpressure peak and then returns to the ambient pressure) [2], with large differences in hearing damage. The spectral analysis also reveals that light weapons contain more high-frequency components that can easily cause injury to the human hearing organ than heavy weapons. Therefore, the safety limits of human hearing organ injury for the two types of weapons are calculated by two equations using $T_A = 1.5$ ms as the boundary between the characteristics of light and heavy weapons.

For most heavy weapons with 1.5 ms $< T_A \leq 100$ ms, the safety limit (in dB) for impulse noise and shock wave damage to the human hearing organ is expressed by:

$$P_{sdB} = 177 - 6lg(T_A * N) \tag{1}$$

where N is the number of shots or explosions.

For light weapons (various guns and small caliber artillery) with 0.25 ms $\leq T_A \leq$ 1.5 ms, the safety limit for impulse noise and shock wave damage to the human hearing organ is expressed by:

$$P_{sdB} = 169 - 8lg(T_A * N) \tag{2}$$

Pressure values exceeding the ambient pressure, usually expressed in dB or Pa, are converted by:

$$P_{dB} = 20lg(P/P_0) \tag{3}$$

where, P_{dB} is the decibel value corresponding to the positive overpressure P. P_0 is the reference pressure level, taking the value of 20μPa.

Results of Time Domain Analysis. The safety limit for impulse noise and shock waves damage to human hearing organ is calculated based on the Eqs. (1) to (3), picking the maximum overpressure peak and T_A. Calculation results are shown in Table 1, and overpressure-time curves with and without protectors are shown in Fig. 4.

Table 1. The safety limit calculation results

	overpressure peak(kPa)	overpressure peak(dB)	T_A(ms)	P_{sdB}(dB)
with protectors	8.26	172.32	2.24	174.90
without protectors	34.30	184.69	0.59	170.86

Table 1 and Fig. 3 show that the parachute gun belongs to the category of light weapons, and the overpressure peak outside the ear (without protectors) is higher than the safety limit. Benefiting from the protective effect of the helmet and earmuffs, the overpressure action time is significantly increased, the overpressure peak in the ear (with protectors) is significantly reduced, which is lower than the safety limit, and the hearing protection meets the protection requirements stipulated in the standard. In addition, the overpressure peaks with and without protectors are higher than 6.9 kPa (170. 7 dB), which is beyond the range of impulse noise (including jet noise, parachute box surge, mechanical noise, etc.) [3], mixed with shock waves.

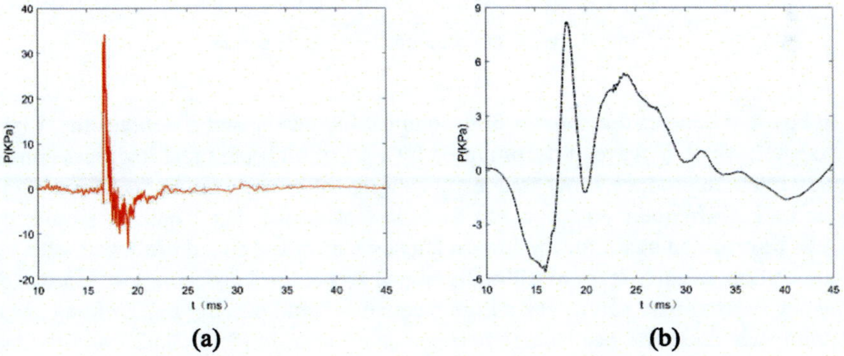

Fig. 3. (a) The overpressure-time curve without protectors, (b) The overpressure-time curve with protectors.

3.2 Time-Frequency Analysis

Wavelet transformations, which are widely used in mathematics and signal processing, are used to obtain the time-frequency characteristics of the overpressure with and without hearing protectors.

The Principle of Wavelet Transformation. The basic principle of wavelet transformation is to decompose the original signal containing different frequency components into multiple non-overlapping frequency components, and then study each component with a resolution that matches its scale [11, 12]. The wavelet transform provides a natural hierarchical structure and multidimensional data representation, so the wavelet multi-scale pyramid algorithm is used to perform multi-resolution decomposition and reconstruction of noisy signals, the principle is shown in Fig. 4.

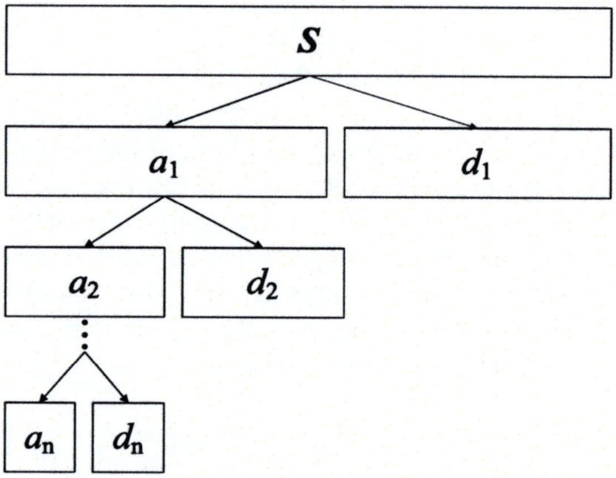

Fig. 4. Wavelet decomposition tree structure

In Fig. 4, n denotes the number of decomposition levels, and ***S*** denotes the original input signal, which is wavelet decomposed by a set of high-pass and low-pass complementary filters to produce a low-frequency component a_1 representing the approximate part and a high-frequency component d_1 representing the detail part. A similar process can be repeated again for the low-frequency part of the signal produced after each decomposition; in each decomposition the signal at that level is decomposed into a low-frequency approximate part a_i and a high-frequency detail part d_i [11]. Define f_s as the sampling frequency, the band range corresponding to a_i is $[0\ (f_s/2)/2^i]$ and the band range corresponding to d_i is $[(f_s/2)/2^i\ (f_s/2)/2^{i-1}]$.

Results of Wavelet Transformation. By using MATLAB, the wavelet transformation is performed on the overpressure data with and without protectors collected during the firing of a seat parachute gun. There are many different types of basis functions that can be selected for the wavelet transform, however, the dB function is mostly used in practical gun firing noise analysis, in addition the decomposition signal scale selection is limited by the length of the sampled data. According to the duration of a noise signal, the sampling frequency and the frequency components, the wavelet function adopted in this paper is dB9, and the wavelet decomposition scale n = 8, and the results are shown in Fig. 5. The frequency range corresponding to each band is shown in Table 2, $\boldsymbol{S} = d_1 + d_2 + d_3 + d_4 + d_5 + d_6 + d_7 + d_8 + a_8$.

Table 2. Frequency range corresponding to each band

Frequency band	Frequency range(kHz)	Frequency band	Frequency range(kHz)
a_8	0~0.25	d_4	4.06~8.13
d_8	0.25~0.51	d_3	8.13~16.25
d_7	0.51~1.02	d_2	16.25~32.50
d_6	1.02~2.03	d_1	32.50~65.00
d_5	2.03~4.06		

The detailed wavelet analysis results are as follows:

(1) a_8 frequency band: This frequency band represents the approximate part of noise signals with and without hearing protectors. As can be seen in Fig. 5, the overpressure-time curve in the approximate part has a significantly different trend from that in d1~d8 frequency band, the overpressure-time curve with hearing protectors follows essentially the same trend as that of the original signal, the peak without hearing protectors is significantly lower than that of original signal, and the pressure peaks with and without hearing protectors are essentially the same.
(2) d_8 frequency band: noise signals of this frequency band with and without hearing protectors exhibit obvious time-varying characteristics. Analysis of its waveform and peak position shows that the main component of this band is low-frequency shock waves, consisting of an initial shock wave and a gunpowder gas shock wave.
(3) d7 frequency band: the noise signal without hearing protectors in this frequency band shows obvious time-varying characteristics, while the time-varying characteristics of the noise signal with hearing protectors is not significant. The overpressure with hearing protectors is approximately equal to zero, and the main component of noise signal without hearing protectors is the low and medium frequency shock waves.
(4) d1~d6 frequency band: the noise signal without hearing protection still shows obvious time-varying characteristics, and the noise signal with hearing protection is approximately equal to zero. In this frequency band, the pressure signal without hearing protection has the sharp sudden change, and it is believed that the noise signal in this frequency band is mainly composed of the medium and high frequency components of shock waves, turbulence noise and interference from the external environment.

In summary, the noise generated by the parachute gun firing is an extremely complex signal with a broadband and continuous spectrum, the sound energy of each frequency band is high. Due to the protective effect of the helmet and earmuffs, the high-frequency component contained in light weapons (parachute guns), which can easily cause damage

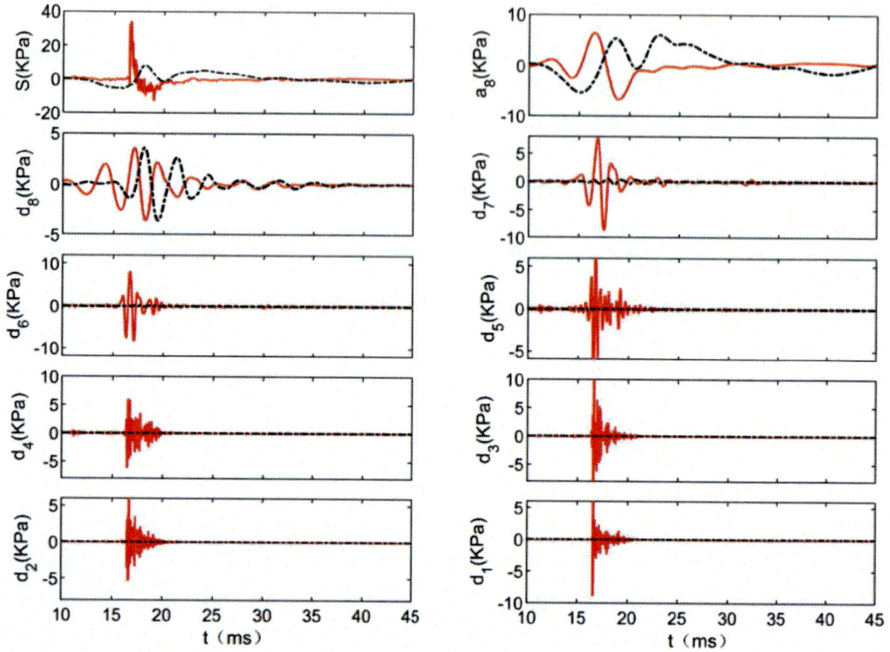

Fig. 5. Wavelet decomposition (red solid line-without protectors, black dotted line-with protectors)

to the human auditory organs, is significantly weakened, and the noise signal with hearing protectors is dominated by the low-frequency component.

4 Conclusion

Pilot hearing protectors are designed primarily for long flight times, and their role in attenuating aircraft steady-state noise and safeguarding call clarity has been fully validated through qualified experiments. However, it is also important to understand the effectiveness of hearing protectors against the complex shock waves and impulse noise encountered during ejection rescue. In this work, a parachute launch experiment was designed to simulate the key actions in the life-saving process of a rocket-assisted ejection seat, the parachute gun firing was completed at high altitude, and the noise signals with and without hearing protectors during the firing process were measured using free-field pressure sensors.

Time-domain analysis and time-frequency analysis based on wavelet transform for noise signals with and without hearing protectors were completed, and the results show that the noise signal generated by a parachute gun belonging to light weapons during firing is extremely complex, with a wide frequency band and high overpressure in all frequency bands; the attenuation effect of hearing protection is uneven across the wide frequency band, and the high-frequency components that easily injure human hearing organs attenuate significantly, basically dropping to zero, while the low-frequency components have no attenuation; with the protection of hearing protectors, the peak of noise signal decreases significantly, and the duration of positive pressure is significantly increased.

This work obtained noise data from the parachute gun firing and evaluated the protective effect of a helmet and earmuffs against shock wave and impulse noise, which is important for the development of better helmets to protect auditory organs. However, real human hearing organs are not sensitive to low frequency sounds, and the next step should consider the characteristics of the acoustic impedance of the real human ear canal to provide further technical support for evaluating pilot protection products.

References

1. Yongli, Z., Tailin, H., Baihe, L., et al.: Detection and extraction of shockwave signal in noisy environments. J. Intell. Fuzzy Syst. **37**(4), 4499–4510 (2019)
2. GJB 2A: Safety limits of impulse noise and blast wave from firing or explosion of conventional weapons and ammunitions on the human auditory organ damage-risk (1996)
3. Zhao, X., Zhou, K., He, L., et al.: Wavelet analysis and numerical simulation of jet noise induced by a large caliber small arms. Acta Armamentarii **40**(11), 2195 (2019)
4. BA, P.: Science and technology of aircraft seat ejection: advanced concepts. Cogent Eng. **9**(1), 2034267 (2022)
5. Ji, X., Zhang, T., Zhang, R., et al.: Attitude and trajectory calculation of ejection seat within the dynamic period. Acta Armamentarii **43**(3), 565 (2022)
6. Huanwen, G., Guodong, L., Yong, C., et al: Study on the influence of explosion impact on pilots in ejection. In: Proceedings of the International Conference Aircraft Utility Systems, China (2022)
7. Fan, Z., Yi, J., Xuewen, H., et al.: Time-frequency analysis of test data from complex noise of rocket engine jet. J. Projectiles, Rockets, Missiles Guidance **32**(3), 145–151 (2012)
8. Joost, O't E., Allen W, Y., Christopher P, B., et al.: Primary blast wave protection in combat helmet design: a historical comparison between present day and World War I. PLoS One **15**(2), e0228802 (2020)
9. Karl, B.: Performance of different types of hearing protectors undergoing high-level impulse noise. Int. J. Occup. Saf. Ergon. **15**(2), 227–240 (2009)
10. Jolyon, C., Dion, K., Ekant, M., et al.: Comparison of hybrid III and THOR 5th percentile female dummies in frontal crash tests. In: Proceedings of IRCOBI Conference (2021)
11. Manel, R., Ali, B.A., Imed Riadh, F., et al.: Wavelet transform application for/in non-stationary time-series analysis: a review. Appl. Sci. **9**(7), 1345 (2019)
12. Hai B, H., Xiao R, H., Ming L, Y., et al.: Identification of vehicle interior noise sources based on wavelet transform and partial coherence analysis. Mech. Syst. Signal Process. **109**, 247–267 (2018)

The High-Temperature Oxidation Behavior of Al Coating on Nickel-Based Single Crystal Superalloy

YanMing Li(✉), QiuShi Wang, Huan Liu, Shan He, ShiJia Liu, and WenWei Tong

Shenyang Engine Research Institute, Aero Engine Corporation of China, Shenyang 110015, China
yanming_linuaa@163.com

Abstract. Al coating was fabricated by chemical vapor deposition on the nickel-based single crystal superalloy. The high-temperature oxidation behavior at 900 °C and 1000 °C for 200 h were detected by scanning electron microscopy, energy disperse spectroscopy and X-ray diffraction. Results indicated that the oxidation kinetics of the alloy substrate and Al coating were all in accord with the parabolic rule. Compared with the alloy substrate, the Al coating manifested lowest oxidation rate and highest activation energy. After oxidation at 900 °C, the alloy substrate and Al coating formed Al_2O_3 protective film, represented some high-temperature oxidation resistance. After oxidation at 1000 °C, the alloy substrate emerged lots of microcracks and holes in the surface layer with serious internal oxidation and nitridation phenomena, resulted in a poor oxidation resistance. The Al_2O_3 protective film of Al coating took on the less surface spallation, more pure and continue, exhibited more excellent high-temperature oxidation resistance.

Keywords: Nickel-based single crystal superalloy · Al coating · Al_2O_3 protective film · high-temperature oxidation resistance

1 Introduction

With the continuous development of high thrust-weight ratio aeroengines and gas turbines, the service environment of turbine blade becomes more and more severe, which proposes higher requirements for the materials of turbine blade. It's very important that the materials need to have high-temperature mechanical properties, oxidation resistance and thermal corrosion resistance [1, 2]. Due to the specific grain orientation and the reduction of grain boundaries, nickel-based single crystal superalloy with perfect high-temperature properties is the preferred material for the manufacture of advanced engine turbine blade [3, 4].

However, turbine blade has been in service for a long time in high-temperature environment, the damage of the substrate material is frequently, because of oxidation phenomenon. Therefore, the surface of turbine blade is generally applied with protective coating, the substrate material ensures the high-temperature mechanical properties, and the protective coating contacts with the oxidation resistance [5–7].

Chinese Society of Aeronautics and Astronautics (Ed.): CASTC 2023, LNME, pp. 58–65, 2024.
https://doi.org/10.1007/978-981-99-8861-7_7

Al coating has the advantages of simple process and short production cycle, and is widely used for high-temperature protection in various fields, But few investigates on the Al coating on the surface of nickel-based single crystal superalloy, and the establishment of oxidation kinetic model and theoretical calculation of the coating uninvolved, which could not characterize the oxidation resistance of the coating completely. In this study, Al coating were prepared on the surface of nickel-based single crystal superalloy by chemical vapor deposition oxidized at 900 °C and 1000 °C for 200 h respectively. The oxidation kinetic curve was drawn by static gain mass method, the oxidation rate were calculated as well as activation energy. The performance of the samples was characterized by X-Ray Diffraction, Scanning Electron Microscope and Energy Dispersive Spectroscopy, which evaluated the high-temperature oxidation resistance.

2 Experimental

Nickel-based single crystal superalloy was used as substrate material with nominal chemical compositions of Cr 6.5~7.5, Co 7.0~8.0, W 4.5~5.0, Al 6.0~6.5, Ta 6.0~7.0, Mo 1.0~1.5, Hf 0.1~0.5, Re 2.5~3.5 and Ni balance (wt.%). Al coating on the surface of alloy substrate was prepared by chemical vapor deposition equipment. Isothermal oxidation tests of the alloy substrate and the coated samples were performed at 900 °C and 1000 °C for 200 h, respectively. The weight changes were measured by an electronic balance (AB204-N) with a precision of 10^{-4} g. The oxidation kinetic curve of the samples were drawn by the method of increasing weight, and the oxidation rate constant and oxidation activation energy were obtained by linear fitting equation. The surface and cross-section morphology of the samples were characterized on a Scanning Electron Microscope (SEM, Sigma 500). The elements distribution of oxide product was determined by Energy Dispersive System (EDS, X-MAM^{N}). The composition of the oxide phases were identified by and X-Ray Diffraction (XRD, D8 focus).

3 Results and Discussion

3.1 Phase Composition of Oxidation Products

Figure 1a presents the XRD patterns of alloy substrate after oxidation. The main product of alloy substrate is Al_2O_3 after oxidation at 900 °C, which indicated that the alloy substrate has some high temperature resistance at this temperature. The oxidation degree of the alloy increases gradually at 1000 °C, and the mixed oxides such as NiO, Cr_2O_3 and Al_2O_3 are formed on the surface, where the mixed oxides can't play a good protective role.

Figure 1b exhibits the XRD patterns of Al coating after oxidation. It is found that Al_2O_3 and γ'-Ni_3Al phases are emerged on the surface of Al coating, and the diffraction peak intensity of Al_2O_3 and γ'-Ni_3Al phases promote gradually with the increase of temperature. The γ'-Ni_3Al phases is transformed from the β-NiAl under the Al_2O_3 film. On the one hand, a large amount of Al are consumed in the oxidation process, which lead to the Al under the oxide film gradually depleted; On the other hand, The formation of γ'-Ni_3Al phases is due to the reduction of Al content where the substrate elements diffuse outward and Al diffuse inward [8].

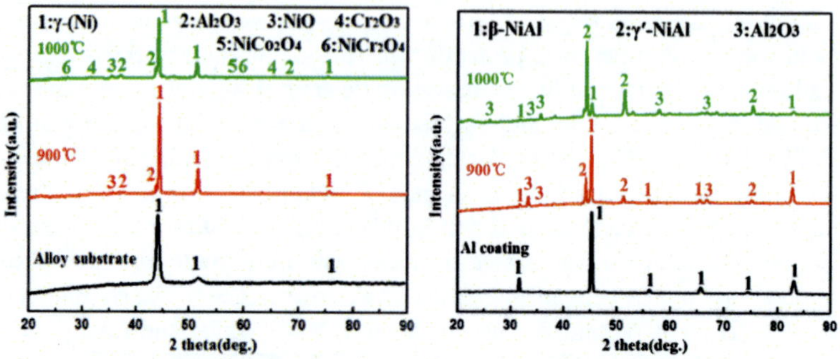

(a) alloy substrate (b) Al coating

Fig. 1. XRD patterns of alloy substrate and Al coating after oxidation

3.2 Oxidation Kinetic Curve

The isothermal oxidation kinetics curves of the alloy substrate and Al coating at 900 °C and 1000 °C for 200 h are shown in Fig. 2. It indicates that the kinetics curves of all the samples are similar basically, which are accord with the parabolic rule. The gain mass and the reaction rate are increased rapidly in the 50 h experimental time, which is due to the formation period of the oxide film. The samples surface in the initial stage of oxidation reaction is full in the oxygen, and the gain mass of oxidation is controlled by the interface reaction [9]. With the extension of experimental time, the gain mass and the reaction rate gradually are tended to be slowly, indicating that dense and protective oxide film has been formed on the surface of the samples, which prevented the further diffusion of oxygen from the outside to the substrate effectively. The increase of oxide gain mass is controlled by the diffusion rate of oxygen in the continuous oxide film, and the reaction rate is slow down relatively [10].

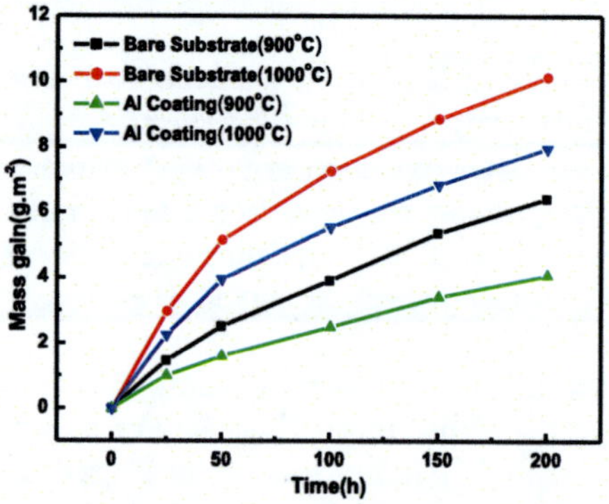

Fig. 2. Isothermal oxidation kinetics curves of alloy substrate and Al coating

3.3 Oxidation Kinetic Model

According to the oxidation kinetics curves of the samples, it can be seen that the gain mass of the oxidation products conform to the parabolic rule of Eq. (1).

$$(\Delta W)^2 = K_{P}t \tag{1}$$

where ΔW is the gain mass per unit area ($g\cdot m^{-2}$); K_P is the parabolic rate constant ($g^2\cdot m^{-4}\cdot h^{-1}$); t is the oxidation time (h).

The gain mass curves of the samples are fitted by Origin8.0 software showed in Table 1. The form of the fitting equation is $Y^2 = A + BX$, Y is the gain mass per unit area, B is the parabolic rate constant, X is the oxidation time. The fitting data results revealed that the oxidation rate constant of Al coating samples is lower than alloy substrate. For instance, the oxidation rate constant of alloy substrate (0.20711 $g^{-2}\cdot m^{-4}\cdot h^{-1}$) is 2.55 times of Al coating (0.08122 $g^{-2}\cdot m^{-4}\cdot h^{-1}$) at 900 °C.

Table 1. Fitting relation between gain mass and oxidation time of the samples.

Sample	Fitting equation
Substrate (900 °C)	$Y^2 = -2.68299 + 0.20711X$
Substrate (1000 °C)	$Y^2 = -0.44766 + 0.52324X$
Al coating (900 °C)	$Y^2 = -1.04033 + 0.08122X$
Al coating (1000 °C)	$Y^2 = -1.19604 + 0.32181X$

The mutual diffusion of material elements and oxygen are inhibited by the existence of the coatings, which reduce the high-temperature oxidation rate of the material, and inevitably lead to the change of oxidation activation energy of the material. According to the Arrhenius relation, the relationship as shown in Eq. (2) is satisfied by the oxidation rate constant K_P and oxidation activation energy Q.

$$K_P = K_0 \cdot \exp(-Q/RT) \tag{2}$$

where K_0 is the model constant; Q is the oxidation activation energy; R is the gas constant; T is the Kelvin temperature.

Respectively bring the K_P value in Table 1 into the equation to calculate the value of K_0 and Q. The calculation results showed the activation energy of alloy substrate was 112316 $kJ\cdot mol^{-1}$, while the Al coating was 170938 $kJ\cdot mol^{-1}$. The surface state of the material was changed by the coating, the activation energy of Al coating was 52% higher than alloy substrate.

3.4 Surface Morphology of Oxidation Products

The surface morphology of alloy substrate after oxidation at 900 °C and 1000 °C are shown in Fig. 3. It was found that intact morphology without microcracks and holes was presented by the oxide film at 900 °C. On the contrary, lots of microcracks and holes presented in the spalling areas are revealed by the surface morphology at 1000 °C. These defects in the oxide film may be caused by the uneven thermal and growth stress between the mixed oxides. Due to the existence of these defects, it would further increase the contact between substrate and air, and led to the higher mass gain. It is obvious that alloy substrate had poor oxidation resistance at 1000 °C.

The surface morphology of Al coating after oxidation at 900 °C and 1000 °C are shown in Fig. 4. It is exhibited that Al coating were uniformly covered with Al_2O_3 protective film. The oxide film was completed relatively, dense and continuous with a small amount of defects such as cavities and pores, which is shown excellent oxidation resistance.

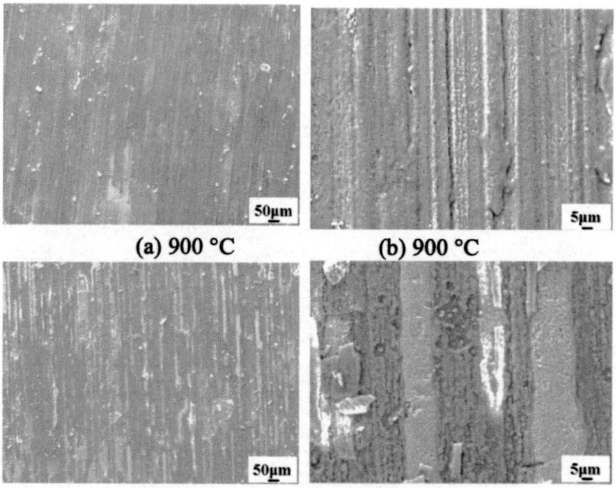

Fig. 3. The surface morphology of alloy substrate after oxidation

3.5 Cross-Section Morphology of Oxidation Products

The cross-section morphology of alloy substrate after oxidation at 900 °C and 1000 °C are shown in Fig. 5. Continuous black areas mainly composed of Al_2O_3 are distributed on the outer surface of alloy substrate at 900 °C, exhibited some oxidation resistance. The inner layer of the alloy substrate is also an oxide with continuous distribution of Al_2O_3, and a large number of black blocks of AlN are generated near the substrate side at 1000 °C too [11]. It is revealed that obvious internal oxidation and nitriding of the alloy substrate is occurred, indicated the reduction of oxidation resistance of the alloy substrate [12].

The cross-section morphology of Al coating after oxidation at 900 °C and 1000 °C are shown in Fig. 6. It is observed that Al coatings could be divided into three layers:

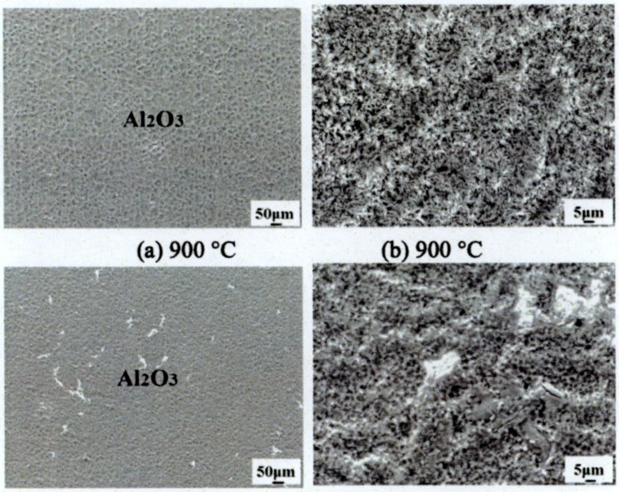

(a) 900 °C (b) 900 °C

(c) 1000 °C (d) 1000 °C

Fig. 4. The surface morphology of Al coating after oxidation

outer layer, middle layer and diffusion layer. The outer layer of the coating was Al_2O_3 protective layer, the middle layer were mainly β-NiAl phase rich in Al and a few γ′-Ni_3Al. There were dense carbide precipitates in the diffusion layer after a long time high-temperature oxidation. At two temperatures, the surface of the coating formed a dense Al_2O_3 protective layer. Compared with 900 °C, the product at 1000 °C contained a large amount of γ′-Ni_3Al phases and little carbides.

After oxidation, Al coating rapidly forms a continuous and dense Al_2O_3 protective layer, which effectively reduce the external diffusion of Al and the internal diffusion of O, and slow down the transformation rate of β-NiAl to γ′-Ni_3Al, showing excellent oxidation resistance. Al_2O_3 protective film is formed on the surface of the coatings after oxidation, and then the Al element in the β-NiAl phase below the Al_2O_3 layer diffuses to the surface and substrate, leading to the transformation from β-NiAl to γ′-Ni_3Al in the coatings. The γ′-Ni_3Al phases which became the diffusion channel of O oxidized with other elements is growing, extending and penetrating through the coatings. With the consumption of Al element, the β-NiAl layer became thinner, while the Al_2O_3 layer and diffusion layer became thicker. At the same time, the elements in the alloy substrate diffused outward, which is easy to form mixed oxides with poor protection effect, resulting in premature failure of the coating [13, 14].

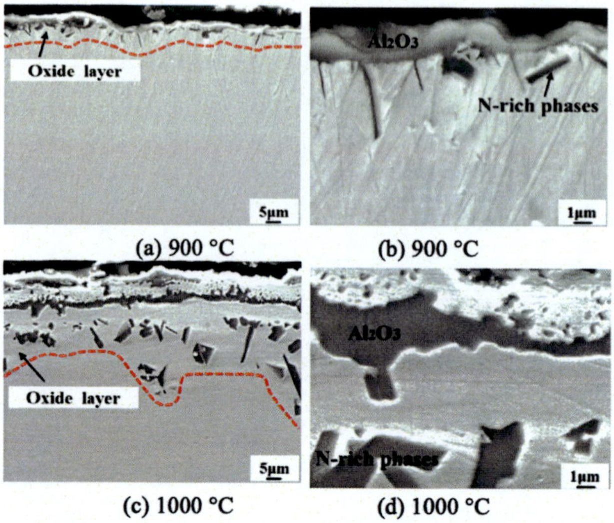

Fig. 5. The cross-section morphology of alloy substrate after oxidation

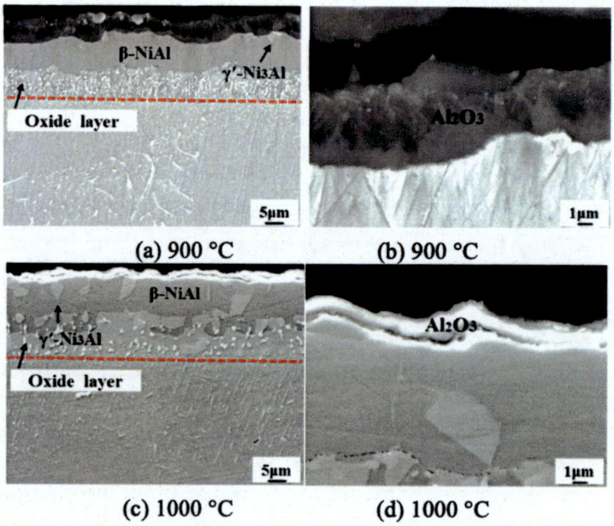

Fig. 6. The cross-section morphology of Al coating after oxidation

4 Conclusions

The oxidation kinetics curves of alloy substrate and Al coating were accord with the parabolic law after oxidation at 900 °C and 1000 °C for 200 h. The mass gain of Al coating was lower than alloy substrate, the activation energy of Al coating was higher than alloy substrate.

Al_2O_3 films were enriched in the oxide layer of the alloy sample, indicating that the alloy had some high-temperature oxidation resistance. However, Al_2O_3 and AlN were formed in the alloy substrate, resulting in the partial reduction of oxidation resistance of the alloy substrate.

The Al_2O_3 protective film of Al coating took on the less surface spallation, more pure and continue, exhibited more excellent high-temperature oxidation resistance. The coatings present a prominent improvement in the high-temperature oxidation resistance compared with alloy substrate.

References

1. Cheng, Y., Zhang, H., Song, L.W., et al.: Effect of Re element on oxidation resistance of Ni3Al-Mo based alloys at 1150 °C. Trans. Nonferrous Met. Soc. China **22**(3), 510–515 (2012)
2. Pyczak, F., Devrient, B., Neuner, F.C., et al.: The influence of different alloying elements on the development of the γ/γ' microstructure of nickel-base superalloys during high-temperature annealing and deformation. Acta Mater. **53**(14), 3879–3891 (2005)
3. Joh, Y., Utada, S., Osawa, M., et al.: Effect of sulfur on creep strength of Ni-base single crystal superalloy, TMS-1700. Mater. Trans. **57**(8), 1305–1308 (2016)
4. Sakaguchi, M., Kouyama, D., Yokoguchi, T., et al.: Effect of crystallographic orientations on stage I fatigue crack propagation in a Ni-base single crystal superalloy. J. Soc. Mater. Sci. Jap. **66**(10), 731–737 (2017)
5. Geoffrey, M., Mick. W.: Turbine blade with variable oxidation resistance coating. EP: EP2354454 A1 (2011)
6. Min, T.K., Si, W.L.: Application of in situ oxidation-resistant coating technology to a home-made 100 kW class gas turbine and its performance analysis. Appl. Therm. Eng. **40**, 304–310 (2012)
7. Movchan, B.A., Malashenko, I.S., Yakovchuk, K.Y., et al.: Two- and three-layer coatings produced by deposition in vacuum for gas turbine blade protection. Surf. Coat. Technol. **67**(1–2), 55–63 (1994)
8. Liu, H., Li, S., Jiang, C.Y., et al.: Oxidation performance and interdiffusion behavior of a Pt-modified aluminide coating with pre-deposition of Ni. Acta Metall. Sin. **12**(32), 1490–1500 (2019)
9. Li, Z.Y., Chen, L.J., Zhang, H.Y., et al.: High-temperature oxidation properties and microstructural evolution of nanostructure Fe-Cr-Al ODS alloys. Materials **14**(3), 526 (2021)
10. Li, Y.Q., Xie, F.Q., Wu, X.Q., et al.: Microstructure and high temperature oxidation resistance of Si-Y co-deposition coatings prepared on TiAl alloy by pack cementation process. Trans. Nonferrous Met. Soc. China **25**(3), 803–810 (2015)
11. Huang, L., Sun, X.F., Guane, H.R., et al.: Oxidation behavior of the single-crystal Ni-base superalloy DD32 in air at 900, 1000, and 1100 °C. Oxid. Met. **65**(3–4), 391–408 (2006)
12. Meng, J.S., Chen, M.X.: Effect of Co on oxidation and hot corrosion behavior of two nickel-based superalloys under Na2SO4–NaCl at 900 °C. Trans. Nonferrous Met. Soc. China **31**(8), 2402–2414 (2021)
13. Jiang, C.Y., Qian, L.Y., Feng, M., et al.: Benefits of Zr addition to oxidation resistance of a single-phase (Ni, Pt) Al coating at 1373K. J. Mater. Sci. Technol. **35**(7), 1334–1344 (2019)
14. Manap, A., Nakano, A., Ogawa, K.: The protectiveness of thermally grown oxides on cold sprayed CoNiCrAlY bond coat in thermal barrier coating. J. Therm. Spray Technol. **21**(3–4), 586–596 (2012)

Vibration Characteristics of Landing Gear Walk Caused by Aircraft Anti-skid Brake

Wei Li, Yan Xie, Junjie Jia(✉), Jing Su, Kai Chang, and Xin Li

Air Industry First Aircraft Design Institute, Xi'an 710089, China
jjj7211@126.com

Abstract. Aircraft anti-skid brake which is usually installed on the wheels of the main landing gear of aircraft, is a key subsystem to ensure the safety of aircraft takeoff and landing. At the state of aircraft taxiing, the anti-skid braking system will cause the landing gear vibrating forward and backward along the aircraft speed direction, such low frequency vibration is also called walking. This paper established the model of aircraft braking system with the help of MATLAB/Simulink software, the data interaction between the model and the multi-body dynamic model of landing gear is conducted through the joint simulation interface, the walking characteristics is then studied. In order to study the vibration influence of the anti-skid control system, a reference rate velocity difference brake control law with bias pressure bias modulated regulation and a slip rate brake control law based on PI were designed respectively. The influence of key control law parameters on walking vibration is then analyzed on dry and wet runways. Simulation results show that for the two different control laws, the influence of proportional stage parameters on walking and braking efficiency is greater than that of integral stage parameters. In addition, the proportional level parameters vary linearly with braking efficiency and buffeting amplitude within a certain range.

Keywords: Aircraft anti-skid brake · Walking · Landing gear · Vibration · Brake control law

1 Introduction

When the aircraft is taxiing on the ground, the friction (bonding force) between the tire and the runway changes, it will cause the landing gear to walk [1]. The traditional anti-skid brake control law is easy to cause the wheels to slip continuously for many times on the wet runway and low-speed brake section, resulting in continuous changes in braking torque, resulting in continuous changes in ground bonding force, causing landing gear steps [2]. In addition, random changes in the friction characteristics of the brakes could also be the cause of the landing gear walking. It could be seen that the mechanical characteristics of tires, external excitation of the runway, brake device, landing gear structure and anti-skid brake system are all related to the longitudinal direction of the landing gear [3, 4]. Walking is a typical low-frequency vibration, and the vibration frequency generally does not exceed 10 Hz. Walking will not only affect the fatigue life

Chinese Society of Aeronautics and Astronautics (Ed.): CASTC 2023, LNME, pp. 66–77, 2024.
https://doi.org/10.1007/978-981-99-8861-7_8

of the landing gear, damage the efficiency of braking, cause discomfort to the on-board, but also cause safety accidents. Therefore, it is necessary to study the characteristics of landing gear step vibration caused by aircraft anti-skid brakes.

For a long time, the modeling and analysis of landing gear steps has been a research hotspot in aircraft ground dynamics.

Balasubramanian et al. develop a parameter-focused landing gear model and analyze the influence of landing gear parameters on walking by applying tire-road friction directly to the wheels [5].

R Lernbeiss et al. simplify the landing gear struts into three-segment connected Euler-Bernoulli beams, and analyze the bending deformation of the struts caused by static tire-road friction [6].

Yin et al. establish a step dynamics model containing six complex equations based on the Lagrangian equation, in which the bending deformation of the landing gear is reduced to a torsion spring damping system [7].

In addition, due to the large number of landing gear and brake system components and the complexity of the force, the method of modeling the landing gear and anti-skid brake system with the help of commercial software is also widely used. Stefania Gualdi develops a multibody dynamic model of the three-point aircraft with an anti-skid brake controller, which enables the analysis of the coupling between the anti-skid brake system and the landing gear system [8]. Prashant et al. develop a model of an aircraft with landing gear and an aircraft skid brake system in multibody dynamics software SIMPACK [9, 10]. Huang establishes a rigid-flex hybrid model considering the flexible deformation of the landing gear struts in the multibody dynamics software ADAMS [11].

Yin et al. carried out a study on the vibration characteristics and parameters of the landing gear step based on the rigid-flexible coupling dynamic analysis model of a half-shaft pillar main landing gear. By modifying the frequency and amplitude of the braking torque, the effect of braking torque on the landing gear step is studied [1]. Zhang et al. analyze the influence of the structural parameters of the landing gear pillar and tire characteristic parameters on the landing gear walking by establishing the mechanical analysis model of each component [12]. Wang et al. propose the brake vibration and landing gear walking were analyzed from the friction characteristics and assembly clearance of the brake device, and some vibration suppression measures [13].

In this paper, the landing gear rigid-flex hybrid model is first established in LMS Virtual.Lab Motion, and then a complete aircraft braking system model is established, and through the establishment of a co-simulation interface, the brake system model established by MATLAB/Simulink and the landing gear rigid-flex hybrid model are exchanged to establish the braking system and landing gear co-simulation model. In addition, this paper also establishes the anti-skid brake control law based on two different architectures: deceleration rate and slip rate, and analyzes the parameter influence of landing gear shake on dry and wet runways. Finally, the ground inertial platform test is carried out using the reference rate-speed differential brake control law adjusted by biased PBM (Pressure-Bias-Modulated) as an example to verify the simulation conclusion.

2 Landing Gear Rigid-Flex Hybrid Model

When studying the coupling vibration of the anti-skid brake and the landing gear structure, the flexibility of the landing gear pillar must be considered. Considering that the parameters such as true longitudinal stiffness and damping are difficult to determine when modeling the landing gear, combined with engineering practice, this paper establishes a rigid-flex hybrid model of the landing gear in the multi-body dynamics software LMS Virtual.Lab.

Taking a single-pillar 6-wheel trolley landing gear as an example, the 3D CAD model is shown in Fig. 1 below. Then we Import the simplified 3D model of the landing gear into LMS Virtual.Lab Motion, and add constraint relations to create a purely rigid landing gear dynamic model.

Due to the large deformation of the landing gear pillars, it is considered to flex the landing gear pillars. The finite element analysis method is used to flex the landing gear pillars, first simplify the three-dimensional model of the landing gear, remove the actuator rod, support rod and other components, and mesh the remaining components. The model is then divided into simple entities that can be mapped and complex entities that cannot be mapped, the mappable entities generate hexahedral mesh elements through 2D mesh mapping, and the non-mapable entities are generated by tetrahedral meshing method, and different types of mesh elements are connected by common faces. The modal analysis of the landing gear structure uses ABAQUS as the post-processing software, and uses the binding contact to realize the constraint relationship between the components, with the surface with thicker mesh and greater stiffness as the main surface, and the surface with finer mesh and less stiffness as the slave surface to establish the contact relationship of multiple connection parts. In order to facilitate the replacement of the flexible body, the multi-point constraint (MPC) is established with the RBE3 flexible unit at the equidistant point on the axis of the cylindrical structure of the inner and outer cylinders of the buffer, and the MPC is established with the RBE2 rigid unit at the fixed hinge point and the rotating sub-center. We replace the main buffer requiring flexible treatment with a flexible body obtained by finite element analysis and add constraints between rigid and flexible bodies.

Fig. 1. 3D model of single-strut, 6-wheeled trolley landing gear

The elastic tire model is imported in Virtual.Lab, which is based on the second-order tension theory, which can analyze the tire force and calculate its rolling characteristics, and finally establish a complete landing gear rigid-flex hybrid model, as shown in Fig. 2 below.

Fig. 2. Rigid-flex hybrid model of single-strut 6-wheeled trolley landing gear

Due to the large deformation of the main buffer pillar after being flexed, the landing gear force during the sliding brake process could not be well simulated by simply adding the body load force, so the other side of the main landing gear model is established by the same method, and the front landing gear module, runway surface module and fuselage model are added to establish a full model based on the landing gear rigid-flex hybrid model, as shown in Fig. 3.

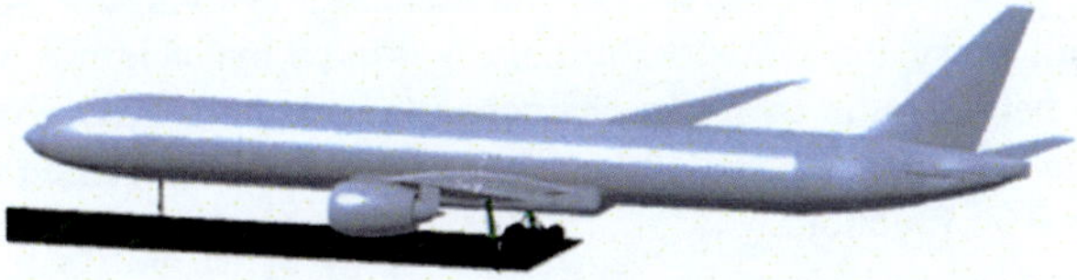

Fig. 3. Full machine dynamics model

3 Model of Aircraft Brake System

Since the control module in Virtual.Lab Motion is relatively simple, it is not conducive to the analysis of brake system control parameters, so we build a brake system model in MATLAB/Simulink.

3.1 Simulation Modeling of Engine Dynamics

According to the braking principle and the wheel dynamics equation, the wheel dynamics equation can be obtained as follows:

$$J\dot{\omega} = F_f \cdot r - T_b = \mu \cdot F_N \cdot r - P_b k_b \tag{1}$$

In the above equation: J is Moment of inertia of the wheels, r is effective radius of the wheel, μ is coefficient of friction between tires and runways, ω is angular velocity of rotation of the wheels. FN is vertical load on the wheels, kb is brake pressure and brake torque conversion coefficient, Pb is brake pressure.

3.2 Model of Brake Device

The brake device is the key component that converts the brake pressure into brake torque, and the brake device model can be represented as:

$$M_b = \begin{cases} 0 & P_b \le P_0 \\ k_2(P_b - P_0) & P_0 < P_b \le \frac{T_1}{k_2} + P_0 \\ \mathrm{T}_1 & \frac{T_1}{k_2} + P_0 < P_b \le P_{bl} \\ \mathrm{T}_2 & P_{bl} < P_b \le P_0 + \frac{T_1}{k_1} \\ k_1(P_b - P_0) & P_0 + \frac{T_1}{k_1} < P_b \end{cases} \quad (2)$$

In the above equation: k1 = Tsm/(Pm-P0), k2 = Tsm/(Px-P0), Tb is the output torque of this cycle, T1 is output torque on the previous cycle, Tsm is maximum braking torque, Pbl is brake pressure in the previous cycle, P0 is minimum brake pressure, Pb is current brake pressure, Px is Maximum hysteresis brake pressure.

3.3 Hydraulic Actuation Section Modeling

During aircraft braking, the hydraulic system realizes pressure control during braking by adjusting the pressure servo valve. For the modeling of hydraulic servo valves, too many factors will make the model extremely complex, but it is not easy to find out the relationship between the dynamic characteristics of the servo valve and the main parameters of the valve. The model of the servo valve can be simplified to the form of a second-order transfer function:

$$G(S) = \frac{{}^{1}/_{\omega_h^2}}{\left(\frac{S^2}{\omega_h^2} + \frac{2\xi_h}{\omega_h}S + 1\right)} \quad (3)$$

In the above equation: wh is the natural frequency of the servo valve, ε is the relative damping coefficient of the servo valve.

According to the experience of the inertia table experiment of the wheel brake system, the hydraulic pipeline of the brake system can be simplified to the first-order inertia link, as shown in the following formula:

$$G(S) = \frac{K}{TS + 1} \quad (4)$$

And K = 1, T = 0.01.

3.4 Model of Friction Between the Wheels and the Runway

Burckhardt model is a semi-empirical tire model based on a large number of tire and road experiments, and the model is expressed as:

$$\mu(\lambda) = C_1(1 - e^{-C_2\lambda}) - C_3\lambda \tag{5}$$

In this equation: C1, C2, C3 are fitting coefficients, λ is wheel slip rate. The relationship between slip rates and bonding coefficients corresponding to several typical pavements is shown in Fig. 4.

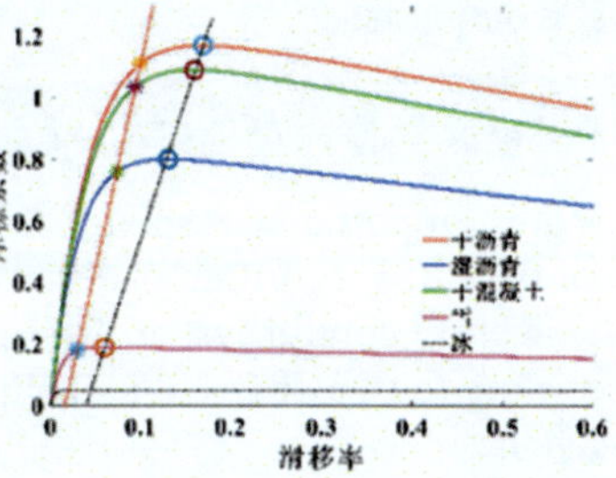

Fig. 4. Burckhardt model friction characteristic curves in different runway states

The slip rate is the ratio of the difference between the speed of the aircraft and the linear speed of the brake wheels to the speed of the aircraft. Among the various factors that affect the bonding coefficient of the brake wheel to the ground, the slip rate is the most influential factor. The slip rate is calculated as:

$$\lambda = \frac{V_P - \omega r}{V_P} = 1 - \frac{\omega r}{V_P} \tag{6}$$

4 Anti-skid Brake Control Law Design

4.1 Reference Rate-Speed Differential Brake Control Law with Bias PBM Adjustment

PBM control law replaces aircraft speed by setting a reference speed, which is updated in real time at each control cycle according to a preset fixed deceleration rate. The reference speed is calculated as:

$$V_r(k) = \begin{cases} V_w(k) & V_r(k-1) - a_r T_s \leq V_w(k) \\ V_r(k-1) - a_r \cdot T_s & V_r(k-1) - a_r T_s \geq V_w(k) \end{cases} \tag{7}$$

In the above equation: Vr (k), Vr (k-1) are the reference speed of the current moment, the previous moment, Vw (k) is current wheel line speed, ar is Preset reference depreciation rate, Ts is the control period of the anti-slip control law.

The PBM algorithm is not a typical closed-loop tracking control algorithm, but a rule-based control algorithm. PBM determines the control output based on the difference between the line speed of the reference speed wheel and the set threshold, when the speed difference is greater than the threshold of the proportional stage and the micro-grading, similar to the typical PID control output, the proportional anti-slip output is:

$$V_p(k) = k_p \times (e(k) - V_{pt}) \tag{8}$$

In this equation: kp is Scale factor, Vpt is proportional response threshold, e(k) = Vr(k)-Vw(k) is the difference between the reference speed and the wheel line speed at the current moment.

The micro-graded anti-slip output is:

$$V_p(k) = k_p \times (e(k) - V_{pt}) \tag{9}$$

kd is micro-grading factor, Vdt is proportional response threshold value.

The pressure deflection of the PBM algorithm plays an important role in regulating the performance of the anti-skid brake control system. The output of the pressure deflection stage is determined according to two different anti-slip thresholds, and the output of the pressure deflection stage is:

$$V_i(k) = \begin{cases} V_i(k-1) + k_{i1}(e(k) + e(k-1) - 2V_{it1}) & 0 \leq e(k) - V_{it1} \leq V_{it2} \\ V_i(k-1) + k_{i1} \times T & e(k) - V_{it1} \geq V_{it2} \\ V_{const} - k_{i3}(n \times T_s)^2 & e(k) - V_{it1} \leq 0 \end{cases} \tag{10}$$

In the above equations, Vi(k) is pressure deflection stage control output, Vit1(k) is shallow non-slip threshold, Vit2(k) is deep non-slip threshold value, ki1 is constant rise factor, ki2 is boost rate, ki3 is discharge acceleration, Vconst is the initial value of the second descent.

Therefore, the final pressure output Pb is the difference between the brake command Pcmd and the anti-skid output:

$$P_b = P_{cmd} - V_p - V_d - V_i \tag{11}$$

4.2 PI-Based Slip Rate Brake Control Law

The slip rate anti-skid brake control law takes the wheel slip rate as the control object. Among the various factors that affect the bonding coefficient of brake wheels to the ground, slip rate is the key factor. Anti-skid braking system based on slip rate generally selects a fixed optimal slip rate as the control target, the selection of the optimal slip rate λopt has an important impact on the braking efficiency, if the optimal slip rate selection is too small, the braking efficiency will be low, which could not make full use of the ground bonding force; If the optimal slip rate is chosen too large, it will lead to frequent slippage. In this paper, 0.2 is selected as the optimal slip rate value, and a slip rate brake control law based on PI is designed.

The slip rate tracking error can be obtained from the estimated or measured slip rate:

$$e = \lambda_{opt} - \lambda \tag{12}$$

PI control is carried out according to the tracking error, and the brake pressure output can be obtained:

$$P_b = k_{p1}e(t) + k_{i1}\int e(t) \tag{13}$$

5 Co-simulation Model of Brake System and Landing Gear

For simulation models established in two different software, MATLAB/Simulink and LMS Virtual.Lab Motion, a co-simulation interface is established to achieve co-simulation. By establishing a control interface for data exchange with MATLAB/Simulink in LMS Virtual.Lab Motion, the virtual prototype model in Virtual.Lab Motion can be integrated into MATLAB/Simulink as a kinetic module to achieve the purpose of connecting with the brake system. MATLAB/Simulink is used as the platform for simulation analysis, and MATLAB will continuously write the results to the results file of Virtual.Lab Motion during the simulation process.

When the two different software of MATLAB/Simulink and LMS Virtual.Lab Motion co-simulate, the data exchange between the models in the two software platforms is completed through the established control interface, and the parameters measured from the LMS Virtual.Lab Motion multibody dynamics model and output to MATLAB/Simulink are the motion parameters during the ground running of the aircraft, and from MATLAB/Simulink The parameters output to the landing gear rigid-flex hybrid model in the subsystem are the forces and torques experienced by the aircraft, including aerodynamic and aerodynamic torques, wheel braking torques, etc.

Based on the above-mentioned established landing gear rigid-flex hybrid model, in Motion, add input control nodes to the rotating pair acting on the braking torque, add output control nodes to the aircraft speed and wheel speed, and set the simulation settlement mode to cosim, you can generate a plant.m file, which can be run in MATLAB to generate a simulation module with braking torque as input and aircraft speed and wheel speed as output, and connect it to the braking system model established above and write it into the control part, you can build a co-simulation model of brakes and landing gear. The internal structure of the plant module in the co-simulation model is shown in Fig. 5.

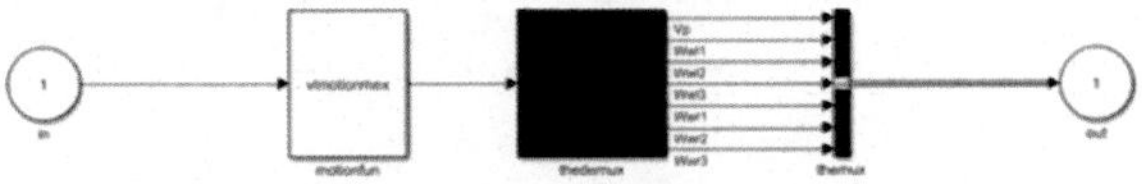

Fig. 5. Co-simulation module

6 Simulation Analysis

6.1 PBM Control Law Parameter Influence Analysis

In order to study the influence of PBM control law parameters on landing gear jitter, the established braking system and landing gear co-simulation model are used for simulation tests, and some key simulation parameter settings in the co-simulation test are shown in Table 1 below.

The wet runway with friction coefficient of 0.3 was selected as the simulation condition, the anti-skid scale factor KP was adjusted, and the landing gear jitter under different anti-skid ratio parameters was compared. Take the values of kp of 0.5, 1.2 and 1.8 respectively, and the landing gear jitter curve for a period of time after the brake is in action is shown in Fig. 6 below.

Table 1. PBM controls the setting of key parameters

Parameter	Value
Equivalent radius of the wheels	0.61 m
Proportional thresholds Vpt	8
Differential thresholds Vdt	7
Points rise threshold Vit1	6
Points rise threshold Vit2	9
Scale level factor Kp	1
Integration level coefficient Ki1	90
Integration level coefficient Ki2	25
Control period Tsc	0.01 s
Brake command PilotCmd	10 Mpa

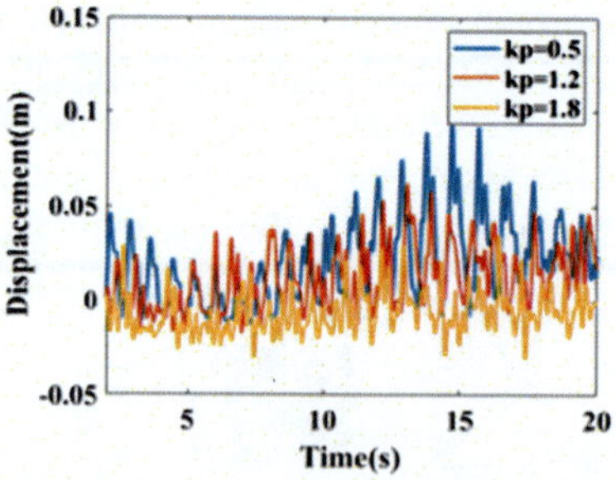

Fig. 6. PBM controls the landing gear jitter curve under different kp

The longitudinal and vertical jitter amplitudes of the landing gear and the braking distance under different scale factors kp are shown in Table 2 below.

Table 2. Simulation results with different kp

kp	0.5	1.2	1.5	1.8
Stride value (m)	0.10	0.07	0.06	0.06
Frame pitch angle amplitude (°)	0.57	0.49	0.49	0.47
Braking distance (m)	1095	996	953	959

From the data obtained in the above chart, it is not difficult to find that when the kp is small (less than 1.5), as the kp increases, the sensitivity of the anti-skid system increases, the landing gear jitter amplitude becomes smaller and the braking distance is shortened; but when the KP is large (less than 1.5), the anti-skid system is too sensitive will not only aggravate the landing gear jitter, but also lead to a longer braking distance.

6.2 PI-Based Influencing Analysis of Slip Rate Control Law Parameters

In order to study the anti-skid effect of the anti-skid brake control law based on slip rate and compare it with the PBM control law, some key simulation parameter settings in the co-simulation test are shown in Table 3 below.

Table 3. Key parameter setting of the control law based on slip rate

Parameter	Value
Optimal slip rate λopt	0.2
Equivalent radius of the wheels	0.61
Scale level factor Kp1	30
Integration level coefficient Ki1	300
Control period Tsc	0.01s
Brake command PilotCmd	10 Mpa
Brake speed Vint	70 m/s

Similarly, the wet runway with friction coefficient of 0.3 was selected as the simulation condition, and the anti-slip scale coefficient kp and integral coefficient ki were adjusted respectively to compare the landing gear jitter under different anti-skid parameters. Start by keeping the value of ki at 100, kp = 20, 30, and 40 were taken for simulation, and the landing gear jitter curves under different kp were obtained as shown in Fig. 7 below.

Similarly, keeping the value of kp at 30, take ki = 100, 200, and 300 respectively for simulation, and obtain the landing gear jitter curve under different ki action as shown in Fig. 8 below.

The braking distance under the above operating conditions is shown in Table 4.

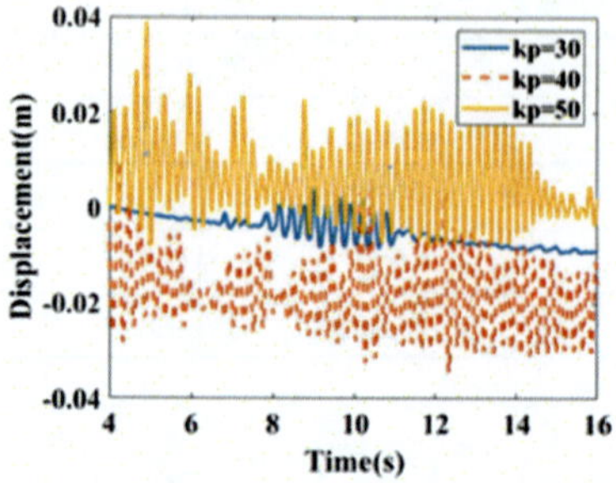

Fig. 7. PI-based slip rate control law landing gear jitter curves under different kp

From the above simulation results, it can be seen that the effect of changing kp on jitter is more obvious than changing ki, and the value of kp and brake efficiency and jitter amplitude show a linear change trend (within a certain range), the smaller the kp, the smaller the jitter amplitude, and the higher the braking efficiency. The effect of changing ki on jitter is not linear, the smaller the ki, the higher the braking efficiency, but the effect of Ki on braking efficiency is smaller.

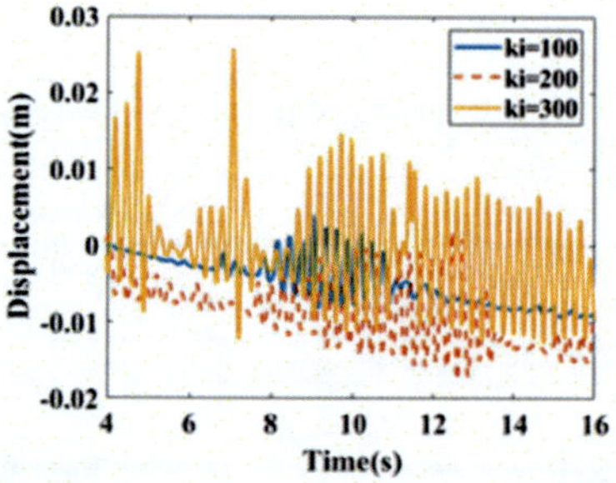

Fig. 8. PI-based slip rate control law different ki landing gear jitter curves

Table 4. Simulation results corresponding to different parameters of PI-based slip rate control law

kp	30	40	50	30	30
ki	100	100	100	200	300
Longitudinal jitter amplitude value (m)	0.0129	0.0442	0.0474	0.0198	0.0385
Frame pitch angle (°)	0.1363	0.3949	0.4829	0.2312	0.3316
Braking distance (m)	1014.70	1051.08	1067.61	1028.32	1037.55

7 Conclusion

In this paper, the landing gear rigid-flex hybrid model is first established in LMS Virtual.Lab Motion, and then a complete aircraft braking system model is established in MATLAB/Simulink, and the braking system and landing gear co-simulation model are established by establishing a co-simulation interface. Two different architectures of anti-skid brake control law with bias PBM adjustment are designed, and the key control parameters of the two control laws are analyzed on the wet runway by using the established co-simulation model of the brake system and landing gear system. Through simulation analysis, it can be found that the effect of changing kp on jitter is more obvious than changing ki for the two control laws, and the value of kp and brake efficiency and jitter amplitude show a linear change trend (within a certain range), the smaller the kp, the smaller the jitter amplitude, and the higher the braking efficiency.

References

1. Yin, Q., Nie, H., Zhang, M., et al.: Study on low-frequency brake-induced vibration characteristics of a half-shaft pillar landing gear. Chin. J. Vibr. Eng. **29**(06), 954–962 (2016). (in Chinese)
2. Denti, E., Fanteria, D.: Analysis and control of the flexible dynamics of landing gear in the presence of antiskid control systems. The Design, Qualification and Maintenance of Vibration-Free Landing Gear **800** (1996)
3. Moseley, D.D., Carter, T.J.: Performance testing on an electrically actuated aircraft braking system. SAE Trans. **10**, 36–67 (1988)
4. Tanner, J.A.: Review of NASA antiskid braking research. SAE Technical Paper (1982)
5. Balasubramanian, K., Farhang, K.: Gear walk instability studies using a vibration model of a reduced scale landing gear system. In: 27th Biennial Mechanisms and Robotics Conference of International Design Engineering Technical Conferences and Computers and Information in Engineering Conference, vol. 5, pp. 243–251 (2002)
6. Lernbeiss, R., Pl¨ochl, M.: Simulation model of an aircraft landing gear considering elastic properties of the shock absorber. Proc. Inst. Mech. Eng., Part K J. Multi-body Dyn. **221**(1), 77–86 (2007)
7. Yin, Q., Jiang, J.Z., Neild, S.A., Nie, H.: Investigation of gear walk suppression while maintaining braking performance in a main landing gear. Aerosp. Sci. Technol. **91**, 122–135 (2019)
8. Gualdi, S., Morandini, M., Ghiringhelli, G.L.: Anti-skid induced aircraft landing gear instability. Aerosp. Sci. Technol. **12**(8), 627–637 (2008)
9. Khapane, P.D.: Simulation of asymmetric landing and typical ground maneuvers for large transport aircraft. Aerosp. Sci. Technol. **7**(8), 611–619 (2003)
10. Khapane, P.D.: Gear walk instability studies using flexible multibody dynamics simulation methods in simpack. Aerosp. Sci. Technol. **10**(1), 19–25 (2006)
11. Huang, M.: Research on loads applied on powered wheels and struts of civil aircraft at touchdown. Proc. Inst. Mech. Eng. C J. Mech. Eng. Sci. **236**(2), 840–851 (2022)
12. Zhang, L., Zhu, P.: Study on the dynamic characteristics of longitudinal jitter of main landing gear. Chin. J. Aeronaut. **03**, 292–296 (1996). (in Chinese)
13. Wang, H., Tian, G., et al.: Simulation analysis of brake vibration of aviation aircraft. Aviat. Precis. Manuf. Technol. **1**, 49–53 (2014). (in Chinese)

Solar Radiation Calculation and Analysis Software for Solar-Powered Aircraft Design Based on Visual Basic and OpenGL

Yueli Li[1,2](✉), Yining Gao[2], and Qi Wang[2]

[1] School of Aeronautics, Northwestern Polytechnical University, Xi'an 710072, China
rockandtcl@163.com
[2] The First Aircraft Institute, AVIC, Xi'an 710089, China

Abstract. This paper develops a design software for the solar radiation calculation and analysis problem in solar-powered aircraft design process, which can carry out various types of aircraft modeling, laying solar battery panels, solar radiation characteristics in any region, any season, any time and any attitude, single-point verification, radiation of all time flight, etc.

Keywords: Solar Radiation · Solar-powered Aircraft · Software

1 Introduction

Net-Zero Aviation has become a major issue in the 21st century. The aircraft based on solar-powered as the main source of energy is one of the most popular configurations [1, 2], and accurate calculation of solar radiation characteristics in any region, any season, any time and any attitude of the vehicle has become the primary task. At present, the typical calculation process is shown as below [3, 4]. It has the following problems: 1) the calculation requires large commercial software, which is cumbersome to operate; 2) the calculation results showed by post-processing, and the real-time is poor; 3) the calculation program needs more control commands input before calculation that could induce error (Fig. 1).

To avoid these problems, some researchers simplify 3D aircraft surface to flat [5, 6] or curved surface [7, 8], which reduces calculation accuracy.

Based on the direct solar radiation model and its atmospheric attenuation correction method in the literature [9], the author uses the OpenGL and Visual basic to develop comprehensive solar radiation analysis software with geometric modeling, solar panel laying, solar radiation characteristics calculation, single-point radiation calculation, and all-day radiation calculation etc.

Chinese Society of Aeronautics and Astronautics (Ed.): CASTC 2023, LNME, pp. 78–85, 2024.
https://doi.org/10.1007/978-981-99-8861-7_9

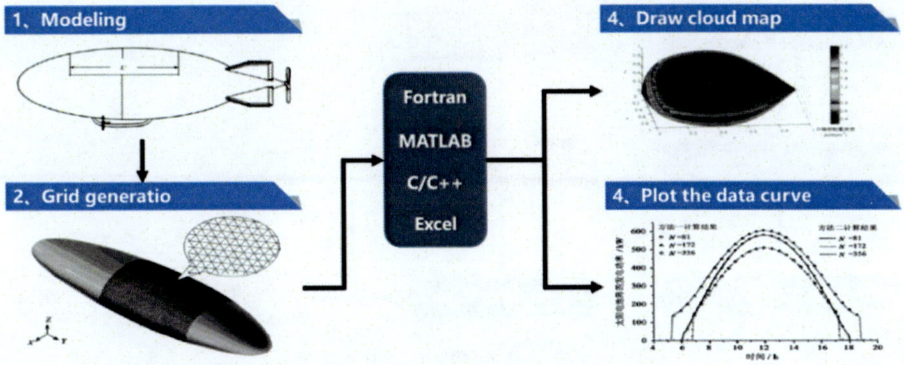

Fig. 1. Traditional calculation processes.

2 Methodology

2.1 Framework

The 3D engine and solar radiation calculation engine is developed with the Visual Basic 6.0 IDE (Integrated Development Environment) and the OpenGL static link library It realizes the construction of aircraft, the laying of solar battery panels, the solution of solar radiation characteristics, the verification of individual calculation, and the analysis of energy during all-day flight (Figs. 2 and 3).

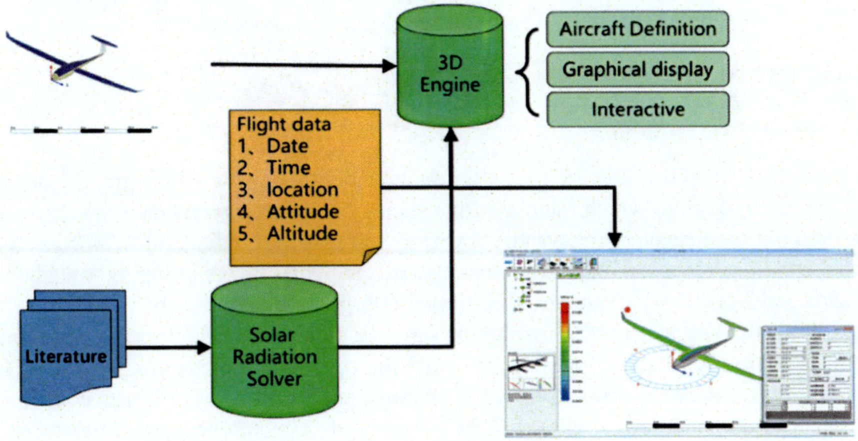

Fig. 2. Frameworks.

2.2 Calculation Process

The process of solar radiation calculation and analysis of the software is as follows:

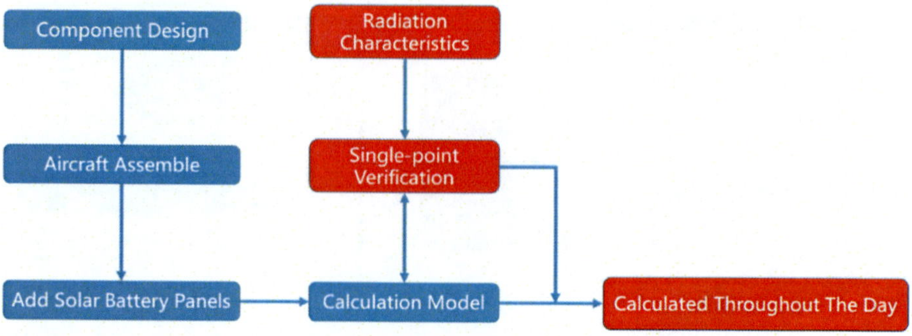

Fig. 3. Calculation process.

a) Aircraft components design;
b) assemble the aircraft according to the relative position of the components,;
c) Laying solar battery panels at the design site to form a computational model;
d) Test the local solar radiation characteristics at that time and whether the total amount of radiation meets the requirements, using the estimated flight position as input;
e) In the case that the total amount of solar radiation in the local area at that time meets the requirements, carry out single-point calculation of aircraft, set different attitudes and orientations, check the panel laying and solar radiation reception, if the panel laying is inappropriate, return to step c) to redesign panel laying, if the flight geographical location is inappropriate, return to step d) to test different geographical location and dates. Then back to the e) step to calculation again;
f) When the more iterative calculation of step e) meets the energy requirements of key flight nodes, begin the all-day computing calculation and analysis.

3 Software Introduction

3.1 General Introduction

Based on the OpenGL standard, the author independently develops a 3D graphic display and operation engine, which has functions such as navigator geometry drawing, scale bar drawing, coordinate axis system drawing, sun position indication drawing, cloud map drawing, orientation indication drawing, etc., supports image cropping, translation, rotation, and pixel copy output. The human-computer interaction has the following features: 1, the aircraft component supports mouse pickup and feedback. 2. The mouse controls zooming, rotating, panning, etc., and the operation logic is similar to CATIA, which is convenient for users to use. 3. The introduction of the scale bar is convenient for users to understand the real size of the model. 4. The sun position information is represented by a red ball, which shows the direction indication, includes east, south, west and north, which is convenient for users to understand the sun position information in the calculation state. 5. Use the date control container to set the flight date, which is convenient and fast. 6. Provides real-time display feedback corresponding to the aircraft azimuth, pitch and yaw settings, and combines with the sun position display to facilitate users to reasonably choose the flight orientation. The software runtime interface is shown as below (Fig. 4):

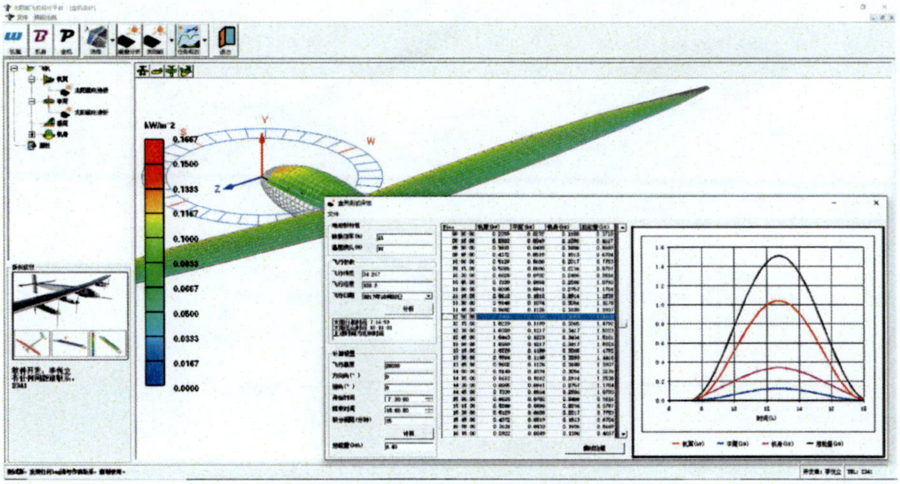

Fig. 4. Software runtime interface.

3.2 Geometric Modeling

Geometric modeling includes wing modeling, fuselage modeling, aircraft assemble and solar battery panel laying (Fig. 5).

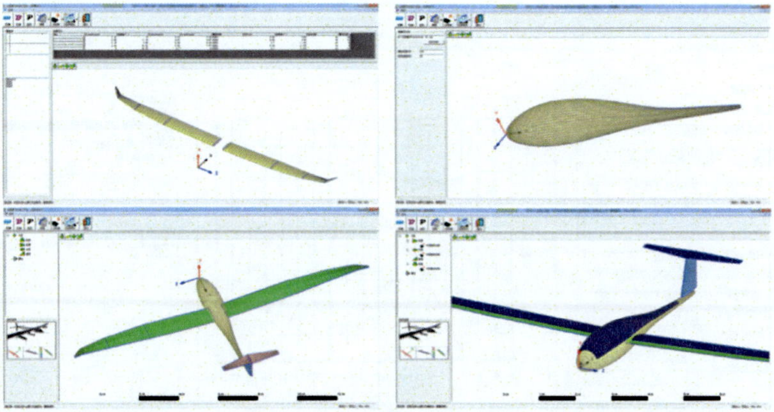

Fig. 5. Modeling process.

The software developed in this paper has a wide range of adaptability to different types of vehicles. Such as airships, aircraft, vehicles, etc. (Fig. 6).

3.3 Radiation Calculation

The solar radiation characteristics calculation form is the calculation settings area on the left, the curve display area in the middle, and the analysis options area on the right. The

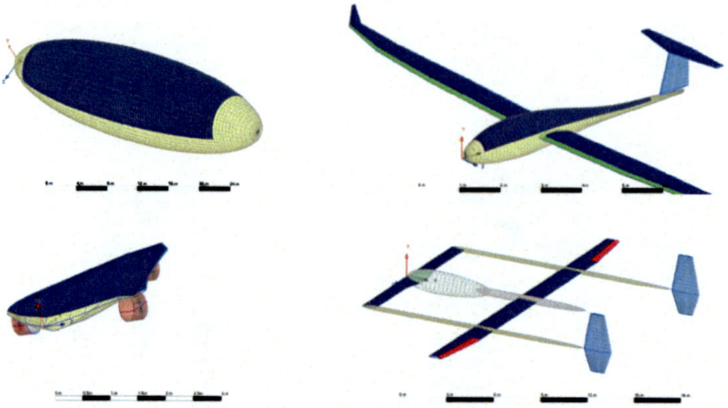

Fig. 6. Various types of vehicles.

calculation setting area can set the longitude and latitude, date, and time of the calculation, and can obtain the true solar time, azimuth angle, and ray vector information at that time. The Analysis options area allows you to calculate the effects of the Earth's orbit, atmospheric attenuation characteristics, altitude angle characteristics, all-day radiation energy characteristics, and the length of time at night at different latitudes. These properties are displayed as curves and can be saved as curve pictures or result data, as below (Fig. 7).

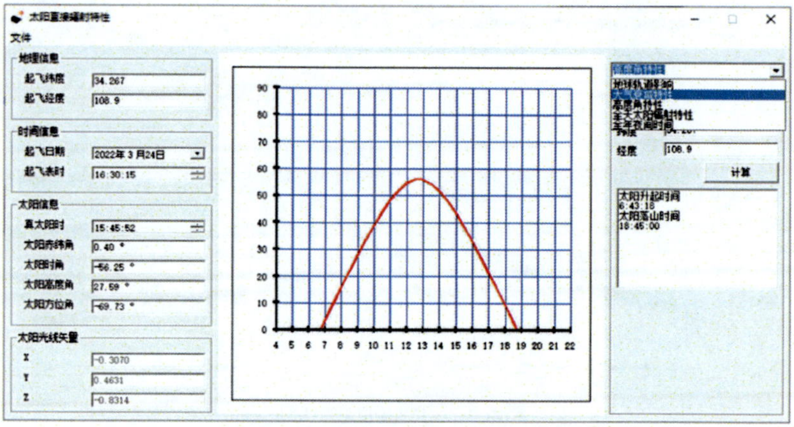

Fig. 7. Radiation characteristics.

After the calculation of the above radiation characteristics, it is possible to choose to fly a solar-powered aircraft in the right latitude and longitude area and in a suitable time period. Before the full-day calculation, the selected flight area needs to be checked for calculation, using a single-point verification module. By changing the flight direction, flight altitude, flight time, flight attitude, etc., the maximum and minimum values of the solar radiation absorption of the calculation model under the current latitude and longitude and date setting conditions are obtained to determine whether the area meets the requirements in the current calculation date conditions, as below (Fig. 8).

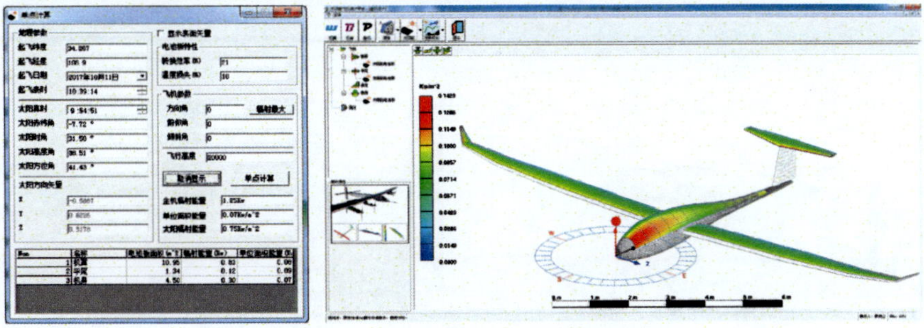

Fig. 8. Single-point verification.

After determining the latitude and longitude of the use area and the flight date, the calculation and analysis of solar radiation throughout the day can be carried out to obtain the solar radiation absorption energy of different components and the total energy of the whole aircraft (Fig. 9).

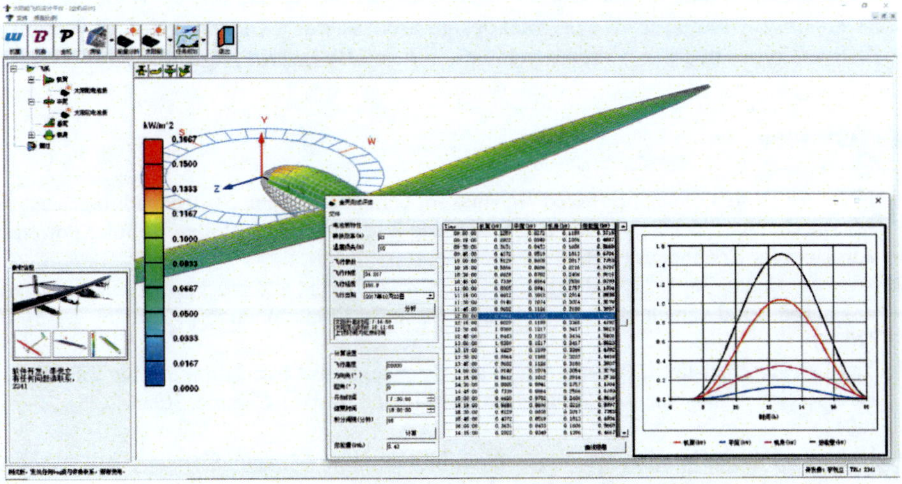

Fig. 9. Calculated throughout the day.

After the calculation is completed, the display area is displayed in real-time in the form of a cloud map to display the difference in solar radiation intensity in different areas, and the software can directly save the display result of the area as an image output (Fig. 10).

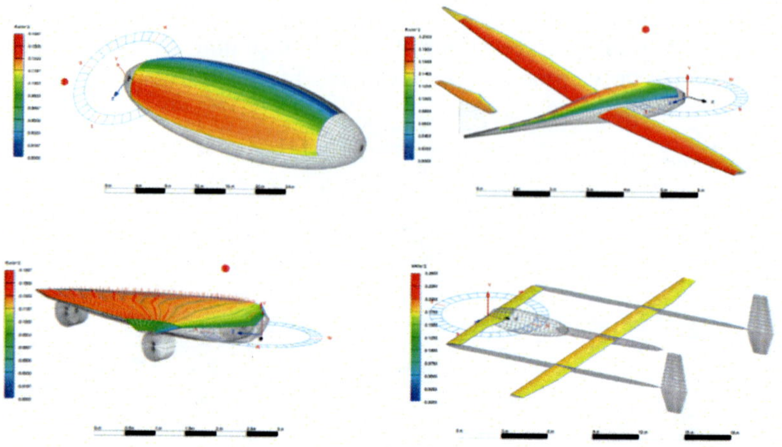

Fig. 10. Cloud map output.

4 Conclusion

Based on the solar radiation model, a solar radiation calculation and analysis software for solar-powered aircraft design based on Visual Basic and OpenGL was developed. The software has powerful geometric modeling, solar battery panel laying, and radiation calculations, and the calculation results support cloud map and curve output. The software greatly simplifies the original design process, has a wide range of adaptability to different types of solar-powered vehicles, and greatly improves the design efficiency.

References

1. Noth, A.: Design of solar powered airplanes for continuous flight. Switzerland: Ingénieuren Microtechnique EcolePoly-technique Fédéralede Lausanne, 2008 Design of Solar Powered Airplanes for Continuous Flight
2. Wen, S., Guang-jia, L., Zhi-sheng, Y., et al.: Current status and prospect of solar-powered unmanned aerial vehicle application in near space (In Chinese). Aerosp. Technol. **1**, 83–90 (2022)
3. Yan-Lei, Z., Zhao-Jie, L., Sheng, W., et al.: Calculation and Simulation of Thin Film Solar Cells Region Arrangement on Stratosphere Airship (in Chinese). Comput. Simul. **26**(8), 71–76 (2011)
4. Hong, S., Baoyin, S., Qiuping, Y.: Study of the solar power system of stratospheric airships (in Chinese). Chin. Space Sci. Technol. **2**, 26–40 (2019)
5. Ma, D.L., Bao, W.Z., Qiao, Y.H.: Study of solar-powered aircraft configuration beneficial to winter flight (in Chinese). Acta Aeronautica et Astrounautica Sinica **35**(6), 1581–1591 (2014)
6. Chan, M., Zhou, Z., Cheng, K., et al.: Exploring the characteristics of power density of tracking PV modules for high-altitude stationary solar-powered airplanes (in Chinese). Acta Aeronautica et Astrounautica Sinica **34**(2), 273–281 (2013)
7. Wei, Z., Qi, S., Yong, L., et al.: Computation and analysis of power generated by the solar cell array of a stratospheric airship (in Chinese). J. Astronaut. **31**(4), 1224–1230 (2010)

8. Zhang, Y., Liu, L.B., Liu, D.X.: Modeling and analysis of solar energy collected by high altitude solar UAV's surface. Appl. Mech. Mater. **644–650**, 234–237 (2014). https://doi.org/10.4028/www.scientific.net/amm.644-650.234
9. Jing, Y., Zhi-zhang, L., Bin, M., et al.: Resources calculation of solar radiation based on matlab (in Chinese). Energy Eng. **01**, 35–38 (2011)

Check for updates

Research on Acceleration Displacement Algorithm in Rocket Sled Test

Ranzhong Yao(✉) and Fusheng Wu

AVIC Aerospace Life-Support Industries, Ltd., Xiangyang 441003, China
Azhong_1@126.com

Abstract. Using rocket sled test to carry out short-range radar acquisition test is an important means to test the acquisition speed and anti-jamming performance of short-range radar at real speed. The radar is placed on the upper part of the rocket sled, and a large deflection will occur when the rocket sled runs at high speed. The vibration sensor is installed at the position of the radar transmitting antenna to calculate the whole swing through integration, so as to ensure that the radar lobe covers the captured body in the sliding process. The purpose of this study is to eliminate the system error through zero compensation, reduce the influence of low-frequency signal on quadratic integration through low-frequency attenuation algorithm, and eliminate the system random error through the adaptation of empirical mode decomposition (EMD) to non-stationary signal. The results show that through the above algorithm, the peak deviation of position data obtained from acceleration integration is less than 5%, which meets the needs of engineering. These innovative algorithm combinations can be applied to other occasions that need to obtain velocity and displacement from acceleration data integration, which is of great significance.

Keywords: Acceleration displacement · Zero value compensation · Low frequency attenuation algorithm · EMD

1 Introduction

The acquisition test of short-range radar by rocket sled test is an important means to test the real-time acquisition speed and anti-jamming performance of short-range radar. The radar is placed on the upper part of the rocket sled. When the rocket sled runs at high speed, it will deflect greatly. During the operation of the rocket sled, it is necessary to ensure that the radar lobe covers the captured object during the sliding process.

Although there are many kinds of equipment and methods to test displacement in theory, these methods still encounter many difficulties in the actual test process. For example, due to the space limitation of the structure, it is sometimes difficult to find the appropriate installation position of the displacement sensor. Even if there is enough space inside the structure to arrange the displacement sensor, the displacement sensor test only obtains the relative displacement between the installation position and the point to be measured. For example, when the test rocket sled is equipped with a radar seeker

Chinese Society of Aeronautics and Astronautics (Ed.): CASTC 2023, LNME, pp. 86–95, 2024.
https://doi.org/10.1007/978-981-99-8861-7_10

for target acquisition, the overall vibration of the rocket sled during operation cannot obtain the displacement of the seeker relative to the orbit [1, 2].

Relatively speaking, the acceleration test is relatively easy. On the one hand, because the acceleration test does not need a static installation position relative to the rocket sled, the acceleration sensor can be directly connected with the rocket sled rigidly; On the other hand, the acceleration sensor is smaller than the displacement sensor and speed sensor, so it does not need a lot of layout space; The above two points make the installation and arrangement of acceleration test in rocket skid very easy. The vibration sensor is installed at the position of the radar transmitting antenna to measure the acceleration signal during operation.

Theoretically, the velocity signal and displacement signal can be obtained by integrating the acceleration signal once and twice. At present, there are mainly two methods for the integration of dynamic acceleration signal, one is to realize it by hardware integration circuit, and the other is to realize it by software integration.

In practical use, the integration accuracy of the hardware integration circuit method is limited by the performance of hardware equipment and the characteristics of the integration circuit itself. The minimum frequency of the acceleration signal to be integrated is required to be high, and generally it cannot be used to measure the motion state signal under the condition of similar impact, which contains more frequency components.

Software integration has the characteristics of diverse algorithms and strong flexibility, and has been widely used. Different software algorithms will bring different errors. In this paper, a new method is proposed: eliminate the system error through zero value compensation, reduce the influence of low-frequency signal on quadratic integration through low-frequency cut-off algorithm, and eliminate the system random error through the adaptation of empirical mode decomposition (EMD) to non-stationary signal [3–5].

2 Eliminating System Error by Zero Acceleration Compensation Method

The zero acceleration compensation method compensates each data to weaken the cumulative effect of error in the integration [6, 7]. The principle diagram of zero acceleration compensation is shown in Fig. 1. It can be seen from Fig. 1 that due to the existence of error, the acceleration area is not equal to the deceleration area. Through compensation, the coordinate axis moves up (into the virtual line segment).

The specific process is as follows:

$$\hat{a}(k) = \begin{cases} 0 & k \in [0, k_1] \\ a(k) - \overline{a_{0-k2}} & k \in [k_1, k_2] \end{cases} \tag{1}$$

$$\hat{a}(k) = \begin{cases} 0 & k \in [0, k_1] \\ \hat{a} - \frac{\overline{a_{0-k3}}}{k2\_k_1}(k - k_1) & k \in [k_1, k_2] \\ 0 & k \in [k_2, k_3] \end{cases} \tag{2}$$

where K1 is the starting point of the motion state, K2 is the end point of the motion state, and 0 and K3 are the starting and ending points of the acceleration sample sequence

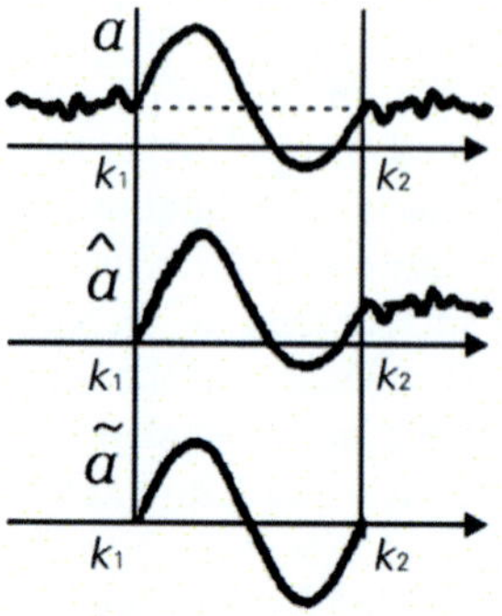

Fig. 1. Schematic diagram of zero acceleration compensation

respectively. a (k) is the acceleration before correction, $\overline{a_{0-k}}$ is the mean value of acceleration in this period, $\widehat{a_k}$ is the acceleration corrected by the acceleration zeroing of the previous static stage. $\widetilde{a_k}$ is the acceleration corrected by zeroing the acceleration in the static phase of the rear end.

Due to the existence of system error, the speed does not become zero after the acceleration and deceleration process. In order to eliminate the system error, zero acceleration compensation is adopted to make the acceleration area equal to the deceleration area, so as to conform to the basic physical law.

2.1 Processing Effect of Zero Acceleration Compensation Method

Figure 2 shows the typical acceleration map of the starting section of the rocket sled, and Fig. 3 shows the results of No.1 and No.2 after matlab programming Shift the waveform before and after correction.

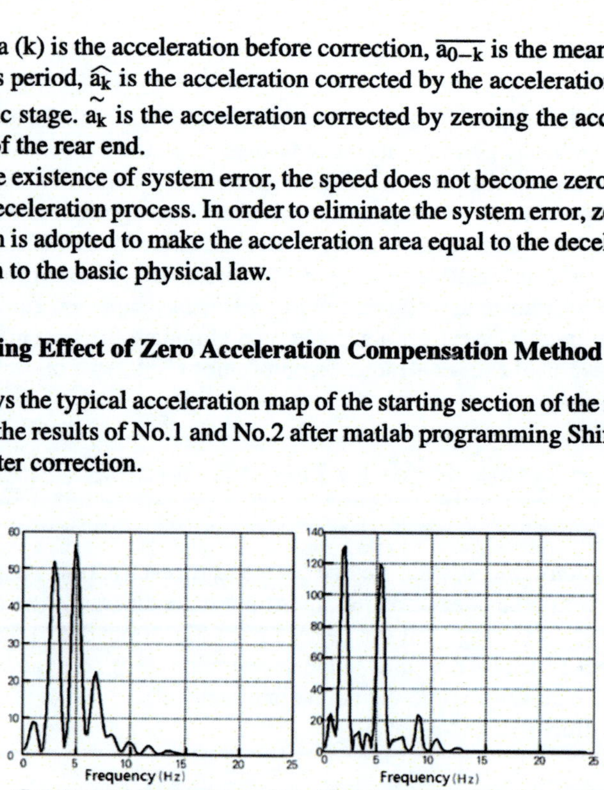

a)Typical rocket sled acceleration map No.1 b)Typical rocket sled acceleration map No.2

Fig. 2. Acceleration curve of typical rocket sled starting section

By comparing the displacement before and after zeroing compensation in Fig. 3, it can be seen that the system error can be obviously eliminated by zero compensation, and the waveform obtained is consistent with the actual situation. After using zero compensation, the displacement offset is greatly reduced.

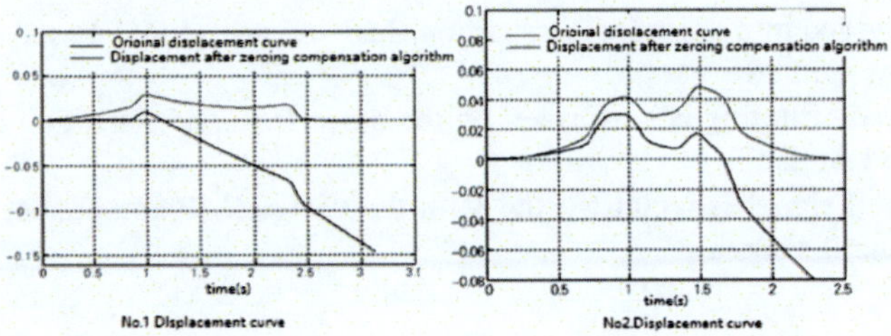

Fig. 3. Displacement comparison before and after zeroing compensation

3 EMD Eliminates Random Errors

EMD algorithm is a new adaptive signal time-frequency analysis method developed in recent years. It extracts the inherent modal function of the signal independently according to the characteristics of the signal itself. It is a method suitable for analyzing nonlinear and non-stationary signals. This method is considered to be a major breakthrough in traditional time-frequency analysis methods such as Fourier analysis and wavelet transform based on the assumption of linear stationarity [8, 9].

Empirical mode decomposition EMD noise reduction is a new signal analysis method proposed by Huang et al. It can decompose the signal into a series of eigenmode functions IMF and residual terms res with high to low frequency. It is different from the traditional Fourier transform, which takes sine function or exponential function as the basis function. Each res component and residual term are extracted from the signal to be decomposed, and has good adaptability.

After EMD decomposition of the original signal, several IMF components and remainder (res) are obtained. The original signal X (t) is expressed as:

$$X(t) = \sum\nolimits_{i=1}^{n} IMF_i(t) + res(t) \tag{3}$$

where: X (t) is the original signal; IMFi(t)is each IMF component; n is the number of IMF components; r (t) is the residual term.

For the reconstruction after signal decomposition, the correlation coefficient threshold method is generally used. The correlation coefficient between IMF components and the original signal is obtained from Eq. 2, and the first minimum value is used as the discrimination threshold. The IMF component corresponding to this point is defined as the transitional IMF. The transitional IMF and the useful IMF are reconstructed to obtain the noise reduction signal after EMD decomposition.

$$r_i(IMF_i, X(t)) = \frac{Cov(IMF_i, X(t))}{\sqrt{Var[IMF_i]Var[X(t)]}} \tag{4}$$

where: IMFi represents the IMF, and X (t) represents the original acceleration signal.

Noise reduction steps are:

(1) perform IMF decomposition on the original signal to obtain several IMF components and a res remainder, as shown in Fig. 4.

(2) IMF is divided into noise IMF, transition IMF and useful IMF based on correlation coefficient method;
(3) IIR low-pass filtering is performed on the noise IMF and transition IMF to obtain the filtered signal;
(4) The filtered signal is reconstructed with the useful IMF component to obtain the noise reduction signal.

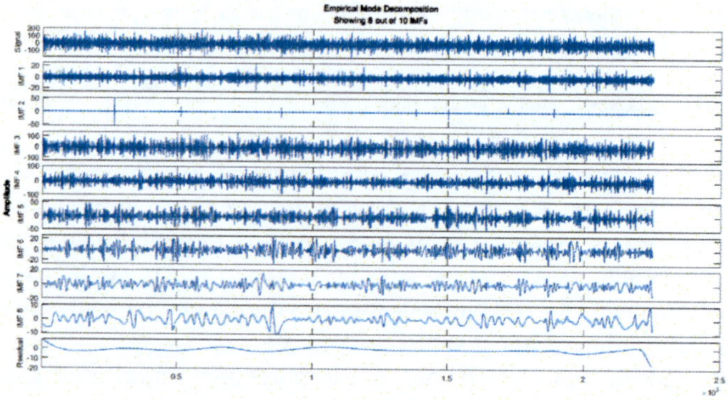

Fig. 4. EMD exploded figure

Calculate the correlation coefficient of each IMF. As shown in Fig. 5, the minimum value appears in the third IMF. Define imf3 as transitional IMF, imf1 and imf2 as noise IMF, and the rest as useful IMF. The transition IMF, the useful IMF and the remainder are reconstructed according to the EMD correlation coefficient reconstruction method.

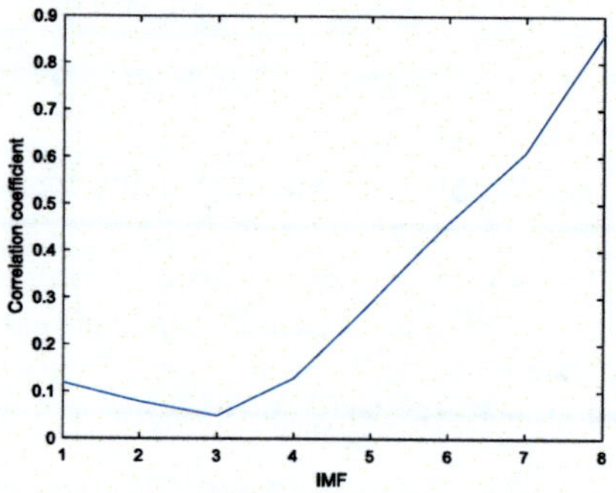

Fig. 5. Correlation coefficient between IMF and curve of each order

Set IMF1 and IMF2 to zero and reconstruct the data. The results of simple low-pass filtering and low-pass filtering combined with EMD noise reduction are compared with Table 1.

Table 1. Comparison of noise reduction results

processing method	Root mean square error(RMSE)	Signal to noise ratio(SNR)	correlation coefficient(R)
Low pass filtering	0.122	25.178	0.984
Low pass filtering + emd noise eduction	0.009	26.355	0.996

4 Frequency Domain Integration Method Based on FFT Transform

In the displacement signal obtained by the twice time-domain integration of acceleration, there is the cumulative error of the quadratic term. This is because the trend term of the measured acceleration signal cannot be completely eliminated. Therefore, the residual small error in the time-domain integration will be cumulatively amplified in the process of the twice integration, and even the displacement signal will be distorted.

For a time-domain discrete acceleration series a (n) with length N, perform discrete Fourier transform.

$$\begin{cases} a(k) = \frac{2}{N}\sum_{1}^{N} a(n)e^{\frac{-2\pi(k-1)(n-1)i}{N}} \\ f(x) = \frac{(n-1)f0}{2N} \end{cases} \tag{5}$$

where f0 is the sampling frequency, a (k) is the complex sequence in the frequency domain after Fourier transform, and f (k) is the corresponding frequency.

Frequency domain integration first Fourier transforms the acceleration signal that needs to be integrated into the frequency domain, and the time-domain integration operation becomes the integration operation of Fourier component coefficients in the frequency domain, and then the integrated time-domain signal is obtained by inverse Fourier transform. Frequency domain integration is directly based on the integral exchange of sine and cosine in the frequency domain, avoiding the cumulative amplification of small errors caused by time domain integration.

The above relationship is expressed by Fourier forward transform and Fourier inverse transform as:

$$\begin{aligned} F(s(t)) &= H_E(\omega)F(\bar{a}(t)) \\ s(t) &= -F^{-1}\left[\frac{1}{\omega^2}F(\bar{a}(t))\right] \end{aligned} \tag{6}$$

where F(x) is the Fourier transform of function x, HE(ω)is the transfer function, $\bar{a}$ (t) Is the measured acceleration signal, s(t) is the integral displacement signal. Under the same frequency component, the amplitude frequency characteristic of the acceleration signal and its represented displacement signal is 1/ ω2, and its plot is shown in Fig. 6.

In the process of acceleration integral displacement, the amplitude of the low-frequency part will be amplified and the high-frequency part will be attenuated. Therefore, to reduce the error of displacement signal, the key is to reduce the error of low-frequency part of acceleration signal [10].

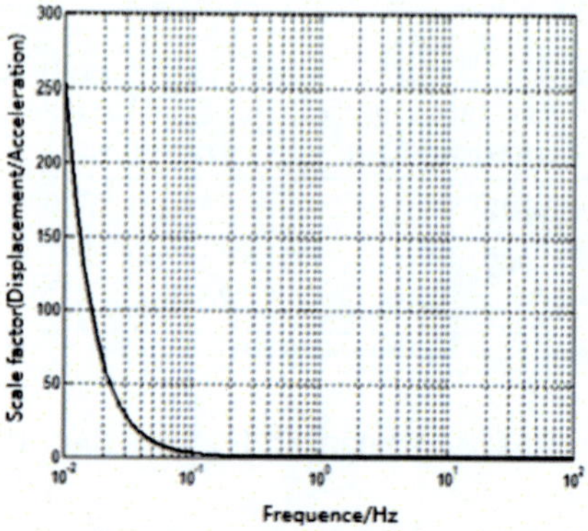

Fig. 6. Signal ratio coefficient curve

In the low-frequency cut-off algorithm, the low-frequency cut-off frequency ft needs to be selected. Assuming that the first peak value of the acceleration signal to be integrated is FS, it is required that FT < FS. Within the whole frequency range, the function $\phi(\omega)$ is defined as:

$$\begin{cases} \phi(\omega) = 0 \ \omega < 2\pi fT \\ \phi(\omega) = 1 \ \omega \geq 2\pi \mathrm{fT} \end{cases} \tag{7}$$

In the low-frequency cut-off algorithm, the signal ratio function $\omega(\phi)$.The acceleration displacement amplitude frequency characteristic is obtained by frequency domain integration, as shown in Fig. 7.

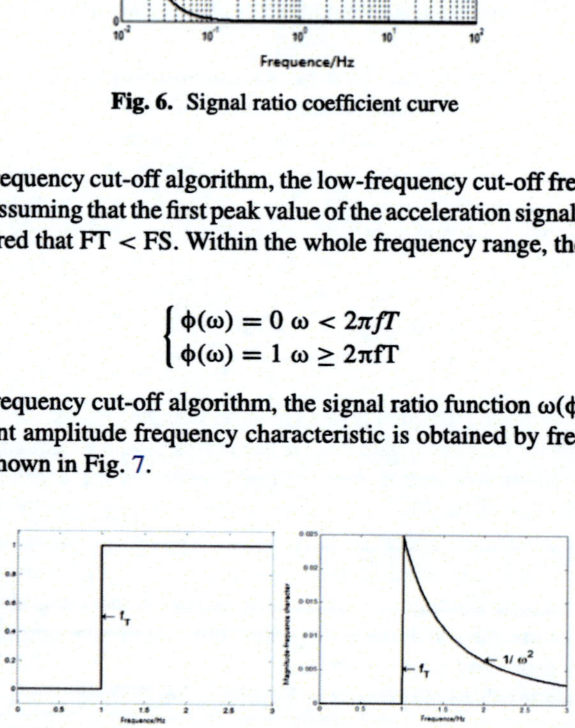

Fig. 7. Low frequency cut off amplitude-frequency characteristics acc-accVS acc-dis

As can be seen from Fig. 7 above, the low-frequency cut-off algorithm selects the low-frequency cut-off frequency ft, and then sets the acceleration signal amplitude below the cut-off frequency to zero. This effectively avoids the existence of low-frequency error components in the displacement signal obtained by integration [11, 12].

5 Laboratory Validation

The validation test is carried out on the shaking table in the engineering laboratory. An acceleration sensor with the model of egcs3-a, the range of 500g and the accuracy of 1% is installed on the shaking table to measure the acceleration value of vibration; Install a model hl-g112-a-c5 on the test bench, with a measurement center distance of 120mm, a measurement range of ± 60 mm, and a resolution of 2.5 μ m. Sampling period 200 μ S laser displacement sensor to measure the displacement of vibration. See Fig. 8 for the test installation diagram. Carry out random frequency sweeping vibration respectively (Fig. 9).

Fig. 8. Sensor installation diagram

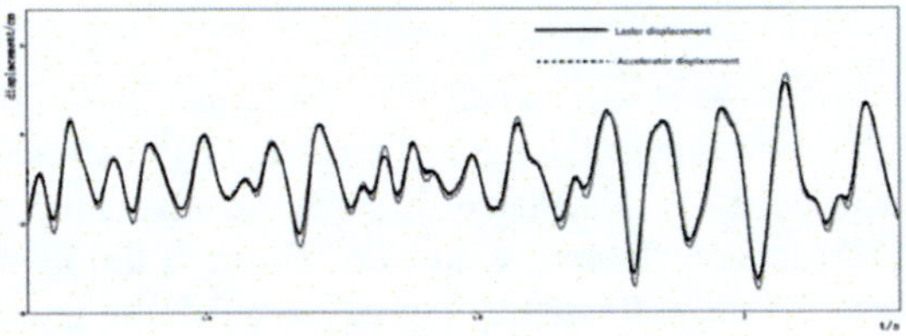

Fig. 9. Comparison diagram between laser displacement sensor and vibration integral

6 Vibration and Displacement of a Rocket Sled

In the real-time acquisition speed and anti-jamming test of a rocket skid short-range radar, an acceleration sensor is installed at the radar, with a running speed of 1.2mach. The vibration data are shown in Figs. 10 and 11.

We can be seen that the displacement data obtained from the acceleration curve through the quadratic integration algorithm is consistent with the actual situation.

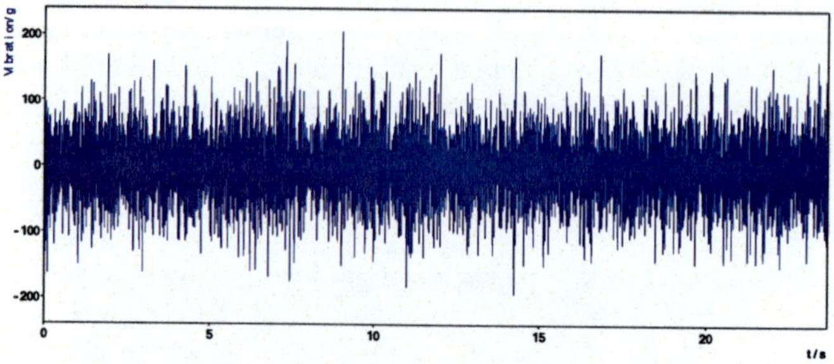

Fig. 10. Radar acceleration curve of rocket sled

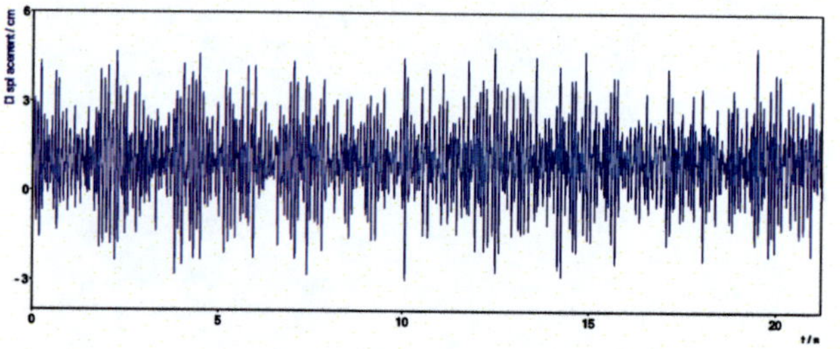

Fig. 11. Radar displacement curve of rocket sled

7 Conclusion

By collecting the acceleration sensor installed on the upper part of the rocket skid, the acceleration sensor data eliminates the system error through zero value compensation, and the system random error is eliminated through the adaptation of empirical mode decomposition (EMD) to non-stationary signals. Through the low-frequency cut-off algorithm, the influence of low-frequency signal on quadratic integration is reduced. It can better measure the left and right displacement amplitude of the rocket sled in operation.

References

1. Guangjun, L., Weidong, J.: Error of train lateral acceleration sensor compensation. Comput. Eng. Appl. **48**(5), 15–17 (2012)
2. Yu, L., Hui, Z., Long, L., et al.: Collection and analysis of acceleration data. Inf. Technol. **12**, 110–116 (2013)
3. Haijiang, L., Shigao, L., Min, L.: EMD and wavelet threshold noise reduction method for gear shift acceleration signal. Noise Vib. Control **38**(2), 198–203 (2018)
4. Tieding, L., Wenlong, Q., Xiaoxing, H., et al.: An EMD noise reduction method to reduce signal-to-noise aliasing. J. Geodesy Geodyn. **40**(2), 111–116 (2020)
5. Lei, D., Min, Z., Wei, Z.: Design and simulation of butterworth low-pass digital filter based on Matlab. Technol. Vis. **21**, 96–97 (2016)
6. Ling, L.Z., Darong, H., Honggang, W.: Application of improved hilbert-huang transform in track irregularities. Control. Eng. **26**(8), 1539–1543 (2019)

7. Li, H., Wang, C., Zhao, D.: Filter bank properties of envelope modified EMD methods. IET Signal Proc. **12**(7), 844–851 (2018). https://doi.org/10.1049/iet-spr.2017.0399
8. Park, K.-T., Kim, S.-H., Park, H.-S., Lee, K.-W., et al.: The determination of bridge displacement using measured acceleration. Eng. Struct. **27**, 371–378 (2005)
9. Park, S.K., Park, H.W., Shin, S.B., Lee, H.S.: Detection of abrupt structural damage induced by an earthquake using time-window technique. Comput. Struct. **86**, 1253–1265 (2008)
10. Hong, Y.H., Kim, H.-K., Lee, H.S.: Reconstruction of dynamic displacement and velocity from measured accelerations using the vibrational statement of aninverse problem. J. Sound Vib. **329**, 4980–5003 (2010)
11. Lee, H.S., Hong, Y.H., Park, H.W.: Design of an FIR filter for the displacement reconstruction using measured acceleration in low-frequency dominant structures. Int. J. Numer. Methods in Eng. **82**(4), 403–434 (2010). https://doi.org/10.1002/nme.2769
12. Hitoshi, K., Masashi, N., Norio, I.: Examination system for accelerometer with parallel link mechanism displacement calculation by integrated. Trans. Japan Soc. Mech. Eng. Ser. C **78**, 27–34 (2012)

Vibration Analysis of Bearing Cage Based on Explicit Dynamics of Complete Machine

Jiang Cong[1,2](✉) and Liu Ding[2]

[1] College of Mechanical and Electronical Engineering, Nanjing University of Aeronautics and Astronautics, Nanjing 210006, China
jclunwen@163.com

[2] Nanjing Mechanical and Hydraulic Engineering Research Center, Aviation Industry Corporation, Nanjing 211106, China

Abstract. In this paper, based on the structural characteristics of air turbine starter itself, a high-precision finite element model of the whole engine is established by adopting the step-by-step modeling and confirmation method of the structural dynamics of the whole engine. Transient dynamics simulation calculation of a certain type of aviation air turbine starter is carried out under normal and abnormal conditions, so as to obtain the characteristics of vibration influence that may be brought by abnormal conditions with assembly deviations. The mechanical state and dynamic characteristics of bearing cages are quantified and compared, and the results show that the deviation of double gear assembly is one of the main reasons for the breakage of turbine end-supported bearing cages, while the change of bearing preload and eccentric assembly in a limited range of turbine angles have relatively little influence.

Keywords: Air turbine starter · Finite element · Turbine · Vibration

1 Introduction

Aviation air turbine starters usually have two working stages: turbine acceleration and deceleration at starting and output shaft rotating with the engine for a long time at the end of starting. These two stages have higher requirements for bearings. The smooth and normal operation of the bearings of the core supporting parts of air turbine starter is also an important guarantee for improving aircraft attendance. When abnormal working conditions occur in bearings, cages are often the first damaged parts in bearings, and the cage rupture is the most serious in its failure mode.

Taking the fracture of bearing cage as an example, such a serious failure mode is usually caused by high temperature, abnormal lubrication, foreign body stuck, etc., and the analysis of the failure is also used by means of ball, raceway, cage wear mark identification, fracture tissue analysis, etc., which fails to take systematic consideration from the bearing working scene. Therefore, it is difficult to confirm the abnormal load of bearings under the condition of assembly deviation of bearings and associated components [1–5], which is often one of the important factors leading to the fracture of bearing cages.

Chinese Society of Aeronautics and Astronautics (Ed.): CASTC 2023, LNME, pp. 96–105, 2024.
https://doi.org/10.1007/978-981-99-8861-7_11

2 Complete Machine Model

In order to determine the influence of starter vibration on turbine support bearings in use, according to the structural characteristics of a certain type of air turbine starter, this paper adopts the strategy of step modeling and confirmation of the overall structure dynamics, and divides the model into rotor and stator subsystems, and then divides the subsystems into several underlying structures. By establishing the finite element model of the underlying structure and using the modal test data to modify the model, then through the thin layer element and friction contact simulation connecting interface, the finite element model and modification of the subsystem are established, and finally the high-precision finite element model of the whole machine is established[6–13]. The specific modeling methods and parameters are not detailed in this paper. The explicit finite element model of the air turbine starter was finally established, as shown in Fig. 1. Explicit finite element model of air turbine starter, with about 164,000 nodes, 382,000 units and 492,000 degrees of freedom. In order to better simulate the bearing contact state, grid encryption was carried out on the raceway of the inner and outer rings of the turbine support bearings.

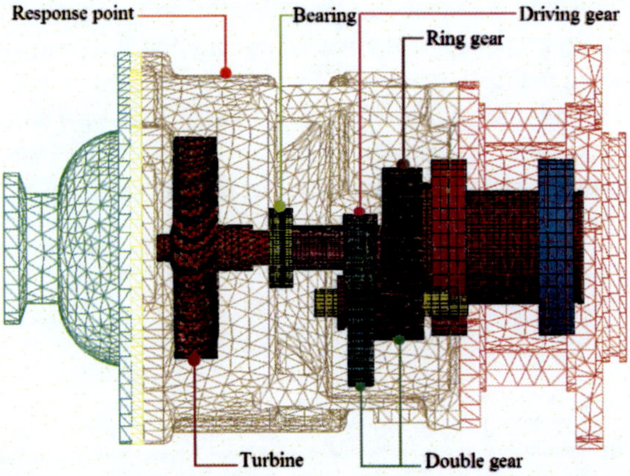

Fig. 1. Explicit finite element model of air turbine starter

At the same time, the rotation frequency of turbine and driving gear is taken as the fundamental frequency, and the meshing frequency and rotation frequency of all gears are calculated based on the meshing transmission ratio of the gears, as shown in Table 1.

Through explicit dynamic analysis and calculation of the whole machine under sudden loading, the response points marked in Fig. 1 were extracted for analysis, so as to determine several factors that have a great influence on the operation of the bearing cage under the vibration condition of the whole machine.

Table 1. Rotation frequency and meshing frequency of gears at all levels

Turbine and drive gear frequency	Frequency of engagement between driving gear and duplex	Double gear frequency	Frequency of engagement between double gear and ring gear	Ring gear and output shaft rotation frequency
1	16	0.36	6.55	0.15

3 Explicit Dynamic Analysis of Sudden Loading Under Normal Conditions

During calculation, unbalance force is applied to the turbine disc core to simulate the sudden load caused by unbalance. Transient dynamics calculation is carried out on the starter under turbine speed conditions of 20000rpm, 40000rpm, 50000rpm, 60000rpm and 80000rpm. Through time-frequency domain analysis of the calculation results, it can be seen that, The vibration frequency of the starter is mainly turbine revolution frequency and its double frequency, meshing frequency and its double frequency of driving gear and double gear, meshing frequency of double gear and ring gear. According to the maximum acceleration response of the whole machine at different speed points in the time domain, as shown in Fig. 2, it can be found that the maximum response of the whole machine near the second critical speed of the turbine (50000rpm) will be larger, and the maximum response at other speed points will increase with the increase of the speed, which conforms to the response characteristics of the test process.

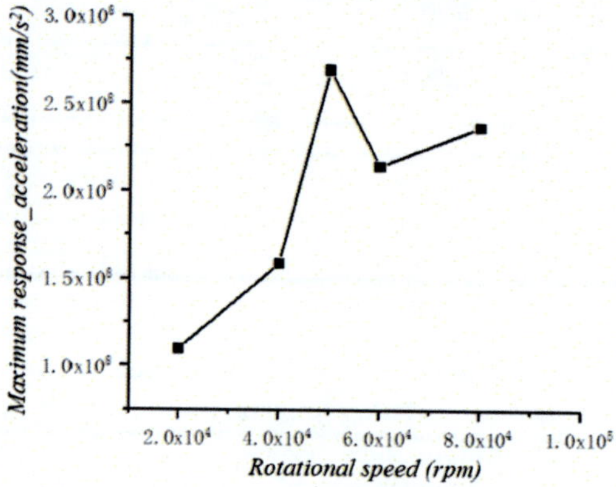

Fig. 2. Comparison of maximum acceleration responses at different rotational speeds

Taking the state of 50000rpm as an example, the displacement amplitude and frequency of the starter under normal state are shown in Fig. 3.

Meanwhile, the maximum axial acceleration of the bearing cage under normal condition can be extracted, as shown in Fig. 4.

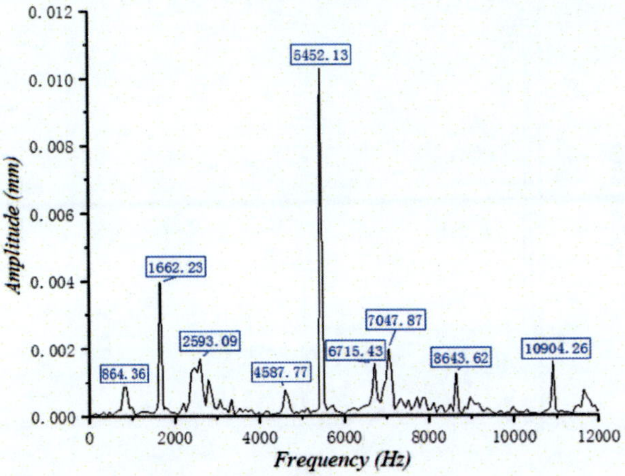

Fig. 3. Displacement spectrum of the turbine at 50000rpm

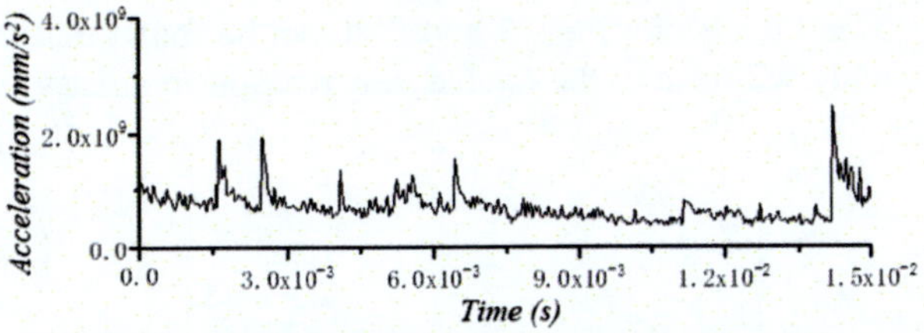

Fig. 4. Maximum axial acceleration of bearing cage under normal condition

Based on the above analysis, the following three assembly factors affecting the turbine transmission frequency, meshing frequency between double gear and drive gear, and ring gear will be focused on when the turbine speed is 5000rpm: The transient dynamics analysis of the whole machine was carried out to obtain the influence characteristics of the abnormal state with assembly deviation caused by different double gear assembly deflection, turbine assembly eccentricity and bearing preload.

4 Explicit Dynamic Analysis of Sudden Load Under Abnormal State

4.1 Double Gear Assembly Deviation State

In Hyper-mesh software, the tiny deflection angle of double gear was used to simulate the assembly deviation state of double gear, and four calculation models with assembly deviation angles of 0.5°, 1°, 1.5° and 2° were established. Finally, the calculation results of four different double gear assembly deviations were obtained and analyzed. Figure 5 shows the displacement amplitude-frequency diagram of double gear assembly deviations of 2°.

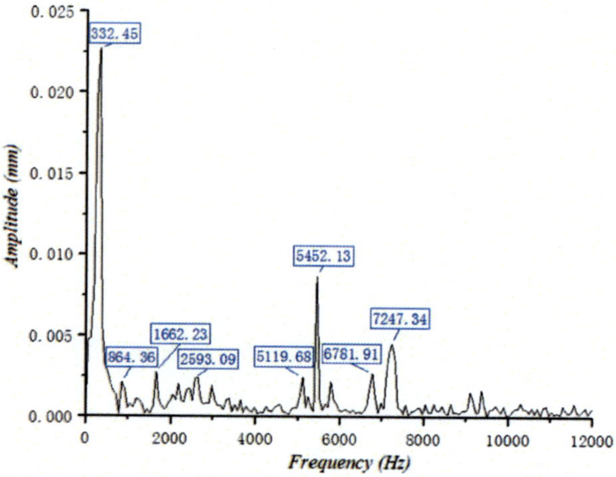

Fig. 5. Spectrum of double gear assembly deviation of 2°

By comparing and analyzing Figs. 3 and 5, it can be found that the deviation of double gear assembly will lead to the double gear rotation frequency on the spectrum diagram.

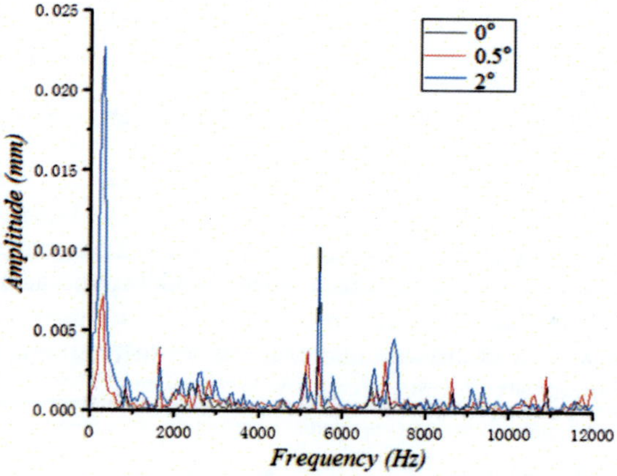

Fig. 6. Angular spectrum comparison of assembly deviation of different double gears

As can be seen from Fig. 6 and Table 2, the overall average vibration amplitude increases with the increase of deviation angle.

The axial acceleration of the bearing cage was extracted and analyzed. Figure 7 below is the comparison of the maximum axial acceleration of the bearing cage under different double gear assembly deviation angles. It can be found that with the increase of double gear assembly deviation angle, the maximum axial acceleration will increase, indicating that the double gear assembly deviation will affect the abnormal vibration of the bearing cage and is one of the reasons for the cage breakage.

Table 2. Average amplitude of assembly deviation angle of different double gears

Eccentric angle	0°	0.5°	1°	1.5°	2°
Mean amplitude(mm)	3.52×10^{-4}	6.86×10^{-4}	9.05×10^{-2}	1.03×10^{-3}	1.13×10^{-3}

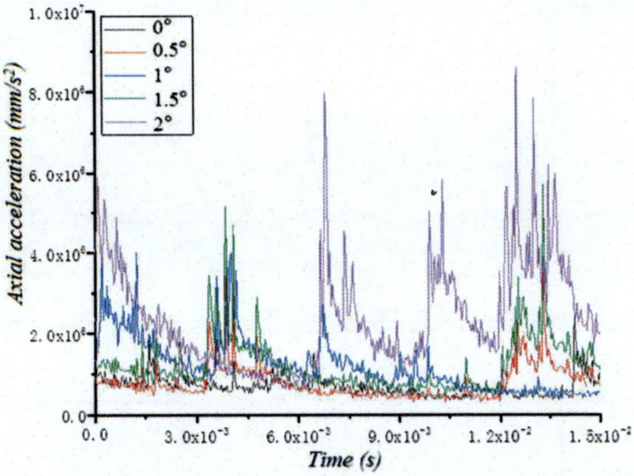

Fig. 7. Maximum axial acceleration of bearing cage under different assembly deviation angles of double gear

4.2 Turbine Assembly Under Eccentric Condition

In the Hyper-mesh software, the turbine rotor was deflected by a tiny angle to simulate the turbine assembly deviation state. Five finite element models of assembly deviation were established with assembly eccentricity angles of 0.02°, 0.04°, 0.06°, 0.08° and 0.1°, respectively. Finally, the calculation results of five different turbine assembly eccentric complete machines were obtained and analyzed. The figure below shows the displacement amplitude-frequency diagram with 0.1° turbine eccentricity.

By comparing Figs. 5 and 8, it can be found that turbine assembly eccentricity will lead to ring gear frequency rotation on the spectrum diagram. As can be seen from Fig. 9 and Table 3, the overall average vibration amplitude increases with the increase of turbine eccentricity angle, but the increase is small compared with the deviation of double gear.

The axial acceleration of the bearing cage was extracted and analyzed. Figure 10 is the comparison diagram of the maximum axial acceleration of the bearing cage under different eccentricity conditions of turbine assembly. It can be found that the maximum axial acceleration of the bearing cage changes little when the turbine has a limited eccentricity angle. The turbine assembly eccentricity is not the main reason for the fracture of bearing cage.

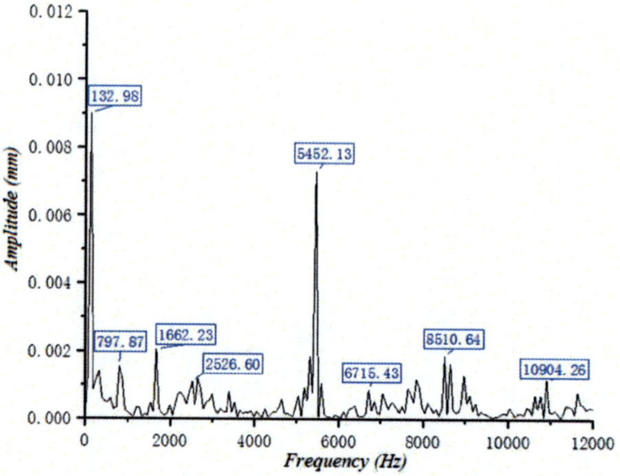

Fig. 8. Spectrum of turbine assembly with eccentricity of 0.1°

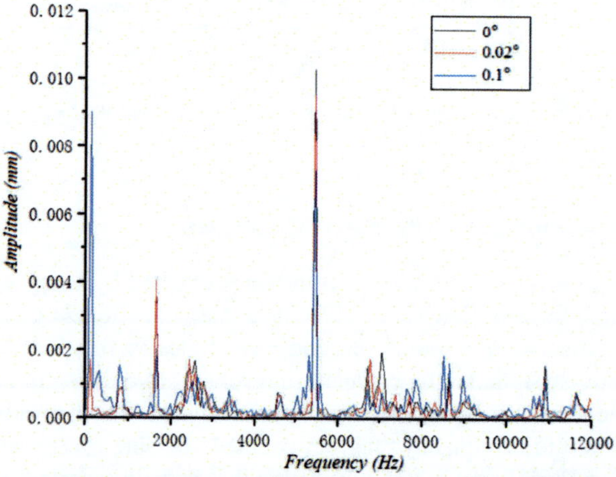

Fig. 9. Spectrum comparison of assembly eccentricity angles of different turbines

Table 3. Average amplitudes of assembly eccentricity angles of different turbines

Eccentric angle	0°	0.02°	0.04°	0.06°	0.08°	0.1°
Mean amplitude(mm)	3.52×10^{-4}	3.6×10^{-4}	3.68×10^{-4}	3.75×10^{-4}	4.42×10^{-4}	4.8×10^{-4}

4.3 Different Bearing Preload States

Under normal condition, bearing preload is 200N. In order to study the influence of bearing preload on the vibration of the whole machine, transient dynamics calculation

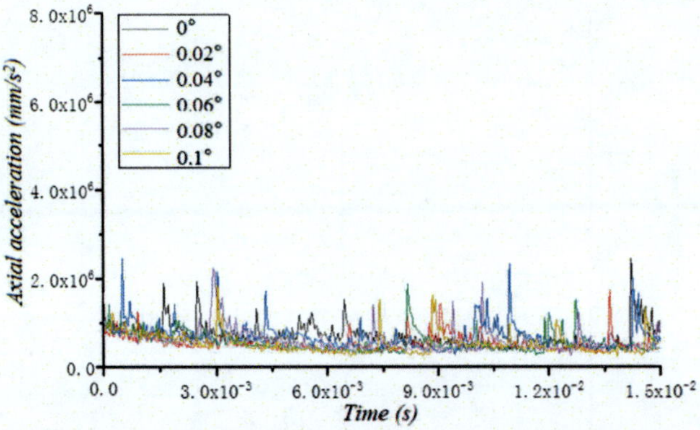

Fig. 10. Maximum axial acceleration of bearing cage under different turbine assembly eccentricities

is carried out on starter with different bearing preload forces. The applied position of bearing preload is shown in Fig. 11.

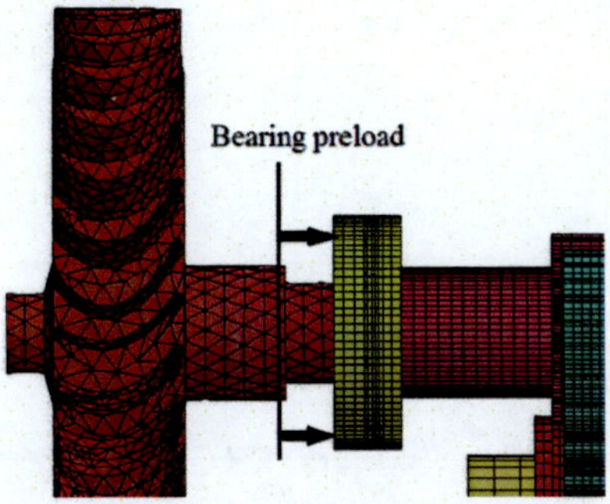

Fig. 11. Finite element model of rotor with bearing preload

Four groups of bearing preloading force (120N, 180N, 260N and 300N) were selected to calculate the transient dynamics of the whole machine. Figure 12 is the displacement spectrum diagram under the condition of 300N preloading force.

Through a comparative analysis of Figs. 3 and 12, it can be found that bearing preload has little influence on the frequency component of the whole machine. Meanwhile, it can be seen from Fig. 13 and Table 4 that bearing preload also has little influence on vibration amplitude.

The axial acceleration of the bearing cage is extracted and analyzed. The finite element calculation results are shown in Fig. 14. It can be found that the maximum axial acceleration of the bearing cage changes little when different bearing preload forces exist, so the bearing preload is not the main cause of the bearing cage breakage when the actual starter works.

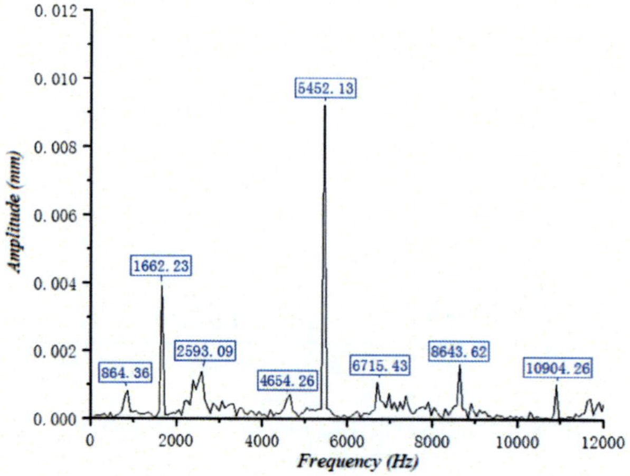

Fig. 12. Displacement spectrum when bearing preload is 300N

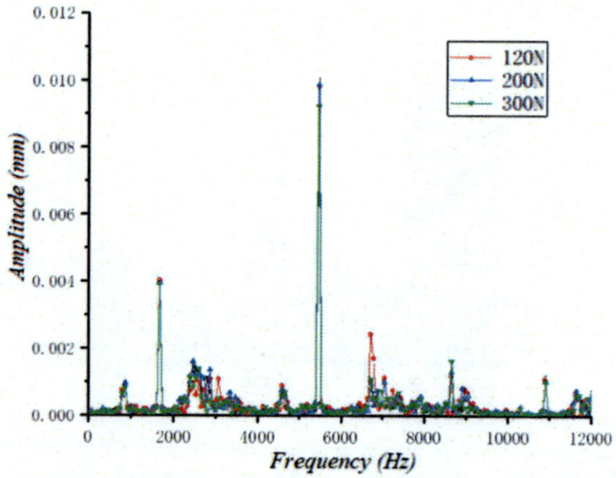

Fig. 13. Spectrum comparison of preloading forces of different bearings

Table 4. Average amplitude under different bearing preloading forces

Preload(N)	120	180	200	260	300
Mean amplitude(mm)	3.52×10^{-4}	3.49×10^{-4}	3.44×10^{-4}	3.50×10^{-4}	3.84×10^{-4}

5 Summarize

In this paper, the maximum axial acceleration of the bearing cage under various states is obtained by calculating the dynamics of the whole machine under normal and abnormal conditions. Through the comparative analysis of the results under normal and abnormal conditions, it can be clearly found that the assembly deviation will make the bearing

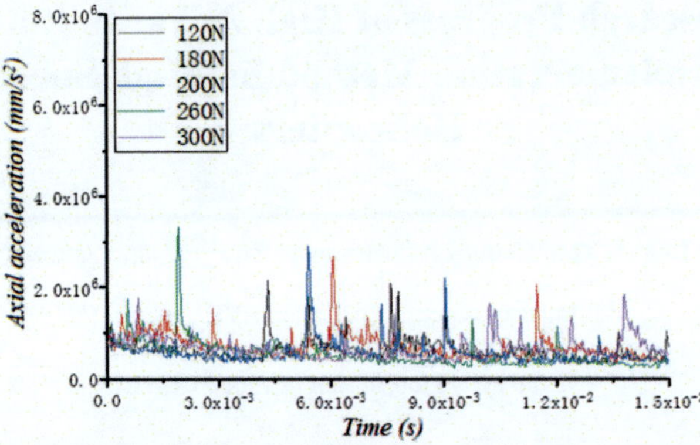

Fig. 14. Maximum axial acceleration of bearing cage under different bearing preload states

under abnormal vibration condition, and the internal force of the bearing may lead to the fatigue crack of the cage. The continuous development of this crack will lead to the fracture of the cage. The deviation of the double gear assembly is one of the main reasons for the fracture of the bearing cage, while the change of the bearing preload and the eccentric assembly in the limited Angle range of the turbine have relatively little influence.

References

1. Meng, G.: Retrospect and prospect of rotor dynamics research. Journal of vibration engineering (01), 5–13 (2002). https://doi.org/10.16385/j.carol-carroll-nki. Issn 1004–4523.2002.01.001
2. Limin, Z., Wang, W., Zhiguang, K.: J. Shenyang Inst. Aeronaut. Technol. **27**(05), 34–37 (2010)
3. Gendai, Z., Chaoping, Z.: J. Vibr. Eng. **25**(4), 365–372 (2012)
4. Chen, X.Q., Xiao, S.F., Liu, X.N.: Finite element modeling and confirmation of structures with uncertain factors. Noise Vibr. Control **33**(5), 26–29 (2013)
5. Qingguo, F., Lingmi, Z., Qintao, G.: Acta aeronautica et astronautica sinica **25**(4), 372–375 (2004)
6. Hao, W., Henghui, Y., Yufang, L.U.: Inf. Syst. Eng. **4**, 38–39 (2014)
7. Chuankai, W., Xin, Y.U.: Sci. Technol. Inf. **30**, 113–114 (2013)
8. Tao, Z.: Performance evaluation method of civil aircraft air starting system. Aeronaut. Sci. Technol. **26**(02), 43–46 (2015)
9. Tingyu, Z.: Aviation Gas Turbine Power Plant, pp. 175–176. Southwest Jiaotong University Press, Chengdu (2004)
10. Hengbin, Z.: Development status and key technology analysis of air turbine starter. Goods and Quality (24), 79 (2017)
11. Xiguang, G., Zhigang, S., Yingdong, S.: Huazhong University of Science and Technology Press, Wuhan, pp. 1–2 (2016)
12. Chaofeng, L., et al.: Basic Theory and Engineering Application of Finite Element Method of Mechanical Structure, pp. 8–10. China Machine Press, Beijing (2020)
13. Guangqi, L.: Numerical Simulation of Aviation Gas Turbine Engine, pp. 1–6. National Defense Industry Press, Beijing (2007)

Research Progress of High Advance Ratio Implementation Method for High Speed Helicopters

Yuqing Luan[1], Yongsheng Liu[1], Hongyan Zhu[1](✉), and Zhirui Wang[2]

[1] Harbin Aircraft Industry (Group) Co. Ltd. of AVIC, Harbin 150066, China
Zhuhy012@avic.com
[2] 95966 PLA Troops, Harbin 150060, China

Abstract. Current requirements for future helicopters proposed by some main investigation projects need helicopters to fly at a higher speed than they commonly achieve while maintaining significant hovering performance. In recent years, interests have increased in the study of high speed helicopters, for example, tiltrotor, compound and coaxial helicopter. They have overcome the flight speed of common helicopters. A main characteristic of high speed helicopters is that they have high advance ratio. The main rotor operates over a wide speed range from 100% in hover to 80% in high speed cruise. Main methods to achieve high advance ratio are variable rotor diameter technique and variable rotor speed technique. The former is realized by change the diameter of rotors. The later is realized by changing the transmission ratio or engine rotating ratio. The paper provides a summary of the research on the variable rotor diameter firstly. There are three kinds of methods to change the rotor diameter. Then, the variable rotor speed technique is described, consisting of the development history, large development plan and implementation challenge. Finally, a prospect of the development for the high advance ratio implementation method in the future is made.

Keywords: Helicopter · High Advance Ratio · Variable Rotor Diameter · Variable Rotor Speed · High Speed

1 Introduction

The helicopter may take off and land vertically in a narrow range. It has good low altitude, speed and maneuverability. Therefore, It is used widely in the military application, such as spying, tracking, supervising, attacking, transporting, and in the civil application, such as earthquake-proof rescuing, first aid at sea, fire fighting, resource exploring. The low performance parameter, such as flight time, range, speed and ceiling, is always a difficult question which puzzles helicopter industry. The special objection of the next generation helicopter is high speed above 407 km/h like the compound helicopter studied by US military in the Future Vertical Lift (FVL) [1, 2]. The successful representation is co-axial rotor and tiltrotor helicopter. High speed is a critical capacity for searching and rescuing. It can improve cost efficiency for civil helicopters. The high speed capability of

Chinese Society of Aeronautics and Astronautics (Ed.): CASTC 2023, LNME, pp. 106–117, 2024.
https://doi.org/10.1007/978-981-99-8861-7_12

conventional helicopters is limited by the compression effect on the advancing blade tip and the reverse flow of stall on the retreading blade. It results in large power requirement and excess vibration in return. The canceling of the compressing on advancing side blades may be realized by reducing blade tip speed. The lowering of blade tip speed can be realized by reducing rotor diameter or rotor speed. The hovering performance is very important for any helicopter. So, a large blade tip speed is necessary right through at the low air speed. Therefore, the variable blade tip speed rotor is an attractive choice. The first object of related investigations is to study the variable rotor diameter and rotor speed, to make it adapt to the high speed helicopter concept. The study progress of variable rotor diameter and rotor speed is reviewed. The aim is to make a technological reserve for the development of high speed helicopters.

2 Variable Rotor Diameter

The design of current helicopter is a compromise between the hovering performance and forward flight performance commonly. This compromise will be a challenge with the high speed and large range becoming important future requirements [3, 4]. A smaller disc load is beneficial for improving the hovering performance, but it will reduce the high speed forward performance. The variable rotor diameter concept was put forward in the 1960s in order to improve the high speed performance for compound and tiltrotor helicopters, so as to reduce the compromise between hovering and high speed cruising properties [5]. A larger rotor diameter is used in the hovering and small speed flight which will reduce the induced power. The rotor diameter is reduced in the high speed forward fight which will avoid compressing effect at high air speed. At the same time, it will reduce the profile power. Thereby, the hovering and high speed performance is made a compromise. The variable diameter concept needs extend and retract every rotor blade. It is a great challenge to design a light variable rotor diameter blade considering damaged tolerance in the large centrifuge load environment. As a result, the variable rotor diameter technology has not been applied in production helicopters. There have some investigations on the variable rotor diameter concept. A review is made for the important experiment and analysis.

Implementation Methods for the variable rotor diameter concept were reviewed by Linden at 1972, comprising the patent after 1913 [6]. The aim was to determine the method which could be applied in the high speed helicopter. Linden described three methods for variable rotor diameter: telescoping blade, folding blade and flexible blade. The so-called telescoping blade was composed of two segments at least. The prospective diameter length was realized by a facility which extended and retracted a segment among them (Fig. 1a). The hinge was incorporated into the blade along the spanwise direction to realize the folding of blade in the flight for the folding blade (Fig. 1b). The folding rotor was designed to unfold as a parachute. The rotor was extended and locked up at startup, but there was no method to retract it. Therefore, it is not suitable for the actively variable rotor. A part of the rotor was needed to be quite soft for the flexible blade. It was retracted by rolling a cylinder (Fig. 1c). Linden judged that the flexible blade was most promising, despite that controlling difficulties existed.

The Telescoping Rotor Aircraft (TRAC) developed by Sikorsky Aircraft Corporation utilized a jackscrew (Fig. 2). The active controlling of the diameter was realized by sliding

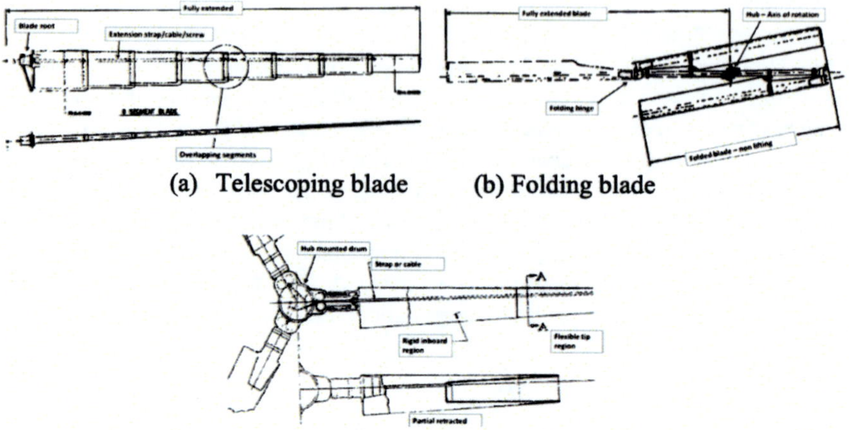

(a) Telescoping blade (b) Folding blade

(c) Flexible blade

Fig. 1. Variable diameter rotor concepts [6].

an outer blade section along a large beam [7, 8]. A one eighth dynamic scaled rotor was tested at 1969, 1970 and 1972. The properties, stress, vibration and aeroelastic stability was measured separately at the most diameter with the flight speed of 277.8 km/h, and at the least diameter with the flight speed of 740.8 km/h. The research shows that the TRAC rotor was capable of improving the properties of the compound and stoppable rotor aircraft, and reducing its vibration level. In the 1920s, this concept was developed into a variable diameter rotor (VDTR). It was used for tiltrotors, and the tunnel test was performed [9]. The test including hovering, transition and cruising validated successfully the feasibility of VDTR for the tiltrotors. The instability was not found in the test. The blade load was acceptable.

The telescoping blade concept has showed the most feasibility for the actual implementation of Variable Diameter Rotor (VDR). Fenny described the history of variable diameter rotors and his research [10]. The VDR rotor developed by Young according to his personal patent was considered as the first full-scale rotor. This rotor arrived at 37.5% powered radius variation at full rotor speed (Fig. 3). The centrifugal force was used to extend the blade and the electric controlling was used to retract the blade for the design concept. The further improvement was made for VDR rotor by Bell Helicopter, until the plan finished at 1971. A good many technique obstacles which prevented VDR rotor from practical applications were pointed out by Fenny. He gave a new design of extending/retracting mechanism with the digital flight controlling and health monitoring in his paper at 2005. It could overcome many obstacles among of them.

A variable diameter mechanism activated by the centrifugal force was developed by Prabhakar and Gandhi aiming at a scaled model rotor [11]. It was adjusted by the spring which joined outer sliding part with the internal fixing part. The variety of the rotor speed changed the rotor radius directly. They found that properties of the rotor were reduced largely comparing with the base-line rotor because that the rotor speed required by rotor extending was high. The effect of rotor radius on the required power

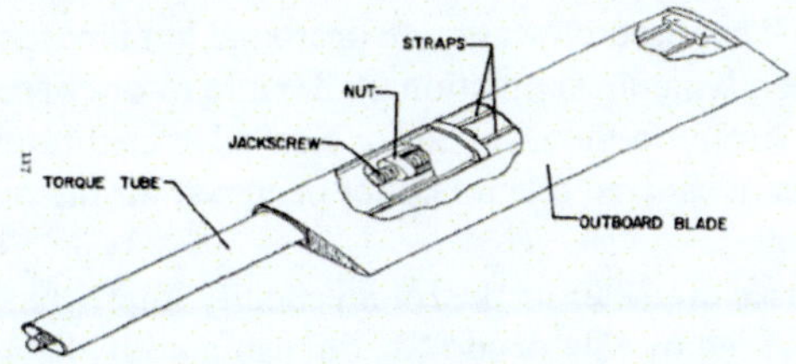

Fig. 2. Schematic of the Sikorsky TRAC rotor [7].

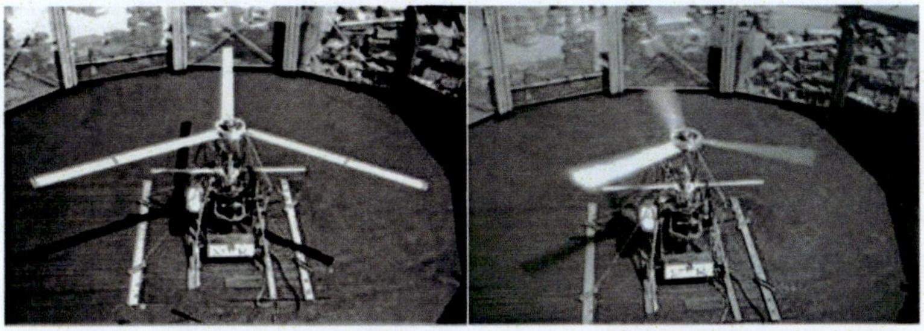

(a) Fully extended (b) Retracted

Fig. 3. Variable Diameter Rotor (VDR) using two telescoping sections, built by Young at Bell [10].

was analyzed by Kang et al. (Fig. 4) [12]. The rotor induced power was dominant in the hover and small speed flight. It is benefit for reducing the induced power when the blade disc area increased. The profile power was increased, but the increment is smaller than the reduction of the induced power. Thereby, the total power was reduced. The rotor profile power reduction was prominent at medium and high flight speed. It was benefit for reducing the required power using a smaller rotor diameter.

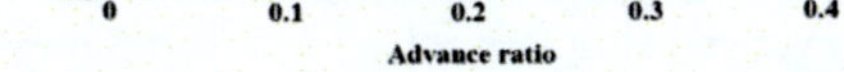

Fig. 4. Comparison of the power required of rotors with different diameters [12].

Lately, the feasibility of the variable rotor radius and speed was evaluated using the analysis method by Mistry et al. [13]. The research showed that the variable rotor radius is more effective for the high thrusting force and high altitude rotor properties. Furthermore, a more improvement may be achieved by combining RPM reducing and rotor radius variation. The research was quite profound but incorporating the simplified blade structure and aerodynamic simulation. A flexible rotor concept was modeled and tested by Sicard and Sirohi on the test bed for a small scaled rotor [14]. The diameter was 18 inches. The result showed that no flutter occurred for the rotor when it unfolded completely.

The effect of variable diameter on the rotor property improvement in different flight environment was discussed by Mistry and Gandhi using utility helicopter as an example [15]. The analysis showed that the maximum flight speed that the helicopter could achieve was reduced, but the effect of power reduction brought by optimized rotor diameter was better and better with the flight altitude increased. The power reduction brought by optimized rotor diameter in the low and high speed flight was larger. The power reduction brought by the optimized rotor diameter was smaller in the medium flight speed (cruising). Especially, the power reduction was quite small in the low altitude flight.

In summary, the variable diameter concept gave feasibility for the improvement of helicopter properties. But, the implement facility of varying radius was completely mechanical in the full scale system.

3 Variable Rotor Speed

Most of helicopters flight at a fixed rotor speed usually. The blade tip speed is 198–228 m/s commonly. But, for high speed helicopters, such as compound helicopters [16, 17], the blade tip speed may reduce in the high speed flight commonly. The rotor speed conversion will be carried out at a fixed forward flight speed. The reason is showed in the Fig. 5 [18]. The large blade tip speed will result in an unacceptable noise level in hover and low speed flight. The rotor thrust is limited and the rotation ability is reduced when the blade tip speed is reduced. The high blade tip speed is limited by compressibility, while the low blade tip speed is limited by stall in hover and low speed flight. However, most helicopters are used in a range of speeds and gross weights, while the properties could be increased through optimizing the rotor speed in every flight status for variable RPM rotorcrafts. The stall limitation could be eliminated or reduced through compound wing and co-axis rotor unloading retreating blade at high speeds. The obvious behalf can be achieved by using variable rotor speed in general helicopters, but the poor efficiency brought by changing engine and transmission speed needs be improved to avoid resonance risk. Over the years, the variable rotor speed concept is important especially for long endurance, high speed or heavy rotorcraft. At the same time, it is considered to improve the property and maneuverability of general helicopters by using medium rotor speed. In the following section, the development history of variable rotor speed is reviewed simply firstly. Then, some important research projects are introduced. Finally, the challenge faced with implementing variable rotor speed is described.

3.1 Development History of Variable Rotor Speed

The effect of blade tip speed on the hovering properties and maximum speed is evaluated by Gustafson and Gessow, as early as 1946 [19]. The research proposed that the blade

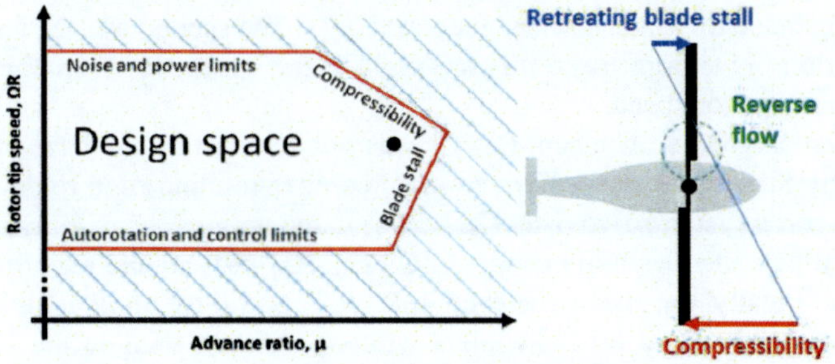

Fig. 5. Typical envelope of the blade tip speed selection [18].

tip speed need be low to the typical value (152.4 m/s) to improve hovering properties. It increased in the forward flight. The maximum speed is improved by the delayed stall. At the time, the maximum speed doesn't reach the transsonic velocity. Therefore, increasing blade tip speed is beneficial for the achieving high speed without the stall of retreating blade. The author proposed that two stage transmissions are required to achieve the two objectives.

A two stage transmission system is installed in the H-5H helicopter at 1953, and the flight test is successful (Fig. 6) [20]. The original objective is to test the retreating blade stall effect when the forward speed increases, which is realized by the variable rotor speed. The engine speed keeps invariable. The 223 RPM large rotor speed and 173 RPM small rotor speed is realized by using high and low variable-speed gear. The normal rotor speed is 194 RPM. The flight experiment shows that two-speed transmission is beneficial for improving high altitude rotorcraft properties.

Fig. 6. Sikorsky H-5 Helicopter from military-aircraft.org.uk.

The XH-51 compound helicopter was developed by Lockheed Corporation at 1963 (Fig. 7). The maximum speed achieved 487 km/h when the rotor was 95.5% normal rotor speed during the flight experiment at 1967. The large advancing blade tip speed effect at the power required was delayed with the gradual reducing of rotor speed [21]. But the pilot reported as well, the stability reduced at small rotor speed. The result of flight experiment showed also that the further expansion of rotor speed and airspeed envelope was limited by two factors. The first was the increasing of vibration level at the conversion from the large rotor speed to medium rotor speed. The reason was that the advancing blade tip Mach number exceeded 0.91. The second was the appearance of rotor oscillation in the high speed forward flight at the conversion from the small rotor speed to medium rotor speed.

The Advancing Blade Concept (ABC) was put forward on the base of rigid rotor technology by Sikorsky at 1964. Two co-axial counter-rotating rigid rotors are used as the main lift force system. Two S-69 (XH-59A) demonstrators were fabricated at 1972 (Fig. 8) to validate the feasibility of the ABC [22]. XH-59A started experimental flight at July, 1973. Finally, the speed achieved 488 km/h, and the ceiling achieved 7770 m. It set a flight speed record for helicopters without assistant wing in the world at that time. The feasibility of the ABC was validated adequately. The rotor speed was 70% of normal rotor speed in the high speed cruising flight. This meant that the high speed advancing blade tip Mach number reduced. Therefore, the power required reduced, and the maximum speed increased. The power turbine speed/rotor speed was adjusted using two speed control sticks. Thereby, the variable rotor speed could be used to optimize the helicopter property or reduce structure resonance. But, the problem such as large weight, large drag force, large vibration, complex mechanical control system and complicated power transmission structure was discovered for the helicopter.

Fig. 7. Lockheed XH-51A from www.aviastar.org

Fig. 8. XH-59A from 1000 air craftphotos.com [22].

The Optimum Speed Rotor (OSR) concept was put forward by Karem [23]. The large varying rotor speed was used to reduce the required power. The OSR concept was used in the A160 UAV (Fig. 9) to improve flight efficiency and reduce power. Thereby, the flying range, property, payload and maximum ceiling increased. At the same time, the fuel consumption and noise reduced. The OSR system flighted in a very large rotor speed range, which will keep the optimized blade load, reduce the power required thereby. The rotor speed range was 60–100% for A160T UAV. The rotor speed was adjusted continually with the advance ratio, weight and air density by OSR to keep the optimized lift force. A160 UAV achieved unprecedented endurance. A 18.7 h record was set up at 2008. A two-speed transmission was used to make the power required being minimum with the variation of the airspeed, weight and altitude. A very high stiffness weight ratio was used in the blade design. Lately, the A-160 project was canceled because of the crash caused by excessive vibration [24, 25].

A variable rotor speed and torsional moment matching system was developed by Airbus Helicopter. It is used in the H145 helicopter. The aim is to reduce noise and improve steering performance, in particular in city. The 96.5–103.5% of normal rotor speed is possible [26]. The similar system is used in the Mi-8 helicopter and its retrofits [27].

The variable rotor speed technology has been applied in some high speed rotorcrafts successfully, and become a key technology [28]. For example, the rotor speed reduces by 20% for Sikorsky X2 technology demonstrator [29], and 20% for Airbus X3 [30], and 18% for V-22 Osprey [31]. The rotor speed of X2 is designed to slow down at high forward flight speed to ensure the Mach number of advancing blade tip being not more than 0.9 [28], in order to avoid the drag compromise in connection with the compressing. The 518 km/h forward speed need rotor speed be reduced to 360 RPM (namely reducing by 20%) at 10,000 foot, ISA, comparing with 446 RPM in hover. Of course, the variety brings difficulty for the design of blade natural frequency.

In conclusion, the application of variable rotor speed has existed for a long while, but it is limited by the real factors such as dynamics problem and implement of driving system. It was used to reduce rotor speed in the forward flight in order to prevent the stall of retreating blades for many rotorcrafts in early days. The objective of current OSR concept is that adjusting rotor speed in a large range for different cruising speed in order to achieve a higher efficiency.

3.2 Large Development Plan for Variable Rotor Speed

The high speed capacity and availability factor promoted is required by all kinds of future advanced rotorcrafts [32]. It is showed in the Fig. 10 [33]. The variable rotor speed is a method which can improve the speed and efficiency of general, compound and tiltrotor helicopters. Currently, the variation of rotor speed is about 15%. It is achieved through reducing the turbine engine speed mostly, which results in the loss of the engine efficiency. The future rotorcraft may need rotor speed reduce by 50%. The subsonic velocity rotorcraft project started at 2012 by NASA, and follow-up Revolutionary Vertical Lift Technology (RVLT) project. Its current work is development and verification of engine technology and design required by the realizing of 50% rotor speed and keep high engine efficiency simultaneously through changing engine speed and fixed ratio transmission system alone.

The current study and development of America, Europe and Russia shows the necessity of variable rotor speed technology for the future rotorcraft [34]. On the one hand, the variable rotor speed increases the efficiency of rotorcrafts, and reduces noise. On the other hand, the departure requirement of rotor speed in hover and faster forward flight may be settled with the variation of the rotor speed. The Heavy Lift Rotorcraft System Investigation project developed by NASA [35] showed the feasibility of the variation of rotor speed on the increase of rotorcraft efficiency. Three different kinds rotorcraft configuration used for passenger transport in the short distance and medium distance flight were investigated. The tiltrotor aircraft was considered to have most likely commercial competitiveness. The transport capacity is 90 passengers, and the air range is 1850 km, and the speed is 155 m/s for the large civil tiltrotor. The variation range of rotor speed required is 50% (Hover, 200 m/s; Cruise, 105 m/s) [36], in order to achieve economic competitiveness.

Fig. 9. A160T UAV in Flight [23].

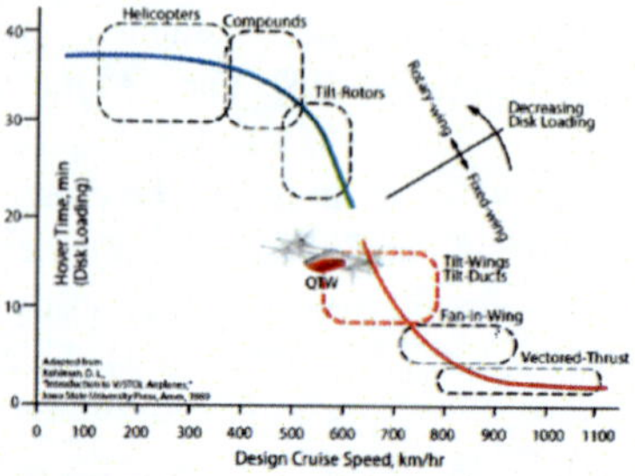

Fig. 10. Features of VTOL Concept [33].

So far, there have been some main rotorcraft research projects, such as the Future Vertical Lift (FVL) project of America and Clean Sky2-Fast Rotorcraft project. The objective of the two projects is to develop high speed rotorcraft with prominent hovering and vertical takeoff and landing capability. These two requirements result in a variable rotor speed. The future project will concentrate on the noise reduction with the variable rotor speed controlling [37]. The objective of the FVL project is to fabricate two kinds of Joint Multi-Role Technology Demonstrator (JMR-TD): one is SB > 1 developed by Sikorsky-Boeing, the other is V-280 developed by Bell and Lockheed Martin [38]. Finally, V-280 is selected by US army. The Clean Sky 2 project will fabricate two demonstrators: LifeRCraft developed by Airbus and the Next Generation Civil Tiltrotor (NGCTR) developed by Lenonardo [37].

3.3 Realization Challenge of Variable Rotor Speed

A constant rotor speed is kept for most helicopters to avoid resonance, because of the complexity, additional weight and depressed efficiency for variable rotor speed realized by a speed changing transmission and throttling engine. These problems will be discussed in short in the follow-up section.

Blade Resonance

One example of UH-60A fanplot is showed in Fig. 11 [39]. It indicates the challenge when the rotor speed is varied. The modal frequency of rotating blade is a function of rotor speed because of the centrifugal stiffness. The careful design is carried out for the blade at the 100% rotor speed to assure the first some order frequencies standing away from the aerodynamic force frequency (Multiple of rotor frequency). There will be a risk of approaching to the resonance mode when the rotor speed is varied. That is to say, one or more blade mode will be activated. Only if there is appropriate damp, otherwise the resonance will result in large deformation, vibration and large dynamic load. And it is a potential hazard. The method to get over the resonance for changing rotor speed may be realized by the advancing blade design and rotor speed controller. The modern manufacture of composite material considers more anti-resonant, innovative, very light

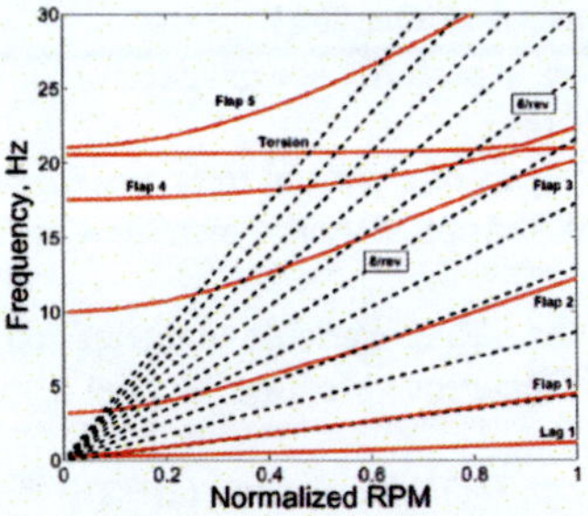

Fig. 11. Fanplot of the UH-60A Black Hawk rotor [39].

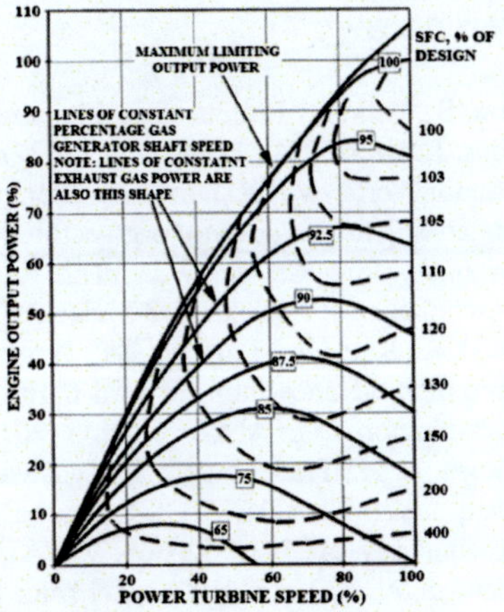

Fig. 12. Typical variation of SFC with turboshaft speed [40].

rigid blade. In the other words, the advanced controller and two stage (or more stage) transmission is selected to avoid the rotor speed approaching to resonance.

Variable Speed Turbines Efficiency

At the present day, turbine engine is designed to use at single output shaft rotating speed peak value. The engine efficiency reduces quickly if the designed rotating speed is deviated. The engine efficiency reduces with the reduction of rotor speed and torsional moment in respect of Specific Fuel Consumption (SFC, the lower the better) (Fig. 12). 40 It is the base line SFC at 100% engine speed and 100% engine output power. If the engine speed is reduced by 50%, and the limiting engine power is retained, then the SFC will increase by 30%. In other words, the full engine speed is kept, but 70% power cost is required, and the SFC only increases by 3–5%. The increase of the SFC for the engine with reduced speed, is able to get over any rotor power increase.

4 Conclusions

In summary, the investigation progress of implement methods for high advance ratio, such as variable rotor diameter and rotor speed is discussed. The following conclusions are drawn.

(1) There are some investigations on variable diameter technology. But, it is not applied in reality because of the complexity for implementation.
(2) The variable rotor speed technology has been applied in the high speed helicopter successfully. It has become a key technology.
(3) The range of variation for rotor speed is about 20% currently. In the future, it will enlarge continually, and is expected to be 50%.

References

1. Im, B., Park, S.: Effects of design parameters on coaxial rotorcraft with propellers on wing tips. J. Aircr. **59**(4), 1104–1111 (2022)
2. Anon: World eVTOL Directory. Electric VTOL News by the Vertical Flight Society [online database]. https://dev.vtol.org/vectored-thrust/. Accessed 12 Oct 2021
3. Han, D., Dong, C., Wei, W.L., et al.: Research progress in performance of adaptive rotor. Acta Aeronautica et Astronautica Sinica **39**(4), 21603 (2018)
4. Feil, R., Rauleder, J., Hajek, M.: Aeromechanics analysis of a coaxial rotor system in hover and high-advance-ratio forward flight. In: 34th AIAA Applied Aerodynamics Conference, Washington, D.C. AIAA Aviation (2016)
5. Segel, R.M., Fradenbrugh, E.A.: Development of the TRAC variable diameter rotor concept. In: AIAA/AHS VTOL Research, Design, and Operations Meeting, George Institute of Technology, pp. 1–10. Reston, VA: AIAA (1969)
6. Linden, A.W.: Variable diameter rotor study, tech. rep., DTIC Document, (1972)
7. Fradenburgh, E.A., Murrill, R.J., Kiely, E.F.: Dynamic model wind tunnel tests of a variable-daimeter, telescoping-blade rotor system (TRAC Rotor), Tech. Rep., USAAMRDL Technical Report 73–32 (1973)
8. Fradenburgh, E.A., Hager, L.N., Keffort, N.F.K.: Evaluation of the TRAC variable diameter rotor: preliminary design of a full-scale rotor and parametric mission analysis comparisons, Tech. Rep., USAAMRDL Technical Report 75–54 (1976)
9. Matuska, D., Dale, A., Lorber, P.: Wind tunnel test of a variable-diameter tiltrotor (VDTR) model, Tech. Rep., NASA CR 177629, National Aeronautics and Space Administration (1994)
10. Fenny, C.A.: Mechanism for varying the diameter of rotors using compound differential rotary transmissions. In: Proceedings of the 61st Annual Forum of the American Helicopter Society. Grapevine, TX (2005)
11. Prabhakar, T., Gandhi, F.: A centrifugal force actuated variable span morphing helicopter rotor. In: Proceedings of the 63rd Annual Forum of the American Helicopter Society, Virginia Beach, VA (2007)
12. Kang, H., Saberi, H., Grandhi, F.: Dynamic blade shape for improved helicopter rotor performance. J. Am. Helicopter Soc. **59**(1), 032008 (2010)
13. Mistry, M., Gandhi F.: Performance improvement with variable rotor span and RPM. In: Proceedings of the 66th Annual Forum of the American Helicopter Society, Phoenix, AZ (2010)
14. Sicard, J., Sirohi, J.: Aeroelastic stability of an extremely flexible, variable- diameter rotor in hover. In: Fifth Decennial AHS Aeromechanics Specialists Conference, San Francisco, CA (2014)
15. Mistry, M., Gandhi, F.: Helicopter performance improvement with variable rotor radius and RPM. J. Am. Helicopter Soc. **59**(4), 17–35 (2014). https://doi.org/10.4050/JAHS.59.042010
16. Han, D., Barakos, G.N.: Transient aeroelastic response of a rotor during rotor speed transition in forward flight. J. Aircr. **59**(4), 1020–1026 (2022)
17. Öhrle, C., Frey, F., Thiemeier, J., et al.: Compound helicopter X3 in high-speed flight: correlation of simulation and flight test. In: Proceedings of the American Helicopter Society 75th Annual Forum, Part 1, American Helicopter Soc. International, Alexandria, VA (2019)
18. Leishman, J.G.: Principles of Helicopter Aerodynamics. Cambridge University Press (2002)
19. Gustafson, F.B., Gessow, A.: Effect of rotor-tip speed on helicopter hovering performance and maximum forward speed, Tech. Rep., NACA ARRL6A16 (1946)
20. Munter, P.L.: Design and flight test of a two-speed helicopter transmission. In: Proceedings of the American Helicopter Society 10th Annual Forum (1954)

21. Lentine, F.P., Groth, W.P., Oglesby, T.H.: Research in maneuverability of the XH-51A compound helicopter, Tech. rep., USAAVLABSTR-68–23 (1968)
22. Jia, Z.Q., Lee, S.: Assessment of turbulence modeling and wake-grid resolution for lift-offset coaxial rotor simulations. In: 2018 Applied Aerodynamics Conference, Atlanta, Georgia AIAA AVIATION Forum (2018)
23. Karem, A.: Hummingbird A160 and optimum speed rotor (OSR). In: AHS International Specialists' Meeting - Unmanned Rotorcraft: Design, Control and Testing, Proceedings, pp. 335–356 (2005)
24. DiOttavio, J., Friedmann, D.: Operational benefits of an optimal, widely variable speed rotor. In: Presented at the American Helicopter Society 66th Annual Forum, Phoenix, AZ (2010)
25. Karem, A.E.: Optimum speed rotor. https://www.google.com/patents/US6007298, US Patent 6, 007, 298 (1999)
26. US Department Of Defence: Eurocopter EC145 UH-72 Lakota Helicopter Flight Manual (2008)
27. MIL Design Bureau: Mi-8-helicopter flight manual (2005)
28. Ward, E.A., Chopra, I., Datta, A.: Design of self-twisting rotor blades for high-speed compound rotorcraft. In: AIAA SciTech Forum, Grapevine, Texas 25th AIAA/AHS Adaptive Structures Conference (2017)
29. Blackwell, R., Millott, T.: Dynamics design characteristics of the Sikorsky X2 technologyTM demonstrator aircraft. In: Proceedings of the American Helicopter Society 64th Annual Forum, Montreal, Canada (2008)
30. Nelms, D.: Speed streak: eurocopter's X3 high-speed hybrid helicopter combines fixed and rotary-wing flying characteristics. Aviat. Week Space Technol. **174**(24) (2012)
31. V-22 Osprey 2010 Guidebook, Tech. Rep. NAVAIR PMA-275 Control Number 10–28, NAVAIR (2010)
32. Suder, K.L., Durbin, K.S., Giel, P., et al.: Variable speed turbine technology development and demonstration. In: Proceeding of the AHS International 74th Annual Forum & Technology Display, Phoenix, Arizona, USA, May 14–17 (2018)
33. Kohlman, D.: Introduction to V/STOL airplanes. Iowa State University Press, Ames, IA (1981)
34. Amri, H., Hartenthaler, K., Weigand, M.: Mass and kinematic analysis of compound split with simulation of the shifting process for variable rotor speed. In: Proceeding of the AHS International 74th Annual Forum &Technology Display, Phoenix, Arizona, USA (2018)
35. Johnson, W., Yamauchi, G.K., Watts, M.E.: NASA heavy lift rotorcraft systems investigation, Tech. report, NASA Ames Research Center and Langley Research Center (2005)
36. Acree, Jr, C.W., Yeo, H., Sinsay, J.D.: Performace optimization of the NASA large civil tiltorotr, Tech. report, NASA Ames Research Center, Moffett Field, California 94035, USA (2008)
37. European Aeronautics Industry. Clean Sky 2 Joint Technical Program (2015)
38. U.S. Army. Future Vertical Lift (FVL) Capability Set 1 Request for Information (2016)
39. Bowen-Davies, G.M.: Performance and loads of variable tip speed rotorcraft at high advance ratios. The Graduate School of the University of Maryland, Doctor of Philosophy (2015)
40. Walsh, P.P., Fletcher, P.: Gas turbine performance. Blackwell Science Ltd and The American Society of Mechanical Engineers (1998)

Progress of Study on Piezoelectric Energy Harvesting Technology and Its Applications in Helicopters

Hongyan Zhu[1](✉), Wei Hu[1], Maolin Sun[2], Zheng Liu[1], Xinhao Yu[1], Kuanyang Xi[1], Bo Liu[1], and Shiyao Liu[1]

[1] AVIC Harbin Aircraft Industry Group Co., Ltd., Harbin 150066, China
zhuhy012@avic.com
[2] 95966 PLA Troops, Harbin 150060, China

Abstract. In order to improve aircraft transport efficiencies, reduce fuel consumption, utilize secondary energies or external energies, three research directions have been determined by the aviation industry in recent years. Energy harvesting technology is one of them. Energy harvesting is the process by which kinetic energy, solar and so on can be converted to usable electric energy. The ultimate objective is to develop self-powered sensors, actuators, or power other small electronic devices. The vibration caused by the engine and rotor blade is a common energy source for helicopters. It may be converted by piezoelectric effect. An advantage of the piezoelectric energy harvesting (PEH) technology is that it can harvest energies in a larger frequency band. The application is simple. Many researchers have utilized the piezoelectric effect to develop vibration energy harvesting devices. The material selection, energy harvesting and storing system, as well as applications of piezoelectric energy harvesting technology in helicopters is introduced. A prospect of the development for its application in aviation in the future is made.

Keywords: Helicopter · EnergyHarvesting · Piezoelectricity · Vibration

1 Introduction

Fuel consumption will increase continually with the development of air transport industry. For example, air transport (passenger and cargo) fuel consumption is 10% of the whole in some developed countries. It is anticipated that the aviation demand will go up steadily, and the civil aviation demand will double at 2032 [1]. The increasing of resource costs will push the rising of aviation transport costs. In order to improve aviation transport efficiencies, three main research subjects have been determined, and one of them is energy harvesting technology. Thereby, the sublevel and exterior energy sources, such as solar, heat, wind and mechanical energy will be utilized [2]. The mechanical energy such as vibration is relatively easier to acquire with respect to other energy sources

This is supported by the Natural Science Foundation of Heilongjiang Province of China (Grant No. JQ2020E006).

Chinese Society of Aeronautics and Astronautics (Ed.): CASTC 2023, LNME, pp. 118–126, 2024.
https://doi.org/10.1007/978-981-99-8861-7_13

of these reproducible energy sources [1]. The energy harvesting demand is increasing because that the application of smart sensors and actuators in aviation, such as structure health monitoring (SHM) and vibration controlling is going up continuously. The smart sensor demand is increasing at 10% annually. The global smart sensor market reaches 378.5hundred million dollars at 2019 [3]. At the same time, the power demand of electronic equipment decreases greatly. For example, the power of a wireless network sensor node is 10^{-3}W [4]. Although the power consumption is reduced, but smart sensors and moveable electronic equipments depend on batteries as energy sources basically. There are some limits such as cyclic replacing or recharging, preventing dimension and weight reduction, polluting environment. Therefore, the research on sensors, radio communications and power generators that harvest energy from the surrounding environment has received more attention. Piezoelectric materials have been used to transform the surrounding vibration energy to electric energy more and more widely. It has been applied on helicopters successively. Material selection, energy harvesting and storage system of piezoelectric harvesting, and its applications on the aviation is introduced, in the hope of providing reference for the aviation intelligentizing.

2 Piezoelectric Energy Harvesting

Energy harvesting is defined as transforming the surrounding environment energy to electric energy, thereby powering electronic equipments. These energies may be captured from mechanical displacement, heat gradient and so on. Some applications are showed in Fig. 1 [5, 6]. There are piezoelectric trees capturing energies from bold wind (Fig. 1(a)), thermoelectricity seats capturing energies from passenger body heat (Fig. 1(b)), solar airplane capturing energies from sunlight (Fig. 1(c)), piezoelectric pavements capturing energies from human movement (Fig. 1(d)), piezoelectric knees capturing energies from

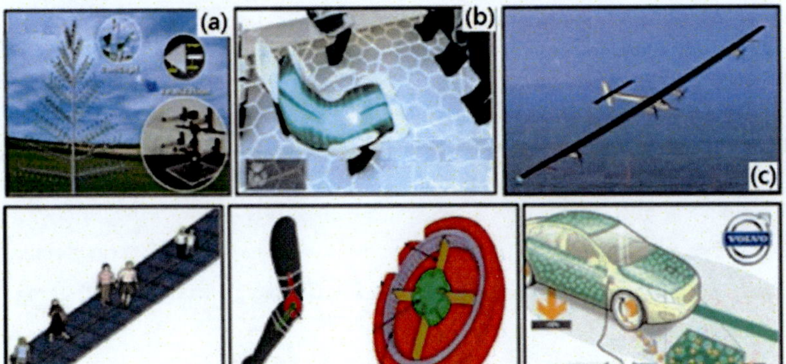

Fig. 1. Energy harvesting applications in aviation and non-aviation: (a) Piezoelectric tree, (b) Thermoelectric seat, (c) Solar airplane, (d) piezoelectric walkway, (e) piezoelectric knees, (f) Car with multifunctional structural panels [5, 6].

knee movement (Fig. 1(e)), cars with multifunctional structural panels, including energy harvesting and storage.

There are four methods by which vibration energies are converted to electric energies, namely, piezoelectric, electrostatic, electromagnetic and magnestostrictive generator. The reason why piezoelectric material is interested is that it is multifunctional. It can be used as sensors, energy harvesting devices and actuators. Moreover, piezoelectric converters may be used to fulfill functions of energy harvesting and actuation simultaneously [7]. There are many piezoelectric materials that can be used for energy harvesting devices, three common materials are Lead Zirconite Titanate (PZT), Macro-Fiber Composite (MFC) and polyvinylidene fluoride (PVDF). PZT is the material used most frequently. It is widely used in piezoelectric micro-electro-mechanicalsystem (MEMS). However, flexible MFC and PVDF are more suitable due to the brittle of PZT. The drawback brought by the flexibility is a reduction of piezoelectric coupling coefficient. A soft wave energy harvester is developed by Kordmahale using piezoelectric-based MFC (Fig. 2) [8]. The final dimensions are 25 cm × 25 cm × 0.4 cm.

Generally, the piezoelectric energy harvesting system is composed of a piezoelectric energy converter, interface circuit and energy storage cell (Fig. 3) [9]. The piezoelectric energy converter is located at the load carrying path usually. It is bonded on the master structure, or embedded in the master structure. It deforms with the master structure. The energy harvesting and storage process can be divided into two steps simply. Firstly, vibration energy is converted to electric energy by the piezoelectric energy converter. Secondly, the electric energy is stored in batteries or capacitors by the interface circuit. Constants required to characterize piezoelectric materials are piezoelectric strain/charge (d), piezoelectric stress/voltage (g), electromechanical coupling coefficient (k). The design method of piezoelectric converter should be considered in order to maximize energy harvesting at smallest coupling loss.

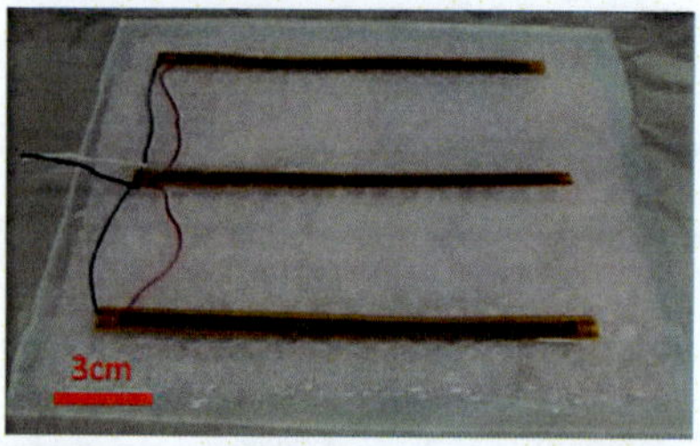

Fig. 2. Final harvester with 3 MFCs [8].

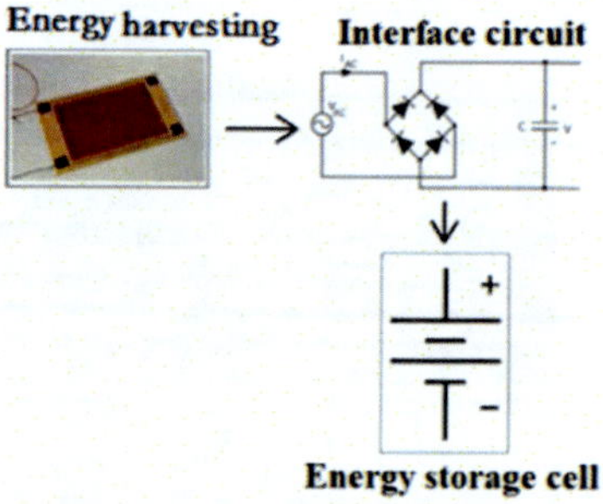

Fig. 3. Energy harvesting and storage system [9].

The piezoelectric vibration energy harvester for the industrialapplications is simple. It is designed as a cantilever with tip mass. The design utilizes a resonance system which is excited by mechanical vibration. A piezoelectric cell is fixed on both sides of a flexible cantilever. The cantilever deforms when the mass moves. Thus, mass movement kinetic energy is converted to electricity by the piezoelectric cell. Apiezoelectric energy harvester product and its harvesting principles shown in Fig. 4 [10–12]. The resonance bandwidth in which piezoelectric harvesting devices operate is narrow. It is depended on the mechanical damping dm. Thereby, the energy harvester is designed as tuning up to the resonance frequency $\Omega = \sqrt{k/m}$. (m is the mass. k is the stiffness.) It is the same as the dominant vibration frequency. The electrical energy harvesting gives feedback to the resonance mechanism by the additional electromechanical damping de. The maximum electrical power harvested by a piezoelectric vibration energy harvesting system may be derived from the equation of motion (1) [10]:

$$m\ddot{x} + d_m\dot{x} + d_e\dot{x} + kx = -m\ddot{z} \quad (1)$$

Conventional piezoelectric energy harvesting systems are all integrated into a master structure. One of the shortcomings is that the mass load effect is often caused. And precious space is exhausted. A self-charged structure (Fig. 5) is developed by Anton et al. to advance functionalities, reduce adverse load effects [13]. The self-charged multi-lamina structure is consisted of active piezo ceramics lamina harvesting energy, membrane battery lamina storing energy, and metal basal lamina in the middle of double piezo configuration. The electric energy is transferred from the piezo ceramics lamina to the membrane battery lamina by a proper regulating circuit. Thereby, the electric energy is harvested and stored by this structure at the same time.

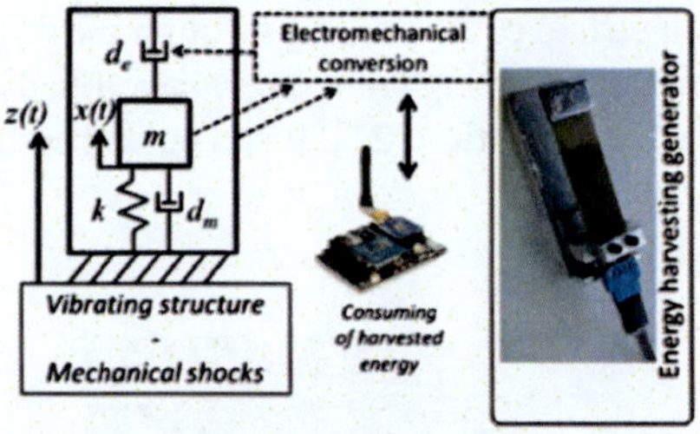

Fig. 4. Piezoelectricvibration energy harvesting principle [10–12].

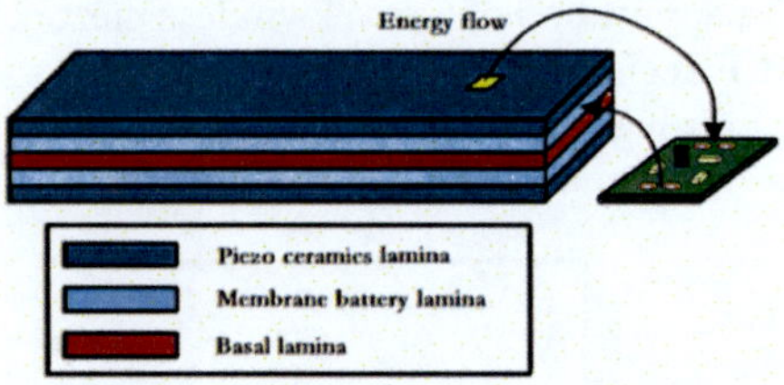

Fig. 5. Self-powered structure [13].

3 Applications of Piezoelectric Energy Harvesting in Helicopters

The main energy of helicopters is chemical or electric. Other sublevel energy may produce, such as heat, vibration, sound, when helicopters flight [1]. These sublevel energies may be used. For example, the sound effect or airflow effect energy converting can be carried out by piezoelectric energy converters [14]. In general, the vibration, flutter aeroelasticity instability caused by the airflow, has become the main excitation in studies [15]. The piezoelectricity based energy harvesting device is small, light and durable. It is adapted to the application in helicopters just right.

The vibration produced by the coupling between aerodynamics and structure is a natural property of helicopters. The vibration is a main limit factor for the maximum forward speed of helicopters. Though the vibration will appear in the all fight envelope, but it is especially significant in the forward flight. For this reason, the helicopter vibration energy harvesting is studied a lot. The peak vibration piezoelectric energy harvesting in the flight of a helicopter is investigated by the laboratory of Zhu et al. [16]. The self-powered smart sensor node which is as big as a credit card is developed. It can produce 240μW electric energy excited at 67 Hz and 0.4 g peak vibration.

The common method characterizing the vibration energy harvester is using the sine wave at resonance frequency to excite harvester and measure the voltage. However, the vibration spectrum is composed of several peaks at different frequencies in practice. Figure 6 shows the actual vibration frequency spectrum of PZLSW-4 helicopter vertical stabilator [17]. It can be seen from the figure that a majority of vibration energies disperse over a frequency band still though there is an amplitude peak at special frequency. It is evident that the energy harvester needs the ability to adapt to the wider vibration frequency band unlike conventional harmonic resonance harvester for the actual applications. Otherwise the whole energy harvesting capability is limited. Therefore, the performance of piezoelectric energy harvesters at vibration with different frequencies is studied by Zhu et al. [17]. Their initial target is the vertical fin of a PZLSW-4 helicopter.

The wireless communication strain sensor powered by piezoelectric energy harvesting is fitted on a helicopter pitch link to measure its load by Arms et al. in the investigation sponsored by US navy [18]. MicroStrain® declares that the flight test of it is successful in 2007 [19]. The piezoelectric vibration energy harvesting powering of the health and usage monitoring system (HUMS) in the helicopter reducer is studied by Agusta Westland [20]. The optimum energy harvesting method is performed for sensors of HUMS in order to reduce the cable distance and prevent electrical cable passing through the joint. The application of a novel wireless piezoeclectric load sensing technology developed by KCF Technologies in rotorcraft HUMS is studied by Loverich [21]. It can be embedded in existing rotorcraft critical load paths (Fig. 7).

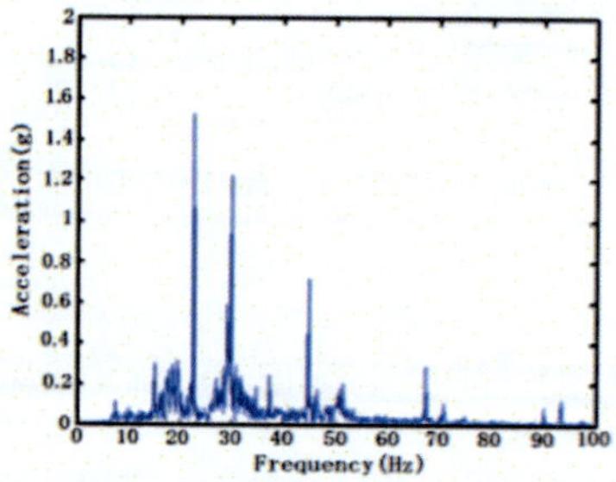

Fig. 6. Vibration frequency for helicopter Vertical stabilizer [17].

Fig. 7. Potential application of KCF's load monitoring on helicopter [21].

A multifunctional composite material for next generation helicopter structures is developed by Harursampath et al. in their research project [9]. Specially, a hybrid composite material called piezoelectricity-battery fiber reinforced composite material (P-BFRC) is studied (Fig. 8). It can be used for helicopter blades. P-BFRC is composed of piezoelectricity fibers and battery fibers arranged in the optimum way. The goal of P-BFRC applied at rotor blades and the region of larger amplitude is to harvest energy from structure vibration using piezoelectric plate. Then, it is stored in the battery of structure itself. In sum, the multifunctional structure system is considered as an effective step toward aviation with more clean and more sustainable development.

The material of commercially available piezoelectric harvesters is ceramic material PZT. But it is becoming less desired when compared to newer single crystal materials (PMN-PT). PMN-32PT single crystal used as an energy harvesting devices in environments typical of helicopter applications is invested by Wozniak et al. [22]. Their initial design resonance frequency is 800 Hz (a common gear mesh frequency in rotorcraft gear boxes). At room temperature, the lab test article (Fig. 9) can produce 25 VRMS at 1.0 g base acceleration and is used continuously for 120 h stably. The overall mass of compact harvester designed by Wozniakis less than 40 g.

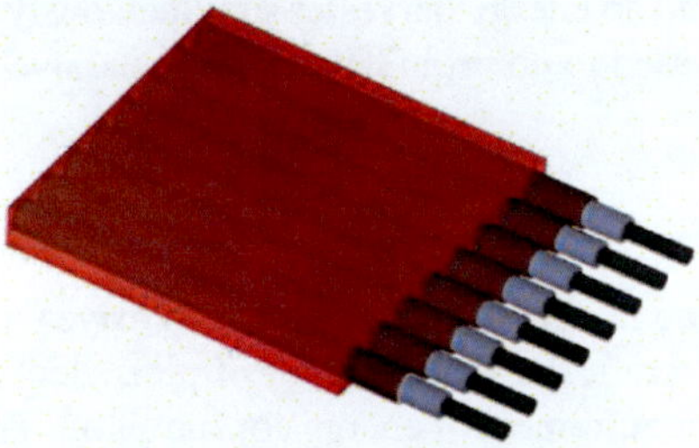

Fig. 8. Piezoelectricity-Battery fiber reinforced composite [9].

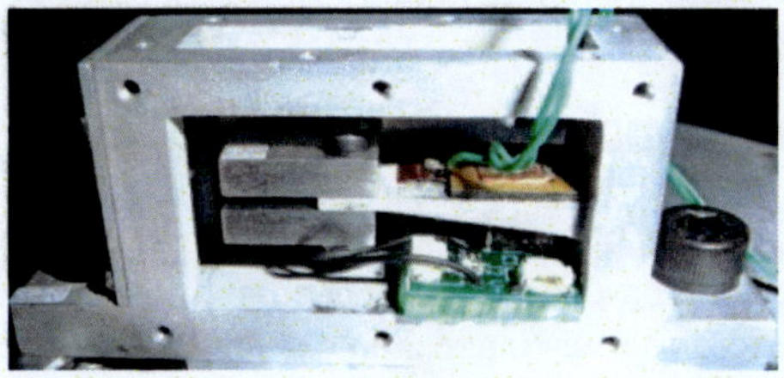

Fig. 9. Compact harvester fully assembled [22].

Currently, the helicopter rotor blade is replaced when the conservative lifetime predicted by analysis arrives. The analysis prediction assumes a worst usage case because that rotor failure is unacceptable. It will make blades being discarded when they can be used several flight hours still. In order to determine the lifetime of blades, the SHM system in blades is required. In addition, with the development of active technologies, the application of active devices embedded in the rotor of helicopters for vibration and noise controlling and performance enhancement is growing rapidly. The power increases from SHM sensor requiring milliwatts to actuator requiring hundreds of watts. Currently, the power required by helicopter blades is fulfilled through slip rings. They are hydraulic or electrical. At the same time, they bring the hub complexity and drag increasing. A method to settle the issue may be the application of a piezoelectric energy harvester embedded in helicopter blades to power the devices. The SHM system in blades and powering methods of the system is studied in the European Clean Sky–GreenRotorcraft project. The piezo electric energy harvesting in the rotor is selected as a method to power the SHM system [23]. A wireless accelerometer powered by the piezoelectric energy harvester is studied by Loverich et al. [24]. The result shows that a 1.25cubic inch wireless sensor and harvester can stream 3.2 kHz vibration data at low vibration level. And the harvester provides 6.4 mW of continuous power to the sensor at 1.0 g and 700 Hz vibration.

The composite blade buried with piezoelectric energy harvesting layer is studied by Vieira et al. [25]. Non-linear analysis on the electromechanical coupling of blade cantilever is carried out. The changing of blade electromechanical properties with the rotary speed increasing is studied. The main aim is to power low energy exhaustion electronic devices such as SHM. The energy harvesting method of self-powered smart blade in a helicopter is given by Bryant et al. [26]. It is called piezoelectric eel. The impetus of the device is provided by aeroelasticity flutter. The maximum power harvested is about 2 mW. It is enough for self-powered sensors. An energy harvesting device fitted in the helicopter blade is developed in the patent of EurocopterDeutcherGmbh [27]. The aim is to power caution light on a helicopter blade tip. In this application, the electric energy is produced by a piezoelectric harvesting device through the outside surface vibration.

Ground resonance (GR) is destructive mechanical instability for helicopters. Nonlinear Energy Sink (NES) is used to control GR passively by attaching on the helicopter blade [28]. NES is also used to control the aeroelastic behaviors of rotor blades [29]. Piezoelectric NES used as a vibration isolator and an energy harvester simultaneously is studied by Xiong et al. [30]. It is hopeful to be used to control GR instability passively.

4 Conclusions and Prospect

The application of piezoelectric energy harvesting in aircraft will be more extensive with the going deep of unmanned and intelligent aircraft, the development of piezoelectric energy harvesting technology, and the reducing requirement of energy consumption. The three issues below will be concerned.

(1) The application of small self-powered electronic equipments such as sensors and actuators powered by piezoelectric vibration energy harvesting.
(2) The study on the piezoelectricity – battery fiber energy harvesting storage composites.
(3) The investigation of self-powered structures, such as piezoelectric energy harvesting skin.

References

1. Erbschloe, D.R., Harrison III, W.E.: A center for aviation system efficiency. In: Proceedings of 14th International Energy Conversion Engineering Conference. Salt Lake City, UT (2016)
2. Thakare, N.S., Thakare, S.S., Shahakar, R.S.: Recent advancement and comparative performance analysis of energy harvesting technique. In: Proceedings of the Second International Conference on Intelligent Computing and Control Systems (ICICCS 2018).IEEE Xplore Compliant Part Number: CFP18K74-ART. ISBN:978–1–5386–2842–3 (2018)
3. http://m.ocn.com.cn/reports/2268zhinengchuanganqi.shtml?origin=baidu_sosuo&renqun_youhua=448401&bd_vid=10136513400854542129
4. Tiwari, R., Buch, N., Garcia, E.: Energy balance for peak detection method in piezoelectric energy harvester. J. Intell. Mater. Syst. Struct. **25**(8), 1024–1035 (2014)
5. Li, S.G., Lipson, H.: Vertical-stalk flapping-leaf generator for wind energy harvesting. In: Proceedings of ASME 2009 Conference on Smart Materials, Adaptive Structures and Intelligent Systems, pp. 611–619, American Society of Mechanical Engineers (2009)
6. Pozzi, M., Zhu, M.: Characterization of a rotary piezoelectric energy harvester based on plucking excitation for knee-joint wearable applications. Smart Mater. Struct. **21**(5), 055004 (2012)

7. Erturk, A., Inman, D.J.: Piezoelectric Energy Harvesting. John Wiley & Sons, Hoboken, NJ, pp. 1–94, 151–385 (2011)
8. Kordmahale, S.B., Do, J., Chang, K.A., et al.: Low cost and piezoelectric based soft wave energy harvester. MRS Advances © 2019 Materials Research Society. https://doi.org/10.1557/adv.2018.675
9. Harursampath, D., Gupta, M., Agrawal, S., et al.: Blending modern multifunctional materials with traditional structures: an approach for a greener and cleaner future. In: Proceedings of the AHS 72nd Annual Forum. West Palm Beach, Florida, USA (2016)
10. Hadas, Z., Janak, L., Smilek, J.: Virtual prototypes of energy harvesting systems for industrial applications. Mech. Syst. Sig. Process. **110**, 152–164 (2018)
11. Mateu, L., Moll, F.: Review of energy harvesting techniques and applications for microelectronics. In: Lopez, J.F., Fernandez, F.V., Lopez-Villegas, J.M., DL Rosa, J.M. (eds). VLSI Circuits Syst. II, Pts 1 2, pp. 359–373 (2005)
12. Hadas, Z., Lan, R.: Modelling and verification of piezoelectric vibration energy harvester. In: Mechatronics 2015, Advanced Mechatronics Solutions, Warsaw, pp. 305–310 (2016)
13. Anton, S.R., Inmany, D.J.: Electromechanical modeling of a multifunctional energy harvesting wing spar. In: Proceedings of the52nd AIAA/ASME/ASCE/AHS/ASC Structures, Structural Dynamics and Materials Conference. Denver, Colorado (2011)
14. DE Marqui, J., R., C.,Erturk, A.: Electroaeroelastic analysis of airfoil-based wind energy harvesting using piezoelectric transduction and electromagnetic induction. J. Intell. Mater. Syst. Struct. **24**(7), 846–854 (2012)
15. Ewere, F., Wang, G., Frendi, A.: Experimental investigation of a bioinspired bluff-Body effecton galloping piezoelectric energy-harvester performance. AIAA J. **56**(3), 1284–1287 (2018)
16. Zhu, D., Beeby, S.P., Tudor, M.J., et al.: A credit cardsized self powered smart sensor node. Sens. Actuators A: Phys. **169**, 317–325 (2011)
17. Zhu, D., Beeby, S., Tudor, M., et al.: Performance of a piezoelectric energy harvester under vibrations taken from helicopter. In: Proceedings of PowerMEMS, pp. 221–224 (2011)
18. Arms, S.W., Townsend, C.P., Churchill, D.L., et al.: Energy harvesting wireless sensors for helicopter damage tracking. In: Proceedings of the AHS International Forum 62. Phoenix, AZ (2006)
19. Arms, S.W., et al.: Tracking pitch link dynamic loads with energy harvesting wireless sensors. In: Proceedings of the American Helicopter Society 63rd Annual Forum. Virginia Beach, VA (2007)
20. Churchill, D.L., Distasi, S., Frattini, T., et al.: Development of a helicopter on-rotor HUM system powered by vibration energy harvesting. In: Proceedings of the 40th European Rotorcraft Forum, pp. 988–1001 (2014)
21. Loverich, J., Frank, J., Szefi, J.: Rotary wing dynamic component structural life tracking with self-powered wireless sensors. In: Proceedings of the American Helicopter Society 66th Annual Forum. Phoenix, Arizona (2010)
22. Wozniak, M.V., Conlon, S.C., Smith, E.C., et al.: Endurance and temperature testing of a PMN-PT single crystal energy harvester for rotorcraft wireless sensor applications. In: Proceedings of the AHS 69th Annual Forum. Phoenix, Arizona (2013)
23. Jong, P.H., Boer, A., Loendersloot, R., et al.: Power harvesting in a helicopter rotor using a piezo stack in the lag damper. J. Intell. Mater. Syst. Struct. **24**(11), 1392–1404 (2012)
24. Loverich, J., Wenner, S., Frank, J., et al.: Energy harvester powered wireless accelerometers. In: Proceedings of the American Helicopter Society 67th Annual Forum. Virgina Beach, VA (2011)

25. Vieira, W.G.R., Nitzsche, F., Marqui Jr, C.D.: Modeling and analysis of piezoelectric energy harvestingfrom helicopter blades. In: Proceedings of the 53rd AIAA/ASME/ASCE/AHS/ASC Structures, Structural Dynamics and Materials Conference, Honolulu, Hawaii, AIAA 2012–1665 (2012)
26. Bryant, M., Fang, A., Garcia, E., et al.: Self-powered smart blade: helicopter blade energy harvesting. In: Ghasemi-Nejhad, M., N. (eds). Proceedings of SPIE 7643, Active and Passive Smart Structures and Integrated Systems 17 (2010)
27. Eurocopter Deutcher Gmbh, Eads Deutcher Gmbh. Energy-self sufficient illumination system for use in helicopter, has energy supply unit designed as energy harvester to generate electrical current from vibrations of rotor blades during operation of rotor blades. German Patent No. DE2020080085517-U1 (2008)
28. Bergeot, B., Bellizzi, S., Cochelin, B.: Passive suppression of helicopter ground resonance using nonlinear energy sinks attached on the helicopter blades. J. Sound Vibr. **392**, 41–55 (2017)
29. Ebrahimzade, N., Dardel, M., Shafaghat, R.: Investigating the aeroelastic behaviors of rotor blades with nonlinear energy sinks. AIAA J. **56**(7), 1–14 (2018)
30. Xiong, L.Y., Tang, L.H., Liu, K.F., et al.: On the use of piezoelectric nonlinear energy sink for vibration isolation and energy harvesting. In: Proceedings of the ASME 2018 Conference on Smart Materials, Adaptive Structures and Intelligent SystemsSMASIS2018.San Antonio, TX, USA (2018)

High Mechanical Properties of Nickel-Based Superalloy with Dispersed Nanoparticles Formed by Laser Metal Deposition

Jiachen Guo(✉), Yong Sun, Yudeng Wang, Chang Ding, Bo Feng, and Jiafu Wang

Shaanxi Key Laboratory of Artificially-Structured Functional Materials and Devices, Air Force Engineering University, Xi'an 710051, Shaanxi, China
gjc_0219@163.com

Abstract. While laser-printed metals do not tend to match the mechanical properties and thermal stability of conventionally-processed metals, incorporating and dispersing nanoparticles in them should enhance their performance. However, this remains difficult to do during laser additive manufacturing. Here, we show that Nickel-based superalloy reinforced by nanoparticles can be deposited via laser melting of nanocomposite powders. Laser deposited nanocomposite materials have high hardness, high Young's modulus and excellent ultimate tensile strength performance in almost all laser printed Nickel-base superalloy. The improved performance is attributed to a high density of well-dispersed nanoparticles, strong interfacial bonding between nanoparticles and matrix, ultrafine grain sizes and dislocation regulation enhancement mechanism.

Keywords: Additive manufacturing · Directed energy deposition · Ni-based superalloys · Nanoparticle · Mechanical properties

1 Introduction

Recently, laser additive manufacturing (LAM) and 3D printing technology have been widely used in national defense, aerospace, automobile, biomedicine and other fields, greatly promoting the innovative development of materials in related fields [1]. In fact, compared with other structural metals, traditional laser printed metals are still at a disadvantage in terms of specific performance improvement and thermal stability. However, some studies have shown that high-density nanoparticles in metals will further enhance their related mechanical properties. It is well known that nano twinned structures are beneficial for the improvement of strength and plasticity of metallic materials [2, 3] and strengthening superhard materials such as boron nitride and diamond [4–6]. Current research shows that nanoparticles dispersed into metals can be used to strengthen metal materials [7–10], and can also play a role in refining grains and preventing solidification cracking [11, 12]. Directed energy deposition (DED), as an advanced LAM technology, has been studied and used in the manufacture of nanoparticle reinforced metals [13–15].

Here we show that Nickel-base superalloy reinforced by dense dispersed nanoparticles can be deposited by laser cladding of Nickel-base nanocomposite powders. Laser

Chinese Society of Aeronautics and Astronautics (Ed.): CASTC 2023, LNME, pp. 127–131, 2024.
https://doi.org/10.1007/978-981-99-8861-7_14

deposited nanocomposite materials have high hardness, high Young's modulus and excellent ultimate tensile strength performance in almost all laser printed Nickel-base superalloy. More importantly, the laser printing technology of such nanoparticle reinforced Nickel-base superalloy can be extended to other materials, which can further break the application performance limitations of related materials.

2 Experimental Details

DD6, the Chinese second-generation nickel-based single crystal superalloy [16], is used as the research object in the present work. Through many experiments, nano particles(up to 10 wt.%) can be slowly added to the initial superalloy powder by ball milling, so as to obtain the metal powder with the outer layer uniformly wrapped with nano particles, as shown in Fig. 1.

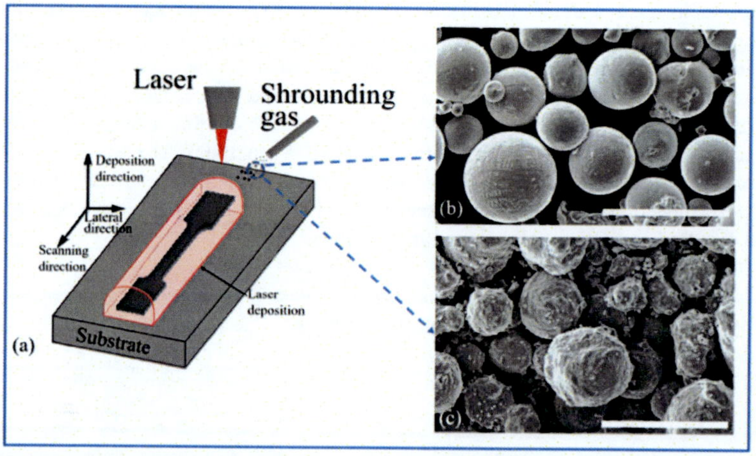

Fig. 1. (a) The schematic diagram of DED experiment, (b) surface morphology of superally powder and (c) superally nanocomposite powder particles. Scale bar, 50 μm in (b) and (c).

As shown in Fig. 1(a), the DED system were performed by using a 1 kW continuous wave Nd: YAG laser. DED processing parameters for the present investigation are laser power 250 W and scanning velocity 5.0 mm/s. The samples for tensile property measurement were cut perpendicular to the deposition direction, and were machined into plate shape, as shown in Fig. 1(a). Hardness and modulus of DEDed samples are measured and analyzed by the nano indentation instrument.

3 Results and Discussion

The morphology of the initial and nano-ZnO decorated metal powder are shown in the SEM images in Fig. 2. Figure 2 (b) (e) are an enlarged view of the local surface of Fig. 2 (a) (d). It can be seen that compared with the smooth surface of the initial powder, uniform ZnO nanoparticles are wrapped and distributed on the surface of Nickel-base nanocomposite powders. These two state powders are used in laser metal deposition experiments, we have obtained two different states of laser deposition, including DEDed and DEDed with nanoparticles samples, as shown in Fig. 2 (c), (f).

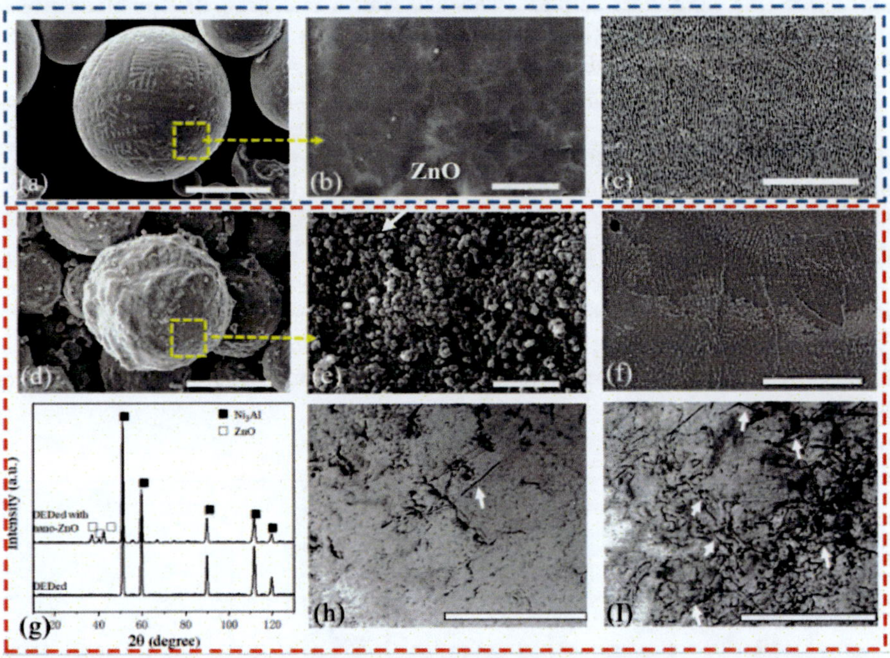

Fig. 2. (a), (d) Surface characteristics of single superally and nanocomposite powders. (b), (e) Magnified image of (a), (d). (c), (f) Cross-sectional SEM images of laser-deposited normal superalloy powder and nanocomposite powder specimens. (g) XRD patterns of the DEDed and the DEDed with nanocomposites samples. (h), (i) High resolution TEM image of dislocation distribution of the normal superalloy powder and nanocomposite powder laser-deposited samples. Scale bar, 20 μm in (a), (d), 500nm in (b), (e), (h), (i) and 300 μm in (c), (f).

By comparing the cross sections of the two laser deposition, the dendrites in the former section grow vertically and uniformly, while there are obvious grain boundaries in the latter, resulting in many fine grains. The XRD patterns of the DEDed and DEDed with nanoparticles components are shown in Fig. 2 (g); composed of face-centred cubic (fcc) Ni_3Al and ZnO phases. Except for the added ZnO phase, there was no apparent difference in the XRD patterns between the two DEDed samples within the limitation of the XRD detection (~1wt%), implying that no formation of new phases occurred or preferred grain orientation (crystallographic texture) developed during the DED.

Furthermore, it can be concluded that the grain refinement caused by ZnO nanoparticles is attributed to two mechanisms. First of all, this particle provides a high-density nucleation site. Once the liquid solidifies during the disturbance process, it will lead to a finer grain size. Second, nanoparticles will prevent the migration of newly formed grains, thus stabilizing the grain size. At the same time, ZnO nanoparticles can be used as grain growth inhibitors in superalloys because they can be used as pinning points to inhibit grain growth during solidification, recrystallization and recovery. This phenomenon has been reported in other nanocomposites [17], indicating that high volume fraction of nanoparticles is very effective for grain growth delay. And the high cooling rate during laser processing can further help to form small grains. Therefore, the combined effects of increased nucleation position, restricted grain growth and high cooling rate in the process of nano reinforced DED lead to the reduction of grain size.

Further research, the dislocation lines in the deposited samples were characterized by TEM and the results are given in Fig. 2 (h) (i). The dislocation lines at DEDed sample are sparse, whereas those at DEDed with nanocomposite are much higher, as indicated by the dislocation line marked by the white arrow in Fig. 2 (i). On this basis, dislocation is another kind of crystalline defect that scientists can play with to design high-performance crystalline materials. According to the well-known Taylor hardening law [18], the strength increases with dislocation density. Pursuing high dislocation density is still an effective way to improve the strength of materials [19]. Coincidentally, nanocomposite laser printing is a way to increase dislocation in alloy structure, as shown in Fig. 2 (i). These dislocations can form tangled dislocation structures, exhibiting higher stability than random dislocations.

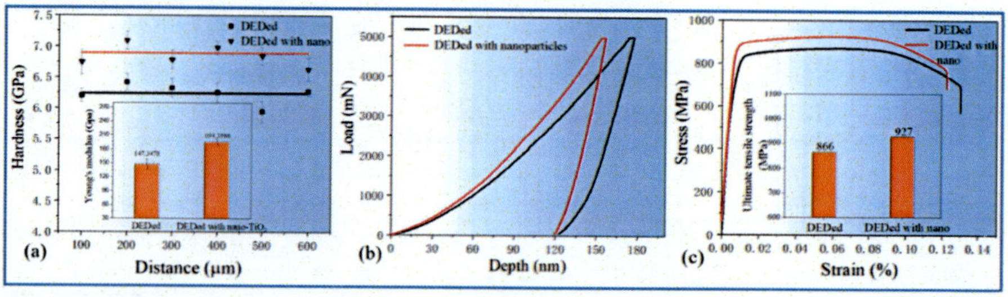

Fig. 3. (a) The hardness as a function of depth and Young's modulus in deposited samples. (b) The load-displacement curve of these two states deposited samples. (c) Tensile curves and room temperature mechanical properties of deposited samples.

For the mechanical properties of the deposited samples, the measured hardness as a function of depth at different samples is presented in Fig. 3(a), in which the data of the Young's modulus are also given for comparison. And the load-displacement curve is also shown in Fig. 2 (b). The value of hardness is approximately 6.24 GPa for the DEDed samples, and it increases to 6.9 GPa in the DEDed with nano. It shows that the hardness of the DEDed with nano deposited area is 10% higher than former. Moreover, the Young's modulus has increased by nearly 32%. It is obvious that compared with ordinary laser deposition, the ultimate tensile strength of DEDed with nano samples has been improved by nearly 7%, as shown in Fig. 3 (c).

Combined with the above research, we believe that the high dislocation density accounts for the excellent mechanical properties by the synergy of dislocation forest hardening, propagation of mobile dislocations and controllable TRIP effect. Therefore, the mechanical properties of materials can be optimized via the manipulation of dislocation density and morphologies.

4 Conclusion

In this study, the processing-microstructure-mechanical properties correlation was established, allowing to draw the following conclusions: The DEDed with nanoparticles has a textureless microstructure, consisting of fine grains and cell structure; The DEDed with nano composites has high tensile strength, excellent ductility and high hardness performance, which were higher than most conventionally fabricated cast and DEDed superalloys. The increase of strength is mainly attributed to the decrease of grain size caused by the combined effects of augmented nucleation sites, restricted grain growth and high cooling rate during DED, as well as the resulting dislocation regulation enhancement mechanism.

Acknowledgments. This work was supported by the Natural Science Basis Research Plan in Shaanxi Province of China (Program No.2023JC-QN-0402).

Declarations of interest. The authors declare that they have no known competing financial interests or personal relationships that could have appeared to influence the work reported in this paper.

References

1. MacDonald, E., Wicker, R.: Science. 353 (2016) aaf2093
2. Lu, L., Shen, Y.F., Chen, X.H., Qian, L.H., Lu, K.: Science **304**, 422–426 (2004)
3. Lu, L., Chen, X., Huang, X., Lu, K.: Science **323**, 607–610 (2009)
4. Tian, Y.J., Xu, B., Yu, D.L., M, Y.M., et al.: Nature. **493**, 385–388 (2013)
5. Huang, Q., Yu, D.L., Xu, B., Hu, W.T., Tian, Y.J.: Nature **510**, 250–253 (2014)
6. Sun, L., et al.: Carbon **127**, 320–328 (2018)
7. Tjong, S.C.: Adv. Eng. Mater. **9**, 639–652 (2007)
8. Pollock, T.M.: Science **328**, 986–987 (2010)
9. Min, Y., Akbulut, M., Kristiansen, K., et al.: Nat. Mater. **7**, 527 (2008)
10. Chen, L.Y., Xu, J.Q., Choi, H., Pozuelo, M., Li, X.C.: Nature **528**, 539 (2015)
11. DeCicco, M.P., Turng, L.S., Li, X., Perepezko, J.H.: Metall. Mater. Trans. A **42**, 2323–2330 (2011)
12. Choi, H., Cho, W.H., et al.: Metall. Mater. Trans. A **44**, 1897–1907 (2013)
13. AlMangour, B., Grzesiak, D., Yang, J.M.: J. Alloy. Compd. **680**, 480–493 (2016)
14. Gu, D., Wang, H., Zhang, G.: Metall. Mater. Trans. A **45**, 464–476 (2014)
15. Song, B., Dong, S., Coddet, C.: Scr. Mater. **75**, 90–93 (2014)
16. Committee, E.: China Aeronautical Materials Handbook: Wrought Superalloy and Cast Superalloy. Standards Press of China, Beijing (2001). (in Chinese)
17. Fattahi, M., Mohammady, M., et al.: J. Mater. Process. Technol. **217**, 21–29 (2015)
18. Taylor, G.I.: Proc. R. Soc. Lond. Ser. A **145**, 362–387 (1934)
19. Zhang, X.D., et al.: Acta Mater. **59**, 3422–3430 (2011)

Array Analysis of Membrane-Type Acoustic Metamaterials Inspired by Spider Web-Inspired Model

Ertai Cao[1], Jun Chen[2(✉)], and Heyuan Huang[1(✉)]

[1] School of Aeronautics, Northwestern Polytechnical University, 127 West Youyi Road, Beilin District, Xi'an, China
huangheyuan@nwpu.edu.cn

[2] Zhuhai Xiang Yi Aviation Technology Company Limited, Lot 32, Free Trade Zone, Zhuhai, China
chenjun@zhftc.com

Abstract. Membrane-type acoustic metamaterials (MAMs) are lightweight materials with excellent low and medium frequency acoustic properties. The typical MAM array process involves using the original single cells as-is. However, practical applicatiossns often require adjusting the size of the cells to fit the actual array area, and few studies have investigated how to determine the appropriate cell size for a given array area. In this study, we selected the spider web model, which has good noise reduction properties at low and medium frequencies between 80–1600 Hz, for array analysis. We investigated the effect of denser arrays on the noise reduction performance (sound transmission loss, STL) of the spider web MAM by comparing the STL of 1 × 1, 2 × 2, and 3 × 3 array models with the same area. Simulation results showed that as the array density increased, the frequency range where the peaks of the STL occurred tended to expand. In the 2 × 2 array model, the free frame contributed to anti-resonance, resulting in a significant high noise reduction region below 700 Hz. This study contributes to the determination of optimal density for practical MAM applications and provides insight into the problem of scale variation in other acoustic metamaterials during the array process.

Keywords: Membrane-type acoustic metamaterial · Noise reduction · Array · Vibration

1 Design

Membrane-type acoustic metamaterials (MAMs) showcase impressive sound control capabilities while retaining a lightweight design [1–4]. Various iterations of these MAMs have been crafted, with particular emphasis on their ability to reduce low-frequency noise. A more comprehensive evaluation can be achieved by investigating the anti-resonance frequency and the sound transmission loss (STL) of the MAM [5–8]. The

J. Chen and H. Huang—Made equivalent contributions to this work.

Chinese Society of Aeronautics and Astronautics (Ed.): CASTC 2023, LNME, pp. 132–137, 2024.
https://doi.org/10.1007/978-981-99-8861-7_15

peak STL frequency can be modified by altering the characteristics of the resonators, thus broadening the sound insulation working bandwidth of MAMs [9–11]. However, in practical applications, it is often necessary to adjust the cell size to match the actual array size, and there are limited studies on how large a single cell should be used for a given array area.

The baseline model for our study adopts the principles of a spider web MAM design [12]. This design not only broadens the sound insulation working bandwidth but also ensures the structure remains lightweight. The membrane shape of the baseline model is circular, and the membrane shape of the baseline model is circular, and for the convenience of arraying, a 0.2 mm thick square polyimide (PI) membrane array is considered for this study in order to facilitate the array. In this study, a square area with a side length of 88.6227 mm is considered, which has the same area as the circular membrane with a 100 mm diameter. Positioned centrally, the EVA resonator, shaped like a cross, measured 40 mm by 4 mm with a depth of 2 mm. The peripheral bands and the cross-like resonators each measured 14 mm by 2 mm, with a depth of 1.8 mm, and they were placed 27 mm away from the midpoint. As shown in Fig. 1, the baseline model, which exhibits good noise reduction effect at low and medium frequencies in the range of 80–1600 Hz, was selected for the array study. By comparing the sound transmission loss of 1 × 1, 2 × 2, and 3 × 3 type array models with the same area, this study probes the influence of dense configurations on the noise mitigation potential of the spider web MAM.

Fig. 1. Exterior view of (a) 1 × 1, (b) 2 × 2 and (c) 3 × 3 type models.

The simulation software COMSOL Multiphysics was used to build a multiphysics field model for the MAM. The software's acoustic-structural interaction module facilitates the determination of the STL and mode. We assume a simulation model of sound field-solid mechanics-sound field. Subsequently, the sound wave of 1Pa passes through the above parts sequentially from a section of input, and is finally completely absorbed by the perfectly matched layer (PML). The surrounding boundary of the MAM remains under fixed constraints, and the boundary of the acoustic field air domain serves as a rigid acoustic field boundary. The interface between the MAM and both the transmitted and incident pressure fields is deemed the acoustic-structural boundary. The numerical model obtained by applying mesh segmentation is shown in Fig. 2.

The STL, as a computational metric, quantifies the decrease in acoustic energy as it progresses through the structure. STL can be depicted by the Eqs. (1).

$$STL = 10\lg\left(\frac{|\langle P_{in}\rangle|}{|\langle P_{out}\rangle|}\right) \tag{1}$$

In this equation, P_{in} signifies the sound pressure at the incident surface, while P_{out} represents the sound pressure right before entering the PML. The symbols < > and | | respectively represent the mean and the modulus.

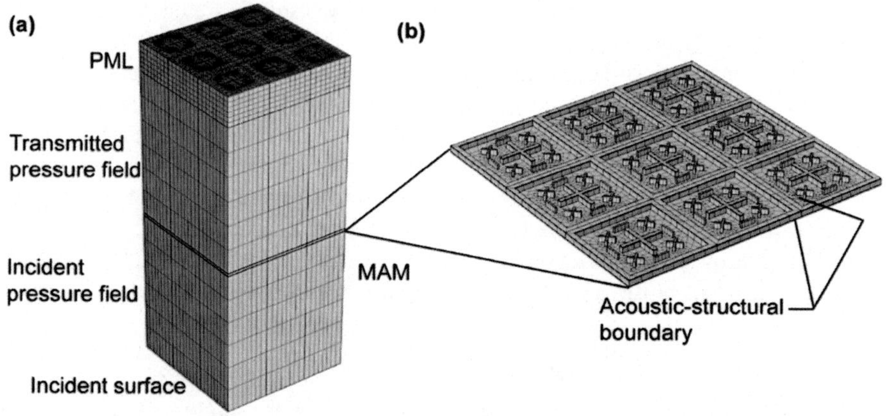

Fig. 2. (a) 3 × 3 type multiphysics mesh model (b) 3 × 3 type MAM mesh model.

2 Results and Discussion

The results of the sound transmission loss (STL) simulations for all models are presented in * MERGEFORMAT Fig. 3. It can be observed that the frequencies of the STL peaks for the 3 × 3 array model can be approximated as an extension of the original 1 × 1 single-cell model, especially for frequencies below 500 Hz for the 1 × 1 single-cell model and below 800 Hz for the 3 × 3 array model. When comparing the STL above 1000 Hz for the 3 × 3 array model and the 2 × 2 array model, there is still a trend of amplifying the frequency where the STL peak is located as the model becomes smaller, but the low frequency of the 2 × 2 array model does not follow this trend. This may be because the frame of the 2 × 2 array model somewhat replaces the role of the cross resonator in the MAM, leading to the appearance of typical low-frequency characteristics of membrane acoustic metamaterials within 700 Hz. Therefore, the 2 × 2 array model can be regarded as a new MAM, where the free frame is also involved in the construction of low-frequency STL peaks, and there is still a tendency for the STL peaks to amplify at frequencies above 700 Hz as the model becomes smaller.

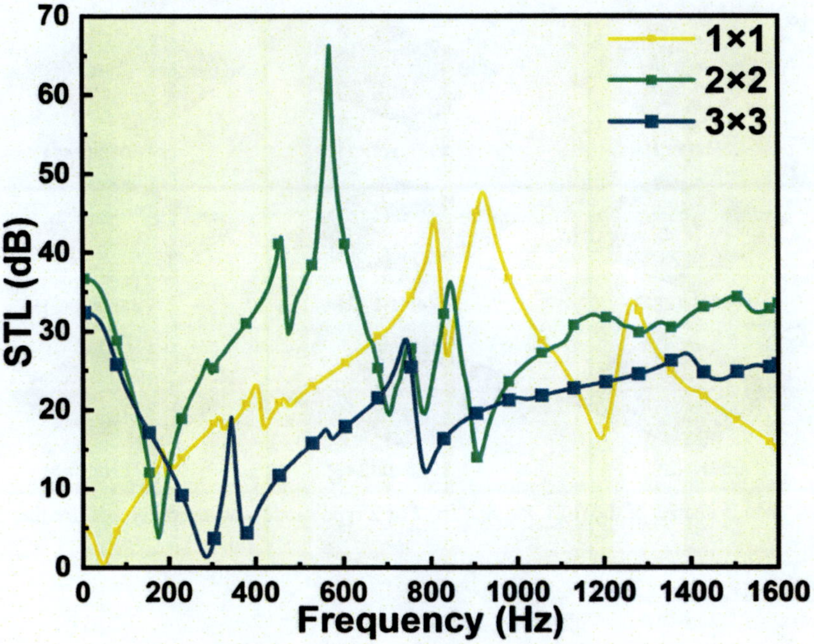

Fig. 3. STL simulation results of MAM for three array approaches.

Figure 4 displays several vibration patterns at typical frequencies for STLs with 1 × 1 and 2 × 2 type array models. All models exhibit similar dipole and quadrupole modes (Dipole mode - 86 Hz (1 × 1), 295 Hz (2 × 2), 60 Hz (3 × 3); Quadrupole mode - 137 Hz (1 × 1), 474 Hz (2 × 2), 1058 Hz (3 × 3)). However, they don't manifest at identical frequencies, and the mode orientations differ across them. For instance, the four dipole modes of the 2 × 2 array model at 295 Hz combine to form a new mode, where the free frame is a crucial component of this mode. This also supports the conclusion that the free frame of the 2 × 2 array model somewhat replaces the role of the cross resonator in the MAM. On the other hand, the 3 × 3 array model is less likely to be involved in the construction of the anti-resonant mode due to its uncentered free frame. In the 3 × 3 array model, only some of the cells often show typical modes instead of all of them, and their amplitudes are not the same, which affects their noise reduction performance. Comparing the modes of the three models near 780 Hz, along with the STL curves, both the 2 × 2 model and the 3 × 3 model exhibit STL peaks, while the 1 × 1 model does not, which explains the correspondence between the modes and the STL results to some extent.

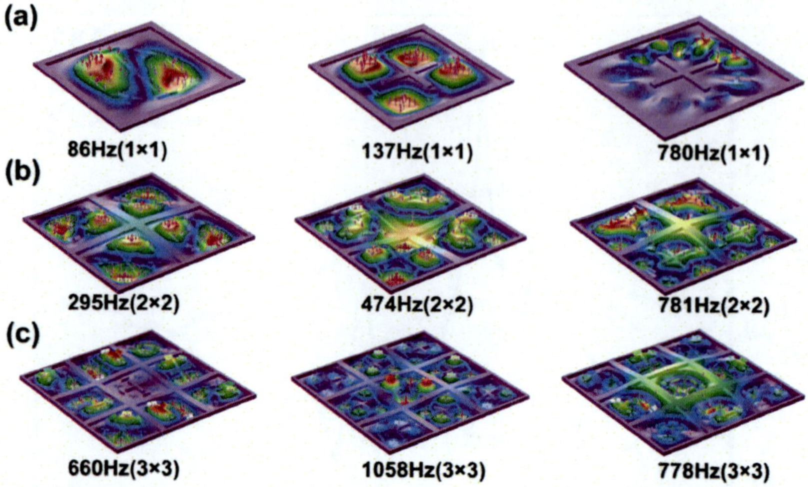

Fig. 4. Mode shapes of the (a) 1 × 1 and (b) 2 × 2 type models at several typical frequencies.

3 Conclusion

This study investigated the effect of denser arrays on the sound insulation performance of the spider web MAM. Simulation results showed that as the array became denser, the frequency range where the sound transmission loss peak occurred tended to expand. In the 2 × 2 array model, the free frame contributed to anti-resonance, resulting in a significant high noise reduction region below 700 Hz, indicating the possibility of multi-stage design. In the 3 × 3 array model, only some of the single cells showed typical modes, and their amplitudes were not the same, which affected their noise reduction performance. These findings will aid in determining optimal density for practical spider web MAM applications and provide insights into the problem of scale changes in other acoustic metamaterials during the array process.

Acknowledgement. This work was jointly supported by the Chinese Aeronautical Establishment Aeronautical Science Foundation (Grant No. 202000150530) and the Fundamental Research Funds for the Central Universities (Grant No. D5000220331).

Funding Statement. The author(s) received no specific funding for this study.

Conflicts of Interest. The authors declare that they have no conflicts of interest to report regarding the present study.

References

1. Yang, Z., Mei, J., Yang, M., Chan, N.H., Sheng, P.: Membrane-type acoustic metamaterial with negative dynamic mass. Phys. Rev. Lett. **101**(20), 204301 (2008)
2. Nassar, H., Yousefzadeh, B., Fleury, R., Ruzzene, M., Alù, A., et al.: Nonreciprocity in acoustic and elastic materials. Nat. Rev. Mater. **5**(9), 667–685 (2020)
3. Wang, Z.W., Zhang, Q., Zhang, K., Hu, G.K.: Tunable digital metamaterial for broadband vibration isolation at low frequency. Adv. Mater. **28**(44), 9857–9861 (2016)
4. Naify, C.J., Chang, C.M., McKnight, G., Nutt, S.: Transmission loss and dynamic response of membrane-type locally resonant acoustic metamaterials. J. Appl. Phys. **108**(11), 114905 (2010)
5. Li, Q.Q., He, Z.C., Li, E., Liu, P., Lin, X.Y., et al.: Design of resonant structures in resin matrix to mitigate the blast wave with a very wide frequency range. Smart Mater. Struct. **29**(4), 045042 (2020)
6. He, Z.C., Xiao, X., Li, E.: Design for structural vibration suppression in laminate acoustic metamaterials. Compos. Part B-Eng. **131**, 237–252 (2017)
7. Liu, F.M., Liu, Z.Y.: Elastic waves scattering without conversion in metamaterials with simultaneous zero indices for longitudinal and transverse waves. Phys. Rev. Lett. **115**(17), 175502 (2015)
8. Zhong, H.B., Gu, Y.H., Bao, B., Wang, Q., Wu, J.H.: 2D underwater acoustic metamaterials incorporating a combination of particle-filled polyurethane and spiral-based local resonance mechanisms. Compos. Struct. **220**, 1–10 (2019)
9. Li, Z.W., Wang, X.D.: Wave propagation in a dual-periodic elastic metamaterial with multiple resonators. Appl. Acoust. **172**, 107582 (2021)
10. Ma, F.Y., Chen, J.Y., Wu, J.H.: Three-dimensional acoustic sub-diffraction focusing by coiled metamaterials with strong absorption. J. Mater. Chem. C. **7**(17), 5131–5138 (2019)
11. Zhou, G.J., Wu, J.H., Lu, K., Tian, X.J., Huang, W., et al.: Broadband low-frequency membrane-type acoustic metamaterials with multi-state anti-resonances. Appl. Acoust. **159**, 107078 (2020)
12. Huang, H.Y., Cao, E.R., Zhao, M.Y., Alamri, S., Li, B.: Spider web-inspired lightweight membrane-type acoustic metamaterials for broadband low-frequency sound isolation. Polymers **13**(7), 1146 (2021)

Cabin Noise Analysis and Noise Reduction Design of the Helicopter with Double-Swept Rotor Blade

Zhongwei Yin[1(✉)], Gang Wang[1], Zhirui Wang[2], Changliang Lin[1], and Yongbin Dang[3]

[1] AVIC Harbin Aircraft Industry Group Co., Ltd., Harbin 150066, China
826455757@qq.com
[2] 95966 Troops of the PLA, Harbin 150060, China
[3] School of Astronautics, Harbin Institute of Technology, Harbin 150001, China

Abstract. The double-swept rotor blade is a new type of blade, and its vortex interference noise level is lower than that of traditional blade. In this paper, the contribution of the cabin noise source of a helicopter is analyzed, and the main gear box vibration and rotor aerodynamic noise are determined to be the main sources. The alternating load at the hub center with double-swept rotor blade is calculated. The vibration load information of the main gear box and the rotor aerodynamic load is predicted. Then the noise level in the cabin is analyzed by the acoustic boundary element method. The result shows that the sound pressure level at passenger's ear is 131dB in the cabin without interior trimming. The skin is determined to be the key structure affecting the cabin noise by the analysis of plane acoustic contribution. After that, four skin structures are designed and their noise reduction effect is analyzed. The metal honeycomb sandwich skin structure has the best noise reduction effect, which can effectively reduce the cabin noise by 10.6%.

Keywords: Cabin noise · Noise reduction · Acoustic boundary element · Double-swept rotor blade · Helicopter

1 Introduction

Although the helicopter has the advantages of vertical take-off and landing, hovering in the air, its cabin noise level is much higher than that of fixed-wing aircraft. The poor acoustic environment in the cabin seriously affects the comfort of passengers on board. With the increasing demand for riding quality, cabin noise reduction has gradually become the key issue in helicopter design. Rotor system noise reduction design and cabin sound insulation design are two effective means of cabin noise reduction [1]. Proper design of the blade tip geometry parameters can not only reduce the alternating aerodynamic force and the airframe vibration level, but also reduce the rotor noise and improve the acoustic environment inside the cabin [2]. Comparing with the traditional blade, The Blue Edge blade designed by Airbus Helicopters optimizes the forward and backward sweep angle of the blade tip, and effectively reduces the blade vortex interference noise [3].

Chinese Society of Aeronautics and Astronautics (Ed.): CASTC 2023, LNME, pp. 138–148, 2024.
https://doi.org/10.1007/978-981-99-8861-7_16

Acoustic simulation analysis can predict the cabin noise level at the beginning of helicopter design, give noise reduction design suggestions, and reduce subsequent improvement and test costs. The acoustic boundary element is a simulation analysis method suitable for super-large acoustic problems. It can complete the acoustic radiation calculation only by extracting the structural surface mesh, and has a fast solution speed [4].

In this present study, a helicopter equipped with double-swept rotor blades is taken as the research object, the hub alternating load and the rotor aerodynamic load on airframe are calculated. On this basis, the noise level in the cabin is analyzed. The noise reduction design is carried out for the key structure, then the noise reduction effect is analyzed.

2 Sound Source Contribution Analysis

Statistical Energy Analysis (SEA) [5] can be used to analyze the noise source, while vibration or sound field information measured by the test is used as input. The main energy transfer path and contribution affecting the cabin acoustic environment can be qualitatively analyzed from the energy point of view.

The contribution of noise source under unit intensity is analyzed by SEA. The structural noise source is set to 1m/s^2 unit vibration acceleration, and the air noise source is set to 1N/m^2 unit sound intensity.

The contribution of noise source is shown in Fig. 1. It can be seen that, the vibration of the main gear box (MGB for short) and rotor aerodynamic noise are the main noise sources in the cabin and their contributions are more than 20%.

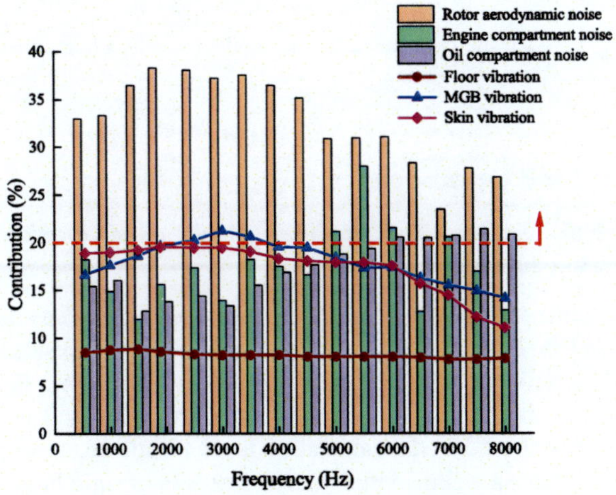

Fig. 1. Sound source contribution.

3 Cabin Noise Analysis Without Interior Trimming

3.1 Acoustic Simulation Model Establishment

Referring to [6], the mass law based on the transmission law is used to equivalent and simplify the glass, cockpit panel, frame and stringer. The acoustic boundary element model composed of platform, floor, skin, floor, windshield and baffle is formed, as shown in Fig. 2. In order to analyze the noise level in the cabin more accurately, six field points conforming to the ISO sound power test standard are defined in the acoustic model. The field points are located at the ears of the pilot and the passenger.

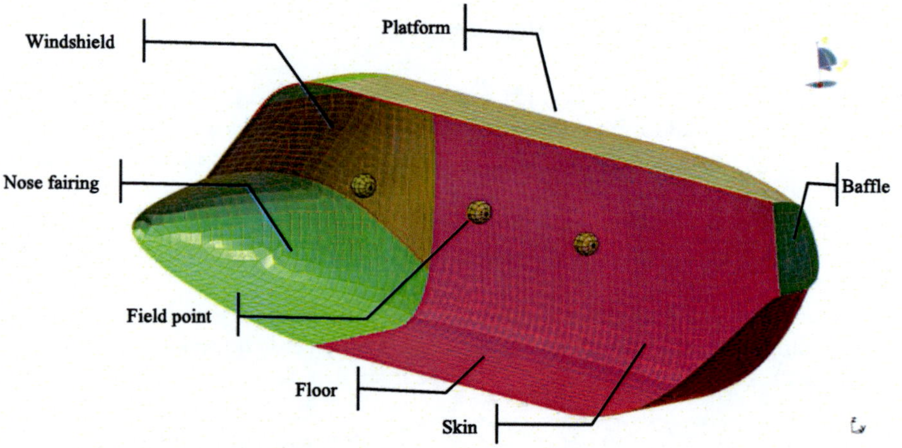

Fig. 2. Acoustic boundary element model (only half is shown).

3.2 Noise Source Load Calculation

MGB Vibration Noise Source. The simplified calculation of vibration noise source is divided into following three steps:

Hub Center Load Calculation. First, it is necessary to calculate the alternating load at the hub center. The load is calculated by the method of [7]. The fuselage aerodynamic data used in the calculation is from the wind tunnel test. The linear density and static stiffness properties of the double-swept rotor blade are calculated by the method of [8]. The structural model used in the calculation is the elastic beam element blade, and the following three factors are considered: a) The freedom of rigid body motion in three directions of blade flapping, lagging and torsion. b) Variation of blade mass and stiffness in spanwise direction. c) The coupling of flapping, lagging and torsion. The aerodynamic models used in the calculation are free wake model, unsteady aerodynamic model and dynamic stall model.

The hub center alternating load at the common horizontal flight speed of 180 km/h is calculated, as shown in Table 1.

Table 1. Alternating load of the hub center at 180 km/h.

Loading direction	Force		Moment	
	Amplitude/N	Phase/°	Amplitude/N·m	Phase/°
Heading	503.6	72.7	216.8	−15.4
Side	416.7	−24.4	186.3	26.2
Vertical	531.6	31.8	812.6	−42.5

MGB Vibration Information Calculation. The MGB vibration information is obtained by dynamic frequency response analysis. First, the mode analysis of the whole helicopter is carried out to find the local mode of the MGB (as shown in Fig. 3.), so as to determine the frequency value (26.13Hz) while the vibration load is applied. Second, the hub center load is applied to the model hub, and the frequency response analysis in free state is carried out. The maximum vibration acceleration value of MGB at 26.13Hz is 0.48g, as shown in Fig. 4.

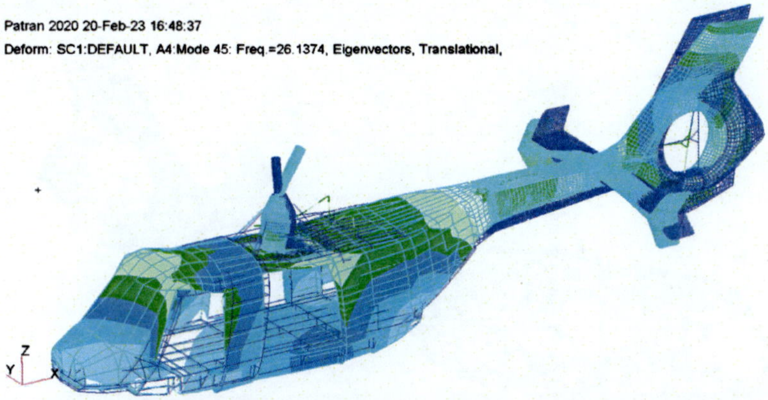

Fig. 3. Vibration frequency of MGB (26.13Hz).

Structural Grid Vibration Velocity Calculation. The cabin detail model is established. Then the grid vibration velocity of this model is calculated caused by the added structural vibration load. The vibration load of the MGB is simplified into a constant amplitude (acceleration) periodic sinusoidal excitation at different frequencies:

$$F = A \sin(ft) \tag{1}$$

where, A is the amplitude, that is, the vibration acceleration. f is the vibration frequency.

The vibration velocity of each grid node is calculated by applying the above periodic vibration load on the cabin structural model. The vibration velocity cloud diagram at a certain moment is shown in Fig. 5. In the calculation of cabin noise, it is necessary to use the method given in [9] to define the data mapping transfer, so as to map the vibration

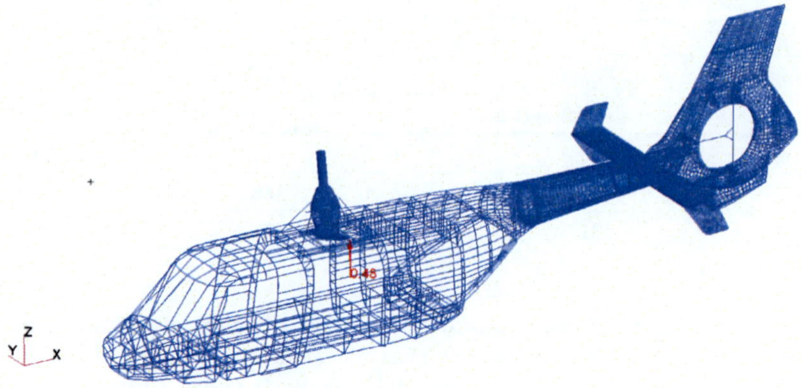

Fig. 4. Vibration acceleration of MGB (0.48g).

information on the structural model to the acoustic model. The vibration velocity cloud diagram of the acoustic model is shown in Fig. 6.

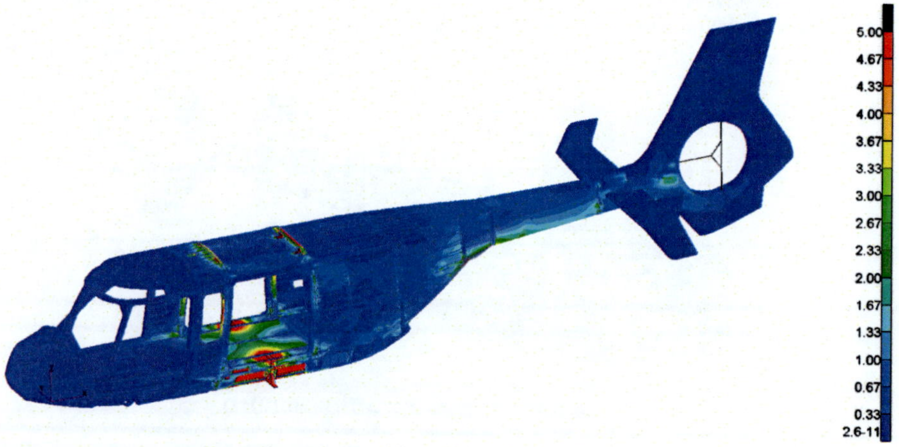

Fig. 5. Vibration velocity cloud diagram of structure model.

Rotor Aerodynamic Noise Source. The calculation of rotor aerodynamic noise load is simplified to a certain extent, only the aerodynamic pressure caused by rotor downwash flow on airframe is calculated. The rotor downwash is set as 10 m/s. The established model is shown in Fig. 7.

First, the steady-state calculation of pressure is carried out. Second, the transient calculation is carried out with the steady-state calculation results as the initial value. The turbulence model selected is large eddy simulation (LES). The calculated aerodynamic pressure on the airframe wall is shown in Fig. 8. Because the calculation results of aerodynamic analysis are time domain, it is necessary to transform it into frequency

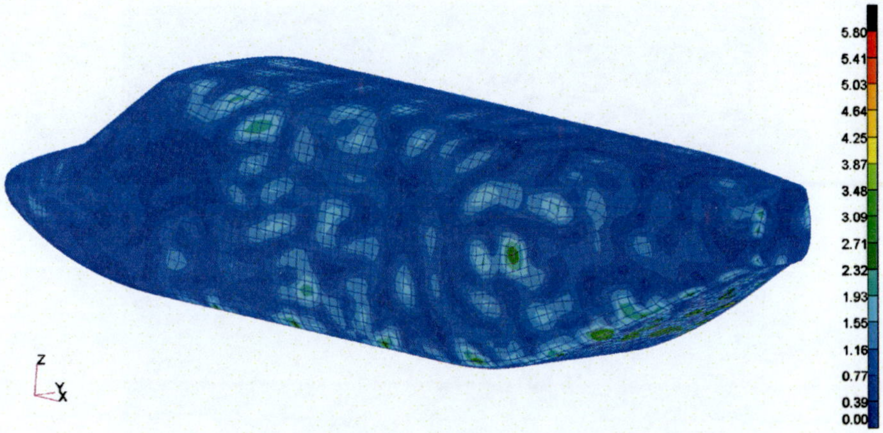

Fig. 6. Vibration velocity cloud diagram of cabin acoustic model.

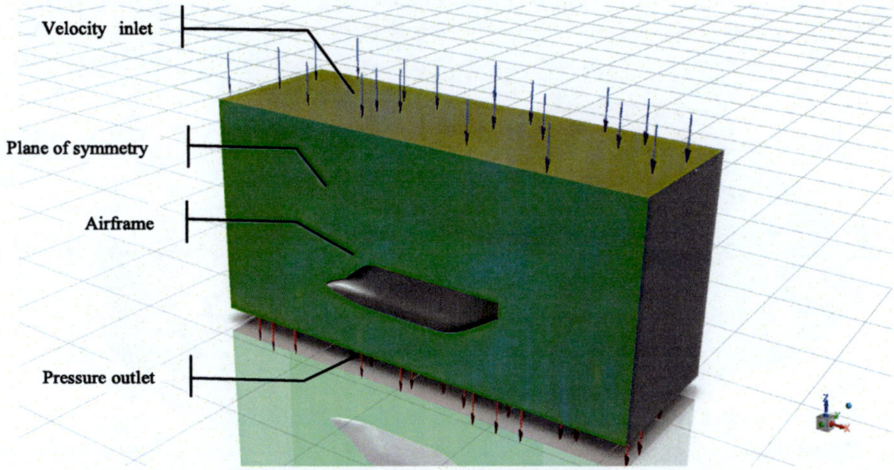

Fig. 7. Aerodynamic analysis model.

domain by fast Fourier transform, and then generate the noise source by Lighthill acoustic analogy.

3.3 Analysis Result

Acoustic Response Analysis. First, the overall sound pressure in the cabin is analyzed, and the sound pressure level distribution cloud map at 2750 Hz is obtained, as shown in Fig. 9. It can be seen that in the range of human ear sensitive frequency, the sound pressure of the passenger cabin is higher than that of the cockpit, reaching 117 dB.

Based on the overall sound field distribution, the sound pressure response at the field points is further calculated, as shown in Fig. 10. It can be seen that the sound pressure at

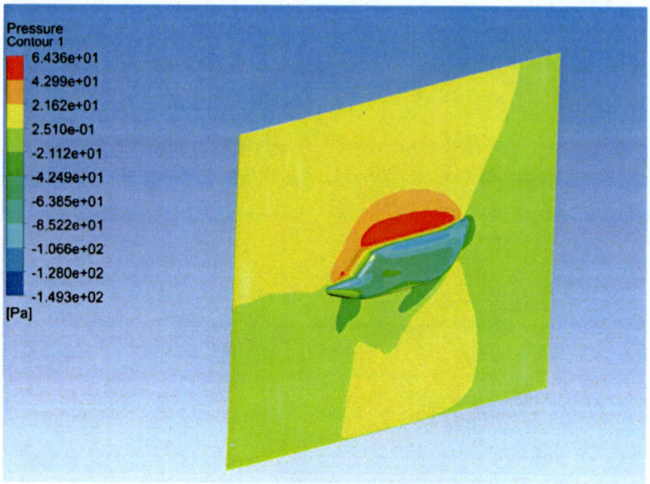

Fig. 8. Pressure on airframe.

Fig. 9. Sound pressure level of the whole cabin at 2750 Hz.

the passenger's ear at 2750 Hz is higher than that of pilot's ear, reaching 131 dB, which is consistent with the overall sound pressure distribution in the cabin.

Panel Contribution Analysis. In order to analysis the influence of each panel on the sound pressure level at the passenger's ear, the plane contribution analysis is carried out. The result is shown in Fig. 11. It can be seen that the contribution of the skin is

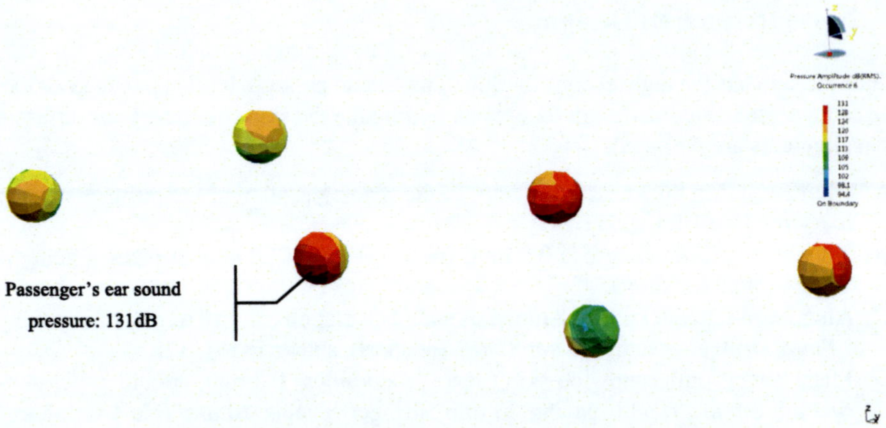

Fig. 10. Sound pressure level of field points at 2750 Hz.

highest, while the skin has weak structure stiffness, which is more likely to cause acoustic vibration coupling with the air in the cabin.

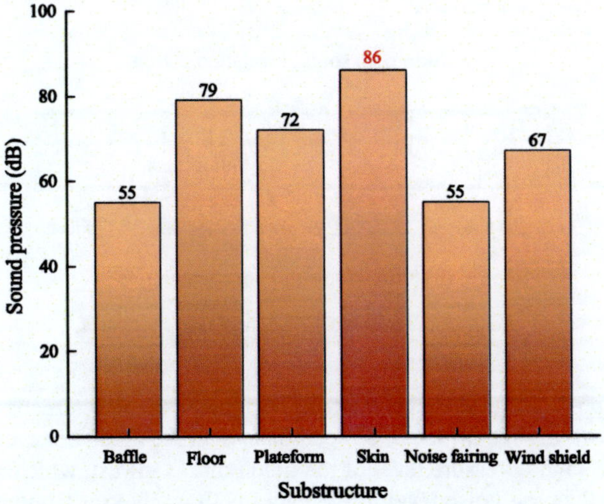

Fig. 11. Sound pressure panel contribution at passenger's ear.

4 Noise Reduction Design

Since the skin has the highest contribution to the sound pressure level at passenger's ear, the cabin noise reduction design focuses on optimizing the skin structure. Four different skin structures are designed:

(A) Aluminum skin thickness is 0.8 mm, frame spacing is 40 mm, and stringer spacing is 40 mm (No.1 for short).
(B) Aluminum skin thickness is 0.8 mm, frame spacing is 20 mm, stringer spacing is 20 mm (No.2 for short).
(C) Aluminum skin thickness 0.8 mm plus sound insulation cotton 8 mm, frame spacing 20 mm, stringer spacing 20 mm (No.3 for short), shown in Fig. 12.
(D) Metal honeycomb sandwich skin (panel is aluminum 0.8 mm, aramid paper honeycomb 8 mm), frame spacing 20 mm, stringer spacing 20 mm (No.4 for short), shown in Fig. 12.

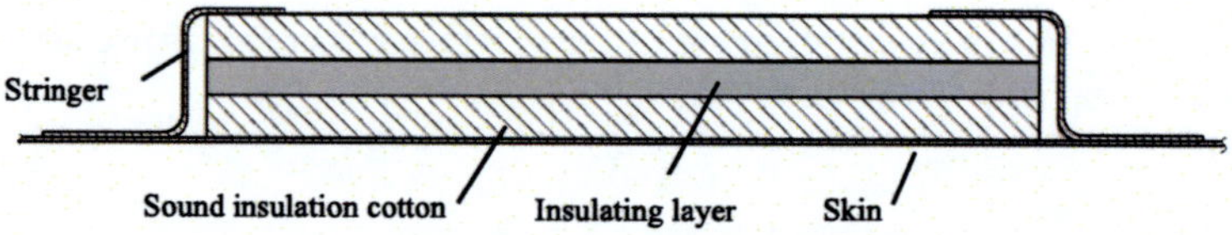

(a) Skin with sound insulation cotton.

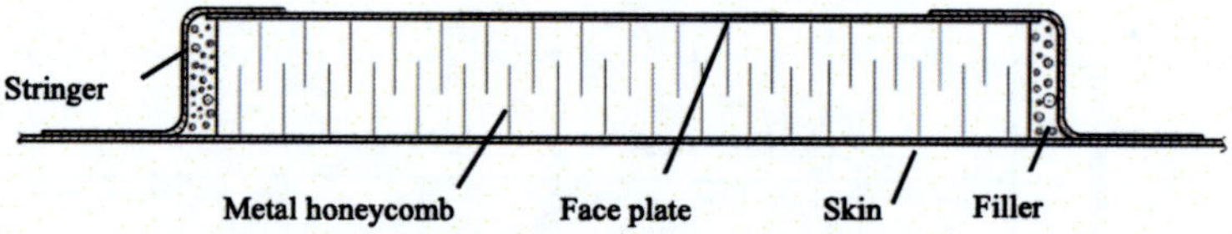

(b) Skin with metal honeycomb.

Fig. 12. No.3 and No.4 designed skin structures.

The sound pressure level under the four designed skin structures is analyzed, and the result is shown in Fig. 13. It can be seen that when the metal honeycomb sandwich skin is used, the sound pressure level of the field point is lowest, which is 117dB. This is 10.6% lower than the noise level in the cabin without interior trimming. The metal honeycomb sandwich skin is the optimal noise reduction structure.

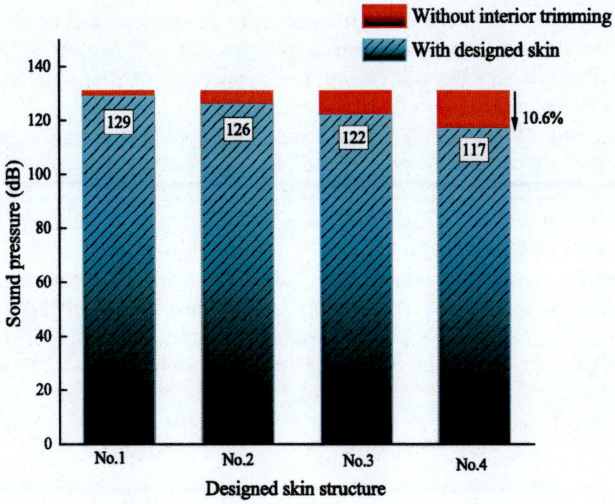

Fig. 13. Noise reduction comparison of designed skin structures.

5 Conclusion

In this paper, the cabin noise and noise reduction design of a helicopter with the double-swept rotor blades are studied by the method of acoustic boundary element simulation. The following conclusions are drawn:

(1) The MGB vibration and the rotor aerodynamic noise are the main noise sources for helicopter cabin, which are determined by the method of SEA.
(2) In the absence of interior trimming, the sound pressure level of passenger cabin is higher than that of cockpit. The maximum sound pressure level at the passenger's ear is 131dB.
(3) The skin is the key structure of cabin noise reduction. Among the designed four skin structures, the metal honeycomb sandwich skin has the best noise reduction effect, which can reduce the cabin noise level by 10.6%.

Acknowledgement. The authors are grateful to the anonymous reviewers for their critical and constructive review of the manuscript. This study was supported by the Natural Science Foundation of Heilongjiang Province of China (No. JQ2020E006).

References

1. Jiang, X.T.: Airplane design manual helicopter design. Aviation Industry Press, Beijing China (2005)
2. Brocklehurst, A., Barakos, G.N.: A review of helicopter rotor blade tip shapes. Prog. Aerosp. Sci. **56**, 35–74 (2013)
3. Wall, A.D., Kebler, C., Pengel, K.: From ERATO basic research to blue edge rotor blade. Journal of the American Helicopter Society **124**, 3103–3121 (2016)

4. Zhan, F.L., Xu, J.W.: Virtual.lab acoustics acoustic simulation calculation from entry and mastery. Northwestern Polytechnical University Press, Xi'an China(2013)
5. Richard, H., Richard, G., De, J.: Theory and application of Statistical Energy Analysis. Butterworth - Heinemann, Newton, MA (1995)
6. Lei, Y., Sheng, M.P., Xiao, H.Y.: Prediction and analysis of helicopter cockpit noise. Journal of Vibration, Measurement & Diagnosis **30**(6), 617–620 (2019)
7. Xia, S.M., Lin, C.L., Zhang, T.L., Cui, Y., Yuan, S.T.: Prediction of hub alternating loads and airframe vibration response in typical flight conditions. Science Technology and Engi-neering **21**(8), 3387–3392 (2021)
8. Xia, S.M., Lin, C.L., Wang, J.L., Zhang, T.L.: Study on dynamic characteristic of double-swept rotor blade. Advances in Aeronautical Science and Engineering **13**(2), 32–37 (2022)
9. Li, Z.G., Zhan, F.L.: Virtual.lab acoustics advanced application examples of acoustics acoustic simulation calculation. National Defense Industry Press, Beijing China(2010)

Research on Influence of Dynamic Balance Deviation on Blade Static Stiffness and Airframe Vibration

Zhongwei Yin[1(✉)], Shangman Xia[1], Yong Wang[2], Changliang Lin[1], and Huaming Wang[3]

[1] AVIC Harbin Aircraft Industry Group Co., Ltd., Harbin 150066, China
826455757@qq.com
[2] 95966 Troops of the PLA, Harbin 150060, China
[3] National Key Laboratory of Science and Technology on Rotorcraft Aeromechanics, Nanjing University of Aeronautics and Astronautics, Nanjing 210016, China

Abstract. During the maintenance and adjustment process of a helicopter blade returning to the factory, the dynamic balance parameters are often deviated and the blades cannot be delivered to users on time. In this paper, the influence of dynamic balance deviation on blade static stiffness properties and airframe vibration level are studied by means of simulation and test. The research shows that the maximum difference of the static stiffness properties between the dynamic balance deviated blade and the qualified one is 7.62%, which appears at the torsional stiffness. The maximum vertical vibration at the key airframe parts of the helicopter equipped with the deviated blade is 0.19g, and the pilot reflects the body feeling well. From the perspective of static stiffness property and airframe vibration level, the dynamic balance deviated blade can be delivered for use.

Keywords: Dynamic balance deviation · Blade · Static stiffness properties · Airframe vibration

1 Introduction

The main blade is an important part of the helicopter upgrade system. The rotating blade provides the necessary lift, lateral force and pitching force to ensure that the helicopter can take off and land vertically, hover and fly forward [1]. The main purpose of the dynamic balance adjustment is to ensure that the blade rotation torque and the deviation of tip rotation trajectory are within the standard range. The qualified dynamic balance check can reduce the vibration level of the helicopter airframe, and the single blade can be exchanged during subsequent use, which is conducive to the blade installation and maintenance [2].

After the main blade is used for a period of time, it often returns to the factory for maintenance due to common faults such as iron-coated rubber opening and polyurethane protective belt aging. The repaired blades often have out-of-tolerance during dynamic

Chinese Society of Aeronautics and Astronautics (Ed.): CASTC 2023, LNME, pp. 149–158, 2024.
https://doi.org/10.1007/978-981-99-8861-7_17

balance inspection, which cannot be delivered to users on time, seriously affecting the maintenance plane process.

In this present study, the influence of dynamic balance deviation on blade static stiffness and airframe vibration level were analyzed by combining simulation analysis with test. The research results can provide important data support for establishing the specified technical standard of repaired blades.

2 Description of Out-of-Tolerance

The repaired main blade is a composite ply structure, including prepreg beam, skin, foam filling block, internal reinforcement beam and other structures. The structure is shown in Fig. 1.

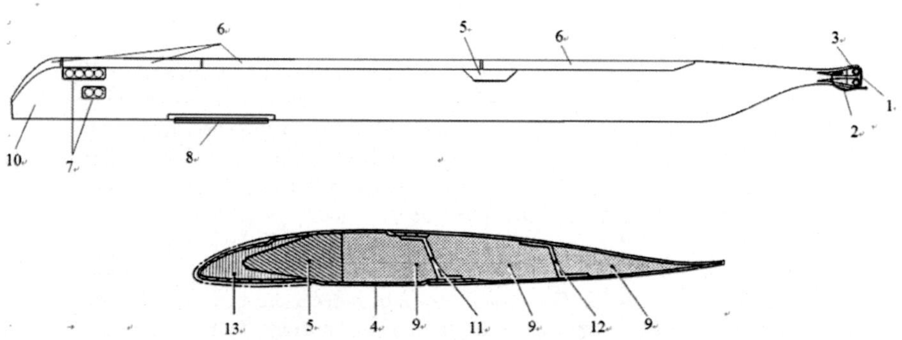

1.Root cover, 2.Attachment filling, 3.Blade root bushings, 4.Skin, 5.Overall counterweight, 6.Leading edge coated iron, 7.Weight bearing pedestal, 8.Adjusting shim, 9.Foam filling block, 10.Blade tip, 11.Front Z-beam, 12. Rear Z-beam, 13.Glass fiber beam

Fig. 1. Blade structure description.

With reference to foreign technical documents and combined with the factory production capacity, the dynamic balance standard of batch production main blades is as follows:

(a) Hinge torque allowable range is ±3 N·m.
(b) Hinge torque slope allowable range is 2.5–4.5.
(c) Vertebral body value allowable range is ±5 mm.
(d) Vertebral slope allowable range is ±5.

In the process of rework, it was found that the main blade generally had moisture absorption and weight gain after a period of use. Some blades also had hinge moment value, vertebral value and other dynamic balance parameters out of tolerance. Table 2 lists the dynamic balance parameters of two repaired blades. No.1 blade is qualified, and No.2 blade is the most deviated one among all the repaired blades.

Table 1. Dynamic balance parameters of partial repaired blades.

No	Parameters									
	Moment value(N•m)				Moment slope	Vertebral value(mm)				Vertebral slope
	1°	2°	3°	4°		1°	2°	3°	4°	
1	1.5	−0.8	−0.5	−0.2	−1.7	0	−2.1	−2.1	−0.7	−0.7
2	23.1	20.6	22.4	19.9	−3.3	−0.2	4.4	13.4	15.9	16.2

3 Static Stiffness Analysis

3.1 Finite Element Simulation

The static stiffness properties of the two blades in Table 1 are calculated by using the method in [3–5]. The results are shown in Fig. 2, Fig. 3 and Fig. 4.

Form the results of simulation analysis, the flap bending stiffness, lag bending stiffness and torsional stiffness curves of No.1 blade and No.2 blade are basically coincident. Since the weight of blade moving chamber needs to be moved while the dynamic balance was adjusted, there is a slight difference in the static stiffness between the No. 1 blade and No. 2 blade. The maximum difference occurs in the torsional stiffness of the blade at 4 m spanwise, which is 6.71%.

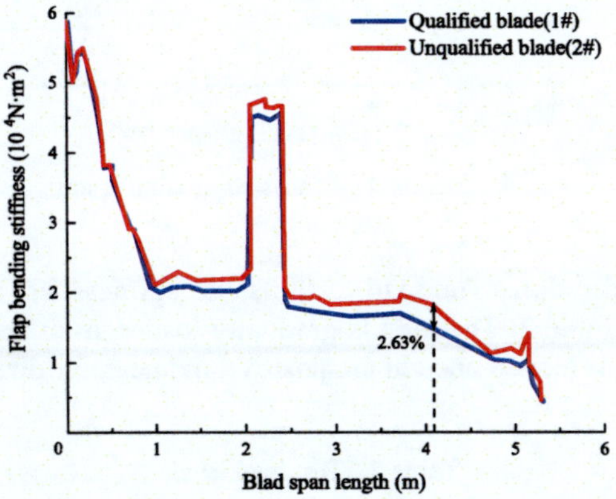

Fig. 2. Flap bending stiffness simulation result.

3.2 Test Measurement

In order to further verify the simulation analysis results, the blade static stiffness test is carried out. The test pieces are the two blades shown in Table 1.

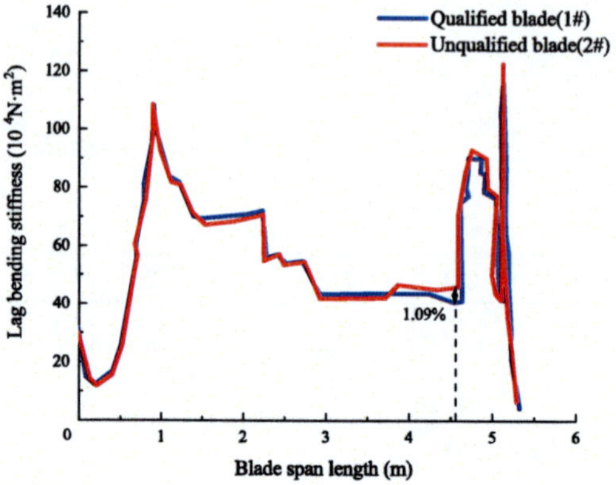

Fig. 3. Lag bending stiffness simulation result.

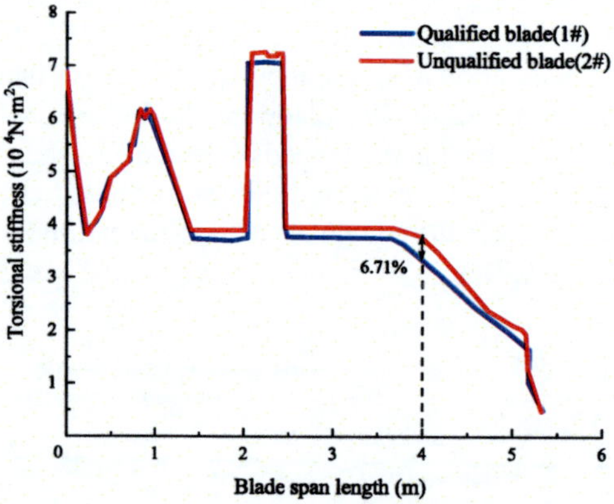

Fig. 4. Torsional stiffness simulation result.

Test Method Introduction. The load of the static stiffness test is divided into four levels, as shown in Table 2. The load is gradually loaded from zero to the maximum value, and each load is loaded once in the positive and negative directions.

Table 2. Test load level.

Load level	Bending stiffness/kg		Torsional stiffness/kg·m	
	Flap	Lag	Ref3690	Ref5137
1	5	10	9	6
2	10	20	18	12
3	15	30	27	18
4	20	40	36	24

For the measurement of flap bending stiffness and lag bending stiffness, resistance strain gauges are used to record the strain values under the external load. The bending stiffness of 10 sections is measured by the full bridge method. The axis of the strain gauge needs to be parallel to the axis of the variable blade pitch.

The torsional stiffness measurement is to measure the relative torsional deformation between different sections with high-precision displacement sensor under the external torque. The vertical displacement of the front and rear edges of each section is measured. A total of 10 sections are measured, of which the first 7 sections are loaded at Ref3690 and the last 3 sections were loaded at Ref5137. In order to ensure that the displacement sensor is perpendicular to the blade span direction, it is necessary to arrange the fixture at each measured section, and the sensor is arranged at the front and rear edges of the fixture.

Figure 5 and Fig. 6 show the measured section sections of bending stiffness and torsional stiffness respectively. Figure 7 shows the experimental conditions.

Fig. 5. Bending stiffness measured sections.

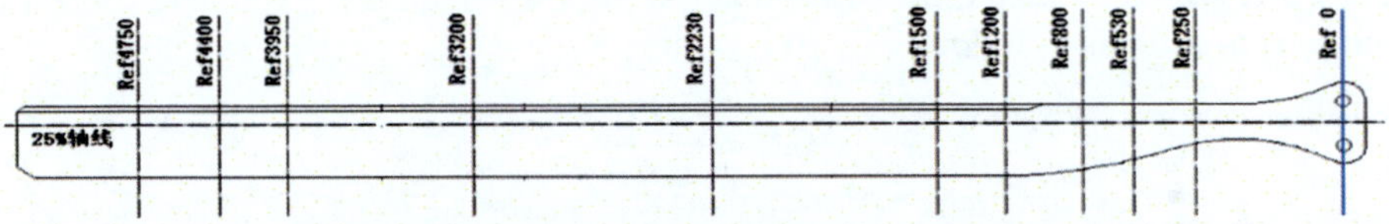

Fig. 6. Torsional stiffness measured sections.

(a) Flap bending stiffness test. (b) Lag bending stiffness test. (c) Torsional stiffness test.

Fig. 7. Experimental conditions.

Test Data Processing. The test data of bending stiffness and torsional stiffness are processed by different formulas.

Bending Stiffness Calculation. According to the classical beam theory, the calculation formula of bending stiffness of beam is as follow:

$$EI = \frac{PLh}{\varepsilon} \tag{1}$$

where L is the distance between the measured section and the loading section. For flap bending stiffness, h is the distance between upper and lower wings. For lag bending stiffness, h is the distance between the strain gauges on the same side. When the upper and lower strain gauges are respectively set, ε is taken as the test value. When the upper and lower strain gauges are two, ε is half of the test value.

Using the above formula to process the test data, the bending stiffness of each section is obtained, as shown in Table 3. The measured section numbers S1 to S10 in the table correspond to S200, S600...S4600 in Fig. 5.

Table 3. Bending stiffness test value ($\times 10^4$ N·m).

No.		Measured Section									
		S1	S2	S3	S4	S5	S60	S7	S8	S9	S10
Flap	1	6.3	3.8	1.9	1.9	2.1	1.6	1.5	1.5	1.1	0.8
	2	6.5	4.1	2.1	1.7	2.2	1.6	1.6	1.6	1.2	0.9
Lag	1	11.3	45.1	73.4	74.5	79.7	59.6	46.3	41.8	51.7	52.0
	2	11.1	45.5	74.4	71.8	78.7	58.0	47.5	45.8	52.6	50.9

Torsional Stiffness Calculation. According to [6], the calculation formula of torsional stiffness of beam is as follow:

$$GJ = \frac{PL}{\Delta\varphi/\Delta\gamma} \tag{2}$$

where L is the distance between the measured section and the loading section. $\Delta\varphi/\Delta\gamma = [(|a_1| + |a_2|)/l_1 - (|b_1| + |b_2|)/l_2]/\Delta$, is the torsion angle per unit length between the measured sections. Δ is the distance between the measured sections. l is the distance between displacement sensors. a and b are the displacement variation of the front and rear edges of the measured section.

Using the above formula to process the test data, the torsional stiffness of each section is obtained, as shown in Table 4. The measured section numbers R1 to R10 in the table correspond to Ref 250, Ref 530...Ref 4750 in Fig. 6.

Table 4. Torsional stiffness test value ($\times 10^4$ N·m).

No.	Measured section									
	R1	R2	R3	R4	R5	R6	R7	R8	R9	R10
1	3.63	3.04	2.44	3.95	3.53	3.62	3.95	3.06	2.50	2.09
2	3.80	2.84	2.48	4.31	3.75	3.70	4.04	3.25	2.72	1.99

Test Data Analysis. The data in Table 3 and Table 4 are sorted into the corresponding static stiffness curves, as shown in Fig. 8, Fig. 9 and Fig. 10. It can be seen that the static stiffness of the two blades is not much different, and the difference is less than 10%. The maximum difference measured in the test also appears in the torsional stiffness, which is 7.62% (at Ref 3950).

The difference of torsional stiffness between the two blades measured by the test is slightly bigger than the simulation. However, considering the measurement accuracy error of the displacement sensor and the selection of the effective value while data processing, it can be considered that the static stiffness property of the dynamic balance deviation is basically the same as that of the qualified blade.

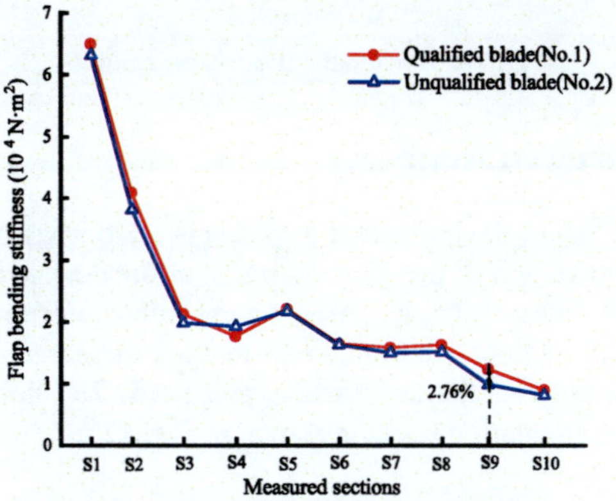

Fig. 8. Flap bending stiffness test results.

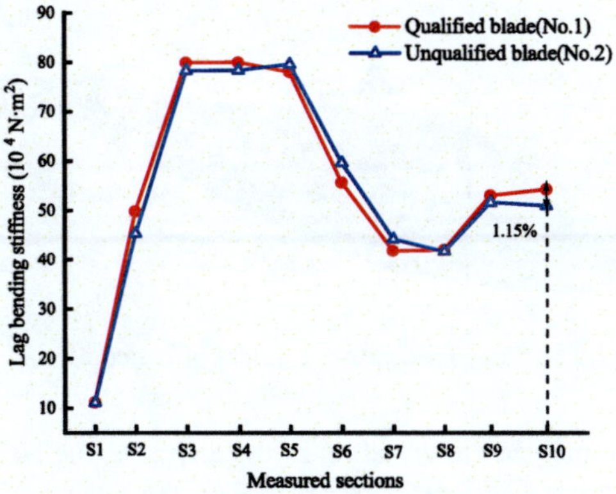

Fig. 9. Lag bending stiffness test results.

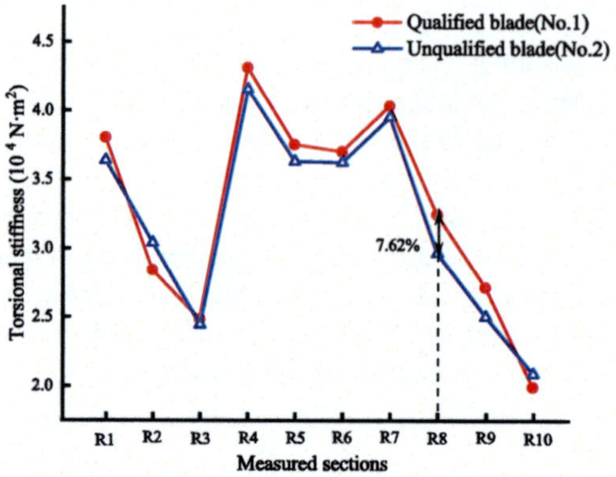

Fig. 10. Torsional stiffness test results.

4 Airframe Vibration Analysis

The finite element model of the helicopter airframe is established by the method of [7]. Since the maximum deviation of the hinge moment of the deviated blades (No. 2 blade in Table 1) is 19 Nm, conservatively, based on the hub center load calculated by the qualified blade, the roll moment and pitch moment are increased by 19 N·m, and then the vertical vibration level of the key parts is predicted. The finite element model is shown in Fig. 11. The simulation result is shown in Fig. 12.

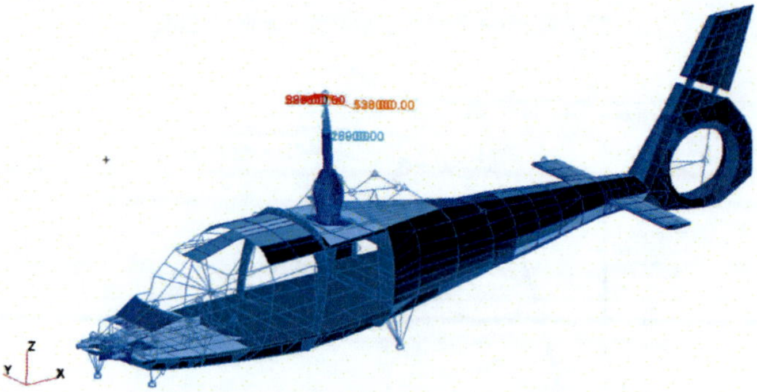

Fig. 11. Finite element model of helicopter.

In order to further analyze the vibration of the airframe, the No.2 blade is equipped on a helicopter with other three qualified ones, then the vibration flight test is carried out. Figure 13 shows the arrangement positions of some vibration sensors. The flight test results are shown in Table 5.

From the simulation analysis and flight test results, the dynamic balance deviation blade will not lead to the increase of airframe vibration, and the vertical vibration value of each key part is all less than 0.2g, which meets the requirements. The pilot also feels well during the flight.

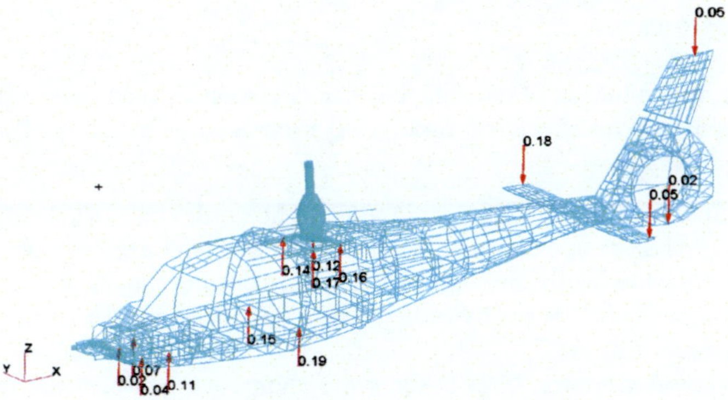

Fig. 12. Vertical vibration level of key parts.

Table 5. Flight test results (g).

Vibration at key parts	Flight speed (km/h)						
	100	120	140	160	180	200	220
Driving pedal	0.14	0.16	0.12	0.12	0.12	0.15	0.17
Controller	0.11	0.18	0.13	0.09	0.10	0.12	0.13
Main reduction platform	0.09	0.16	0.12	0.12	0.08	0.14	0.13
Driver floor	0.18	0.19	0.18	0.16	0.16	0.21	0.22
Tail	0.05	0.07	0.07	0.10	0.11	0.09	0.12
Engine	0.11	0.17	0.19	0.18	0.18	0.21	0.23

(a) Driving pedal. b) Controller. (c) Engine.

Fig. 13. Arrangement positions of some vibration sensors.

5 Conclusions

In this paper, the influence of dynamic balance deviation on blade static stiffness and airframe vibration level is studied through experiment and simulation. The following conclusions are drawn:

(1) The maximum difference of the static stiffness properties between the dynamic balance deviated blade and the qualified blade is 7.62%, which appears at the torsional stiffness. Considering the measurement accuracy error of the displacement sensors, it can be considered that the dynamic balance deviation will not lead to the change of blade static stiffness.
(2) The maximum vibration level at the key airframe parts after using the dynamic balance deviated blade is 0.19g, which is obtained by flight test. This value meets the standard requirements. At the same time, the pilot reflects the body feeling well.

Acknowledgement. The authors are grateful to the anonymous reviewers for their critical and constructive review of the manuscript. This study was supported by the Foundation of National Key Laboratory of Rotorcraft Aeromechanics (No.61422202112).

References

1. Wu, X.M.: Engineering design of helicopter dynamic. Aviation Industry Press, Beijing (2017)
2. Zeng, C., Lin, C.L., Yang K., Zhang, T.L., Fan, G.X., Wen, Y.H.: Study on adjustment technique of helicopter dynamic balance deviated blade. In: 17th China Aviation Measurement and Control Technology Annual Conference, pp. 117–121. Measurement & Control Technology, China (2020)
3. Xia, S.M., Lin, C.L., Wang, J.L., Zhang, T.L.: Study on dynamic characteristic of double-swept rotor blade. Adv. Aeronaut. Sci. Eng. **13**(2), 32–37 (2022)
4. Gao, L., Xia, S.M., Xie, Z.Y.: Method for calculating profile characteristic of composite material blade. China, CN107220458B, 2020.12.29
5. Xia, S.M., Zhang, T.L., Wang, H.M.: Calculation and analysis for cross section rigidity of the composite material flexible beam based on VABS. In: 20th China Helicopter Annual Conference, pp. 477–481. Chinese Society of Aeronautics and Astronautics, China (2010)
6. Zhang, J.Y., Yu, L.C., Xu, C.J., Han, J.: The test technical research on torsional stiffness of the helicopter blades. New Technol. **13**(2), 131–134 (2022)
7. Xia, S.M., Lin, C.L., Zhang, T.L., Cui, Y., Yuan, S.T.: Prediction of hub alternating loads and airframe vibration response in typical flight conditions. Sci. Technol. Eng. **21**(8), 3387–3392 (2021)

Design of Sliding Mode Variable Structure Control System for Resonant Fiber Optic Gyroscope

Fan Wu[1,2](✉), Jun Li[2], Shiqi Lan[2], Bo Yan[2], and Yazhou Yue[2]

[1] School of Aerospace, Northwestern Polytechnical University, Xi'an 710072, Shaanxi, China
wufan_franck@163.com

[2] AVIC Xi'an Flight Automatic Control Research Institute, Xi'an 710076, Shaanxi, China

Abstract. The paper designs a sliding mode variable structure control (VSC) system for resonant fiber optic gyroscope (RFOG). We have derived the state equation of the control system, given the design process of sliding mode variable structure control system, compared the step response of the control system under different control laws by simulation. The sliding mode variable structure control method is applied to the RFOG control system for the first time, the system static testing is performed using different control laws. The result shows that the VSC method can be well applied to the RFOG closed-loop frequency locking control system, compared with the traditional PID control method, the frequency locking accuracy of the RFOG control system is improved by about 18%. It shows that the VSC control method has better closed loop control performance.

Keywords: resonant fiber optic gyroscope · sliding mode · variable structure control

1 Introduction

The demand of high-precision and small bulk inertial gyroscope is more and more urgent nowadays. Resonant fiber optic gyroscope (RFOG) which combines the advantages of laser gyroscope (RLG) and interferometric fiber optic gyroscope (IFOG) measures the rotation rate by detecting laser resonant frequency shifts between the clockwise (CW) and counterclockwise (CCW) directions in ring resonator cavity. By using the optical resonance principle and increasing the number of loops, the detection sensitivity and signal-to-noise ratio of Sagnac effect are improved. Theoretically, it can meet the demand of high precision and small bulk at the same time. In addition, the RFOG has a series of advantages such as high dynamic response, large dynamic range, anti-electromagnetic disturbance, stable scale factor and free from acceleration. It is one of the important development directions to high precision and small bulk inertial gyroscopes [1–3].

In recent years, researchers have proposed a lot of control methods, such as frequency out-of-lock control [4], closed-loop based on square wave quadrature demodulation [5], reciprocal modulation and demodulation technique [6, 7], high-frequency modulation

Chinese Society of Aeronautics and Astronautics (Ed.): CASTC 2023, LNME, pp. 159–168, 2024.
https://doi.org/10.1007/978-981-99-8861-7_18

closed-loop control technique [8] to improve the performance of resonant fiber optic gyroscopes, these methods have effectively reduced the mode-hopping noise and the demodulation curve fluctuation in the closed-loop control system, suppressed the residual amplitude modulation (RAM) noise and relative intensity noise (RIN) in the optical system. However, there is still a significant gap between the actual and theoretical accuracy of the resonant fiber optic gyroscope [9, 10]. The key technology of RFOG is to lock the central frequency of the laser at the resonant frequency of the FRR stably. Therefore, researchers have proposed servo control technologies such as frequency locking control technology based on digital PI controller [11], frequency locking control technology based on fuzzy PI control [12], etc. But the frequency locking performance of the closed-loop control system still needs to be improved. Improving the frequency locking accuracy through optimization of the servo controller is crucial to improving the performance of the resonant fiber optic gyroscope.

In the design of the RFOG laser frequency closed-loop locking control system, there is a contradiction between excellent control performance and high anti-disturbance ability. When we design a traditional PI closed-loop control system using the excellent control parameters, the external disturbance introduced into the control system such as the impact, vibration, mode hopping or rotation, will cause the laser frequency unlock and make the gyroscope system lose its function. The external disturbance will change the internal state function of the control system, that is, the structure of the RFOG system has different forms in practical work. Theoretically, the sliding mode variable structure control (VSC) system has ultra-high control accuracy and complete adaptability to external disturbance of the system [13, 14]. In this paper, the sliding mode variable structure control method is adopted for the laser frequency locking closed-loop control system of the RFOG. Through the design, simulation, parameter optimization and engineering application of the sliding mode VSC system, it is proved that the method can be well applied to the laser frequency locking control system of the RFOG, and the frequency locking accuracy is improved.

2 The System Design and Simulation

The laser frequency of the CW and CCW directions are stabilized at the resonance peak of the ring resonator cavity by modulating and demodulating the light field. When the gyroscope rotates, the laser frequency of the CW and CCW directions will produce frequency shifts due to the Sagnac effect. Then, the rotation rate is obtained by detecting frequency shifts [15, 16]. The Schematic diagram of system structure is shown in Fig. 1. The Laser is a 1550 nm wavelength semiconductor laser and the Y-branch is a dual-arm phase modulator dividing the laser into two beams. The laser enters the CW and CCW directions through the coupler respectively. PD1 and PD2 are photodetectors where the photoelectric conversion are carried out. LPF1 is used to filter the signal to obtain the error signal of the control system, the error signal controls the laser through the servo controller. Then, the laser frequency closed-loop locking control system is formed.

For research purposes, the transfer function of the laser frequency closed-loop locking control system structure is simplified as which is shown in Fig. 2. The input is the frequency deviation e which is caused by the gyroscope rotation. The sliding mode variable structure control servo controller is represented as VSC. The u is the control signal

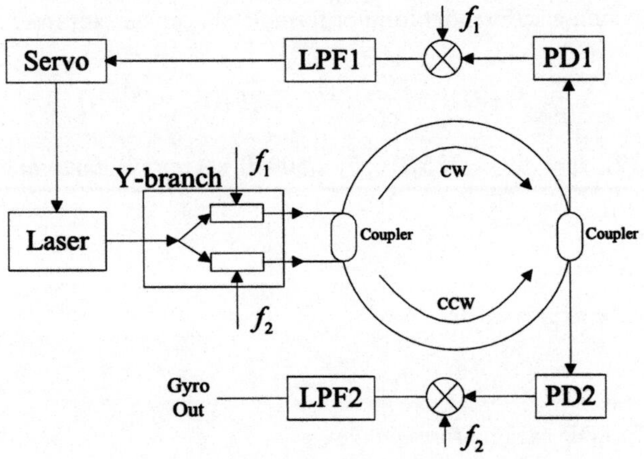

Fig. 1. Schematic diagram of system structure

integrated by the laser, the integration process can be expressed as $1/s$. The frequency deviation of the optical Sagnac resonance system is simplified to proportional transfer function with the coefficient K, in which the external disturbance to the optical Sagnac resonance system is expressed as f_M. Due to the existence of the external disturbance, the transfer function of the control system will change in practical work. In photoelectric conversion process, the signal will be subjected to AC, which can be equivalent to high pass filtering expressed by HPF function. In demodulation process, the signal will be subjected to low-pass filtering, which can be expressed by the low-pass filtering LPF function. Resonant optical frequency bias setting is expressed by f_{bias} and the resonant optical frequency bias error is expressed by f_{out}.

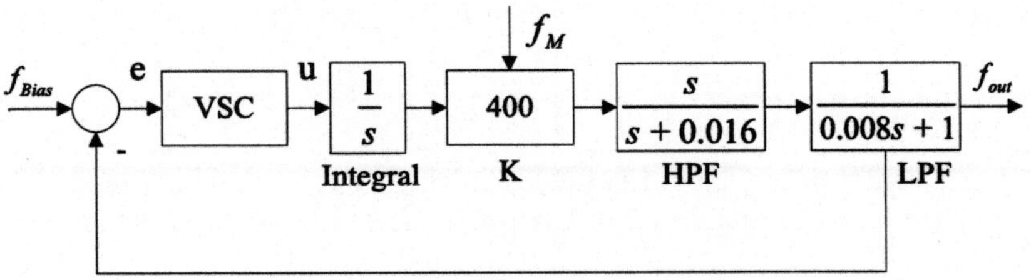

Fig. 2. Control system block diagram

According to the control system block diagram shown in Fig. 2, we can get the transfer function of the RFOG control system.

$$\begin{aligned} G(s) &= \frac{1}{s} \cdot 400 \cdot \frac{s}{s + 0.016} \cdot \frac{1}{0.008s + 1} \\ &= \frac{50000}{s^2 + 125.016s + 2} \end{aligned} \tag{1}$$

The corresponding differential form of formula (1) can be expressed as

$$\frac{d^2}{d^2 t} f_{out}(t) + a_2 \frac{d}{dt} f_{out}(t) + a_1 f_{out}(t) = b_1 u(t) \tag{2}$$

In formula (2), $a_1 = 2$, $a_2 = 125.016$, $b_1 = 50000$, set the system error e in the control system.

$$e = f_{bias} - f_{out} \tag{3}$$

Formula (2) can be expressed as

$$f_{out} = f_{bias} - e \tag{4}$$

Substitute formula (4) into formula (2)

$$\frac{d^2}{d^2 t} e + a_2 \frac{d}{dt} e + a_1 e = -b_1 u \tag{5}$$

Set state variable $x_1 = e, x_2 = \dot{x}_1 = \dot{e}$, we can get the description of the system state space as

$$\begin{cases} \dot{x}_1 = x_2 \\ \dot{x}_2 = -a_2 x_2 - a_1 x_1 - b_1 u \\ x_1 = e \end{cases} \tag{6}$$

Hence, the equation of state system can be expressed by formula (7)

$$\begin{cases} \begin{bmatrix} \dot{x}_1 \\ \dot{x}_2 \end{bmatrix} = \begin{bmatrix} 0 & 1 \\ -a_1 & -a_2 \end{bmatrix} \begin{bmatrix} x_1 \\ x_2 \end{bmatrix} + \begin{bmatrix} 0 \\ -b_1 \end{bmatrix} u \\ e = [1\ 0] \begin{bmatrix} x_1 \\ x_2 \end{bmatrix} \end{cases} \tag{7}$$

From the coefficient matrix of state equation of the RFOG control system, it is easy to get that the control system is controllable and observable, so the switching function can be set

$$s = c_1 x_1 + x_2, c_1 > 0 \tag{8}$$

In the equation, s is the sliding surface, c_1 is the parameter that determining the slope of the sliding surface, when the c_1 is set an appropriate value, the sliding mode variable structure control system is gradually stable and has good quality.

In the design of sliding mode variable structure control system, it is also necessary to select appropriate switching function as the VSC system control law, which including constant switching, proportional switching, function switching, etc. Among them, the variable structure control law of proportional switching function is relatively simple in engineering and has good control performance. Therefore, we design the VSC control law of the RFOG system using the proportional switching function.

$$u = \psi_1 x_1,\ \psi_1 = \begin{cases} \alpha,\ x_1 s > 0 \\ \beta,\ x_1 s < 0 \end{cases} \tag{9}$$

It can be seen from formula (9) that the design of VSC control law based on proportional switching sliding mode surface is actually the design of the optimal values of c_1, α, β, the parameters have to satisfy the arrival condition showing in formula (10).

$$\dot{s} s < -\delta \tag{10}$$

δ is any small positive number, we set $\Psi_1 = \alpha, e > 0$, the Eq. (11) can be get according to the formula (5)

$$p^2 + a_2 p + (a_1 + b_1 \alpha) = 0 \tag{11}$$

To ensure the formula (11) has non-negative real roots, we substitute the system state space description formula (6), the switching function formula (8) and the proportional switching function variable structure control law formula (9) into arrival condition formula (10), then the following inequalities can be obtained

$$\begin{cases} \alpha > \dfrac{a_2^2 - 4a_1}{4b_1} \\ \alpha > \dfrac{c_1(a_2 - c_1) - a_2}{b_1} \\ \beta < \dfrac{c_1(a_2 - c_1) - a_2}{b_1} \\ c_1 > 0 \end{cases} \tag{12}$$

Thus, if the parameters c_1, α, β satisfy the inequality (12), the system state can enter the sliding mode surface in limited time, and the closed-loop control system is asymptotically stable.

We import the theoretical model to the Simulink software, establish the sliding mode variable structure closed-loop control system mode, and optimize the control parameters by applying step response. The Simulink simulation model is shown in Fig. 3. The sampling rate is set to 10 MHz, the frequency offset f_{bias} is set to system step signal, the VSC model is the variable structure controller. The signal generator is used to simulate the external disturbance of the RFOG closed-loop control system, system response is obtained in the scope.

The sub-model of sliding mode VSC controller is shown in Fig. 4, where the parameters c_1, α, β are the control law parameters to be optimized. Gain1 is the conversion coefficient from laser frequency to control code and Gain2 is the conversion coefficient from control code to laser frequency. The Switch is the switching module that determines which channel is the output.

Through the iterative simulation, the optimal parameters that satisfying formula (12) are obtained, which are $c_1 = 14$, $\alpha = 30$, $\beta = 6$. The 766 Hz frequency deviation is

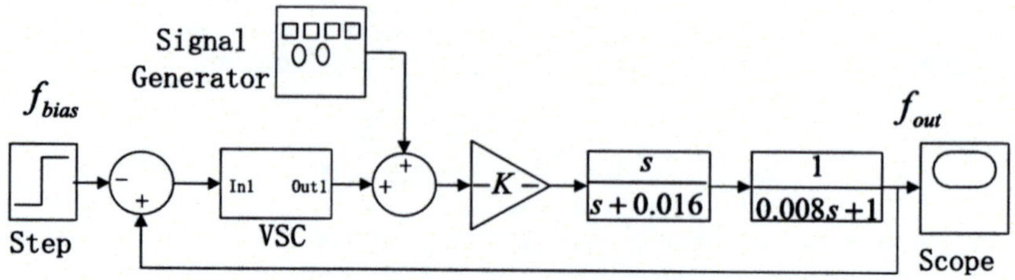

Fig. 3. System simulation model

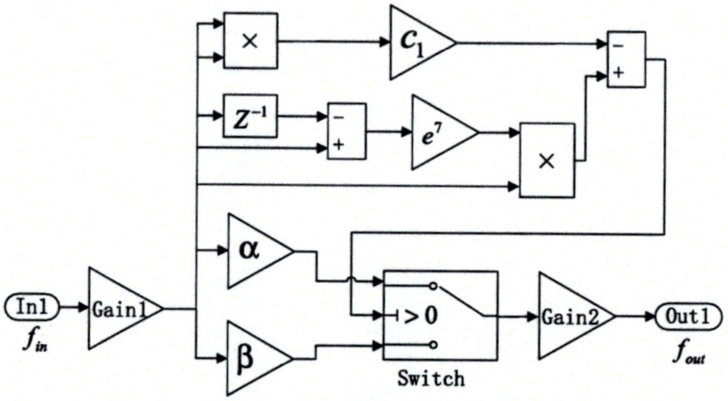

Fig. 4. The sub model of sliding mode VSC controller

equivalent to the rotating speed of 1 °/s for the optical Sagnac ring resonator of diameter d = 10 cm. Therefore, the step signal 766 Hz is applied to the system to simulate when the gyro starts rotating at the rotating speed of 1 °/s. Comparing the system response using VSC control law and traditional PID control law, the results are shown in Fig. 5. From the simulation results, it can be seen that the system step response time using VSC control law is about 20 ms, and the system step response time using PID control law is about 60 ms. The response time of VSC control law is about 3 times less than that of PID control law. Therefore, The sliding mode VSC system has better response performance and better control performance in theory.

When the optical Sagnac resonance system is affected by external disturbance such as vibration, temperature, mode hopping or gyro rotation, the resonant laser frequency in the optical resonance system will change. Because the 75 Hz is equivalent to the rotating speed of 0.083 °/s, which is the maximum random walk value of our system, we assume that the optical system is subject to vibration disturbance with frequency of 4 Hz and frequency amplitude of 75 Hz. The system signal generator in Fig. 3 can apply a square wave signal to simulate with the frequency of 4 Hz and the frequency amplitude of 75 Hz. Then, we get the demodulate the error response in the Scope. The anti-disturbance ability of the closed-loop system using VSC control law and traditional PID control law is compared. The simulation results are shown in Fig. 6. The closed-loop system with VSC control law can recover the system stability faster when the system is disturbed by external disturbance, so its anti-disturbance ability is better.

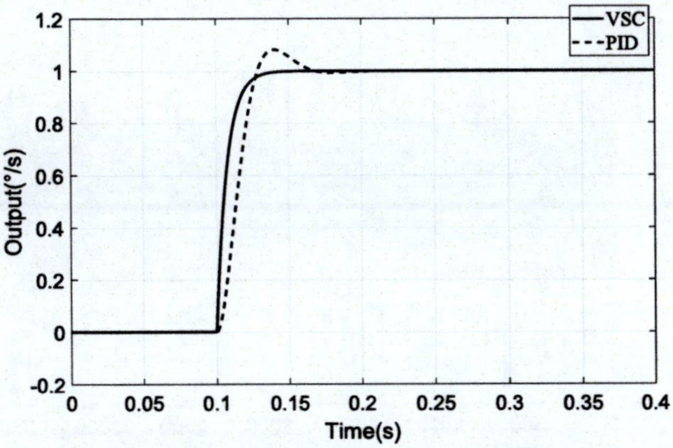

Fig. 5. Step response under different control laws

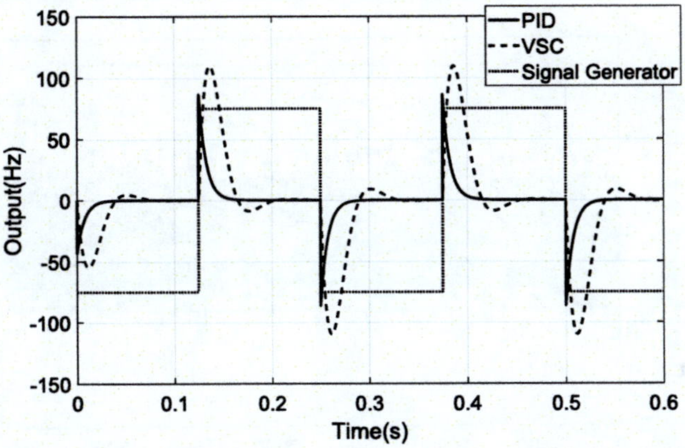

Fig. 6. System demodulation error response under different control laws

3 System Test

The PID control system and the sliding mode VSC system are applied to the closed-loop control of the RFOG respectively. The wavelength of semiconductor laser used in the experiment is 1550 nm and the line-width is 3 kHz. The optical resonator is made of 20 m long polarization fiber with a diameter of 10 cm and a fineness of 15.6. The modulation frequency used in the laser closed-loop control is 191 kHz, and the gyro output loop is 197 kHz. Keeping other parameters unchanged, the RFOG system is tested respectively at room temperature for the time 3600 s. The demodulation error data of the PID control law control system and the VSC control law control system are collected, as shown in Fig. 7 and Fig. 8.

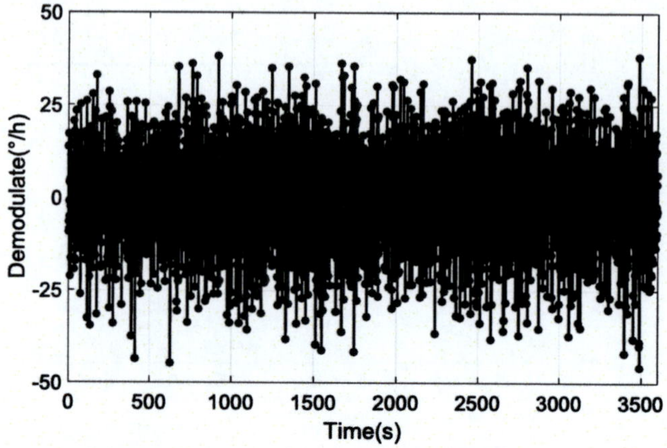

Fig. 7. Data of system demodulation error under PID control law

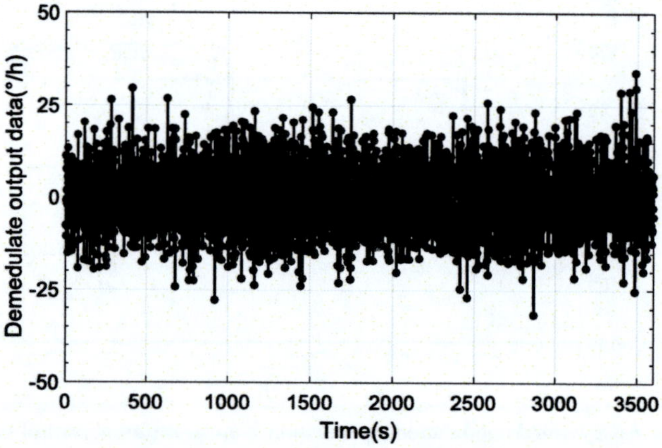

Fig. 8. Data of system demodulation error under VSC control law

The data are analyzed by Allan variance analysis. The results are shown in Fig. 9 and Fig. 10. The Allan variance noise level of the system demodulation error under the PID control law is better than that of the VSC control law. Under the 100 s integration time, the frequency locking accuracy of the PID control law control system is 0.3215°/h, and the frequency locking accuracy of the VSC control law control system is 0.2624°/h. Obviously, the sliding mode VSC system improves the frequency locking accuracy of the RFOG.

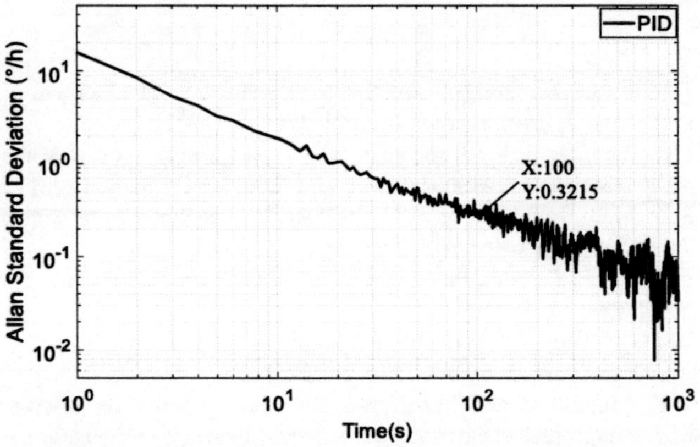

Fig. 9. Allan variance of system demodulation error under PID control law

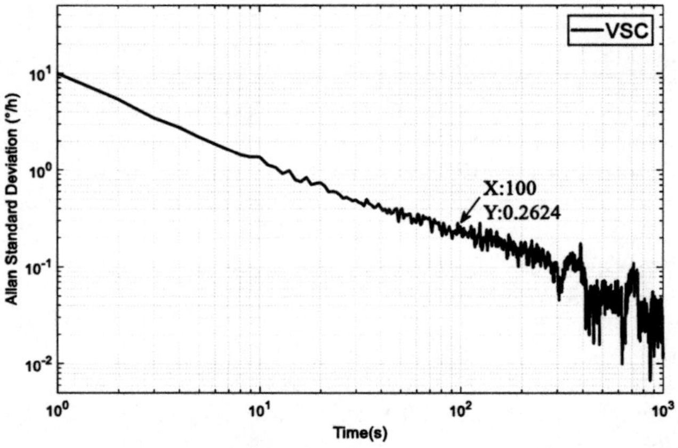

Fig. 10. Allan variance of system demodulation error under VSC control law

4 Conclusion

In this paper, we propose a sliding mode variable structure control method of the RFOG. Through design and simulation of the control system, a better system step response and demodulation error response are obtained, which prove that the sliding mode VSC control method has better anti-disturbance ability. We apply the sliding mode VSC control method to the RFOG system. The test results show that the sliding mode VSC control method can be well applied to the RFOG laser frequency closed-loop locking control system. Compared with the traditional PID control method, the frequency locking accuracy is improved by about 18%. It shows that the VSC control method has better closed loop control performance.

References

1. Shang, K., Lei, M., Li, H.: Design, manufacturing and future development of the integrated fiber optic gyroscope. J. Chin. Inertial Technol. **29**(4), 8 (2021)
2. Sanders, G.A., Taranta, A.A., Narayanan, C., et al.: Hollow-core resonator fiber optic gyroscope using nodeless anti-resonant fiber. Opt. Lett. **46**(1), 46–49 (2021)
3. Lai, Y.H., Suh, M.G., Lu, Y.K., et al.: Earth rotation measured by a chip-scale ring laser gyroscope. Nat. Photonics **14**(6), 345–349 (2020)
4. Lei, M., Yu, H., Fang, Y., et al.: Frequency tracking out-of-lock control in a resonator fiber-optic gyro. Appl. Opt. **59**(4), 923–928 (2020)
5. Ying, D., Liu, Q., Wang, Z., et al.: Closed-loop RFOG based on square wave quadrature demodulation. Optics Commun. **493**, 127015 (2021)
6. Li, H., Lin, Y., Liu, L., et al.: Signal processing improvement of passive resonant fiber optic gyroscope using a reciprocal modulation-demodulation technique. Opt. Express **28**(12), 18103–18111 (2020)
7. Liu, L., Liu, S., Hu, J., et al.: Resonant fiber optic gyroscope using a reciprocal modulation and double demodulation technique. Opt. Express **30**(7), 12192–12203 (2022)
8. Liu, S., Liu, L., Hu, J., Liu, Q., Ma, H., He, Z.: Reduction of relative intensity noise in a broadband source-driven RFOG using a high-frequency modulation technique. Opt. Lett. **47**(19), 5100–5103 (2022)
9. Wang, Z., Wang, G., Wang, Z., et al.: Incoherence suppression method of optical noises in a resonant fiber optic gyro based on the circularly polarized light propagation mechanism in a resonator. Opt. Lett. **46**(13), 3191–3194 (2021)
10. Wang, Z., Wang, G., Kumar, S., Marques, C.: Recent advancements in resonant fiber optic gyro-A Review. IEEE Sensors J. **22**, 18240–18252 (2022)
11. Yang, Z.: Studieson Optical Noises and Signal Detection Technology in Resonator Fiber Optic Gyro. Zhejiang University, Hangzhou (2010)
12. Suling, L.I., Shuhua, B.A.I.: Laser frequency tracking and locking technology based on fuzzy PI control. J. Nanchang Univ. (Sci. Ed.) **44**(5), 492 (2020)
13. Zinober, A.S.I.: An introduction to sliding mode variable structure control. In: Variable Structure and Lyapunov Control, pp. 1–22. Springer, Heidelberg (1994). https://doi.org/10.1007/BFb0033676
14. Xu, D., Gao, Y., Tan, Z., et al.: Study on sliding mode variable structure control for gyro stabilized system. In: 2010 8th World Congress on Intelligent Control and Automation, pp. 3860–3863. IEEE (2010)
15. Ying, D., Xie, T., Wang, Z., et al.: A closed-loop RFOG based on digital serrodyne and sine modulations with two LiNbO3 phase modulators. Optics Commun. **452**, 151–157 (2019)
16. He, M., Wang, G., Long, W., Gao, W.: Modeling and simulation of double closed-loop resonant fiber optical gyro. Opt. Optoelectron. Technol. **18**(6), 98 (2020)

Aeroelastic Stability for Straight and Double Swept Rotor Blades in Forward Flight

Shuangman Xia[1](✉), Dingfeng Ma[2], Changliang Lin[1], Jinliang Wang[1], and Shengtao Yuan[1]

[1] Aircraft Design and Research Institute, AVIC Harbin Aircraft Industry Group Co., Ltd., Harbin 150066, China
xia19840229@163.com
[2] 95966 Troops of the PLA, Harbin 150060, China

Abstract. Double-swept blades can effectively reduce helicopter rotor blade vortex interference noise. However, they can lead to the deviation of the aerodynamic center and center of gravity of the blade section relative to the pitch axis, and this can pose a risk to aeroelastic stability during high-speed forward flight. Therefore, it is necessary to investigate this issue in depth. Based on the straight blade of a star flexible rotor, a double-swept blade structure was designed and the dynamic characteristics and aeroelastic stability of the straight blade and double-swept blade were compared and analyzed. The effects of parameters such as sweep-forward angle, sweep-backward angle, sweep-forward station, and sweep-backward station on aeroelastic stability were studied. The analysis results showed that the double-swept structure, compared to the straight blade, can reduce the third flap frequency and first torsional frequency of the blade. Additionally, the double-swept blade can improve aeroelastic stability during forward flight when the double-swept blade beam is strengthened. Increasing the sweep-backward angle is beneficial to aeroelastic stability. However, when the sweep-backward station is closer to the blade tip, the aeroelastic stability becomes worse.

Keywords: Helicopter · Rotor Blade · Double-Swept · Aeroelastic Stability

1 Introduction

With the continuous development and wide application of helicopter technology, helicopter vibration and noise levels have become important indicators for measuring their quality and performance. Further reducing noise and vibration has become the main objective of modern helicopter development [1, 2]. Relevant research has confirmed that rotor blade design, especially the blade tip structure, has a great impact on the overall performance, vibration, and acoustics of the helicopter. The representative structure is the double-swept blade, which can significantly improve helicopter performance and effectively reduce rotor noise [3, 4].

In the 1990s, the French National Aeronautics and Astronautics Office and the German Aerospace Center developed a "Blue Blade" as part of the ERATO (Rotor Aerodynamic Noise Optimization Project) project. Its purpose was to design a new rotor

Chinese Society of Aeronautics and Astronautics (Ed.): CASTC 2023, LNME, pp. 169–185, 2024.
https://doi.org/10.1007/978-981-99-8861-7_19

blade to reduce vortex interference noise. Compared with traditional blades, the blade structure was designed with forward-sweep-backward, which is called a double-swept blade. Flight test results showed that in the case of low-speed drop, the BVI noise of the "Blue Blade" was reduced by an average of 3–4 dB compared to the traditional rotor blade, which has a very good noise reduction effect [5, 6].

Foreign researchers have conducted a lot of analysis and experimental research on double-swept blades. Filippi [7, 8] established a multi-dimensional finite element dynamic analysis model of double-swept blades and verified them through experiments. At the same time, they studied the establishment methods of dynamic and static finite element models of double-swept composite blades. Truong [9] also established a three-dimensional dynamic finite element model of double-swept blades and researched the dynamic characteristics of the blades. They compared and analyzed the analysis results of the three-dimensional model and one-dimensional model. Thomas [10] carried out an experimental study on the aeroelastic stability of two blades with the same material and different structural forms, and analyzed the changes of the first-order damping of lag under different forward flight velocities, collective and pre-cone angles. Kumar [11, 12] used a comprehensive dynamic load analysis tool to analyze the aeroelastic response of double-swept blades in hover, and considered that the double-swept structure will increase the lag load and torsional torque. They studied the influence of rotor blade structure parameters on the dynamic characteristics and load of the rotor on the rotor tower. Arnold [13] used numerical analysis methods to analyze the aeroelastic stability of the helicopter in hover. They found that due to the influence of the double-swept structure of the double-swept rotor blade, the blade torsion mode was easily coupled with the flapping bending mode.

Zhang Junhao [14] established a rotor/fuselage coupling system model based on double-swept blades and studied the effects of geometric parameters such as forward-swept angle and backward-sweep angle on ground resonance and air resonance with bearingless rotor helicopters in detail. Yang Weidong and Deng Jinghui [15] established a new rotor aeroelastic model with a new blade tip and found that the blade sweep tip introduced strong bending-torsion coupling, and the first-order lag stability decreased with the increase of the tip sweep angle. Xia Shuangman and Lin Changliang [16] analyzed the influence of the double-swept structure on the dynamic characteristics of the blade and found that the double-swept structure has little influence on the low-order modes of the blade.In China, research on the double-swept blade structure has mainly focused on its influence on helicopter mechanical stability and the single-swept blade on rotor aeroelastic stability, but there have been few studies on the rotor aeroelastic stability of forward-backward swept blade structures.

Therefore, this paper establishes a comprehensive analysis model of rotor aeroelastic coupling dynamics using a star flexible rotor as an example. The validity of the model is verified through blade natural frequency testing. Based on this, the differences in dynamic characteristics and aeroelastic stability between straight blades and double-swept blades are analyzed, and the influence of the forward-backward swept blade structure on rotor aeroelastic stability is studied. The effects of forward-swept blade structure parameters, such as forward-swept angle, swept-back angle, forward-swept station, and backward-swept station, on aeroelastic stability are also revealed.

2 Aeroelastic Dynamic Model

2.1 Aeroelastic Coupling Dynamics Modeling

In order to describe the motion of the blade, the aeroelastic coupling dynamic equation of the rotor is obtained using Hamilton's principle. The equation is expressed as follows:

$$\int_{t_1}^{t_2} [\delta(U - T) - \delta W]dt = 0 \tag{1}$$

where δU is the variation of the strain energy of the elastic blade established according to the strain-displacement field, δT is the variation of the kinetic energy of the blade established according to the motion field, and δW is the variation of the external force work acting on the blade, including aerodynamics and other constrained external forces.

The total kinetic energy, potential energy, and external force work of the rotor aeroelastic coupling system can be expressed by the following equation:

$$T = \sum_{k=1}^{NB} \sum_{i=1}^{NE} T_{ki}^b \tag{2}$$

$$E = \sum_{k=1}^{NB} \sum_{i=1}^{NE} E_{ki}^b \tag{3}$$

$$W = \sum_{k=1}^{NB} \sum_{i=1}^{NE} W_{ki}^b \tag{4}$$

In the formula, the kinetic energy, elastic potential energy, and external force work corresponding to the i beam section of the k blade are represented by T_{ki}^b, E_{ki}^b, and W_{ki}^b respectively. Here, NB is the total number of blades and NE is the total number of units divided by each blade. The nonlinear dynamic equation of the k blade can be obtained by inserting the above formula into the Hamilton's principle formula.

$$Q_{ki}^T(\ddot{q}_{ki}, \dot{q}_{ki}, q_{ki}, t) + Q_{ki}^E(q_{ki}) + Q_{ki}^A(\dot{q}_{ki}, q_{ki}, t) = 0 \tag{5}$$

The aeroelastic dynamics perturbation equation of the rotor can be expressed as follows:

$$M\left(\hat{q}(t)\right)\Delta\ddot{q} + C\left(\hat{\dot{q}}(t), \hat{q}(t), t\right)\Delta\dot{q} + K\left(\hat{\dot{q}}(t), \hat{q}(t), t\right)\Delta q = F \tag{6}$$

The equations of rotor dynamics motion can be expressed as second-order differential equations:

$$M\ddot{q}(t) + C\dot{q}(t) + Kq(t) = f(t) \tag{7}$$

In the formula, $f(t)$ the column vector represents the external force acting on the system, and the matrix M, C, K represents the mass, damping, and stiffness of the system.

2.2 Blade Natural Frequency Calculation

By ignoring the damping and external force terms in formula (7), the free vibration equation expression of the rotor dynamics is obtained:

$$M\ddot{q}(t) + Kq(t) = 0 \tag{8}$$

The solution of the generalized coordinates q is used to describe the motion of the system, which is assumed to be in the following form:

$$q = \Phi \sin \omega (t - t_0) \tag{9}$$

Here, Φ is an n-order vector, ω is the vibration frequency of the vector Φ, t is a time variable, and t_0 is a time constant determined by the initial conditions. If Eq. (9) is brought into (8), the generalized eigenvalue problem is obtained:

$$K\Phi - \omega^2 M\Phi = 0 \tag{10}$$

The eigenvectors Φ and eigenvalues ω can be determined by solving the above equations. According to the order of eigenvalues, n eigensolutions (ω_n, Φ_n) (n = 1, 2, 3……n) can be obtained., ω_n represents the n natural frequencies of the rotor dynamic system, and Φ_n represents the natural vibration modes of the corresponding n natural frequencies.

2.3 Rotor Aeroelastic Stability Solution

In this paper, the eigenvalue method is used to solve the rotor aeroelastic stability, and the dynamic instability of the system is judged according to the real part of the eigenvalue.

If the external force term of the equation of motion (7) is ignored, the dynamic free vibration equation of the rotor with damping term is expressed as:

$$M\ddot{q}(t) + C\dot{q}(t) + Kq(t) = 0 \tag{11}$$

FLOQUET describes its solution in the form of

$$\tilde{q}(t) = B(t)\left\{C_k e^{P_k t}\right\} \ (k = 1, 2, \cdots, 2N) \tag{12}$$

$B(t)$ is a $2N \times 2N$ matrix, which is regarded T as a period, there is $B(T) = B(0)$, from which we can get:

$$\tilde{q}(0) = B(0)\{C_k\} \tag{13}$$

$$\tilde{q}(T) = B(T)\left\{C_k e^{P_k T}\right\} = B(0)\left\{C_k e^{P_k T}\right\} \tag{14}$$

It can also be expressed as:

$$\tilde{q}(T) = \underbrace{\left[\tilde{q}^{(1)}\ \tilde{q}^{(2)} \cdots \tilde{q}^{(2N)}\right]}_{[Q]} \begin{Bmatrix} \tilde{q}_1(0) \\ \tilde{q}_2(0) \\ \vdots \\ \tilde{q}_{2N}(0) \end{Bmatrix} \tag{15}$$

$[Q]$ is called FLOQUET transitive matrix.

$\lambda_K = e^{p_k t}$ are the eigenvalues of the FLOQUET transfer matrix $[Q]$, which can be obtained by QR (orthogonal trigonometric) decomposition method, and then can be written as

$$P_k = \frac{1}{T}\ln\lambda_k = \alpha_k + i\omega_k \tag{16}$$

According to the above formula, the real part and the imaginary part can be given:

$$\alpha_k = \frac{1}{T}\ln|\lambda_k| = \frac{1}{2T}\ln\left[(\lambda_k)_R^2 + (\lambda_k)_I^2\right] \tag{17}$$

$$\omega_k = \frac{1}{T}\tan^{-1}\frac{(\lambda_k)_I}{(\lambda_k)_R} \tag{18}$$

The real part represents the damping of the system,and reflects the growth and convergence of the response. When $\alpha_k < 0\,(|\lambda_k| < 1)$ indicates the system is stable, the imaginary part ω_k reflects the system frequency.

3 Blade Structure

3.1 Blade Root Constraint

The blades are connected to the rotor shaft through a star flexible hub,which mainly includes central star,clamp,blade pin,variable pitch horn, elastomer bearing and frequency matcher,as shown in Fig. 1.

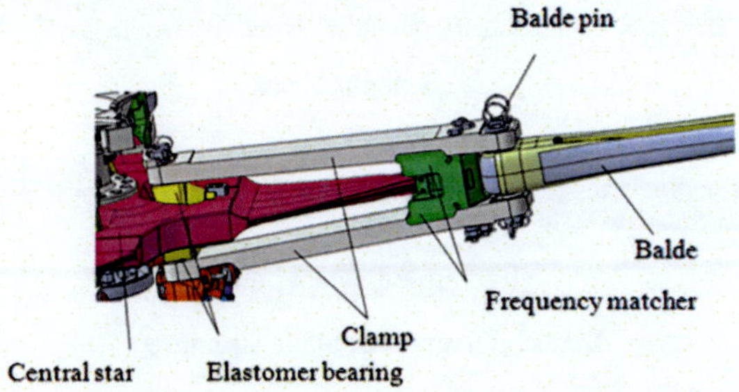

Fig. 1. Schematic diagram of main rotor hub

During the flap motion of the blade, the star flexible arm is flexible in the flapping direction, so the blade and a pair of clamps move up and down around the center of the elastomer bearing, and the elastomer bearing produces shear deformation around the center. Meanwhile, the star flexible arm produces upper and lower bending deformation. The elastomer bearing can be regarded as a flapping hinge with elastic constraints.

The stiffness of the star flexible arm in the lag direction is much larger than in the flap direction. Therefore, when the blade and a pair of clamps move around the center of the elastomer bearing, the elastomer bearing produces shear deformation, and the frequency matcher produces lower stiffness than the star flexible arm, which not only provides lag damping but also adds elastic constraints to the root of the blade, and the natural frequency of blade lag can be adjusted. In this way, the elastomer bearing can be recognized as a lag hinge with elastic constraints.

It can be seen from the above that the star flexible hub is an articulated rotor with elastic constraints at the hinge. According to the structure and function of the star flexible hub, the root constraint disposal mode is as follows: in the flap direction, the clamp stiffness and frequency matcher compression stiffness provide constraint stiffness, which is paralleled with the constraint stiffness provided by the elastomer bearing and the flexible arm. Due to the stiffness of the clamp in the lag direction, the constraint stiffness is obtained by the frequency matcher and the flexible arm in series, and then in parallel with the elastomer bearing. In the pitch direction, the constraint stiffness is provided by the elastomer bearing.

3.2 Blade Structure Parameters

In order to study the influence of double-swept blade aeroelastic stability, the reference straight blade of H425 helicopter's rotor blade is used as a basis, and the double-swept structure is applied, as shown in Fig. 2. Figure 2a represents the reference straight blade, while Fig. 2b represents the double-swept blade. To enhance the torsional stiffness of the double-swept blade, a strengthening prepreg zone is added along the blade beam direction. The other structural parameters of the double-swept blade are basically the same as those of the reference straight blade.

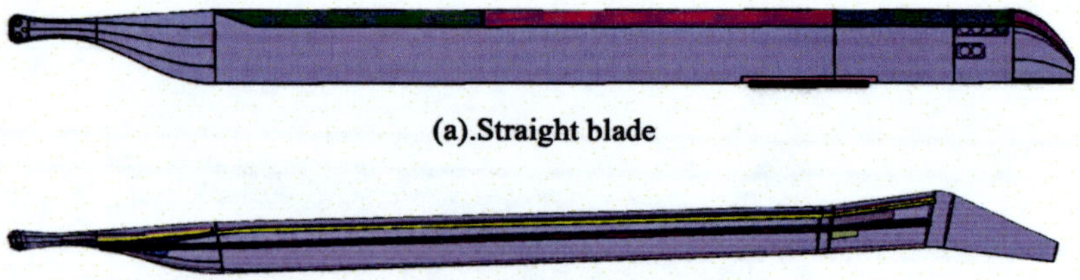

(a).Straight blade

(b).doubled-swept blade

Fig. 2. Diagram of blade structure

The radius of the double-swept blade is 6 m. The sweep-forward angle is 7.7° at the forward station located 4.8 m from the blade radius, and the sweep-backward angle is 30° at the forward station located 5.4 m from the blade radius. The required calculation data is shown in Table 1.

Table 1. Rotor parameters

Parameters	numerical value
Number of blade	4
Rotor speed	360 r/min
Rotor radius	6 m
Hub radius	0.69 m
Radial station of hinges	0.23 m
Blade chord	0.385 m
Blade twist angle	−12°
Precone angle	−4.5°
Droop angle	−2°
Sweep-forward angle	7.7°
Sweep-backward angle	30°
Sweep-forward station	4.8 m
Sweep-backward station	5.4 m

4 Verification of Calculation Method

4.1 Calculation and Test of Blade Scetion Characteristics

In the process of solid dynamics modeling of blades, two-dimensional linear section characteristic analysis and one-dimensional engineering beam are simplified. The blade section characteristics mainly describe a series of characteristics by beam characteristics, such as linear density, flap stiffness, lag stiffness, torsional stiffness, section center of gravity, section rigid center and moment of inertia, which are the basis for rotor natural frequency calculation and aeroelastic stability analysis. To verify the accuracy of the method for calculating the blade section characteristics in this paper, full-size composite blades were selected for calculation and compared with the test results. The calculation and test methods of composite blade section stiffness are mature, and the details are not repeated in this paper. Figure 3 shows the blade section stiffness test, and Figs. 4, 5 and 6 show the comparison between the calculated results and the test results.

The calculation and test methods of composite blade scetion stiffness are mature, and the details will not be repeated in this paper.

Figure 3 shows the blade scetion stiffness test, and Figs. 4, 5 and 6 shows the comparison between the calculated results and the test results.

Fig. 3. Test of blade sections stiffness

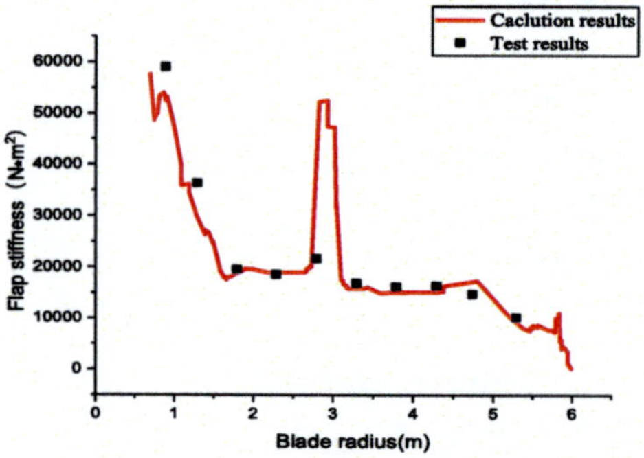

Fig. 4. Flap stiffness of blade

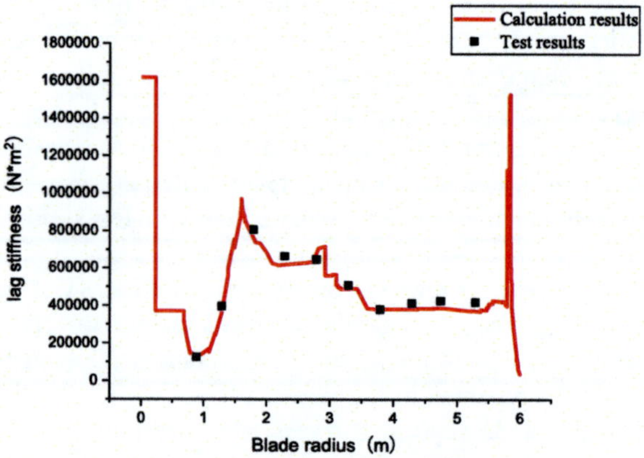

Fig. 5. Lag stiffness of blade

It can be seen from Figs. 4, 5 and 6 that the calculated results are basically consistent with the test results, and the error is controlled within 10%. This indicates that the calculation results are reliable.

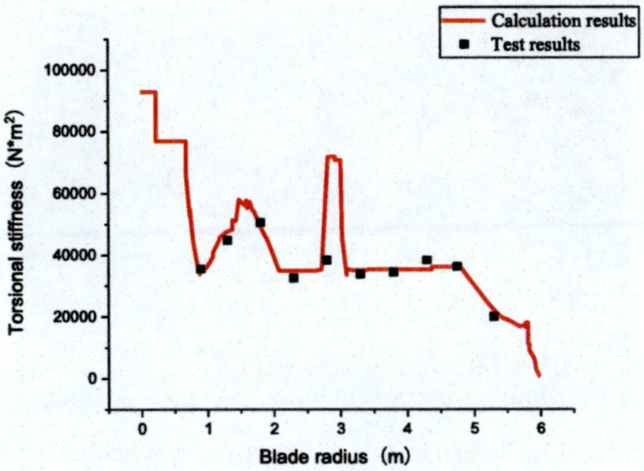

Fig. 6. Torsional stiffness of blade

4.2 Calculation and Test of Blade Natural Frequence

When it is not possible to carry out the rotating dynamic test of the blade, the static test of the blade is an effective method to verify the calculation model. In this paper, the root fixed support method and the suspension method are adopted to carry out the natural frequency test according to the different constraint modes of the blade. Figure 7 shows the root fixed test, and Fig. 8 shows the suspension test. Table 2 and Table 3 show the comparison between the calculated results and the test results of two different test states.

Fig. 7. Modal test of the fixed blade

From Tables 2 and 3, it can be found that the calculated results are in good agreement with the test results, and the error is controlled within 5%. The calculation accuracy meets the requirements of engineering design.

Fig. 8. Modal test of the suspension blade

Table 2. Calculation and test results of the fixed blade

Mode	Calcultion results	Test results	Error%
1st Flap	1.16	1.19	1.97
2st Fap	5.92	6.14	2.73
1nd Lag	4.39	4.28	4.33
2rd Lag	25.25	25.76	2.08
1st Torsional	38.01	36.89	3.03

Table 3. Calculation and test results of the suspension blade

Mode	Calcultion results	Test results	Error%
1st Flap	5.41	5.65	4.2
2st Fap	15.36	16	4.17
1nd Lag	30.57	31.05	3.37
2rd Lag	95.25	94.85	0.62
1st Torsional	72.47	74.4	2.66

5 Analysis Results of Double Swept Blades

5.1 Modal Analysis Results of Blades

The unique structure of the double-swept blade causes the aerodynamic center and center of gravity of the blade section to the pitch axis offset, which makes the design of dynamic characteristics for forward-backward swept blades very challenging. The analysis of the modal characteristics of the blade can not only better control the aeroelastic stability of the rotor but also provide technical support for the optimal design of blade structural characteristics and mass distribution to achieve the ideal frequency distribution.

Figure 9 and Fig. 10 show the variation of the natural frequencies with rotational speed for the reference straight blade and double-swept blade, respectively. In the figure, F represents the blade flap frequency, L represents the blade lag frequency, T represents the blade torsional frequency, and the number represents the frequency order, such as 1L representing the first-order blade lag frequency.

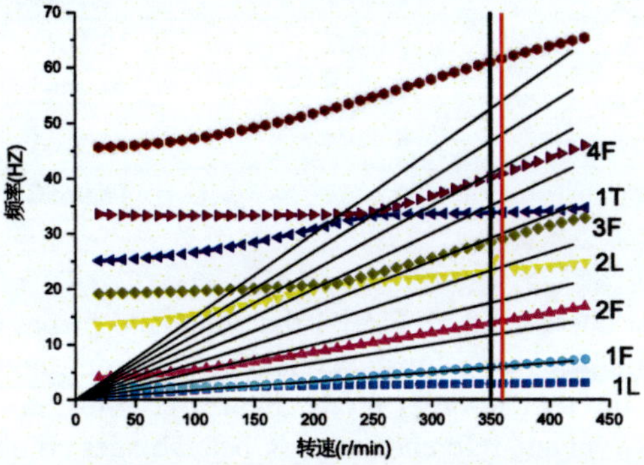

Fig. 9. Fan diagram of frequencies versus rotor speed

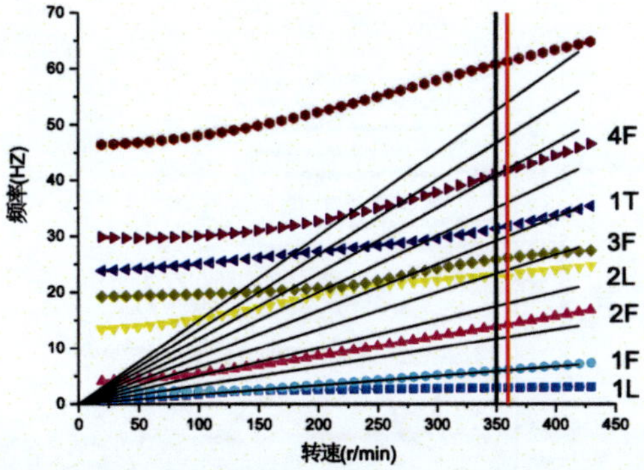

Fig. 10. Fan diagram of frequencies versus rotor speed

It seems evident from Fig. 9 and Fig. 10 that the double-swept blade structure has a significant impact on the first-order torsion and third-order flap frequency. Furthermore, Table 4 presents the natural frequencies of the two types of blades at the operating speed of 360 r/min ($\Omega = 6$ Hz).

It can be observed that, at this speed, the first-order frequencies of the lag and flap of the double-swept blade are nearly identical to the reference straight blade, which significantly affects the third-order flap frequency and the first-order torsional frequency. This is due to the decrease in torsional stiffness of the tip section with the double-sweep design.

Figures 11, 12 and 13 display the mode shapes of the double-swept blade and reference straight blade, where r denotes the blade section's spanwise station, and R is the

Table 4. Frequency calculation results(unit: Ω)

Mode	Norm blade	Double-swept blade
1st lag mode frequency	0.5012	0.5013
1st flap mode frequency	1.0360	1.0360
2nd flap mode frequency	2.3995	2.4020
2nd lag mode frequency	3.8007	3.8012
3rd flap mode frequency	4.8627	4.3722
1st torsional mode frequency	5.6485	5.3163
4ur flap mode frequency	6.8341	6.9650

blade radius. It can be observed from Fig. 11 that the mode shapes of the double-swept and reference straight blades are almost identical at low-order modes, but they change significantly beyond the second-order mode. Figure 12 illustrates that the first-order mode of the double-swept and reference straight beam blades coincides, but the trough position of the second-order lag mode changes, likely caused by the prepreg layer's modification. Figure 13 depicts the torsional mode shape of the blade, showing that there is little difference between the double-swept blade and reference straight blade.

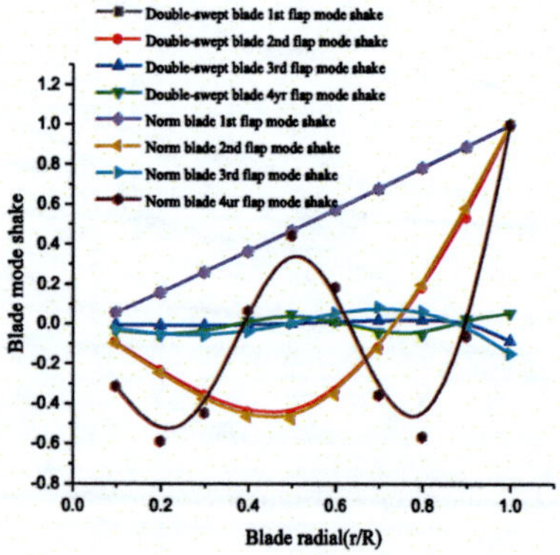

Fig. 11. The flap mode shapes diagram of blade

5.2 Results of Aeroelastic Stability Analysis

The aeroelastic stability of the rotor with double-swept blades is analyzed and compared to that of the reference straight blades. The calculation state is shown in Table 5.

Figure 14 shows the results of aeroelastic stability analysis of double swept blade and reference straight blade, from the results, under the condition of 6° collective pitch, with the increase of forward flight velocity, the first-order damping of lag increases,

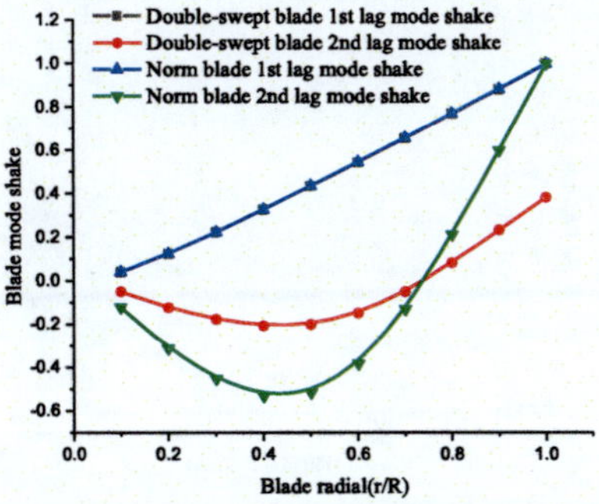

Fig. 12. The lag mode shapes diagram of blade

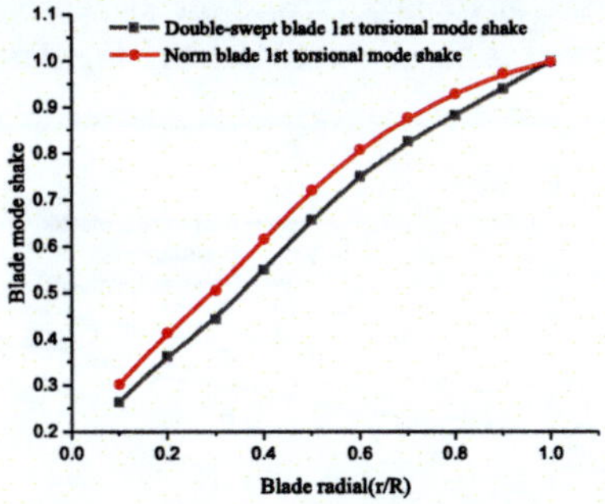

Fig. 13. The torsional mode shapes diagram of blade

Table 5. Calculation state

Calculation Parameters	Calculation range
Collective (°)	6
Rotor speed(r/min)	360 r/min
Flight speed(m/s)	0, 10, 20, 30, 40, 50, 60, 70, 80

the stability margin increases. From the comparison results, the stability margin of the double-swept blade is slightly better than that of the reference straight blade, this is due to the increase of prepreg along the beam direction of double-swept blade, which improves the torsional stiffness of blades.

5.3 Influence of Double-Swept Structure Parameters

The effects of the forward-swept angle, backward-swept angle, forward-swept position, and backward-swept position on the aeroelastic stability of double-swept blades are analyzed.

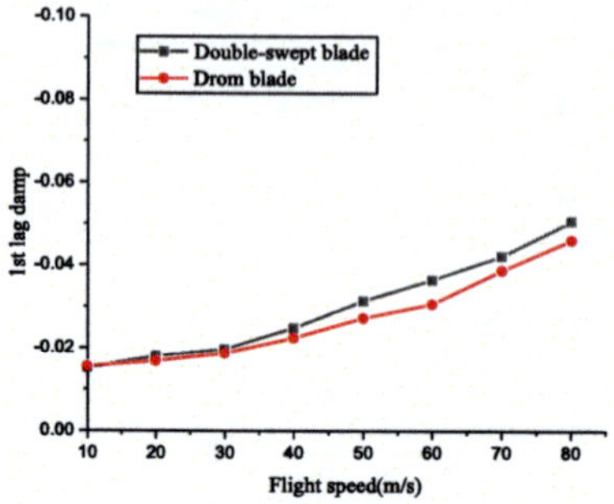

Fig. 14. Diagram of 1st lag mode damping versus speed

When the sweep-forward angle is 7.7°, and the sweep-backward angle is 20°, 30°, and 40°, respectively, the variation of the first-order lag damping with flight speed is shown in Fig. 15.

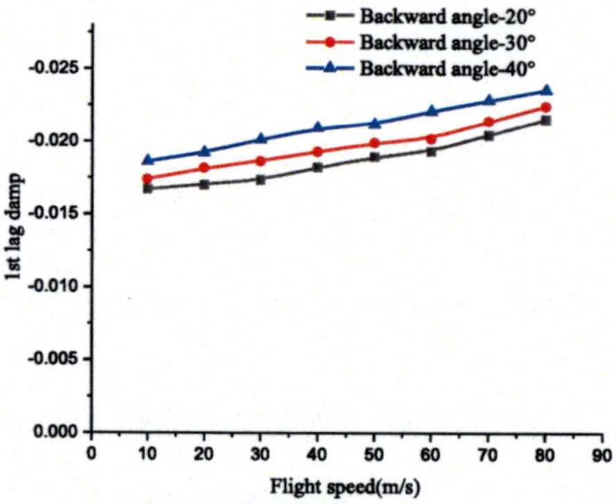

Fig. 15. Diagram of 1st lag mode damping versus speed

It can be seen that as the sweep-backward angle increases, the first-order lag damping also increases, indicating that increasing the sweep-backward angle is beneficial for aeroelastic stability.

When the sweep-backward angle is 30°, and the sweep-forward angle is 6.7°, 7.7°, and 8.7°, respectively, with the sweep-forward position and sweep station position unchanged, the variation of the first-order lag damping with flight speed is shown in Fig. 16.

It can be considered that the forward-swept angle has little influence on aeroelastic stability because the first-order lag damping of the three forward-swept angles changes only slightly.

When the current sweep-forward station position is constant, and the sweep station position is 0.83R, 0.9R, and 0.93R respectively, with the sweep angle and sweep angle unchanged, the variation of the first-order lag damping with flight speed is shown in Fig. 17.

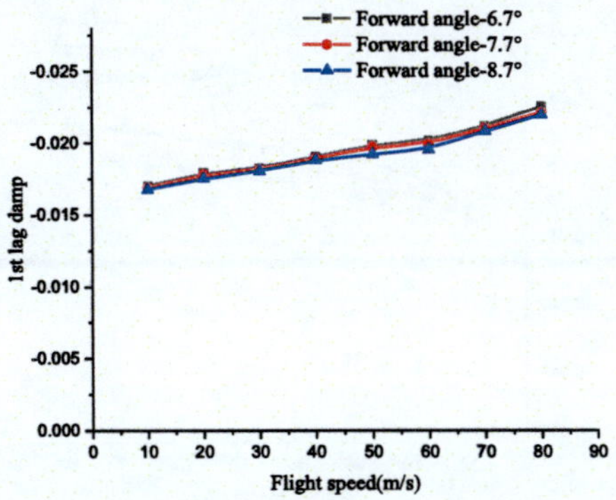

Fig. 16. Diagram of 1st lag mode damping versus speed

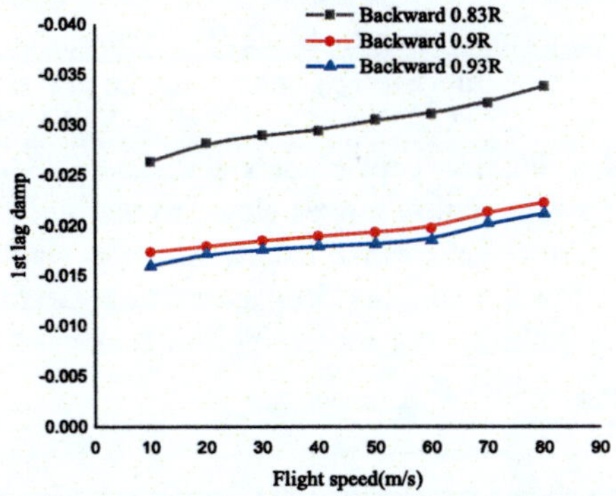

Fig. 17. Fan diagram of 1st lag mode damping versus speed

It can be seen that as the backward swept station position moves closer to the blade tip, the first-order lag damping decreases, indicating that a position closer to the tip is more disadvantageous for aeroelastic stability.

When the backward swept station position is constant, and the sweep-forward station position is 0.7R, 0.8R, and 0.83R, respectively, with the sweep angle and sweep angle unchanged, the variation of the first-order lag damping with flight speed is shown in Fig. 18.

It can be seen that as the forward swept station moves backward, the first-order lag damping increases, indicating that a backward movement is beneficial for aeroelastic stability.

6 Conclusions

(1) Compared with a conventional straight blade, the double-swept structure reduces the third-order flap and the first-order torsional natural frequency, with little influence on the blade's lower natural frequency.

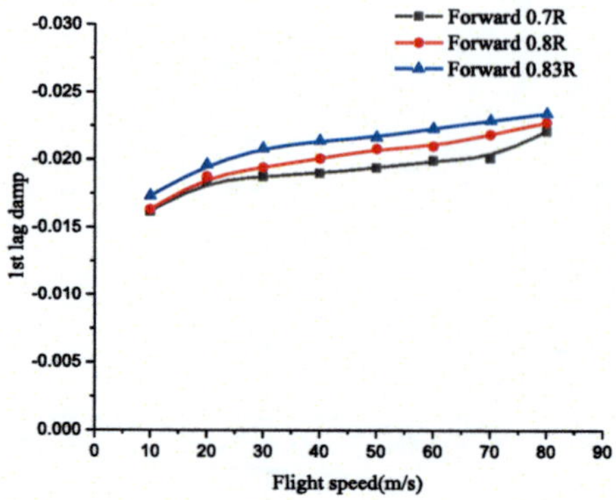

Fig. 18. Diagram of 1st lag mode damping versus speed

(2) Compared to a conventional straight blade, the aeroelastic stability margin of a double-swept blade can be improved by enhancing the overall stiffness design of the rotor.
(3) Analysis results of a double-swept blade's aeroelastic stability parameters show that as the forward-swept station moves closer to the blade tip, the lag first-order damping increases. Conversely, as the backward-swept station moves closer to the blade tip, the lag first-order damping decreases. The forward-swept angle has little effect on aeroelastic stability, but increasing the backward-swept angle results in greater first-order lag damping.

Acknowledgement. The authors are grateful to Liu zheng for providing blade structure data. They also thank the anonymous reviewers for their critical and constructive review of the manuscript. This study was supported by the Natural Science Foundation of Heilongjiang Province of China (No. JQ2020E006).

References

1. Ward, E., Chopra, I., Datta, A.: Rotation frequency driven extension torsion coupled self-twisting rotor blades. J. Aircr. **55**, 1929–1941 (2018)
2. Milluzzo, J.I., Drayton, S.: Particle Image velocimetry measurements and performance analysis of rotors with slotted blades. J. Aircraft **55**, 2401–2413 (2018)
3. Brocklehurst, A., Barakos, G.N.: A review of helicopter rotor blade tip shapes. Prog. Aerosp. Sci. **56**, 35–74 (2013)
4. van Der Wall, B.G., et al.: From aeroacoustics basic research to a modern low-noise rotor blade. J. Am. Helic. Soc. **62**(4), 1–16 (2017)
5. Rauch, P., Gervais, M., Cranga, P., et al.: Blue edge: the design, development and testing of a new blade concept. In: Proceedings of the 67th Annual Forum of the American Helicopter Society, pp. 542–555 (2011)
6. Van Der Wall, B., Kebler, C., Pengel, K., et al.: From ERATO basic research to the blue edge rotor blade. In: Proceedings of 72th Annual Forum of the American Helicopter Society, pp. 3103–3121 (2016)

7. Filippi, M., Zappino, E., Carrera, E.: Multidimensional models for double-swept helicopter blades. AIAA J. **57**, 2609–2616 (2019)
8. Filippi, M., Zappino, E., Carrera, E., Castanié, B.: Effective static and dynamic finite element modeling of a double swept composite rotor blade. J. Am. Helic. Soc. **65**(3), 1–12 (2020)
9. Yeo, H., Truong, K., Ormiston, R.A.: Comparison of one-dimensional and three-dimensional structural dynamics modeling of advanced geometry blades. J. Aircr. **51**, 226–235 (2014)
10. Maier, T.H., Sharpe, D.L., Anita, I.A.: Aeroelastic stability for straight and swept-tip rotor blades in hover and forward flight. In: Proceedings of 55th Annual Forum of American Helicopter Society, pp. 1031–1047 (1999)
11. Kumar, M., Rao, L.: Aeroelastic analysis of double- swept rotor blade in hovering conditions. In: Proceedings of 17th International Forum on Aeroelastic and Structural Dynamics, pp. 1318–1331 (2017)
12. Kumar, M.R., Venkatesan, C.: Effects of blade configuration parameters on helicopter rotor structural dynamics and whirl tower loads. Aeron. J. **120**(1224), 271–290 (2016)
13. Jurgen, A.: Using multibody dynamics for flutter assessment of an isolated rotor with innovative blade layout. In: Proceedings of 72th Annual Forum of American Helicopter Society (2016)
14. Zhang, J., Xia, P., Chopra, I.: Aeromechanical stability of a bearingless rotor helicopter with double-swept blades. J. Aircraft **58**(2), 244–252 (2021)
15. Yang, W.D., Deng, J.H.: Aeroelastic stability analysis of helicopter rotor blade with swept tips. J. Nanjing Univ. Aeronaut. Astronaut. **35**(3), 248–252 (2003). (in Chinese)
16. Xia, S.M., Lin, C.L., Wang, J.L., et al.: Study on dynamic characteristics of double-swept rotor blade. Adv. Aeronaut. Sci. Eng. **13**(2), 32–37 (2022). (in Chinese)

Analysis and Testing of Rotating Blade Natural Frequency

Shuangman Xia[1](✉), Renzhong Wang[2], Changliang Lin[1], Haibin Xu[1], and Huaming Wang[3]

[1] Aircraft Design and Research Institute, AVIC Harbin Aircraft Industry Group Co., Ltd., Harbin 150066, China
xia19840229@163.com
[2] 95966 Troops of the PLA, Harbin 150060, China
[3] National Key Laboratory of Science and Technology on Rotorcraft Aeromechanics, Nanjing University of Aeronautics and Astronautics, Nanjing 210016, China

Abstract. Reasonable allocation of the natural frequencies of a blade can help to avoid resonance or excessive vibration within the helicopter's working speed range, thus reducing and controlling the vibration level. For composite blades, the motion equation of the blade was deduced using the Hamilton principle to obtain the natural frequency and natural vibration mode of the blade. Full-scale blade natural frequency tests were carried out on the rotor testing device, and the analysis results were in good agreement with the test results. Based on this, the influence of rotor stiffness parameters on the natural frequency of the blade was studied. The results show that the first lag frequency of the blade increases with the increase of the elastic bearing and blade lag stiffness, which has little influence on the flap frequencies and torsional frequency.

Keywords: Helicopter · Rotor Blade · Composite · Frequency · DynamicTest

1 Introduction

As one of the key components of a helicopter, rotor blades bear centrifugal force, flap moment, lag moment, and torsional moment generated by rotation. Rotating blades can generate lift, lateral force, and pitching force of the helicopter, allowing it to take off, land vertically, hover, and fly forward by controlling collective pitch and periodic pitch [1, 2]. To achieve efficient vertical flight, the rotor is mostly designed with a large diameter and low disk load, and the rotor blade is an elastic structure with a slender and large aspect ratio. Research shows that the natural frequencies of the rotating blade are significantly higher than those of non-rotating blades, due to the large centrifugal force stiffness added when the blade rotates [3, 4].

With the progress of aerospace technology, and the advantages of composite materials in strength, durability, and weight, composite materials are increasingly used in blades. Designers can adjust the characteristics of the helicopter rotor blade structure by selecting different composite materials and ply angles to meet the requirements of rotor blade

Chinese Society of Aeronautics and Astronautics (Ed.): CASTC 2023, LNME, pp. 186–197, 2024.
https://doi.org/10.1007/978-981-99-8861-7_20

dynamics [5, 6]. However, the composite blade is affected by transverse shear deformation, geometric nonlinearity, section warping, and various elastic couplings during operation, which present significant challenges to the research of composite blades [7, 8]. Therefore, it is necessary to accurately and effectively predict the dynamic characteristics of composite blades and carry out dynamic tests to ensure reliable design.

In recent decades, domestic and foreign researchers have conducted extensive work on dynamic characteristics analysis and rotating blade tests. The calculation methods for blade natural frequencies have become mature, and the solutions mainly include the assumed mode method and the concentrated mass method. The assumed mode method is a generalized coordinate approximation method, and the linear sum of finite assumed modes approximately describes the vibration of the elastic body, obtaining the approximate value of the blade's natural frequency using the energy principle. The lumped mass method includes the transfer matrix method and finite element method. With the continuous development of computer technology, the finite element method has become the mainstream method to calculate blade natural frequencies. Kee and Shin [9] stablished a three-dimensional composite blade dynamic model using the finite element method and conducted blade dynamic analysis. The finite element model considered transverse shear deformation, Coriolis effect, and anisotropic elastic coupling. Lee [10]and Kee [11]completed vibration analysis of the rotating pre-twisted plate with the assumption of small displacement. Heo and Bauchau [12] used shell elements to establish the blade dynamic model with sweep angle and droop angle and adopted a domain decomposition algorithm to reduce computational complexity. Kang [13] introduced the geometrically accurate shell element and applied it to rotor blade modeling. Zhou Yun [14] calculated the natural frequencies of the model blade with different damping using the second-order nonlinear elastic beam model and conducted natural frequency tests of rotating blades. Huang Jun [15] conducted calculations of the rotor blade's natural frequency and carried out the first test of the blade's natural frequency with rotation in China. Shang Lina [16] used the variational asymptotic method to determine the arbitrary warping of the beam profile and established the geometrically accurate nonlinear motion equation of the beam using the generalized Hamilton principle. The research results show that section warping has a significant impact on the natural frequencies of composite beams.

In this paper, the dynamic equation of the rotor blade is established based on the Hamilton principle, and the natural frequency of the rotating blade is obtained. To verify the accuracy and reliability of the calculation results, a full-scale composite rotating blade natural frequency test is carried out. On this basis, the influence of blade stiffness change on the natural frequency is studied.

2 Blade Motion Equation

In order to describe the motion of the rotor blade, the equations of motion are derived using Hamilton's principle. The equation is expressed as follows:

$$\int_{t_1}^{t_2} [\delta(U - T) - \delta \mathrm{W}]dt = 0 \tag{1}$$

δU is the variation of the elastic blade strain energy, δT is the variation of the elastic blad kinetic energy, and δW is the variation of the external force work acting on the blade, include aerodynamics and other constrained external forces.

The rotor dynamic motion equation can be expressed as a second-order differential equation:

$$M\ddot{q}(t) + C\dot{q}(t) + Kq(t) = f(t) \tag{2}$$

where $f(t)$ is the external force column vector acting on the system, and M, C, K are the mass and stiffness matrices of the system, respectively.

Neglecting damping and external force, the free vibration equation of the rotor dynamic system can be obtained.

$$M\ddot{q}(t) + Kq(t) = 0 \tag{3}$$

The generalized coordinates q are used to describe the motion of the system, which are assumed to be in the following form:

$$q = \Phi \sin \omega(t - t_0) \tag{4}$$

where Φ is an n-order vector, ω is the vibration frequency of vector Φ, t is a time variable, and t_0 is a time constant determined by the initial conditions. If Eq. (4) is substituted into (3), a generalized eigenvalue problem can be obtained:

$$K\Phi - \omega^2 M\Phi = 0 \tag{5}$$

The eigenvectors Φ and eigenvalues ω can be determined by solving the above equations.According to the order of eigenvalues, n eigenolutions (ω_n, Φ_n) (n = 1, 2, 3……n)can be obtained, where ω_n represents the natural frequencies of the rotor dynamic system and Φ_n represents the natural vibration modes of the corresponding natural frequencies.

3 Rotor Structure

The rotor system includes the hub and blades, as shown in Fig. 1. The hub is a star flexible hub, which is an articulated rotor with a flaphinge and elastic constraints. The hub is equipped with frequency matchers, which are viscoelastic dampers widely used in helicopters and provide elastic constraints at the blade root. The frequency matcher bears shear deformation and has stiffness and damping effects in the lag direction. The stiffness characteristics of the frequency matcher have a decisive impact on the rotor blade lag natural frequency, and its damping characteristics have a decisive influence on rotor flutter, helicopter air resonance, and helicopter ground resonance stability.

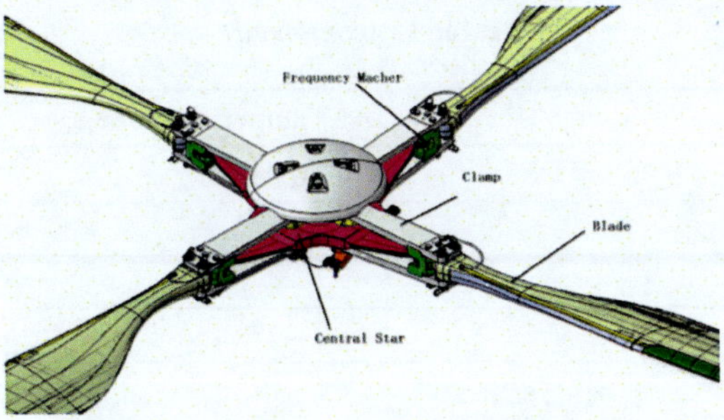

Fig. 1. Schematic diagram of main rotor

The blade is a composite laminated structure with a C-beam, front and rear Z-beams, and box-shaped pieces formed with the skin to withstand torque. The internal cavity is filled with foam (Fig. 2).

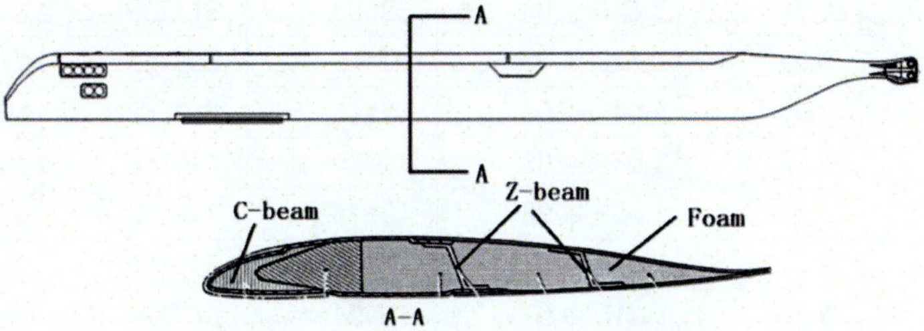

Fig. 2. Diagram of blade structure

The blade structure data for calculation is shown in Table 1, and typical section data of the blade is shown in Table 2.

Table 1. Rotor parameters

Parameters	numerical value
Number of blade	4
Rotor speed	360 r/min
Rotor radius	6 m
Hub radius	0.69 m
Radial station ofhinges	0.23 m

(continued)

Table 1. (*continued*)

Parameters	numerical value
Blade chord	0.385 m
Blade twist angle	−12°
Precone angle	−4.5°
Droop angle	−2°

Table 2. Rotor section property

Radius m	MASS Kg/m	EA $N \cdot m^2$	EIY $N \cdot m^2$	EIZ $N \cdot m^2$	GJ $N \cdot m^2$
0	48.23	1.98E + 08	186347	1623030	93020
0.2	48.23	1.98E + 08	186347	1623030	93020
0.7	50.74	1.62E + 08	58083	315233	65413
1.3	7.09	1.22E + 08	27365	603782	51534
1.6	7.09	1.04E + 08	17846	857148	56955
2.1	6.13	1.03E + 08	19084	619186	34924
3.1	6.17	1.00E + 08	17574	512002	35436
3.4	6.0	9.65E + 07	16186	493506	35290
4.5	5.77	9.41E + 07	16924	388994	36247
5.4	6.06	9.31E + 07	8075	382159	20361
6.0	1.45	1.20E + 07	284	38063	864

4 Calculation Results

The natural frequencies of the rotor blade are calculated at the working speed, and the collective pitch is 0°. The natural frequencies of the blade are shown in the figures in Fig. 3. The horizontal axis in the diagram is the rotor speed, and the vertical axis is the frequency. It can be seen from the figure that the natural frequency of the blade increases with the increase of rotor speed, which indicates that the centrifugal force stiffness of the blade increases with the increase of rotor speed. However, the blade lag frequency does not change significantly with the speed because the blade lag stiffness is large, and the effect of centrifugal force in the lag direction is smaller than that in the flap direction. However, the first-order flap frequency increases significantly with the increase of rotating speed because the centrifugal force in the flap direction increases with the increase of rotating speed.

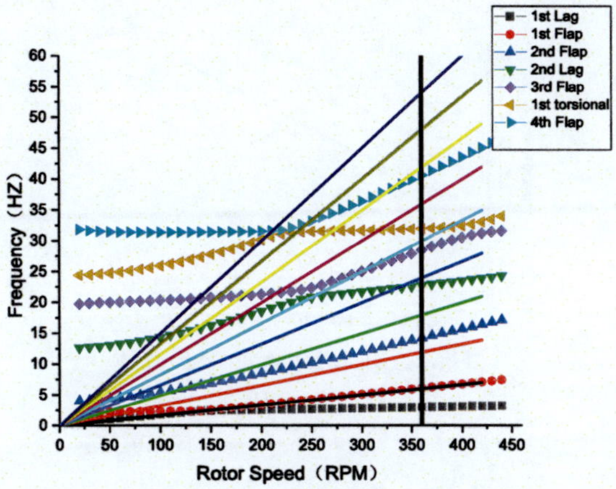

Fig. 3. Rotor resonance diagram

It can be observed from Fig. 3 that the blade's natural frequencies have remained within a certain safety range with respect to the frequency of the exciting force, which is in accordance with the dynamic design requirements.

Figures 4, 5, and 6 show the blade's flap mode, lag mode, and torsional mode, respectively. It can be seen from these figures that the first-order vibration modes of lag, flap, and torsion are similar to rigid body motion. These vibration modes provide a basis for blade frequency modulation and reference for the rotating blade's natural frequency test patch position.

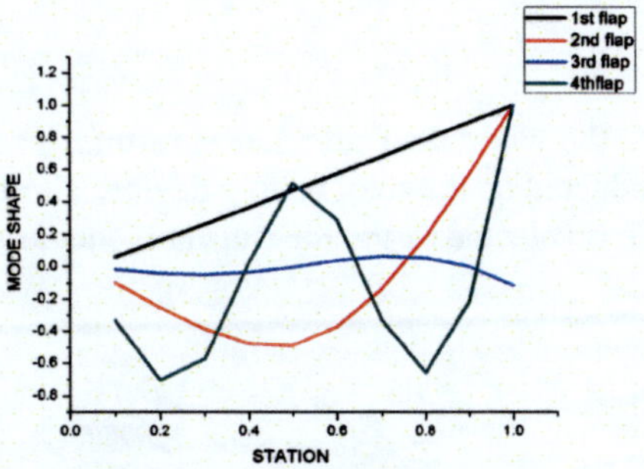

Fig. 4. Blade flap mode shape

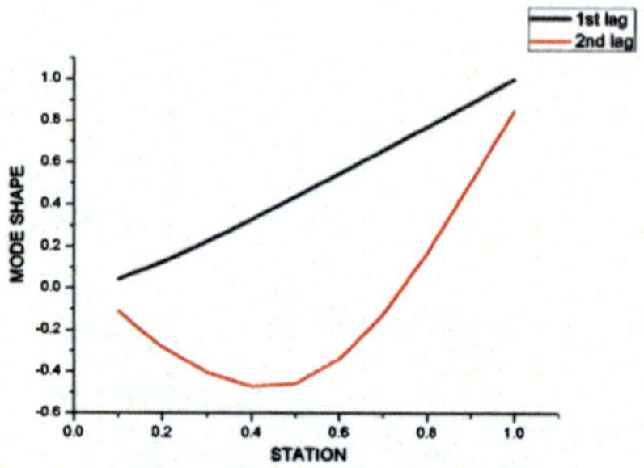

Fig. 5. Blade lag mode shape

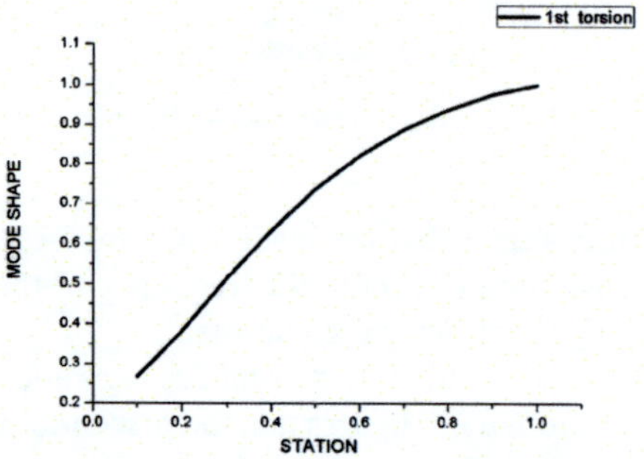

Fig. 6. Blade torsional mode shape

5 Natural Frequency Test of Rotating Blade

5.1 Test Method

To verify the accuracy of the theoretical calculation results, a natural frequency test of the rotating blades is conducted on the rotor testing device. Three blades are installed on the device, and one blade is selected as the measuring blade, as shown in Figs. 7 and 8.

Fig. 7. Rotor testing device

Fig. 8. Blade for test

According to the blade's vibration mode analysis in the previous section, strain gauges are affixed to 10 sections of the blade along the span direction. For each section, a group of flap strain gauges and a group of torsional strain gauges are affixed to the 1/4 chord of the section. Two groups of lag strain gauges are affixed to the front and rear edges of the blade. The primary deformation direction of the flap strain gauges and the lag strain gauges is along the blade span direction. The strain gauge affixing scheme is illustrated in Fig. 9.

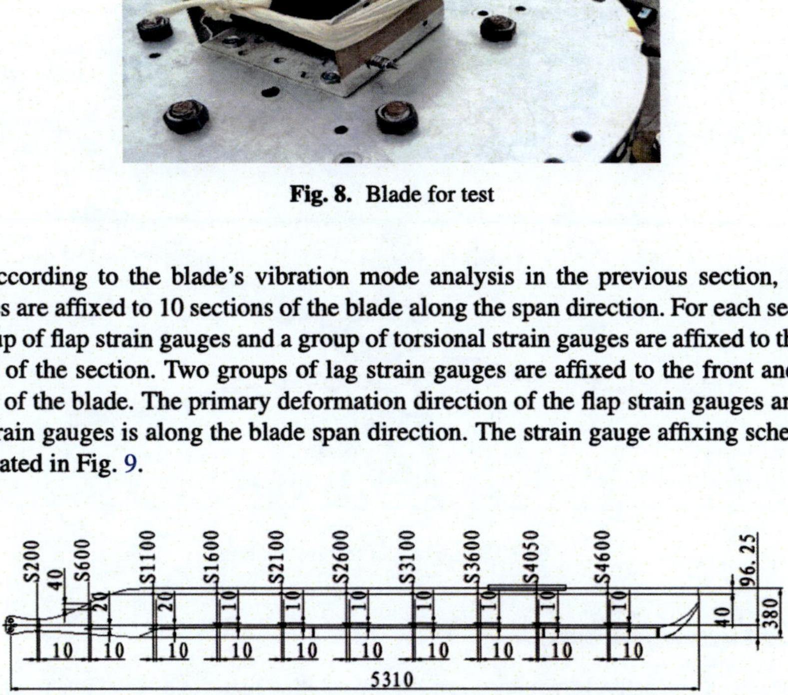

Fig. 9. Diagram of blade sticking strain gauge

The natural frequencies of the blades are measured at different speeds, as listed in Table 3.

An actuator is used for excitation during the test, and it is installed between the bench and the moving tilter. The actuator excites the fixed tilter, and the excitation force is transmitted to the moving tilter through the fixed tilter. The pitch motion of the blade is generated by the pitch link, after which the blade generates flap motion and lag motion. The acquisition device is installed at the center of the rotor hub, as shown in Fig. 10.

Table 3. Speed measurement range

Number	Speed(RPM)
1	120
2	235
3	295
4	320
5	325
6	350
7	360
8	380

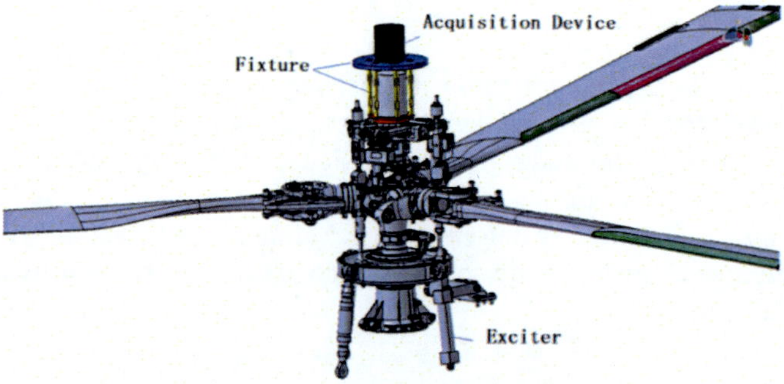

Fig. 10. Diagram of test installation

5.2 Test Results and Analysis

The results of the natural frequency test for the rotating blade at a collective pitch of 0° are shown in Table 4. The test mainly identified the first, second, and third order natural frequencies of the flap, the first and second order natural frequencies of lag, and the first torsion order natural frequency.

Table 5 shows the comparison between the test results and the calculated results at the working speed of 360RPM.

As shown in Table 5, the calculated results are in good agreement with the test results, with errors controlled within 5%. This proves that the analysis model is reliable.

Table 4. Blade mode frequencies

RPM	Mode frequency(HZ)						
	1st flap	2nd flap	3rd flap	4th flap	1st Lag	2rd Lag	1st torsioal
120	2.0	6	15.7	27.4	2.0	19.0	37.2
235	3.9	9.5	19.6	29.7	2.7	19.6	38.3
295	4.9	11.7	22.6	32.0	3.2	20.1	38.5
320	5.3	12.8	24.5	34.7	3.3	20.7	39.0
325	5.4	12.9	24.6	34.8	3.4	20.8	39.1
350	5.8	13.6	26.4	38.3	3.6	21.7	39.4
360	6.0	14.0	27.2	39.3	3.8	22.2	39.6
380	6.3	14.8	28.4	41.4	3.8	23.2	39.9

Table 5. Blade mode frequencies of calculation and test(HZ)

Mode	Calcultion results	Test results	Error%
1st Lag	3.71	3.8	2%
1st flap	6.22	6	4%
2nd flap	14.29	14	2%
2rd Lag	22.8	22.2	3%
3rd flap	28.52	27.2	5%
1st torsional	39.06	39.6	1.30%
4th flap	40.66	39.3	3%

6 Parameter Analysis

In the blade dynamic design process, blade stiffness can be altered due to manufacturing deviations. Therefore, it is necessary to analyze the influence of blade stiffness on the blade's natural frequency. Based on the characteristics of the star flexible hub, changes in elastic bearing stiffness and blade lag stiffness are chosen as parameters for analysis.

Table 6 shows the blade natural frequencies when the stiffness of the elastomer bearing is changed by 20%, and the rotor speed is 360 RPM.

The calculation results show that when the stiffness of the elastomer bearing changes, the blade's flap and torsional frequencies remain mostly unchanged. It mainly affects the first and second order lag frequencies of the blade. When the lag stiffness of the elastomer bearing increases by 20%, the frequency increases by 5%. When the lag stiffness of the elastomer bearing decreases by 20%, the frequency decreases by 4%. The same trend is observed for the second order lag frequency of the blade. When the lag stiffness of the elastomer bearing increases by 20%, the frequency increases by 1.1%, and when it decreases by 20%, the frequency decreases by 1.3%.

Table 6. Results comparison(HZ)

Mode	Initial	Stiffness reduced by 20%	Stiffness increased by 20%
1st Lag	3	2.88	3.18
1st flap	6.22	6.22	6.22
2nd flap	14.29	14.28	14.34
2rd Lag	22.8	22.5	23.04
3rd flap	28.5	28.5	28.5
1st torsional	39.06	39.06	39.06
4th flap	40.66	40.62	40.62

Table 7 shows the blade natural frequencies when the blade lag stiffness changes.

Table 7. Results comparison(HZ)

Mode	Initial	Stiffness reduced by 20%	Stiffness increased by 20%
1st Lag	3	2.98	2.98
1st flap	6.22	6.22	6.22
2nd flap	14.29	14.28	14.28
2rd Lag	22.8	21.6	23.88
3rd flap	28.5	28.5	28.5
1st torsional	39.06	39.06	39.06
4th flap	40.66	40.2	40.2

Table 7 shows the blade natural frequencies when the blade lag stiffness is changed. It can be observed that when the blade lag stiffness changes, the blade's flap and torsional frequencies remain mostly unchanged. It mainly affects the second order lag frequency of the blade. When the blade lag stiffness increases by 20%, the frequency increases by 5%, and when it decreases by 20%, the frequency decreases by 5%.

7 Conclusions

1) The rotor blade dynamic analysis model is reliable, with errors between the calculated results and test results controlled within 5%.
2) The method of automatic tilter periodic pitch excitation can quickly identify the required blade frequency.
3) The stiffness of the elastomer bearing has a significant influence on the first order frequency of the lag, and the blade lag stiffness has an influence on the second order lag frequency of the blade, with a minor effect on the flap and torsional frequencies.

Acknowledgement. The authors are grateful to Xinhao Yu and Chunlan Zhao for providing blade structure data and test pictures. They also thank the anonymous reviewers for their critical and constructive review of the manuscript. This study was supported by the Foundation of National Key Laboratory of Rotorcraft Aeromechanics (No. 61422202112).

References

1. Dong, H.: Helicopter Structural Dynamics, pp. 1–7. Science Press, Beijing (2020). (Chinese)
2. Insititute of science and technology,Ministry of aerospace industry. Helicopter dynamics manual. Aviation industry press, Beijing (1991). (Chinese)
3. Sun, Z., Xiao, Q., Xu, G.: Helicopter Strength. Aviation Industry Press, Beijing (1990). (in Chinese)
4. Wu, X.: Engineering Design of Helicopter Dynamics, pp. 1–35. Aviation Industry Press, Beijing (2017). (Chinese)
5. Bauchau, O.A.: A beam theory for anisotropic materials. J. Appl. Mech. **52**(2), 416–422 (1985)
6. Epps, J.J., Chandra, R.: The natural frequencies of rotating composite beams with tip sweep. J. Am. Helic. Soc. **41**(1), 29–36 (1996)
7. Cesnik, C.E.S., Hodges, D.H.: VABS: a new concept for composite rotor blade cross-sectional modeling. J. Am. Helic. Soc. **42**(1), 27–38 (1997)
8. Hodges, D.H.: Nonlinear Composite Beam Theory. AIAA, Reston (2006)
9. Kee, Y., Shin, S.: Structural dynamic modeling for rotating blades using three-dimensional finite elements. J. Am. Helic. Soc. **29**, 1607–1618 (2015)
10. Lee, S.J., Han, S.E.: Free vibration analysis of plates and shells with a nine-node assumed natural degenerated shell element. J. Sound Vibr. **241**(4), 605–633 (2001)
11. Kee, Y.J., Kim, J.H.: Vibration characteristics of initially twisted rotating shell type composite blades. Compos. Struct. **64**(2), 151–159 (2004)
12. Heo, S.S., Bauchau, O.A.: A novel parallel algorithm for rotor dynamics simulation. In: 70th AHS Forum, Montreal, Canada (2014)
13. Kang, H., Chang, C., Saberi, H.: Assessment of beam and shell elements for modeling rotorcraft blades. J. Aircraft **51**(2), 520–531 (2014)
14. Zhou, Y., Hu, H.: Identification of helicopter rotating blade frequency based on transfer function. Helic. Techn. **3**, 001–007 (2011). (Chinese)
15. Huang, J., Deng, J., Han, J.: Frequency tests of helicopter rotating rotor blade. J. Nanjing Univ. Aeronaut. Astronaut. **33**(5), 410–415 (2001). (Chinese)
16. Shang, L., Xia, P.: Geometrically exact non-linear modeling and non-classical effect influence of composite beams. Acta Materiae Compositae Sinica **35**(1), 218–228 (2018). (Chinese)

Check for updates

A Synthesis Decision Model for Level of Repair of Civil Aircraft Based on SVIAE-MEE Method

Yuxin Wang, Wanyi Deng, Xiaowei Wen, and Yiqiang Wang(✉)

Civil Aviation University of China, Tianjin 300300, China
Wyq80@126.com

Abstract. The level of repair analysis plays a key role for aircraft to achieve satisfactory life-cycle cost, in which the maintenance capability, time, cost and other factors should be considered comprehensively. This paper presents a synthesis decision model for aircraft repair level based on the matter-element extension (MEE) method. According to the characteristics of maintenance support system of civil aircraft, both the economic and non-economic factors were considered as inputs for the model. The alternative repair programs were ranked by the MEE method in view of their comprehensive performance. The set value iteration (SVI) combined with anti-entropy (AE) approach was used to eliminate subjective bias during the weight assignment. It is finally shown via demonstration on an example problem that the proposed model is effective and reasonable providing a novel way to evaluate the optimal repair level in the maintenance supporting system of civil aircraft.

Keywords: Level of Repair Analysis · Maintenance · Matter-Element Extension · Combination Weighting

1 Introduction

The level of repair analysis (LORA) is one of the important parts of civil aircraft maintenance engineering analysis, which is intended to determine the optimal maintenance solution by considering the maintenance capability, time, cost, etc., comprehensively. The real aircraft systems are usually designed in a modular way characterized by multiple indentures representing the hierarchy of components. Added to that is the existence of supporting facilities and depots in which the repair action can be performed, as well as the hierarchy relationship that exists between these repair locations which is usually known as multiple-echelon repair network [1]. The multiple-echelon repair network coupled with multiple indentures of components constitutes the complex maintenance supporting system of civil aircraft.

In recent years, several efforts have been made by government agencies, universities and research centers worldwide for the implementation of LORA [2–4]. In 2014, the Society of Automotive Engineers has published the SAE AS1390 standard based on the US military standard MIL-STD-1390 [5, 6]. Besides these standards, many researchers have also proposed various integer programming models to solve the LORA problem

Chinese Society of Aeronautics and Astronautics (Ed.): CASTC 2023, LNME, pp. 198–216, 2024.
https://doi.org/10.1007/978-981-99-8861-7_21

[7–9]. Meanwhile, spare parts are critical resources of a support system which play an important role in operating missions. LORA and spare parts stocks are typically optimized individually to obtain the best solution within their individual constraints. Many researchers have neglected the relation-ship between spare parts stocks and maintenance, which leads to the non-optimal cost of the fleet system. Therefore, most efforts have been transferred to the joint optimization method of spare parts stocks and level of repair analysis [10–12].

However, most previous literature has focused on the economic cost factors, while the non-economic constraints such as repair time, safety, etc., have not been receiving much attention [13]. There are limited number of studies in the literature dealing with the comprehensive evaluation of the LORA problem. Due to the fuzzy nature of the comprehensive index system of LORA, deviations between the scores from scholars and experts and the real situation are inevitable. In view of incompleteness of in-formation, there have been various fuzzy methods for complex systems in recent years of which is the matter-element extension (MEE) method [14–17]. It is based on the matter-element theory using the ex-tension set relevance to solve the incompatibility between the multiple objectives and constraints. The principle of the MEE theory is to divide the object under evaluation into different grades, then the results are obtained based on the correlation between the object and different grades. It is adept at quantitative processing of data with multiple factors and has been applied in many fields, such as environmental evaluation, quality evaluation and sustainable development. Compared with traditional evaluation methods, the model construction is relatively simple, and the evaluation results are specific and explicit.

Therefore, in this paper, we propose a synthesis decision model where both the economic and non-economic factors are considered simultaneously by using the MEE method to obtain an optimal maintenance solution.

2 Constructing the Index System of LORA

This study constructed an evaluation index system including a total of 9 economic indicators and 20 non-economic indicators, as shown in Fig. 1. The comprehensive evaluation index system is characterized with a four-layer hierarchy, viz. The first layer of object (A), the secondary layer of attributes (B1–B2), the third of criteria (C1–C7), and the bottom of factors(D1–D29).

The object reflects the suitability of maintenance alternative plans for the requirements in total, which can be divided into two attributes, i.e., economic and non-economic. When selecting the criteria, this study improved and added some from the relevant literature, standards, and the actual situation, al-lowing for factors that affected the criteria during the implementation of maintenance plans [18].

3 Combination Weighting Based on SVI-AE

The determination of weights is critical to the comprehensive evaluation where the final result heavily relies on the weights' rationality and operability. The existing subjective weighting methods are susceptible to experts' experience and arbitrariness, while the

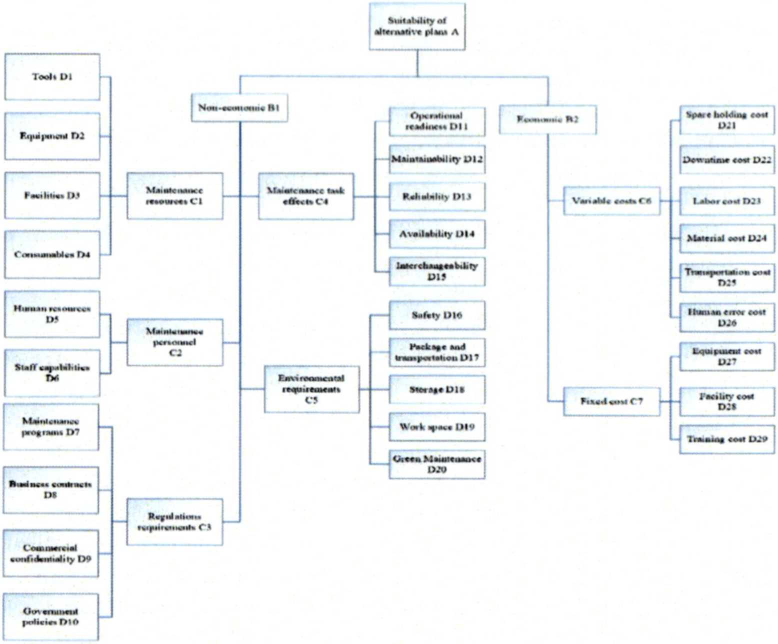

Fig. 1. The comprehensive evaluation index system of level of repair of civil aircraft

objective ones are prone to be sensitive to indicators leading to invalid results. Therefore, on the basis of the shortage of the present subjective and objective weighting methods, an optimal combination weighting method is proposed based on the quad-ratic programming where the subjective weights are derived from the Set Valued Iteration method (SVI) and the objective ones from the Anti Entropy method (AE).

3.1 Calculation of Subjective Weights by SVI

The inconsistency analysis of judgements must be performed when using analytic hierarchy process (AHP) to determine subjective weights. However, it is hard to achieve a consistent judgement matrix due to the limitations of knowledge and preference difference of subject matter experts. In contrast, the SVI method is easier to conduct and more suitable to human's thinking ways by using the degree of coverage of indicators to determine weights. Hence, due to the large amount of LORA indicators and subsequent cumbersome consistency check in AHP, the SVI method is adopted to determine weights.

Supposing the index set as $X = \{x_1, x_2, \cdots, x_p\}$, the expert set $L = \{l_1, l_2, \cdots, l_q\}$, the subject weights are determined as follows,

1) Select the positive integer $g(1 \leq g \leq p)$ as the primary subset capacity. The set X composed of the same level of indicators is ranked according to the importance degree. The first g, $2g$, …, sg indicators constitute the subsets by iteration of s steps. The selection is stopped when $sg + r = p$ $(1 \leq r \leq g)$. The first ξg $(1 \leq \xi \leq s)$ indicators

selected by the expert l_k $(1 \le k \le q)$ can be recorded as

$$X_{\xi g}^{l_k} = \left\{x_1^{l_k}, x_2^{l_k}, \ldots, x_{\xi g}^{l_k}\right\} \subset X \tag{1}$$

2) Count the degree of coverage of the indicator x_j by the indicative function $u_{\xi k}(x_j)$ $(1 \le j \le p)$

$$u_{\xi k}(x_j) = \begin{cases} 1, x_j \in X_{\xi k}^{l_k} \\ 0, x_j \notin X_{\xi k}^{l_k} \end{cases} \tag{2}$$

3) Calculate the weights by the statistical processing as

$$f(x_j) = \sum\nolimits_{k=1}^{q} \sum\nolimits_{\xi=1}^{s+1} u_{\xi k}(x_j) \tag{3}$$

Then the weight v_j is determined by the normalization as:

$$v_j = \begin{cases} \frac{f(x_j)}{\sum_{j=1}^{p} f(x_j)}, x_j \text{ is chosen} \\ \frac{f(x_j)+\frac{1}{2p}}{\sum_{j=1}^{p}\left(f(x_j)+\frac{1}{2p}\right)}, x_j \text{ is not chosen} \end{cases} \tag{4}$$

The final subjective weight vector is $\boldsymbol{V} = (v_1, v_2, \ldots, v_n)'$.

3.2 Calculation of Objective Weights by AE

The conception of entropy was originally proposed by the physicist Rudolph Clausius. It was used to describe the irreversible phenomena of motion in thermodynamics. The entropy weighting is a method using the value of information entropy to calculate the weight of each index according to its variation degree. However, the entropy weight would be zero in case the index values are 0 or 1. Consequently significant original information would be omitted. Thus, the anti-entropy weighting method is used in this study to avoid the sensitivity to the index values and is proved more suitable for the real LORA evaluation.

The anti-entropy is defined as follows,

$$R_j = -\sum_{i=1}^{m} r_{ij}^* \ln\left(1 - r_{ij}^*\right) \tag{5}$$

where R_j is the anti-entropy, i the number of alternative plan, $i = 1, 2, \ldots, m$; j the number of indicator $j = 1, 2, \ldots, n$. $r_{ij}^* = c_{ij} / \sum\limits_{i=1}^{m} c_{ij}$, c_{ij} is the indicator value constituting the evaluation matrix $\boldsymbol{C}$.

The objective weight u_j is calculated as follows,

$$u_j = R_{ij} / \sum_{j=1}^{n} R_{ij} \tag{6}$$

The final objective weight vector is $\boldsymbol{U} = (u_1, u_2, \ldots, u_n)'$..

3.3 Combination Weighting by Quadratic Programming

The objective weighting method does not take the importance degree of indicators into account thus leading to poor interpretability of weights. Therefore, it should be combined with subjective method and not be used solo. The combination weighting methods take account of the advantages of both methods, and reflect the importance of indicators, as well as the data information to improve the credibility of evaluation results. The classic combination methods include addition as well as multiplication methods, level difference maximum and so on. The former two methods lack good explainability, while the others could not avoid the bias by the experts. Therefore, the quadratic programming combination method is proposed with the minimum goal of deviation of objective and subjective weights (U, V), as shown in Eq. (7),

$$\begin{aligned} \min Q(\boldsymbol{W}) &= [\boldsymbol{Y}(\boldsymbol{U}-\boldsymbol{W})]^T[\boldsymbol{Y}(\boldsymbol{U}-\boldsymbol{W})] + [\boldsymbol{Y}(\boldsymbol{V}-\boldsymbol{W})]^T[\boldsymbol{Y}(\boldsymbol{V}-\boldsymbol{W})] \\ &= 4\left[\frac{1}{2}\boldsymbol{W}^T\left(\boldsymbol{Y}^T\boldsymbol{Y}\right)\boldsymbol{W} - \left(\frac{1}{2}\boldsymbol{Y}^T\boldsymbol{Y}\boldsymbol{U} + \frac{1}{2}\boldsymbol{Y}^T\boldsymbol{Y}\boldsymbol{V}\right)^T\boldsymbol{W}\right] + \varphi \\ &\text{s.t. } \boldsymbol{\alpha}\boldsymbol{W} = 1,\ \boldsymbol{W} \ge 0 \end{aligned} \tag{7}$$

where $\boldsymbol{W}$ is the synthesis weight column vector, $\boldsymbol{Y}$ is the row vector of indicators' value of the alternative plan, $\boldsymbol{\alpha}$ is row vector of all-ones, φ is the constant.

According to the Lagrange multiplier method, the Lagrange function can be constructed as

$$f(w,\lambda) = f(w) + \lambda(\alpha w - 1) \tag{8}$$

where w is the column vector, λ is the Lagrange multiplier.

The necessary conditions for the extrema of multi-variable function are

$$\begin{cases} f_w' + \lambda\alpha = 0 \\ \alpha w - 1 = 0 \end{cases} \tag{9}$$

where $f_{w'}$ is the derivative of $f(w,\lambda)$ for each component of w.

Therefore, after substitution one can get

$$\boldsymbol{Y}^T\boldsymbol{Y}\boldsymbol{W} - \left(\frac{1}{2}\boldsymbol{Y}^T\boldsymbol{Y}\boldsymbol{U} + \frac{1}{2}\boldsymbol{Y}^T\boldsymbol{Y}\boldsymbol{V}\right) + \boldsymbol{\alpha}^T\lambda = 0 \tag{10}$$

It can be transformed as

$$\begin{aligned} &\begin{bmatrix} \boldsymbol{Y}^T\boldsymbol{Y} & \boldsymbol{\alpha}^T \\ \boldsymbol{\alpha} & 0 \end{bmatrix}\begin{bmatrix} \boldsymbol{W} \\ \lambda \end{bmatrix} = \begin{bmatrix} \frac{1}{2}\boldsymbol{Y}^T\boldsymbol{Y}\boldsymbol{U} + \frac{1}{2}\boldsymbol{Y}^T\boldsymbol{Y}\boldsymbol{V} \\ 1 \end{bmatrix} \\ &\text{s.t. } \boldsymbol{W} \ge 0 \end{aligned} \tag{11}$$

The final optimal weight can be obtained as

$$\left.\begin{array}{c} W = -FL + D \\ L = \frac{1}{2}Y^T YU + \frac{1}{2}Y^T YV \\ F = \left(Y^T Y\right)^{-1}\alpha^T\left(\alpha\left(Y^T Y\right)^{-1}\alpha^T\right)^{-1}\alpha\left(Y^T Y\right)^{-1} - \left(Y^T Y\right)^{-1} \\ D = \left(Y^T Y\right)^{-1}\alpha^T\left(\alpha\left(Y^T Y\right)^{-1}\alpha^T\right)^{-1} \end{array}\right\} \tag{12}$$

The overall flow by using improved SVI-AE combination weighting method is shown in Fig. 2.

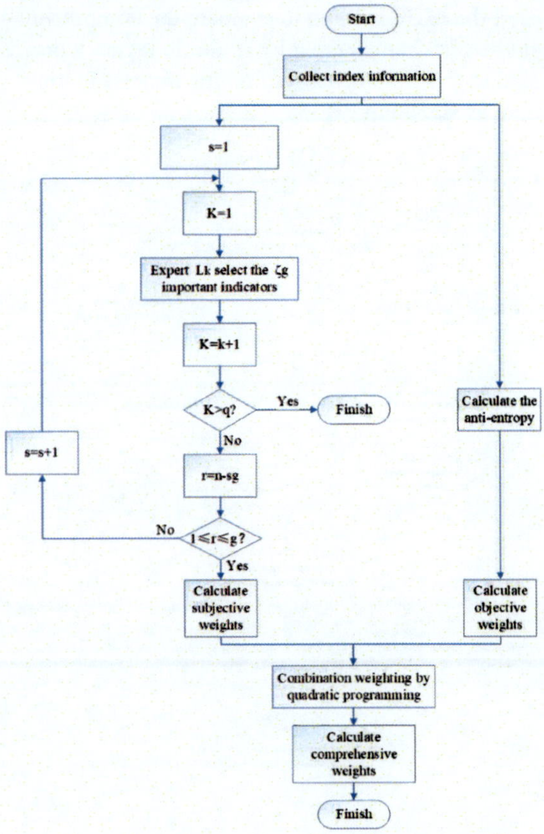

Fig. 2. Improved process of combination weighting

4 Establishing the Synthesis Decision Model of LORA

The goal of the synthesis decision model of LORA is to select the solution with the top score between the feasible alternatives based on the economic and non-economic factors. The decision function *D* is

$$D = f(F, I, E, W) \tag{13}$$

where *F* is the feasible alternatives, *I* the comprehensive index system, *E* the set of expert score information, *W* the weights of indicators.

As the synthesis decision of the LORA is a nonlinear event with complex influencing factors, and the MEE method is good at assessing incompatible complex and multi-factor issues, this study used the MEE method to evaluate the comprehensive LORA system. Since the system studied is characterized by multiple layers whereas the classic MEE method is only available to single-layer system, the process is improved in this section for multi-layer system as shown in Fig. 3.

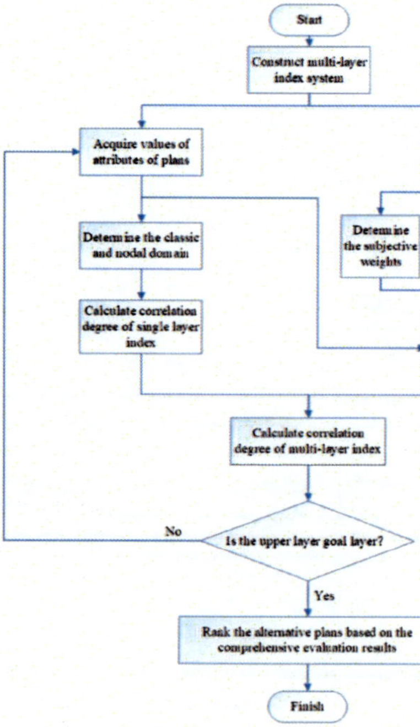

Fig. 3. The synthesis decision flow of repair level of civil aircraft based on SVIAE-MEE

4.1 Calculation of Correlation Degree of Single Layer Index

The matter element is the basic unit of the model, incorporating the relationship between quality and quantity of the objective matter. Specifically, a matter-element, R, is a triple of alternatives N, attributes C, and values V, i.e., $R = (N, C, V)$, as shown below.

$$R = (N, C, V) = \begin{bmatrix} R_1 \\ R_2 \\ \cdots \\ R_n \end{bmatrix} = \begin{bmatrix} N & c_1 & v_1 \\ & c_2 & v_2 \\ & \vdots & \vdots \\ & c_n & v_n \end{bmatrix} \tag{14}$$

where $C = [c_1, c_2, \ldots, c_n]$ is the vector of characteristics, $V = [v_1, v_2, \ldots, v_n]$ the corresponding values of C.

The distance between the evaluated matter-element X and the domain can be expressed as

$$\rho(X, X_o) = \left| X - \frac{a+b}{2} \right| - \frac{1}{2}(b-a) \tag{15}$$

where $\rho(x, x_o)$ is the distance, X_o the domain, i.e. $X_o = (a, b)$.

The process for calculating the correlation degree of single layer index is as follows:

(1) Before carrying out the calculation, the characteristic indexes were divided into five grades, as shown in Table 1 according to the influence of each index, also the value range of each grade was determined. The basis for formulating the grading standards and the boundary values for each grade are contingent on the maintenance plans for civil aircraft, criteria for LORA evaluations, the analysis and judgment of the historical data, academic achievements related to LORA models and professional recommendations from scholars and experts.

Table 1. Grading of LORA indexes

Grade	excellent	good	medium	poor	very poor
code	Q_1	Q_2	Q_3	Q_4	Q_5

(2) Determine the classical domain of the matter-element, which is shown as below

$$R_{ej} = \left(N_{ej}, C, V_{eij}\right) = \begin{bmatrix} N_{ej} & c_1 & v_{e1j} \\ & c_2 & v_{e2j} \\ & \vdots & \vdots \\ & c_n & v_{enj} \end{bmatrix} = \begin{bmatrix} N_{ej} & c_1 & \langle a_{e1j}, b_{e1j} \rangle \\ & c_2 & \langle a_{e2j}, b_{e2j} \rangle \\ & \vdots & \vdots \\ & c_n & \langle a_{enj}, b_{enj} \rangle \end{bmatrix} \tag{16}$$

where R_{ej} is the j-th matter element, v_{eij} the classical domain with the upper limit a_{eij} and the lower limit b_{eij}.

(3) Determine the nodal domain of the matter-element which is shown as follows

$$R_p = \left(N_p, C, V_{pi}\right) = \begin{bmatrix} N_p & c_1 & v_{p1} \\ & c_2 & v_{p2} \\ & \vdots & \vdots \\ & c_n & v_{pn} \end{bmatrix} = \begin{bmatrix} N_p & c_1 & \langle a_{p1}, b_{p1}\rangle \\ & c_2 & \langle a_{p2}, b_{p2}\rangle \\ & \vdots & \vdots \\ & c_n & \langle a_{pn}, b_{pn}\rangle \end{bmatrix} \tag{17}$$

where N_p represents the entirety of grades, v_{pi} the nodal domain with the upper limit a_{pi} and the lower limit b_{pi}, $v_{eij} \subset v_{pi}$.

(4) Determine the matter-element to be comprehensively evaluated which can be represented using the matter-element as follows:

$$R_o = (N_o, C, V) = \begin{bmatrix} N_0 & c_1 & v_1 \\ & c_2 & v_2 \\ & \vdots & \vdots \\ & c_n & v_n \end{bmatrix} = \begin{bmatrix} N_o & c_1 & \langle a_1, b_1\rangle \\ & c_2 & \langle a_2, b_2\rangle \\ & \vdots & \vdots \\ & c_n & \langle a_n, b_n\rangle \end{bmatrix} \tag{18}$$

where R_O is the matter-element to be evaluated, $c_1, c_2, \ldots, c_n$ represent the indicators in the index system, and $v_1, v_2, \ldots, v_n$ are the values of the corresponding indicators.

(5) The value of the correlation function related to the *j*-th grade corresponding to the *i*-th index is obtained as follows

$$k_j(v_i) = \begin{cases} -\dfrac{\rho\left(v_i, v_{eij}\right)}{\left|v_{eij}\right|} & v_i \in v_{eij} \\ \dfrac{\rho\left(v_i, v_{eij}\right)}{\rho\left(v_i, v_{pi}\right) - \rho\left(v_i, v_{eij}\right)} & v_i \notin v_{eij} \end{cases} \tag{19}$$

where $\left|v_{eij}\right| = \left|b_{eij} - a_{eij}\right|$ denoting the range of classical domain of the *i*-th index related to the *j*-th grade, $\rho\left(v_i, v_{eij}\right)$ and $\rho\left(v_i, v_{pi}\right)$ represent the distances between v_i and its classical domain v_{eij}, and its nodal domain v_{pi}, respectively.

The comprehensive degree of correlation in grade j can be calculated based on the weights by the SVI-AE strategy as follows,

$$k_j(P_o) = \sum_i w_{ij} k_j(v_i) \tag{20}$$

Since the degree of correlation cannot be directly used for evaluating the grade of alternatives, the comprehensive evaluation grade could be indicated by the evaluation eigenvalue as follows,

$$K_j(N_i) = \frac{k_j(N_i) - \min k_j(N_i)}{\max k_j(N_i) - \min k_j(N_i)} \tag{21}$$

$$k^* = \frac{\sum_j j \times K_j(N_i)}{\sum_j K_j(N_i)} \tag{22}$$

where k^* is the grade eigenvalue that explains which adjacent grade the matter-element is closer to. Through k^*, the extent to which the matter-element is in the certain grade can be measured.

4.2 Calculation of Multi-Layer Index Correlation Degree

Supposing the index system is of p layers, the layer of correlation and evaluation eigenvalue of the j-th grade for the p-th layer indicators can be achieved by the preceding process. Subsequently the eigenvalue $k^*(P)$ of all alternatives are taken to constitute the matter element matrix of the p-1 layer. Allowing for the weights of the indicators of the p-1 layer, the degree of correlation and evaluation eigenvalue for the p-1 layer indicators can be derived thus creating the matter element matrix of the p-2 layer. Accordingly, the final evaluation eigenvalue of all alternatives to the first (object) layer of indicators could be obtained from the bottom up by iterations. By combining the subjective judgements with the objective calculations, the final decision result is more effective and reasonable.

5 Case Study

A cockpit voice recorder is considered for the application of synthesis decision in our LORA model, which consists of a Flight Data Acquisition Unit (FDAU), Underwater Locator Beacon (ULB), and Crash-Survivable Memory Unit (CSMU) as shown in Fig. 4. It is representative of high value aviation materials with complicated maintenance technologies and expensive costs. It is thus necessary to screen out the best alternative maintenance plan. There are four alternatives (S_1, S_2, S_3, S_4) to be ranked after economic and non-economic analysis to determine the optimal solution, where S_1, S_2 are the plans to be conducted on line level, and S_3, S_4 to be conducted on shop level.

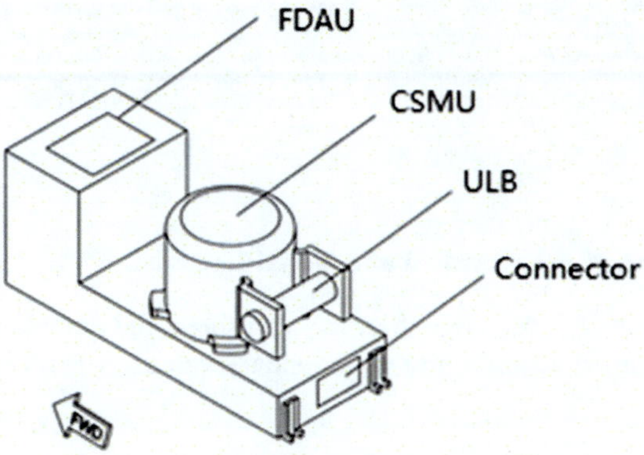

Fig. 4. Illustration of a cockpit voice recorder

5.1 Grading and Boundary Values

According to the actual data and advice from experts, five grades and the boundary values for each grade are determined. There are four alternative maintenance plans and the values of the bottom layer indicators were collected for each plan. For example, the variable cost indicator C_6 of each plan are shown in Table 2, and the boundary values for each grade are shown in Table 3.

Table 2. Values of variable cost indicator of each plan

Plan	Values					
	D_{22}	D_{23}	D_{24}	D_{25}	D_{26}	D_{27}
S1	1.20 × 106	1.47×10^6	8.70×10^5	4.42×10^7	53.40 × 103	75.60×10^4
S2	1.18×10^6	0.98×10^6	6.30×10^5	5.30×10^7	36.92×10^3	64.43×10^4
S_3	1.56×10^6	0.86×10^6	9.30×10^5	6.42×10^7	36.25×10^3	27.00×10^4
S4	1.48×10^6	0.52×10^6	8.80×10^5	4.96×10^7	74.00×10^3	46.00×10^4

Table 3. Boundary values of five grades for the variable cost indicator

Boundary values	Indicators					
	D_{22}	D_{23}	D_{24}	D_{25}	D_{26}	D_{27}
Q1	$[1.50 \times 10^6, 1.40 \times 10^6)$	$[1.50 \times 10^6, 1.30 \times 10^6)$	$[10.00 \times 10^5,$ 9.9.00 $\times 10^5)$	$[7.00 \times 10^7, 6.40 \times 10^7)$	$[80.00 \times 10^3, 70.00 \times 10^3)$	$[70.00 \times 10^4, 60.00 \times 10^4)$
Q_2	$[1.40 \times 10^6, 1.30 \times 10^6)$	$[1.30 \times 10^6, 1.10 \times 10^6)$	$[9.00 \times 10^5, 8.00 \times 10^5)$	$[6.40 \times 10^7, 5.80 \times 10^7)$	$[70.00 \times 10^3, 60.00 \times 10^3)$	$[60.00 \times 10^4, 50.00 \times 10^4)$
Q_3	$[1.30 \times 10^6, 1.20 \times 10^6)$	$[1.10 \times 10^6, 0.90 \times 10^6)$	$[8.00 \times 10^5, 7.00 \times 10^5)$	$[5.80 \times 10^7, 5.20 \times 10^7)$	$[60.00 \times 10^3, 50.00 \times 10^3)$	$[50.00 \times 10^4, 40.00 \times 10^4)$
Q_4	$[1.20 \times 10^6, 1.10 \times 10^6)$	$[0.90 \times 10^6, 0.70 \times 10^6)$	$[7.00 \times 10^5, 6.00 \times 10^5)$	$[5.20 \times 10^7, 4.60 \times 10^7)$	$[50.00 \times 10^3, 40.00 \times 10^3)$	$[40.00 \times 10^4, 30.00 \times 10^4)$
Q_5	$[1.10 \times 10^6, 1.00 \times 10^6]$	$[0.70 \times 10^6, 0.50 \times 10^6)$	$[6.00 \times 10^5, 5.00 \times 10^5]$	$[4.60 \times 10^7, 4.00 \times 10^7]$	$[40.00 \times 10^3, 30.00 \times 10^3]$	$[30.00 \times 10^4, 20.00 \times 10^4]$

5.2 Determination of the Classical and Nodal Domains

The classical domain R_{o1}, R_{o2}, R_{o3}, R_{o4}, R_{o5} and the nodal domain R_p for the indicator C_6 were determined according to the boundary values of grades, as shown below, respectively.

$$R_{o1} = \begin{bmatrix} N_{o1} & D_{22} & \langle 1.40 \times 10^6, 1.50 \times 10^6 \rangle \\ & D_{23} & \langle 1.30 \times 10^6, 1.50 \times 10^6 \rangle \\ & D_{24} & \langle 9.00 \times 10^5, 10.00 \times 10^5 \rangle \\ & D_{25} & \langle 6.40 \times 10^7, 7.00 \times 10^7 \rangle \\ & D_{26} & \langle 70.00 \times 10^3, 80.00 \times 10^3 \rangle \\ & D_{27} & \langle 60.00 \times 10^4, 70.00 \times 10^4 \rangle \end{bmatrix},$$

$$R_{o2} = \begin{bmatrix} N_{o2} & D_{22} & \langle 1.30 \times 10^6, 1.40 \times 10^6 \rangle \\ & D_{23} & \langle 1.10 \times 10^6, 1.30 \times 10^6 \rangle \\ & D_{24} & \langle 8.00 \times 10^5, 9.00 \times 10^5 \rangle \\ & D_{25} & \langle 5.80 \times 10^7, 6.40 \times 10^7 \rangle \\ & D_{26} & \langle 60.00 \times 10^3, 70.00 \times 10^3 \rangle \\ & D_{27} & \langle 50.00 \times 10^4, 60.00 \times 10^4 \rangle \end{bmatrix},$$

$$R_{o3} = \begin{bmatrix} N_{o3} & D_{22} & \langle 1.20 \times 10^6, 1.30 \times 10^6 \rangle \\ & D_{23} & \langle 0.90 \times 10^6, 1.10 \times 10^6 \rangle \\ & D_{24} & \langle 7.00 \times 10^5, 8.00 \times 10^5 \rangle \\ & D_{25} & \langle 5.20 \times 10^7, 5.80 \times 10^7 \rangle \\ & D_{26} & \langle 50.00 \times 10^3, 60.00 \times 10^3 \rangle \\ & D_{27} & \langle 40.00 \times 10^4, 50.00 \times 10^4 \rangle \end{bmatrix},$$

$$R_{o4} = \begin{bmatrix} N_{o4} & D_{22} & \langle 1.10 \times 10^6, 1.20 \times 10^6 \rangle \\ & D_{23} & \langle 0.70 \times 10^6, 0.90 \times 10^6 \rangle \\ & D_{24} & \langle 6.00 \times 10^5, 7.00 \times 10^5 \rangle \\ & D_{25} & \langle 4.60 \times 10^7, 5.20 \times 10^7 \rangle \\ & D_{26} & \langle 40.00 \times 10^3, 50.00 \times 10^3 \rangle \\ & D_{27} & \langle 30.00 \times 10^4, 40.00 \times 10^4 \rangle \end{bmatrix},$$

$$R_{o5} = \begin{bmatrix} N_{o5} & D_{22} & \langle 1.00 \times 10^6, 1.10 \times 10^6 \rangle \\ & D_{23} & \langle 0.50 \times 10^6, 0.70 \times 10^6 \rangle \\ & D_{24} & \langle 5.00 \times 10^5, 6.00 \times 10^5 \rangle \\ & D_{25} & \langle 4.00 \times 10^7, 4.60 \times 10^7 \rangle \\ & D_{26} & \langle 30.00 \times 10^3, 40.00 \times 10^3 \rangle \\ & D_{27} & \langle 20.00 \times 10^4, 30.00 \times 10^4 \rangle \end{bmatrix},$$

$$R_{P} = \begin{bmatrix} N_{P} & D_{22} & \langle 1.00 \times 10^6, 1.50 \times 10^6 \rangle \\ & D_{23} & \langle 0.50 \times 10^6, 1.50 \times 10^6 \rangle \\ & D_{24} & \langle 5.00 \times 10^5, 10.00 \times 10^5 \rangle \\ & D_{25} & \langle 4.00 \times 10^7, 7.00 \times 10^7 \rangle \\ & D_{26} & \langle 30.00 \times 10^3, 80.00 \times 10^3 \rangle \\ & D_{27} & \langle 20.00 \times 10^4, 70.00 \times 10^4 \rangle \end{bmatrix}.$$

The matter element matrix R_o was established to be evaluated based on the attribute values of alternatives.

$$R_o = \begin{bmatrix} N_o & D_{22} & 1.20 \times 10^6 & 1.18 \times 10^6 & 1.56 \times 10^6 & 1.48 \times 10^6 \\ & D_{23} & 1.47 \times 10^6 & 0.98 \times 10^6 & 0.86 \times 10^6 & 0.52 \times 10^6 \\ & D_{24} & 8.70 \times 10^5 & 6.30 \times 10^5 & 9.30 \times 10^5 & 8.80 \times 10^5 \\ & D_{25} & 4.42 \times 10^7 & 5.30 \times 10^7 & 6.42 \times 10^7 & 4.96 \times 10^7 \\ & D_{26} & 53.40 \times 10^3 & 36.92 \times 10^3 & 36.25 \times 10^3 & 74.00 \times 10^3 \\ & D_{27} & 75.60 \times 10^4 & 64.43 \times 10^4 & 27.00 \times 10^4 & 46.00 \times 10^4 \end{bmatrix}$$

5.3 Combination Weighting of Indicators

Take the indicator set of C_4 with five indicators namely $X = \{D_{12}, D_{13}, D_{14}, D_{15}, D_{16}\}$ as an example. Five experts were selected to assess the top important indicators according to the actual maintenance activities. The assessment result is shown in Table 4.

Table 4. Importance rating results of indicators by experts

Experts	Iterations		
l_1	$D_{16}D_{12}$	$D_{15}D_{13}$	D_{14}
l_2	$D_{16}D_{12}$	$D_{13}D_{15}$	D_{14}
l_3	$D_{16}D_{15}$	$D_{12}D_{14}$	D_{13}
l_4	$D_{16}D_{12}$	$D_{15}D_{14}$	D_{13}
l_5	$D_{16}D_{15}$	$D_{13}D_{12}$	D_{14}

The coverage situation of each indicator during the assessment is shown in Table 5.

Table 5. Coverage situation of indicators

Indicator	Statistical coverage	Rate of coverage
D_{12}	$4+4+3+4+2$	$(4+4+3+4+2)/25$
D_{13}	$2+3+1+1+3$	$(2+3+1+1+3)/25$
D_{14}	$1+1+2+2+1$	$(1+1+2+2+1)/25$
D_{15}	$3+2+4+3+4$	$(3+2+4+3+4)/25$
D_{16}	$5+5+5+5+5$	$(5+5+5+5+5)/25$

The weights of D_{12} to D_{16} for C_4 can thus be calculated as $V_{C_4} = [0.23, 0.12, 0.09, 0.22, 0.34]$. Similarly, all the subjective weights of indicators can be acquired. The objective weights and the final combinative weights of indicators of D-layer to the C-layer are shown in Table 6.

It can be seen that the combination weights are not totally between the subjective and objective weights. This is because the combination weights by the quadratic programming are not only related to the subjective and objective weights, but also the specific values of indicators of each alternative plan.

Table 6. Weights of indicators of D-layer to the C-layer

Layer	Subjective weights	Objective weights	Combinative weights
Factors	$V_{C_1} = [0.20, 0.20, 0.20, 0.07, 0.33]$	$U_{C_1} = [0.20, 0.18, 0.18, 0.22, 0.22]$	$W_{C_1} = [0.35, 0.18, 0.06, 0.14, 0.27]$
	$V_{C_2} = [0.53, 0.47]$	$U_{C_2} = [0.48, 0.52]$	$W_{C_2} = [0.51, 0.49]$
	$V_{C_3} = [0.10, 0.20, 0.30, 0.40]$	$U_{C_3} = [0.25, 0.25, 0.25, 0.25]$	$W_{C_3} = [0.28, 0.26, 0.14, 0.32]$
	$V_{C_4} = [0.23, 0.12, 0.09, 0.22, 0.34]$	$U_{C_4} = [0.22, 0.19, 0.15, 0.21, 0.23]$	$W_{C_4} = [0.19, 0.21, 0.15, 0.20, 0.25]$
	$V_{C_5} = [0.34, 0.25, 0.13, 0.09, 0.19]$	$U_{C_5} = [0.23, 0.22, 0.17, 0.16, 0.22]$	$W_{C_5} = [0.34, 0.16, 0.10, 0.17, 0.23]$
	$V_{C_6} = \begin{bmatrix} 0.27, 0.18, 0.14, \\ 0.09, 0.06, 0.26 \end{bmatrix}$	$U_{C_6} = \begin{bmatrix} 0.18, 0.13, 0.16, \\ 0.17, 0.19, 0.17 \end{bmatrix}$	$W_{C_6} = \begin{bmatrix} 0.14, 0.12, 0.15, \\ 0.11, 0.31, 0.17 \end{bmatrix}$
	$V_{C_7} = [0.29, 0.37, 0.16, 0.18]$	$U_{C_7} = [0.26, 0.27, 0.24, 0.23]$	$W_{C_7} = [0.27, 0.32, 0.20, 0.21]$

5.4 Calculating the Correlation Degree of Indicators

The correlation degree of each alternative plan to the D-layer indicators were calculated from the bottom up. Taking the indicator D_{22} in C_6 for example, the correlation degree of S_1 plan to grade Q_1 can be calculated as follows,

$$\rho\left(v_{D_{22}}, v_{o1}\right) = \left|1.2 \times 10^6 - \frac{1}{2}\left(1.5 \times 10^6 + 1.4 \times 10^6\right)\right|$$

$$-\frac{1}{2}\left(1.4 \times 10^6 - 1.5 \times 10^6\right) = 3$$

$$\rho\left(v_{D_{22}}, v_{p1}\right) = \left|1.2 \times 10^6 - \frac{1}{2}\left(1.5 \times 10^6 + 1.0 \times 10^6\right)\right|$$

$$-\frac{1}{2}\left(1.5 \times 10^6 - 1.0 \times 10^6\right) = -2$$

$$k_1\left(v_{D_{22}}\right) = \frac{\rho(v_{D22}, v_{o1})}{\rho\left(v_{D_{22}}, v_{pi}\right) - \rho\left(v_{D_{22}}, v_{o1}\right)} = -0.6$$

Accordingly, all the correlation degree of S_1 plan to D_{22}–D_{27} are shown in Table 7.

Table 7. Correlation degree of S_1 plan to indicator D_{22}–D_{27}

Indicator	Q_1	Q_2	Q_3	Q_4	Q_5
D_{22}	−0.6000	−0.5000	−0.3333	−0.3333	−0.5000
D_{23}	−0.8500	−0.9250	−0.9500	−0.9625	−0.9700
D_{24}	−0.5000	−0.3500	−0.5667	−0.6750	−0.7400
D_{25}	−0.8600	−0.8250	−0.7667	−0.6500	−0.5000
D_{26}	−0.5320	−0.4150	−0.2200	−0.3641	−0.5000
D_{27}	−1.5600	−1.2800	−1.1867	−1.1400	−1.1120

Based on the combination weights W_{C6} of D_{22}–D_{27}, the correlation degree of indicator C_6 of S_1 plan to each grade is [−0.7912, −0.6759, −0.6049, −0.6458, −0.6997]. The final evaluation eigenvalue of indicator C_6 is 3.3956. Similarly, all the correlation degrees and the eigenvalues of all the alternative plan to the indicator C_6 is shown in Table 8.

Table 8. Correlation degrees and eigenvalues of all alternative plans to C_6

Plan	Q_1	Q_2	Q_3	Q_4	Q_5	Eigenvalues
	−0.7912	−0.6759	−0.6049	−0.6458	−0.6997	3.3956
S_2	−0.6758	−0.6552	−0.5727	−0.5294	−0.5678	3.9042
S_3	−0.6200	−0.6484	−0.6681	−0.7119	−0.7712	2.1613
S_4	−0.8468	−0.8026	−0.7791	−0.7258	−0.6717	4.0467

The eigenvalues shown in Table 8 make up the evaluation data of indicator C_6. Similarly, the evaluation data of indicator C_7 can be calculated to make up the matter element matrix

$$R_o = \begin{bmatrix} N_o & C_6 & 3.3956 & 3.9042 & 2.1613 & 4.0467 \\ & C_7 & 3.0046 & 2.8906 & 2.4259 & 4.0909 \end{bmatrix}$$

Based on the combination weights of C_6 and C_7, the evaluation data and the weight of indicator B_2 can be also calculated, which are similar to the indicator B_1. All the weights of layers of attributes and criteria are shown in Table 9. Finally, the eigenvalues of each alternative plan to the top layer of object A can be obtained which are used as reference to sort alternatives. All the evaluation results of each plan to each layer are shown in Table 10. The radar chart of each plan is shown in Fig. 5.

5.5 Discussion

It can be seen from Table 9 that in terms of sorting order, $S_4 > S_2 > S_1 > S_3$ according to its superiority and inferiority. The disparity of evaluation results between the non-economic indicators is less and the evaluation grading of non-economic indicators has

Table 9. Weights of layers of attributes and criteria

Layer	Subjective weights	Objective weights	Combinative weights
Attribute	$V_A = [0.53, 0.47]$	$U_A = [0.57, 0.43]$	$W_A = [0.55, 0.45]$
Criteria	$V_{B_1} =$ $[0.19, 0.08, 0.31, 0.26, 0.16]$	$U_{B_1} =$ $[0.22, 0.16, 0.21, 0.21, 0.20]$	$W_{B_1} =$ $[0.28, 0.06, 0.18, 0.20, 0.28]$
	$V_{B_2} = [0.60, 0.40]$	$U_{B_2} = [0.53, 0.47]$	$W_{B_2} = [0.57, 0.43]$

Table 10. Evaluation results of each plan to each layer

Index layer	Index code	S_1	S_2	S_3	S_4
Object	A	1.9008	2.1349	1.6982	2.4741
Attributes	B_1	3.7712	3.8931	3.5740	3.9789
	B_2	2.0929	2.5620	1.1919	3.4353
Criteria	C_1	4.4348	4.4954	4.3416	4.6087
	C_2	4.0232	4.1818	4.3980	4.2376
	C_3	4.6087	4.6087	4.4741	4.6087
	C_4	4.3840	4.3374	3.9001	4.2950
	C_5	4.2028	4.3939	4.2335	4.4451
	C_6	3.3956	3.9042	2.1613	4.0467
	C_7	3.0046	2.8906	2.4259	4.0909

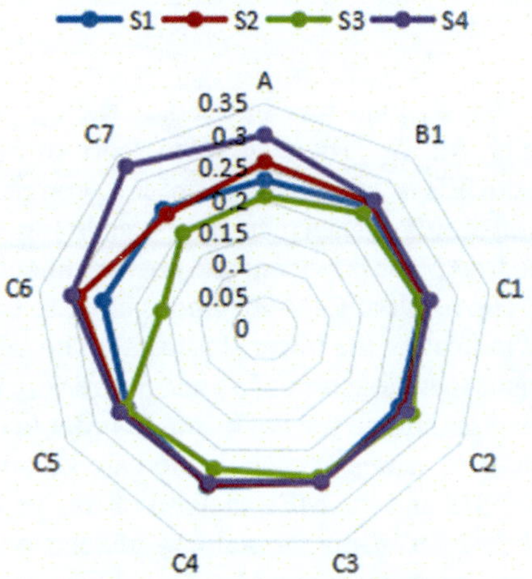

Fig. 5. Radar chart of four plans

a greater impact on the final decision according to the combination weights. The plan S_3 is dominant in terms of indicator maintenance personnel requirements, etc. However, it is in inferior position of grade Q_2 due to the small weight of C_2 and the influence of indicators of variable and fixed costs. Taking account of the indicators of attribute layer, the plan S_1 is superior to S_3, but the costs of downtime, transportation, labor and maintenance are higher thus leading to an inferior grade to S_2 and S_4. Although the plan S_2 is predominant in view of costs of spare holding, consumables and transportation, the inferiority of non-economic indicators such as supportability and human resources finally result in the inferior grade to S_4. Therefore, the plan S_4 ranks the first in terms of comprehensive evaluation out of four plans.

The results of several commonly used evaluation methods were also compared to verify the rationality of the present sorting order, which include Technique for Order Preference by Similarity to Ideal Solution (TOPSIS), Fuzzy Comprehensive Evaluation (FSE), and Grey Relation Analysis (GRA). All the normalized results and sorting orders are shown in Table 11.

Table 11. Normalized results and sorting orders by various methods

Plan	Evaluation methods			
	SVIAE-TOPSIS	SVIAE-FSE	SVIAE-GRA	SVIAE-MEE
S_1	0.2337	0.1982	0.2314	0.2316
S_2	0.2645	0.2663	0.2591	0.2601
S_3	0.1795	0.2049	0.2144	0.2069
S_4	0.3223	0.3314	0.2951	0.3014
Order	$S_4 > S_2 > S_1 > S_3$	$S_4 > S_2 > S_3 > S_1$	$S_4 > S_2 > S_1 > S_3$	$S_4 > S_2 > S_1 > S_3$

It can be concluded that S_4 plan was selected as the optimal one by all the four methods, inferior to which was S_2. The evaluation score of S_1 is higher than that of S_3 by TOPSIS, GRA and MEE in this study. In order to compare the sensitivity of the methods, all the evaluation scores of each plan are shown in Fig. 6. It can be seen that the overall trend between the four plans is evident and appears linearly by the three methods except the FSE method. The result of GRA method is close to the MEE method studied but has a flatter gradient indicating the lower sensitivity. The inflection point shown by FSE on the S_1 plan indicates that the scores of S_1 and S_3 are very close and the sensitivity of fuzzy processing is poor leading to the deviation. In order to compare the stability of the methods, all the evaluation scores of each method are shown in Fig. 7. It is obvious that the scores of the TOPSIS and the MEE method of this paper are widely scattered and distributed uniformly, which ensure the stability of decision. In contrast, the scores of GRA and FSE are very close which could lead to rank reversal. Therefore, the MEE method in this study demonstrates great superiorities in terms of both sensitivity and stability of decision.

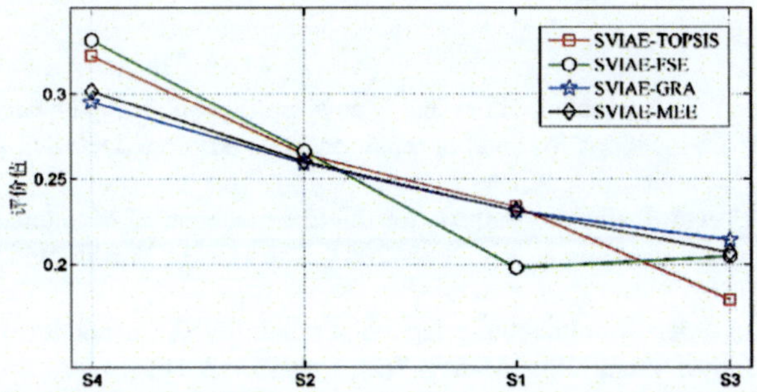

Fig. 6. Normalized evaluation results of various methods to four plans

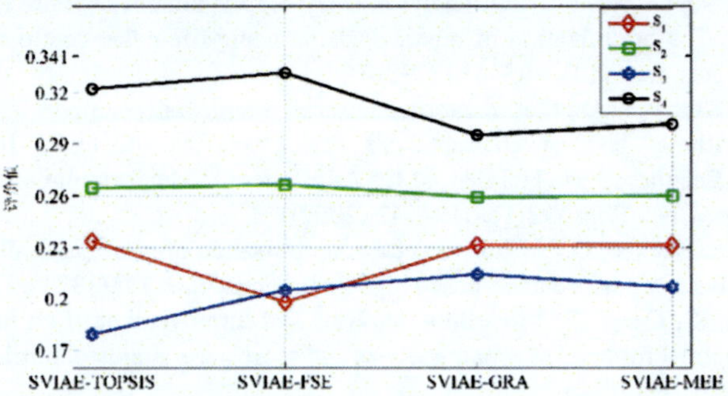

Fig. 7. Normalized evaluation results of four plans with various methods

6 Conclusions

A synthesis decision model of level of repair for civil aircraft based on the SVIAE-MEE combination strategy was established, which was targeted at the applicability of alternative plans taking account of economic and non-economic factors comprehensively including maintenance requirements, personnel, regulations, environments, variable and fixed costs, etc. An example of the cockpit voice recorder was studied. It was found that the rank order of the present method is consistent with those of TOPSIS and GRA methods thus indicating its credibility. The rank reversal maybe occurs by the GRA and FSE methods due to close and indiscernible results of them. The widely-scattered and uniformly distributed evaluation values of the MEE method ensure the stability of decision results. Consequently, the proposed model provides a new perspective to quantitatively evaluate the maintenance alternative plans and has significant benefits in the configuration of maintenance support system of civil aircraft.

Acknowledgments. This research is supported by the National Natural Science Foundation of China (Grant No. U2033209). Many thanks for their help to our work.

References

1. Bicakci, I., Ic, Y.T., Karasakal, E., et al.: A new multi-echelon repair network model with multiple upstream locations for level of repair analysis problem. Def. Sci. J. **71**(6), 762–771 (2021)
2. Marino, L.C.: Level of Repair Analysis for the Enhancement of Maintenance Resources in Vessel Life Cycle Sustainment, pp. 2038–2043. The George Washington University, USA (2018)
3. Rawat, M., Lad, B.K.: Simultaneous selection of reliability design and level of repair for fleet systems. Int. J. Syst. Assur. Eng. Manag. **7**(2), 138–155 (2016)
4. Rawat, M., Lad, B.K., Duffuaa, S., et al.: An integrated strategy for fleet maintenance planning. J. Qual. Maint. Eng. **23**, 457–479 (2017)
5. Level of Repair Analysis: AS1390. Warrendale: SAE International (2014)
6. MIL-STD-1390D. Level of repair analysis. USA: Department of Defense (1993)
7. Barros, L.L.: The optimization of repair decision using life-cycle cost parameters. IMA J. Math. Appl. Bus. Ind. **9**(4), 403–413 (1998)
8. Saranga, H., Kumar, U.D.: Optimization of aircraft maintenance/support infrastructure using genetic algorithms—level of repair analysis. Ann. Oper. Res. **143**(1), 91–106 (2006)
9. Basten, R.J., Schutten, J.M., Heijden, M.C.: An efficient model formulation for level of repair analysis. Ann. Oper. Res. **172**(1), 119–142 (2009)
10. Malyemez, C., Baykoc, O.F.: Multi-objective optimisation of spare parts allocation and level of repair analysis in performance-based logistics. Aeronaut. J. **126**(1301), 1210–1221 (2022)
11. Liu, W., Liu, K., Deng, T.: Modelling, analysis and improvement of an integrated chance-constrained model for level of repair analysis and spare parts supply control. Int. J. Prod. Res. **58**(10), 3090–3109 (2020)
12. Wang, R., Chen, G., Wu, J.: Joint Optimization method of spare parts stocks and level of repair analysis considering the multiple failure modes. Applied Science **11**, 7254 (2021)
13. Bicakci, I., Ic, Y.T., Karasakal, E.: A Multi-objective mathematical model for level of repair analysis with lead times and multi-transportation modes. Int. J. Inf. Technol. Decis. Mak. **21**(1), 423–440 (2022)
14. Zhao, Y., Xue, X., Huang, Y.: Evaluating comprehensive carrying capacity of coastal area using the matter-element exten-sion method: a case study in Fujian province of China. Ocean Coast. Manag. **214**, 105902 (2021)
15. Shan, C., Dong, Z., Debao, L.: Study on river health assessment based on a fuzzy matter-element extension model. Eco-logical Indicators **127**, 107742 (2021)
16. Liu, S., Li, W.: Indicator's sensitivity analysis for environmental engineering geological patterns caused by underground coal mining with integrating variable weight theory and improved matter-element extension model. Sci. Total Environ. **686**, 606–618 (2019)
17. Wang, Y., Li, F., Yang, J., et al.: Demand response evaluation of RIES based on improved matter element extension model. Energy **212**, 118121 (2020)
18. Mao, Y., Xu, T.: Research of 4M1E's effect on engineering quality based on structural equation model. Syst. Eng. Procedia **1**, 213–220 (2011)

Analysis of Heterogeneous Cooperative Unmanned System with System Engineering

Linggong Zhao(✉), Zhijun Liu, Jian Zhou, Long Zhou, and Qian Xiang

Chinese Aeronautical Establishment, Beijing, China
zhaolinggong@163.com

Abstract. The collaborative application of heterogeneous unmanned systems is one of the key development directions in the unmanned battlefield and civil field in the future. However, it involves multiple different platform types, huge scale, and complex interdisciplinary issues, and there is still a certain distance from large-scale applications. This paper uses the system engineering method to analyze the process of collaborative application of heterogeneous unmanned systems and sort out some directions of future development and some key technologies that need to be breakthrough. The process is divided into four parts: demand acquisition, demand analysis, architecture analysis, and platform design. The system engineering method is suitable for the design and analysis of complex systems and conforms to the characteristics of unmanned systems. It can effectively reduce repeated iterations in the design and development process of heterogeneous unmanned system collaborative applications, and improve inheritance.

Keywords: Systems Engineering · Unmanned Systems · Overview

1 Introduction

With the upgrading of equipment and the changes in battlefield form, the future war is bound to be a multi-dimensional offensive and defensive war of "one area with multiple layers, and air-ground integration" [1], and the form of confrontation will also be upgraded from equipment confrontation to inter-system confrontation. The development of intelligent control technology promotes the application of unmanned aerial vehicle (UAV), unmanned underwater vehicle (UUV), unmanned surface vehicle (USV), unmanned ground vehicles (UGV), and other new unmanned systems (collectively referred to as UxV) "battle groups" in the future battlefield to achieve cross-domain cooperative joint operations [2], and the formation of systems and collaborative use can maximize equipment efficiency, becoming the subversive weapons and equipment in the future war [3].

Heterogeneous unmanned systems are generally divided into structural heterogeneity and functional heterogeneity: structural heterogeneity is caused by the constraints of the unmanned platform itself, resulting in the different space and scope of unmanned equipment to perform tasks, such as the difference in space movement and load capacity or energy and power constraints; Functionally heterogeneous payloads carried by

Chinese Society of Aeronautics and Astronautics (Ed.): CASTC 2023, LNME, pp. 217–230, 2024.
https://doi.org/10.1007/978-981-99-8861-7_22

unmanned platforms lead to different types of tasks that unmanned equipment can perform [4].

Cross-domain collaboration of heterogeneous clusters can give full play to the advantages of different unmanned systems, realize the complementation of cluster intelligence functions and energy efficiency multiplication through information sharing and integration, behavior interaction and coordination, and task collaboration and cooperation, and improve the ability to cope with complex environments and missions [5], for example, multi-UAV-UAV air-ground cooperative reconnaissance [6], multi-UAV-UAV air-ground cooperative patrol [7], etc. At present, the research on formation control and encirclement control of heterogeneous cluster systems is still in its infancy, and the relevant research results are few [8].

The research on the collaborative application of heterogeneous unmanned systems involves more knowledge breadth, time length, management scale, etc. than the research content of the original single platform or homogeneous unmanned aerial vehicle single task. The analysis method of system engineering can effectively improve research efficiency and reduce the number of iterations. The system engineering idea was introduced into China by Academician Qian Xuesen, and combined with the engineering practice in China, formed the system engineering idea with Chinese characteristics, guiding the development of aerospace and even China's various industries [8]. With the rise of model-driven system development methods, especially in the software field, people combine model-driven with system engineering and propose model-based system engineering (MBSE) [9]. Combined with the MBSE method, realize modular modeling and further accelerate the development progress of new collaborative application mode.

2 Overview of Unmanned System

In the field of unmanned systems, a lot of research work has been carried out or is being carried out at home and abroad.

2.1 Unmanned Aerial Vehicle

With the continuous improvement of UAV's intelligence, autonomy, and information acquisition ability, UAV's operational application has evolved from single combat to cluster combat and cooperative combat, thus promoting the change of UAV's operational application mode. UAVs have become an indispensable and important force in joint operations under the background of the new era [10]. There are many relevant reports and literature materials on cluster UAV projects at home and abroad, such as cooperative operation (CODE), Gremlins UAV colony, Flock-93 colony, Russia's 'Молния' swarm drone; Israel's artificial intelligence drone swarm; German Ghost Play, etc. (Fig. 1).

2.2 Unmanned Surface Vehicle

Unmanned ships/boats have the advantages of autonomous navigation, long endurance, and strong load capacity, and have been applied and promoted in many aspects in recent years, such as maritime safety and induction minesweeping, territorial sea surveillance,

Fig. 1. Research of Unmanned Aerial Vehicle

marine environmental monitoring, marine biological research, coastal surveying, and mapping, etc. [11]. The US military will equip the underwater unmanned squadron (UUVRON-1) with autonomous underwater robots in the coastal combat space, 'Snake-Head' large unmanned underwater vehicles, and Orca ultra large unmanned underwater vehicles. The sea wolf class attack nuclear submarine "Jimmy Carter" is classified as belonging to the underwater squadron to conduct submarine spy reconnaissance activities [12]. Russia's 'deepwater force' has tested and developed a variety of small deepwater nuclear power plants, the special nuclear submarine "Moscow suburb", the nuclear submarine "Belgorod" and a variety of unmanned underwater vehicles. In the ASW exercise conducted by the UK on the northern coast of Scotland, the persistent mobile detection network composed of four-wave glider USVs successfully detected and tracked a forward unmanned underwater vehicle and diesel-powered submarine for the first time, proving that the system can perform anti-submarine missions [12] (Fig. 2).

Fig. 2. Research of Unmanned Surface Vehicle

2.3 Unmanned Vehicle

In 2022, the United States Army conducted a comprehensive test of a complete unmanned combat system, including heavy, medium, and light unmanned combat vehicles and technology control vehicles, including the medium-sized unmanned combat vehicle "Rough Tooth Saw" M5, the heavy unmanned combat vehicle modified from M113, and the unmanned combat vehicle mission authorization technology control vehicle modified from M2 "Bradley".

Russia has equipped five sets of "Uranus" - 9 armored unmanned combat vehicles with unmanned combat vehicles and armored forces, while all engineering units of the Navy have also equipped "Uranus" - 6 mine clearing unmanned vehicles for clearing landmines along the coast and on land. Both of these unmanned vehicles have undergone actual combat testing on the Syrian battlefield. The British Army's new generation of "Boxer Dog" series of wheeled armored combat vehicles have been put into mass production, including infantry transport vehicles, command and control vehicles, field ambulances, and special transport vehicles.

China's UV-T10 track-type unmanned combat vehicle has a certain off-road capability, but its height is not limited to surmounting too large obstacles. The sharp-claw type 1 unmanned combat vehicle can perform close reconnaissance, detection, and surveillance tasks for class or platoon units; The highly mobile Longma 2 unmanned off-road vehicle can meet the needs of unmanned ground operations and material transportation of the Army. Raytheon Intelligent Unmanned Combat Vehicle has a certain stealth and low infrared design. It is equipped with omnidirectional photoelectric probes around the body, which can monitor the surrounding conditions in all directions. The maximum speed is 120 km/h, and the driving is very smooth. It can even be replaced with triangular rubber tracks to enhance off-road performance (Fig. 3).

Fig. 3. Research of Unmanned Vehicle

2.4 Unmanned System

Unmanned systems can be divided into manned/unmanned cooperative systems, air-ground cooperative systems, and air-surface cooperative systems. In addition, more collaborative application models are being explored.

In the aspect of manned/unmanned cooperation, relying on the advanced system architecture, based on network communication, on the development of intelligent autonomous capability, and with the support of advanced combat management, manned equipment, and unmanned aerial vehicles are interconnected and closely coordinated to perform tasks [10]. The future unmanned war is not an autonomous war with unmanned equipment, but the manned front is gradually "backward". Unmanned troops and equipment are engaged with the enemy on the front line of the battlefield, and human beings participate and control "remotely" [13]. The United States, the United Kingdom, Australia, and other military powers are vigorously promoting the "loyal wingman" project, and have carried out relevant demonstration and verification work. The Autonomous Control System (ACS) being developed by the US Air Force's "Skyborg" project will become the "brain" of the "loyal wingman" project, enabling the UAV to independently perform complex combat tasks and achieve cooperative operations with the UAV [14]. The air-launched UAV can effectively shorten the attack distance by air-launched UAV, keep the manned platform away from the threat of the enemy, and improve its survivability, which may become an important way for future UAV/UAV formation to cooperate (Fig. 4).

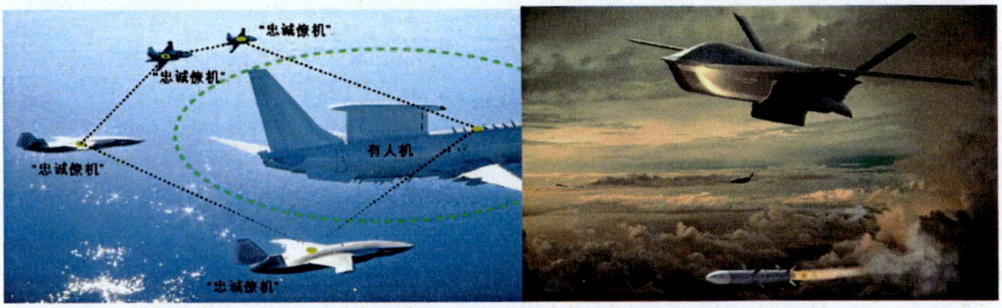

Fig. 4. Manned/unmanned Cooperation and The Air-launched UAV

In terms of air-ground cooperation, the ground-air cooperative unmanned system has the advantages of multi-dimensional perception, global response, and efficient mobility. It can independently adapt to a complex dynamic environment and cooperate in operations, completing complex tasks such as cross-regional cooperative reconnaissance, air-ground integrated attack, and multi-target operations. It is an important tool for the transformation of combat form to non-contact, asymmetric, and zero casualties [15]. For example, NASA has developed the "Persevere" Mars rover with a miniature unmanned helicopter, which is the first time in human history to launch a ground-air cooperation platform on an extraterrestrial planet. At present, the platform has successfully landed on Mars [15].

In terms of air-surface cooperation, the two different unmanned platforms, unmanned vehicles, and unmanned aerial vehicles have strong "heterogeneous" characteristics in terms of behavior mode, operational domain, operational mission, and intelligence level [15]. The organic combination and cooperative control of the unmanned vehicle and unmanned aerial vehicle can improve the time sensitivity, range, and duration of sea-air detection and strike against medium and small targets, form a sea-air integrated reconnaissance, surveillance, and strike system, and efficiently complete the operational tasks of alert surveillance, cooperative detection, and sea-air strike. For example, the United States and France have successfully carried out the unmanned aerial vehicle landing test, thus opening the prelude to the autonomous landing of unmanned aerial vehicles [10, 16]; In 2017, the annual meeting of the American Naval Surface Ship Association released the concept map of the air-sea cooperative system, indicating that the United States Navy will be equipped with a single ship carrying multiple unmanned aerial vehicles to carry out combat missions in the future; The French Ship Construction Agency (DCNS) successfully achieved the cooperative operation of three unmanned systems (UAV, unmanned ship and unmanned underwater vehicle) in May 2017, which is the first time in Europe [10] (Fig. 5).

Fig. 5. The Projects of Air-surface Cooperation

3 Analysis of Heterogeneous Unmanned System by System Engineering Method

3.1 Brief Description of System Engineering

The design of unmanned systems covers multiple motion platforms and the coupling and cross-linking between platforms. The complexity of systems is not only the accumulation of platforms but also the cross-linking between platforms and systems, as well as task scheduling. The workload is huge, requiring complex system design, development, and testing, which is very suitable for the application of system engineering methods. As an interdisciplinary method and approach to guide system development, system engineering's core value is to plan and constrain the development process of general systems, It effectively reduces the development cost and risk of the system.

System engineering integrates multiple professional technologies, and through repeated iterations of analysis, integration, testing, and evaluation, forms a system that meets system requirements and is gradually optimized overall. According to NASA's definition of the scope of system engineering, these development processes include the technology development process and the technology management process, in which the technology development process constitutes the life cycle stage model of the system [17]. The more classical system engineering methodology is Hall's system engineering methodology and Qian Xuesen's integrated methodology. Hall's system engineering methodology divides system engineering into three dimensions - time dimension, logic dimension, and professional dimension.

The time dimension is a rough structure classification. For a specific project, from the beginning of the planning work to the system update, its whole process can be divided into the following seven stages: planning stage, design stage, development stage (system development), production stage, installation stage, operation stage, and update stage. The logical dimension is the classification of fine structure. Each stage of the time dimension can be divided into several logical steps to show the fine structure of system engineering, that is, setting out problems, determining objectives, system synthesis, system analysis, system evaluation, decision-making, and implementation. The professional wiki is based on many disciplines involved in system engineering. Based on the two-dimensional structure, add the third dimension - the professional dimension. 'This dimension can be measured by the degree of normalization or mathematical structure" [18].

Qian Xuesen's comprehensive integration methodology elevates the perspective of consideration to "the System of Systems (SoS)" [19], which is a high degree of integration on all kinds of integration as well as on all kinds of integration. Comprehensive integration is corresponding, opposite, and complementary to reductionism, and the comprehensive integration workshop system is the specific application of this method. The seminar hall system is composed of three parts: the machine system, expert system and knowledge system formed by the integration and integration of modern high-tech with a computer as the core, and a highly intelligent human-machine integration system, which has the functions of knowledge and information collection, storage, transfer, analysis and integration, and can also generate new knowledge and new applications.

3.2 Research Ideas for System Engineering

The development process of system engineering is the process of deepening the understanding of the system, which needs to follow the repeated process of analysis, time, re-analysis and re-practice. The system engineering process generally includes task analysis, function analysis, design synthesis and verification, etc. The process will have different adjustments in combination with different system engineering (Fig. 6).

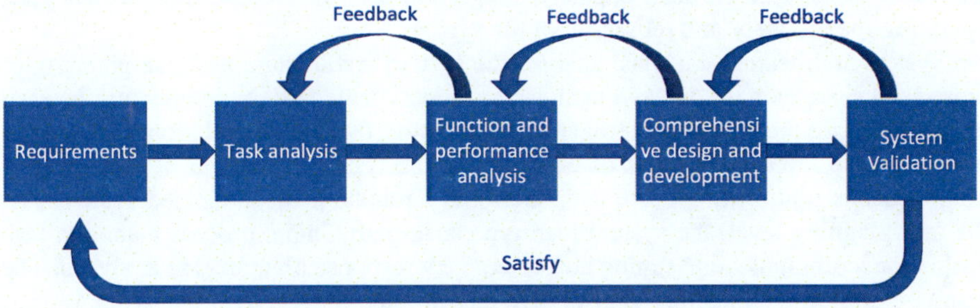

Fig. 6. The Process of System Engineering

Task analysis is to clarify the requirements of the task and the objectives and constraints of the work, to put forward requirements for the function and performance of the system, and then further analyze and decompose to obtain a comprehensive description of the system function, as well as the logical relationship between the function and performance. According to the function and performance analysis, the optimized system's physical structure that can meet the requirements is developed. The purpose of verification is to confirm that the physical structure of the designed system at all levels meets the requirements of the system, that is, the requirements of the task.

3.3 Unmanned System Application

From the perspective of system engineering, the process of establishing a collaborative application system for unmanned systems is generally divided into four parts: demand acquisition, demand analysis, system architecture analysis, and platform design analysis.

Each part is divided into several small items, as shown in the figure. From top to bottom is the process of requirement decomposition. The process of requirement realization is to obtain the executable content at the bottom layer through layer-by-layer analysis, and then reverse layer-by-layer integration.

Requirements acquisition: There are three main requirements acquisition methods: top-level document, application scenario design and new equipment application. The top-level document is that the higher authorities take the lead in guiding the research institutions and departments to conduct pre-research utilizing documents or policies for the consideration of the future; Application scenario design refers to the design of future use scenarios by research institutions or institutions based on future practical applications to guide the development direction of unmanned systems; New equipment application is aimed at putting into use new equipment with subversive or significant capability improvement, and designing a series of customized unmanned system collaborative application modes around the equipment.

Requirements analysis: according to the requirements obtained from various channels, the mission capability and functions of the unmanned system are analyzed from the two aspects of mission requirements and functional requirements, such as the realization of reconnaissance and early warning, armed strike, material transportation, communication relay, surveying and mapping, battlefield minesweeping, underwater anti-submarine and other tasks, and the load capacity, range and time indicators, use survival rate, operating speed Safety and reliability of the system.

System architecture analysis: analyze the system architecture and task planning of unmanned systems with demand analysis as input. At the system architecture level, it mainly analyzes the types of heterogeneous platforms, the docking relationship between platforms, the input and output of information, the type and number of each type of homogeneous platform, and the deduction and simulation of unmanned systems. At the task planning level, the system load type, task path Guiha, process planning, task evaluation, communication timing and emergency response planning are analyzed. The results of system architecture analysis directly provide input requirements for various platform designs.

Platform design analysis: according to the design requirements provided by the unmanned system architecture, the platform design shall be carried out according to the platform type, such as an unmanned aerial vehicle, unmanned boat/ship and unmanned vehicle, including the conventional process design content of each type of platform, such as the overall platform design, subsystem design, subsystem development, integration, testing, acceptance and maintenance support. According to their characteristics, each type of platform also includes some other design points, such as the aerodynamic design of the UAV shape, stealth design, etc. (Fig. 7).

The design and development of an unmanned system is a process of repeated iteration and continuous approximation. It is necessary to feed back the research results of the previous step or several steps according to the specific research content and results of each step and adjust the overall plan in time to ensure the overall coordination and effectiveness of the system. To reduce the repetition of the design process, an expert system should be added to the work of demand acquisition, demand analysis, and system

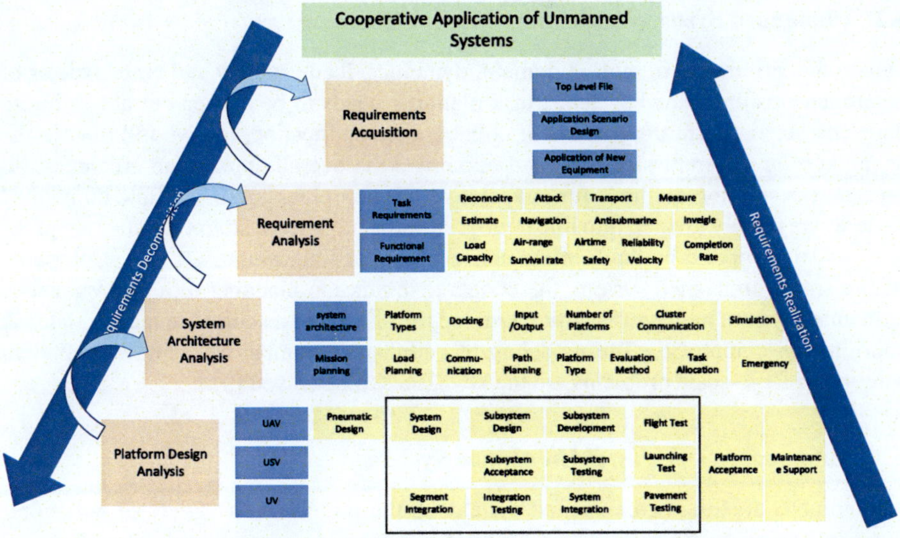

Fig. 7. Unmanned System Application of System Engineering

architecture analysis, to integrate the actual situation of various disciplines and take into account the realizability of the requirements for the platform.

4 Analysis of the Development Direction of Cooperative Application of Unmanned Systems

4.1 Strengthen Top-Level Planning and Construction

Top-level design refers to macro, overall and overall long-term planning and design. Based on the operational tasks and typical operational scenarios, the new system architecture design and integration methodology are adopted. Based on the existing operational system, the UAV and manned equipment are effectively and dynamically combined to generate new combat forces to achieve system reconstruction and energy enhancement. The cooperative operation of heterogeneous unmanned systems should have the reconfigurable ability, adaptive ability, and expandable ability to accelerate the formation of new qualitative operational capabilities such as situational awareness diversification, decision-making multi-polarization, attack diversification, variable combat methods, and damage resistance, improve operational effectiveness and task completion rate, and provide for the subsequent formation of cross-domain coordination, distributed, open Support is provided by a dynamically collaborative and highly autonomous composable joint combat system [9]. The distributed interconnection mechanism supporting the intelligent reconfigurable multiple battle kill chain [20–22] is built by adopting the system method oriented to logical unit design and integration.

4.2 Unmanned System Cooperative Mission Planning

Due to the differences in motion domain, dynamics, load capacity and other aspects of heterogeneous unmanned systems, task planning needs to be carried out according to their characteristics in the process of collaborative application, to give full play to the advantages of different platforms and maximize the overall application efficiency. At present, the solution methods for cooperative mission planning of multiple unmanned systems are mainly divided into linear programming model, constraint satisfaction problem model, heuristic method, intelligent evolutionary algorithm, etc. [3]. Compared with a single unmanned system, the cooperative mission planning of a heterogeneous unmanned system has significantly increased the task type, resources, quantity and task coordination complexity. The modeling of cooperative combat tasks and the efficient solution of large-scale problems are the key issues to be solved [3].

4.3 Autonomous Docking of Unmanned System

Autonomous docking of unmanned systems is the premise of a variety of collaborative application modes, such as airborne air-launched UAV, unmanned vehicle-carried UAV, unmanned boat and unmanned aerial vehicle [11]/unmanned underwater vehicle, unmanned boat and unmanned vehicle, etc. With the increase of the number and scale of unmanned systems, the control mode of humans in the loop gradually becomes invalid, and the precise docking and autonomous docking capabilities between different unmanned platforms become increasingly important, while the platform motion mode, motion range, load The difference between task planning brings different degrees of interference to autonomous docking. The autonomous docking mode, docking interface, and other contents of the unmanned system need to be designed and reflected in the mission planning.

4.4 Self-Planning Ability of Sports

The biggest feature of the unmanned system is unmanned. It requires the unmanned cooperative system to have a certain ability of motion trajectory planning, to detect the position and movement of the equipment in the system network in real-time, as well as the obstacle information in the external environment, to plan its action path in real-time, and to achieve the maximum action effect with the minimum energy price. The application of autonomous learning technologies such as artificial intelligence is the premise to realizing autonomous planning of unmanned equipment movement. There have been many research results on this frontier technology, but there is still a long way to go from a practical application or large-scale application.

4.5 Reasonable Allocation of Computing Resources

The volume, load, power consumption and bandwidth of unmanned equipment are extremely limited. To process massive computing data, such as environment, motion, autonomous planning, collaborative tasks, etc., computing resources are seriously insufficient. Therefore, it is necessary to reasonably allocate the limited airborne computing capacity and communication bandwidth. Some computing capacity or training requirements can be placed on the ground station, and the ground large computer can perform fast calculations, and only upload the results; It can also reasonably allocate parallel computing resources and computing time sequence in multi-machine tasks, call computing resources of heterogeneous cluster unmanned equipment, and compute by priority. It requires computing resource planning and communication bandwidth to provide a strong guarantee.

5 Key Technologies of Unmanned System Cooperation

5.1 Homogeneous Unmanned System Cluster Decision-Making and Planning

The decision and planning of homogeneous unmanned swarm systems is the premise of the cooperative application of heterogeneous unmanned systems. Homogeneous unmanned equipment has consistency or complementarity in scope, motion characteristics, functional characteristics, and other aspects, and can cooperate in mission planning and cluster decision-making to obtain a solution that maximizes global benefits. Many research institutions and scholars have participated in the research on the methods of cluster decision-making and task planning of homogeneous unmanned systems.

With the increase in the number and scale of swarm systems and the exponential increase in the difficulty of collaboration, there are inflection point effects in system implementation, cluster management, collaborative control, and other technologies. At the same time, the communication distance and bandwidth of the unmanned system are limited, and the design of the communication network topology is constrained, which makes it difficult to achieve large-capacity inter-computer information distribution and arbitrary topology transmission, further compressing the optimal performance of the large cluster system in theory. Therefore, large clusters and communication constraints put forward stringent requirements for architecture design, collaborative planning and decision-making, and swarm flight control.

The application of artificial intelligence technology will greatly improve the autonomy of unmanned systems, and further improve the efficiency and effectiveness of cooperation between systems. Autonomy includes four key technologies, namely artificial intelligence/machine learning, improving efficiency and effectiveness, trust and weaponization [17].

5.2 Cooperative Real-Time Control of Heterogeneous Systems

The movement ability and dynamic characteristics of cross-domain platforms are significantly different. For example, air platforms and underwater platforms can move in three-dimensional space, while ground and surface platforms can only move in two-dimensional space. In addition, the dynamics and kinematics forms of different platforms are also significantly different, which is closely related to the feasibility of coordination behavior, making it necessary to build more complex models and consider more environment/task constraints when studying the problem of cross-domain platform collaborative control, further improving the difficulty of research on collaborative control theory and technology [4].

Based on the control algorithm, network security and communication technology, the collaborative real-time control of heterogeneous systems studies the technologies of coordinated command, mixed formation and trusted control of the cross-domain unmanned system, and improves the robustness and interoperability of heterogeneous unmanned systems. Specifically, the collaborative control of heterogeneous systems mainly includes three key technologies: collaborative command and task allocation, hybrid formation and formation reorganization, and human-in-loop trusted control [15].

5.3 Seamless Integration of Different Situational Information

Situational awareness is a technical challenge that unmanned systems must face. For collaborative awareness, the key to information sharing is to uniformly express and integrate the perception information obtained by different unmanned systems. For cross-domain platforms, the information available from different platforms has significant differences in perception perspective, data type, data scale, noise level and other aspects. In addition, the environment in which cross-domain platform systems operate is usually characterized by strong dynamics and high complexity, which makes the seamless fusion of perception information one of the difficulties in the coordination and consistency of heterogeneous unmanned systems [4].

The cross-domain cooperative unmanned system carries multimodal sensors to obtain its surrounding environment data and aggregates the information through collaborative sensing technology to achieve "one aircraft discovery, the whole network is known", and improve the system's distributed situation awareness ability. It mainly includes three key technologies such as cross-modal understanding and cognition, target tracking and intention reasoning, and distributed information sharing and interaction [15].

5.4 Real-Time Behavior Optimization Under Multiple Constraints

The environmental information obtained by different platforms in cross-domain collaboration includes large-scale "macro" and local fine "micro" environmental information, which includes both environmental information and task and target information. The multi-dimensional characteristics of situational constraints are obvious, which will provide more environmental information for each platform in the collaboration system and provide a basis for achieving global optimal decision-making. However, The behavior optimization decision problem under multi-dimensional constraints is easy to cause problems such as local minimum and even non-solvable (feasibility) and real-time [4].

6 Summary

Unmanned systems have been paid attention to and developed in many aspects from beginning to end. With the diversification of war forms and the extension of the battlefield front, the development and application of unmanned equipment have received unprecedented attention. The cooperative application of heterogeneous unmanned system clusters, represented by unmanned aerial vehicles, unmanned vehicles, and unmanned ships, also brings many problems. The motion domain, kinematics and mission capabilities of various unmanned platforms are different. Therefore, many key technologies need to be broken through, such as the decision-making and planning of unmanned systems, collaborative real-time control of heterogeneous systems, a fusion of different situational information, and behavior optimization under multiple constraints.

For complex systems, system engineering methods have broad application prospects. The system engineering method is used to analyze the design and application of heterogeneous unmanned systems, starting from the acquisition of top-level requirements, decomposing the requirements layer by layer, gradually landing the requirements to the executable level, and then gradually integrating and synthesizing to obtain heterogeneous systems that meet the top-level requirements. The application of the system engineering analysis method can strengthen the role of top-level design and reduce repeated iterations in the design process.

Based on the preliminary analysis of the system engineering method, the development direction of the future collaborative application of unmanned systems is sorted out, such as strengthening the top-level planning and construction, strengthening the collaborative task planning, autonomous docking of unmanned systems, the self-planning ability of motion and reasonable allocation of computing resources.

In the future, model-based system engineering can be used for the analysis of the cooperative application of unmanned systems, so that the system design work has better inheritance and iteration.

References

1. Zhang, Y., Zhang, Y., Qiu, M., et al.: Research on the application of ground-air unmanned platform cooperative operation. Fire Command Control **46**(5), 6 (2021)
2. Hodicky, J., Prochazka, D.: Challenges in the implementation of autonomous systems into the battlefield. In: International Conference on Military Technologies, pp. 743–747 (2017)
3. Ma, S., Ma, Y.: Mission planning method for cooperative operation of heterogeneous unmanned system group. Command Control Simul. (2), 7 (2019)
4. Fan, B., Zhao, G., Bo, Y., Wu, X.: Multi-target air-ground heterogeneous unmanned system cooperative task allocation method. J. Military Eng. **44**(06), 1564–1575 (2023)
5. He, Y., Qin, T., Wang, N.: Cross-domain collaboration: new trend of unmanned system technology development and application. Unmanned Syst. Technol. **4**(4), 13 (2021)
6. Chen, J., Zhang, X., Xin, B., et al.: Coordination between unmanned aerial and ground vehicles: a taxonomy and optimization perspective. IEEE Trans. Cybern. **46**(4), 959–972 (2016)
7. Lindemuth, M., Murphy, R., Steimle, E., et al.: Sea robot-assisted inspection. IEEE Robot. Autom. Mag. **18**(2), 96–107 (2011)
8. Wei, Z., Weng, Z., Hua, Y., et al.: Heterogeneous unmanned cluster formation - encirclement tracking control under switching topology. J. Aviat. **44**(2), 16 (2011)
9. Zhang, P., Huang, B., Ju, H.: MBSE: development direction of system engineering. Sci. Technol. Herald **38**(21), 6 (2020)
10. Zhang, X., Sun, Z., Wu, L., et al.: Top concept thinking of intelligent cooperative operation of future manned and unmanned aerial vehicles. Unmanned Syst. Technol. (2021)
11. Xu, X., Duan, H., Zeng, Z., et al.: Research progress of unmanned aerial vehicle/unmanned vehicle cooperative control. Air Weap. **27**(6), 6 (2020)

12. Zhang, H., Jing, D., Wang, X., et al.: Current situation and trend of overseas marine unmanned systems research. In: Proceedings of the Fifth Underwater Unmanned Systems Technology Summit Forum, vol. 5 (2022). https://doi.org/10.26914/c.cnkihy.2022.054572
13. Fu, C., Li, T., Jiang, Y.: Research on manned/unmanned underwater cooperative warfare. In: Proceedings of the Fifth Underwater Unmanned Systems Technology Summit Forum, vol. 5 (2022). https://doi.org/10.26914/c.cnkihy.2022.054601
14. https://www.woiwrj.com/wurenjibaike/46593/
15. Fu, M., Yang, Y., Yue, Y., et al.: Overview of ground-air cooperative unmanned system. Def. Sci. Technol. **42**(3), 8 (2021)
16. Li, L., Zhang, X., Wang, H., et al.: Integrated cooperative operation architecture of heterogeneous unmanned equipment at sea. Ship Sci. Technol. **41**(23), 4 (2019)
17. Wang J, Liu B, Dong Q.: Analysis of the United States department of defense's integrated roadmap for unmanned systems. Air Missile (5), 4 (2019)
18. Wang, Z.: Introduction to System Engineering, 4th edn, pp. 97–103. Electronic Industry Press, Beijing (2012)
19. Wang, Z.: Introduction to System Engineering, 4th edn, pp. 108–113. Electronic Industry Press, Beijing (2012)
20. Cheater, J.C.: Accelerating the kill chain via future unmanned aircraft. In: Air War College: Center for Strategy and Technology (2007)
21. Hahn, A., Thomas, R.K., Lozano, I., et al.: A multi-layered andkill-chain based security analysis framework for cyber-physical systems. Int. J. Crit. Infrastruct. Prot. **11**, 39–50 (2015)
22. Bloye, B.A.: Optimizing the air-to-ground kill chain for time-sensitive targets. Monterey. Naval Postgraduate School, USA (2009)

Coupling Analysis of the Factors Influencing the Man-Machine Efficacy Based on System Dynamics

Qiuqi Wu, Youchao Sun(✉), Yifan Wang, Weicheng Liu, and Zichang Chen

Nanjing University of Aeronautics and Astronautics, Nanjing, China
{wuqiuqi,sunyc,wang_yf,liuweicheng,chenzichang}@nuaa.edu.cn

Abstract. The future intelligent cockpit's complex components make it challenging to evaluate the efficacy of man-machine cooperative tasks. This paper uses system dynamics and coupling theory to construct a system dynamics model and quantify the effects of equipment, human, and the environment to reveal the evolution law of man-machine efficacy. The coupling relationship and the factors influencing man-machine efficacy are first determined. The construction of the system dynamics model is the second step. Then, using the Delphi method, the expert weights and the improved interval-valued intuitionistic fuzzy—analytic hierarchy process method are combined to calculate the weights of each factor. The coupling degree of the factors is calculated using the coupling degree model and the N-K model. When the equations have been input, the simulations are then performed. The findings show that changes in the degree of coupling have the following effects: during the first seven units of time, the trend of the change in man-machine efficacy value is essentially flat, and from the seventh unit of time on, the change in man-machine efficacy value exhibits significant performance.

Keywords: Man-machine Efficacy · Aircraft Cockpits · System Dynamics

1 Introduction

It is generally believed that future intelligent cockpits will evolve in a way that emphasizes multimodal intelligent man-machine interaction (MMI) with the help of decision support systems. Man-machine efficacy (MME) in this paper therefore refers to the effectiveness and qualities of the human and intelligent cockpit's work, relying on the man-machine interface for information exchange [1]. Through reasonable function allocation and cooperation, the human and the decision support system carry out the respective tasks they are good at and ultimately complete the task. Compared to conventional aircraft cockpits, the collaborative operational tasks in intelligent cockpits are more highly automated and time-sensitive, especially when it comes to control information input techniques and dynamic changes in man-machine control. It is unknown whether the factors of conventional aircraft cockpits still apply to the man-machine collaboration (MMC) in future cockpits [2].

Chinese Society of Aeronautics and Astronautics (Ed.): CASTC 2023, LNME, pp. 231–240, 2024.
https://doi.org/10.1007/978-981-99-8861-7_23

Over the years, scholars have done lots of research using different methods and points of view to figure out how different things affect MME. Human factors, the environment, the equipment, the layout, and the strategy are frequently considered. Human reliability analysis, which quantifies the effect of individual factors such as fatigue level [3] and cognitive patterns [4], is frequently used in the analysis of these factors. Ismail et al. [5]'s laboratory simulations of various light, temperature, and humidity conditions showed a clear correlation between MME and environmental parameters. Tsarouhas [6] investigated the impact of optimizing automated production processes by conducting an analysis of the equipment. Based on six major categories of downtime losses, he proposed a framework for improving automated systems. Song et al. [7] suggested a design strategy for improving the man-machine environment of military cockpits based on deep learning while studying the impact of layout design. Zhang et al. [8] developed an abstraction method to provide a solution for multimodal intelligent MMI modeling in aircraft cockpits to improve MME for the MMI resource matching and scheduling issue using system dynamics (SD). The review reveals that different methods were used to explore the impact of various influencing factors (IFs) on MME, ranging from general modeling methods to physiological experiments.

Yoo et al. [9] developed an IF architecture for airport passenger security screening based on human resources, facilities, and responsibility structure. They suggested adjustments that should be made to improve the chances of finding contraband during security checks. Cui et al. [10] built an influence system for civil aviation security efficiency with nine IFs and used a panel regression model to evaluate the degree of the effect. Li et al. [11] built a hierarchical structure for evaluating the aircraft service quality, developed a rating model using fuzzy hierarchical analysis and the binary fuzzy language method, and found factors that needed improvement. By incorporating decision testing and explanatory structural models, the key factors were found.

The rest of this paper is divided into five parts. Section 2 describes in detail the principles for screening the IFs and the CD calculation method. Section 3 outlines the SD model constructed based on the IF system. Section 4 describes the calculation method for each data point in the SD parameter equation. Section 5 runs the SD system and discusses the sensitivity of the CD. Finally, Sect. 6 concludes the paper.

2 Construction of the Multi-factor Coupled Model of MME

2.1 Establishment of the Index System of the IFs

This paper analyzes three categories of factors that affect MME: equipment, human, and environment, based on the aircraft operation process. Hierarchy of factors was subjected to the first round of expert deliberation using the Delphi method. Ultimately, 3 primary factors, 7 secondary factors, and 26 tertiary factors were selected (see Fig. 1).

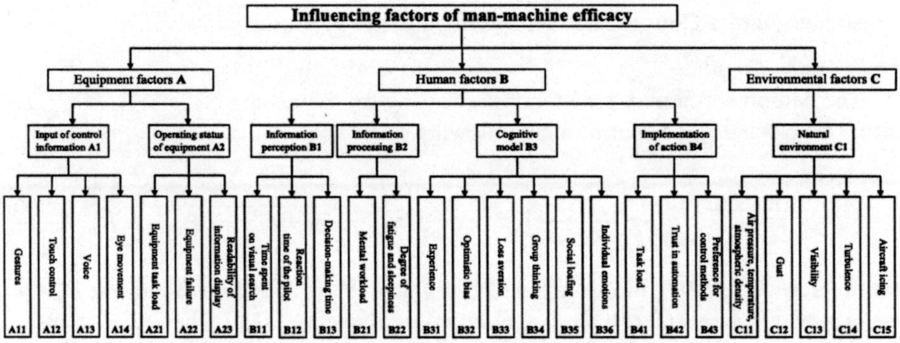

Fig. 1. Hierarchy diagram of intelligent cockpit IFs.

2.2 Coupling Analysis of IFs

Construction of a Coupled Model Based on the CD Model

Construction of Efficacy Function

Depending on whether the factor has a positive effect on MME, the efficacy function can be calculated according to the following formula:

$$U_{ij} = \begin{cases} (X_{ij} - B_{ij})/(A_{ij} - B_{ij}), \ U_{ij} \text{ has positive effect} \\ (A_{ij} - X_{ij})/(A_{ij} - B_{ij}), \ U_{ij} \text{ has negative effect} \end{cases} \tag{1}$$

where, Uij is the efficacy coefficient, and $0 \leq Uij \leq 1$. The order parameters' upper and lower limits for the entire system in steady state are represented by Aij and Bij, respectively, and the expert rating for each factor is represented by X_{ij}.

Uij indicates the degree of agreement with the target. A linear weighting method can be used to determine each factor's contribution to the system.

$$U_i = \sum_{j=1}^{m} \lambda_{ij} U_{ij}, \quad \sum_{j=1}^{m} \lambda_{ij} = 1 \tag{2}$$

where, Ui represents the actual contribution efficacy of each subsystem to the over-all system, and λij represents the weight of the j th indicator of subsystem i.

Calculation of Coupling Degree

The following equation can be used to calculate the CD model of multiple factors:

$$C_m = m\left\{(U_1, U_2, U_3 \ldots U_m)/\left[\prod U_i + U_j\right]\right\}^{1/m} \tag{3}$$

where Cm represents the degree of coupling, m represents the number of factors, and Ui represents the factor's actual contribution efficacy.

Construction of a Coupled Model Based on the N-K Model

Multi-factor Coupling

The equipment-human-environment coupling can be expressed as $T(A, B, C)$, and the CD is calculated as shown in the following equation:

$$T(A, B, C) = \sum_h \sum_i \sum_j \left[P_{hij} \cdot \log_2 \left(\frac{P_{hij}}{P_h *** \cdot P*_i ** \cdot P **_j *} \right) \right] \tag{4}$$

where, A, B, C refers to the coupling of equipment, human, and environment; $P_h ***$, $P*_i **$, and $P **_j *$ refer to the probability of influencing MME when subsystems are in the corresponding states of h, i, and j; and the values of h, i, and j are $\{0, 1\}$.

3 Construction of the SD Model

Stocks and flows have an algebraic relationship that is as follows:

$$Stock(t) = \int_{t_0}^{t} [inflow(t) - outflow(t)]dt + Stock(t_0) \tag{5}$$

where, t_0 is the initial time, and t is the present time.

The flow diagram of the SD model regarding each subsystem, and the total human-equipment-environment system, is shown in Fig. 2.

4 Determination of Parameter Equation of SD Model

4.1 Calculation of the Weight of the IFs

Table 1 shows the results after adding the values and calculating the percentage.

4.2 Calculation of the CD of IFs

Figure 3(a) displays the values of the two-factor coupling, which are determined individually using the N-K model. The CD secondary factors is calculated in Fig. 3(b).

4.3 Determination of the Final Parametric Equation

The SD parameter equations related to equipment factors are given in detail in Table 2.

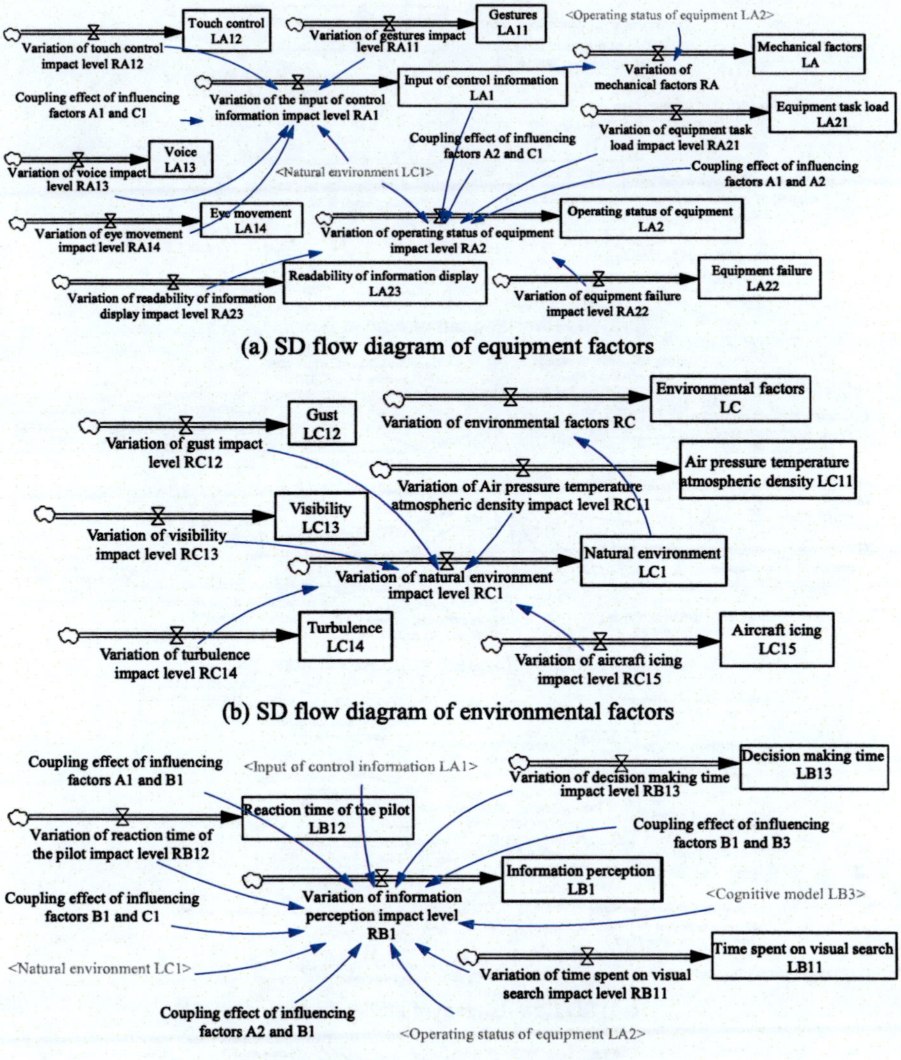

(a) SD flow diagram of equipment factors

(b) SD flow diagram of environmental factors

(c-1) SD flow diagram of human factors

Fig. 2. SD flow diagram of the total system and each subsystem.

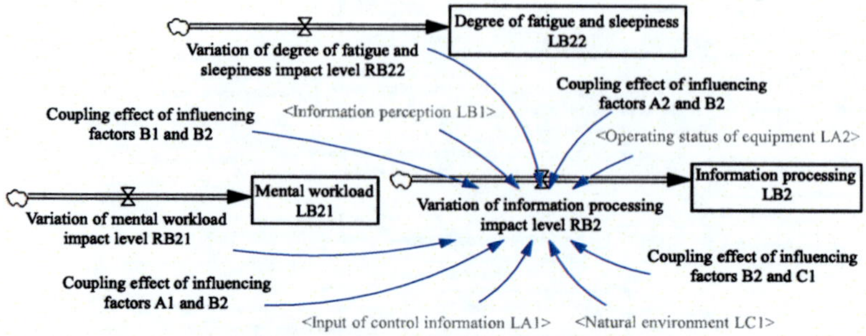

(c-2) SD flow diagram of human factors

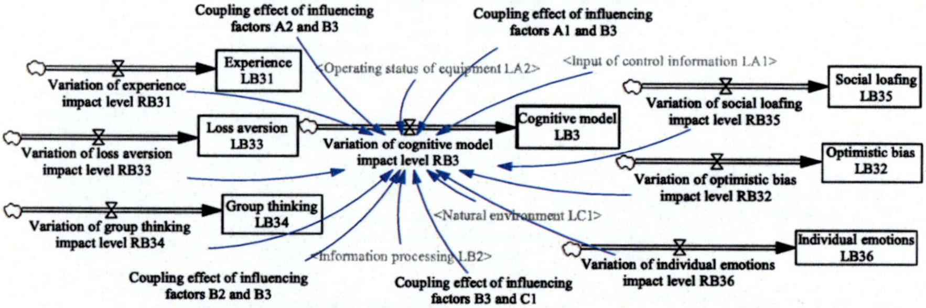

(c-3) SD flow diagram of human factors

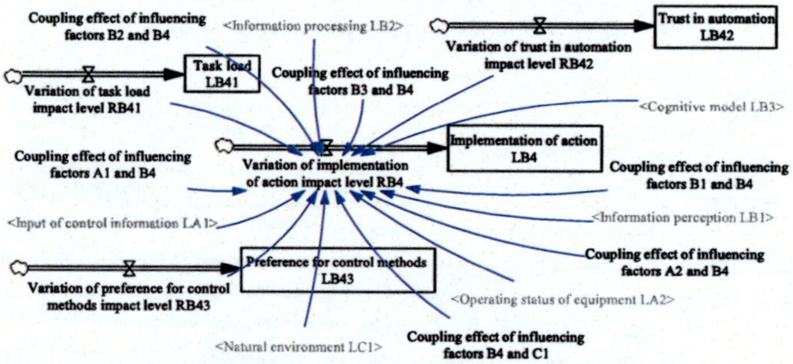

(c-4) SD flow diagram of human factors

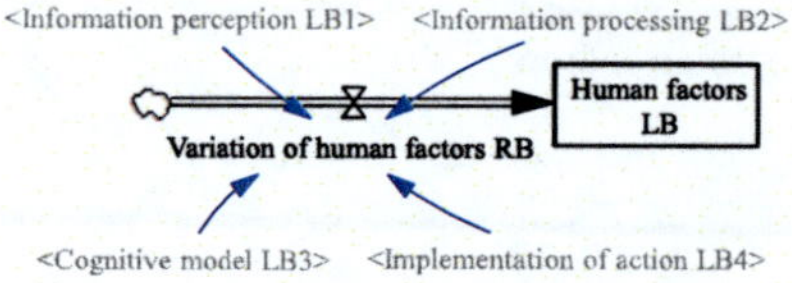

(c-5) SD flow diagram of human factors

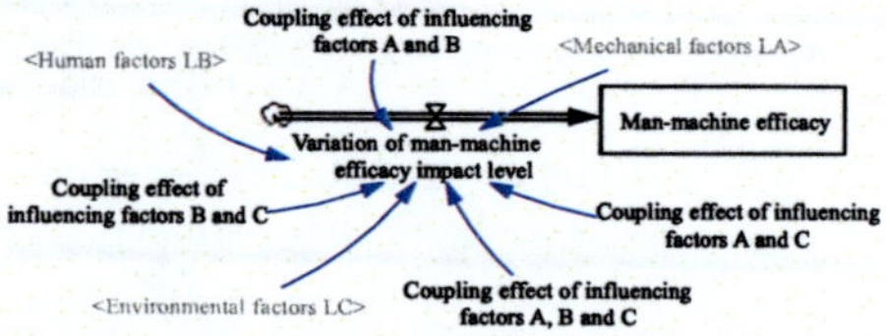

(d) SD flow diagram of the total system

Fig. 2. (*continued*)

Table 1. Weight of IFs of the MME.

Factor	**A11**	**A12**	**A13**	**A14**	**A21**	**A22**	**A23**	**B11**
Weight	0.0329	0.0414	0.0395	0.0362	0.0554	0.0551	0.0551	0.0561
B12	**B13**	**B21**	**B22**	**B31**	**B32**	**B33**	**B34**	**B35**
0.0547	0.0547	0.0764	0.0744	0.0210	0.0177	0.0177	0.0138	0.0114
B36	**B41**	**B42**	**B43**	**C11**	**C12**	**C13**	**C14**	**C15**
0.0141	0.0573	0.0560	0.0561	0.0191	0.0189	0.0260	0.0200	0.0191

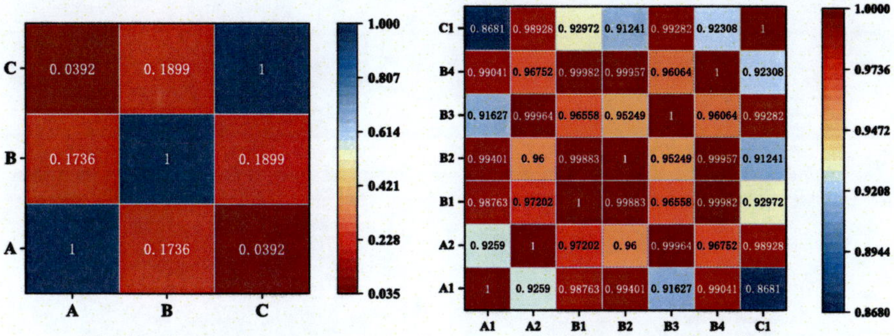

(a) CD of primary factors (b) CD of secondary factors

Fig. 3. CD of primary and secondary factors.

Table 2. Parametric equations for system dynamics related to equipment factors.

Variable	Type	Equation
LA14	State variable	LA14 = INTEG("RA14",1)
LA1	State variable	LA1 = INTEG("RA1",4)
LA	State variable	LA = INTEG("RA",7)
LA12	State variable	LA12 = INTEG("RA12",1)
LA13	State variable	LA13 = INTEG("RA13",1)
RA14	Rate variable	RA14 = 0.0362
RA11	Rate variable	RA11 = 0.0329
RA1	Rate variable	RA1 = 0.2193*RA11 + 0.276*RA12 + 0.2633*RA13 + 0.2413*RA14 + C[A1-C1]*LC1
RA12	Rate variable	RA12 = 0.0414
RA13	Rate variable	RA13 = 0.0395

5 Simulation Results and Discussion

5.1 Results of the Secondary Factors

The simulation was run, as shown in Fig. 4.

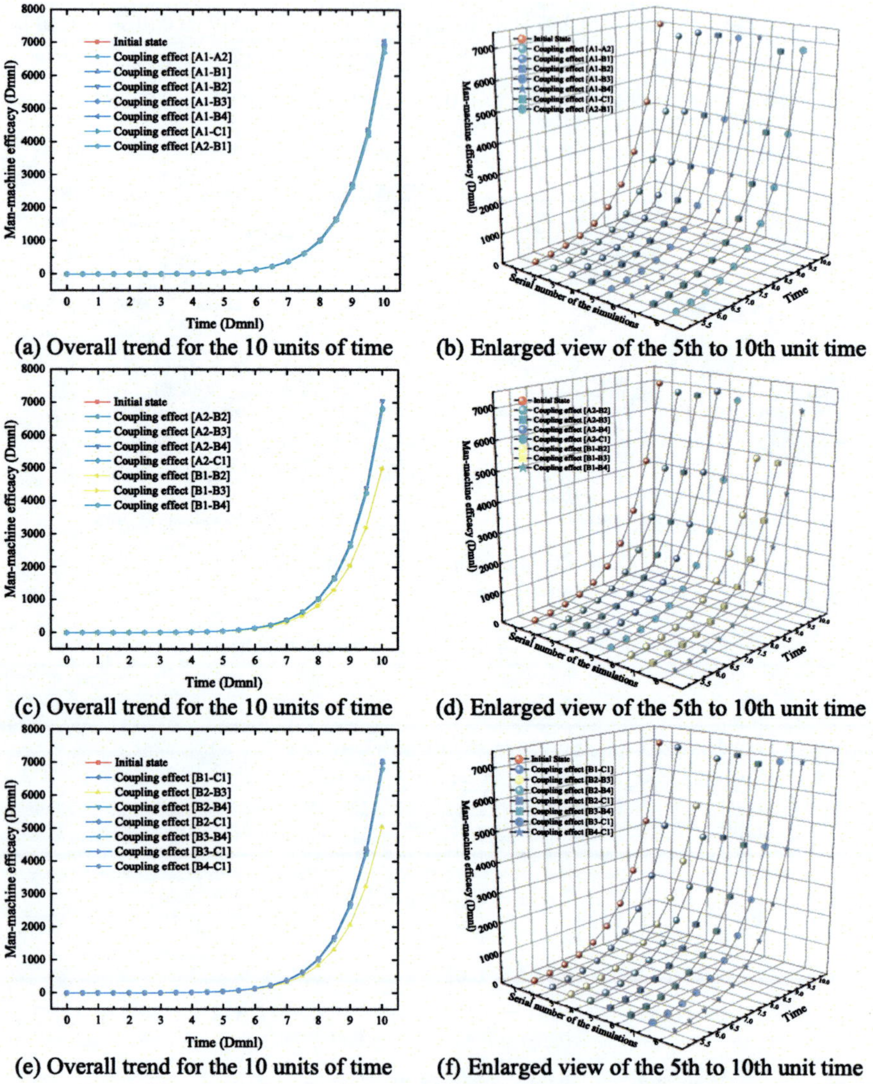

(a) Overall trend for the 10 units of time (b) Enlarged view of the 5th to 10th unit time

(c) Overall trend for the 10 units of time (d) Enlarged view of the 5th to 10th unit time

(e) Overall trend for the 10 units of time (f) Enlarged view of the 5th to 10th unit time

Fig. 4. The effect of changes in the CD of secondary factors on MME.

Information perception-information processing, information processing-cognitive model, and information perception-cognitive model were the three coupling factors that deviated most from the system's initial value.

5.2 Results of the Tertiary Factors

Figure 5 displays the simulation findings.

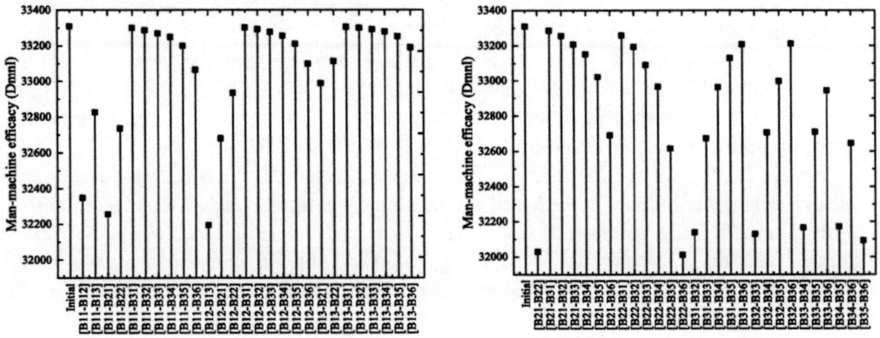

Fig. 5. The effect of changes in the CD of tertiary factors on MME.

Mental workload—fatigue and sleepiness level; fatigue and sleepiness level—individual emotions; and social loafing—individual emotions are the three sub-coupling factors that deviate most from the system's initial value.

6 Conclusion

All simulation results follow the same pattern of change: for the first seven units of time, the trend of the change in the efficacy value is comparatively flat, but starting with the eighth unit of time, the performance of the change is noticeable.

The multi-factor coupling of equipment-human-environment, the two-factor coupling of information perception-information processing, information processing-cognitive model, and information perception-cognitive model in the secondary factor, and the two-factor coupling of mental workload-fatigue and sleepiness level, fatigue and sleepiness level-individual emotions, and social loafing-individual emotions in response to the simulation results should all be taken into consideration.

The findings can serve as a theoretical foundation for the restriction of flight time as well as a guide for the early detection of abnormal situations to ensure flight safety.

Acknowledgments. This work was financially supported by Joint Fund of National Natural Science Foundation of China and Civil Aviation Administration of China (No. U2033202, U1333119), National Natural Science Foundation of China (No.52172387) and Fundamental Research Funds for the Central Universities (ILA220321A22, 56XBC22006).

References

1. Jipp, M., Ackerman, P.L.: Impact of higher levels of automation on performance and situation awareness. J. Cogn. Eng. Decis. Mak. **10**, 138–166 (2016)

2. Wang, L., Jamieson, G.A., Hollands, J.G.: Trust and reliance on an automated combat identification system. Human Fact. J. Human Fact. Ergon. Soc.. **51**, 281–291 (2009)
3. Myszewski, J.M.: Mathematical model of the occurrence of human error in manufacturing processes. Qual. Reliab. Eng. Int. **26**, 845–851 (2010)
4. Kosmowski, K.T.: Human factors and cognitive engineering in functional safety analysis. Adv. Intell. Syst. Comput. **635**, 434–448 (2017)
5. Ismail, A.R., Nizam, C.M., Haniff, M.H.M., Deros, B.M.: The impact of workers productivity under simulated environmental factor by taguchi analysis. APCBEE Proc. **10**, 263–268 (2014)
6. Tsarouhas, P.H.: Overall equipment effectiveness (OEE) evaluation for an automated ice cream production line. Int. J. Prod. Perf. Manag. **69**, 1009–1032 (2019)
7. Song, Z., Su, S., Yan, M., Wang, B.: Research on man-machine environment optimization design strategy of military cabin based on deep learning. https://ieeexplore.ieee.org/abstract/document/9969753. Accessed 04 Mar 2023
8. Zhang, X., Sun, Y., Zhang, Y.: Evolutionary game and collaboration mechanism of human-computer interaction for future intelligent aircraft cockpit based on system dynamics. IEEE Trans. Human-Mach. Syst. **52**, 87–98 (2022)
9. Yoo, K.E., Choi, Y.C.: Analytic hierarchy process approach for identifying relative importance of factors to improve passenger security checks at airports. J. Air Transp. Manag. **12**, 135–214 (2006)
10. Cui, Q., Li, Y.: The change trend and influencing factors of civil aviation safety efficiency: the case of Chinese airline companies. Saf. Sci. **75**, 56–63 (2015)
11. Li, W., Yu, S., Pei, H., Zhao, C., Tian, B.: A hybrid approach based on fuzzy AHP and 2-tuple fuzzy linguistic method for evaluation in-flight service quality. J. Air Transp. Manag. **60**, 49–64 (2017)

Path Planning and Optimization of a UAV-Based Relay Communication System with Co-channel Interference

Wenqiang Zhao(✉), Zhan Zhang, Jianfeng Feng, and Xin Jiang

COMAC Beijing Aircraft Technology Research Institute, Beijing 102211, China
zhaowenqiang@coamc.cc

Abstract. Unmanned aerial vehicle (UAV) based relay communication is an important technical scheme for long-distance wireless communications. However, the UAV-based relay communication system is susceptible to co-channel interference. In this paper, to improve the link reliability of the UAV-based relay communication system, a path planning method was proposed for the UAV. First, a two-hop decode-and-forward (DF) relay communication system model was presented, in which the UAV acted as a relay platform between a mobile access point (AP) and a fixed base station (BS). Then an analytical expression of the outage probability of the system was derived. Based on the criterion of minimizing the outage probability, the problem of jointly optimizing the UAV's heading and flight speed was formulated and addressed by the Nelder-Mead simplex algorithm. The simulation results demonstrated the validity of this proposed path planning method.

Keywords: Unmanned Aerial Vehicle · Relay Communication · Path Planning · Co-channel Interference · Decode-and-Forward

1 Introduction

Unmanned aerial vehicle (UAV) based relay communication is an important technical scheme for wide-area wireless communications. Compared to conventional fixed relay communication schemes, UAV-based relay communication has several advantages such as wide-area communication coverage, quick deployment, flexibility of relay node locations, and lower maintenance costs. Hence, UAV-based relay communication has been widely used in many military and civilian applications [1–3]. However, due to the mobile characteristics of the relay node, energy constraints of UAVs, and complex working environments, the UAV-based relay communication system also has problems. Some of these problems are that the path planning of a UAV-based relay node is complicated, the transmission schemes of a UAV-based relay communication system are complex, and UAV-based relay network topology varies, etc. [4–6]. Therefore, it is of great significance to conduct research on the key technologies of a UAV-based relay communication system.

Chinese Society of Aeronautics and Astronautics (Ed.): CASTC 2023, LNME, pp. 241–252, 2024.
https://doi.org/10.1007/978-981-99-8861-7_24

To date, UAV-based relay communication systems have been heavily researched. This research has mainly been concentrated in the following areas: air-to-ground channel modeling of a UAV-based relay communication system [7–10], the transmission scheme of a UAV-based relay communication system [11, 12], the path planning of the UAV acting as a relay [13–18], and the optimization of the UAV-based relay network [19–21]. Researchers have tried various ideas to solve the path-planning problem of UAVs for a multiple-access wireless channel. Zhan [13] developed a heading optimization algorithm to maximize the ergodic normalized transmission rate of the links between the ground nodes and the relay. Jiang [14] further proposed a UAV path planning method by dynamically adjusting the UAV's heading based on the criterion of maximizing the average sum rate of the ground-to-air uplink. In order to solve the path-planning problem of UAVs for a wireless broadcast channel, Yin [15] proposed a UAV position optimization method based on the criteria of minimizing the outage probability of the links from the source to the user nodes. Choi [16] further proposed a maneuvering control scheme to maximize the minimum ergodic link capacity. To solve the path-planning problem of UAVs for point-to-point wireless communications, Ouyang [17] proposed a UAV relay scheme with joint transmitter and receiver beamforming, and developed a UAV path planning method based upon the criterion of maximizing the output signal-to-noise ratio. Nevertheless, the proposed scheme required that both the transmitter of the access point and receiver of the base station have the knowledge of channel fading information. Using the UAV relay transmission scheme with space-time block coding proposed by [18], Haitao proposed a UAV heading optimization method based on maximizing the ergodic capacity of the dual-hop link. However, in practical military and civilian applications of UAV-based relay communication systems, the UAV relay node is often exposed to intentional jamming or radio interference [22]. Typical research for the path planning of UAV-based relay communication systems do not consider the effects of co-channel interference. Thus, the following are the primary objectives of this study.

1) Study the effect of co-channel interference on the link performance of a UAV-based relay communication system.
2) Give a path planning method for UAV-based relay communication systems with co-channel interference.

In this paper, a scenario where the point-to-point UAV-based relay communication system with co-channel interference at the UAV-based relay node was considered. The UAV uses the DF protocol to relay messages, and all the nodes within the system where equipped with a single antenna. The system performance was then quantified by deriving the outage probability of the system, and it was found that the UAV heading and flight speed were the main influencing factors for the outage probability. Thus, a path-planning problem to optimize the UAV heading angle and flight speed based upon the criterion of minimizing the outage probability was developed. By using the variable substitution method, the constrained UAV heading angle and flight speed optimization problem was transformed into an unconstrained optimization problem that was addressed by the Nelder-Mead simplex algorithm. The main contributions included the following 3 aspects.

1) A heading control algorithm was developed for the UAV-based relay communication system with co-channel interference.
2) The closed-form expression of the outage probability of the UAV-based relay communication was obtained by theoretical derivation.
3) The outage probability performance of the system was proven to be significantly improved by optimizing the flight path of the UAV.

This paper follows the following format. Section 2 describes the system model, the signal model, and the channel model. In addition, the instantaneous signal-to-interference-plus noise-ratio (SINR) at the UAV and BS is obtained. In Sect. 3, the outage probability for the AP-UAV link and the UAV-BS link is analyzed. In addition, the path-planning problem for the UAV is formulated, and then solved using the Nelder-Mead simplex algorithm. Section 4 presents the simulation results that demonstrate the validity of the path-planning algorithm proposed. The conclusion is summarized in Sect. 5.

2 System Description

2.1 System Model

The model of a UAV-based relay communication system with co-channel interference at the relay is presented in Fig. 1. The UAV-based relay communication system consists of a fixed ground base station (BS), a fixed-wing UAV acting as an airborne relay node, a mobile access point (AP), and K fixed deployed co-channel interferers, where I_k represents the k-th interferer. It is assumed that there is no direct communication link between the AP and the BS. Thus, a UAV relay node is needed to realize the communication between the AP and the BS. Although the information transmission between the BS and the AP is essentially bidirectional, this paper only pays respect to the AP-UAV-BS link. It is further assumed that there is no direct communication link between the interferers and the BS. This implies that the interfering signals are continuously transmitted by the interferers, which means only the UAV node receiver is affected by the interferers. For simplicity, it is assumed that all the nodes within the system are equipped with a single antenna.

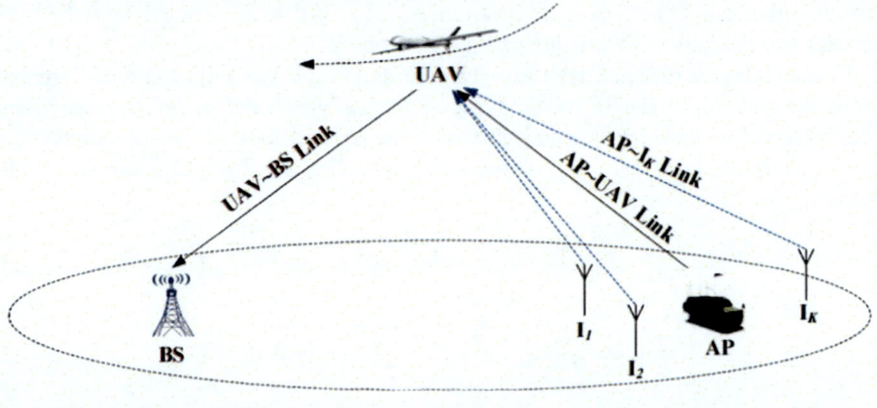

Fig. 1. The model of a UAV-based relay communication system with co-channel interference at the relay.

2.2 Signal Model

At time t, the communication from the AP to the BS takes place in 2 continuous time slots. In the first time slot, the AP sends its signal to the UAV. The received signals at the UAV can be expressed as

$$r_{\mathrm{U},t} = \sqrt{P_{\mathrm{A}}} h_{\mathrm{AU},t} s_t + n_{\mathrm{AU},t} + \sum_{k=1}^{K} \sqrt{P_k} h_{k,t} w_{k,t} \tag{1}$$

where P_A represents the transmitting power at the AP, $h_{\mathrm{AU},t}$ is the channel fading coefficient between the AP and the UAV, and s_t is the transmitted signal with 0 mean and variance 1. $n_{\mathrm{AU},t}$ is the received additive white Gaussian noise (AWGN) with 0 mean and variance σ_{A}^2. K is the is the number of interferers, P_k represents the interfering power at interferer I_k, $h_{k,t}$ is the channel fading coefficient between the interferer I_k and UAV, and $w_{k,t}$ represents the transmitted signal from the interferer I_k with 0 mean and variance 1.

In the second time slot, the Decode-and-forward (DF) protocol is adopted to relay the message to the BS. Accordingly, the received signals at the BS can be written as

$$r_{\mathrm{B},t} = \sqrt{P_{\mathrm{U}}} h_{\mathrm{UB},t} x_t + n_{\mathrm{UB},t} \tag{2}$$

where P_{U} represents the transmit power at the UAV, $h_{\mathrm{UB},t}$ denotes the channel fading coefficient between the UAV and the BS, x_t is a transmitted symbol signal by the UAV with 0 mean and variance 1, and $n_{\mathrm{UB},t}$ is the received AWGN with 0 mean and variance σ_{U}^2.

2.3 Channel Model

The channel-fading coefficient $h_{X,t}(X \in \{\mathrm{AU}, k, \mathrm{UB}\})$ with large-scale path loss related to the distance between each node and the UAV is [13]

$$h_{X,t} = \frac{g_{X,t}}{d_{X,t}^{\alpha}} (X \in \{\mathrm{AU}, k, \mathrm{UB}\}) \tag{3}$$

where $g_{X,t}$ is the small-scale fading coefficient, which is a normalized complex Gaussian random variable with zero mean and variance 1, $d_{X,t}$ is the distance between the UAV and each ground node, and α is the path loss exponent.

To determine the distance between each node and the UAV, a 3-dimensional Cartesian coordinate system with the BS and the I_k located at $(x_{\mathrm{B}}, y_{\mathrm{B}}, 0)$ and $(x_k, y_k, 0)$ respectively is considered. The time-varying coordinates of the AP and the UAV can be expressed as $(x_{\mathrm{A},t}, y_{\mathrm{A},t}, 0)$ and $(x_{\mathrm{U},t}, y_{\mathrm{U},t}, h_t)$, respectively, thus allowing $d_{\mathrm{AU},t}$, $d_{\mathrm{UB},t}$ and $d_{k,t}$ to be separately calculated by

$$d_{\mathrm{AU},t} = \sqrt{(x_{\mathrm{U},t} - x_{\mathrm{A},t})^2 + (y_{\mathrm{U},t} - y_{\mathrm{A},t})^2 + h_t^2} \tag{4}$$

$$d_{\mathrm{UB},t} = \sqrt{(x_{\mathrm{U},t} - x_{\mathrm{B}})^2 + (y_{\mathrm{U},t} - y_{\mathrm{B}})^2 + h_t^2} \tag{5}$$

$$d_{k,t} = \sqrt{\left(x_{\mathrm{U},t} - x_k\right)^2 + \left(y_{\mathrm{U},t} - y_k\right)^2 + h_t^2} \tag{6}$$

We further assume that the UAV flying at a fixed altitude, the UAV dynamics are governed by the following discrete-time model [23]:

$$\begin{cases} x_{\mathrm{U},t} = x_{\mathrm{U},t-\Delta t} + v_{\mathrm{U},t}\Delta t \cos\delta_t \\ y_{\mathrm{U},t} = y_{U,t-\Delta t} + v_{\mathrm{U},t}\Delta t \sin\delta_t \end{cases} \tag{7}$$

where $v_{\mathrm{U},t}$ is the UAV's flight speed at time t, which is constrained as $v_{\min} \le v_{\mathrm{U},t} \le v_{\max}$. Δt is the time interval representing the UAV's heading update rate, $\left(x_{\mathrm{U},t}, y_{\mathrm{U},t}\right)$ denotes the UAV's x-y coordinate at time t, $\left(x_{\mathrm{U},t-\Delta t}, y_{\mathrm{U},t-\Delta t}\right)$ denotes the UAV's x-y coordinate at time $t - \Delta t$, and δ_t is the UAV's heading at time t. Considering that the UAV's heading in every moment is constrained by the turning radius of the UAV, δ_t satisfies $\delta_{t-\Delta t} - \delta_{\max} \le \delta_t \le \delta_{t-\Delta t} + \delta_{\max}$ where $\delta_{\max}$ is the maximum change in the UAV's heading. By substituting Eq. (7) into Eqs. (4), (5) and (6), Eqs. (4), (5) and (6) are further simplified as

$$d_{\mathrm{AU},t} = \left(a_{\mathrm{A},t} + b_{\mathrm{A},t}\cos\delta_t + c_{\mathrm{A},t}\sin\delta_t\right)^{\frac{1}{2}} \tag{8}$$

where

$$\begin{cases} a_{\mathrm{A},t} = \left(x_{\mathrm{U},t-\Delta t} - x_{A,t}\right)^2 + \left(y_{\mathrm{U},t-\Delta t} - y_{A,t}\right)^2 + h^2 + v_{\mathrm{U},t}^2\Delta t^2 \\ b_{\mathrm{A},t} = 2v_{\mathrm{U},t}\left(x_{\mathrm{U},t-\Delta t} - x_{\mathrm{A},t}\right)\Delta t \\ c_{\mathrm{A},t} = 2v_{\mathrm{U},t}\left(y_{\mathrm{U},t-\Delta t} - y_{\mathrm{A},t}\right)\Delta t \end{cases} \tag{9}$$

$$d_{M,t} = \left(a_{M,t} + b_{M,t}\cos\delta_t + c_{M,t}\sin\delta_t\right)^{\frac{1}{2}} \tag{10}$$

where $M \in \{\mathrm{UB}, k\}$, and

$$\begin{cases} a_{M,t} = \left(x_{\mathrm{U},t-\Delta t} - x_M\right)^2 + \left(y_{\mathrm{U},t-\Delta t} - y_M\right)^2 + h^2 + v_{\mathrm{U},t}^2\Delta t^2 \\ b_{M,t} = 2v_{\mathrm{U},t}\left(x_{\mathrm{U},t-\Delta t} - x_M\right)\Delta t \\ c_{M,t} = 2v_{\mathrm{U},t}\left(y_{\mathrm{U},t-\Delta t} - y_{M,t}\right)\Delta t \end{cases} \tag{11}$$

Assuming that the UAV can predict the AP position $\left(\hat{x}_{\mathrm{A},t}, \hat{y}_{\mathrm{A},t}, 0\right)$ using a Kalman filter at time $t - \Delta t$, the heading and flight speed of the UAV at time t become the only variables of $d_{\mathrm{AU},t}$ and $d_{M,t}$.

2.4 Instantaneous Received SINR at the UAV and the BS

Using the signal models given in Eqs. (1) and (2), the instantaneous received SINR at the UAV and the BS can be separately calculated by

$$\gamma_{\mathrm{U},t} = \frac{\rho_{\mathrm{A}} d_{\mathrm{AU},t}^{-2\alpha} \cdot \left|g_{\mathrm{AU},t}\right|^2}{1 + \sum\limits_{k=1}^{K} \rho_k d_{k,t}^{-2\alpha} \cdot \left|g_{k,t}\right|^2} \tag{12}$$

$$\gamma_{\mathrm{B},t} = \rho_{\mathrm{U}} d_{\mathrm{UB},t}^{-2\alpha} \cdot \left|g_{\mathrm{UB},t}\right|^2 \tag{13}$$

where $\rho_{\mathrm{A}} = P_{\mathrm{A}} / \sigma_{\mathrm{A}}^2$, $\rho_k = P_k / \sigma_{\mathrm{A}}^2$ and $\rho_{\mathrm{U}} = P_{\mathrm{U}} / \sigma_{\mathrm{U}}^2$ represents the average SNR. It is observed that the small-scale and large-scale fading coefficients are the main influencing factors of the instantaneous SINR in Eqs. (12) and (13) at time t.

3 Path Planning Method for UAV

The instantaneous received SINR in Eqs. (12) and (13) are constantly changing due to the position change of the UAV and the AP. Therefore, to ensure the transmission reliability of the UAV-based relay communication system with the co-channel interference, it is necessary to plan the flight path of the UAV. Moreover, the outage probability is an important performance indicator to measure the transmission reliability of wireless communication systems. Thus, a path planning method is presented that jointly optimizes the heading and flight speed of the UAV based on the criterion of minimizing the outage probability of the UAV-based relay communication system.

At time t, the outage probability of the system can be expressed as

$$P_{\text{out},t} = 1 - \left(1 - P_{\gamma_{\text{U}},t}\right)\left(1 - P_{\gamma_{\text{B}},t}\right) \tag{14}$$

where $\Pr\{\cdot\}$ denotes the probability, $P_{\gamma_{\text{U}},t}$ is the outage probability of the AP-UAV link, and $P_{\gamma_{\text{B}},t}$ is the outage probability of the UAV-BS link.

3.1 Outage Probability of the AP-UAV Link

At time t, the outage probability of the AP-UAV link can be written as

$$P_{\gamma_{\text{U}},t} = \Pr\left(\gamma_{\text{U},t} \le \gamma_{\text{T}}\right) = F_{\gamma_{\text{U},t}}(\gamma_{\text{T}}) \tag{15}$$

where γ_{T} is a threshold value, $F_{\gamma_{\text{U},t}}(x)$ denotes the cumulative distribution function (CDF) of $\gamma_{\text{U},t}$ [24].

Since $g_{\text{AU},t} \sim \mathcal{CN}(0,1)$ and $g_{k,t} \sim \mathcal{CN}(0,1)$, $\left|g_{\text{AU},t}\right|^2$ and $\left|g_{k,t}\right|^2$ are two independent exponential random variables. Equation (15) can be further expressed as

$$F_{\gamma_{\text{U},t}}(\gamma_T) = 1 - e^{-\frac{\gamma_T}{\rho_{\text{A}} d_{\text{AU},t}^{-2\alpha}}} \sum_{i=1}^{\alpha(\mathbf{R})} \sum_{j=1}^{\tau_i(\mathbf{R})} \chi_{i,j}(\mathbf{R}) \times \left(\frac{\gamma_{\text{T}} R_{[i]}}{\rho_{\text{A}} d_{\text{AU},t}^{-2\alpha}} + 1\right)^{-j} \tag{16}$$

where $\mathbf{R} = diag\left(R_1, R_2, \cdots, R_{k,} \cdots R_K\right)$ while $R_k = \rho_k d_{k,t}^{-2\alpha}$, $\alpha(\mathbf{R})$ is the number of distinct diagonal elements of $\mathbf{R}$, $R_{[1]} > R_{[2]} > \cdots > R_{[\alpha(\mathbf{V})]}$ are distinct diagonal elements in decreasing order, $\tau_i(\mathbf{R})$ represents the multiplicity of $R_{[i]}$, and is the (i, j) - th characteristic coefficient of $\mathbf{R}$.

3.2 Outage Probability of the UAV-BS Link

At time t, the outage probability of the UAV-BS link can be written as

$$P_{\gamma_{\text{B}},t} = \Pr\left(\gamma_{\text{B},t} \le \gamma_{\text{T}}\right) = F_{\gamma_{\text{B},t}}(\gamma_{\text{T}}) \tag{17}$$

where $F_{\gamma_{\text{B},t}}(x)$ denotes the CDF of $\gamma_{\text{B},t}$.

Since $g_{UB,t} \sim \mathcal{CN}(0,1)$, $\left|g_{\text{UB},t}\right|^2$ is an exponential random variable. Hence, the CDF of $\gamma_{\text{B},t}$ is given by

$$F_{\gamma_{\text{B}},t} = 1 - e^{-\frac{\gamma_{\text{T}}}{\rho_{\text{U}} d_{\text{UB},t}^{-2\alpha}}} \tag{18}$$

3.3 UAV Path Planning Mathematical Problem

Substituting Eqs. (16) and (18) into Eq. (14) obtains the following result

$$P_{\mathrm{out},t} = 1 - e^{-\left(\frac{1}{\rho_{\mathrm{A}} d_{\mathrm{AU},t}^{-2\alpha}} + \frac{1}{\rho_{\mathrm{U}} d_{\mathrm{UB},t}^{-2\alpha}}\right)\gamma_{\mathrm{T}}} \times \sum_{i=1}^{\alpha(\mathbf{R})} \sum_{j=1}^{\tau_i(\mathbf{R})} \chi_{i,j}(\mathbf{R}) \left(\frac{\gamma_{\mathrm{T}} R_{[i]}}{\rho_{\mathrm{A}} d_{\mathrm{AU},t}^{-2\alpha}} + 1 \right)^{-j} \tag{19}$$

Equation (19) shows that the large-scale channel fading coefficients $\left(d_{\mathrm{AU},t}, d_{\mathrm{UB},t}, d_{k,t}\right)$ are the main influencing factors of the outage probability of the system at time t, Furthermore, according to the Eqs. (8) and (10), the outage probability $P_{\mathrm{out},t}$ is essentially a function of the heading δ_t and speed $v_{\mathrm{U},t}$ of the UAV.

Based on the criterion of minimizing the outage probability, the UAV's heading and flight speed optimization problem can thus be formulated as

$$\begin{aligned} \{\delta_t^{\mathrm{opt}}, v_{\mathrm{U},t}^{\mathrm{opt}}\} &= \arg \min_{\delta_t, v_{\mathrm{U},t}} P_{\mathrm{out},t} \\ s.t. \quad & \delta_t^{\mathrm{l}} \le \delta_t \le \delta_t^{\mathrm{u}} \\ & v_{\min} \le v_{\mathrm{U},t} \le v_{\max} \end{aligned} \tag{20}$$

where $\delta_t^{\mathrm{l}} = \delta_{t-\Delta t} - \delta_{\max}$, $\delta_t^{\mathrm{u}} = \delta_{t-\Delta t} + \delta_{\max}$, δ_t^{opt} is the optimal heading angle, and $v_{\mathrm{U},t}^{\mathrm{opt}}$ is the optimal flight speed.

3.4 Solution of the Path Planning Problem

Equation (20) is a 2-dimensional nonlinear optimization problem with boundary constraints. The conventional method to solve similar problems requires the gradient of the objective function. However, the expression of $P_{\mathrm{out},t}$ is complex enough that it is difficult to take the required derivative of the UAV's heading or speed to acquire the gradient. To overcome this difficulty, the problem of Eq. (20) is first transformed into an unconstrained optimization problem. Then, the Nelder-Mead simplex algorithm [25] is applied to solve this unconstrained optimization problem. The advantage of this proposed method is that it does not use numerical or analytic gradients.

To transform the problem of Eq. (20) into an unconstrained optimization problem, a method of variable equivalent transformation is introduced. It is assumed that a variable x is constrained as

$$x^{\mathrm{l}} \le x \le x^{\mathrm{u}} \tag{21}$$

where x^{l} is the lower bound of x, and x^{u} is the upper bound of x. By introducing a new variable $y \in R$, the constrained variable x can thus be equivalently transformed as

$$x = x^{\mathrm{l}} + \left(x^{\mathrm{u}} - x^{\mathrm{l}}\right) \cdot (\sin(y) + 1)/2 \tag{22}$$

Note that it can be proven that Eq. (22) satisfies Eq. (21). Applying Eq. (22), the heading and the speed of the UAV can also be equivalently transformed as

$$\delta_t = \delta_t^l + \left(\delta_t^u - \delta_t^l\right) \cdot \left(\sin\left(\hat{\delta}_t\right) + 1\right) \Big/ 2 \tag{23}$$

and

$$v_{U,t} = v_{\min} + (v_{\max} - v_{\min}) \cdot \left(\sin\left(\hat{v}_{U,t}\right) + 1\right)/2 \tag{24}$$

where $\hat{\delta}_t \in R$ and $\hat{v}_{U,t} \in R$. By substituting Eqs. (23) and (24) into Eq. (20), the problem of Eq. (20) can be rewritten as

$$\{\hat{\delta}_t^{\text{opt}}, \hat{v}_{U,t}^{\text{opt}}\} = \arg \min_{\hat{\delta}_t, \hat{v}_{U,t}} \hat{P}_{\text{out},t} \tag{25}$$

where $\hat{P}_{\text{out},t}$ is a function of $\hat{\delta}_t$ and $\hat{v}_{U,t}$. The problem of Eq. (25) can now be solved using the Nelder-Mead simplex algorithm. Finally, substituting $\hat{\delta}_t^{\text{opt}}$ and $\hat{v}_{U,t}^{\text{opt}}$ into Eqs. (22) and (23) respectively, the optimal heading angle and flight speed can be obtained.

4 Numerical Results

4.1 Simulation Settings

This section provides simulation results to validate the proposed path planning method. The simulation system consisted of a fixed ground BS, a mobile UAV, a mobile AP, and one fixed ground interferer. All the nodes in the system had a single antenna, and the path loss exponent was set as $\alpha = 1$ [14], The simulation time of the system was set as 300 s, and the time interval to update the UAV's heading angle was $\Delta t = 1$ s. The simulation parameters for the AP were an initial position of (3500, 1000, 0) m, a speed of 20 m/s, and a transmitting power of $\rho_A = P_A/\sigma_A^2 = 70$ dB. In order to make the simulation more realistic, it was assumed that the AP made a sharp turn with $\delta_{110}/\delta_{109} = -1.8856$. The simulation parameters for the UAV were an initial position of (500, 250, 350) m, and altitude of $h = 350$ m. The speed constraints of the UAV were $(v_{\min}, v_{\max}) = (10, 30)$ m/s, and the maximum heading angle change was set to $\delta_{\max} = 15^{\circ}$. The UAV's transmitting power was also set to $\rho_U = P_U/\sigma_U^2 = 70$ dB. The simulation parameters for the interferer were an initial positions of (3000, 1500, 0) m, and transmitting power of $\rho_1 = P_1/\sigma_A^2 = 75$ dB, 80 dB or 85 dB depending on the case considered.

4.2 Optimal Path and System Performance

Figure 2 shows the optimal path of the UAV with an interferer. The transmitting power of the interferer was 75 dB. The solid blue curve denotes the movement trajectories of the AP. The red and solid curve represents the optimal path of the UAV, which was calculated by the proposed method. The black dashed curve was obtained by using the exhaustive search method. From Fig. 2, it can be seen that the optimal path of the UAV and the exhaustive search method are nearly identical.

Figure 3 shows the optimal path of the UAV in 2 different scenarios. One scenario consists of the system with one interferer, while the other scenario has no interferer. The initial positions of the AP, the UAV, the BS, and the interferer are the same as used in Fig. 2. The black and solid curve denotes the optimal path of the UAV without an

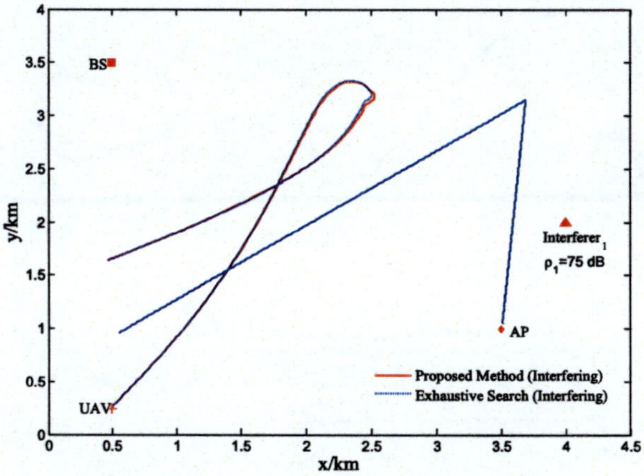

Fig. 2. Optimal path of the UAV with one interferer.

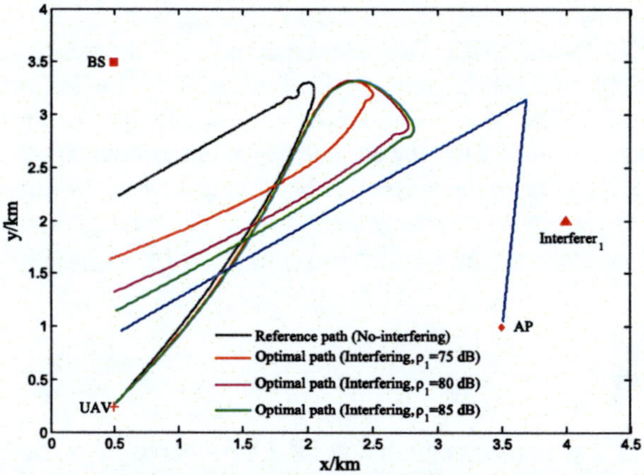

Fig. 3. Comparison of the optimal path of the UAV in 2 different scenarios: the UAV-based relay communication system with and without one interferer.

interferer. For comparison to the optimal path of the UAV with one interferer, it was used as a reference path. The solid curves colored as red, pink, and green were the optimal paths of the UAV with one interferer, and the transmitting power at the interferer was set as 75 dB, 80 dB, or 85 dB. The following results were observed from Fig. 3. Compared to the scenario without an interferer, the optimal path of the UAV was affected by the interference power of the interferer node. The greater the interference power, the closer the optimal relay position of the UAV was to the AP. This can be explained by as the interference power was increased, to improve the transmission reliability of AP-UAV link, the optimal relay position of the UAV moved closer to the AP node.

Figure 4 shows the outage probability curve versus simulation time for the UAV-based relay communication system with and without one interferer. The interfering

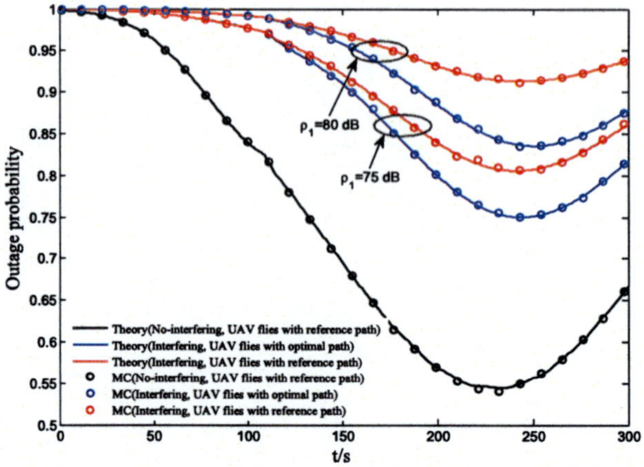

Fig. 4. Comparison of the outage probability performance for the UAV-based relay communication system with and without one interferer.

power was set to 75 dB and 80 dB. The curves marked as 'o' are from Monte Carlo (MC) simulations, and the solid curves were obtained by Eq. (20). The following results were observed from Fig. 4. The Monte Carlo simulation results matched perfectly with the theoretical results, which validates the correctness of the theoretical analysis. Compared to the outage performance of the system without an interferer, the outage performance of the system with one interferer greatly deteriorated. These results show that the outage performance of the system was significantly improved by the proposed optimization method.

5 Conclusion

In this paper, a UAV path planning method for a UAV-based relay communication system with co-channel interference was presented. Simulation results demonstrated the correctness and effectiveness of the proposed method. From the research, the following conclusions were obtained.

1) The outage probability performance of the UAV-based relay communication system was susceptible to co-channel interference.
2) The outage probability performance of the UAV-based relay communication system with co-channel interference was improved by using the proposed path planning method.

References

1. Frew, E.W., Brown, T.X.: Airborne communication networks for small unmanned aircraft systems. Proc. IEEE **96**(12), 2008–2027 (2008)

2. Hayat, S., Yanmaz, E., Muzaffar, R.: Survey on unmanned aerial vehicle networks for civil applications: a communications viewpoint. IEEE Commun. Surv. Tutor. **18**(4), 2624–2661 (2016). Fourthquarter
3. Orfanus, D., de Freitas, E.P., Eliassen, F.: Self-organization as a supporting paradigm for military UAV relay networks. IEEE Commun. Lett. **20**(4), 804–807 (2016)
4. Zeng, Y., Zhang, R., Lim, T.J.: Wireless communications with unmanned aerial vehicles: opportunities and challenges. IEEE Commun. Magaz. **54**(5), 36–42 (2016)
5. Gupta, L., Jain, R., Vaszkun, G.: Survey of important issues in UAV communication networks. IEEE Commun. Surv. Tutor. **18**(2), 1123–1152 (2016). Secondquarter
6. Saleem, Y., Rehmani, M.H., Zeadally, S.: Integration of cognitive radio technology with unmanned aerial vehicles: issues, opportunities, and future research challenges. J. Netw. Comput. Appl. **50**, 15–31 (2015)
7. Willink, T.J., Squires, G.W., Colman, K., Muccio, M.T.: Measurement and characterization of low-altitude air-to-ground MIMO channels. IEEE Trans. Vehicul. Technol. **65**(4), 2637–2648 (2016)
8. Matolak, D.W., Sun, R.: Unmanned aircraft systems: air-ground channel characterization for future applications. IEEE Vehicul. Technol. Magaz. **10**(2), 79–85 (2015)
9. Matolak, D.W., Sun, R.: Air-ground channel characterization for unmanned aircraft systems: the near-urban environment. In: Proceedings of the MILCOM 2015–2015 IEEE Military Communications Conference, Tampa, pp. 1656–1660 (2015)
10. Simunek, M., Fontán, F.P., Pechac, P.: The UAV low elevation propagation channel in urban areas: statistical analysis and time-series generator. IEEE Trans. Antennas Propagat. **61**(7), 3850–3858 (2013)
11. Ono, F., Ochiai, H., Miura, R.: A wireless relay network based on unmanned aircraft system with rate optimization. IEEE Trans. Wirel. Commun. **15**(11), 7699–7708 (2016)
12. Palat, R.C., Annamalau, A., Reed, J.R.: Cooperative relaying for ad-hoc ground networks using swarm UAVs. In: Proceedings of the MILCOM 2005–2005 IEEE Military Communications Conference, Atlantic, vol. 3, pp. 1588–1594 (2005)
13. Zhan, P., Yu, K., Swindlehurst, A.L.: Wireless relay communications with unmanned aerial vehicles: performance and optimization. IEEE Trans. Aerosp. Electron. Syst. **47**(3), 2068–2085 (2011)
14. Jiang, F., Swindlehurst, A.L.: Optimization of UAV heading for the ground-to-air uplink. IEEE J. Select. Areas Commun. **30**(5), 993–1005 (2012)
15. Jin, Y., Zhang, Y.D., Chalise, B.K.: Joint optimization of relay position and power allocation in cooperative broadcast wireless networks. In: Proceedings of the 2012 IEEE International Conference on Acoustics, Speech and Signal Processing (ICASSP), Kyoto, pp. 2493–2496 (2012)
16. Choi, D.H., Jung, B.H., Sung, D.K.: Low-complexity maneuvering control of a UAV-based relay without location information of mobile ground nodes. In: Proceedings of the 2014 IEEE Symposium on Computers and Communications (ISCC), Funchal, pp. 1–6 (2014)
17. Ouyang, J., et al.: Optimization of beamforming and path planning for UAV-assisted wireless relay networks. Chin. J. Aeronaut. **24**(2), 313–320 (2014)
18. Liu, H.T., et al.: Path planning method for UAV relay communication system with space-time block coding. Acta Aeronaut. Astronaut. Sinica **38**(9) (2017)
19. Han, Z., Swindlehurst, A.L., Liu, K.J.R.: Optimization of MANET connectivity via smart deployment/movement of unmanned air vehicles. IEEE Trans. Vehicul. Technol. **58**(7), 3533–3546 (2009)
20. Kim, S., et al.: Coordinated trajectory planning for efficient communication relay using multiple UAVs. Control Eng. Pract. **29**(19), 42–49 (2014)
21. Koulali, S., Sabir, E., Taleb, T., Azizi, M.: A green strategic activity scheduling for UAV networks: a sub-modular game perspective. IEEE Commun. Magaz. **54**(5), 58–64 (2016)

22. Dong, Y., He, C., Wang, Z., Zhang, L.: Radio map assisted path planning for UAV anti-jamming communications. IEEE Signal Process. Lett. **29**, 607–611 (2022)
23. Dubins, L.E.: On curves of minimal length with a constraint on average curvature, and with prescribed initial and terminal positions and tangents. Am. J. Math. **79**(3), 497–516 (1957)
24. Proakis, J.G.: Digital Communications. McGraw-Hill, Columbus (2001)
25. Lagarias, J.C., Reeds, J.A., Wright, M.H., Wright, P.E.: Convergence properties of the Nelder-Mead simplex method in low dimensions. SIAM J. Optimiz. **9**(1), 112–147 (1998)

Failure Analysis of Splitter Based on Grey Correlation

Lianhe Zhang and Chengyu Ju(✉)

Chinese Flight Test Establishment, Yanliang District, Xi'an 710089, China
juchengyu1221@foxmail.com

Abstract. As an important part of the power supply system of civil aircraft, the wire splitter undertakes important functions. This paper uses fault tree analysis to give typical failure modes for the wire splitter, and then uses grey correlation analysis to evaluate the importance of each failure mode. Finally, corresponding suggestions are given for the evaluation results. The method is of great significance to improve product reliability.

Keywords: Line Splitter · Failure Mode · Grey Correlation · Preventive Measure

1 Introduction

The power supply system of civil aircraft is an important system. The function of the power supply system is to provide the electrical energy that meets the predetermined design requirements for each electrical system or equipment of the aircraft, including sufficient capacity and reliable power transmission and distribution, and ensure that the power supply quality meets the requirements. According to the different types of main power supply, the aircraft power supply system has gone through the development process of low-voltage DC, constant speed constant frequency AC, variable speed constant frequency AC, high-voltage DC and variable frequency AC. At present, the internationally advanced main power supply configurations are variable-frequency AC and high-voltage DC. The power supply system has the ability of fault-tolerant power supply and automatic load management. The distributed power distribution system is adopted to realize the nearby transmission, distribution, control, protection and management of aircraft electrical energy through the application of bus and computer control technology [1–3].

As an important part of the aircraft power supply system, the splitter is responsible for the power distribution management function [4], which can achieve one in two and one in three. This paper studies the splitter selected for a civil aircraft.

The splitter is composed of a feedback module and a composite guide rail. The feedback module can realize the internal confluence of 6 to 20 contacts in a small area, and can form multiple combinations of wiring, which improves the product's environmental and vibration resistance characteristics. The termination module uses crimp socket contacts, which can distribute large current to small size contacts with the bus installed inside the module. Figure 1 shows the feedback module [5].

Chinese Society of Aeronautics and Astronautics (Ed.): CASTC 2023, LNME, pp. 253–260, 2024.
https://doi.org/10.1007/978-981-99-8861-7_25

The composite guide rail realizes the installation and fixation of the feedback module. The composite guide rail is lighter than the aluminum alloy guide rail in weight. The guide rail material is resistant to high temperature and corrosion, and the feedback module can be released from the guide rail and taken out with a simple tool [6, 7]. The guide rail has a positive locking device, which can make the module reliably fixed in the guide rail. Figure 2 shows the composite guide rail.

Fig. 1. Feedback module

Fig. 2. Composite guide rail

2 Failure Mode Analysis

According to the composition, main functions, working principle, technical requirements, etc., the failure mode analysis is carried out. After analysis, the main failure modes of the splitter include the following types:

1) The internal contact of the feedback module is poor. As an important component of the splitter, it is required to ensure good contact with the link within the specified time. In the process of use, due to the ambient temperature, improper assembly and connection, etc., it is easy to cause internal looseness and damage, resulting in poor contact, etc., and ultimately resulting in the distributor failure;
2) The internal insulation of the feedback module is poor (leakage, low insulation resistance, etc.). Poor insulation conditions such as internal leakage and low insulation resistance of the feedback module will cause damage to components and lead to the failure of normal use of the splitter;
3) The internal contact of the feedback module is instantaneous. The internal structure of the feedback module, such as spring structure and components, is vulnerable to impact stress such as vibration, which will lead to instant power failure and affect the normal operation of the product;
4) The fixing between the feedback module and the guide rail is loose or loose. The use of the distributor may encounter vibration or other impact stress. If the feedback module and the guide rail are not firmly installed, the contact point, external structure and even the cable of the distributor may be damaged.

A failure tree is established for the failure of the wire splitter, and its failure tree analysis results are shown in Fig. 3.

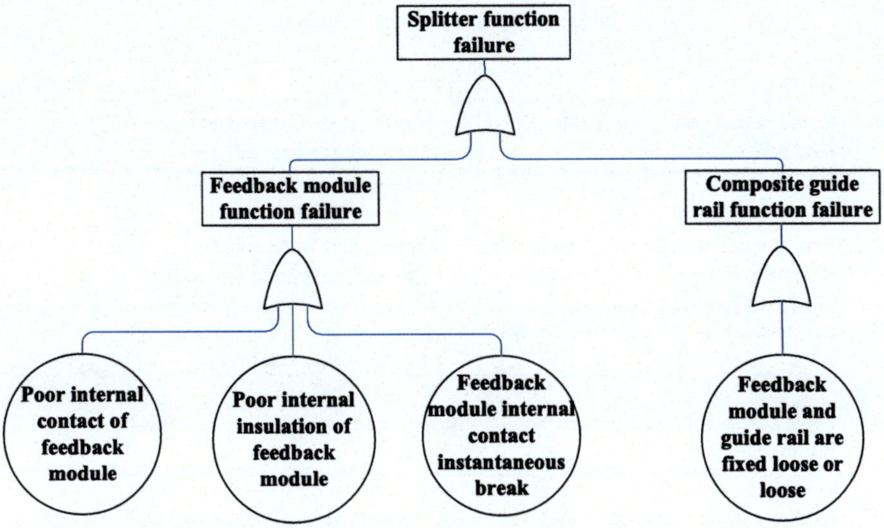

Fig. 3. Fault Tree of Splitter

3 Failure Cause Analysis

By analyzing the causes of the above failure modes one by one, the main causes of failure include the following:

1) The design is unreasonable. Unreasonable design material selection and structural design may easily lead to the inherent deficiency of the product and constitute the hidden danger of product failure;
2) Poor machining accuracy and dimensional consistency. Poor machining accuracy and dimensional consistency can easily lead to redundant objects in the feedback module, resulting in changes in contact resistance, poor internal contact, abnormal heating, ablation, short circuit and other phenomena;
3) There are errors in assembly. There are errors in assembly or unreasonable structure matching. Under the action of environmental stress and other external stresses, it is easy to break, insulation damage and other failures.

4 Failure Mechanism Analysis

According to the above failure modes and possible failure reasons, the failure mechanism analysis is carried out. See Table 1 for specific analysis results.

Table 1. Failure mechanism analysis

No.	Failure Mode	Failure Mechanism	Remarks
1	Poor internal contact of feedback module	1. Dust deposition on contact surface 2. Contact corrosion 3. Contact reed profit relaxation 4. Spark and arc ablation	
2	Poor internal insulation performance of feedback module (leakage, low insulation resistance, etc.)	1. There is dust on the contact surface; 2. Moisture absorption on the contact surface forms ionic conductive channels 3. Material aging and burning carbonization	
3	Feedback module internal contact instantaneous break	Spring structure and components are affected by vibration	
4	Looseness of fixation between feedback module and guide rail	The matching between feedback module and guide rail is not reasonable	

5 Fault Mode Analysis Based on Grey Correlation

The failure mode and effect analysis (FMEA) method has been widely used and recognized in aviation, nuclear energy, chemical and manufacturing industries. The traditional FMEA method uses the risk priority index (RPN) to rank the risks in order to make full use of the limited resources in the prevention and maintenance of the most dangerous failure modes. The RPN value is obtained by multiplying the occurrence probability (O), severity (S) and detection degree (D) of the fault. The value range of the three decision factors is 1 to 10. The higher the RPN value, the more important the failure mode is, and higher priority should be given. See Table 2 for details.

The priority method for evaluating failure modes using grey correlation is as follows:

1) Normalize the index value of the original decision matrix and convert it to a number in the [0,1] interval. Since all indicators are cost type, the normalization formula is:

$$b_{ij} = \frac{1/r_{ij}}{\sqrt{\sum_{i=1}^{n} (1/r_{ij})^2}} (i = 1, 2, \ldots, n; j = 1, 2, 3) \tag{1}$$

Then the normalization matrix can be obtained as follows:

$$U = \begin{bmatrix} u_{11} & u_{12} & u_{13} \\ u_{21} & u_{22} & u_{23} \\ \vdots & \vdots & \vdots \\ u_{n1} & u_{n2} & u_{n3} \end{bmatrix} \tag{2}$$

2) Determine the hazard ideal failure mode xo:

$$x_0 = \max_{1 \leq i \leq n} u_{ij} (j = 1, 2, 3) \tag{3}$$

Table 2. Meaning of fuzzy language evaluation terms

result(RPN = S*O*D)	Severity(S)	Occurrence(O)	Detection degree(D)
Very low	The system operates normally, and the failure has little impact on the system function;	Very low probability of failure	The failure can be basically determined to be detected
low	Some parts of the system operate normally, and the failure will degrade some system functions;	Low probability of failure	High probability of failure being detected
commonly	The system operates abnormally, and the failure will affect the system function;	Failure occurs occasionally	The probability of failure being detected is medium
high	The system operation is seriously abnormal, and the failure has a serious impact on the system function;	High probability of failure	Low probability of failure being detected
Very high	The system stops running, and the failure causes serious damage to the system;	Very high probability of failure	The probability of failure being detected is extremely low

3) Calculation of correlation coefficient:

Take the hazard ideal failure mode xo = [u01, u02, u03] as the comparison standard, and xi = ui1, ui2, ui3 (i = 1,2,…, n) as the comparison set, then

$$\begin{aligned} \Delta_{ij} &= \left|u_{0j} - u_{ij}\right| (i = 1, 2, \ldots, n; j = 1, 2, 3) \\ \Delta_{\max} &= \max_i \max_j \left|u_{0j} - u_{ij}\right|, \\ \Delta_{\min} &= \min_i \min_j \left|u_{0j} - u_{ij}\right| \end{aligned} \tag{4}$$

The calculated correlation coefficient is:

$$\gamma_{ij} = \frac{\Delta_{\min} + \zeta \Delta_{\max}}{\Delta_{ij} + \zeta \Delta_{\max}} (i = 1, 2, \ldots, n; j = 1, 2, 3) \tag{5}$$

where: $\zeta \in [0, 1]$ according to experience, generally $\zeta \leq 0.5$, the correlation coefficient matrix G can be obtained

$$G = \begin{bmatrix} r_{11} & r_{12} & r_{13} \\ r_{21} & r_{22} & r_{23} \\ \vdots & \vdots & \vdots \\ r_{n1} & r_{n2} & r_{n3} \end{bmatrix} \quad (6)$$

4) Calculate the hazard grey correlation degree:

$$\delta = \omega G^T \quad (7)$$

where: $\delta = [\delta_1, \delta_2, \ldots, \delta_n]$ Represents the hazard grey correlation degree matrix, and $\delta = [\omega_1, \omega_2, \omega_3][\gamma_{i1}, \gamma_{i2}, \gamma_{i3}]^T$

5) The hazard of each failure mode is ranked according to the grey correlation degree δ of hazard, and the failure mode corresponding to the smallest δ is the failure mode with the greatest hazard.

Taking the failure mode of the wire splitter in this paper as an example, the evaluation results using the expert scoring method are shown in Table 3.

Table 3. Failure Mode Assessment of Splitter

No	Failure Mode	S(A1)	O(A2)	D(A3)	Rpn
1	Poor internal contact of feedback module	5	6	4	120
2	Poor internal insulation performance of feedback module (leakage, low insulation resistance, etc.)	6	6	3	108
3	Feedback module internal contact instantaneous break	7	4	4	112
4	Looseness of fixation between feedback module and guide rail	8	3	5	120

The harmfulness of these three failure modes is weighed by the traditional RPN method, but the RPN values are relatively close and difficult to compare. The grey correlation is used to evaluate.

1) Establish decision matrix:

$$R = \begin{bmatrix} 5\ 6\ 4 \\ 6\ 6\ 3 \\ 7\ 4\ 4 \\ 8\ 3\ 5 \end{bmatrix} \quad (8)$$

2) Normalize the decision matrix with equation, and obtain

$$U = \begin{bmatrix} 0.6207\ 0.3482\ 0.4757 \\ 0.5173\ 0.3482\ 0.6343 \\ 0.4434\ 0.5222\ 0.4757 \\ 0.3880\ 0.6963\ 0.3806 \end{bmatrix} \quad (9)$$

3) Use formula (3) to determine the hazard ideal failure mode x0:

$$x_0 = \left[0.6207\ 0.6963\ 0.6343\right] \tag{10}$$

4) Determine the correlation coefficient matrix G (where $\zeta = 0.5$):

$$G = \begin{bmatrix} 1 & 0.3333 & 0.5232 \\ 0.5717 & 0.3333 & 1 \\ 0.4954 & 1 & 0.5232 \\ 0.4279 & 0.4999 & 0.4247 \end{bmatrix} \tag{11}$$

5) According to the designer's design intent, preferences and expert opinions, the weight of each index can be determined as $\omega = [0.5,0.3,0.2]$, weigh which failure mode has the greatest hazard. Use formula (7) to calculate the hazard grey correlation degree:

$$\delta = [0.7046, 0.5858, 0.6523, 0.4489] \tag{12}$$

Then the sequence of hazard of failure mode can be obtained: x4 > x2 > x3 > x1, and failure mode 4 is the most hazardous.

6 Preventive Measures and Requirements

According to the above analysis, the relevant preventive measures and suggestions for the line splitter are as follows:

1) Give priority to products that meet the requirements of relevant standards and technical indicators to improve the reliability of products;
2) Implement the assembly process control and operation management, optimize the assembly process, ensure the installation effect, and ensure the fixation and reliability between the feedback module and the guide rail;
3) Strengthen the product acceptance inspection, effectively screen the products whose performance meets the requirements but whose reliability is defective before installation, prevent the defective products from being installed and use, and improve the reliability of the products.

References

1. Bowles, J.B.: An assessment of RPN prioritization in a failure modes effects and criticality analysis. In: Proceeding Annual Reliability and Maintainability Symposium Tampa, pp. 380–386 (2003)
2. Sankar, N.R., Prabhu, B.S.: Modified approach for prioritization of failures in a system failure mode and effects analysis. Int. J. Qual. Reliab. Manag. **18**(3), 324–335 (2001)
3. Pillay, A., Wang, J.: Modified failure mode and effects analysis using approximate reasoning. Reliab. Eng. Syst. Saf. **79**(1), 69–85 (2003)
4. Chang, C.L., Liu, P.H., Wei, C.C.: Failure mode and effect analysis and grey theory. Integr. Manuf. Syst. **4**(12), 211–216 (2001)

5. Dong, Y.: The lower confidence limit of reliability for exponential life data with varying censoring time. AC-TA Math. Appl. Sinica **29**(1), 61–67 (2006)
6. Vidal, L.A., Marle, F., Bocquet, J.C.: Using a Delphi process and the analytic hierarchy process (AHP) to evaluate the complexity of projects. Exp. Syst. Appl. **38**(5), 5388–5405 (2011)
7. Zhang, J., Peng, A.: Decision-making of slicing scheme in fused deposition modeling process based on analytical hierarchical process. Nanjing Univ. Aeronaut. Astronaut. **27**(2), 125–130 (2010)

Quantitative Evaluation Method for Aircraft Maintainability

Chengyu Ju[1(✉)], Na Zhao[2], and Kun Sun[1]

[1] Chinese Flight Test Establishment, Yanliang District, Xi'an 710089, China
juchengyu@139.com
[2] Shaanxi Yuchen Aviation Equipment Co., Ltd., Yanliang District, Xi'an 710089, China

Abstract. Maintainability test and evaluation is of great significance to improve aircraft maintenance efficiency and support level. However, at present, maintainability evaluation involves many qualitative measurement indicators such as accessibility, interchangeability, maintenance human quality engineering, error prevention and maintenance safety. However, qualitative indicators rely too much on human experience, resulting in great subjective differences and difficult to form systematic and objective comments. Aiming at this problem, this paper designs a maintainability verification form, which comprehensively represents the importance of the problem by using the frequency and impact of the problem, so as to reduce the human subjectivity.

Keywords: Maintainability · Quantification · Occurrence Frequency · Influence Degree · Accessibility

1 Introduction

Maintainability is the ability to maintain or restore the product to the specified state under the specified conditions and within the specified time according to the specified procedures and methods. It is a general quality characteristic reflecting the simplicity, convenience, technical requirements and low maintenance cost of the product. Maintainability test and evaluation refers to the work process of testing the design of products at all levels by means of verification, verification and evaluation, supplemented by modeling and simulation demonstration verification methods [1, 2], so as to confirm the compliance of products with Maintainability requirements, so as to ensure that they meet the design requirements.

The flight test stage is the last pass to carry out aircraft maintainability test, which is of great significance to improve the aircraft maintainability level. Large scale tests have been carried out at home and abroad. The logistics support test and evaluation (LT & E) carried out by the U.S. military is an important part of maintainability test and evaluation. Taking F-22 "Raptor" as an example, the U.S. Air Force purchases special test aircraft for 9-week logistics support test and evaluation. During the five months during the test, the logistics support test and evaluation team verified 734 maintenance operations and 686 related maintenance procedures, and completed 44 support equipment operations and

Chinese Society of Aeronautics and Astronautics (Ed.): CASTC 2023, LNME, pp. 261–266, 2024.
https://doi.org/10.1007/978-981-99-8861-7_26

human quality engineering operations [3]. Through the test, the implementation effect of maintenance procedures is improved, unnecessary support equipment is eliminated, the requirements for tools and support steps are reduced, and then the requirements for maintenance manpower are reduced [4, 5].

Therefore, maintainability test and evaluation are of great significance to improve aircraft maintenance efficiency and support level. However, at present, maintainability evaluation involves many qualitative measurement indicators such as accessibility, interchangeability, maintenance human quality engineering, error prevention and maintenance safety [6–8]. However, qualitative indicators rely too much on human experience, resulting in great subjective differences and difficult to form systematic and objective comments. Aiming at this problem, this paper comprehensively represents the importance of the problem by using the occurrence frequency and influence degree of the problem, so as to reduce the human subjectivity [9, 10].

2 The Construction of Maintenance Verification Principle

Aircraft maintainability verification is generally carried out according to the items in Table 1. The following describes accessibility and error prevention in detail.

Table 1. Maintenance verification project

Serial number	Inspection items
1	Safety redundancy measures inspection
2	System alarm prompt function check
3	Inspection for the setting of safety signs on the body surface
4	Error prevention, interchangeability, accessibility, maintenance human engineering inspection
5	Inspection on the configuration of safety protection facilities at dangerous points of maintenance operation
6	Safety warning instructions of random technical data

2.1 Accessibility Verification Principle

The principles of accessibility verification generally include the following contents:

1) The system, equipment and parts of the aircraft shall be reasonably arranged according to the failure frequency, difficulty of adjustment, disassembly time, weight, installation characteristics, etc. as far as possible, so as to avoid large-scale disassembly and unloading during maintenance and troubleshooting;
2) Ensure good accessibility of equipment with high failure frequency and frequent preventive maintenance;
3) Try to arrange the connector and switch at the position with good accessibility, and the equipment test points shall be arranged outside the equipment;
4) Ensure good accessibility of all points requiring lubrication and maintenance.

2.2 Error Proof Verification Principle

The principles of error prevention verification generally include the following contents:

1) The system and equipment shall prevent errors in connection, assembly, installation and cover, so as to ensure that even if there are operation errors, they can be found immediately;
2) Special measures of "misplaced installation" shall be adopted for important equipment or parts;
3) During the installation of left and right parts, the system performance shall not be adversely affected due to the exchange of positions, otherwise obvious marks or error proofing measures shall be set.

2.3 Verification Form Design

There are many aircraft maintainability verification parts. Take the aircraft equipment cabin as an example. Various system equipment are installed in the equipment cabin.

Table 2. Design of verification form for aircraft equipment cabin

NO	e.g. (1)	e.g. (2)
Equipment cabin name	1#equipment cabin	2#equipment cabin
Corresponding cover	XXX	XXX
Equipment name	XXX	XXX
Reasonable pipeline/line layout A yes B no	A	B
Clear equipment identification	A	B
Layout A Centralized layout B Double layer layout C Decentralized layout D Single layer layout	B	A
Fixed A Rollock B Screw fixed C Slot type	B	B
Approach mode A Direct approach B Work ladder required	B	A
Is there any problem threatening personal or equipment safety when disassembling and assembling the equipment A yes B no	B	A
Maintenance convenience A convenient B More convenient C Inconvenient	C	A
Existing problems	XXX	XXX

Therefore, it is necessary to check the rationality of the layout of various equipment/parts and pipelines/lines in the cabin, the installation mode, identification, accessibility, error prevention measures of the equipment, and whether there are problems threatening personal or equipment safety (such as high-voltage electricity and high-voltage gas) when disassembling and assembling the equipment. See Table 2 for the design of typical verification form [11, 12].

3 Quantitative Evaluation of Maintainability Problems

The maintainability problems found in the verification process shall be graded and evaluated according to the impact degree (including major defects, serious defects and general defects) and occurrence frequency (frequent, likely, accidental and rare). Finally, the comprehensive evaluation of the impact index shall be carried out. The problems with impact index 1–3 must be changed.

3.1 Frequency of Occurrence

According to the frequency of occurrence, it is divided into four types: always, frequently, sometimes and Rarely. See Table 3 for specific quantitative classification criteria.

Table 3. Frequency of maintainability problems

Serial number	Frequency of occurrence	Division criteria
1	always	At least once every 5 sorties
2	frequently	At least once every 50 sorties
3	sometimes	At least once every 500 sorties
4	Rarely	At least once every 1000 sorties

3.2 Degree of Influence

According to the degree of influence, it is divided into three categories: major defects, serious defects and general defects. See Table 4 for details.

3.3 Comprehensive Evaluation

The impact index of maintainability problems is graded according to the impact degree and occurrence frequency. See Table 5 for details. From level 1 to level 12, level 1 is the most serious level, and then decreases successively. The acceptance principle is shown in Table 6.

Table 4. Impact degree of maintainability problems

Serial number	Degree of influence	meaning
1	Major defects	It led to a long time of suspension
2	serious defects	The maintenance time is prolonged by 1 * MTTR to 4 * MTTR
3	General defects	Cause the maintenance time to be extended by 0–1 * MTTR

Note: MTTR refers to the average maintenance time of aircraft

Table 5. Determination of problem impact index

Frequency level	Impact level		
	Major defects	Serious defects	General defects
always	1	3	7
frequently	2	5	9
sometimes	4	6	11
Rarely	8	10	12

Table 6. Acceptance principles for maintainability issues of different impact indexes

Impact index	Acceptance principle
1–3	Unacceptable/must be changed
4–8	Dissatisfied/very need to change
9–12	Dissatisfaction/change required

4 Classic Case

During the disassembly and assembly of a civil aircraft engine, it was found that the accessibility of the air starter bleed pipeline was poor and it was difficult to disassemble. Compared with the same type of engine, it took 2 h, which was a serious defect. However, the engine generally needs to be disassembled every 400 sorties, which was accidental. According to Table 4, the comprehensive evaluation result is grade 6, which belongs to the item of maintainability that needs to be changed very much.

5 Conclusion

This paper systematically combs the Verification Principles of maintainability evaluation, constructs the risk parameters and evaluation methods based on quantitative measurement of maintainability problems, reduces the difference of human subjective evaluation, and is of great significance to guide the follow-up maintainability evaluation.

References

1. Zhang, B.: Foreign New Generation Fighter Integrated Support Program Practice. Aviation Industry Press, Beijing
2. Chabal, C., Megard, C., Sibille, L.: EMM-3D: A virtual environment for evaluating maintainability from CAD models (2022)
3. Zhou, D., et al.: Maintenance accessibility evaluation method based on D-H model and comfort in a virtual maintenance environment. Int. J. Adv. Manufact. Technol. **121**(1), 1429–1442 (2022)
4. Wang, R., et al.: Failure criticality evaluation of ship propeller shaft system based on Fuzzy FMECA method. J. Phys. Conf. Ser. **1820**(1), 012117 (2021)
5. Feng, H., Chen, L., Zhao, S.: Simulation model of aggregated equipment combat unit's mission maintainability in the mobile mission. In: International Conference on Frontiers of Manufacturing and Design Science (2012)
6. Yang, Z., et al.: Aircraft survivability tradeoff design considering combat capability. Syst. Eng. Electron. Technol. **36**(1), 90–94 (2014)
7. Hou, M., Li, S., Li, S.: Evaluation of aircraft combat injury emergency repair based on grey correlation analytic hierarchy process. Electro Optic Control **6**, 68–71 (2006)
8. FAR25. Airworthiness Standards: Transport Category Airplanes. FAA, Washington, D.C. (2007)
9. Xiao, C., Li, D.: Reliability research on airborne dual redundancy of electrical wiring interconnection system. In: 2018 11th International Symposium on Computational Intelligence and Design (ISCID), vol. 2. IEEE (2018)
10. Wymore, A.W.: Model-Based Systems Engineering, vol. 3. CRC Press (2018)
11. Lei, Y., Zuo, M.J., Hoseini, M.: Rotating machinery fault detection using EEMD and bispectrum. In: Proceedings of the ASME 2009 International Design Engineering Technical Conferences and Computers and Information in Engineering Conference IDETC/CIE 2009, San Diego, pp. 1–6 (2009)
12. Lu, Z., Sun, Y.: Research on maintainability evaluation model based on fuzzy theory. Chin. J. Aeronaut. **20**, 402–407 (2007)

Aircraft Reliability Evaluation Method Based on Multi-mission Profile

Tao Ma(✉), Chengyu Ju, and Kun Sun

Chinese Flight Test Establishment, Yanliang District, Xi'an 710089, China
2224837594@qq.com

Abstract. Aircraft mission reliability is an important indicator to measure its operational capability, which is of great significance to the aircraft's ability to perform tasks. This paper establishes a mission reliability evaluation method based on operational requirements for the three main factors that affect flight altitude, speed and acceleration, and calculates the flight profiles of multiple mission profiles to obtain the aircraft's mission reliability, At the same time, the influence of the difference of flight profile on the reliability is considered, which has guiding significance for the reliability evaluation of subsequent aircraft missions.

Keywords: Reliability Assessment · Mission Profile · Flight Parameters · Flight Test

1 Introduction

The mission reliability of the aircraft is one of the important evaluation and verification contents for the type design finalization/identification flight test, and is of great significance for the evaluation of equipment operational applicability and operational effectiveness [1, 2]. According to the relevant national military standard, if a special mission profile is arranged for flight test in the design finalization flight test, the evaluation shall be conducted according to formula (1); If a special mission profile is not arranged for flight test in the finalization flight test, but is only evaluated in combination with other mission profiles [4, 5], the evaluation calculation shall be carried out according to formula (2):

$$\hat{P}_{MC} = \frac{S}{S + N} \tag{1}$$

$$\hat{P}_{MC} = e^{-\frac{\Delta t}{T_{BCF}}} \tag{2}$$

where $\hat{P}_{MC}$ is the point estimate of the mission success probability; S is the number of successful tasks; N is the number of task failures; Δt specify the task time; T_{BCF} is the mean time between serious failures.

At present, the special mission profile flight test is not arranged in the model flight test, but is conducted simultaneously in combination with the performance flight test.

Chinese Society of Aeronautics and Astronautics (Ed.): CASTC 2023, LNME, pp. 267–274, 2024.
https://doi.org/10.1007/978-981-99-8861-7_27

Therefore, formula (2) is generally used for calculation. According to the definition of GJB 451A [6], "mission reliability refers to the ability of the product to complete the specified functions within the specified mission profile", "the probability measure of mission reliability is mission reliability". It can be seen that mission reliability is closely related to the mission profile performed by the aircraft. The same aircraft performs different tasks, and its reliability also varies. Therefore, mission reliability modeling and evaluation based on flight profile becomes one of the ways to solve this problem [7, 8]. However, this method does not thoroughly analyze various factors affecting the mission reliability of combat aircraft (including environment, service profile, stress, etc.), which is not conducive to finding the main factors affecting the reliability from the perspective of failure mechanism and taking targeted effective measures. Let the success times X of n independent tests obey the binomial distribution, and the probability of success r times r is [9]:

$$R(X = r) = \binom{n}{r}\theta^{r}(1-\theta)^{n-r} \tag{3}$$

Starting from the definition of mission reliability, this paper analyzes and determines the factors that have a great impact on aircraft mission reliability, such as air pressure, altitude and overload, and on this basis, introduces the reduction coefficient method, and establishes an aircraft mission reliability evaluation method based on flight parameters.

2 Analysis of Factors Affecting Aircraft Mission Reliability

The reliability level of aircraft and aviation equipment is closely related to its use environment. When performing different tasks, the environmental stress of the combat aircraft is different, and the load it bears is also quite different, which may be affected by the stress of temperature, vibration, overload, low pressure and humidity at the same time. In fact, it is the combined effect of these stresses that ultimately affects the reliability of the aircraft. However, in practical engineering, due to the different failure mechanisms caused by different stress loads and the complex coupling effect between various stresses, it is difficult to combine them with aircraft reliability from a physical point of view. In order to simplify the problem so as to achieve engineering feasibility, according to relevant domestic and foreign data and actual engineering experience, several environmental stresses, such as altitude, speed and acceleration, which have a great impact on the reliability of aircraft equipment, can be mainly considered.

From the perspective of flight parameters, it can be seen that the impact of flight altitude, speed and acceleration on the mission reliability of aircraft is represented by air pressure, vibration and overload. These three factors will have a great impact on the internal equipment of the aircraft, thus also affecting the mission reliability. When designing the reliability verification profile, the above three kinds of load conditions of the aircraft during the actual execution of the mission can be simulated, and then the verification profile matching the actual load conditions of the aircraft can be developed.

3 Mission Reliability Assessment Model Based on Use Requirements

3.1 Height and Speed

Height and speed will cause dynamic pressure of airborne products, and dynamic pressure will cause vibration. Note that the dynamic pressure is q(Pa), the height is h(km), and the speed is v(m/s). The relationship between dynamic pressure and height and velocity basically conforms to Bernoulli's theorem under ideal gas state, and its formula is as follows:

$$q = \frac{1}{2}\rho v_{\mathrm{TAS}}^2 \tag{4}$$

where: ρ is the atmospheric density, unit: kg/m^3; v_{TAS} is the real air velocity, unit: m/s.

Various parameters in the atmosphere, such as density, temperature and pressure, change with the altitude, so it is necessary to consider the pressure altitude of flight during calculation. MIL-STD-810F [10] pointed out that for the product air data system, to calculate the real air velocity, it is necessary to accurately know the local atmospheric characteristics. If the instrument used to indicate the speed has corrected the effect of air density on it, the gauge speed is the equivalent air speed at this time, so the dynamic pressure formula of Eq. (4) can be rewritten as follows:

$$q = \frac{1}{2}\rho_n v_{\mathrm{EAS}}^2 \tag{5}$$

where ρ_n is the atmospheric density at sea level, unit:kg/m^3; v_{EAS} is the equivalent air velocity, in. Because ρ_n is fixed, the dynamic pressure is only related to v_{EAS}. GJB899A-2009 specifies calculation products the formula of vibration value w is

$$w = kq^2/22.925 \tag{6}$$

where w is the power spectral density, unit: $\left(\mathrm{m/s^2}\right)^2$/HZ; q is the dynamic pressure, unit: Pa, k is the position coefficient, and its value can be found in Table B.3.5-6A of GJB899A-2009.

Guangbin Yang of Ford Motor pointed out that vibration can be used for accelerated fatigue failure test of mechanical and electronic products, and deduced the relationship between failure time L and vibration as follows:

$$L = \frac{A}{w^B} \tag{7}$$

where w is the power spectral density, unit: $\left(\mathrm{m/s^2}\right)^2$/HZ; A, B is a constant. For aircraft equipment under random vibration, B is usually 3–5.

By substituting Formula (5) and Formula (6) into Formula (7), we can get

$$L = \left(\frac{91.7}{k}\right)^B \frac{A}{\rho_n^{2B} v_{\mathrm{EAS}}^{4B}} \tag{8}$$

For complex systems such as aircraft equipment, it can be assumed that the time between failures follows exponential distribution. At this time, the failure rate of aircraft equipment can be expressed as

$$\lambda = \left(\frac{k}{91.7}\right)^B \frac{\rho_n^{2B} v_{\mathrm{EAS}}^{4B}}{A} \tag{9}$$

where, ρ_n and A is constant; k is the coefficient to be estimated. Let $\beta_1 = \left(\frac{k}{91.7}\right)^B \frac{\rho_n^{2B}}{A}$, formula (9) be simplified as $\lambda = \beta_1 v_{\mathrm{EAS}}^{4B}$,Where β_1 is the coefficient to be estimated.

3.2 Acceleration

The relationship between MTBF and acceleration of aircraft equipment is as follows

$$L = \frac{C}{a^D} \tag{10}$$

Which a represents the longitudinal acceleration of the product. Assuming that the time between failures of aircraft equipment follows exponential distribution, the failure rate can be expressed as

$$\lambda = \frac{a^D}{C} \tag{11}$$

If $\beta_2 = \frac{1}{C}$, $\beta_3 = D$, then formula (11) can be expressed as

$$\lambda = \beta_2 a^{\beta_3} \tag{12}$$

Considering the influence of altitude, speed and acceleration on the failure rate of aircraft equipment, the failure rate model can be obtained by adding the two damage models

$$\lambda = \lambda_1 + \lambda_2 = \beta_1 v_{\mathrm{EAS}}^{4B} + \beta_2 a^{\beta_3} \tag{13}$$

Unlike common fault data, the fault data of aircraft equipment is the fault data accumulated after it executes multiple flight profiles within a given calendar time. It is assumed that within the calendar time t_s, m product has performed many different flight missions. Note that the total flight time t_{jf} of the j product. Within the flight time t_{jf}, the flight parameters are $Y(t) = \left(v_j(t), a_j(t)\right)$, and the relationship between the aircraft equipment failure rate and flight parameters is

$$\lambda_j(t) = \beta_1 v_j(t)^{4B} + \beta_2 a_j(t)^{\beta_3} \quad 0 \le t \le t_{jf} \tag{14}$$

Assuming that the number of failures of j product equipment in the total flight time t_{jf} is f_j, there are

$$\int_0^{t_{j/j}} \lambda_j(t)dt = f_j + \varepsilon_j \quad j = 1, 2, \cdots, m \tag{15}$$

Bring formula (14) into the above formula

$$\int_0^{t_{ff}} \left(\beta_1 v_j(t)^{4B} + \beta_2 a_j(t)^{\beta_3}\right) dt = f_j + \varepsilon_j \quad j = 1, 2, \cdots, m \tag{16}$$

If the flight time, flight parameters $\boldsymbol{Y} = (\boldsymbol{V}, \boldsymbol{A})$ and fault number $\boldsymbol{F}$ of the product are known, when $m > 3$, then the least squares estimation of the unknown parameters can be obtained according to Gauss-Markov theorem:

$$\hat{\boldsymbol{\beta}} = \left(\boldsymbol{H}^{\mathrm{T}}\boldsymbol{H}\right)^{-1}\boldsymbol{H}^{\mathrm{T}}\boldsymbol{F} \tag{17}$$

If $\boldsymbol{\varepsilon}$ the distribution is normal, there is $\boldsymbol{\beta} \approx N\left(\hat{\boldsymbol{\beta}}, \hat{\sigma}^2(\boldsymbol{H}^{\mathrm{T}}\boldsymbol{H})^{-1}\right)$, where $\hat{\sigma}^2 = \frac{\|\boldsymbol{F}-\boldsymbol{H}\hat{\boldsymbol{\beta}}\|_2^2}{(m-3)}$ is the unbiased estimate of $\sigma^2 \cdot \|\boldsymbol{F} - \boldsymbol{H}\hat{\boldsymbol{\beta}}\|_2^2 = \sum_{j=1}^{m}\left(\sum_{i=1}^{n_j}\left(\beta_1 v_{ji}^{4B} + \beta_2 a_{ji}^{\beta_3}\right)\Delta t - f_j\right)^2$ Given the confidence level $1 - \alpha$, the available confidence interval β_i is $\left[\hat{\beta}_i - z_{1-\alpha/2}\hat{\sigma}\sqrt{c_i}, \hat{\beta}_i + z_{1-\alpha/2}\hat{\sigma}\sqrt{c_i}\right]$, where c_i is the ith diagonal element of $\left(\boldsymbol{H}^{\mathrm{T}}\boldsymbol{H}\right)^{-1}$, and $z_{1-\alpha/2}$ is the quantile of the standard normal distribution $1 - \alpha/2$. Given the mission time t and flight parameters $\boldsymbol{Y} = (\boldsymbol{V}, \boldsymbol{A})$ during this period, the mission reliability of the product equipment can be obtained:

$$R(t|\boldsymbol{Y}) = \exp\left(-\sum_{i=1}^{n}\left(\hat{\beta}_1 v_i^{4B} + \hat{\beta}_2 a_i^3\right)\Delta t\right) \tag{18}$$

If $n = 300t$, $D = (d_2, d_3) = \left(\sum_{i=1}^{n} v_i^{4B}\Delta t, \sum_{i=1}^{n} a_i^3 \Delta t\right)$, according to $\hat{\lambda}(t) = d_2\hat{\beta}_1 + d_3\hat{\beta}_2$. $\boldsymbol{\beta} \sim N\left(\hat{\boldsymbol{\beta}}, \hat{\sigma}^2(\boldsymbol{H}^{\mathrm{T}}\boldsymbol{H})^{-1}\right)$ It can be seen that $\lambda(t) \sim \left(\hat{\lambda}(t), \sigma^2 w\right)$. Given the confidence level $1 - \alpha$, the available confidence interval is:

$$\left[\hat{\lambda}(t) - z_{1-\alpha/2}\hat{\sigma}\sqrt{w}, \hat{\lambda}(t) + z_{1-\alpha/2}\hat{\sigma}\sqrt{w}\right] \tag{19}$$

From this, the lower confidence limit of mission reliability of product equipment can be obtained:

$$R_L(t|\boldsymbol{Y}) = \exp\left(-\left(\hat{\lambda}(t) + z_\alpha\hat{\sigma}\sqrt{w}\right)\right) \tag{20}$$

4 Mission Reliability Assessment of a Certain Type of Aircraft

Take the flight parameter monitoring data of an aircraft as an example. Data sampling interval $\Delta t = 1$ s. Because the magnitude difference of each element in the same matrix is too large, it will significantly increase the error. Therefore, in order to reduce this error, the data shall be processed as follows: when B = 3, the speed is divided by 100, and the acceleration remains unchanged; When B = 4, the speed is divided by 100, and the acceleration is multiplied by 10; When B = 5, the speed is divided by 100, and the acceleration is multiplied by 100. The following data can be obtained for subsequent calculation

$$H_{B=3} = \begin{pmatrix} 37296450 & 29853379 \\ 49278485 & 13693141 \\ 994333577 & 25127439 \\ 2975905328 & 28209537 \end{pmatrix}, H_{B=4} = \begin{pmatrix} 562125673 & 29853379000 \\ 810658963 & 13693141000 \\ 234353407693 & 25127439000 \\ 706961576912 & 28209537000 \end{pmatrix},$$
$$H_{B=5} = \begin{pmatrix} 10392318521 & 29853379000000 \\ 17053858735 & 13693141000000 \\ 58231261512140 & 25127439000000 \\ 175024680680470 & 28209537000000 \end{pmatrix}, F = \begin{pmatrix} 0 \\ 0 \\ 1 \\ 2 \end{pmatrix} \tag{21}$$

According to formula (17), when B is equal to 3, 4 and 5, the estimated value is

$$\hat{\beta}_{B-3} = 10^{-9}\begin{pmatrix} 0.662 \\ 3.841 \end{pmatrix}, \hat{\beta}_{B-4} = 10^{-12}\begin{pmatrix} 2.746 \\ 4.836 \end{pmatrix}, \hat{\beta}_{B-5} = 10^{-14}\begin{pmatrix} 1.109 \\ 0.485 \end{pmatrix} \tag{22}$$

Because $\hat{\sigma}^2 = \|\boldsymbol{F} - \boldsymbol{H}\hat{\boldsymbol{\beta}}\|_2^2/(m-3)$, the result is $\sigma_{B=3} = 0.2950, \sigma_{B=4} = 0.2820, \sigma_{B=5} = 0.2790$. It can be obtained $\hat{R}_{B=3} = \hat{R}_{B=4} = \hat{R}_{B=5} = 0.9959$ by substituting the collected flight parameters into formula (19). Similarly, according to $D = (d_2, d_3) = \left(\sum_{i=1}^{n} v_i^{4B}\Delta t, \sum_{i=1}^{n} a_i^3 \Delta t\right)$ it can be calculated from

$$\begin{aligned} D_{B-3} &= (1357644.254, 842945.3613) \\ D_{B-4} &= (10303366.7, 842945361.3) \\ D_{B-5} &= \left(92450257.31, 8.42945 \times 10^{11}\right) \end{aligned} \tag{23}$$

It is solved by substitution $\hat{\lambda}(t) = d_2\hat{\beta}_1 + d_3\hat{\beta}_2$, Calculated $\hat{\lambda}_{B-3} = \hat{\lambda}_{B-4} = \hat{\lambda}_{B-5} = 0.0041$.then $\omega_{B=3} = 7.2826 \times 10^{-4}$, $\omega_{B=4} = 7.2757 \times 10^{-4}$, $\omega_{B=5} = 7.2764 \times 10^{-4}$ According to formula (20), when the confidence level is 0.8, the corresponding confidence interval is as follows:

a) When B = 3, the confidence interval of is;
b) When B = 4, the confidence interval is;
c) When B = 5, the confidence interval is.

According to formula (20), when the confidence level is 0.8, the lower limit of reliability is respectively

$$R_{L(B-3)} = 0.9892; \quad R_{L(B-4)} = 0.9895; \quad R_{L(B-5)} = 0.9896 \tag{24}$$

See Table 1, Table 2 and Table 3 for specific calculation results.

Table 1. Point estimation of failure rate * time and reliability based on flight parameter

Value of B	Failure rate * time	Reliability
3	0.0041	0.9959
4	0.0041	0.9959
5	0.0041	0.9959

Table 2. Interval estimation of failure rate * time based on flight parameter method

Confidence	B		
	3	4	5
0.9	0.0041 ± 0.0131	0.0041 ± 0.0125	0.0041 ± 0.0124
0.8	0.0041 ± 0.0102	0.0041 ± 0.0097	0.0041 ± 0.0096
0.7	0.0041 ± 0.0083	0.0041 ± 0.0079	0.0041 ± 0.0078
0.6	0.0041 ± 0.0067	0.0041 ± 0.0064	0.0041 ± 0.0063

Table 3. Lower limit of reliability based on flight parameter method

Confidence	B		
	3	4	5
0.9	0.9858	0.9862	0.9864
0.8	0.9892	0.9895	0.9896
0.7	0.9917	0.9919	0.9920
0.6	0.9939	0.9940	0.9940

It can be seen that the mission reliability assessment method based on flight parameters proposed in this paper fundamentally considers the impact of the difference of flight profile on reliability, and analyzes the various factors affecting mission reliability from the perspective of failure mechanism, and can give the inter-area estimation results of mission reliability and the evaluation results of unilateral lower confidence limit.

5 Conclusion

Based on the analysis of many factors that affect the mission reliability of aviation equipment in the flight test phase and combined with the use requirements, this paper determines the altitude, speed and acceleration as the main factors affecting the mission reliability of aircraft, and introduces the conversion coefficient to convert the flight time and fault data under different mission profiles, and establishes the mission reliability evaluation model based on the use requirements flight parameters, It provides a reference for the follow-up research on reliability test verification and virtual simulation technology based on failure mechanism from a microscopic perspective.

References

1. Manual. Aircraft Design Manual Part 20: Reliability and Maintainability Design, pp. 414-418. Aviation Industry Press, Beijing (1999)
2. Feng, Q., et al.: Importance measure-based phased mission reliability and UAV number optimization for swarm. Reliab. Eng. Syst. Safety **223**, 108478 (2022)
3. Sohn, S.Y., Yoon, K.B., Chang, I.S.: Random effects model for the reliability management of modules of a fighter aircraft. Reliab. Eng. Syst. Saf. **91**(4), 433–437 (2006)
4. Xing, L., Meshkat, L., Donohue, S.: An efficient approach for the reliability analysis of phased-mission systems with dependent failures (2022)
5. Liu, Z., et al.: Mission reliability assessment method for battle-plane based on flight profile. J. Beijing Univ. Aronaut. Astronaut. **38**(1), 59–63 (2012). (in Chinese)
6. GJB451A-2005. Reliability Maintainability Supportability Terminology. General Assembly Electronic Information Base Department, Beijing (2009)
7. Hong, D., et al.: Environment referring factor based on proportional hazards model. J. Beijing Univ. Aeronaut. Astronaut. **36**(4), 443–446 (2010). (in Chinese)
8. Du, H., Junhai, C., Xsiying, H.: Research on task oriented reliability simulation evaluation and maintenance strategy optimization of degraded system. In: Asian Simulation Conference. Springer, Singapore (2022)
9. Zhang, X., Wang, W., Zhang, W.: Mission reliability analysis of flexible manufacturing cells considering component reliability, task demand, and product quality. Int. J. Adv. Manufact. Technol. **124**(11), 4403–4414 (2023)
10. MIL-STD-810F. Test method standard for environmental engineering considerations and laboratory tests

Experimental Investigation of Sound Absorption Properties of the Damped Sonic Black Holes

Lihua Peng and Qibo Mao(✉)

School of Aircraft Engineering, Nanchang Hangkong University, Nanchang 330063, China
qbmao@nchu.edu.cn

Abstract. The Sonic Black Holes (SBH) absorber consists a cylindrical cavity with a set of rigid thin rings. The inner diameters of the rings are gradually decreasing. The sound speed in SBH can be progressively slowed down, then the effective sound absorption performance can be obtained. In order to further improve the acoustic performance of the SBH, the sponges are used as sound-absorbing material in SBH. The sponges of 2 mm thickness are arranged at the bottom and the different ranges of the inner wall of the SBH. The absorption coefficients of the SBH before and after the sponge arrangement are experimentally investigated. The experimental results reveal that the SBH structure is different from the traditional vibration-based acoustic black hole (ABH) structure. The sound absorption performance cannot be improved effectively when the sponge is placed at the bottom of the SBH. However, when the sponge is placed on the inner wall around the bottom of the SBH, the sound absorption performance can be improved markedly.

Keywords: Sonic black hole · Sound absorbing material · Damping distribution · Absorption coefficient

1 Introduction

With the rapid development of the aviation industry, noise pollution in the aviation field is becoming more and more serious, and aircraft noise has also attracted more and more attention. In the past decade, acoustic black hole (ABH) structure has attracted increasing attention for passive aircraft vibration control. Mironov [1] and Krylov [2] first proposed the concept of one-dimensional ABH, a beam which thickness decreases according to $h(x) = \varepsilon x^n$ $(n \geq 2)$ and the bending waves propagating over the structure will be concentrated at the tip of the beam. It is noted that ABH as a new wave manipulation technology that can achieve the aggregation and dissipation of wave energy through gradient changes in structural size [3, 4]. Further research found that the ABH structure can effectively achieve energy convergence even if there is a certain thickness at the end of the structure and the efficient consumption of wave energy can be achieved by placing the damping material on the end of the structure [5–7].

In recent years, the concept of ABH has begun to be expanded and applied in the field of noise control. The sonic black hole (SBH) structure was proposed and theoretically and experimentally demonstrated it has the sound insulation effect of low-frequency

Chinese Society of Aeronautics and Astronautics (Ed.): CASTC 2023, LNME, pp. 275–281, 2024.
https://doi.org/10.1007/978-981-99-8861-7_28

broadband by experiments [8, 9]. It is noted that the damping material arranged at the end of the ABH can obtain the vibration reduction of the structure. In recent years, scholars have found that the SBH structure is different from the traditional ABH structure for vibration control. The sound absorption performance cannot be improved effectively when the sponge is placed at the bottom of the SBH [10–12].

Notice that previous studies have mainly focused on the effect of the changes of SBH size on sound absorption performance [13, 14]. In this paper, the sponges are used as sound-absorbing material. The influence of the arrangement position of the sponge on the sound absorption performance in the SBH is studied experimentally and provide the guidance for practical engineering applications.

2 Theoretical Analysis

The structure of SBH used in this paper is shown in Fig. 1.

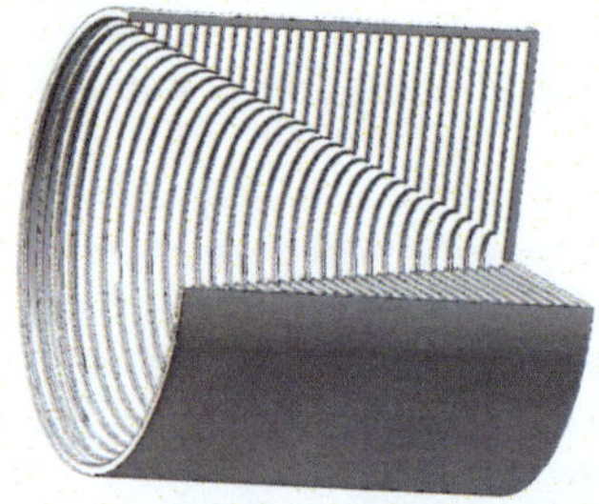

(a) Three dimensional images of SBH

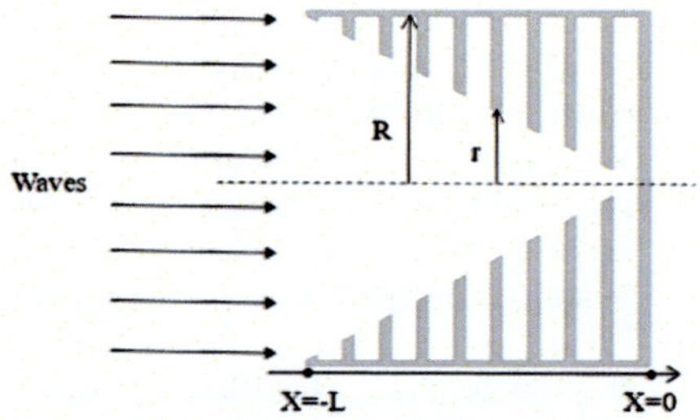

(b) Two dimensional images of SBH

Fig. 1. Conceptual view of the SBH structure

The sound waves is incoming the SBH structure at x = −L in Fig. 1(b). Based on linearized Euler's formula and continuity condition, the sound wave propagation in the SBH can be expressed.

$$\frac{1}{c^2}\frac{\partial^2 p(x,t)}{\partial t^2} + \frac{2Y\rho_0}{r}\frac{\partial p(x,t)}{\partial t} = \frac{\partial^2 p(x,t)}{\partial x^2} + \frac{\partial p(x,t)}{\partial x}(\ln S)'_X \tag{1}$$

where $S = \pi r^2$ is the area of the hole of the ring, r is the waveguide radius, ρ_0 is the density of air, c is the speed of sound, x is the coordinate of the corresponding positions of the structure.

Assuming that the relationship between sound pressure and time is given as $p(x,t) = p(x)e^{-j\omega t}$. The radius of the inner rings in the SBH exhibit a linearly decay, such as

$$r(x) = -\frac{R}{L}x \tag{2}$$

where R is the radius of SBH, L is the length of SBH.

The wall admittance of the SBH can be obtained from Ref. [13].

$$Y(x) = -j\frac{k_0}{Z_0}\frac{R^2 - r^2}{2r} \tag{3}$$

where $Z_0 = \rho_0 c$ is the characteristic impedance of air and $k_0 = \omega/c$ is the wavenumber.

Substituting (2) and (3) into (1), we get

$$\frac{\partial^2 p(x)}{\partial x^2} + \frac{2}{x}\frac{\partial p(x)}{\partial x} + \frac{k_0^2 L^2}{x^2} p(x) = 0 \tag{4}$$

The general solution to (4) is given by

$$p(x) = c_+ e^{\frac{1}{2}\left[-1+\sqrt{1-(2k_0L)^2}\right]\ln x} + c\_ e^{\frac{1}{2}\left[-1-\sqrt{1-(2k_0L)^2}\right]\ln x} \tag{5}$$

where C_+ and C_- stand for real constants.

The local wavenumber $k(x)$ can be obtained from (5)

$$k(x) = \frac{1}{2x}\sqrt{(2k_0L)^2 - 1} \tag{6}$$

The group velocity of the sound waves in the structure is

$$c_g = \left(\frac{\partial k}{\partial \omega}\right)^{-1} = \frac{c_0^2 x}{2L^2\omega}\sqrt{(2k_0L)^2 - 1} \tag{7}$$

From (7) and Fig. 1(b), it can be found that the time for the sound wave from the inlet of the ABH to arrive at the ABH end-wall can be expressed as

$$T = \lim_{l \to 0+} \int_{-L}^{-l} \frac{dx}{c_g} = \lim_{l \to 0+} \frac{L^2\omega}{2c_0^2\sqrt{(2k_0L)^2 - 1}} \ln\left(\frac{L}{l}\right) \to \infty \tag{8}$$

From the above analysis, it can be found that, ideally, it takes infinite time for the sound waves to propagate from x = −L to x = 0 as shown in Fig. 1. Similar to the one-dimensional ABH structure that realizes the vibration reduction, the structure shown in Fig. 1 can be called SBH.

3 Experimental and Discussion

According to the analysis in Sect. 2, ideally, SBH has excellent sound absorption performance. But in the actual manufacturing process, the thickness of rings can-not be equal to the ideal state of zero, and the incoming sound waves can-not be fully absorbed by SBH. Therefore, sponge is arranged in SBH as a sound absorbing material to improve the sound absorption performance of SBH.

The SBH model shown in Fig. 1(a) is fabricated by 3D printing. Figure 2 shows the picture of the sample. The length and radius of the sample are 94 mm and 49.5 mm, respectively. There are 30 rings with thickness 0.7 mm in sample, the distance between the neighbouring rings is 3 mm. The experimental model of SBH as shown in Fig. 2 is obtained by 3D printing.The sound absorption coefficient of the SBH model is measured by impedance tube as shown in Fig. 3. For comparison, the numerical finite element results calculated by COMSOL are also presented in Fig. 4. Both the numerical results and the experimental results show the SBH has a good broadband sound absorption performance.

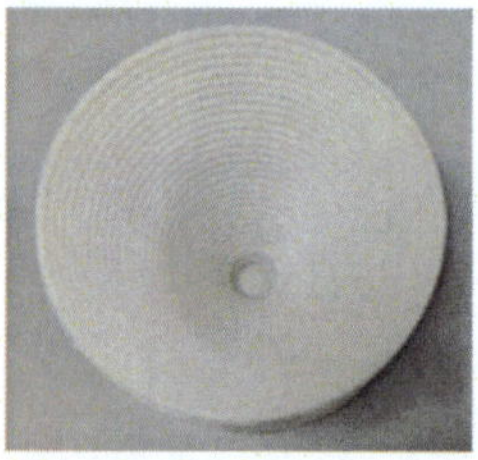

Fig. 2. 3D-printed SBH model

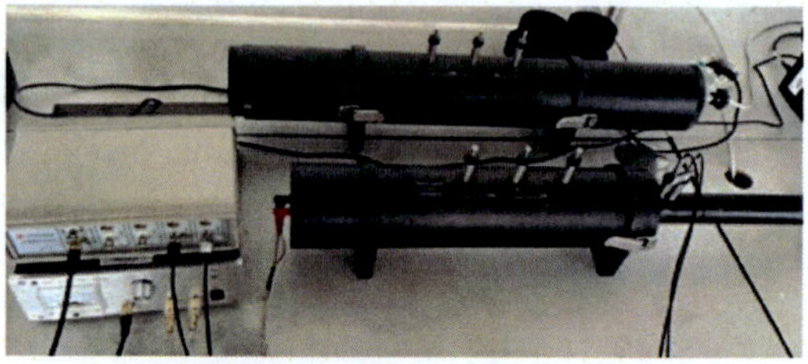

Fig. 3. Experimental equipment

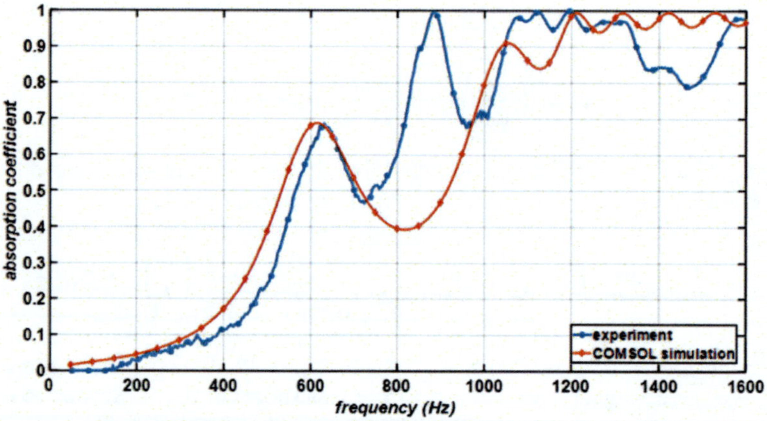

Fig. 4. Comparison between measured absorption coefficient of SBH and simulation calculation

The sound pressure level distributions of SBH at 200 Hz, 600 Hz, 1100 Hz, and 1600 Hz are shown in Fig. 5. Clearly, the sound pressure level in SBH keeps the same at 200 Hz, it means that no ABH effect at this frequency. However, the sound pressure level at the bottom of SBH has been significantly increased at 600 Hz, 1100 Hz and 1600 Hz.

In order to further enhance the SBH sound absorption characteristic, sponges with 2 mm thickness are glued on the bottom and the different ranges of the inner wall of the SBH, as presented in Fig. 6. The heights of the glued sponge are the ABH sample height, 2/3 of the SBH height or 1/3 of the SBH height, respectively.

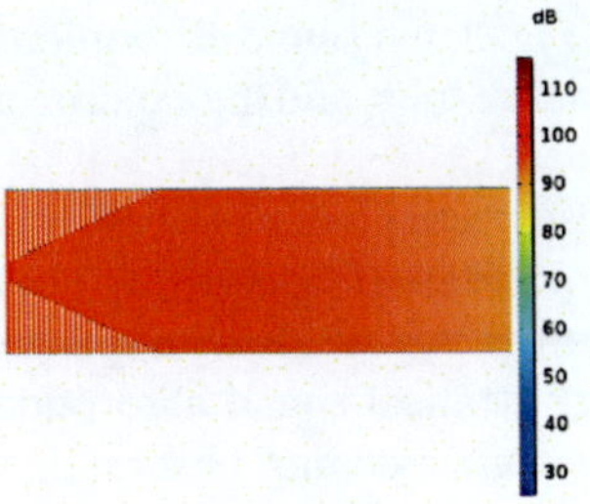

(a) 200Hz

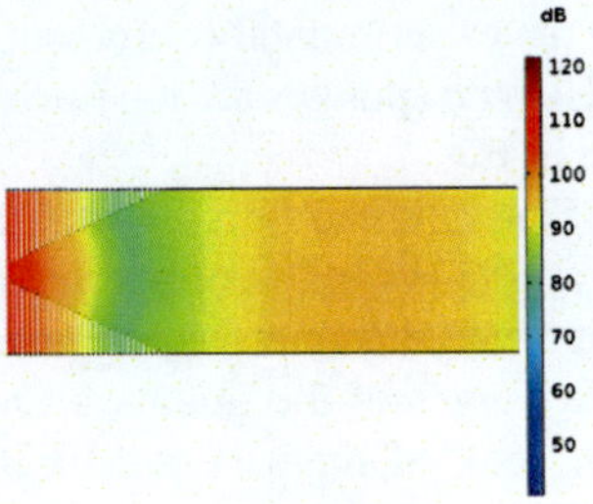

(B)600Hz

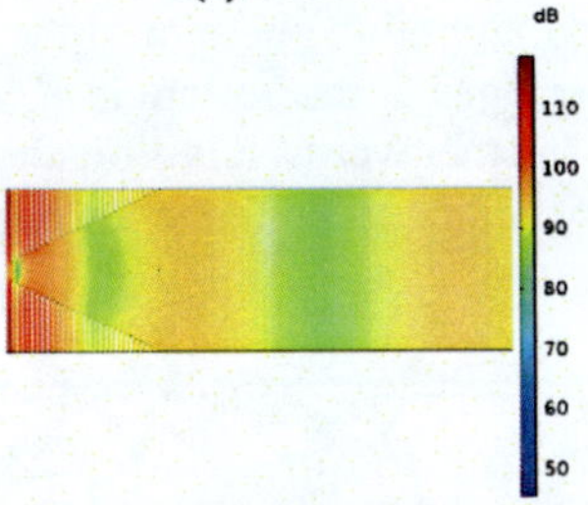

(c)1100Hz

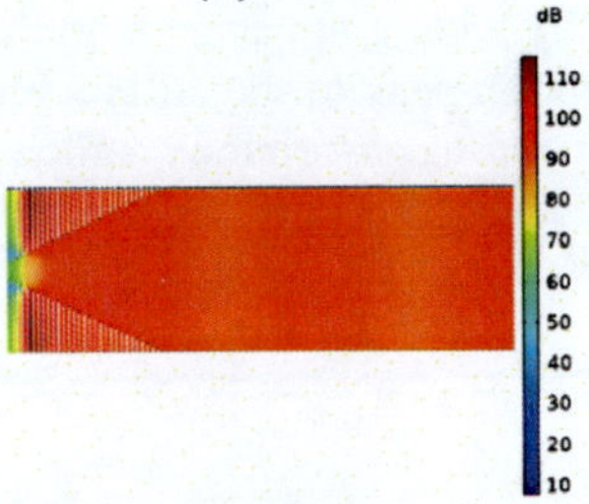

(d)1600Hz

Fig. 5. The sound pressure level distribution at different frequencies of the SBH

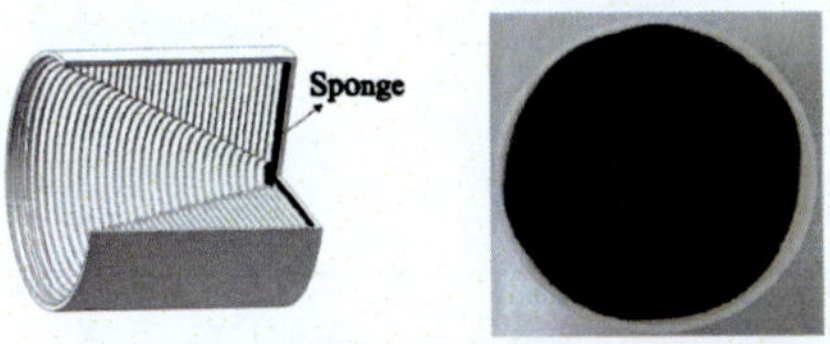

(a) The sponge is arranged at the bottom of the SBH

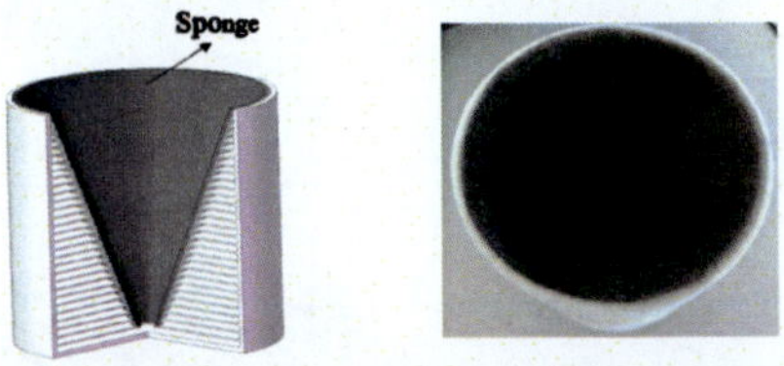

(b) The sponge with the same height as the structure is arranged on the inner wall of the SBH

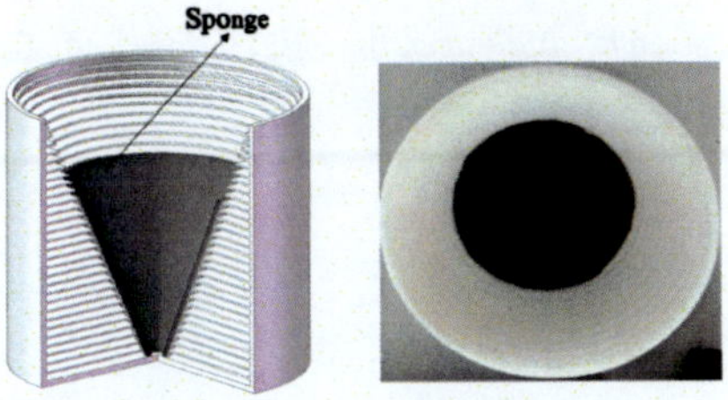

(c) The sponge with 2/3 of the structure height is arranged on the inner wall of the SBH

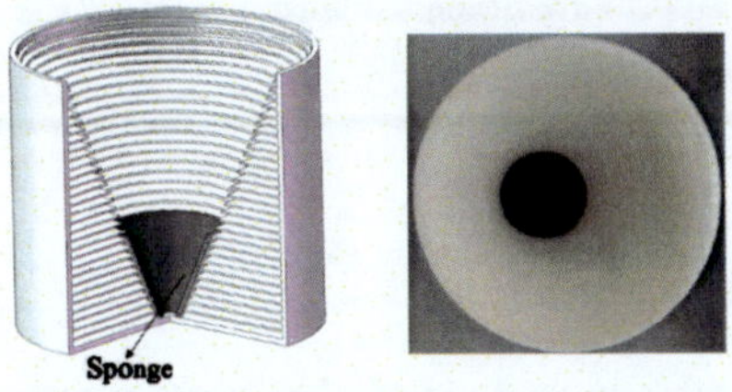

(d) The sponge with 1/3 of the structure height is arranged on the inner wall of the SBH

Fig. 6. The sponges of 2 mm thickness are arranged at the bottom and the inner wall of the SBH

Using the experimental equipment shown in Fig. 3, the sound absorption coefficient of the SBHs with sponges are measured by impedance tube, and the experimental results are shown in Fig. 7.

From Fig. 7, it can be seen that the first absorption peak of the SBH without the sponge is at 633 Hz and the corresponding absorption coefficient is 0.7. The absorption coefficient of ABH is improved by arranging sponges. All ABHs with sponges arranged at different ranges of the inner wall have almost identical sound absorption. The first absorption peak of them is at 610 Hz and the corresponding absorption coefficient is 0.92. Obviously, the sound absorption performance of the SBH can be effectively increased by adding the sponge on the inner wall around the bottom of the SBH. The first absorption peak of the ABH with sponge arranged at the bottom is at 648 Hz and the corresponding absorption coefficient is 0.86. In other words, it is difficult to improve the sound absorption performance of the SBH by adding a sponge to the bottom of the ABH.

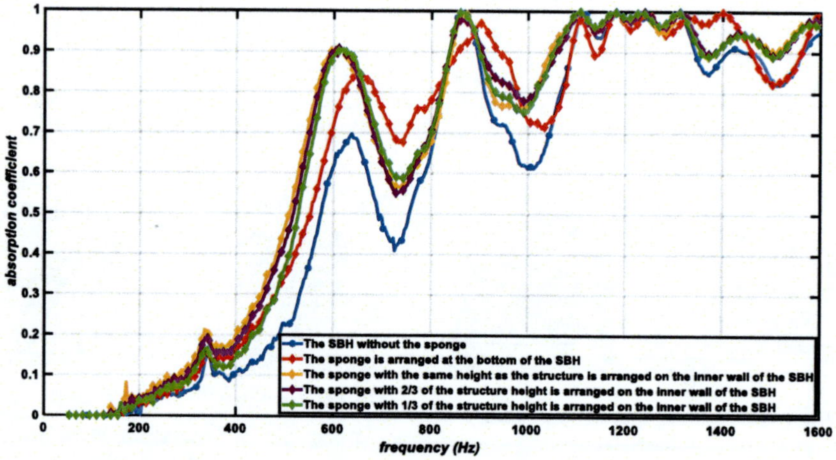

Fig. 7. Experimental measurements of the absorption coefficient of the SBH before and after the sponge arrangement

4 Conclusions

In this study, the sound absorption performance of SBH is demonstrated based on experiments and COMSOL simulation. In order to further improve the sound absorption performance of the SBH, arranging the sponges in the bottom and different ranges of the inner wall of the ABH we designed. The influence of the arrangement position of the sponge on the sound absorption performance in the SBH is studied experimentally. Comparing to the traditional SBH, the first absorption peak of the experimental absorption coefficient for the SBHs with sponges on the inner wall can be increased by 0.22. Furthermore, it is found that the absorption frequency band can be lowered 23 Hz. However, if the sponge is placed on the bottom of the SBH, the first absorption peak is only improved about 0.16. It means that the sound absorption performance of the SBH can be effectively increased by adding the sponge on the inner wall around the SBH bottom. It is surprised that the sound absorption can not be effectively improved by adding sponge on the SBH bottom. It means that optimizing the sponge location in the SBH may achieve a better sound absorption effect, further theoretical and experimental investigations for SBH with damping material should be carried out to achieve higher values of sound absorption coefficients.

Acknowledgement. We would like to express our gratitude to the funding support from the National Natural Science Foundation of China (No. 12364058 and No. 51975266) and the Aeronautical Science Foundation of China (No. 2020Z073056001).

References

1. Mironov, M.A.: Propagation of a flexural wave in a plate whose thickness decreases smoothly to zero in a finite interval. Soviet Phys. Acoust. **34**(3), 318–319 (1988)
2. Krylov, V.V.: Conditions for validity of the geometrical-acoustics approximation in application to waves in an acute-angle solid wedge. Soviet Phys. Acoust. **35**, 176–180 (1989)
3. Huang, W., et al.: Energy focusing effect of two-dimensional acoustic black hole on flexural waves. J. Vib. Shock **36**(09), 51–57+92 (2017)
4. Gao, N.S., et al.: Progress and applications of acoustic black holes. Sci. Bull. **67**(12), 1203–1213 (2022)
5. Krylov, V.V., Tilman, F.J.B.S.: Acoustic 'black holes' for flexural waves as effective vibration dampers. J. Sound Vib. **274**, 605–619 (2004)
6. Bowyer, E.P., et al.: Effect of geometrical and material imperfections on damping flexural vibrations in plates with attached wedges of power law profile. Appl. Acoust. **73**, 514–523 (2012)
7. Kralovic, V., Krylov, V.: Damping of flexural vibrations in tapered rods of power-law profile: experimental studies. Appl. Acoust. **76**, 359–365 (2014)
8. Mironov, M.A., Pislyakov, V.V.: One-dimensional acoustic waves in retarding structures with propagation velocity tending to zero. Acoust. Phys. **48**(3), 347–352 (2002)
9. Mi, Y.Z., et al.: Wave trapping by acoustic black hole: simultaneous reduction of sound reflection and transmission. Appl. Phys. Lett. **118**(11), 114101 (2021)
10. Umnova, O., et al.: Multiple resonances in lossy acoustic black holes-theory and experiment. J. Sound Vib. **543**, 117377 (2023)
11. Červenka, M., Bednařík, M.: On the role of resonance and thermoviscous losses in an implementation of "'acoustic black hole'" for sound absorption in air. Wave Motion **114**, 103039 (2022)
12. Mousavi, A., Berggren, M., Wadbro, E.: How the waveguide acoustic black hole works: a study of possible damping mechanisms. Acoust. Soc. Am. **151**(6), 4279–4290 (2022)
13. Guasch, O., Arnela, M., Sánchez-Martín, P.: Transfer matrices to characterize linear and quadratic acoustic black holes in duct terminations. J. Sound Vib. **395**, 65–79 (2017)
14. Ouahabi, A.A.E., Krylov, V.V., O'Boy, D.: Experimental investigation of the acoustic black hole for sound absorption in air. In: The 22nd International Congress on Sound and Vibration, Florence (2015)

Aerodynamic Modeling and Analysis of Receiver Under the Disturbance of Tanker's Wake Flow

Zhou Wangyi[1(✉)], Zang Xiheng[1], Qiao Lei[2], Jiang Chao[1], and Zhu Xueyao[1]

[1] AVIC Xi'an Flight Automatic Control Research Institute, Xi'an 710065, China
1208210734@qq.com

[2] School of Aerodynamics, Northwestern Polytechnical University, Xi'an 710072, China

Abstract. Aiming at the wing-mounted and hose-drogue refueling platform of tanker, the disturbance of the wake flow field of tanker on the aerodynamic characteristics of the receiver is studied. Based on the patched mesh technology, the mesh of the tanker, the hose-drogue and the receiver are generated respectively and then patched together to reduce the difficulty of mesh generation. By solving the Reynold Averaged Navier-Stokes (RANS) equation, the numerical simulation of the entire flow field of refueling is carried out on the receiver under the influence of the wake flow field of tanker and its engine jet, and the increments of the aerodynamic force and moment coefficient along the flow wise, span wise and height direction of the receiver is obtained. The influence of the wake of tanker on the trim control of the receiver is studied. The finite element method (FEM) is adopted to analyze the aerodynamic loading of the hose-drogue assembly, and the stress and deformation of each element are obtained. Based on the numerical simulation technology of the coupling solution of CFD and FEM, the equilibrium position of the hose-drogue assembly under the influence of the wake field of the tanker is calculated.

Keywords: Aerial Refueling · Wake Flow of Tanker · Aerodynamic Characteristics of Receiver · Hose-Drogue Assembly

1 Introduction

Aerial refueling technology can effectively increase the range of the receiver, enhance the combat radius, and prolong the flight time, thereby greatly improving the combat effectiveness of the receiver [1]. When the tanker flies, the surrounding air is strongly disturbed, mainly caused by the jet flow of the engine of tanker, the turbulent flow of boundary layer and tail vortex of the wing. The influence of the engine jet is small, and the influence of the oil tanker tail vortex on the oil engine is more significant [2]. The unsteady, nonlinear and strongly disturbed wake of the tanker changes the attitude, position and stability of the receiver, which in turn affects the success rate of aerial refueling and the flight safety of the receiver.

Some research has been carried out on the modeling of the wake flow field of the tanker and hose-drogue assembly at home and abroad. Jewell [3] and Rossow [4] first proposed a mathematical model of the wake flow field of the tanker. Bloy [5] uses

Chinese Society of Aeronautics and Astronautics (Ed.): CASTC 2023, LNME, pp. 282–292, 2024.
https://doi.org/10.1007/978-981-99-8861-7_29

Horseshoe Vortex Model (HVM), Vortex Lattice Model (VLM), Wake Vortex Sheets (FVS) and Roll-up Vortex Models (RVM) to set up the wake flow field model of the tanker and analyze its influence on the receiver. Liu Jiaolong [6] combined the Computational Fluid Dynamics(CFD) method and the equivalent aerodynamic effect method to conduct a modeling study of the aerodynamic effects of the receiver under the disturbance of the wake. Jackson [7] adopted the Euler equation to simulate the refueling process of the KC-135 tanker and the Learjet business jet, and analyzed the effect of the wake on the lift and drag characteristics of the receiver. YUE Kuizhi [8] simulated the wake and engine jet flow of the tanker by solving the NS equation, and analyzed the characteristics of the wake flow field, but didn't analyze its influence on the tanker. Wang Wei [9] established a variable-length hose-drogue model, and analyzed its stability and dynamic characteristics. Zhu & Meguid [10] considered large torsion and deformation of the hose based on the elastic beam theory. Kamman [11] applied the multi-body dynamics theory to divide the hose into multiple finite segments, and each segment of the hose is connected by hinges to realize the modeling of the hose.

In this paper, based on the wing-mounted and hose-drogue refueling platform, the wake flow field and the engine jet of the tanker are numerically simulated by solving the N-S equation, and the changes of aerodynamic force and torque of the receiver at different flight positions under the disturbance of the wake flow are analyzed. Comparing the trimmed states of the receiver with the wake flow and without the wake provides a basis for the design of the control law of the receiver. Starting from the analysis of aerodynamic loading of the hose, the hose-drogue assembly is discretized by FEM, and the deformation and displacement of each element are solved to calculate and analyze the equilibrium position of the overall hose-drogue assembly in wake flow of tanker.

2 Numerical Method

2.1 CFD Numerical Simulation Method

As shown in Eq. (1), the governing equation used in this work is the integral form of three dimensional unsteady Navier-Stokes equation [12].

$$\frac{\partial}{\partial t}\iiint_{\Omega} \boldsymbol{U} dV + \iint_{\partial\Omega} \boldsymbol{F}(\boldsymbol{U})\cdot \boldsymbol{n} dS = \iint_{\partial\Omega} \boldsymbol{G}(\boldsymbol{U})\cdot \boldsymbol{n} dS \tag{1}$$

where $\boldsymbol{U} = (\rho, \rho u, \rho v, \rho w, \rho e)^T$ is the conservative variable, $\mathbf{F}(\mathbf{U})$ and $\mathbf{G}(\mathbf{U})$ refer to inviscid flux and viscous flux respectively, Ω is the control volume, $\partial\Omega$ is the boundary of control volume, dS refers to the surface integral element and $\mathbf{n}$ is the outer normal vector of grid element boundary.

The turbulence model used in this work is $k-\omega$ SST model [13] which shown in Eq. (2) and Eq. (3).

$$\frac{\partial(\rho k)}{\partial t} + u_i\frac{\partial(\rho k)}{\partial x_i} = P_k - \beta_k \rho k\omega + \frac{\partial}{\partial x_i}\left[\left(\mu_l + \frac{\mu_t}{\sigma_k}\right)\frac{\partial k}{\partial x_i}\right] \tag{2}$$

$$\frac{\partial(\rho\omega)}{\partial t} + \mu_i\frac{\partial(\rho\omega)}{\partial x_i} = C_\omega P_\omega - \beta_\omega \rho\omega^2 + \frac{\partial}{\partial x_i}\left[\left(\mu_l + \frac{\mu_t}{\sigma_\omega}\right)\frac{\partial \omega}{\partial x_i}\right] + 2\rho(1-F_1)\frac{1}{\sigma_{\omega 2}}\frac{1}{\omega}\frac{\partial k}{\partial x_i}\frac{\partial \omega}{\partial x_i} \tag{3}$$

where μ_t is the eddy viscous coefficient and can be evaluated by Eq. (4). P_k and P_ω refer to turbulence generation term and can be calculated by Eq. (5) and Eq. (6).

$$\mu_t = \min\left[\frac{\rho k}{\omega}, \frac{a_1 \rho k}{\Omega F_2}\right] \tag{4}$$

$$P_k = \mu_t\left[\left(\frac{\partial u_i}{\partial x_j}\right)\left(\frac{\partial u_i}{\partial x_j} + \frac{\partial u_j}{\partial x_i}\right) - \frac{2}{3}\left(\frac{\partial u_k}{\partial x_k}\right)^2\right] \tag{5}$$

$$P_\omega = \rho\left[\left(\frac{\partial u_i}{\partial x_j}\right)\left(\frac{\partial u_i}{\partial x_j} + \frac{\partial u_j}{\partial x_i}\right) - \frac{2}{3}\left(\frac{\partial u_k}{\partial x_k}\right)^2\right] \tag{6}$$

2.2 Engine Jet Flow Numerical Method

The principle to solve the fluid-structure coupling problem is to satisfy the coupling of liquid and solid by the kinematic equilibrium equation and dynamic equilibrium equation, respectively:

$$\boldsymbol{v}_f \cdot \boldsymbol{n}_f = -\boldsymbol{v}_s \cdot \boldsymbol{n}_s \tag{7}$$

$$\boldsymbol{P}_f \cdot \boldsymbol{n}_f = \boldsymbol{\sigma}_s \cdot \boldsymbol{n}_s \tag{8}$$

In the formula, $\boldsymbol{v}_f$ and $\boldsymbol{v}_s$ represent the velocity vectors of the fluid and the solid; $\boldsymbol{n}_f$ and $\boldsymbol{n}_s$ represent the unit outer normal vector of the fluid and solid boundary, and the directions of the two are opposite; $\boldsymbol{P}_f$ and $\boldsymbol{\sigma}_s$ represent the stress tensors of the fluid and the solid.

The displacement of the node on the fluid-structure interaction interface is determined by the kinematic coupling condition. By solving the N-S equations in arbitrary Lagrange-Euler (ALE) coordinates, the flow field information such as pressure and velocity at the fluid nodes can be obtained. Integrating the pressure at the fluid nodes at the coupled interface can solve the force acting on the solid for structural kinematics and dynamic variables. Figure 1 shows a schematic diagram of the fluid-structure interaction solution process.

2.3 Engine Jet Flow Numerical Method

The state of the far field flow at infinity is determined by the flight state of the tanker, and Ma, ρ, U, p, A and $\dot{M}$ represent Mach number, air density, normal velocity of boundary, static pressure, area and mass flow rate. The subscript *inf* indicates the values at the far field, and the subscript *fan* indicates the value at the inlet of fan.

$$U_{inf}\,\rho_{inf}A_{inf} = U_{fan}\rho_{fan}A_{fan} = \dot{M} \tag{9}$$

$$\left(\frac{A_{inf}}{A_{fan}}\right)^2 = \frac{Ma_{fan}^2}{Ma_{inf}^2}\left(\frac{1+\frac{\gamma-1}{2}Ma_{inf}^2}{1+\frac{\gamma-1}{2}Ma_{fan}^2}\right)^{\frac{\gamma+1}{\gamma-1}} \tag{10}$$

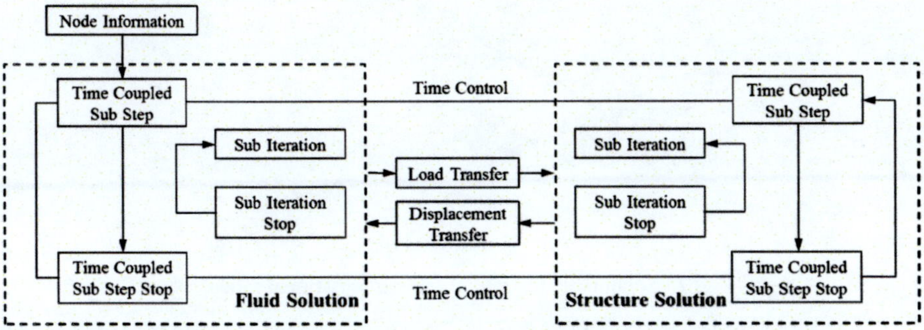

Fig. 1. Schematic diagram of the process of fluid-structure interaction solution.

$$p_{fan} = p_{inf} \frac{p_{fan}}{p_{inf}} = p_{inf} \frac{p_{fan}}{p_0} \frac{p_0}{p_{inf}} = p_{inf} \left(\frac{1 + \frac{\gamma-1}{2} Ma_{inf}^2}{1 + \frac{\gamma-1}{2} Ma_{fan}^2} \right)^{\frac{\gamma}{\gamma-1}} \tag{11}$$

where γ is the specific heat ratio of air, p_0 is the total pressure of the far field flow. The engine bypass ratio R_B, the total temperature and total pressure at the outlet are both taken as known quantities. The mass flow at the outlet of engine can be divided into two parts by bypass ratio, where the mass flow rate $\dot{M}_{EF}$ at the outlet of fan is:

$$\dot{M}_{EF} = \frac{R_B}{1 + R_B} \dot{M} = \rho_{EF} A_{EF} U_{EF} \tag{12}$$

Through the above formula, all the state parameters of the inlet and outlet of the engine can be obtained [14].

3 Simulation and Analysis

3.1 Computational Model

The computational model is shown in Fig. 2, including the tanker, the hose-drogue body and the receiver. According to the symmetrical condition, the tanker is a half model, and the hose-drogue body and the receiver are full model. The origin of the reference coordinate is located at the aerodynamic center of the tanker, the X axis is parallel to the fuselage axis pointing to the empennage, the Y axis is vertical to the plane of symmetry of the tanker pointing to the right wing, and the Z axis is vertical to OXY plane pointing to upward. The main parameters of the computational model are shown in Table 1.

3.2 Computational Mesh

If the overall mesh generation of tanker, the hose-drogue body and the receiver is performed will greatly increase the difficulty, and sharply increase the amount of mesh and the calculation cost. In order to solve the problem, the patched grid technology is adopted, that is, the computational model is divided into three separate parts, and generates the

Fig. 2. Schematic diagram of the computational model.

Table 1. Main parameters of the computational model.

Parameter	Value
Wingspan of tanker (m)	49.478
Mean aerodynamic chord of tanker (m)	6.42
Reference area of tanker (m^2)	144
Length of hose (m)	18
Outer diameter of hose (m)	0.06
Bottom diameter of drogue (m)	0.4
Wingspan of receiver (m)	12.8
Mean aerodynamic chord of receiver (m)	5.046
Reference area of receiver (m^2)	55.596

three separate meshes respectively, and the interpolation of the flow field information among the mesh is realized through the interface. Figure 3 show the surface mesh of the tanker and its engine nacelle, the hose-drogue body, and the receiver, respectively. The mesh quantity of the tanker is 25 million, the mesh quantity of the hose-drogue body is 5 million, and the mesh quantity of the receiver is 20 million. The flight condition for CFD computation is 0.4 Ma and 2000 m.

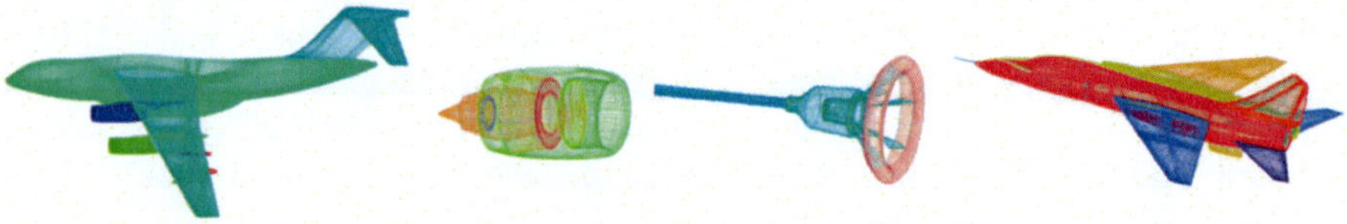

(a) Tanker. (b) Engine nacelle. (c) Hose-drogue body. (d) Receiver.

Fig. 3. Surface mesh of the computational model.

3.3 Characteristics of the Wake Flow Field

Figure 4 is the contour diagrams of the induced angle of attack, induced sideslip angle and relative velocity of the local flow field in the $Z = -3.641$ m plane, respectively. There appears peak and valley value of the induced angle of attack $\Delta\ \alpha$ near the position of drogue. The reason is that the influence of the drogue on its surrounding flow field makes the streamline deflect to the positive direction of Y axis at the position before flowing over the drogue. At different Y-axis position, with the increase of the Y coordinate, the absolute value of the negative induced angle of attack quickly increases to more than 1°. After flowing over the drogue, the streamline deflect to the negative direction of Y axis, and the induced angle of attack first increases to a peak value of about 0.5°. The induced sideslip angle $\Delta\ \beta$ is within the range of 5 m on both sides of symmetry plane of the drogue. Starting from the position of the drogue, as the Y coordinate increases, the absolute value of the negative sideslip angle at the corresponding position shows a decreasing trend. When the coordinate value of Y is greater than 40 m, the absolute value of the negative sideslip angle reduces to below 0.06°, which is only about 5% of the peak value in the region. The change of relative velocity $\Delta\ U_M$ is quite small, which is less than 0.4% of the far field flow velocity. In the vicinity of tanker's engine, the flow field is affected by the jet flow, resulting in a maximum relative velocity of 5 m/s.

(a) Induced angle of attack. (b) Induced sideslip angle. (c) Relative velocity.

Fig. 4. Distribution contour in the $Z = -3.641$ m plane

3.4 Aerodynamic Characteristics Analysis of the Receiver

Based on the suspending point of hose mounted on the wing of the tanker, the refueling channel is designed, where the aerodynamic influence of the wake flow field on the receiver is considered. The refueling channel is displayed in Fig. 5. The center line of the refueling channel is at the position of Y = 20 m in the span direction. The flow direction ranges from the position of X = 30 m to X = 65 m, including all region during the refueling process. The spanwise width of the refueling channel is designed to be 6 m away from the centerline. The center in height direction of the refueling channel is located at 4 m below the wing pod, and the height is 6 m in total.

The influence of wake flow is mainly reflected in the effect of the wing tip vortex rotation, which affects the longitudinal and lateral aerodynamic characteristics of the receiver. Figure 6 to Fig. 8 show the variation curves of aerodynamic force and torque coefficient increments caused by the wake flow of tanker ranging from −2°–16° angle of attack along the flow direction, spanwise direction and height direction. When the receiver flies in the region of refueling channel, it suffers aerodynamic changes compared to flying in the far field flow, so additional deflection of control surface is required to rebalance the variation to maintain the relative position of flight in refueling process (Fig. 7).

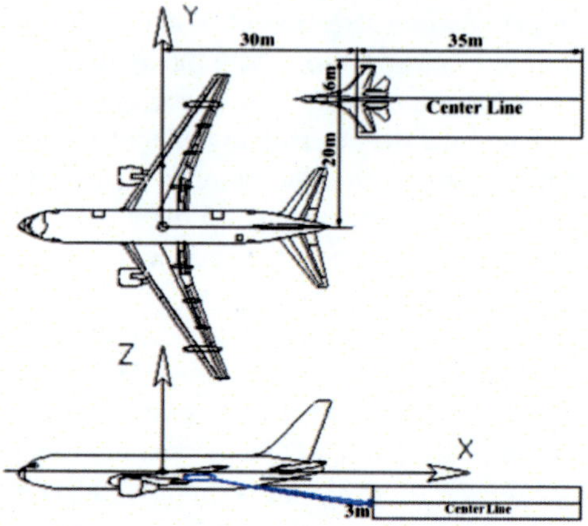

Fig. 5. Schematic diagram of the refueling channel.

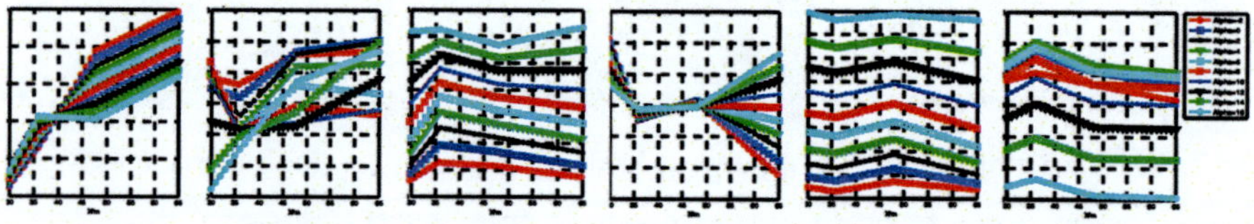

(a) CL (b) Cd (c) Cy (d) Cm (e) Cl (f) Cn

Fig. 6. Aerodynamic coefficient increments of the receiver along flow direction.

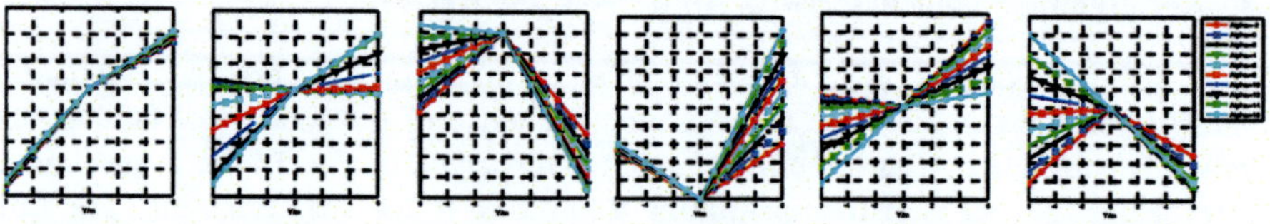

(a) CL (b) Cd (c) Cy (d) Cm (e) Cl (f) Cn

Fig. 7. Aerodynamic coefficient increments of the receiver along spanwise direction.

3.5 Influence of the Wake Flow on the Trim of the Receiver

Table 2 demonstrates the trimmed variables with the wake flow and without the wake flow in typical flight condition at 2000 m altitude and 0.4 Mach number. When there

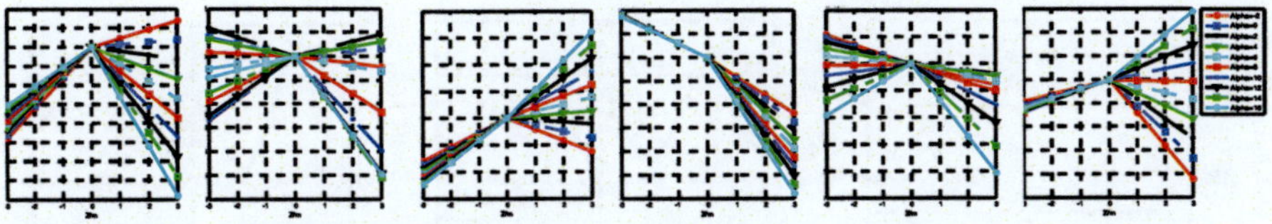

(a) CL (b) Cd (c) Cy (d) Cm (e) Cl (f) Cn

Fig. 8. Aerodynamic coefficient increments of the receiver along height direction.

exists the wake flow, additional deflection of aileron and rudder is required to keep the lateral balance. When the distance between the probe of receiver and the hose decreases from 6 m to 0 m along the center line of the refueling channel, the trimmed angle of attack, the trimmed deflection of the aileron and rudder gradually increase, and the trimmed deflection of the elevator and the throttle gradually decrease.

Figure 9 shows the comparison of the trimmed state variables of the receiver with the input of the trimmed deflection of control surfaces and throttle within 10 s simulation. It can be concluded that the trimmed state variables of the receiver goes to divergence which cannot meet the stability requirements of flight during aerial refueling with the influence of the wake flow. Therefore, the trimmed input variables of the receiver in the wake flow are supposed to be calculated for the feed forward control law design of the receiver.

Table 2. Comparison of the trimmed variables of the receiver with and without wake flow.

Situation / Variables	No wake flow	The distance between the probe of receiver and the hose with wake flow						
		6 m	5 m	4m	3 m	2 m	1 m	0 m
V(m/s)	133.0	133.0	133.0	133.0	133.0	133.0	133.0	133.0
α(°)	1.06	1.08	1.09	1.10	1.12	1.13	1.15	1.17
β(°)	0	0	0	0	0	0	0	0
p(°/ s)	0	0	0	0	0	0	0	0
q(°/ s)	0	0	0	0	0	0	0	0
r(°/ s)	0	0	0	0	0	0	0	0
X(m)	0	0	0	0	0	0	0	0
H(m)	2000	2000	2000	2000	2000	2000	2000	2000
Z(m)	0	0	0	0	0	0	0	0
De(°)	0.53	0.47	0.45	0.44	0.43	0.42	0.41	0.40
Da(°)	0	-1.45	-1.42	-1.44	-1.47	-1.50	-1.53	-1.56
Dr(°)	0	2.06	2.07	2.08	2.09	2.10	2.11	2.12
TLA(%)	41.90	43.56	43.56	43.55	43.53	43.52	43.50	43.48

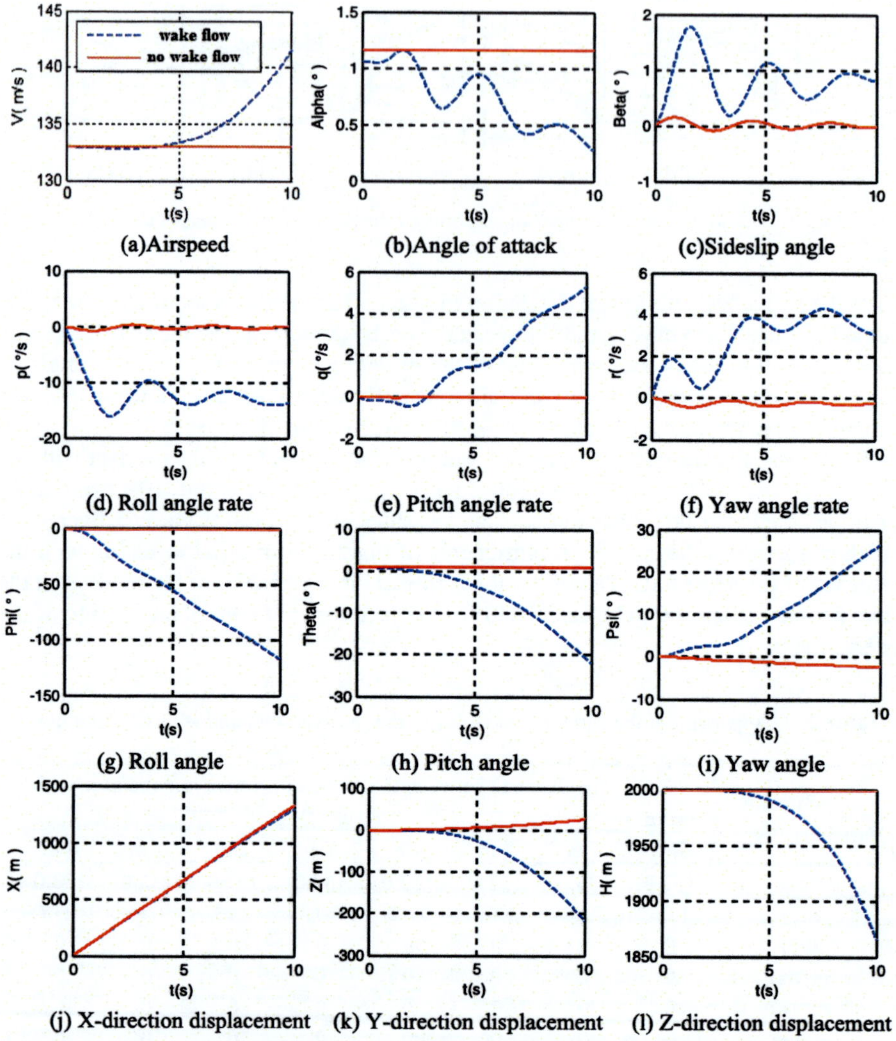

Fig. 9. Comparison of trimmed state variables of the receiver with and without wake flow.

3.6 Equilibrium Position of the Hose-Drogue Assembly

Through the numerical simulation of fluid-structure coupling, the equilibrium position of the hose-drogue assembly whose length varies from 15 m to 22 m is obtained when the flight altitude is 2000 m and the Mach number is 0.4. Figure 10 shows equilibrium position of hose at each length. Table 3 shows the coordinates of the circle center of the bottom circle of the drogue. As the length of the hose increases, the sinking amount of the drogue gradually increases, and the lateral distance gradually moves closer to the fuselage.

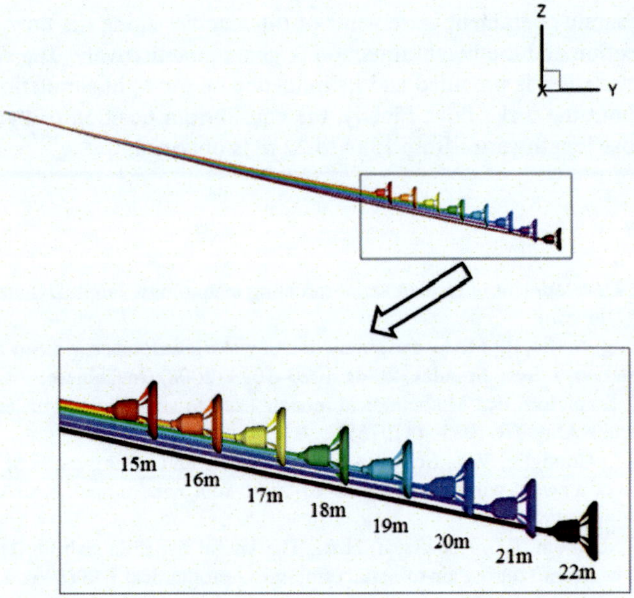

Fig. 10. The equilibrium position of hose at different length.

Table 3. The coordinates of the center of the bottom circle of the drogue.

Length of hose (m)	Lateral position X (m)	Horizontal position Y (m)	Sink position Z (m)
15	−0.156	14.686	−2.521
16	−0.213	15.637	−2.882
17	−0.258	16.585	−3.248
18	−0.302	17.553	−3.506
19	−0.344	18.534	−3.696
20	−0.411	19.504	−3.941
21	−0.457	20.465	−4.226
22	−0.485	21.421	−4.541

4 Conclusion

Based on the patched mesh technology, the CFD numerical simulation of the entire flow field of the tanker, the hose-drogue body and the receiver is carried out considering the jet flow of tanker's engine. Through the numerical simulation method of fluid-structure coupling, the aerodynamic forces acting on the hose-drogue assembly is obtained, and the deformation of the hose is calculated by the FEM to obtain different hose lengths. Firstly, the characteristics of the wake flow field is analyzed, and then the variation of the aerodynamic coefficient increments of the receiver along the flow direction, the spanwise direction and the height direction is given quantitatively. The comparison of the trimmed variables is provided and its influence on the fight simulation is analyzed with and without the wake flow. Finally, the equilibrium position of the hose-drogue assembly whose length varies from 15 m to 22 m is obtained.

References

1. Yuyan, H.: Development of foreign aerial refueling technology. Flight Dynam. **12**(4), 14–22 (1994). (in Chinese)
2. Li, H., Huang, A., Jia, R.: Model and simulation of wake interference in air-to-aerial refueling flight simulation. J. Syst. Simulat., **20**(4), 1020–1023 (2008) (in Chinese)
3. Jewell, W., Stapleford, R.: Mathematical models used to simulate aircraft encounters with wake vortice. NASA-TR-1035–4 (1975)
4. Rossow, V.J., Corsiglia, V.R., Schwind, R.G.: Velocity and rolling-moment measurements in the wake of a swept-wing model in the 40-by-80-foot wind tunnel. NASA-TM-X-62414 (1975)
5. Bloy, A.W., Lamont, P.J., Abu-Assaf, H.A.: The lateral dynamic stability and control of a large receiver aircraft during air-to-aerial refueling. Aeronautical J. **90**(896), 237–243 (1986)
6. Jiaolong, L., Jianping, X., Xinmin, D.: Research on aerodynamic effect modeling of receiver based on numerical simulation of wake vortex. Flight Dynam. **32**(5), 389–393 (2014). (in Chinese)
7. Jackson, D.K., Tyler, C., Blake, W.B.: Computational analysis of air-to-aerial refueling. AIAA-2007–4289 (2007)
8. Kuizhi, Y., Liangliang, C., Tieying, Z.: Numerical simulation of aerodynamic influence of an aircraft on the hose-refueling system during aerial refueling operations. Aerosp. Sci. Technol. **49**, 34–40 (2016)
9. Wei, W., Xizang, L., Peng, W.: Dynamics of hose-drogue refueling systems during coupling. Flight Dynam. **31**(2), 181–182 (2013)
10. Zhu, Z.H., Meguid, S.A.: Elastodynamic analysis of aerial refueling hose using curved beam element. AIAA J. (S0001–1452), **44**(6), 1317–1324 (2006)
11. Kamman, J.W,, Hustopn, R.L.: Multibody dynamics modeling of variable length towed and tethered cable system. Multibody Syst. Dynam. (S1384–5640), **5**(3), 211–221 (2001)
12. Teman, R.: Navier-Stokes equations : theory and numerical analysis (1977)
13. Menter, F.R., Rumsey, C.L.: Assessment of two-equation turbulence models for transonic flows. In: AIAA Fluid Dynamics Conference (1994). https://doi.org/10.2514/6.1994-2343
14. Lei, Q., Junqiang, B., Jun, H.: Interference effects of wing-mounted high bypass ratio pod with engine power. Acta Aerodynamica Sinica **32**(4), 433–438 (2014)

Effects of Film Holes Position on Anti-icing Characteristics of Aero-engine Inlet Adjustable Blade

Xinwei Jiang(✉), Huan Gong, Yundan Li, Qi Jia, and Miao Li

AECC Shenyang Engine Research Institute, Shenyang, China
690509242@qq.com

Abstract. In order to obtain the influence of film hole position on the anti-icing characteristics of aero-engine inlet adjustable blade, numerical simulation and experimental investigation was carried out for different film hole structure of the blade. The performance of different film hole position (include the trailing edge, the side wall and the rotating shaft of strut) and blowing ratio on the heating effects of the adjustable blade surface were analyzed, and experimental verification was carried out under ice wind tunnel conditions. The results show that with the increasing of blowing ratio, the temperature of adjustable blade increases gradually. The surface with film holes located at the trailing edge has the highest temperature, which is 315 K. Under the same working conditions of mainstream and hot air, the trailing edge inclined hole structure has the smallest ice thickness among the 3 structures. Based on the calculation and experimental results for anti-icing of different positions film hole structures on the adjustable blade, the inclined film holes at the trailing edge has better heating and icing protection effect.

Keywords: film heating · blowing ratio · anti-icing

1 Introduction

The inlet guide vane (IGV) is located at the entrance of aero-engine. Ice can accumulate on the surface of adjustable blade and affect flight safety when the engine operates under icing conditions. Nowadays, Hot Air Anti-Icing System has been widely applied to protect engine inlet components from icing. Numerical simulations and experimental studies have been widely carried out to obtain the icing mechanism [1–4]. The water droplets impingement characteristics of IGV was investigated [5–8], and the distribution of the anti-icing thermal load on the surface was obtained [9].Numerical simulation results show that the inner side of the adjustable part is also a key area requiring anti-icing [10]. It can be seen that previous studies on the icing prevention of engine IGV mainly focused on the water droplet impingement and icing characteristics of the fixed strut part, while relatively few studies have been conducted on the adjustable blade part. Usually, there are film holes on the strut to provide protection to the adjustable blade by blowing and heating.

Chinese Society of Aeronautics and Astronautics (Ed.): CASTC 2023, LNME, pp. 293–299, 2024.
https://doi.org/10.1007/978-981-99-8861-7_30

Based on the above research background, this paper has carried out numerical simulation research on the film heating of adjustable blades, established different film structure calculation models, and focused on the influence of the film hole position on the heating temperature of the adjustable blade surface. And experimental verification under ice wind tunnel conditions has been conducted. The research results will be helpful to design the film anti-icing structure of the adjustable blade and improve the anti-icing ability of aero-engine.

2 Numerical Simulation Methods

In order to study the effects of film hole location on the ice formation of adjustable blades, models with different film holes location were established in the case of adjustable blades having a deflection angle of 30°. The structures are shown in Fig. 1. The film holes are located on the side wall, the trailing edge and rotating shaft of the strut respectively. The hot air film holes are all cylindrical, with a diameter of D = 1mm and a spacing of 2 times the diameter.

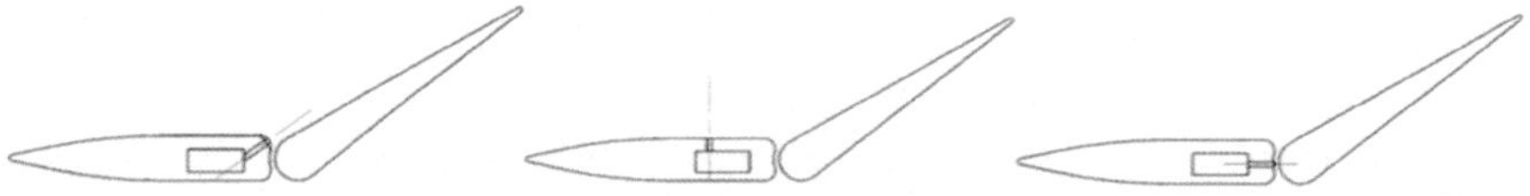

(a)Film holes at trailing edge (b)Film holes at side wall (c)Film holes at the rotating shaft

Fig. 1. Film hole Structures at different positions

The commercial software CFX was used for calculation. The turbulence model was Shear Stress Transport (SST) model, main equations are as follows:

$$\frac{\partial}{\partial t}(\rho k) + \frac{\partial}{\partial x_i}(\rho k u_i) = \frac{\partial}{\partial x_j}\left(\Gamma_k \frac{\partial k}{\partial x_j}\right) + G_k - Y_k + S_k \tag{1}$$

$$\frac{\partial}{\partial t}(\rho \omega) + \frac{\partial}{\partial x_i}(\rho \omega u_i) = \frac{\partial}{\partial x_j}\left(\Gamma_k \frac{\partial \omega}{\partial x_j}\right) + G_\omega - Y_\omega + S_\omega \tag{2}$$

where G_k represents the kinetic energy of turbulence, $G\omega$ express ω equation, $\Gamma\omega$, $\Gamma\omega$ represents the effective diffusion term, Y_k, $Y\omega$ represents the divergence term, $D\omega$ represents the orthogonal divergence term, S_k and $S\omega$ is customized for users.

The blowing ratio (*M*) is defined as follows:

$$M = \frac{\rho_h u_h}{\rho_g u_g} \tag{3}$$

ρ_h and u_h are the density and velocity of the hot air flow, ρ_g and u_g are the density and velocity of the mainstream.

Given the velocity boundary of the fluid domain inlet, which is 45m/s, and the inlet temperature is 263K.Given the pressure at the outlet, which is calculated based on the actual aerodynamic parameters of the engine. The blowing ratio (*M*) of the secondary flow hot air is range of 1.0 to 2.0.

The overall and local grids used for the calculation are shown in Fig. 2. Figure 3 shows the trend of pressure distribution on the surface of adjustable blades with different grid numbers. When the grid size reaches 1.5 million or more, the pressure variation difference on the wall surface of adjustable blades is very small, and the y + is lower than 3.5. Therefore, in the calculation, the mesh size is selected in the numbers of 1.5 to 2 million.

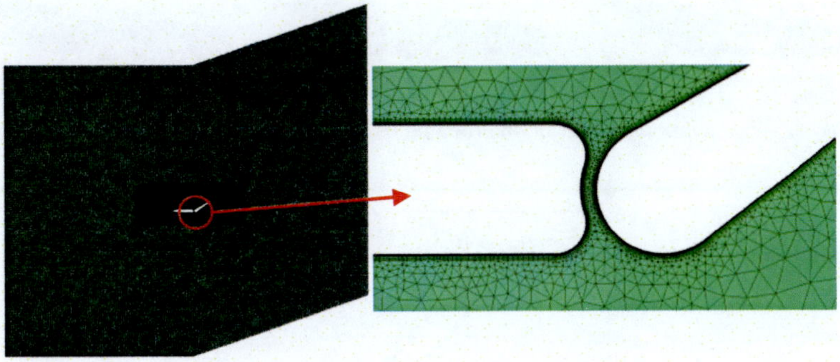

Fig. 2. Grid division of computing domain

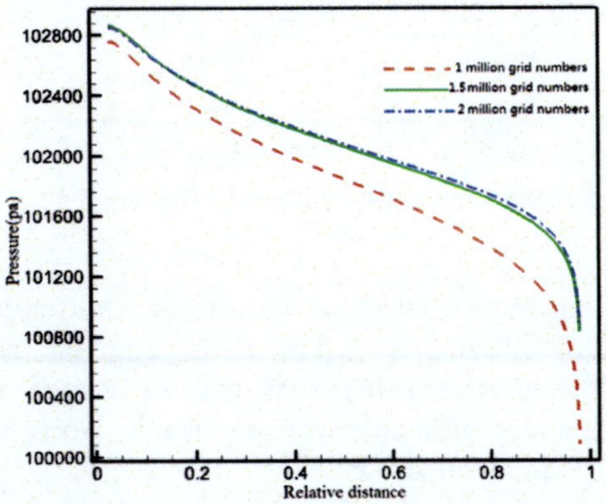

Fig. 3. Comparison of wall pressure with different numbers of grids

3 Numerical Results of Heating Effects

As shown in Figs. 4 and 5, the heating effect comparison of the film in trailing edge under different blowing ratios is shown. According to the cloud images, the wall temperature at the front of the adjustable blade surface is relatively high, while the wall temperature at the trailing edge is relatively low. An obvious high temperature region is formed on the adjustable blade surface area directly opposite the air film holes. The surface temperature of the adjustable blade gradually decreases along the flow direction. With the increase of blowing ratio, the average surface temperature of the adjustable blade gradually increases. The surface wall temperature of the adjustable blade increases by an average of about 20 k when the blowing ratio increases from 1.0 to 2.0.

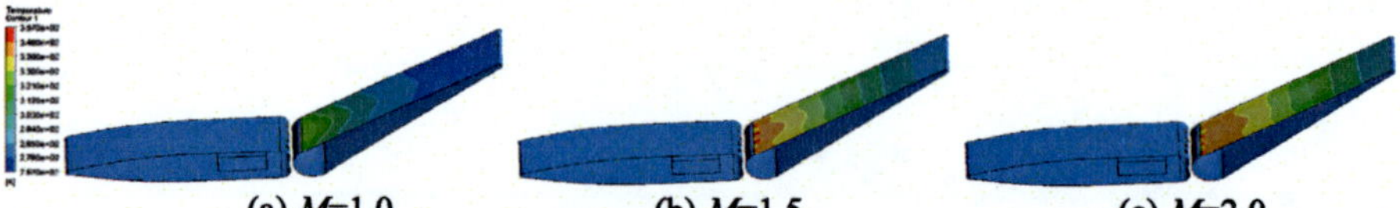

Fig. 4. Wall temperature of the inclined holes structures at the trailing edge of strut

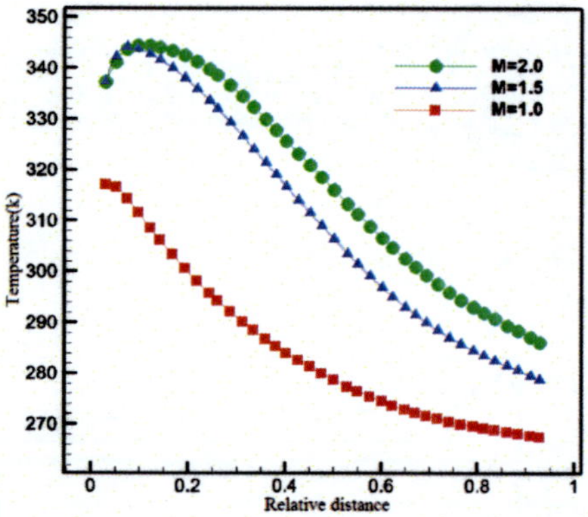

Fig. 5. Wall temperature curves of the inclined holes structure at the trailing edge

Figure 6 shows a comparison curve of the surface wall temperature of adjustable blades with different film hole structures. The trailing edge inclined film hole structure has the best heating effect on adjustable blades, with an average surface temperature of about 315 k. The rotating shaft film hole structure has the worst heating effect, with an average wall temperature of about 269 k.

In order to reveal the influence of different air film structures on the heating characteristics of adjustable blades further, the flow characteristics of different structures are compared and analyzed, as shown in Fig. 7. For the trailing edge inclined hole structure, all secondary flow hot air can better attach to the blade surface, and has a higher outflow speed, while less diffusion to the mainstream, forming a better protective effect on the blade concave surface. For the side wall film hole structure, due to the influence of the position and direction of the film outflow, its normal momentum is relatively larger, resulting in more mixing into the mainstream. At the same time, due to the reduced momentum after reaching the adjustable blade, a part of the hot gas is pumped to the convex surface of the blade, resulting in a reduced protective effect on the adjustable

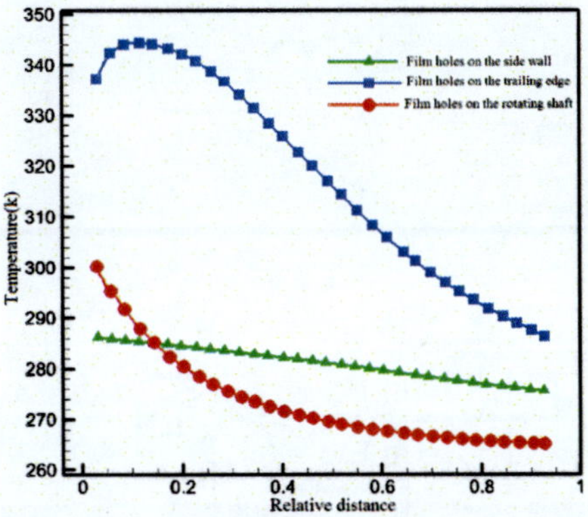

Fig. 6. Comparison of heating effect of different film hole structure

blade. For the rotating shaft film hole structure, due to its relatively close distance to the blade convex side, most of the hot air is pumped to the convex surface due to the suction effect, and only a small portion attach to the surface of the blade concave side, resulting in a waste of hot air so its heating effect is the worst .

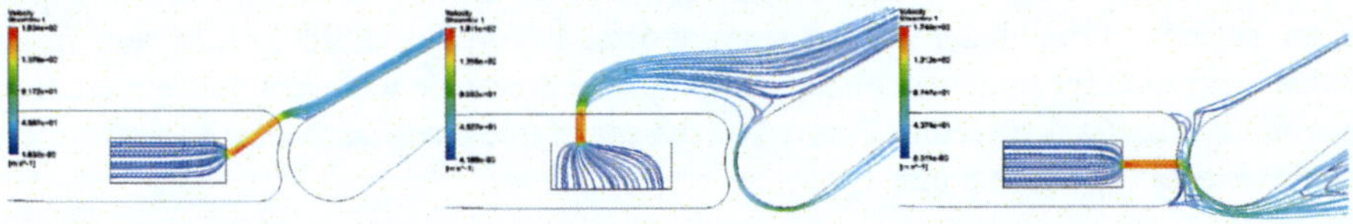

(a)Film holes on trailing edge (b)Film holes on side wall (c)Film holes on the rotating shaft

Fig. 7. Comparison of streamline of different film hole structure

4 Experimental Results of Ice Wind Tunnel

In order to further verify the anti-icing effect of different position air film structures on adjustable blades, preliminary anti-icing effect tests on different structure test pieces were conducted in an ice wind tunnel. The structure of the experimental system is shown in Fig. 8.

The facility is composed of eleven parts, including wind tunnel, air dynamic system, cooling system, atomization water spray system, vacuum system, secondary flow system, circulating water system, softening water system, corner heat system, controlling system, viewing system and corresponding auxiliary systems. The icing wind tunnel is located in AECC Shenyang Engine Research Institute, which is mainly used for engine inlet cones, struts and other inlet components tests and verifications.

Figure 9 shows the photos of the experimental results for different structures under the same icing conditions. It can be seen that adjustable blades with different structures all accumulate with a certain thickness ice. However, the amount of ice on the surface

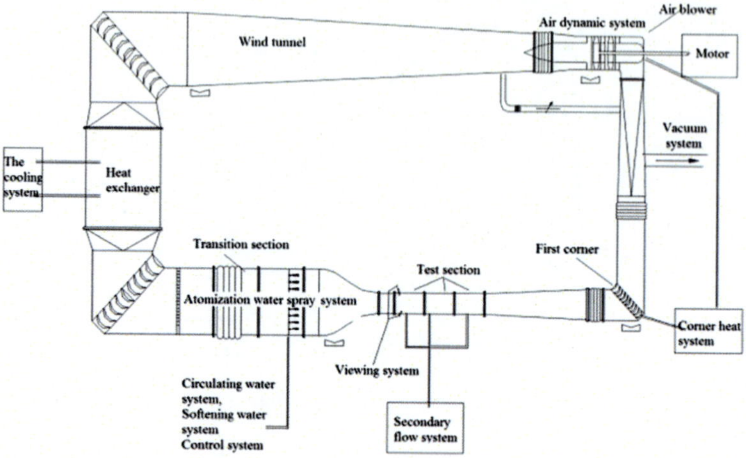

Fig. 8. Experimental systems

of different structure is different. The side wall film hole structure shows significant ice accumulation, with a maximum ice thickness of about 5 mm. The trailing edge inclined hole structure has the smallest ice thickness, and there are obvious marks of film impact on the surface of the blade. The ice amount directly facing the film hole is very small, with a maximum ice thickness about 3mm. Finally, for blade with film hole structure on the rotating shaft, severe icing occurred on the adjustable blade surface, with a maximum icing thickness of about 8 mm.

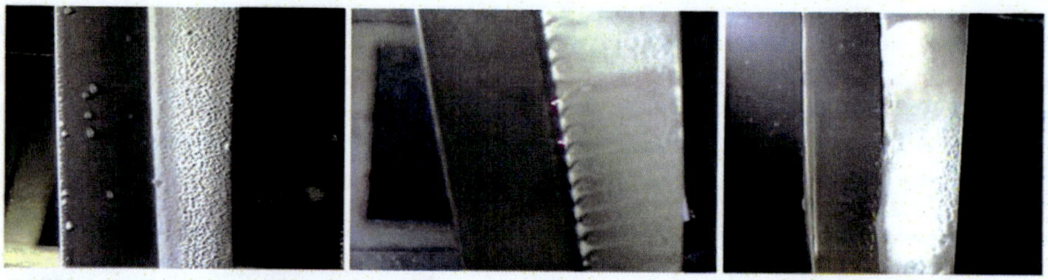

(a)Film holes at the side wall (b)Films hole at the trailing edge (c)Film holes at the rotating shaft

Fig. 9. Icing test results of different film hole structures

Comparing the ant-icing experimental results of adjustable blades with three structures, the trailing edge inclined hole structure has the best protective effect on adjustable blades, which is also consistent with the calculated conclusions of heating effects. It is recommended to consider both the location of the hot air film hole and the realizable film structure during engineering design. When the hot air temperature and flow rates are constant, refer to the performance of the film hole position in this paper to achieve effective icing protection for adjustable blades.

5 Conclusions

In this study, numerical simulations and experiments were carried out to acquire the anti-icing performances of inlet adjustable blade with different film hole structures. The following conclusions can be obtained.

(1) As the blowing ratio increases, the heating effect of different film structures on adjustable blades increases. When the blowing ratio is increased from 1.0 to 2.0, the maximum wall temperature of adjustable blades with different structures increase by about 20 K.
(2) The trailing edge inclined hole structure has a good heating effect, with a maximum average wall temperature of about 315 K.
(3) A portion of the hot air coming out of the film holes will be sucked to the convex surface of the blade. The lower the blowing ratio, the more obvious the suction effect is. When the blowing ratio is 2.0, the trailing edge inclined hole structure hot air can fully attach to the blade surface.
(4) According to the ice wind tunnel verification results, the trailing edge inclined hole structure has the best icing protection ability to adjustable blades, consistent with the calculated conclusions.

References

1. Sang, L., Eric, L.: Simulation of icing on a cascade of stator blades. J. Propulsion Power **24**(6) (2008)
2. Papadakis, M., Yeong, H.W., Wong, S.C., et al.: Comparison of experimental and computational ice shapes for an engine inlet. In: AIAA Atmospheric and Space Environments Conference, Toronto (2010)
3. Pellissier, M., Habashi, W., Pueyo, A.: Design optimization of hot-air anti-Icing systems by FENSAP-ICE. In: 48th AIAA Aerospace Sciences Meeting Including the New Horizons Forum and Aerospace Exposition, Orlando, Florida (2010)
4. Guido, S., Baruzzi, M.O., et al.: Multi-physics simulation of aircraft ice protection system and ice impact with turbofan blades. In: 8th AIAA Atmospheric and Space Environments Conference, Washington, D.C (2016)
5. Yang, J., Guo, W., Lou, D.: Numerical analysis of trajectories of water droplets on the anti-icing vane of an engine. Gas Turbine Exper. Res. **24**(1), 19–24 (2011)
6. Liu, H., Yang, J.: Effect of film gap outflow on water droplets impingement characteristics of the anti-icing vane. Gas Turbine Exper. Res. **30**(2), 51–57 (2017)
7. Ke, P., Zhang, Y., Yu, G., et al.: Influence of exterior hot-film on droplet impingement characteristics over aero-engine inlet strut. J. Aerospace Power **32**(3), 621–629 (2017)
8. Zhao, J., Wu, D.: Effects of regulated angles to droplet impingement characteristics of aero-engine strut. Aeronau. Comput. Tech. **43**(4), 61–64 (2013)
9. Xu, L., Zhou, J., Zhang, D., et al.: Numerical simulation of anti-icing heat load for strut of aero-engine inlet. Aeroengine **35**(4), 21–24 (2009)
10. Yaguo, L., Zhenxia, L., Lifen, Z., et al.: Numerical simulation of ice accretions on trailing adjusted strut. Aeronau. Comput. Tech. **41**(6), 58–65 (2011)

Research on Cooling Test Technology of Civil Helicopter Power Plant and Lubricating Oil System

Chunzhuang Zhao[1](✉), Yong Wang[2], Jiaxin Gao[1], Tengyuan Liu [1], and Lei Wang [1]

[1] Harbin Aircraft Industry Group Co., Ltd., Harbin 150066, China
zhaochunzhuang@126.com

[2] Troops of the Chinese People's Liberation Army, Harbin, China

Abstract. In view of the airworthiness review requirements of civil aviation regulations on the cooling capacity assessment of civil helicopters, the cooling test technology of civil helicopter power plant and lubricating oil system is studied. The test data under the planned flight subjects are obtained by adding temperature sensors and signals on the helicopter. The measured temperature data is corrected to the temperature correction data under the upper limit of the helicopter's operating temperature according to the formula, and then the temperature correction data is compared with the temperature limit of relevant accessories and materials to obtain the cooling capacity assessment results. The cooling test technology of civil helicopter power plant and lubricating oil system is studied, which provides a reference for the cooling test of various types of helicopters.

Keywords: helicopter · power plant · lubricating oil · cooling test · test

1 Introduction

In order to improve flight safety, modern helicopters are equipped with two or more engines. After one engine fails and shuts down, the remaining engines can still provide sufficient power for the helicopter [1]. The helicopter lubricating oil system [2] (also known as lubrication system) is mainly distributed in the engine, main reducer, tail reducer and other parts, providing cooling, lubrication and cleaning functions for these parts. The accessories and manufacturing materials of the power unit and the lubricating oil system need to be kept in normal operation within a certain temperature range. The temperature is maintained mainly through the lubricating oil system, in addition to the way of directly radiating heat to the outside.

The assessment of the cooling function of the power unit and lubricating oil system of the newly developed aircraft is the flight test subject of the civil aviation review [3]. The civil aviation review is based on three points of assessment: first, the use of engine power and flight subject planning; The second is the definition of temperature "stability" [4–5]: during the cooling test, the outside atmospheric temperature during the test should be at least 38 °C in principle (except for the devices used in winter), and the temperature

Chinese Society of Aeronautics and Astronautics (Ed.): CASTC 2023, LNME, pp. 300–305, 2024.
https://doi.org/10.1007/978-981-99-8861-7_31

is considered "stable" when the temperature change rate is lower than 1.1 °C per minute; Third, after the temperature is stable, it shall comply with the operating temperature limit and material temperature limit of the accessories. In theory, the cooling function of the power plant and the lubricating oil system will be affected by different ambient air temperatures. Regardless of the ambient air temperature, the measured temperature should be corrected to the upper limit of the helicopter after obtaining the temperature test data, so that the helicopter cooling function can still work normally when the ambient air temperature is the upper limit of the helicopter, When the ambient air temperature is low, it will have a negative impact on the cooling function assessment, and when the ambient air temperature is high, it will have a positive impact on the cooling function assessment.

2 Overview and Index Analysis of Helicopter Power Unit and Lubricating Oil System

The power unit of a helicopter is usually a turboshaft engine [6]. The turboshaft engine is designed according to the unit, including five units, including accessory gearbox/power output shaft, compressor, gas generator, power turbine and reduction gearbox [7–8]. The helicopter's lubricating oil system is mainly distributed at the engine, main reducer and tail reducer. The lubricating oil in the lubricating oil system is equipped with temperature sensors at the booster pump engine and main reducer to monitor the lubricating oil temperature, which are the left engine lubricating oil temperature [9], the right engine lubricating oil temperature and the main reducer lubricating oil temperature.

The temperature limits of typical accessories can be obtained from the engine installation manual, and the limits of engine oil temperature and main reducer oil temperature can be obtained from the helicopter flight manual.

3 Test Method Design and Test Planning

The cooling test includes two parts: ground test and flight test. The ground test is used as the preliminary test of flight test to initially confirm the cooling capacity of the helicopter. If the cooling capacity cannot meet the standard in the ground test, the test shall be repeated after adjustment until it meets the standard.

The ground test mainly checks the measured value of the stable temperature and the measured value of the maximum temperature in the whole process of the engine running at idle power and flight (without lift) power for a long time under the ground driving condition, and the temperature change trend of each temperature measuring point and lubricating oil temperature within a period of time after the ground driving and shutdown.

The flight test mainly checks the measured value of stable temperature and the measured value of the highest temperature in the whole process of the helicopter under various typical flight conditions, different takeoff weights and various engine power conditions. For twin-engine helicopters, the engine working state is divided into twin-engine working and single-engine working. The typical takeoff weight is divided into twin-engine maximum takeoff weight and single-engine maximum takeoff weight. Considering the engine working state and typical takeoff weight, the main test subjects are planned as follows:

(1) In the state of maximum takeoff weight of two engines, the helicopter hovers with ground effect in the state of maximum continuous power of two engines until the temperature measurement points or oil temperature is stable. Affected by the external atmospheric environment, if the helicopter cannot hover at the maximum continuous power of two engines under the maximum takeoff weight, the helicopter's takeoff weight shall be reduced until it can hover. The reduced helicopter weight can be achieved by reducing the counterweight or the fuel consumption of helicopter takeoff.
(2) The maximum takeoff weight state of a single engine requires the continuity of helicopter action. The helicopter first hovers with ground effect for a short time, then climbs to the height of 1000 m with the maximum continuous power of two engines, and then changes to the simulated single engine state (one engine is idling, the other engine is flying), and continues to climb at a climb rate of no less than 150 feet/min in the simulated single engine state until the climb rate is less than 150 feet/min Or the climb height reaches 3000 m, or the climb time reaches 30 min. If any of the above three conditions is met, the climb will be exited, and the level flight will be changed to the maximum continuous power of a single engine until the temperature measurement points or the oil temperature is stable.

4 Temperature Test Modification Technology

The temperature sensor (K-type thermocouple) is installed at the selected measuring point on the engine, and the KAM-500 data collector produced by Ireland's ACRA CONTROLLTD company is installed on the floor of the engine room. The bus drawn from the engine and the added temperature sensor are connected to the KAM-500 data collector through cables.

5 Method of Data Processing

Since the cooling test of the power plant is to assess the cooling system's ability to ensure that the power plant and its accessories and cowling are within the temperature limit under various flight conditions and use conditions of the helicopter, it is necessary to measure the external atmospheric temperature during the test (which must be less than the upper limit of the helicopter's use temperature), The measured temperature obtained at each measuring point during the cooling test of the power plant is corrected to the corrected temperature when the external atmospheric temperature is the upper limit of the helicopter's operating temperature (the upper limit of the operating temperature is related to the flight altitude, which can be obtained by referring to the altitude-temperature envelope (flight envelope) in the restricted section of the helicopter's "flight manual"). For example, if the upper limit of helicopter operating temperature is 50 °C and the external atmospheric temperature is 30 °C during the test, the measured temperature at each measuring point should be corrected to the corrected temperature when the external atmospheric temperature is 50 °C.

Assuming that the measured temperature at a measuring point is T1, the corrected temperature is T2, the external atmospheric temperature is T0, and the standard atmospheric pressure altitude is Hp, the calculation method for correcting it to the external atmospheric temperature is the upper limit of the helicopter's operating temperature is as follows:

(1) Given the standard atmospheric pressure altitude Hp, there are two methods to calculate the upper limit of temperature Tp of the helicopter at this flight altitude, one is obtained by consulting the altitude-temperature envelope (flight envelope), and the other is obtained by calculating the altitude-temperature relationship. Two calculation methods are described in detail below.

Method 1: According to the altitude-temperature envelope (flight envelope), the upper limit of helicopter operating temperature at this flight altitude is Tp. Taking a helicopter as an example, Tp is calculated as follows:

The use temperature of the helicopter is linear with the flight altitude. The two points (10.4 °C, 20000 ft) and (50 °C, 0 ft) on the line are known. Assuming that Tp (°C) = a * Hp (ft) + b, the two points are brought into the binary linear equation, and a = -0.00198, b = 50 is obtained. Thus, Tp (°C) = 50–0.00198 * Hp (ft) is obtained.

Method 2: The known law of temperature change with height is as follows: for every 1000m increase in height, the temperature decreases by 6.5 °C until it drops to - 56.5 °C, above which the temperature is considered to be constant. Assuming Tp (°C) = a * Hp (ft) + b, from the height increase of 1000m (3281ft) and the temperature decrease of 6.5 °C, a = 6.5 °C/1000 = 0.0065 °C/m or a = 6.5 °C/3281 = 0.00198 °C/ft; If the upper limit of helicopter operating temperature is 50 °C at an altitude of 0 m, then b = 0. To sum up, Tp (°C) = 50–0.00198 * Hp (ft) is obtained.

(2) The calculation method of the upper temperature limit Tp of the helicopter at a fixed flight altitude is obtained from (1). The correction range of the measured temperature is set at one time of the difference between Tp and the outside atmospheric temperature T0 (one time is already set too large, that is, the larger the correction range, the greater the error will be. Therefore, try to select the weather with higher temperature for the test, which can reduce the calculation correction range and the correction error). The calculation formula of the corrected temperature can be obtained as follows:

$$\begin{aligned} T2\,(°C) &= T1\,(°C) + 1.0^{*}[Tp\,(°C) - T0\,(°C)] \\ &= T1\,(°C) + 1.0^{*}[50\,(°C) - 0.0065^{*}Hp(m) - T0\,(°C)] \\ &= T1 + \;(°C)1.0^{*}[50\,(°C) - 0.00198^{*}Hp(ft) - T0\,(°C)] \end{aligned} \quad (1)$$

where, T2 is the corrected temperature, T1 is the measured temperature, Tp is the upper limit of the helicopter operating temperature at a certain flight altitude, Hp is the standard atmospheric pressure altitude (absolute atmospheric pressure altitude), and T0 is the external atmospheric temperature during the test.

6 Application of Cooling Test Technology for Power Unit and Lubricating Oil System of AHelicopter

During the cooling test of the power unit and lubricating oil system of a helicopter, the flight altitude, flight speed, external atmospheric temperature, main reducer lubricating oil temperature, left engine lubricating oil temperature, right engine lubricating oil temperature and engine status parameters (left/right engine torque, gas turbine speed, power turbine speed, turbine temperature, etc.) were obtained by means of signal extraction on the helicopter, The measured temperature values of typical temperature measuring points on the engine (10 measuring points, 5 for each side of the engine, torque matching box, P3 sensor, high-energy igniter, T45 matching box, N2 sensor) were obtained by adding temperature sensors.

The correction of the measured value of temperature is calculated in two ways. One is the subject with fixed altitude (for example, hover subject, level flight subject, and fixed flight altitude is considered as fixed external atmospheric temperature). This temperature correction can first determine the stable temperature or maximum temperature of the measured value, and then only correct the temperature. The other is the subject of altitude change (for example, climb subject, the change of flight altitude will bring about the change of external atmospheric temperature). For the correction of such temperature, the temperature must be corrected in the whole process, and then the stable temperature or maximum temperature shall be confirmed in the correction value.

The test results are as follows:

(1) In the ground test, the measured values of temperature at each measuring point and the measured values of lubricating oil temperature were obtained. Due to the fixed height, the highest temperature in the measured values was directly confirmed, and only the temperature was corrected to obtain the temperature correction value. The temperature correction value was compared with the temperature limit of accessories and materials, and both were within the temperature limit range.
(2) During the flight test, the measured values of temperature at each measuring point and the measured values of lubricating oil temperature were obtained. In the subjects with fixed altitude (hover and level flight), the highest temperature in the measured values was directly confirmed, and only the temperature was corrected to obtain the temperature correction value; In the subject of height change (climbing), first correct the measured value of the whole process temperature, then confirm the maximum temperature in the corrected value, and compare the temperature correction value with the temperature limit of accessories and materials, which are within the temperature limit.
(3) It can be seen from (1) and (2) that the cooling function of the helicopter is good.

7 Conclusion

Through the research on the cooling test technology of civil helicopter power unit and lubricating oil system, the following conclusions are obtained:

(1) The civil aviation review and assessment of helicopter cooling function is mainly based on three requirements: firstly, flight subjects; secondly, stability requirements; and thirdly, comparison of corrected data with temperature limits of accessories and materials.
(2) The ground test of the helicopter cooling test mainly plans the long-term ground driving and the analysis of the temperature change after shutdown. The flight test mainly plans the double-engine hover with the maximum takeoff weight and the double-engine hover with the maximum takeoff weight of the single engine, the double-engine climb, the single-engine climb and the long-term stable level flight with the maximum continuous power of the single engine.
(3) The measured temperature data shall be calculated according to the formula to obtain the temperature correction data under the upper limit of the helicopter's operating temperature. The corrected temperature is related to the upper limit of the helicopter's operating temperature, the barometric height, and the external atmospheric temperature. The temperature correction data shall be compared with the temperature limit of accessories and materials.
(4) The application of cooling test technology on a helicopter is introduced, which provides a reference for cooling test of other aircraft.

References

1. Liu, T., Kuang, C., Wang, Q., et al.: Research on helicopter power system design. China Sci. Technol. Inform. **11**, 41–45 (2019)
2. Wei, X.: Prent situation and future analysis of aeroengine lubricating oil system. Internal Combustion Engine Parts **12**, 63–64 (2019)
3. Shi, S., Lv, W., Zhang, H.: Optimal design on aerodynamic performance of cooling hole intakes in engine bay cooling system of helicopter. J. Nanjing Univ. Aeronau. & Astronau. **02**, 45–60 (2020)
4. CCAR-29-R2, Airworthiness Regulations for Transport Rotorcraft (2017)
5. AC 29–2C, Consultation Notice on Airworthiness Standards of Transport-type Rotorcraft (2002)
6. Zhi, W., Ye, T.: Development requirements of general aviation and helicopter engines. Aerospace Power **06**, 17–20 (2019)
7. Zou, W., Zheng, X.: Development trends of aero turboshaft engines. J. Aerospace Power **12**, 2577–2587 (2019)
8. Ge, N.: Development and technical trend of turbo shaft engine. J Nanjing Univ. Aeronau. & Astronau. **02**, 145–156 (2018)
9. Zhang, X.: Fault analysis and improvement verification of engine oiu=l temperature entering yellow zone of a helicopter. Mech. Electrical Inform. **05**, 42–43 (2020)

Space and Temporal Distribution Characteristics of Lightning in Aircraft Flight Areas in Sichuan and Gansu Regions

Wei Yu(✉), Tao Tang, and Yue Wu

AVIC Chengdu Aircraft Industrial Group Co., Ltd., Chengdu 610000, China
474735325@qq.com

Abstract. Lightning is one of the disastrous weather phenomena affecting the safety of aviation aircraft. It is very necessary to study the spatial and temporal distribution characteristics of lightning in the flight airspace. Based on the lightning observation data provided by the National Lightning Monitoring Network, we analyzed the time and space distribution of lightning in Sichuan and Gansu areas from 2010 to 2019. The results showed that the monthly frequency of lightning presented a single-peak feature, which was mainly concentrated in April-September, with the highest occurrence month from June to August, and the peak of lightning occurrence appeared in July, with 3.61 million times. Lightning rarely occurs from October to March, but it occurs every month. The frequency of lightning occurrence is the most in summer, accounting for 68% of the whole year, spring and autumn account for 19% and 13% respectively; winter is the least, less than 1%. The frequency of lightning occurrence has obvious annual variation characteristics, and the frequency of annual occurrence in high frequency years can reach more than twice in low frequency years. The spatial distribution of lightning in the region has an obvious large value center, which is located in Luzhou, the southern part of the basin. Generally speaking, the frequency of lightning occurrence varies greatly among regions. The frequency of lightning in Sichuan Basin is higher than Panxi area, and the western Sichuan Plateau is the least. The frequency of lightning activity is relatively consistent with the heavy precipitation in the region, and the precipitation from the western to the southern Sichuan Basin is relatively large, which also corresponds to the most active area of lightning, which has a good consistency.

Keywords: Lightning · Space-time distribution

1 Introduction

Lightning is a strong discharge phenomenon between clouds and clouds, and between clouds and the ground. It will release powerful energy in the process of emergence, and then cause casualties, electronic system damage, fire and explosion accidents, which cause significant economic losses and social impact. Worldwide, more than 10,000 people are killed or injured by lightning every year, causing economic losses of more than $1

Chinese Society of Aeronautics and Astronautics (Ed.): CASTC 2023, LNME, pp. 306–313, 2024.
https://doi.org/10.1007/978-981-99-8861-7_32

billion. As an important part of thunderstorms, lightning is an important factor threatening flight safety in current aviation activities. Thunderstorms not only bring harm to the aerial flight of aviation aircraft, but also can cause the failure of radio communication and electronic equipment of aircraft. Therefore, the study of lightning is particularly important. Thunderstorms are often accompanied by lightning, strong winds, heavy precipitation, downburst and other dangerous weather. In terms of meteorology, the research on strong wind and heavy precipitation has been relatively mature. Due to the extremely short occurrence time, scattered occurrence range, data collection difficulty and other factors, the research on lightning is also relatively limited at present. It is generally believed in China that lightning activity mostly occurs in coastal areas, mountainous areas, mesoscale convection areas and convergence areas of ITCZ. Ma Ming et al. made statistics on the spatial distribution of lightning in China, and divided four active lightning areas in China, from high to low: South China, Southwest China, East China and North China. It can be seen that there are obvious differences in the distribution characteristics of lightning in different regions of China. On the whole, lightning is more frequent in the east and less in the west, more in the south and less in the north, and the coastal areas are higher than the inland. However, the terrain of Sichuan includes plateaus, mountains and plains, and the landform is complex. The dynamic uplift of mountains and the evening rain in Sichuan make the characteristics of lightning activity different from those in most areas of China.

According to the study of lightning activity in different regions of China, the lightning location data has about 2 h advance compared with radar data during hail and thunderstorm; lightning mostly occurs in densely populated cities, tall buildings and hardened ground, which is related to air pollution and urban heat island effect. However, due to the difficulty in collecting lightning data, most scholars' research on lightning is limited to flash. The research on the spatial distribution of lightning is limited in the collection point of electricity.

With the rapid development of economy and science and technology in southwest China, the aviation industry has put forward higher requirements for meteorological service, especially thunderstorm weather services. When aviation aircraft fly in the lightning area, the strong turbulence in the cloud will cause severe turbulence to the aircraft, and the plane will even be in a state of inoperability, increasing the difficulty of the pilot's operation. The strong thunderstorm electric field in lightning can seriously interfere with or damage the aircraft radio communication or other flight electronic instruments. Aircraft flight zones in Sichuan and Gansu are located on the east side of the Qinghai-Tibet plateau. Due to the dynamic and thermal benefits of high terrain of the plateau, the climate on the east side of the plateau is greatly affected. The spatial and temporal characteristics of lightning occurrence in aircraft flight zones in Sichuan, Gansu area are also different from those in other areas. Due to the close connection between lightning and convection, lightning activity can be used as an effective index to study convection activity in the region. Compared with radar and radar observation, lightning detection network is less affected by terrain, strong real-time and wide coverage, which may play a unique role in the monitoring of convection activity in the aircraft flight area of Sichuan and Gansu. In order to provide better aviation meteorological support, the use of lightning location data to understand the distribution of lightning weather phenomenon in the aircraft flight area of Sichuan and Gansu can improve the ability of aviation weather forecasters to forecast thunderstorms to a certain extent. It can provide the basis for the control department to direct the aircraft to avoid the thunderstorm area, and help the aviation weather forecaster to establish the awareness of lightning climate scenario in operation.

2 Data and Methods

The scope of this study is 26°N-34°N, 100°E-109°E, including Sichuan Basin, the eastern part of the Western Sichuan Plateau, Panxi region, and the southern part of Gansu province. The lightning location data is derived from the cloud-ground lightning observation data provided by the National Lightning Monitoring Network. The lightning data is from 2010 to 2019, which can cover the research area better. The lightning location information includes lightning occurrence time, lightning intensity, lightning waveform slope, lightning occurrence position, etc. The data accuracy is better than 300 m, the detection efficiency is greater than 80%, and the data time resolution is day by day.

This paper adopts mathematical statistical method to analyze and study the frequency of lightning activity, the seasonal, monthly and daily variation characteristics of lightning occurrence, and the spatial distribution characteristics of lightning activity in the aircraft flight area in Sichuan and Gansu.

3 Temporal Variation Characteristic of Lightning Activity

3.1 Monthly Variation of Lightning Frequency in the Flight Areas of Aircraft in Sichuan and Gansu Regions

From 2010 to 2019, the frequency of lightning in aircraft flight areas in Sichuan and Gansu was calculated monthly. As shown in Fig. 1, lightning in the region presented a single-peak characteristic, mainly in April to September. The high occurrence months were from June to August, and the peak of lightning occurrence appeared in July, with 3.61 million times. From June to August, the air is abundant in water vapor, and the warm and humid updraft is a necessary condition to generate lightning. Meanwhile, the strong convection in summer is conducive to the enhancement of unstable energy, and the final formation of lightning. Lightning rarely occurs from October to March of the next year in the winter half year, but lightning occurs every month. This is because the temperature in the winter half year is low, the atmospheric heat is not obvious, and the convection is not strong. However, when the strong cold air moves south, the phenomenon of lightning will also appear. The lightning frequency decreases rapidly in October in September, probably because the temperature drops quickly in October, and the water vapor in the atmosphere decreases obviously.

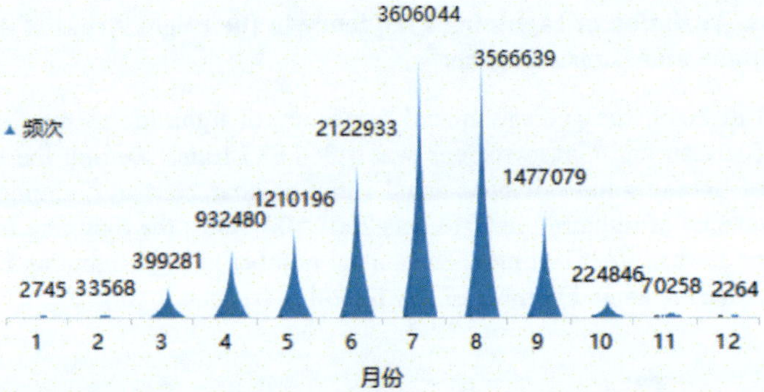

Fig. 1. Monthly frequency of lightning in the flight area from 2010 to 2019

3.2 Seasonal Variation of Lightning Frequency in the Flight Areas of Aircraft in Sichuan and Gansu Regions

Figure 2 shows the seasonal distribution of lightning occurrence frequency in the flight area. Lightning occurrence in summer is the most, accounting for 68% of the year, followed by spring for 19%, then autumn for 13%, and winter for the least. Summer is the main season of high incidence of lightning in aircraft flight areas in Sichuan and Gansu regions, while there are few lightning activities in winter. This is because the temperature is generally low in winter, and the atmosphere is dry, the convection is weak, and the triboelectrification of atmospheric motion is weak, so it is the period with the least lightning frequency.

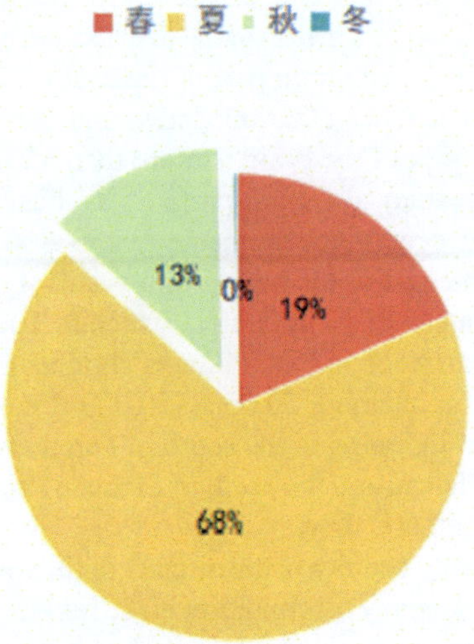

Fig. 2. Seasonal distribution of lightning occurrence in the flight area from 2010 to 2019

3.3 Annual Variation of Lightning Frequency in the Flight Areas of Aircraft in Sichuan and Gansu Regions

From 2010 to 2019, the average annual frequency of lightning in the flight area of aircraft in Sichuan and Gansu regions was 1,364,833 times. During the decade, the frequency of lightning has obvious annual variation characteristics, among which the highest frequency of lightning in 2013 was 2005,500 times; the lightning frequency in 2015 was the lowest, 982579 times. The annual number of high lightning frequency is more than twice the annual number of low lightning frequency (Fig. 3).

Fig. 3. Annual variation of lightning occurrence frequency in the flight area from 2010 to 2019

4 Spatial Distribution of Lightning in Aircraft Flight Areas in Sichuan and Gansu Regions

As shown in Fig. 4, within the flight area of aircraft in the Sichuan and Gansu regions, there are obvious high-value centers in the spatial distribution of lightning.The high-value center is located in Luzhou, the southern part of the basin. The maximum frequency of the high-value center area reaches 500,000 times, and the frequency of lightning in most areas in the southern part of the basin is more than 350,000 times. The frequency of lightning is from the western to central parts of the basin, the central and western parts of Chongqing, and the occurrence frequency is 250–350,000 times; the occurrence frequency of lightning in the rest of the basin,but it is still significantly higher than that in the western Sichuan Plateau and Panxi area, which is 100,000–250,000 times. The frequency of lightning occurrence in Panxi area is significantly higher than that in the western Sichuan Plateau, maintaining 50,000–150,000 times, among which there is a relatively high incidence of lightning in the south of Panxi area, with a total of 250,000 times. The West Sichuan Plateau and the north of Sichuan Province has a low frequency of lightning activity of 1–100,000 times.

The frequency of lightning activity in the aircraft flight area of Sichuan and Gansu is relatively consistent with the heavy precipitation in the region, and the precipitation from the west to the south of Sichuan Basin is relatively large, which also corresponds to the most active area of lightning, with a good consistency. Moreover, lightning is more active in the transition zone of complex terrain, which is mainly due to the topography, daily heating, underlying surface conditions and the water vapor content in the atmosphere and other jointly determined. In the transition zone between the basin and the mountainous area, when the water vapor in the basin is sufficient and the water vapor moves to the mountainous area, it is conducive to generating electricity in the cloud and forming thunderstorms due to the effect of topographic uplift. (Fig. 5)

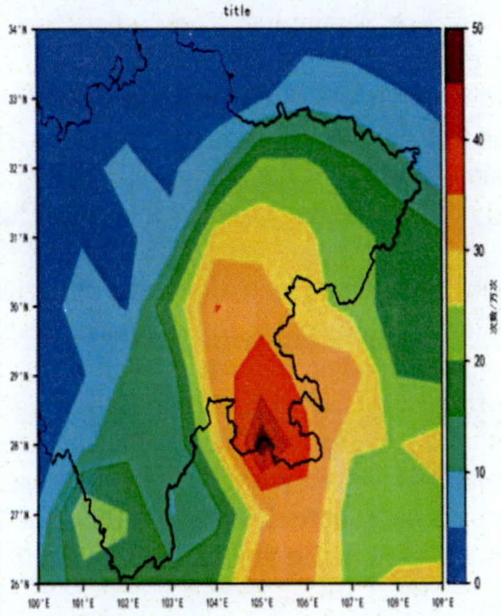

Fig. 4. Spatial distribution of lightning occurrence in the flight area from 2010 to 2019

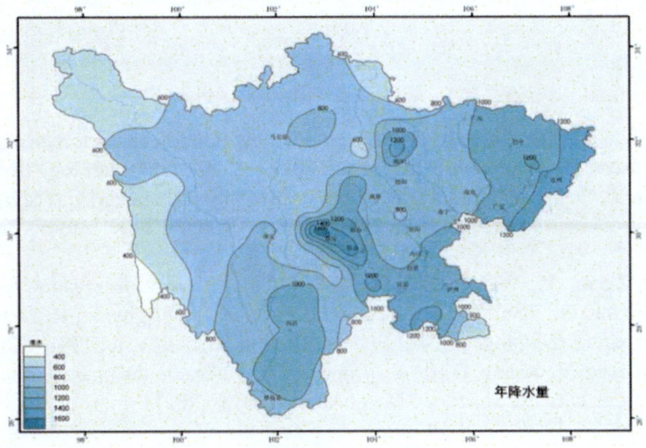

Fig. 5. Spatial distribution of annual precipitation in Sichuan region

5 Conclusion

Based on the analysis of the spatial and temporal distribution characteristics of lightning in aircraft flight areas in Sichuan and Gansu from 2010 to 2019, the main conclusions are as follows:

The monthly occurrence frequency of lightning showed unimodal characteristics, mainly concentrated in April to September. The high occurrence months were from June to August, and the peak of lightning appeared in July, with 3.61 million times. Lightning rarely occurred from October to March, but lightning appeared every month.

The seasonal variation distribution of lightning occurrence frequency: the most lightning in summer, accounting for 68% of the whole year, spring and autumn account for 19% and 13% respectively; winter is the least, less than 1%. The frequency of lightning occurrence has obvious annual variation characteristics, and the frequency of annual occurrence in high frequency year can reach more than twice in low frequency year.

There is an obvious large value center in the spatial distribution of lightning in the region, and the large value center is located in Luzhou, the south of the basin. Generally speaking, the frequency of lightning occurrence in Sichuan Basin is more than that in Panxi region, and the western Sichuan Plateau is the least. The frequency of lightning activity is relatively consistent with the heavy precipitation in the region, and the precipitation from the western to the southern Sichuan Basin is relatively large, which also corresponds to the most active area of lightning, and has a good consistency.

Fund Project. Research Project of Meteorological Support and Automatic Warning Technology of Dangerous Weather in Flight Airspace and Airline (Technology (21) (1) (B11) (0022)); Science and Technology Development Fund Project of Sichuan Provincial Key Laboratory of Rainstorm,Drought and Flood Disasters in Plateau and Basin (SCQXKJQN2020032).

References

1. Wang, K., Zhou, L., Jum X., et al.: Spatio-temporal distribution characteristics of lightning in the Yangtze River Delta region of China from 2011 to 2019. J. Meteorology Environ (2021)
2. Liu, H., Chu, Y., Wang, L.: Spatial and temporal distribution characteristics of lightning in the terminal control area of Chengdu from 2015 to 2021. Chengdu Univ. Inform. Eng. (037–003) (2022)
3. Zhou, K.H., Zheng, Y., Wang, T.: Using NWP and multi-source observation data
4. Zhang, R., Zhang, G., Wang, Y., et al.: Preliminary analysis of lightning characteristics in the northeastern part of the Qinghai-Tibet Plateau. Plateau Meteorological Syst. **32**(3), 9 (2013)
5. Zhang, W., Zhang, Y., Zheng, D., et al.: Summary of studies on the characteristics of tropical cyclone lightning activity. Shandong Meteorol (041–003) (2021)
6. Xie, D., Wang, X., Li, P., et al.: Comparative analysis of lightning activity in the Yunnan-Guizhou Plateau and the Qinghai-Tibetan Plateau. Plateau Meteorol. Meteorol. **42**(S01), 6 (2022)
7. Chai, J., Wang, X., Xiao, W.: Seasonal comparison of lightning activities in the five provinces in central China. In: Promote the Innovation of Meteorological Science and Technology, Improve the Ability of Disaster Prevention and Mitigation and Climate Change"—— Collection of the 7th Academic Exchange meeting of Jiangsu Meteorological Society (2011)

8. Wang, X., Jin, W., Wang, Y., et al.: Characteristics of spatial and temporal changes of thunderstorms and background analysis of severe annual atmospheric circulation in Dalian. J. Meteorol. Environ. **35**(2), 8 (2019)
9. Gong, C., Zeng, S., Wang, J., et al.: Analysis of the climate characteristics of thunderstorms in China in the last 30 years. Plateau Meteorol. Syst. **32**(5), 8 (2013)

Research on the Bidirectional Sealing Performance of a Fixed Ball Valve

Yajie Xu(✉), Rui Guo, Yan Xiao, and Shulin Li

AVIC Jincheng Nanjing Engineering Institute of Aircraft System, 211100 Nanjing, China
15295751363@163.com

Abstract. In this paper, the bidirectional sealing performance of a fixed ball valve in the closed state is studied through the theoretical calculation and finite element simulation. Based on the theoretical calculation model of sealing performance, the results prove that the sealing structure can ensure the bidirectional sealing performance without leakage. At the same time, the contact at the sealing surface is simulated and calculated using Workbench software. The deformation and contact specific pressure distribution on the entire contact surface are obtained. The results show that the deformation gradually increases from the center of the sealing surface to both sides, and the contact specific pressure gradually decreases from the center of the sealing surface to both sides. Reliable sealing circular-bands are formed at the center of the sealing surface in both directions. By reasonably establishing the model, setting boundaries and loading conditions, the simulation calculations are efficient and can obtain more realistic contact specific pressure distributions.

Keywords: Fix Ball Valve · Bidirectional Sealing Performance · Contact Specific Pressure · Sealing Circular-band

1 Introduction

As a shut-off valve, ball valve has the advantages of simple principle, high reliability, good sealing performance and low flow resistance [1], and is widely used in water supply, drainage, oil or gas transportation, aerospace and other fields [2]. Sealing performance is the key to ball valve design, and the sealing surfaces must be tightly fit without gaps to prevent leakage of fluid media through the gaps. The sealing performance is affected by many factors, including the quality of sealing pair, surface roughness, material characteristics, structural dimensions and physical properties of the fluid media [3]. It is of great significance to study the sealing performance during the design stage of ball valve, which is conducive to reducing the risk of leakage during actual operation [4, 5]. The sealing problem is theoretically the contact problem between the valve ball and the valve seat. Based on reasonable assumptions, the sealing performance of ball valves can be calculated by the theoretical calculation model. However, in fact, the contact between the valve ball and the valve seat contains complex nonlinear problems, which makes it difficult to obtain the contact state through the theoretical calculation. Therefore, using finite element simulation to analyze the contact problem of ball valves is a more economical and effective method.

Chinese Society of Aeronautics and Astronautics (Ed.): CASTC 2023, LNME, pp. 314–323, 2024.
https://doi.org/10.1007/978-981-99-8861-7_33

In the aircraft fuel system, the central wing fuel tank is often equipped with a cross-feed shutoff valve to control the on-off of the cross-feed pipeline in the event of an imbalance alarm for fuel in the left and right fuel tanks. Both sides of such valves may be subjected to the force of fluid media, so the sealing performance in both directions needs to be considered in the closed position. Currently, there have been many studies on the sealing performance of ball valves with different structures [6–10], but there is no research on the bidirectional sealing performance. In this paper, the bidirectional sealing performance of a fixed ball valve in the closed position is studied by theoretical calculation and finite element simulation.

2 Theoretical Model

2.1 Structural Model

The designed fixed ball valve is shown in Fig. 1. When the valve is closed, the valve seat is in close contact with the valve ball under the dual effects of fluid force and spring preload, ensuring the bidirectional sealing performance when there is fluid force on the left or right side.

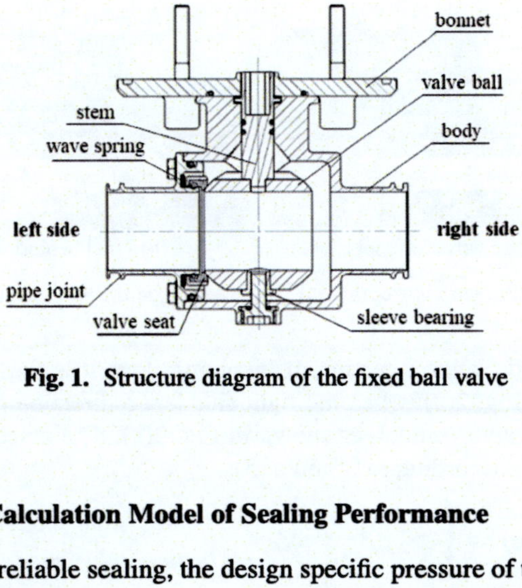

Fig. 1. Structure diagram of the fixed ball valve

2.2 Theoretical Calculation Model of Sealing Performance

In order to achieve reliable sealing, the design specific pressure of the ball valve cannot be lower than the required specific pressure, nor can it exceed the allowable specific pressure of the material:

$$q_b < q < [q] \quad (1)$$

where q_b is required specific pressure; $[q]$ is allowable specific pressure, $[q] = 18\text{MPa}$ [7] because the material of the valve seat in this paper is PTFE; q is design specific pressure. Therefore, the design specific pressure in this paper must meet the following condition to ensure sealing:

$$0.49\,\text{MPa} < q < 18\,\text{MPa} \tag{2}$$

The design specific pressure q can be calculated based on the structural model:

$$q = \frac{4Q}{\pi\left(D_2^2 - D_1^2\right)} \tag{3}$$

where Q is the sealing force, which is the resultant force acting on the sealing surface in the direction of fluid flow; D_1 and D_2 are the inner and outer diameters of the contact surface between the valve seat and the valve ball, respectively. Q can be calculated by the following formula:

$$Q = Q_{MJ} - Q_J + Q_1 - Q_2 \tag{4}$$

where Q_{MJ} is the force of hydrostatic pressure on the sealing surface; Q_J is the force of hydrostatic pressure in the clearance of the sealing surface; Q_1 is the preload acting on the valve seat by the wave spring; Q_2 is the friction force of the valve seat sliding. When the fluid pressure acts on different sides in the closed state of the ball valve, the area of thrust surface is different and the corresponding Q_{MJ} is also different, while the values of Q_J, Q_1 and Q_2 are not affected by the direction of incoming flow.

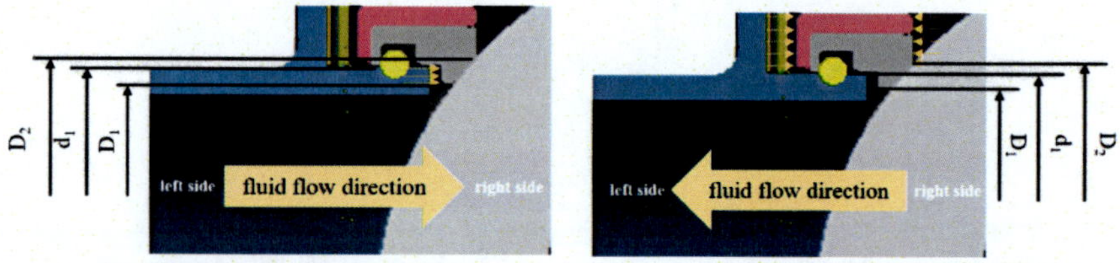

(a) "seal before valve" model (b) "seal behind valve" model

Fig. 2. The partially enlarged view of the fluid force acting on the valve seat in the closed position

When the left side is the direction of incoming flow, it is the "seal before valve" model, as shown in Fig. 2(a); when the right side is the direction of incoming flow, it is the "seal behind valve" model, as shown in Fig. 2(b). The calculations of Q_{MJ} are performed separately according to formula (5):

$$Q_{MJ} = \begin{cases} \frac{\pi}{4}P\left(d_1^2 - D_1^2\right)''\textit{seal before valve}''\textit{model} \\ \frac{\pi}{4}P\left(D_2^2 - d_1^2\right)''\textit{seal behind valve}''\textit{model} \end{cases} \tag{5}$$

where d_1 is the outer diameter of the valve seat bracket (part of the pipe joint).

Substituting the design dimension parameters into the above formulas, the design specific pressures are obtained as follows:

$$q = \begin{cases} 1.77MPa\,''\textit{seal before valve}''\textit{model} \\ 1.73MPa\,''\textit{seal behind valve}''\textit{model} \end{cases} \tag{6}$$

Equation (6) satisfies the sealing condition of Eq. (2), and the bidirectional sealing performance of the ball valve in this paper is verified through theoretical calculation.

3 Finite Element Simulation Model

3.1 Model Introduction

The ball valve model is reasonably simplified, focusing on the analysis of the sealing pair formed by the valve ball and the valve seat, and taking into account the friction between the valve seat and the pipe joint. The simplified model is shown in Fig. 3.

1-valve seat; 2-simplified pipe joint; 3-valve ball

Fig. 3. Simplified contact model of the ball valve

PTFE has more obvious plastic characteristics than rubber, and often exhibits BKIN characteristics. In the finite element analysis, the BKIN material model is used for simulation. The main mechanical performance parameters of the material are shown in Table 1. The materials of the valve ball and pipe joint are aluminum alloy. Set the contact pairs according to Table 2, and select the augmented-Lagrange contact algorithm to perform modified iterative calculation of the contact surface to improve the accuracy of contact calculation.

Table 1. The main mechanical performance parameters of PTFE[11]

Performance parameters	Density	Elastic modulus	Yield strength	Tangent modulus	Poisson's ratio
Values	2100 kg/m^3	0.4 GPa	18 MPa	120 MPa	0.35

The upper and lower cylindrical holes of the valve ball are reinforced with fixed support constraints to ensure that the valve ball will not deflect under the action of fluid force. The outer wall surface and sides of the valve seat are applied with radial and tangential constraints so that the valve seat can only move freely in the axial direction. A preload force of 2MPa is applied to the valve seat, and a fluid force of 0.41 MPa is applied to the surfaces of the valve ball and valve seat. The boundary conditions and load settings in both directions are shown in Fig. 4.

Table 2. Contact pairs settings

Contact pairs	Contact states	Contact bodies	Target bodies
valve ball and valve seat	frictional contact (friction coefficient = 0.1)	valve seat	valve ball
valve seat and pipe joint	frictional contact (friction coefficient = 0.15)	valve seat	pipe joint

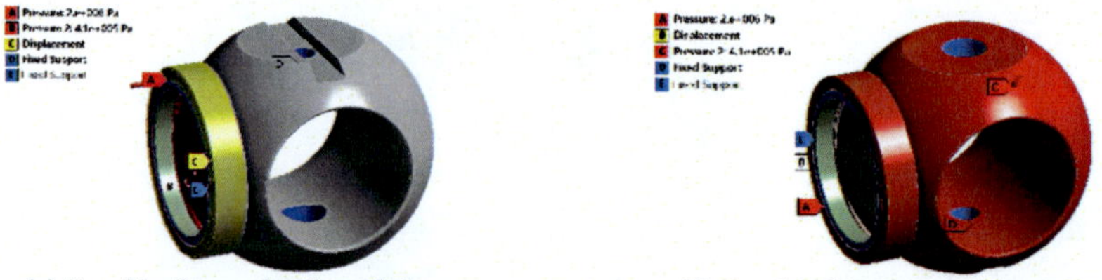

(a) "seal before valve" model (b) "seal behind valve" model

Fig. 4. Boundary conditions and load settings in both directions

3.2 Simulation Results and Analysis

The stress cloud chart of the ball valve contact model ("seal before valve" model) is shown in Fig. 5. The upper and lower parts of the valve ball are limited by the stem and sleeve bearing which can easily cause stress concentrations at right angle positions. Thus the position with the greatest stress is the root of the contact position between the valve ball and the stem.

Fig. 5. Stress cloud chart of the ball valve contact model ("seal before valve" model)

The deformation and stress distribution cloud charts on the contact surface of the "seal before valve" model are extracted, as shown in Fig. 6. The deformation and stress distribution on the contact surface are consistent. Although the constraints on the upper and lower ends of the valve ball are not exactly the same, the deformation approximately presents the law of symmetry between the upper and lower ends. Because the pressure bearing surface at the opening position of the valve ball is the thinnest position, the maximum deformation occurs here. The deformation on the contact surface of the valve seat gradually increases from the center of the sealing surface to both sides, and the deformation value of the outer circle of the contact surface is the largest. The maximum deformation on the contact surface of the valve seat and the valve ball are 0.0256 mm and 0.00192 mm, respectively. The difference in deformation will cause the actual sealing specific pressure of the contact surface to be inconsistent with the design sealing specific pressure. The stress gradually decreases along the center of the sealing surface to both sides. The maximum stresses of the valve seat and the valve ball are 3.34 MPa and 3.38 MPa, respectively, which both meet the strength requirements of the material.

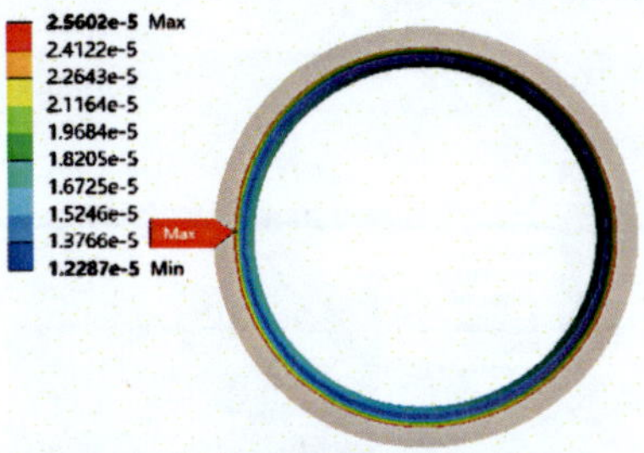

(a) deformation cloud chart of the valve seat

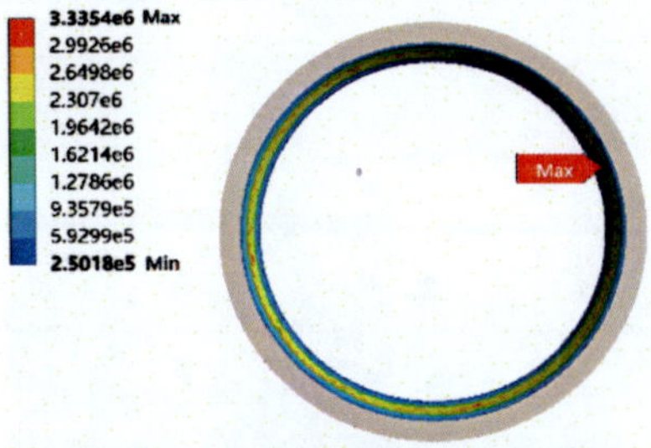

(b) stress cloud chart of the valve seat

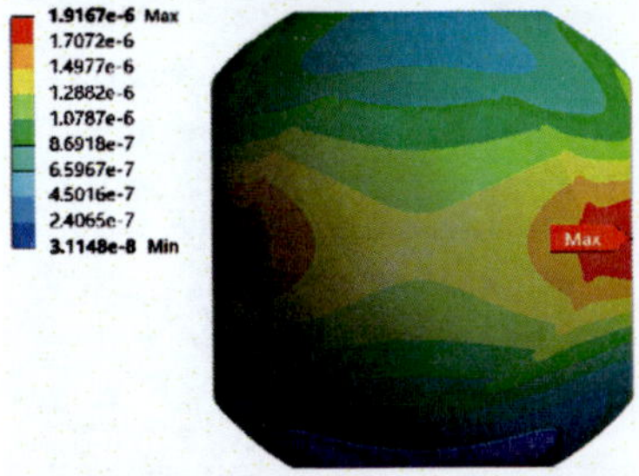

(c) deformation cloud chart of the valve ball

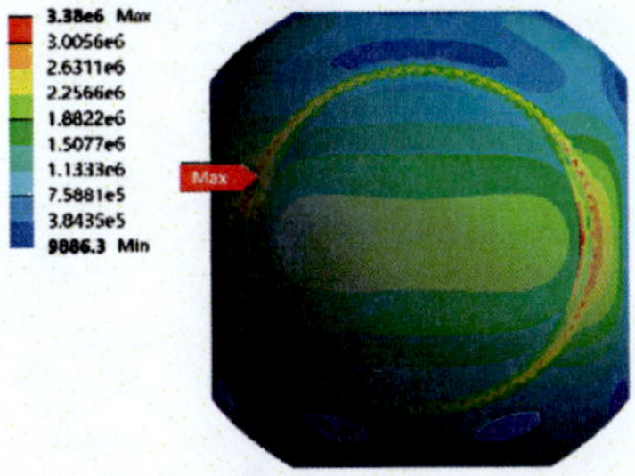

(d) stress cloud chart of the valve ball

Fig. 6. Deformation and stress cloud charts on the contact surface of the "seal before valve" model

Due to the difference in deformation between valve ball and valve seat, a change in the sealing specific pressure can occur at the sealing surface. Extract the contact state, openness, contact specific pressure and sliding distance of the contact pair for sealing evaluation and analysis. As can be seen from Fig. 7:

a) Contact state: "Near" are non-contact areas that only appear in small annular zones on the inside and outside of the sealing surface. The contact state of the central area is mainly characterized by "sliding" or "sticking". There is no through-opening (yellow area) on the sealing surface, which means that no macroscopic leakage will occur.
b) Openness: Most of the gaps between the two contact surfaces of the sealing pair are 0, indicating that there is no opening in the sealing surface relative to the initial assembly. In particular, a sealing circular-band is formed at the center which ensures

sealing performance. In this way, even if there is a certain gap between the inside and outside of the contact surface, there is no risk of leakage.

c) Contact specific pressure: Contact specific pressure gradually decreases from the center of the contact surface to both sides, contrary to the trend of deformation change. This is due to the difference in deformation, resulting in gaps in the contact of the sealing surfaces. The sealing specific pressure is the smallest at the maximum deformation. The maximum specific pressure reaches 5.19 MPa, which is greater than the necessary specific pressure for sealing. A sealing circular-band is formed at the center of the contact surface to ensure sealing.
d) Sliding distance: The sliding distance at the center of the contact surface can make the valve ball and valve seat in contact and achieve sealing.

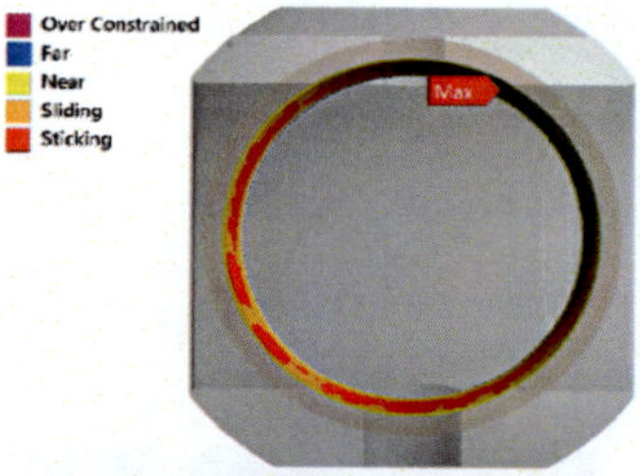

(a) cloud chart of contact state

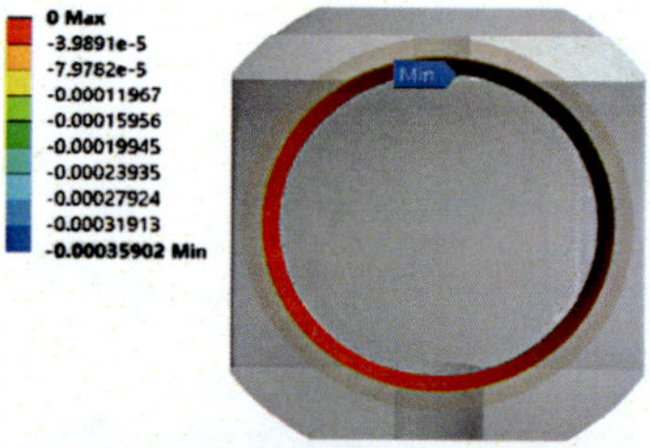

(b) cloud chart of openness

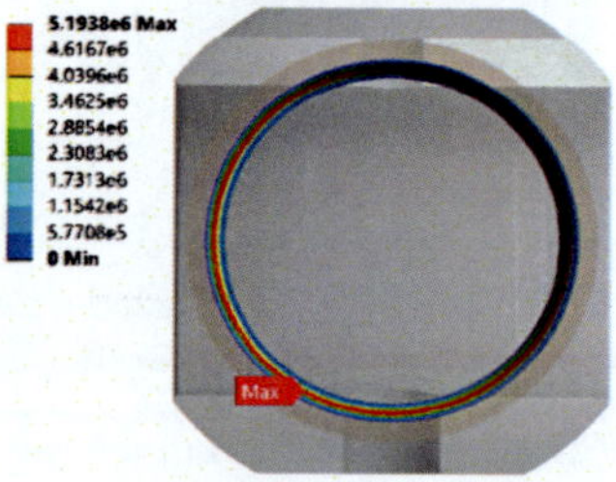

(c) cloud chart of contact specific pressure

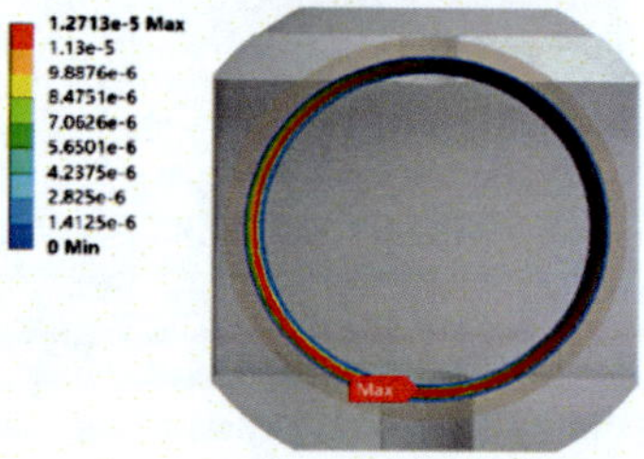

(d) cloud chart of sliding distance

Fig. 7. Contact state, openness, contact specific pressure and sliding distance cloud charts of the "seal before valve" model

The deformation and stress distribution on the contact surface of the "seal behind valve" model are shown in Fig. 8. It can be seen from the figures that the deformation and stress distribution on the contact surface are consistent. The deformation of the valve ball and valve seat is different, and the stresses all meet the strength requirements of the material.

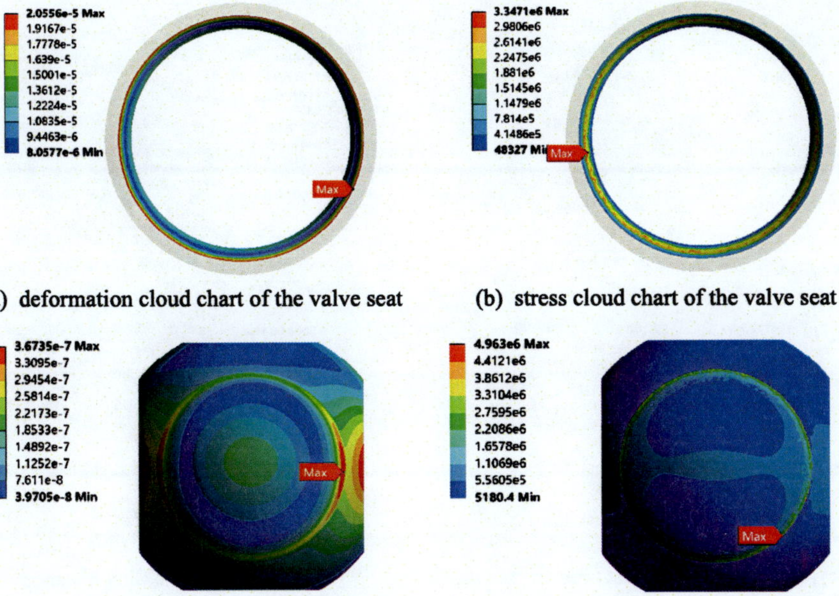

(a) deformation cloud chart of the valve seat (b) stress cloud chart of the valve seat

(c) deformation cloud chart of the valve ball (d) stress cloud chart of the valve ball

Fig. 8. Deformation and stress cloud charts on the contact surface of the "seal behind valve" model

The contact state, openness, contact specific pressure and sliding distance data of the contact pair show that the "seal behind valve" model also forms a reliable seal on the center ring belt of the sealing surface (Fig. 9).

In this section, the contact problem of the ball valve sealing structure is simulated and the results also prove the bidirectional sealing performance. However, the theoretical solutions calculated in Sect. 2.2 are relatively smaller compared to the simulation results, because the specific pressure of the sealing surface presents a distribution law that gradually decreases from the center to both sides. And the theoretical model assumes that the specific pressure of the seal acts uniformly on the entire contact surface, so the theoretical values are smaller.

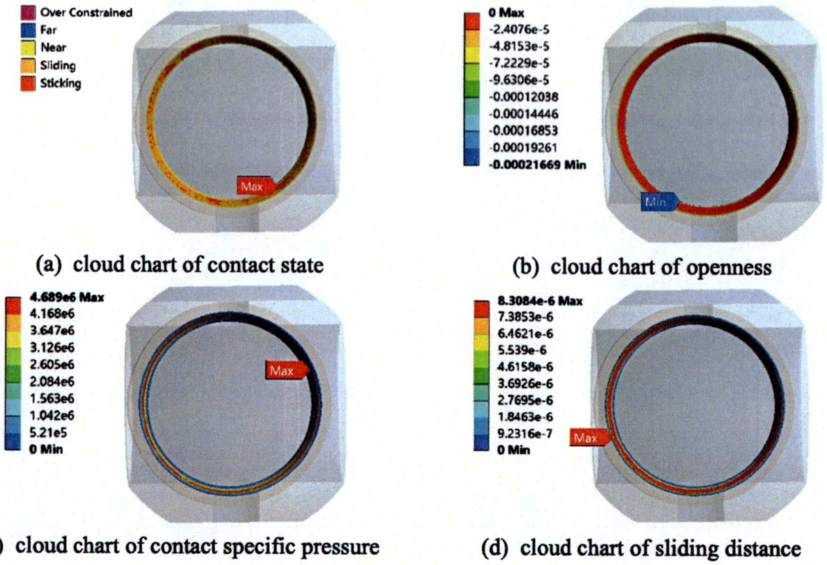

Fig. 9. Contact state, openness, contact specific pressure and sliding distance cloud charts of the "seal behind valve" model

4 Conclusion

In this paper, the bidirectional sealing performance of a fixed ball valve in the closed state is studied by theoretical calculation and finite element simulation. The following conclusions can be obtained:

a) Based on the theoretical calculation model of sealing performance, the theoretical calculation results prove that the sealing structure of the ball valve in this paper can ensure the bidirectional sealing.
b) Based on Workbench software, the nonlinear contact problem between the valve ball and the valve seat is calculated by finite element simulation, and the stress distribution on the entire contact surface is intuitively obtained. The contact specific pressures gradually decrease from the center of the contact surface to both sides, forming reliable sealing circular-bands. The simulation results also verify the bidirectional sealing performance.
c) The advantage of theoretical calculation is that the calculation amount is small, while the disadvantage is that the calculation difficulty will increase for complex sealing structures. The theoretical calculation value is based on some ideal assumptions, which may differ from the actual use. The finite element simulation calculation method in this paper greatly improves the calculation efficiency and can obtain a more realistic contact specific pressure distribution, which can further become an important tool in the design of ball valve sealing structure.

References

1. Qiang Zhu, F.: Discussion on the classification and characteristics of ball valves. Theoretical Research in Urban Construction **8**, 1–6 (2014). (in Chinese)
2. Jinxiao Li, F.: Numerical simulation and analysis of the contact stress of the sealing surface of a fixed ball valve during the opening and closing process. Lanzhou University of Technology (2018). (in Chinese)
3. Dingrui Yang, F.: Research and optimization on sealing performance of hard seal fixed ball valve. Lanzhou University of Technology (2021). (in Chinese)
4. Hui Gao, F., Xiaohui Hou, S.: Sealing principle and leakage analysis of pipeline ball valve. Oil Gas Storage Trans. **25**(3), 57–60 (2006). (in Chinese)
5. Huayou Zhang, F., Zerong Yan, S., Yuanfang Chen, T.: Design and selection of ball valves. Beijing Science and Technology Publishing, Beijing (1994). (in Chinese)
6. Xing Liu, F., Jing Mei, S., Li Chen, T.: Analysis of the structure and sealing characteristics of the GQ47H-DN370 type orbiting ball valve seat. Heilongjiang Sci. **12**(24), 22–28 (2021). (in Chinese)
7. Yufeng Wang, F.: Strength stiffness and friction characteristics analysis and research on the opening and closing process of high temperature and high pressure ball valve. Lanzhou University of Technology (2016). (in Chinese)
8. Jun Wei, F.: Sealing performance analysis and optimization of high-pressure metal hard sealing fixed ball valve. Nanchang Hangkong University (2017). (in Chinese)
9. Min Jiang, F.: Simulation analysis and experimental research for sealing characteristics of positive seal ball valve. Shanghai Jiaotong University (2015). (in Chinese)
10. Min Chen, F., Wencheng Tang, S., Yifang Zhang, T.: Finite element analysis of hyperelastic contact problem in valve sealing structure. China Mech. Eng. **18**(15), 1773–1775 (2007). (in Chinese)
11. Yifang Zhang, F., Bin Wu S., Jianxin Wang, T.: Study on sealing characteristics of ball valve. Valve **5**, 14–17 (2015). (in Chinese)

Research on Testing and Evaluation Methods of Infrared Radiation Characteristics of Air Targets

Yi Wang(✉), Hao Wang, Zhen Wang, and Xue Wang

Chinese Flight Test Establishment, Xi'an, Shaanxi Province, China
478964608@qq.com

Abstract. For the problems of implementation difficulty, high cost and low efficient in obtaining the distribution of the infrared radiation characteristics of the dynamic target, the test system and dynamic test implementation method of air targets' infrared radiation characteristics were introduced in this paper systematically. At the same time, the data processing methods were described in detail for different test equipment. The effectiveness and correctness of the proposed method were verified by flight test. The results show that the infrared characteristic distribution data of the target in the all visual angles can be obtained through the proposed method efficiently and the accuracy were reliable. The method proposed in this paper can support the acquisition of infrared characteristic data of different types of air targets.

Keywords: Infrared Radiation · Dynamic Test · Air Targets

1 Foreword

The infrared radiation characteristics of air targets are important indicators to characterize the infrared stealth ability of targets, and the key indicators to measure the comprehensive performance of military aircraft. Especially, with the rapid development of military technologies, such as detection, anti-detection, countermeasures, and anti-countermeasures, provided serious threats to the battlefield viability of military aircraft. Therefore, carrying out the research of infrared radiation characteristics testing and evaluation of air targets, obtaining the infrared characteristic distribution and adopting various infrared stealth methods to reduce the probability of detecting and attacking aerial targets have become an inevitable trend in the development of weapons and equipment.

The infrared radiation characteristics of air targets can be obtained by ground-to-air test and air-to-air test. For these two methods, the air-to-air test method obtains targets features in more visual angles, which are helpful for refined evaluation of target infrared characteristics. Air-to-air test measures obtains the infrared radiation characteristics of aerial targets in real flight conditions by loading test equipment on an aerial measurement platform and implement test by airborne companion flight with the target. Through data processing and calculation, the infrared stealth effect of stealth aircraft in real

Chinese Society of Aeronautics and Astronautics (Ed.): CASTC 2023, LNME, pp. 324–331, 2024.
https://doi.org/10.1007/978-981-99-8861-7_34

environments can be evaluated [1]. Due to the importance of infrared characteristics in the survivability of equipment, many countries attach great importance to test and evaluate the infrared radiation characteristics of air targets and have established a large number of infrared test systems. Take the United States for example, a number of test systems have been developed to obtain aircraft infrared characteristics, mainly focusing on two major military bases, the Eglin Air Force Base and the Naval Air Warfare Center. The test method is that using airborne external infrared measurement pods to form an aerial formation with the targets. In addition, the United States also uses multiple large aircraft to establish airborne target optical characteristics laboratories to collect and measure the optical characteristics of air targets and backgrounds [2, 3]. A large amount of testing, evaluation, and simulation modeling work has also been carried out in China in terms of infrared radiation characteristics testing of air targets, the dynamic test mainly conducted in the form of air-to-air or ground-to-air method [4].

Air-to-air infrared radiation characteristics test is difficult to carry out due to its high cost, implementation difficulty, and significant constraints from the environment and test equipment. Aiming at the above issues, this article focuses on the test and evaluation methods of air-to-air infrared radiation characteristics, supporting the acquisition and evaluation of infrared characteristics data of air targets.

2 Analysis of Infrared Radiation Characteristics of Aircraft

During the flight in air, the intrinsic infrared radiation sources of the aircraft mainly include two parts, one is the infrared radiation from the engine exhaust nozzle and exhaust flame, which is mainly concentrated in the medium wave band, and the other is the infrared radiation from the fuselage skin, which is mainly concentrated in the long wave band [5].

In real environment, the infrared radiation characteristics of an aircraft are inevitably affected by background infrared radiation, mainly manifested as the following:

- The reflection of solar radiation from the fuselage.
- The reflection of sky (including clouds) radiation from the fuselage.
- The reflection of ground radiation from the fuselage.

The infrared radiation characteristic obtained in air is the result of a combination of various factors which are difficult to analyze in isolation, especially the radiation of the fuselage skin [6, 7]. On the other hand, the infrared radiation of an aircraft is closely related to parameters such as engine operating conditions, flight altitude, speed, skin emissivity, meteorological conditions, and detection angle [8, 9]. In order to analyze the infrared radiation characteristics of aircraft, it can be divided into two parts: (1) The infrared radiation characteristics analysis of the fuselage skin. (2) The infrared radiation characteristics analysis of engine exhaust nozzle and exhaust flame. In the process of aerial dynamic test and evaluation, it is also necessary to carry out aerial dynamic test of targets based on visual angles.

3 Infrared Radiation Characteristics Test of Aerial Targets

3.1 Test System

This section focuses on air-to-air infrared radiation characteristics test. In this method, the aerial tested target and the aircraft equipped with infrared test equipment fly in formation to complete the infrared radiation characteristics test of the target in different visual angles. In the process of dynamic infrared characteristic testing, different aircrafts mount corresponding infrared test equipment to form different airborne dynamic infrared test systems.

There are mainly two types of mounting methods:

1. Pod type air-to-air dynamic infrared test system
Utilize the external mounting capabilities of existing combat aircraft to design a dedicated pod equipped with infrared thermal imagers, spectral radiometers, and visible light cameras to conduct testing on the infrared radiation brightness, radiation intensity, spectral distribution, and other characteristics of the tested target.
2. Loaded type air-to-air dynamic infrared test system
This type of system mainly utilizes the high loading capacity of aircraft such as transport aircraft, and is equipped with testing equipment such as thermal imagers, spectral radiometers, and visible light cameras to conduct testing on the infrared radiation brightness, radiation intensity, spectral distribution, and other characteristics of the tested target.

In the process of testing the infrared radiation characteristics of air targets, the setting of test equipment parameters is determined based on the status of test target. Generally, the forward and lateral areas of target are mainly the fuselage skin, with relatively low temperatures. The radiation bands are mainly concentrated in the long wave band (8 μm–14 μm). The backward mainly includes the engine nozzle and tail flame, with high temperature, and the radiation band is mainly concentrated in the medium wave band (3 μm–5 μm). According to the test visual angles to set the parameters of the spectral radiometer and infrared thermal imager in the test system, and then begin the infrared radiation characteristics test of air targets.

3.2 Air-to-Air Infrared Test Method

During the air-to-air dynamic infrared radiation characteristic test, the aerial tested target and the aircraft equipped with infrared test equipment fly in formation, and the test distance is controlled within 1000m. Changing the formation situation and the installation status of the test equipment according to different visual directions, the 360° circumferential infrared radiation characteristics of the target aircraft can be obtained.

The specific test plan is as the following:
1. The tested target and the test aircraft equipped with infrared test equipment are flying in formation at the specified flight altitude and speed.
2. Control the direction of the field of view of the test equipment, keeping the tested target fully appears in the center of the field of view, and adjusting the formation distance to maximize the proportion of the target in the field of view of the test equipment.

3. Change the relative positions between the tested target and aircraft equipped with infrared test equipment from the horizontal and vertical sections, while the test equipment keeps track of the target from beginning to end, and complete the infrared radiation characteristics test of the target at other visual angles. The flight trajectory of the aircraft equipped with infrared test equipment around the tested target is shown in Fig. 1.
4. Transform the field of view direction of the equipment to test the infrared radiation characteristics of the target in different directions. The measurement distance is required to be as close as possible to eliminate the impact of the atmosphere on infrared attenuation over the test distance.

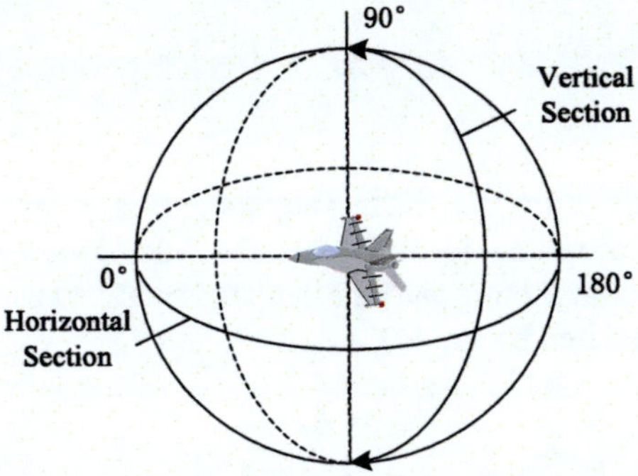

Fig. 1. The circling fly path for testing omnidirectional infrared radiation characteristics of a target.

4 Data Processing

The infrared image, spectral radiation intensity, and radiation intensity of the target at different visual angles can be obtained through the air-to-air dynamic infrared testing. According to the different detection principles of spectral radiometers and infrared imaging system, the data processing methods are different.

4.1 Data Processing for Infrared Spectroradiometer

When using an infrared spectral radiometer to obtain the infrared radiation characteristics of a target, the signal obtained is the sum of the radiation of the target and the background, which needs to be deducted to obtain the target radiation intensity. Target radiation brightness L_t is:

$$L_t = L_{tb} - k \cdot L_{bg} \tag{1}$$

where,

L_t is the target radiation brightness, $W{\cdot}m^{-2}{\cdot}Sr^{-1}$.

L_{tb} is the radiation brightness of target and background, $W{\cdot}m^{-2}{\cdot}Sr^{-1}$.

k is the proportion coefficient of the background to the field of view.

L_{bg} is the background radiation brightness, W·m^{-2}·Sr^{-1}.

k can be obtained by Eq. (2).

$$k = \frac{(A_0 - A_t)}{A_0} \tag{2}$$

where,

A_t is the target radiation area, m^2.

A_0 is the field of view area at the test distance (target), m^2.

A_0 can be obtained by Eq. (3).

$$A_0 = \pi l^2 tg^2\left(\frac{\theta}{2}\right) \tag{3}$$

where,

l is the distance between target and test equipment, m.

θ is the field angle of test equipment (when the field of view is circular), (°).

The target radiation brightness can be calculated by Eq. (1), and the target radiation intensity I_t can be calculated by Eq. (4):

$$I_t = A_o \cdot L_t \tag{4}$$

where, I_t is the radiation intensity of targe, W · Sr-1.

The target radiation intensity calculated by Eq. (4) can be corrected using Eq. (5) for atmospheric absorption to obtain the true radiation value of target.

$$I_{te} = \frac{I_t}{\tau_a} \tag{5}$$

where,

I_{te} is the total effective radiation intensity of the target, W · Sr^{-1}.

τ_a is the atmospheric transmittance.

4.2 Data Processing for Infrared Imaging System.

When process data obtained by infrared imaging system, it is necessary to divide the image into isothermal regions, then give the average radiation temperature and radiation area of the isothermal region i (i=1,2,3,... n). The radiation brightness of isothermal zone i can be calculated by Eq. (6) and Eq. (7).

$$M(\lambda T) = \frac{2\pi \cdot c^2 \cdot h}{\lambda^5} \frac{1}{e^{\frac{hc}{\lambda \cdot k \cdot T}} - 1} \tag{6}$$

where,

$M(\lambda T)$ is the blackbody spectral radiation emission at target equivalent absolute temperature T, W · m^{-2}· μm^{-1}.

h is Planck constant, the value is 6.626 × 10^{-34}, J·s.

k is Boltzmann constant, the value is 1.38×10^{-23}, $W \cdot s \cdot K^{-1}$.

The radiation brightness of the target in the isothermal region at the radiation temperature can be calculated by Eq. (7).

$$L_t = \frac{1}{\pi} \int_{\lambda_1}^{\lambda_2} M(\lambda, T) d\lambda \tag{7}$$

where,

L_t is the radiation brightness in the isothermal region at the target radiation temperature in the band of $\lambda_1 \sim \lambda_2$, $W \cdot m^{-2} \cdot Sr^{-1}$.

The radiation intensity of the target isothermal zone i can be obtained by Eq. (8).

$$I(i) = A_i \cdot L(i) \tag{8}$$

where,

$I(i)$ is the radiation intensity of the target isothermal region i within the working wavelength range of the equipment, $W \cdot Sr^{-1}$.

$L(i)$ is the radiant brightness of the target isothermal zone i, $W \cdot m^{-2} \cdot Sr^{-1}$.

The total radiation intensity I_t of the target can be obtained by Eq. (9).

$$I_t = \sum_{i=1}^{n} I(i) \tag{9}$$

where,

I_t is the total radiation intensity of the target within the working wavelength range of the equipment, $W \cdot Sr^{-1}$.

The total effective radiation intensity I_{te} of the target through atmospheric absorption can be corrected by Eq. (10).

$$I_{te} = \frac{I_t}{\tau_a} \tag{10}$$

In the above Equation, τ_a is the atmospheric transmittance of the test path within the working wavelength range ($\lambda_1 \sim \lambda_2$) of the infrared imaging system.

5 Results and Analysis

According to the above test method, when conducting infrared radiation characteristic testing for a certain type of aircraft, the aerial tested target and the aircraft equipped with infrared test equipment fly in formation and change their positions in the agreed formation manner, and the target radiation intensity in different visual angles can be obtained within 6 to 8 effective sorties. The visual angles distribution of the tested target within a certain sortie is shown in Fig. 2 and Fig. 3. In terms of test accuracy, the calculated error is within the range of 15% to 20%, which meets the requirements of dynamic test after correction.

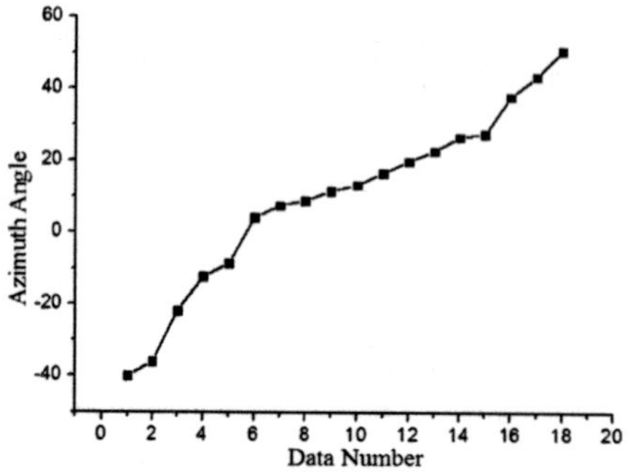

Fig. 2. Distribution of azimuth angle

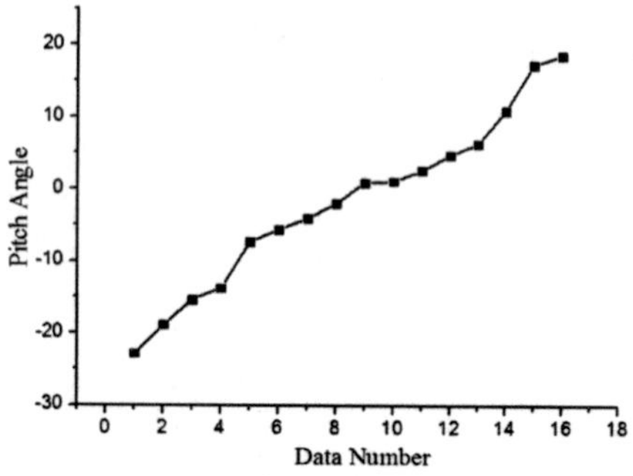

Fig. 3. Distribution of pitch angle

For aerial targets, the projected area of the aircraft changes from large to small and then increases when the pitch angle changes from large to small, the projected area of the aircraft reaches the minimum at the pitch angle of 0°. So when the aircraft's azimuth angle is fixed and the pitch changes from large to small, the radiation intensity first decreases and then increases, with the radiation intensity being the smallest near the pitch angle of 0°. The reason why the radiation intensity at upper view is greater than that at lower view is that when viewed upward, the radiation energy at the lower surface of the aircraft accounts for the majority, while when viewed downward, the radiation energy at the upper surface of the aircraft accounts for the majority. The temperature at the lower surface of the aircraft is greater than the temperature at the upper surface of the aircraft, which means that the radiation energy at the lower surface is greater than the radiation energy at the upper surface.

When the pitch angle is fixed and the azimuth angle increases from 0° to 90°, the projected area of the aircraft continues to increase. Therefore, the radiation intensity increases with the increase of the projected area.

6 Conclusion

The test methods for dynamic infrared radiation characteristics of air targets and the data processing methods for different test equipment were described in detail in this article. Through flight tests, the effectiveness and correction of the test and evaluation methods proposed for infrared radiation characteristics of air targets are verified. The dynamic test method for infrared radiation characteristics proposed in this article can be used to obtain infrared characteristics data of different types of aerial targets, and also support the identification and evaluation of targets' infrared characteristics.

References

1. Chaozhe, T.O.N.G., Zhongxiang, T.O.N.G., Yongzhi, L.I.U., et al.: Evaluation method of stealth aircraft's infrared radiation measurement. Infrared Laser Eng. **41**(11), 2891–2897 (2012)
2. Cline, J., Behm, D., Kidd, K., et al.: F/A-22IR signature flight test model validation. Aircraft Surv. **3**, 9–12 (2003)

3. Lizhu, K.A.N.G., Jinsong, Z.H.A.O., Zhenhua, L.I., et al.: The overview of the research work developments on infrared signature of aircrafts. Infrared Technol. **39**(2), 105–124 (2017)
4. Yongmin, X.I.E., Weidong, S.H.E.N., Sihong, S.O.N.G., et al.: Effectiveness evaluation of infrared steal based on the contrast of target and background infrared radiation. Infrared Technol. **33**(2), 113–115 (2011)
5. Xue, C.: Research on infrared radiation energy measurement method based on skin. shenyang university, Shenyang (2014)
6. Shuheng, C., Niyao, Z., Ning, C., et al.: Test of infrared radiation characteristic of aircraft and study on launch strategy for infrared decoy flare. Infrared Technol. **43**(10), 949–953 (2021)
7. Yao, K., Fan, X., Wang, Y.: Research on testing technology of aircraft. Measure Contr. Technol. **39**(4), 6–9 (2020)
8. Ren, L., Zhang, J., Wang, X., et al.: Analysis of steal properties on IR radiation suppressor embed inside helicopter rear airframe. Infrared Laser Eng. **40**(11), 2091–2097 (2011)
9. Yan, L., Feng, W.: Research on Infrared radiation characteristics modeling method of ship-to-air missile's typical flight target. Ship Electron. Eng. **41**(9), 186–189 (2021)

Multi-objective Optimization of a Simulating Specimen Design of a Turbine Rotor

Tianzi Zhang[1], Yixiong Liu[1,2](✉), Qing Du[1], and Yunwu Wu[1]

[1] AECC Shenyang Engine Research Institute, Shenyang 110015, China
Yixiong.liu5021@gmail.com

[2] Northwestern Polytechnical University, Xi'an 710072, China

Abstract. This paper aims to perform an optimization of the simulating specimen to search for the optimal specimen design. A design flow chart was adopted and the basic guidelines of the simulating design for the turbine rotor were introduced. The main principle of simulating specimen design is to ensure that the maximum stress and stress gradient at the notch of the specimen are equivalent to the critical part of the component. Then, the genetic algorithm (GA) optimizer capable of optimizing multi-objective was developed combining the ANSYS APDL code. It was found that the vent hole of the turbine front baffle was the critical part. Stress analysis of each individual generated by the GA optimizer was performed and the maximum stress of the specimen was almost the same as the component, reaching 1024MPa. The stress gradient was also compared and showed good consistence with the component. Afterwards, the LCF test of the specimen was carried out and the specimen encountered failure after 150572 cycles. It is 2.18% and 3.36% lower than the predicted results of the specimen and component respectively. It offers a good reference for the LCF life prediction of the turbine rotor.

Keywords: Genetic Algorithm · Stress Gradient · Maximum Stress · LCF test · LCF life

1 Introduction

The low cycle fatigue (LCF) life of aeroengine rotor is a crucial design requirement for aeroengine. It is closely related to the structural integrity and reliability as well as flight safety. The rotor slot, bolt hole, ventilation hole, balancing hole and flange are the most concerned parts that occupy high stress concentration. In order to satisfy the civil aviation regulation, the LFC life prediction and test need to be carried out to verify the rotor LCF life. However, the fatigue tests of actual components are time-consuming and expensive, making it difficult to achieve the test objectives. To be specific, it is highly dependent on the material dispersity and manufacturing quality. The LCF life might vary in a large range and cannot be used to estimate the engine life. Besides, the number of the test samples is limited due to the economic cost. Therefore, it is more helpful to utilize the rotor LCF test specimen to study the fatigue life and material dispersity of the turbine rotor. The aim is to design the specimen that employing the same stress gradient as the real component and investigate the LFC life of the specimen.

Chinese Society of Aeronautics and Astronautics (Ed.): CASTC 2023, LNME, pp. 332–343, 2024.
https://doi.org/10.1007/978-981-99-8861-7_35

Numerous researchers have carried out studies on the LCF specimen design and test. At first, the design principle is to ensure that the loading and deformation of the maximum stress point of the specimen equal to the real component [8]. Meanwhile, the structure and shape of the specimen are similar to the simulated component. Nevertheless, with the improvement of fatigue theory and design requirements, the stress gradient and stress concentration are both considered in the simulating specimen design. Zhao [2] proposed a design criterion for engine simulating specimen based on consistency in the strain range at the maximum stress point and the coefficient of strain distribution. Liu [1] designed a compressor tenon slot simulating specimen based on the consistency in the distribution of the first principal stress. To be specific, the stress distribution in the range of 6mm within the maximum principal stress maintaining the same level with the simulated component. Similarly, Lu [4] performed the simulating specimen design considering the stress gradient along the 3 smm of the maximum principal stress. Yang [5] simulated the fatigue life of a compressor rotor using the tensile strain energy theory. Zheng [6] optimized the simulating specimen for the bolt hole of a high-pressure turbine of a turbofan engine based on the fatigue life prediction model of tensile strain energy.

Recently, optimization software and method are implemented in the simulating specimen design to accurately simulating stress gradient and maximum stress. Kuang [7] utilized the Design Xplorer parameter optimization module of ANSYS WORKBENCH to design a LCF specimen considering the maximum stress point and stress gradient distribution.

This paper presents a multi-objective parameter optimization design method for simulation parts based on a genetic algorithm for typical parts of aircraft engine components. A simulation part for the air vent hole of a certain engine model's low-pressure turbine front baffle was designed, and experimental validation was conducted.

2 Methodology

2.1 Design Process

The design flow chart of the paper is depicted in Fig. 1. The first step is to perform the stress analysis of the component to obtain the stress gradient and the maximum stress. Afterwards, the LCF life analysis would be carried out and the critical location limiting the LCF life would be confirmed. It is crucial to achieve the stress distribution and the maximum stress of the critical area, providing a good reference for the simulating specimen design. As for simulating specimen optimization, a parametric model has to be established to consider all the influential factors. Then, the stress analysis is performed using the genetic algorithm to approaching the optimal design. The APDL code would be generated for each iteration and calculate the stress of every individual. The design objectives are to obtain the same stress distribution of real component and the specimen. The main reason is to accurately estimated the LCF life of the specimen maintains and predict the real LCF life of the component. Therefore, it highlight the importance of ensuring that both the stress gradient within the critical area and the maximum stress would be simulated concurrently.

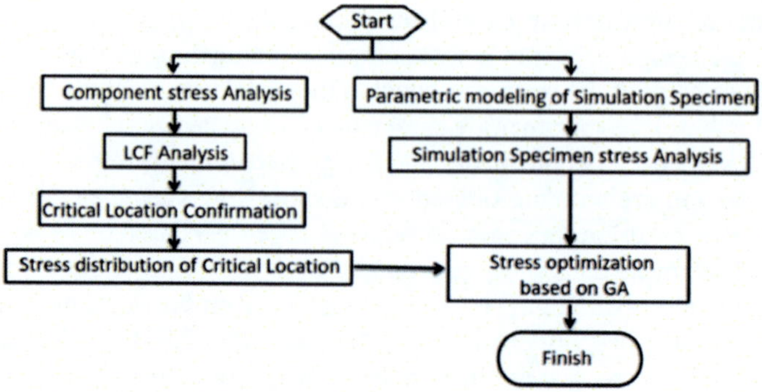

Fig. 1. Simulating specimen design flow chart

2.2 LCF Life Prediction Method

Neuber Method

The LCF life of the turbine rotor is calculated based on the local stress-strain method. It is developed according to the Neuber equation. The fatigue cycle life of the component is predicted implementing the Manson-Coffin equation. It should be noted that if the material data is not available for all cases, the LCF life would be predicted according to the slope method.

The modified Manson-Coffin equation is shown in (1). $\Delta\varepsilon$ is the range of strain while the ε'_f is the fatigue ductility coefficient. σ_m and σ'_f are average stress and fatigue strength coefficient respectively. N_f means the cycle life. E is the elastic modulus. b and c are fatigue strength exponent and fatigue ductility exponent.

$$\frac{\Delta\varepsilon}{2} = \frac{\sigma'_f - \sigma_m}{E}(2N_f)^b + \varepsilon'_f(2N_f)^c \tag{1}$$

The cyclic stress-strain equation is shown in (2):

$$\frac{\Delta\varepsilon}{2} = \frac{\Delta\sigma}{2E} + (\frac{\Delta\sigma}{2K'})^{1/n'} \tag{2}$$

where,

K' is the coefficient of cyclic strength.

$\Delta\sigma$ is the range of stress.

n' is the cyclic strain hardening exponent.

The modified Neuber equation is written as (3):

$$\Delta\sigma \cdot \Delta\varepsilon = \frac{K_f^2(\Delta s)^2}{E} \tag{3}$$

where,

K_f is the effective stress concentration coefficient.

The primary assumption is that the life in logarithm form is linear to the stress concentration coefficient, as presented in (4).

$$K_f = a\,\lg(N) + b \tag{4}$$

When K_f equals 3, the effective stress concentration coefficient is achieved by interpolation shown in (5):

$$K_f' = \frac{(K_f - 1)(K_t - 1)}{2} + 1 \tag{5}$$

Slope Method.
As it is not possible to access all the test data of the LFC life, the slope method is available to estimate the cycle life, as presented in (6).

$$\Delta\varepsilon = \left[\ln\frac{1}{1-\psi(t,T)}\right]^{0.6} N_f^{-0.6} + \frac{3.5[\sigma_{sr}(t,T) - \sigma_m]}{E(T)} N_f^{-0.12} \tag{6}$$

where,

$\Delta\varepsilon$ is the strain range of the critical part considering the stress concentration.

$\sigma_{sr}(t,T)$ is the ultimate stress limit corresponding to the maximum temperature and total duration of transition state during the service life.

$\psi(t,T)$——is Reduction of area of the material corresponding to the maximum temperature and total duration of the transition state during the life cycle.

$E(T)$ is elastic modulus;

σ_m is the static component of the cycle.

2.3 Genetic Algorithm Method

In this scenario, the multi-objective GA method is implemented to perform the simulating specimen optimization, which would achieve the maximum stress and stress gradient concurrently within the design space. It offers an in-depth understanding into the influences of the critical parameters on the stress distribution.

During the optimization, GA tool was applied to search for the optimum simulating specimen within the design space. The GA optimizer was developed by Liu [8, 9] based on the NSGA II [10]. It would generate the design variables according to the mutation, selection and crossover operators. Afterwards, the ANSYS software would be called to perform the finite element analysis for each individual. It is widely accepted that the GA operator settings would affect the optimal results. Therefore, the verification process was carried out to guarantee a stable solution. The crossover fraction, mutation fraction, population size and maximum generation were selected as 0.7, 0.7, 100, 300. By confirming the design variables, design constraints and design objectives, the optimizer would run ANSYS to obtained the desired values and compare them with the design objectives. The optimal solution is achieved when the fitness is quite small, indicating the calculated maximums stress and stress gradient are close to the targets. It should be noted that the initial values of the optimizer are highly dependent on the designer's experience and should be carefully designed to achieve the optimum solution at the lowest cost.

3 Simulating Specimen Design

3.1 Design Principle

In this scenario, a non-standard simulating specimen was designed to simulate the stress distribution of the turbine rotor. The main principle is to ensure that the maximum stress and stress gradient at the critical parts are equivalent to the component. Meanwhile, the material, manufacturing process, surface quality and shot blasting should maintain the same levels.

The basic shape of the simulating specimen is shown in Fig. 2. Obviously, the stress concentration is mainly affected by the notch radius and notch depth. The stress concentration and maximum stress of the simulating specimen are obtained by adjusting the notch radius and depth.

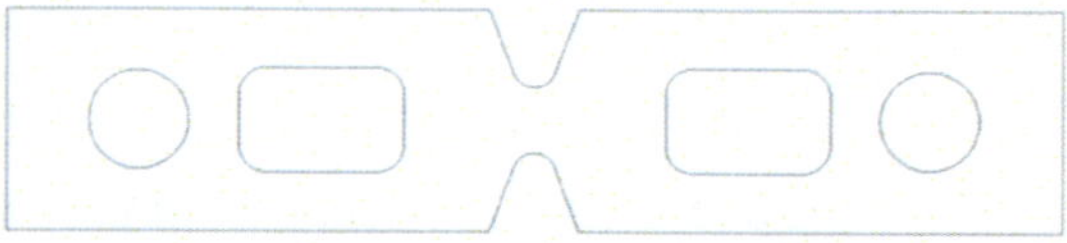

Fig. 2. Demonstration of simulating specimen

The basic guidelines for material selection, shape design, and loading simulation of the specimen are listed as follows:

a. Keep the material type and fatigue performance consistent with the component. Besides, the forging direction should also be considered. It would be better to take the sample of the specimen from the real component to maintain the maintain the mechanical property.
b. The temperature of the specimen needs to be the same as the component. It is essential for the LCF life design since temperature impacts the stress and cycle life simultaneously.
c. The strain range $\Delta\varepsilon$ or stress range $\Delta\sigma$ at the maximum stress point equals that of the component.
d. The geometric shape of the stress concentration area is similar to that of the component. The strain distribution influencing factor life model only considers the effects of strain distribution on the virtual crack in the main direction (generally the direction with the largest gradient), but not includes other directions. It would be better to design a similar shape of the specimen to mitigate the stress differences.

3.2 Design Space Exploration

In order to fully explore the design space and achieve the optimal simulating specimen design, the GA method is implemented in the design optimization. As presented in Fig. 3, 17 parameters are included as the design variables. The definitions, lower bounds and upper bounds of the variables are listed in Table 1. The most concerned parameters are the notch radius and notch depth as they are closely related to the critical area.

All the variables are generated by the GA optimizer and inputted to the ANSYS APDL code. Afterwards, the stress analysis would be carried out to calculate the stress gradient and maximum stress. Then, the obtained stress gradient and maximum stress of the specimen are compared with the component. The fitness is the stress difference between the specimen and component, as presented in (7). The maximum stress point and the stress within 2 mm are considered. In summary, the fitness is desired to be close to zero to precisely simulate the component stress distribution (Table 2).

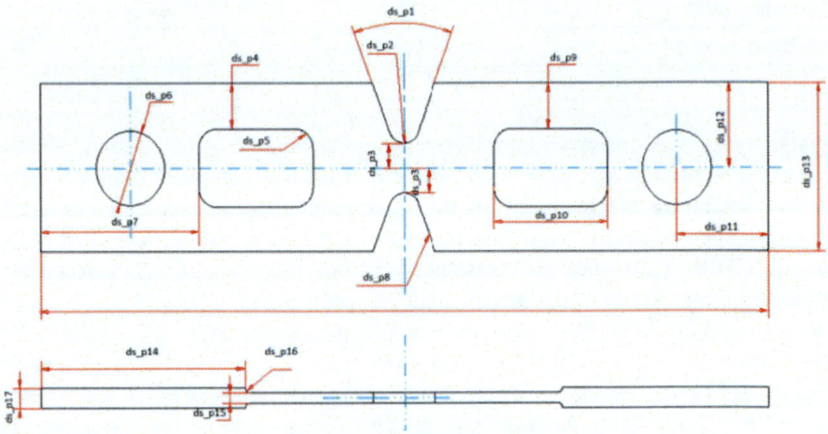

Fig. 3. Design Variables

Table 1. Design variables

Variables	Variables meaning	Lower bound	Upper bound
P1	Notch angle	30	50
P2	Notch radius	2.5 mm	4 mm
P9	Notch depth	10 mm	13 mm
P15	Specimen thickness	2 mm	6 mm

$$Fitness = \frac{Ssp - Sco}{Sco} \tag{7}$$

3.3 Simulating Specimen Optimization

Turbine Rotor Finite Element Results.
During the optimization, the ANSYS APDL code was executed for each individual

Table 2. Results comparisons

Items	Vent hole of the front baffle	
	Specimen	Component
Maximum stress point(MPa)	1024	1024
Average stress (MPa)	744	761
Stress concentration coefficient	1.38	1.35
Notch radius(mm)	3.4	
Notch depth(mm)	11.9	

generated by the GA optimizer and the stress was calculated. As mentioned above, the maximum generation is set as 300 with 100 individuals for each generation. Before moving to the simulating specimen design, the component stress distribution was obtained. The most dangerous part is confirmed to be the vent hole of the turbine front baffle by simulating all the steady-state and transient conditions. It becomes the critical component constraining the LCF life of the turbine and the aeroengine.

As depicted in Fig. 4, the maximum stress locates at the upper side of the vent hole of the front baffle, reaching 1024 MPa. An obvious stress concentration area is observed around the vent hole. The stress within 2 mm of the maximum stress is 1024 MPa, 751 MPa, 685 MPa, 585 MPa. Approaching the calculated stress of the component is the desired target for the simulating specimen.

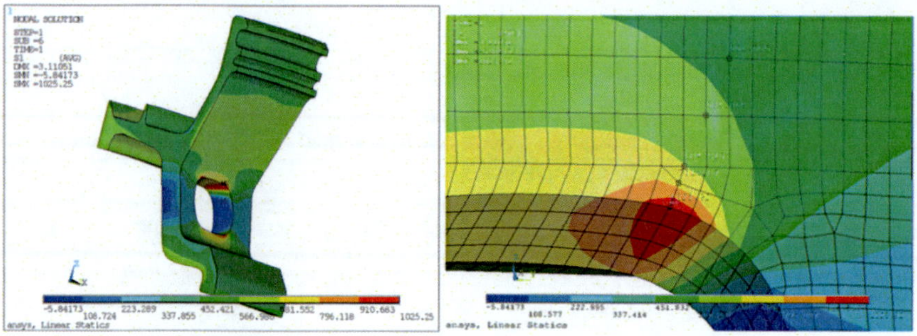

Fig. 4. Stress gradient and maximum stress of the component

Optimization Results.

After crossover, mutation and selection, the genes of the elite individual are passed down to the next generations. The optimal solution of the simulating specimen is presented in Fig. 5. It is clear that the maximum stress of the simulating specimen is almost the same as the component, reaching 1024 MPa. Meanwhile, the stress gradient of the specimen is very similar to the component. It should be noted that the element size of the model is about 0.5 mm, indicating that the distance of each point along the critical path is almost 0.5 mm. It is helpful for the results comparisons.

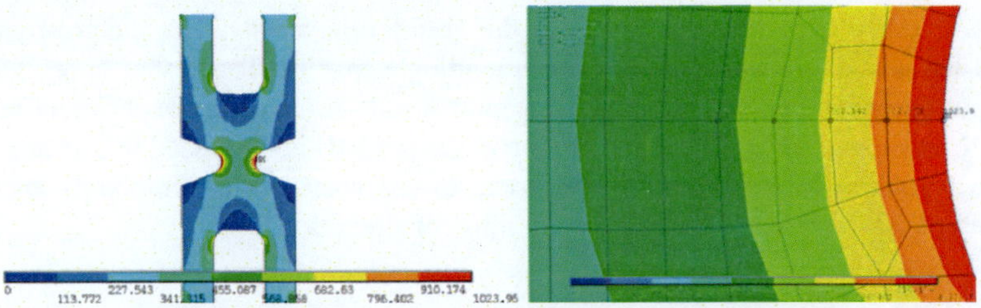

Fig. 5. Stress gradient and maximum stress of the specimen

The stress gradient between the specimen and the component are compared in Fig. 7. The overall trend shows that the stress of the specimen is consistent with the component along the critical path. However, when the point is 2 mm away from the maximum stress point, the stress difference reaches 6.23%. It is acceptable since the crack always initiated from the maximum stress point (Fig.6).

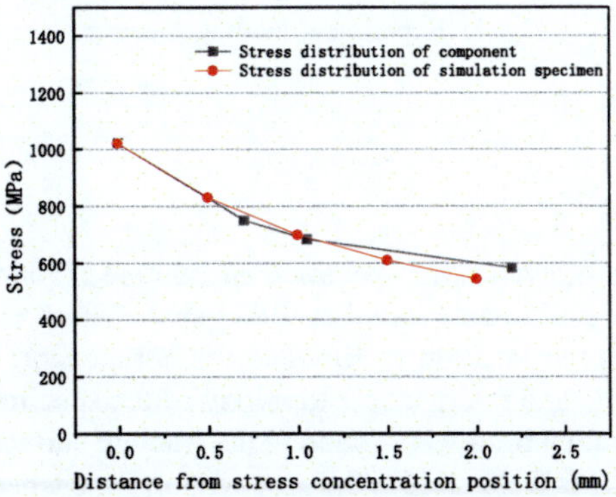

Fig. 6. Stress gradient comparisons

As shown in Table 3, the average stress of the specimen is 744 MPa, which is 2.23% lower than the component. The primary reason is that the stress far away from the maximum stress point is lower. Consequently, the stress concentration coefficient of the specimen is higher than the component. Nevertheless, it would not impact the results of the estimated LCF life.

4 Experimental Verification

4.1 Experiment Condition

The LCF test was carried out to verify the simulation results. The electro-hydraulic fatigue rig was employed with a 10 Hz loading capacity. The tested specimen was heated to an equivalent temperature. The total tensile force of the experiment reached 23000N, as shown in Fig. 7. The tensile force was loaded and unloaded in 0.1s and lasted until the rupture of the simulating specimen. In this scenario, 10 simulating specimen were tested taking the material dispersity into consideration.

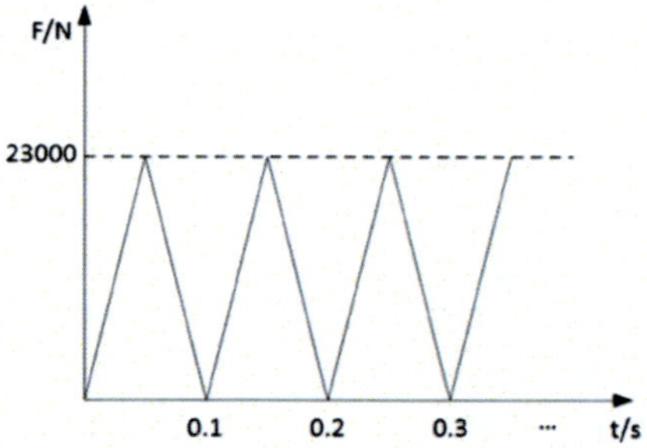

Fig. 7. Experiment loading spectrum

4.2 Failure Analysis

After a certain amount of cycles, the specimen encountered failure, as shown in Fig. 8. The fracture of all the specimen located at the U-shape notch area, as the red arrows indicated. It is the desired location to simulate the component. Figure 9 and Fig. 10 present the macroscopic and microscopic topology of the specimen fracture. Apparent fatigue striations and radial lines were found in the fracture surface, indicating a fatigue failure. The fatigue stripes were clear at the propagation zone whereas the instantaneous fracture zone exhibited a ductile dimple morphology. Besides, no defects and other surface quality issues were found.

4.3 Results Comparisons

Finally, the LFC life of the simulation results and tested results are compared in Table 4. It is found that the predicted LCF life of the specimen and component are quite close, with the difference within 1.21%. Meanwhile, the calculated LCF life of the simulating specimen is 2.18% higher than the tested results. It means that the LCF life of the vent hole of the turbine front baffle is approximately 150572 cycles. The relative error of the LCF life between the tested specimen and the real component is about 3.36%.

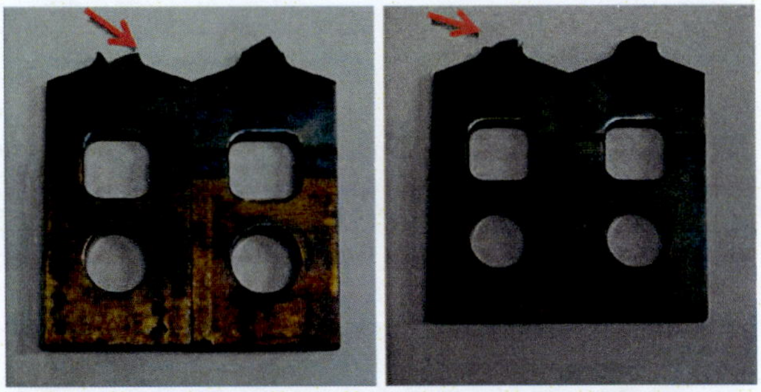

Fig. 8. Location of the fracture of the specimen

Fig. 9. Macroscopic topology of the fracture

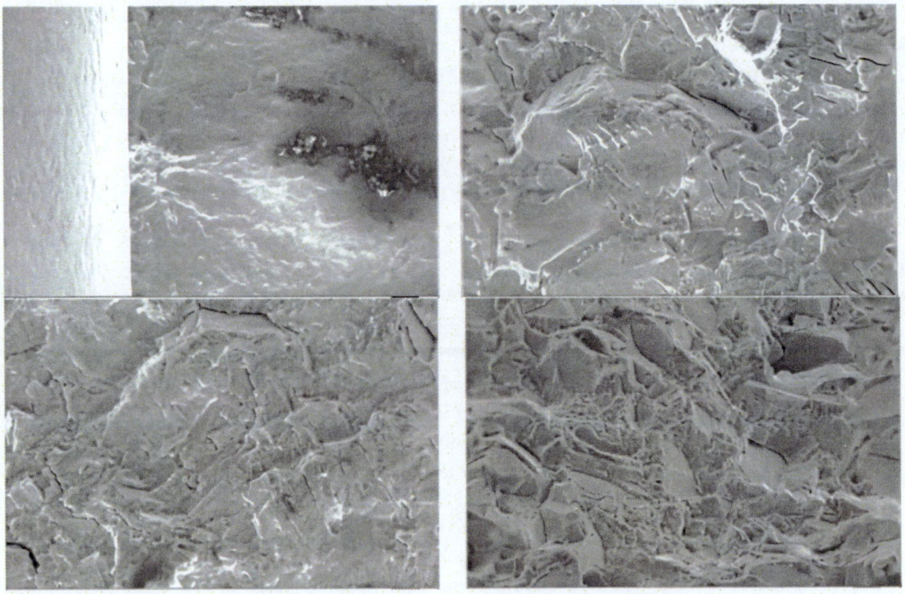

Fig. 10. Microscopic topology of the fracture

Table 3. LCF life cycles comparison

Item	Vent hole of the front baffle		
	Component simulation	Specimen simulation	Specimen tested
Cycle life	155810	153927	150572

5 Conclusions

This paper presents a design optimization of the simulating specimen of a turbine rotor. 4 parameters were varied during the optimization to achieve the same stress gradient and maximum stress as the real component. Afterwards, the LCF test was carried out and the LFC life of the simulation results and tested results were compared. The main outcomes are listed as follows:

The main principle of simulating specimen design is to ensure that the maximum stress and stress gradient at the notch of the specimen are equivalent to the critical part of the component. Furthermore, the material, manufacturing process, surface quality and shot blasting should maintain the same levels.
The stress gradient and maximum stress optimized by the GA method are close to the real component. Besides, the LCF test results indicate that the specimen failed at the maximum stress point. It highlights the importance of implementing optimization algorithm in specimen design.
The LCF life of the simulation results and the tested results is within 2.18% of the simulating specimen. Meanwhile, the difference of the estimated LCF life of the specimen and component lies in 1.21%. It provides a good reference and economic way to utilize simulating specimen to predict the LCF life of the turbine rotor.

References

1. You, M., He, X.Y., Xie, L.Y.: Theory and method of engine disc simulation technology. J. Mech. Design **02**, 62–64 (2007). (in Chinese)
2. Zhao, F.X.,Yang, X.Y.:Design method for low cycle fatigue simulation test pieces of engine components. J.Gas Turbine Experiment and Research (**02**), 50–52(2003). (in Chinese)
3. Liu, T.Y., Geng, R., Zhang, J.F.: Design of simulated specimen for low-cycle fatigue of turbine engine disk. J. Aerosp. Power **23**(1), 32–39 (2008). (in Chinese)
4. Lu, S., Wang, C.G., Chen, J.: Design method for disc simulator with arbitrary maximum stress gradient path. J. J. Aerodyn. **25**(09), 2000–2005 (2010). (in Chinese)
5. Yang, X.Y., Dong, L.W., Geng, Z.X., Liu, X.L., Zhao, F.X.: Design and experimentation of simulation specimen for aero-engine compressor disk slot used in low cycle fatigue test. Aerosp. Power **23**(10), 1829–1834 (2008). (in Chinese)
6. Zheng, X.M., Sun, Y.T., Yang, X.Y., Xiao, X.H., Liao, X.J.: Design of low cycle fatigue simulating specimen for bolt holes of a turbofan engine high pressure turbine. J. Aerosp. Power **33**(10), 2351–2358 (2018). (in Chinese)
7. Kuang, C.Y., Liu, Y.F.: Design of simulated specimen for low-cycle fatigue of aircraft engine titanium alloy disk. Equip. Manufact. Technol. **01**, 36–40 (2020). (in Chinese)

8. Liu, Y.X., Nalianda, D., Mo, D., et al. Multi-objective optimization of a three-shaft high bypass ratio engine for EIS2050. In: Proceedings of Global Power and Propulsion Society, 0113 (2021)
9. Liu, Y.X.: Optimization of hybrid electric propulsion system. MSc Thesis, Cranfield University
10. Song L. NGPM -- A NSGA-II Program in Matlabv1.4 (https://www.mathworks.com/matlabcentral/fileexchange/31166-ngpm-a-nsga-ii-program-in-matlab-v1-4), MATLAB Central File Exchange. Retrieved December 13, 2020

Check for updates

A Method for Tensile Strength Prediction of Composite Laminates

Houbing Wang(✉), Huafeng Xiong, Limin Chen, Xinxiang Li, and Shengchun Yang

National Key Laboratory of Strength and Structural Integrity, Aircraft Strength Research Institute of China, Xi'an 710065, China
wanghoby@sina.com

Abstract. From the perspective of engineering application, a method to predict the tensile strength of the carbon fiber reinforced polymer (CFRP) was proposed using classical laminate theory and the theory of function of a complex variable. In the method, the influence of defects in the manufacturing process of composite laminates on the tensile strength was considered. The tensile strength of arbitrarily lay-up composite laminate with the same composite material and the same molding process was predicted by the modulus and strength of unidirectional lamina and ultimate strength of a laminate. The method was used to calculate the tensile strength of 5 kinds of composite material systems and 25 types of laminates. The predictions of the present approach were compared to test results. The research indicated that the calculated results were in good agreement with experimental results. Two assumptions about defect treatment of composite laminates can be established. Owing to the current strength theories of composite materials were relatively complicated, the study provided a simple and efficient way to predict the tensile strength of composite laminates for composite structure design.

Keywords: Composite · Laminate · Tensile · Strength · Defect

1 Introduction

The strength theory of metal materials mainly refers to failure criteria. The first strength theory [1] is the maximum tensile stress criterion, which assumes that the material will be damaged when the maximum tensile stress reaches the ultimate stress related to the material properties. The second strength theory [2] is the maximum elongation linear strain criterion, which assumes that brittle fracture occurs when the maximum elongation linear strain reaches the ultimate stain related to the material properties. The third strength theory [3] is the maximum shear stress criterion, which assumes that the materials yield when the maximum shear stress reaches the ultimate shear stress. The fourth strength theory [4] is a criterion for the specific energy of deformation, which assumes that materials yield when the specific energy of deformation reaches the ultimate value related to material properties.

Since the application of composite materials in aerospace structures in the 1960s, their strength theories have been widely concerned and incessantly developed. Currently,

Chinese Society of Aeronautics and Astronautics (Ed.): CASTC 2023, LNME, pp. 344–357, 2024.
https://doi.org/10.1007/978-981-99-8861-7_36

there are many famous strength theories [5–7] for composite materials. The British Engineering and Natural Science Research Council and the Society of Mechanical Engineers had jointly evaluated and ranked the strength theories of composite materials around the world [8, 9]. Among them, the top ranked theories were Zinoviev theory [10], Bogetti theory [11], Puck theory [12], Cuntze theory [13], Liu-Tsai theory [14] and Huang theory [15]. Due to the anisotropy of composite materials and the diversity and complexity of failure modes, the strength theories of composite materials are much more complex than that of metal materials. In a word, these strength theories of composite materials mainly included four aspects: constitutive equations, stress calculations, failure criteria, and degradation criteria. In the constitutive equations, linear elastic constitutive equations (such as Zinoviev theory, Liu-Tsai theory) and nonlinear elastic constitutive equations (such as Bogetti theory, Puck theory, Cuntze theory, and Huang theory) were used. In nonlinear elastic constitutive equations, the nonlinear deformation effects of composite materials only in certain directions were considered (such as Bogetti theory, Puck theory, Cuntze theory), or classical elastic-plastic constitutive theories (such as Huang theory) was directly used. In stress calculations the effects of mechanical loads were studied in some theories (Zinoviev theory, Bogetti theory, Puck theory, Cuntze theory, and Liu-Tsai theory of macroscopic theory), while the effects of thermal stress and residual stress were studied in others (such as Huang theory). The failure criteria include: the maximum stress criterion [16], the maximum strain criterion [17], Tsai-Hill criterion [18], Tsai-Wu criterion [19], Hoffman criterion [20], Hashin criterion [21, 22], Puck criterion [12], LaRC criterion [22, 23] etc. There are currently multiple degradation criterions. Various degradation criteria were adopted by Zinoviev theory, Bogetti theory, Puck theory, Cuntze theory, Liu-Tsai theory, and Huang theory.

Due to the complexity of the strength theory of composite materials, it is difficult to calculate the ultimate strength of composite structures. In the design of aircraft, the failure strength of the first ply is mainly used as basis for designing composite laminates according to strength requirements, which is also one of the reasons for the conservative design of composite structures and the excessive weight of structures [24]. During the manufacturing and processing of composite laminates, various defects are inevitably introduced, such as voids, bubbles, inclusions, resin starvation, resin richness, wrinkles, delamination, and microcracks, which can affect the ultimate strength of composite laminates. Currently, the main strength theories of composite materials, such as Zinoviev theory, Bogetti theory, Puck theory, Cuntze theory, Liu-Tsai theory, and Huang theory, did not contain the impact of various forms of defects on the ultimate strength of composite laminates.

From the perspective of engineering applications, this article studied the defects in composite laminates on their ultimate strength. Using the LBPF (Load Bearing Ply Failure) criterion [25], classical laminate theory, and complex variable function theory, an efficient and rapid approach for the tensile strength prediction of composite laminates was proposed and established.

2 Prediction Method

2.1 Defect Hypotheses

The manufacturing methods of composite structures are different from metal structures, and the types of defects are also different. The main internal defects of composite laminate include: voids, bubbles, inclusions, resin starvation, resin richness, wrinkles, delamination, microcracks, etc. [24, 26], as shown in Fig. 1. The various defects can lead to different degradation in the mechanical properties of composite laminates.

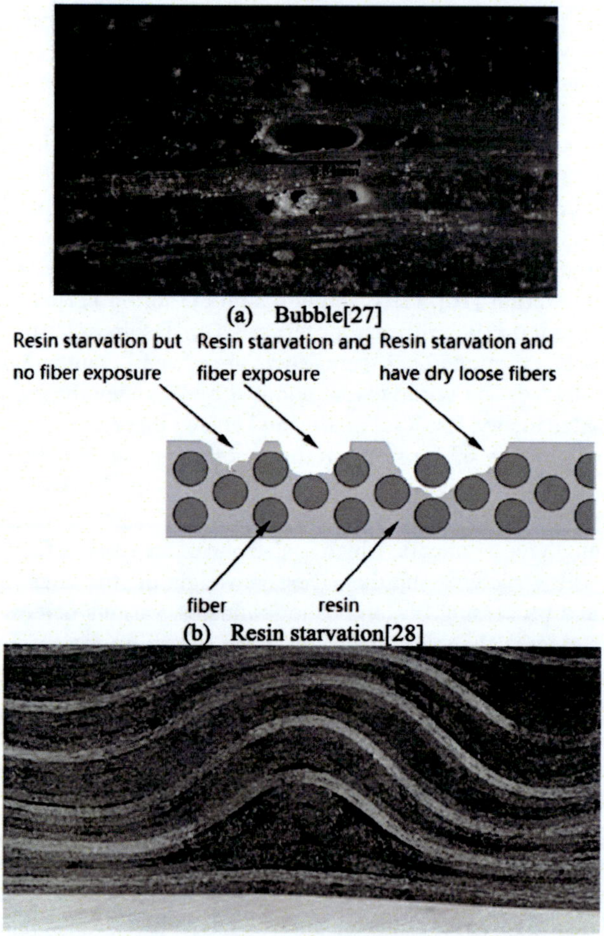

(a) Bubble[27]

(b) Resin starvation[28]

(c) Resin richness, wrinkles, delamination, microcracks[29]

Fig. 1. Defects in composite laminates

In the evaluation method for tensile allowable values of composite laminates, an open-hole with a diameter of 6 mm (inch diameter of 0.25 in) was used to cover various defects and damages during the manufacturing process. In literature [31], discrete source damage was simplified as an open-hole when calculating the residual strength of composite panels containing discrete source damage, and in literature [32], crack was treated as open-hole when calculating the residual strength of composite laminates containing crack. Inspired by these, various internal defects and damages in composite laminates were unified and treated as an open-hole.

In the application of composite laminates, mechanical performance tests of unidirectional plates are usually conducted for each material system, and the mechanical performance parameters of unidirectional plates can be easily obtained. Due to the wide variety of laminates (selecting different stacking sequences and number of plies, theoretically there can be infinite lay-ups for each composite material), usually typical lay-ups are selected for mechanical property testing, so the mechanical property data of laminates with different lay-ups are often limited. In the manufacturing process of composite laminates, Lay-up can introduce more defects in laminates than in laminas. In summary, in order to simplify the calculation and facilitate engineering applications, it was assumed that the lamina was a perfect structure without manufacturing defects, and the laminates with lay-ups were a structure with defects. Since the manufacturing defects of composite laminates are closely related to the molding process, it was assumed that the laminates had the same defects compare to laminas with the same composite material and the same molding process.

2.2 LBPF Criterion

The LBPF criterion was a failure criterion for calculating the residual strength of composite plates with defects (or damages, geometric shapes could be circular, elliptical or crack) under tensile or compressive loads, and proposed by Chen Puhui et al. LBPF criterion [25] assumed that when the average stress (average value was shown in Fig. 2) of the 0° ply adjacent to the defect (or damage) of a composite laminate reaches the ultimate strength of a unidirectional plate (0° ply) within a certain range(the length is l_0) perpendicular to the load direction (x direction was shown in Fig. 2), the plate would fail. The characteristic length (shown in Fig. 2) was a constant that was only related to the composite material system and load type, and was independent of the lay-ups and hole size of the laminate. The characteristic length could be determined by testing a laminate with an open-hole. The expression of the LBPF criterion is shown in Eq. (1).

$$\frac{1}{l_0}\int \sigma_y^0(x,0)dx = X \tag{1}$$

where, $\sigma^0{}_y$ was the stress component in the direction y of the 0° ply, X was the tensile ultimate strength or compressive ultimate strength of the 0° ply.

According to the classical laminate theory and the theory of functions of a complex variable, the tensile strength of a composite laminate with an open-hole can be deduced from Eq. (1), as shown in Eq. (2) below [25].

$$\sigma_{\mathrm{tu}} = \frac{l_0 X}{\beta_1 F_1 + \beta_2 F_2 + \beta_2 F_3} Y \tag{2}$$

where

$$\begin{aligned} \beta_1 &= Q_{11}a_{11} + Q_{12}a_{12} \\ \beta_2 &= Q_{11}a_{12} + Q_{12}a_{22} \\ \beta_3 &= Q_{11}a_{16} + Q_{12}a_{26} \end{aligned} \tag{3a}$$

$$\begin{aligned} \beta_1 &= Q_{11}a_{11} + Q_{12}a_{12} \\ \beta_2 &= Q_{11}a_{12} + Q_{12}a_{22} \\ \beta_3 &= Q_{11}a_{16} + Q_{12}a_{26} \end{aligned} \tag{3b}$$

$$\begin{aligned} \beta_1 &= Q_{11}a_{11} + Q_{12}a_{12} \\ \beta_2 &= Q_{11}a_{12} + Q_{12}a_{22} \\ \beta_3 &= Q_{11}a_{16} + Q_{12}a_{26} \end{aligned} \tag{3c}$$

$$Y = \frac{3(1 - 2r/W)}{2 + (1 - 2r/W)^3} \tag{3d}$$

whereσ_{tu} was the tensile strength of the composite laminate. $\beta_i (i = 1, 2, 3)$ was the intermediate variable, a_{ij} $(i 、 j = 1, 2, 6)$ were the compliance component of the composite laminate with 1 and 2 parallel and transverse to the loading direction respectively, and Q_{ij} $(i 、 j = 1, 2, 6)$ is the stiffness coefficient. F_1 、 F_2 、 F_3 are also intermediate variables, r being the radius of the opening hole (defect), and x being the coordinate variable. Y is the finite width correction coefficient for composite laminates, and $\mu_j (j = 1, 2)$ were the roots of the characteristic Eq. (4) (one for each pair of conjugate roots).

$$a_{22}\mu^4 - 2a_{26}\mu^3 + (2a_{12} + a_{66})\mu^2 - 2a_{16}\mu + a_{11} = 0 \tag{4}$$

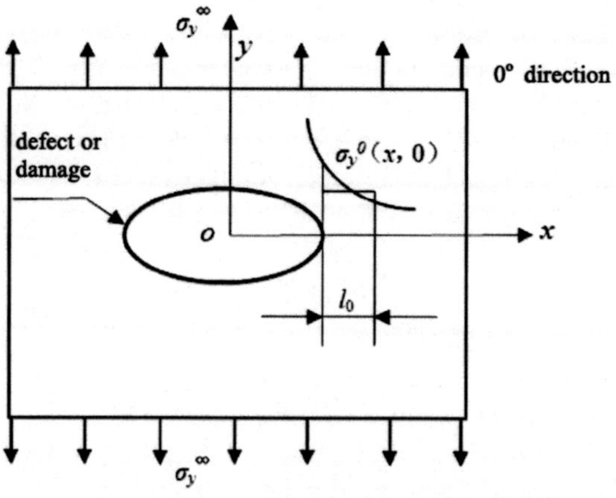

Fig. 2. Definition of the characteristic distances for LBPF Criterion

2.3 Computational Method

The process for calculating the tensile strength of any series of laminate with arbitrary lay-up was shown in Fig. 3 using the mechanical property parameters of a unidirectional composite plate and a laminate. Substituted the parameters (modulus, strength, characteristic length, etc.) of a unidirectional composite plate and the parameters(each ply orientation angle, tensile strength, etc.) of a laminate with the same composite material and the same molding process into Eq. (2), and used the LBPF criterion the defect size (open-hole radius) of the laminate was obtained through inverse calculation. The tensile strength of the laminate can be calculated by substituting the parameters of the unidirectional plate, the defect size, and the parameters of the laminate with arbitrary lay-up (including geometric parameters, each ply orientation angle) into Eq. (2) again.

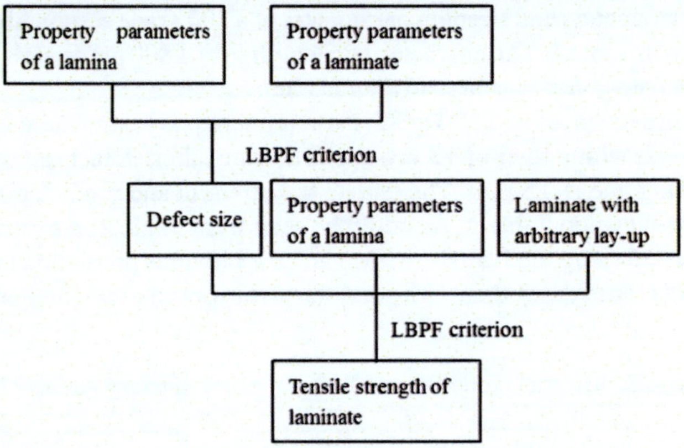

Fig. 3. Flow chart of tensile strength of composite laminate

3 Method Verification

3.1 Test Method

The composite laminate test was conducted in accordance with ASTM D3039 "Standard test method for tensile properties of polymer matrix composite materials" [33]. The specimen was shown in Fig. 4. In order to ensure that the test process did not cause damage to the clamping section of the specimens (both ends of the specimen were clamped in the grips of the testing machine during the test), the tabs were affixed to both ends of the specimens (there were no tabs in Fig. 4). There were a total of 5 specimens per group.

The tensile strength of the plate σ_{tu} (MPa) can be obtained from Eq. (5).

$$\sigma_{tu} = \frac{P^{max}}{A} \tag{5}$$

where P^{max} was the maximum load before failure, N, A (=25δ) was the cross-sectional area of laminate, mm^2.

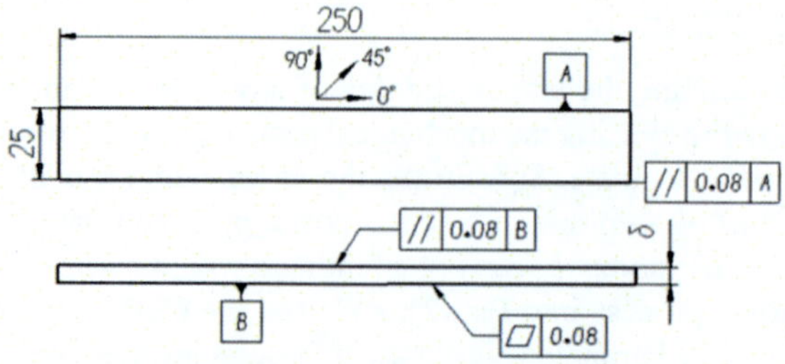

Fig. 4. Tensile specimen drawing

3.2 Experimental Result of T800 Grade Carbon Fiber/Epoxy Composites

The modulus parameters and strength parameters of T800 grade carbon fiber/epoxy lamina were shown in Table 1. The characteristic length l_0 of T800 grade carbon fiber/epoxy composite laminate under tensile load was 1.3 mm.

Ply stacking sequences of T800 grade carbon fiber/epoxy composite laminates were shown in Table 2, with a total of 13 lay-ups. The equivalent defect radius of laminates to unidirectional plate is 0.5 mm. The tensile test strength and predicted strength of the laminates were shown in Table 2. As could be seen from Table 2, the error between the strength calculated using the LBPF criterion and the previous prediction method and the test strength was mostly less than 10%, and the error for only two laminates was 11%.

Table 1. Material properties of T800 grade carbon/epoxy lamina [34]

Parameters	Value	Note
E_1(GPa)	179	Longitudinal tensile modulus
E_2(GPa)	8.11	Transverse tensile modulus
V_{12}	0.317	Poisson ratio
G_{12}(GPa)	4.14	In-plane shear modulus
t(mm)	0.191	Thickness of ply
X_t(MPa)	3102	Longitudinal tensile strength
Y_t(MPa)	68.6	Transverse tensile strength
S_{xy}(MPa)	168	In-plane shear strength

3.3 Experimental Result of T300 Grade Carbon Fiber/Epoxy Composite 1 Laminates

The modulus parameters and strength parameters T300 grade carbon fiber/epoxy composite 1 lamina were shown in Table 3. The characteristic length l_0 of the T300 grade carbon fiber/epoxy composite 1 laminates under tensile load was 1.3 mm.

Table 2. Comparisons of experimental and predicted strength of T800 grade carbon/epoxy laminates [35]

No. of lay-up	Lay-up	Experimental strength	Predicted strength	Relative error
1	$[45/-45/0/90]_S$	910	942	4%
2	$[45/-45/0/90]_{3S}$	899	942	5%
3	$[(45/-45/0/90)_3/0/(90/0/-45/45)_3]$	938	1005	7%
4	$[45/-45/0/90/45/-45/0/-45/45/-45/-45/45/90/45/0]_S$	754	840	11%
5	$[45/-45/0/90]6_S$	901	942	5%
6	[45/-45/0/0/90/0/45/-45/0]s	1430	1356	5%
7	[45/-45/0/0/0/90/0/45/-45/0]s	1387	1470	6%
8	$[45/-45/0/90/45/-45/0/0/90/0/0/45/-45/0]_S$	1288	1321	3%
9	$[45/-45/0/90/0/45/-45/0/90/45/-45/0/0/45/-45/0]_S$	1088	1213	11%
10	$[45/-45/0/0/90/0/45/-45/0/0/90/45/-45/0/0/45/-45/0]_S$	1310	1356	4%
11	[45/0/0/-45/0/90/0/0/90/0/-45/0/0/45]	1635	1602	2%
12	[45/0/0/-45/0/90/0/0/90/0/-45/0/0/45]s	1614	1602	1%
13	[45/0/0/-45/0/90/0/0/90/0/-45/0/0/45/0/0]s	1673	1711	2%

Ply stacking sequence of T300 grade carbon fiber/epoxy composite 1 laminates were shown in Table 4, with a total of 4 lay-ups. The equivalent defect radius of laminates to unidirectional plate was 0.02 mm. The tensile test strength and predicted strength of the laminates were shown in Table 4. As could be seen from Table 4, the error between the strength calculated using the LBPF criterion and the previous prediction method and the test strength was mostly less than 5%.

Table 3. Material properties of T300 grade carbon/epoxy composite 1 lamina [36]

Parameters	Value	Note
E_1(GPa)	135	Longitudinal tensile modulus
E_2(GPa)	10.4	Transverse tensile modulus
V_{12}	0.287	Poisson ratio
G_{12}(GPa)	6.14	In-plane shear modulus
t(mm)	0.125	Thickness of ply
X_t(MPa)	1956	Longitudinal tensile strength
Y_t(MPa)	84.5	Transverse tensile strength
S_{xy}(MPa)	168	In-plane shear strength

Table 4. Comparisons of experimental and predicted strength of T300 grade carbon/epoxy composite 1 laminates [37]

No. of lay-up	Lay-up	Experimental strength	Predicted strength	Relative error
1	$[45/0/-45/90/0]_{2s}$	962	975	1%
2	$[45/0/-45/0/-45/0/45/90/45/0]_s$	961	971	1%
3	$[45/-45/0/-45/90/45/0]_s$	785	802	2%
4	$[45/-45/0/-45/90/45/0/45/0/-45]_s$	862	818	5%

3.4 Experimental Result of T300 Grade Carbon Fiber/Epoxy Composite 2 Laminates

The modulus parameters and strength parameters of T300 grade carbon fiber/epoxy composite 2 lamina were shown in Table 5. The characteristic length l_0 of the T300 grade carbon fiber/epoxy composite 2 laminates under tensile load was 1.897 mm.

Ply stacking sequence of T300 grade carbon fiber/epoxy composite 2 laminates were shown in Table 6, with a total of 4 lay-ups. The equivalent defect radius of laminates to unidirectional plate was 0.02 mm. The tensile test strength and predicted strength of the laminates were shown in Table 6. As could be seen from Table 6, the error between the strength calculated using the LBPF criterion and the previous prediction method and the test strength was mostly less than 10%, and the error for only one laminates was 11%.

Table 5. Material properties of T300 grade carbon/epoxy composite 2 lamina [38]

Parameters	Value	Note
E_1(GPa)	138	Longitudinal tensile modulus
E_2(GPa)	9.32	Transverse tensile modulus
V_{12}	0.31	Poisson ratio
G_{12}(GPa)	5.96	In-plane shear modulus
t(mm)	0.12	Thickness of ply
X_t(MPa)	1821	Longitudinal tensile strength
Y_t(MPa)	89.3	Transverse tensile strength
S_{xy}(MPa)	111	In-plane shear strength

3.5 Experimental Result of T300 Carbon Grade Fiber/Epoxy Composite 3 Laminates

The modulus parameters and strength parameters of T300 carbon fiber/epoxy composite 3 lamina were shown in Table 7. The characteristic length l_0 of T300 grade carbon fiber/epoxy composite 3 laminates under tensile load was 1.3 mm.

Table 6. Comparisons of experimental and predicted strength of T300 grade carbon/epoxy composite 2 laminates [38]

No. of lay-up	Lay-up	Experimental strength	Predicted strength	Relative error
1	$[45/0/\text{-}45/90/0]_{2s}$	972	929	4%
2	$[45/0/\text{-}45/0/45/90/\text{-}45/0/45/0]_{s}$	1049	931	11%
3	$[45/\text{-}45/0/45/\text{-}45/90/45/0/\text{-}45/45]_{s}$	648	631	3%
4	$[45/0/\text{-}45/90]_{2s}$	719	708	2%

Ply stacking sequence of T300 grade carbon fiber/epoxy composite 3 laminates were shown in Table 8, with a total of 4 lay-ups. The equivalent defect radius of laminates to unidirectional plate was 0.14 mm. The tensile test strength and predicted strength of the laminates were shown in Table 8. As could be seen from Table 8, the error between the strength calculated using the LBPF criterion and the previous prediction method and the test strength was mostly less than 10.

Table 7. Material properties of T300 grade carbon/epoxy composite 3 lamina [38]

Parameters	Value	Note
E_1(GPa)	135	Longitudinal tensile modulus
E_2(GPa)	10.4	Transverse tensile modulus
V_{12}	0.287	Poisson ratio
G_{12}(GPa)	6.14	In-plane shear modulus
t(mm)	0.125	Thickness of ply
X_t(MPa)	1956	Longitudinal tensile strength
Y_t(MPa)	84.5	Transverse tensile strength
S_{xy}(MPa)	168	In-plane shear strength

3.6 Experimental Result of T300 Grade Carbon Fiber/Epoxy Composite 4 Laminates

The modulus parameters and strength parameters of T300 grade carbon fiber/epoxy composite 4 lamina were shown in Table 9. The characteristic length l_0 of T300 grade carbon fiber/epoxy composite 4 laminates under tensile load was 1.1 mm.

Ply stacking sequence of T300 grade carbon fiber/epoxy composite 4 laminates were shown in Table 10, with a total of 4 layers. The equivalent defect radius of laminates to unidirectional plate was 0.1 mm. The tensile test strength and predicted strength of the laminates were shown in Table 10. As could be seen from Table 10, the error between

Table 8. Comparisons of experimental and predicted strength of T300 grade carbon/epoxy composite 3 laminates [38]

No. of lay-up	Lay-up	Experimental strength	Predicted strength	Relative error
1	$[45/0/\text{-}45/90/0]_{2s}$	962	975	1%
2	$[45/0/\text{-}45/0/\text{-}45/0/45/90/45/0]_s$	961	971	1%
3	$[45/\text{-}45/0/\text{-}45/90/45/0]_s$	785	802	2%
4	$[45/\text{-}45/0/\text{-}45/90/45/0/45/0/\text{-}45]_s$	862	818	5%

the strength calculated using the LBPF criterion and the previous prediction method and the test strength was mostly less than 10.

Table 9. Material properties of T300 grade carbon/epoxy composite 4 lamina [38]

Parameters	Value	Note
E_1(GPa)	137	Longitudinal tensile modulus
E_2(GPa)	9.73	Transverse tensile modulus
V_{12}	0.303	Poisson ratio
G_{12}(GPa)	5.39	In-plane shear modulus
t(mm)	0.12	Thickness of ply
X_t(MPa)	1865	Longitudinal tensile strength
Y_t(MPa)	74.8	Transverse tensile strength
S_{xy}(MPa)	120	In-plane shear strength

Table 10. Comparisons of experimental and predicted strength of T300 grade carbon/epoxy composite 4 laminates [38]

No. of lay-up	Lay-up	Experimental strength	Predicted strength	Relative error
1	$[45/0/\text{-}45/90/0]_{2s}$	897	892	1%
2	$[45/0/\text{-}45/0/45/90/\text{-}45/0/45/0]_s$	897	895	0%
3	$[45/\text{-}45/0/45/\text{-}45/90/45/0/\text{-}45/45]_s$	564	607	8%
4	$[45/0/\text{-}45/90]_{2s}$	637	682	7%

3.7 Analysis of Experimental Results

According to the test results of 5 kinds of composite material systems in use and 25 types of laminates, it can be seen that the predicted strength is in good agreement with the experimental strength, and the error of most results is within 10%. For the same composite material, the same manufacturing process, and different lay-ups, the difference in predicted strength error is also small, indicating that it is feasible using the same open-hole to simulate the defects of laminate of different lay-ups with the same composite material, the same manufacturing process.

4 Conclusion

The defects such as voids, bubbles, inclusions, resin starvation, resin richness, wrinkles, delamination, and microcracks generated during the manufacturing process of composite laminates can be equivalent to an open-hole when calculating tensile strength. The defects of laminates made of the same composite material, the same manufacturing process, and different lay-ups can be simulated using the same open-hole. The tensile strength of composite laminates calculated based on LBPF criterion, classical laminate theory, and the theory of functions of a complex variable is in good agreement with experimental results.

References

1. Gou, W.X.: Mechanics of Materials[M]. Beijing: Science Press, 2005: 297–300 (in Chinese)
2. De, S.V.: Memoire sur l'establissement des equations differentielles des mouvements interieurs operes dans les corps solides ductiles au dela des limites ou l'e´lasticite´ pourrait les ramener a leur premier e´tat[J]. Comptes Rendus hebdomadaires des Seances de l'Academie des Sciences **70**, 473–480 (1870)
3. Tresca, H.: Sur I'ecoulement des corps solids soumis a defortes pression. Comptes Rendus hebdomadaires des Seances del'Academie des Sciences **59**, 754–758 (1864)
4. Von, M.R.: Mechanik der festen Ko¨rper im plastisch deformablen Zustand, Nachrichten von der Ko¨niglichen Gesellschaft der wissenschaften zu Go¨ettinger. Mathematisch-physikalische Klasse **52**, 582–592 (1913)
5. Echaabi, J., Trochu, F., Gauvin, R.: Review of failure criteria of fibrous composite materials. Polym. Compos. **17**(6), 786–798 (1996)
6. Sutherland, L.S., Soares, C.G.: Review of probabilistic models of the strength of composite materials. Reliab. Eng. Syst Safety **56**, 183–196 (1997)
7. Orifici, A.C., Herszberg, I., Thomson, R.S.: Review of methodologies for composite material modeling incorporating failure. Compos. Struct. **86**, 194–210 (2008)
8. Schulte, K.: World wide failure exercise on failure prediction in composites. Compos. Sci. Technol. **62**, 1479 (2002)
9. Hinton, M.J., Kaddour, A.S., Soden, P.D.: A further assessment of the predictive capabilities of current failure theories for composite laminates: comparison with experimental evidence. Compos. Sci. Technol. **64**, 549–588 (2004)
10. Zinoviev, P.A., Grigoriev, S.V., Lebedeva, O.V., et al.: The strength of multilayered composites under a plane-stress state. Compos. Sci. Technol. **58**(7), 1209–1223 (1998)
11. Bogetti, T.A., Hoppel, C.P.R., Harik, V.M., et al.: Predicting the nonlinear response and progressive failure of composite laminates. Compos. Sci. Technol. **64**, 329–342 (2004)
12. Puck, A., Schurmann, H.: Failure analysis of FRP laminates by means of physically based phenomenological models. Compos. Sci. Technol. **58**(7), 1045–1067 (1998)

13. Cuntze, R.G., Freund, A.: The predictive capability of failure mode concept-based strength criteria for multidirectional laminates. Compos. Sci. Technol. **64**, 343–377 (2004)
14. Liu, K., Tsai, S.W.: A progressive quadratic failure criterion for a laminate. Compos. Sci. Technol. **58**(7), 1023–1032 (1998)
15. Huang, Z.M.: A bridging model prediction of the ultimate strength of composite laminates subjected to biaxial loads. Compos. Sci. Technol. **64**, 395–448 (2004)
16. Sun, C.T., Quinn, B.J., Tao, J., et al.: Comparative evaluation of failure analysis methods for composite laminates. DOT/FAA/AR-95/109[R]. Washington D. C. : NASA, (1996)
17. Tsai, S.W.: Strength characteristics of composite materials. NASA/CR-224[R]. Washington D. C. : NASA, (1965)
18. Tsai, S.W., Wu, E.M.: A general theory of strength for anisotropic materials[J]. J. Compos. Mater. **5**(1), 58–80 (1971)
19. Hoffman, O.: The brittle strength of orthotropic materials. J. Compos. Material. **1**(2), 200–206 (1967)
20. Hashin Z, Rotema. A fatigue failure criterion for fiber reinforced materials[J]. J. Compos. Material. **7**(4), 448–464 (1973)
21. Hashin, Z.: Failure criteria for unidirectional fiber composites. J. Appl. Mech. **47**(2), 329–334 (1980)
22. Davila, C.G., Camanho, P.P.: Failure criteria for FRP laminates in plane stress. NASA/TM-2003–212663[R]. Washington D. C. : NASA (2003)
23. Pinho, S.T., Davila, C.G., Camanho, P.P., et al. Failure models and criteria for FRP under in-plane or three-dimensional stress states including shear non-linearity. NASA/TM-2005–213530[R]. Washington D. C. : NASA (2005)
24. Chinese Aeronautical Establishment: Handbook of composites design [M], pp. 157–172. Aviation Industry Press, Beijing (1990)
25. Chen, P.H., Shen, Z., Wang, J.Y.: A new method for compression after impact strength prediction of composite laminates. J. Compos. Mater. **35**(20), 589–610 (2001)
26. Chinese Aeronautical Establishment: Handbook of composites structure design [M], pp. 72–73. Aviation Industry Press, Beijing (2001)
27. Yang, B.: Research on numerical simulation and defect prediction for the filling process of resin transfer molding[D]. Harbin: Harbin Institute of Technology, pp. 106–110 (2015)
28. Hu, D.Y.: Root cause and preventive actions on rein starvation defects of composite parts. Aircr. Des. **37**(5), 34–37 (2017)
29. Johnson, K.J., Butler, R., Loukaides, E.G., et al.: Stacking sequence selection for defect-free forming of uni-directional ply laminates. Compos. Sci. Technol. **2019**(171), 34–43 (2018)
30. ASTM Standard D5766. Standard test method for open-hole tensile strength of polymer matrix composite laminates[S]. ASTM Internation, West Conshohocken, PA, 2018. http://www.satm.org
31. Wang, H.B., Wang, X.H., Lin, G.W., et al.: Residual strength evaluation of curved hat-stiffened composite panels with discrete-source damage. Acta Aeronautica et Astronautica Sinica **43**(6), 525899 (2022)
32. Wang, H.B., Deng, F.C., Wei, H.Y., et al.: Residual strength evaluation of laminates with open hole or crack under tensile load. Composites Science and Engineering(Accepted)
33. ASTM Standard D3039. Standard test method for tensile properties of polymer matrix composite materials[S]. ASTM Internation, West Conshohocken, PA, 2017. http://www.satm.org
34. Xu, F.: Test report on basic properties of low pressure curing of XXX composite material system[R]. AVIC/00–623S-2019–113–0053, 2019
35. Sun, H.: Test report on design allowable value of C919 composite wing research and development project[R]. AVIC/ Y12WX25–60SY002, 2015

36. Xiong, H.F.: Test report on mechanical properties of XXX composite material[R]. AVIC/ 623S-201223–021 (2012)
37. Xiong, H.F.: Test report on allowable value of composite structure[R]. AVIC/ 623S-201813–95 (2018)
38. Xiao, J.: Test report on mechanical properties of domestic carbon fiber XXX composite material[R]. AVIC/ 623S-200801–12 (2008)

Dynamic Response Analysis of Oscillating Blade Gust Generator System and Wind Tunnel

Jinge Yu[1,2](✉), Guoning Huang[3], Libao Wang[1], Dongqiang Zhao[3], Nan Liu[1], Siyu Wu[1,2], and Pengfei Zhang[1,2]

[1] AVIC The Aerodynamics Research Institute, Harbin 150001, China
yujinge2005@163.com
[2] Aerodynamics for Low Speed and High Reynolds Number, Harbin 150001, China
[3] AVIC The First Aircraft Institute, Xi'an 710089, China

Abstract. Aiming at the aerodynamic and structural coupling characteristics between the oscillating blade gust generator system and the wind tunnel body in large low-speed wind tunnel, a coupling analysis system including rigid/elastic gust generator blades, elastic wind tunnel body and flow field aerodynamic force is established. The fluid-solid coupling method is used to analyze the influence of gust generator motion on wind tunnel vibration and gust flow field. The main frequency characteristics of the horizontal and vertical motion direction deformation of the generator blade are obtained. The aerodynamic and structural coupling design and analysis method of gust generator-tunnel system in large low-speed wind tunnel is established, which provides an analysis method for the design of gust generator in large low-speed wind tunnel.

Keywords: oscillating blade type · gust generator · wind tunnel body · fluid-solid coupling

1 Introduction

Gust is a strong deterministic wind disturbance in the atmosphere. When the aircraft encounters gust, the body will produce additional unsteady aerodynamic force and torque, which will adversely affect the flight performance of the aircraft [1, 2].

Modern civil aircraft, such as large passenger aircraft, emphasizes economy, comfort, safety and reliability, requires higher aerodynamic efficiency and lower structural weight, usually adopts large aspect ratio wing, uses a large number of composite materials on structural materials, and the aircraft wing has greater flexibility, so the aircraft is more sensitive to gust response. Gust load, especially vertical discrete gust load, often becomes the most serious flight load, which has a serious impact on the comfort, safety, economy and reliability of large passenger aircraft [3]. When the aircraft encounters the unsteady aerodynamic load caused by the low and medium intensity gust, it will cause the turbulence of the aircraft, bring tension to passengers, reduce the comfort of civil aircraft, and even cause personal injury of passengers [4].

Chinese Society of Aeronautics and Astronautics (Ed.): CASTC 2023, LNME, pp. 358–371, 2024.
https://doi.org/10.1007/978-981-99-8861-7_37

Domestic and foreign civil aircraft airworthiness columns have made clear provisions for gust load. Chinese Airworthiness Standards for Transport Aircraft (CCAR-25-R4) and American Aviation Administration Regulations (FAR-25) provide corresponding provisions for gust load design criteria [5].

Therefore, accurate prediction of gust load is an important work that must be carried out in aircraft design stage, and wind tunnel test is the most important and indispensable means. The development of large civil aircraft in China has put forward requirements for the development of advanced gust generator technology in large low-speed wind tunnel. However, the large wind tunnel gust generator-wind tunnel -flow field constitutes a complex coupling system. The gust generator produces discrete gusts with different frequencies and amplitudes by oscillating blades. If designed improperly, it is easy to cause coupling between the systems. Due to the aerodynamic and structural coupling, the gust generator vibrates with the wind tunnel body, and even the aeroelastic instability occurs. Therefore, in the design of gust generator, it is necessary to carry out the aerodynamic and structural coupling analysis of gust generator- wind tunnel -flow field to study the influence of generator- wind tunnel coupling effect on gust flow field.

In this paper, the nested grid technology is used to establish the fluid calculation grid of the blade of gust generator and the structure of wind tunnel. The blade and shaft of gust generator are the foreground grid, and the interior of wind tunnel is the background grid. The numerical simulation of gust flow field formed by blade deflection is carried out by combining the nested grid technology and rigid dynamic grid technology [6]. The loose coupling algorithm is used to realize the CFD/CSD time domain coupling solution [7]. The dynamic mesh technology is used to realize the elastic deformation of the wall boundary on the foreground grid where the gust generator blade is located. The CFD/CSD data exchange and the dynamic mesh are integrated through the radial basis function (Radial Basis Function, RBF) method. Finally, through CFD/CSD time domain coupling solution, the structural dynamic response of gust generator and wind tunnel and the gust flow field change in the test section are calculated, and the influence of structural dynamic response characteristics and elastic effect of gust generator on gust flow field quality is analyzed.

2 Overall Scheme of Gust Generator

The gust generator is installed in the closed test section of FL-10 wind tunnel of AVIC The Aerodynamics Research Institute. The size of the test section is 8 m (width) × 6 m (height). The generator is mainly composed of oscillating blade, hydraulic motor, driving end, following end and vertical support, as shown in Fig. 1. The blades of the generator are four groups, each of which is composed of three units. The blade airfoil is NACA0015, the span is 6 m, the chord length is 0.5 m, and the blade spacing is 0.7 m.

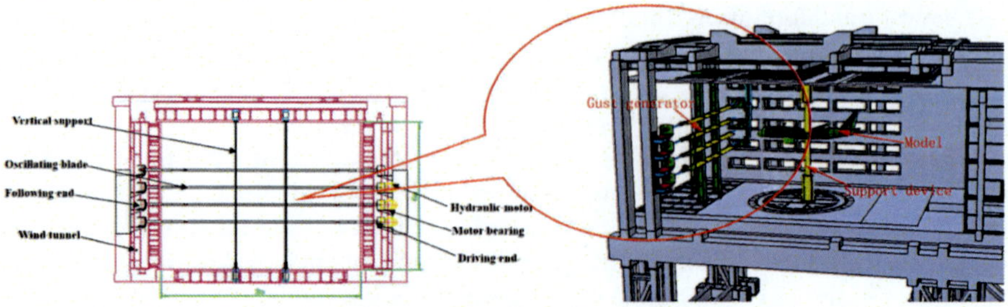

Fig. 1. General scheme of gust generator

3 Model and Calculation Method

3.1 Structure Calculation Model

3.1.1 Structural Modeling

The finite element analysis model includes the wind tunnel body and gust generator, as shown in Fig. 2. The natural nodes are basically used in meshing, and the analysis is not merged. The analysis nodes are taken on the theoretical shape, and the internal structure is taken on the intersection point constructed by the longitudinal and lateral main load. After analyzing the stress characteristics of gust generator structure, it is necessary to determine the element selection of finite element model in the process of structural discretization, because the selection of model element type needs to reflect the stress characteristics of actual structural members, and the connection between elements also needs to correctly reflect the force transmission path of the structure.

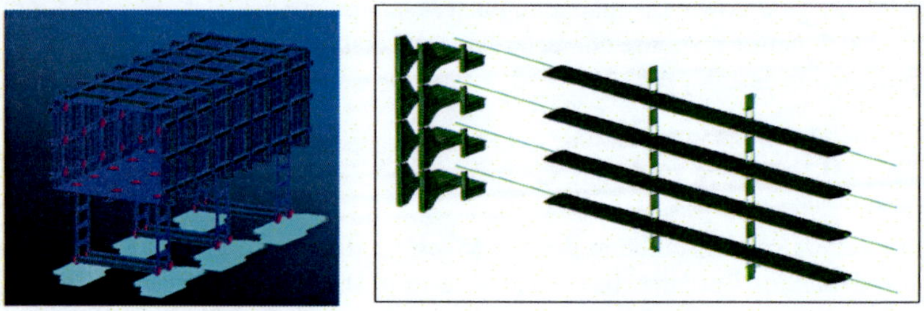

(a) wind tunnel body (b) gust generator

Fig. 2. Structure finite element model

3.1.2 Structural Response Analysis

Structural dynamic equations of motion are as follows:

$$\boldsymbol{M}\ddot{\boldsymbol{q}}(t) + \boldsymbol{C}\dot{\boldsymbol{q}}(t) + \boldsymbol{K}\boldsymbol{q}(t) = \boldsymbol{Q}(t) \quad (1)$$

where, $\boldsymbol{M}$、$\boldsymbol{C}$、$\boldsymbol{K}$ and are respectively structural mass matrix, damping matrix and stiffness matrix, $\boldsymbol{Q}(t)$ which are all external forces borne by the object. $\ddot{\boldsymbol{q}}(t)$、$\dot{\boldsymbol{q}}(t)$ and $\boldsymbol{q}(t)$ the acceleration, velocity and displacement of the moving object.

3.1.3 Model Modal

The first 11 order modals of wind tunnel body and generator are calculated. The first-order mode is shown in Fig. 3. The first-order modal frequencies of the wind tunnel body and generator are 1.53 Hz and 23.5 Hz, indicating that the lowest-order frequency of the generator is higher than its maximum operating frequency of 15 Hz, and it will not resonance occurs.

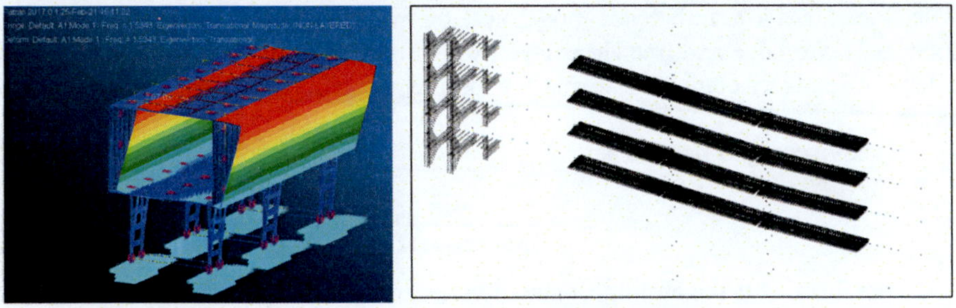

(a) wind tunnel body (b) gust generator

Fig. 3. The first-order modes of wind tunnel and generator

3.2 Fluid Calculation Model

3.2.1 Flow Control Equation

The fluid control equation used in this paper is Navier-Stokes (N-S), and its three-dimensional rectangular coordinate system can be written as:

$$\frac{\partial \boldsymbol{Q}}{\partial t}+\frac{\partial}{\partial x}(\boldsymbol{F}_1+\boldsymbol{G}_1)+\frac{\partial}{\partial y}(\boldsymbol{F}_2+\boldsymbol{G}_2)+\frac{\partial}{\partial z}(\boldsymbol{F}_3+\boldsymbol{G}_3)=\dot{\boldsymbol{S}} \tag{2}$$

where, $\boldsymbol{Q}$ is the dependent variable vector, $\boldsymbol{F}_1$、$\boldsymbol{F}_2$ and $\boldsymbol{F}_3$ is the non-viscous flux vector, $\boldsymbol{G}_1$、$\boldsymbol{G}_2$ and $\boldsymbol{G}_3$ is the viscous flux vector, and $\dot{\boldsymbol{S}}$ is the source term vector. The expression is as follows:

$$\boldsymbol{Q}=\begin{pmatrix} e \\ \rho \\ \rho u \\ \rho v \\ \rho w \end{pmatrix} \boldsymbol{F}_1=\begin{pmatrix} (e+p)u \\ \rho u \\ \rho u^2+p \\ \rho vu \\ \rho wu \end{pmatrix} \boldsymbol{F}_2=\begin{pmatrix} (e+p)v \\ \rho v \\ \rho uv \\ \rho v^2+p \\ \rho wv \end{pmatrix} \boldsymbol{F}_3=\begin{pmatrix} (e+p)w \\ \rho w \\ \rho uw \\ \rho vw \\ \rho w^2+p \end{pmatrix}$$

$$G_1 = \begin{pmatrix} \dot{q}_x - u\tau_{xx} - v\tau_{xy} - w\tau_{xz} \\ 0 \\ -\tau_{xx} \\ -\tau_{xy} \\ -\tau_{xz} \end{pmatrix} G_2 = \begin{pmatrix} \dot{q}_y - u\tau_{yx} - v\tau_{yy} - w\tau_{yz} \\ 0 \\ -\tau_{yx} \\ -\tau_{yy} \\ -\tau_{yz} \end{pmatrix} G_3 = \begin{pmatrix} \dot{q}_z - u\tau_{zx} - v\tau_{zy} - w\tau_{zz} \\ 0 \\ -\tau_{zx} \\ -\tau_{zy} \\ -\tau_{zz} \end{pmatrix}$$

where e is the total energy, ρ is the density, p is the pressure, u, v, and w are the velocity components in the x, y, and z directions. The $\dot{\mathbf{S}}$ source term vector includes the Euler discrete term model, the P-1 radiation model, and the porous equations.

3.2.2 Turbulence Model

SST (Shear-Stress Transport) k-ω shear stress transport k-ω turbulence model uses Wilcox standard k-ω turbulence model near the wall and k-ε turbulence model at the boundary layer edge and free shear layer, which transits through a mixed function. In order to effectively combine k-ω and k-ε turbulence models, they are unified into k-ω forms.

Vortex viscosity is defined as:

$$\nu_t = \frac{a_1 k}{\max(a_1 \omega, \Omega F_2)} \tag{3}$$

In the form, Ω is the absolute value of vorticity, $a_1 = 0.31$, F_2 is a mixed function, expressed as:

$$F_2 = \tanh\left[\max\left(2\frac{\sqrt{k}}{0.99\omega y}, \frac{500\mu}{\rho y^2 \omega}\right)\right]^2 \tag{4}$$

The form of ν_t solves the transport of turbulent shear stress in inverse pressure gradient boundary layer. k and ω are obtained by the corresponding turbulent model transport equation.

3.2.3 Wall Function

The viscous flux at the wall is given as follows:

$$\tau_w = \begin{cases} \frac{\kappa^* \rho_w \sqrt{k_1}(U_{c1} - U_{t_w})}{\ln(E y_1^*)}, & y_1^* > y_v^* \\ \frac{\mu_w (U_{c1} - U_{t_w})}{y_1^*}, & y_1^* \le y_v^* \end{cases} \tag{5}$$

where ρ is density, k is turbulent kinetic energy, μ is dynamic molecular viscosity, and U_{t_w} is tangential flow component of wall velocity vector. $\kappa^* = c_\mu^{1/4}\kappa$, $c_\mu = 0.09$, $\kappa = 0.41, E = 8.8, y_v^* = 11.2$.

$$y^* = \frac{c_\mu^{1/4} \rho_w y \sqrt{k}}{\mu_w} \tag{6}$$

where y is the local coordinate perpendicular to the wall, and U_c is the deformation of U_t.

3.2.4 Computational Grid

The blade of gust generator needs to rotate sinusoidally around its rotating shaft, and the coordinate definition is shown in Fig. 4.

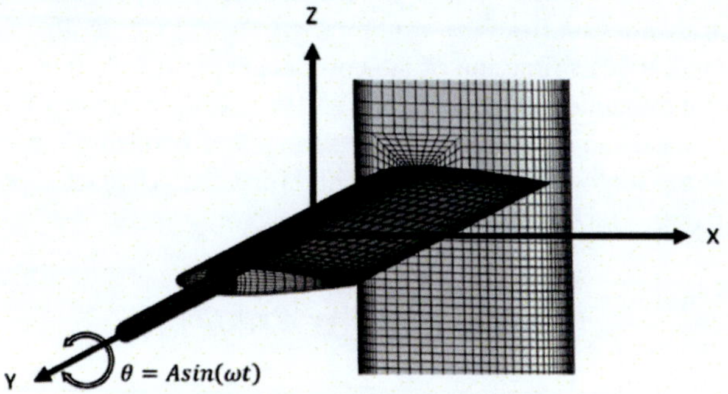

Fig. 4. Definition of coordinates

Through the topological relationship of geometric shape, hexahedral meshes are drawn, with a total of 15 million meshes and a mesh quality of more than 0.3. Appropriate grid encryption is carried out in the part where the flow parameters change dramatically. In order to minimize the numerical dissipation in the gust test section, the grid encryption is carried out in the direction of gust velocity variation (Z direction). Due to the wall function method, the actual surface Y + does not need to be strictly controlled at a certain value, Y + range from 30 to 100. It can be seen from the figure that the drawn computational grid can meet the requirements from orthogonality, grid scale, and density change (Fig. 5).

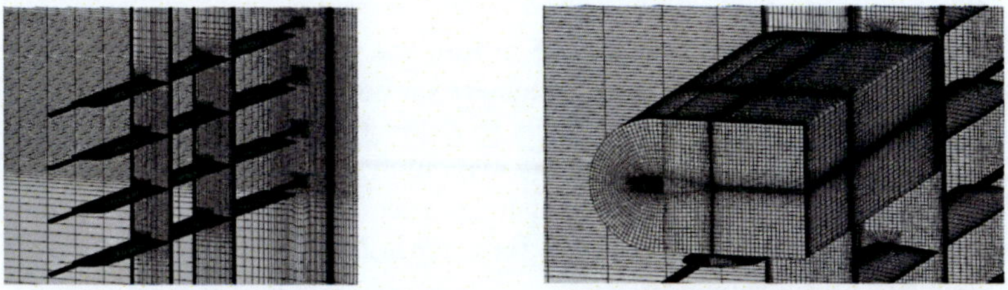

(a) Surface mesh of shafts, blades, supporting structures (b) External nested grids of shaft and blade

Fig. 5. Grid division

3.3 Pneumatic/Structural Data Exchange

In this paper, radial basis function (RBF) is used for data transmission and grid deformation of spatial flow field. The basic method is as follows. Firstly, the interpolation relationship between the structural grid and the aerodynamic grid is constructed according to the structural vibration mode. Secondly, the structural vibration mode is interpolated to the center point of the aerodynamic grid on the object surface by using the radial basis interpolation function. Then, the generalized force, generalized displacement and structural deformation are solved. Finally, the structural deformation is interpolated to the whole flow field space grid by using the radial basis interpolation function, so as to realize the data transmission from the structural grid to the whole flow field grid.

Because the radial basis function of the compact support function is consistent with the requirements of dynamic mesh calculation and fluid-solid coupling data transmission, and the solution matrix in the calculation is sparse and banded distribution, which is conducive to solving large-scale problems. Therefore, the compact support function is selected in this paper. The basic form is:

$$F(\mathbf{r}) = \sum_{i=1}^{N} w_i \varphi\left(\|\mathbf{r} - \mathbf{r}_i\| / r\right) \tag{7}$$

Let variable ξ be the ratio of the distance from the center to the tight radius, namely, $\xi = \|\mathbf{r} - \mathbf{r}_i\| / r$ Previous studies have shown that the quality of the updated mesh calculated by CP C2 function is the highest, and the ability to adapt to large deformation is stronger. Therefore, this paper selects the CP C2 compactly supported RBF basis function, and its specific expression is as follows:

$$\varphi(\xi) = \begin{cases} (1-\xi)^4(4\xi+1)\ , \xi \le 1 \\ 0 \qquad\qquad\qquad\ , \xi > 1 \end{cases} \tag{8}$$

4 Calculation Results and Analysis

4.1 Calculation Model and Working Condition

The generator blade is located at 7.6 m from the inlet upstream of the flow field in the test section, and 27 observation points are set downstream of the gust generator shaft to monitor the change of gust velocity in Z direction. The calculation results show that the change of time step will affect the maximum gust velocity, and has no effect on the frequency, shape and phase of the generated gust. Therefore, considering the calculation accuracy and efficiency, the unsteady time step is 2.5×10^{-4}s. The blade rotation motion equation of gust generator is:

$$\theta = \theta_{\max} \sin(2\pi f t) \tag{9}$$

where $\theta_{\max}$ is the maximum rotation angle of the blade, and f is the rotation frequency of the blade.

There are five calculation conditions, this paper only to the worse conditions $\theta_{\max} = 10°$、$f = 10\text{Hz}$、wind speed 60 m/s as an example to illustrate.

4.2 Simulation Analysis of Gust Generator

Figure 6 shows the variation curve of the maximum overall deformation (three directions) of the gust generator with time. It can be seen from the figure that the maximum deformation of all blades is less than 3 mm in a rotating cycle. The deformation along the Z direction is the largest, and presents a non-sinusoidal periodic change. This is because the maximum deformation of all blades is counted here, and the coordinates of the maximum deformation at each moment are not necessarily the same. The deformation along the X direction is almost greater than 0, and the period is half of the deformation along the Z direction. This is because the force required for the deformation in the X direction is air resistance, and the component of the resistance in the X direction is always positive. The deformation along the Y direction fluctuates around 0, which can be ignored compared with the other two directions.

Figure 7 is the time-varying curve of blade deformation at X = 0.3347 m, Y = 0 m, Z = 1.4 m, 0.7 m, −0.7 m. It can be seen from the figure that when the deformation monitoring points are fixed at a certain position of the structure, the time-history curve of blade deformation is obviously cyclical. The three monitoring points are on the axis of the wind tunnel, from top to bottom. It can be seen from Fig. 7 (a) and Fig. 7 (b) that the closer to the axis of the wind tunnel, the larger the peak value of the deformation in the X direction is, and the smaller the peak value of the deformation in the Z direction is, and the peak values are not more than 1.5 mm. From Fig. 7 (a) and Fig. 7 (c), it can be seen that due to the hole wall effect, when the blade is close to the wind tunnel wall, the Z direction deformation of the blade is asymmetric about 0m. And the more close the blade along the Z direction, the more negative the Z direction deformation.

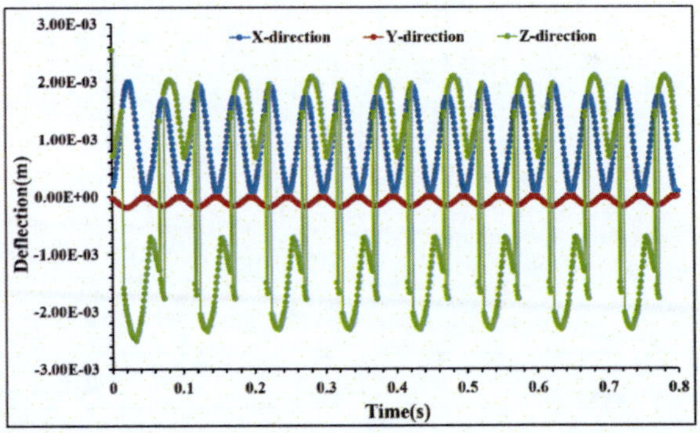

Fig. 6. Time history of the maximum deformation of the gust generator (three directions)

The time-domain curves in the above three cases are transformed by FFT to obtain the frequency characteristics of deformation in X and Z directions. It can be seen from Fig. 8 (a) that the main frequency of deformation in the Z direction is the same as the deflection frequency of the gust generator blade, which is 10 Hz, indicating that the elastic deformation of the blade is mainly due to the periodic variation of the pressure on the blade, which is a typical harmonic forced vibration. It can be seen from Fig. 8 (b) that the main frequency of the deformation in the X direction is 20 Hz. The reason for

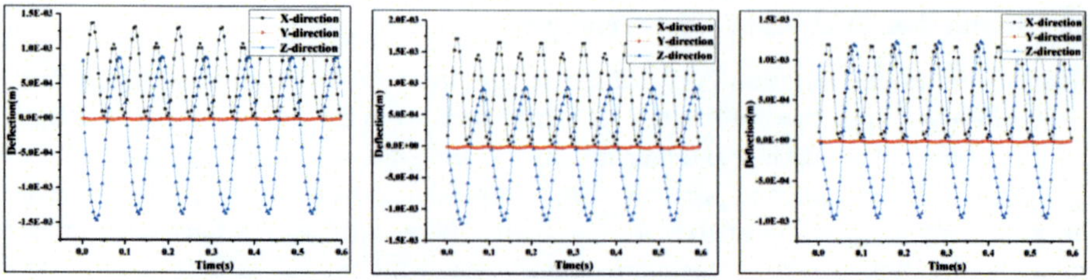

(a) Z=1.4m (b) Z=0.7m (c) Z= -0.7m

Fig. 7. Time history of gust generator deformation (X = 0.3347 m, Y = 0 m)

this phenomenon is that the excitation force is aerodynamic resistance, and the direction of the force in the 1/4 cycle and 3/4 cycle is X positive direction, which is equivalent to the absolute value of the sine, so the frequency is twice the excitation frequency. It can also be found from Fig. 8 (a) and Fig. 8 (b) that there are half-frequency and double-frequency situations. To explain the reasons for this phenomenon, it is necessary to further study the motion mechanism of gust generator. As shown in Fig. 9, the blade of gust generator deflects upward and downward sinusoidally. In the process of upward deflection, the fluid on the lower surface of the blade is separated and the separated vortex is generated. When the vortex develops to a sufficient extent, the shedding phenomenon occurs, and the vortex shedding will produce a main frequency and several subharmonic and frequency doubling resonant components.

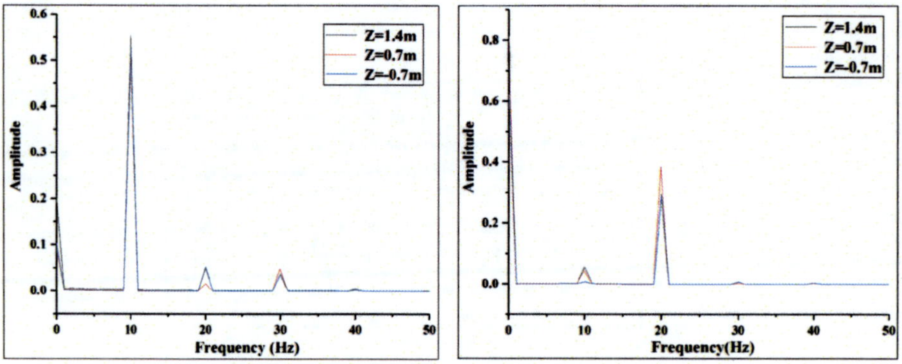

(a) X direction (b) Z direction

Fig. 8. Spectrum of gust generator deformation

Figure 10 shows the gust profile shape generated by gust generators with rigid and elastic structures. It can be seen from the figure that since the deformation of the elastic gust generator is very small (less than 1.5 mm) in the working process, the influence on the generated gust profile is very small (slightly reducing the gust amplitude), which can be basically ignored.

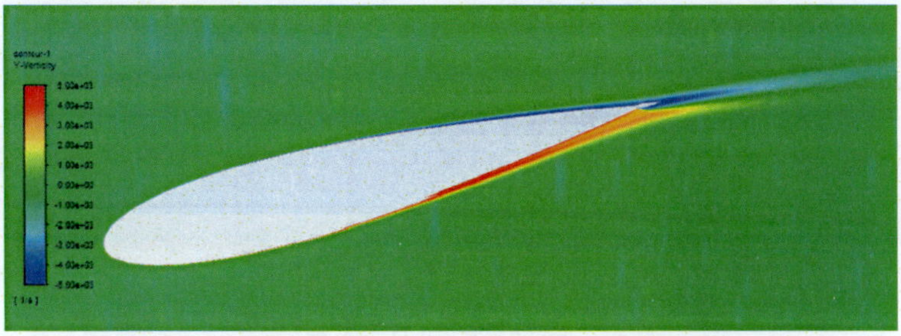

Fig. 9. Vortex nephogram of blade section of gust generator

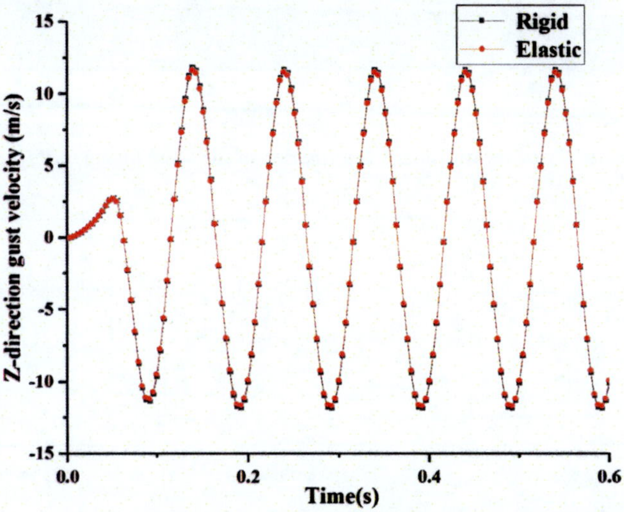

Fig. 10. Rigid\Elastic gust generator gust profile shape

4.3 Wind Tunnel Simulation Analysis

Figure 11 shows the variation curve of the maximum overall deformation (three directions) of the wind tunnel with time. It can be seen that the maximum deformation of the wind tunnel wall is less than 0.15 mm in a blade rotation cycle. The deformation along Z direction is the largest and presents non-sinusoidal periodic variation. The deformation along the X direction is almost no less than 0 m, because the force required for the deformation in the X direction is air resistance, and the component of the resistance in the X direction is always positive.

Figure 12 shows the variation curves of blade deformation with time at the positions of X = 0.8 m, 3.4 m, 7.3 m, Y = 0 m, Z = –2.65 m. It can be seen from the figure that the three observation points are on the central axis at the bottom of the wind tunnel, from forward to backward. Comparing the three figures, it can be seen that the maximum deformation of wind tunnel wall is at 3.4 m behind the blade.

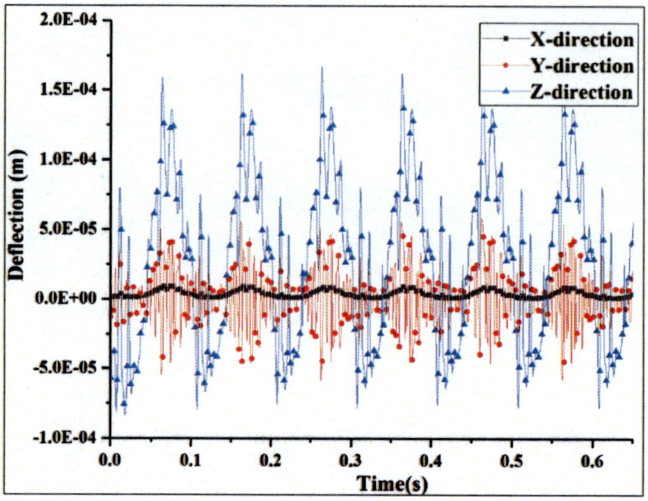

Fig. 11. Time history of the maximum deformation of the wind tunnel wall (three directions)

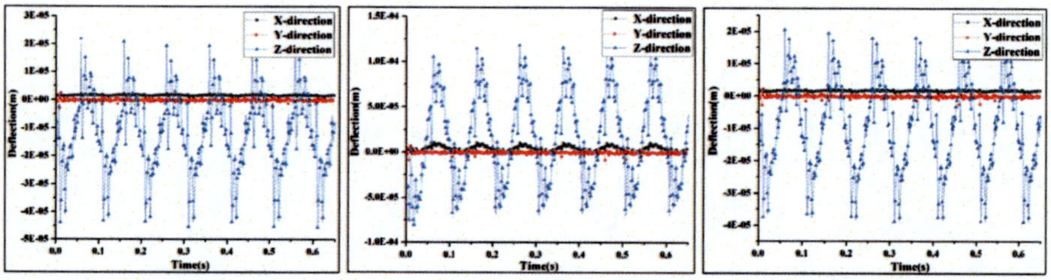

(a) x=0.8 m (b) x=3.4 m (c) x=7.3 m

Fig. 12. X coordinate change, Y = 0 m, z = –2.65 m position blade deformation versus time curve

The time-domain curves in the above three cases are transformed by FFT to obtain the frequency characteristics of deformation in X and Z directions. It can be seen from Fig. 13 that the blade deflection frequency of gust generator is 10 Hz, and the main frequency of the wind tunnel wall deformation in X and Z directions is 10 Hz, and there is a double frequency phenomenon.

Figure 14 shows the forces borne by the wall in different directions. It can be seen from the figure that the aerodynamic force on the upper and lower walls mainly comes from the Z direction, and the peak value is about 3 × 104 N. The aerodynamic force on the left and right walls mainly comes from the Y direction, and the peak value is about 2 × 104 N. This is also the cause of the maximum deformation in the Z direction.

4.4 Generator-Wind Tunnel Coupling Simulation Analysis

Under the condition of no wind speed, the influence of inertial force of gust generator on the wind tunnel body is studied, without considering the influence of gust aerodynamic

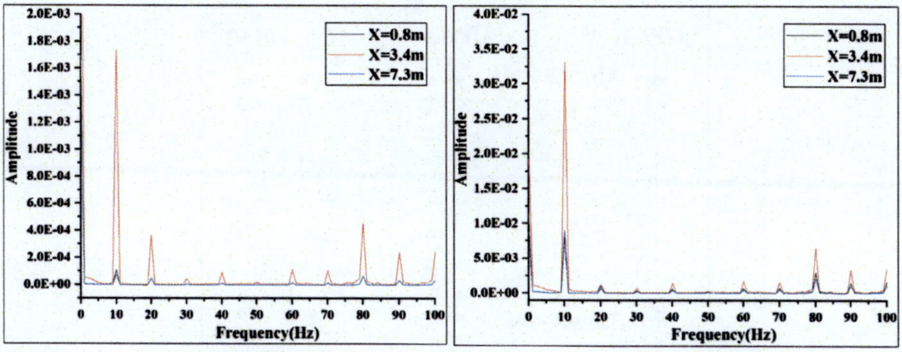

(a) X direction (b) Z direction

Fig. 13. Spectrum of wall deformation in wind tunnel

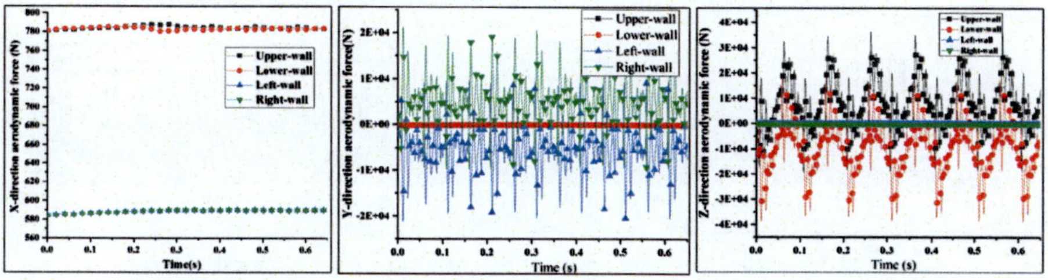

(a) X direction (b) Y direction (c) Z direction

Fig. 14. The time history of the force on the wall in different directions

force on the blade and the tunnel wall. Taking the blade swing frequency of $\theta_{max} = 10°$ and $f = 10$ Hz under the condition of bad working condition of the generator as an example, an acceleration sensor is set up to monitor the upper and lower walls of the wind tunnel. The location of the sensor is shown in Fig. 15.

The vibration of the wind tunnel caused by the generator movement is analyzed by the accelerometer arranged on the wall of the tunnel, and only the monitoring point 1 of the upper wall with large vibration is taken as an example. As can be seen from Fig. 16, according to the fact that when the oscillating frequency of the generator blade is 10 Hz, the frequency response of each position of the wind tunnel is mainly the excitation frequency of the gust generator and its frequency doubling under the inertial force of the generator oscillation. There is no excitation dispersion or other dangerous response of the elastic mode of the wind tunnel and the blade.

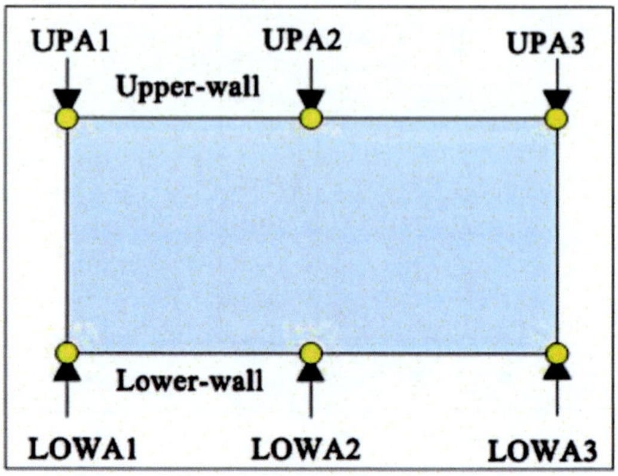

Fig.15. Acceleration monitoring points of upper and lower tunnel walls

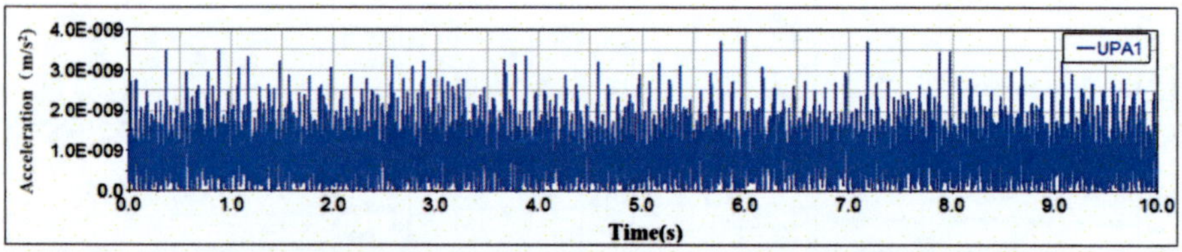

(a) Acceleration time domain curve of No.1 monitoring point in upper tunnel wall

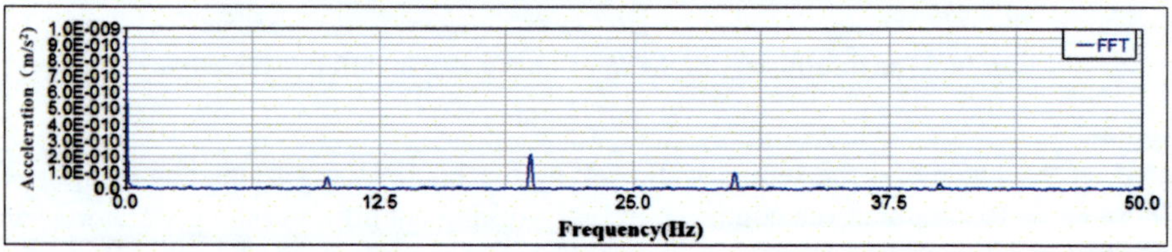

(b) Acceleration frequency curve of No. 1 monitoring point on upper wall

Fig. 16. Acceleration monitoring data curve of No. 1 monitoring point in upper tunnel wall

5 Conclusion

According to the elastic dynamic model of gust generator, the gust flow field quality and structural load considering structural elastic effect are calculated. The conclusions are as follows:

(1) The elastic deformation of gust generator blade is not more than 3 mm. The main frequency of elastic deformation in Z direction is the frequency of blade deflection, which is a typical forced oscillation. The main frequency of elastic deformation in X direction is twice that in Z direction.
(2) Due to the small deformation of the elastic gust generator in the working process, it has little effect on the gust profile (slightly reducing the gust amplitude), which can be ignored.

(3) The elastic deformation of wind tunnel wall does not exceed 0.15 mm. The main frequency of elastic deformation is the frequency of blade deflection, and there is an obvious frequency doubling phenomenon. The maximum deformation of wind tunnel wall is located in the middle part of the upper and lower wall.

References

1. Yang, X.M., Liu, N., Guo, C.P., et al.: A survey of aeroelastic wind tunnel test technology of flight vehicle. Acta Aerodynamica Sinica **36**(6), 995–1008 (2018)
2. Li, Y.: Study on vertical gust generation technology in Low Speed Wind Tunnel[R].Tian jin Civil Aviation University of China (2019)
3. Wang, L., Shang, L., Zhao, C.J., Dai, X.J.: Study on the discrete gust dynamic load and reponses of tactics missiles. J. Dyn. Contr. **13** (3), 194–198 (2015)
4. Hu, H.Y., Zhao, Y.H., Huang, R.: Study on aerodynamica analysis and control of aircraft structures. Chin .J. Theor. Appl. Mech. **48** (1),1–27 (2016)
5. Yang, S.B., Wang, L.B.: Analysis of airworthiness regulations of gust and turbulence loads for civil aircraft. Civil Aircraft Design Res. **02**, 49–53 (2016)
6. Xu, M., Chen, S.L.: CFD /CSD coupling calculation research. Chinese J. Appl. Mech. **21**(2), 33–36 (2004)
7. Xu, M., An, X.M., Chen, S.L.: CFD /CSD coupling numerical computational methodology. Acta Aeronautica et Astronautica Sinica **27**(1), 33–37 (2006)

Integrated Guidance and Control Design of Wide-Area Hypersonic Vehicle Based on Dynamic Inversion

Ying Wang, Guangwen Li(✉), Shaobo Zhai, Haizheng Zhang, Jingyi Wang, and Qiuling Jia

School of Automation, Northwestern Polytechnical University, Xian 710072, Shaanxi, China
liguangwen@nwpu.edu.cn

Abstract. The hypersonic vehicles (HV) represented by the combined powered aerospace vehicle have the characteristics of strong coupling, uncertain aerodynamic parameters, and many control constraints during the ascent phase. The guidance system of HV couple with its trajectory control system strongly and the trajectory optimization and guidance system design process are interactive and complicated. In this paper, a three-dimensional integrated guidance and control algorithm based on dynamic inversion is proposed. The attitude loop controller and guidance controller are designed to establish a mapping relationship between the longitudinal and lateral channel commands and the control surface commands, height and lateral position control of HV can be controlled accurately. The simulation results show that the integrated guidance and control method of HV for the wide-area flight can track reference commands precisely.

Keywords: Hypersonic vehicle · integrated guidance and control · dynamic inversion control

1 Introduction

Hypersonic vehicle (HV) has the characteristics of strong coupling, fast time-varying, and strong uncertainty, which has a significant impact on navigation guidance and control system. The guidance and control technology for HV has become a research hotspot [1–3]. During wide-range flight, the aircraft undergoes rapid state changes, and the uncertainty of the flight environment and aerodynamic models is significantly increased, while unknown disturbances become more apparent. Traditional control methods often separate guidance and control systems in design, which not only fails to optimize the overall control performance of the system to the greatest extent but also cannot meet the control performance requirements of HV under wide-range conditions [4, 5]. Therefore, the design of its guidance and control system needs to consider the coupling characteristics of the guidance and control loops and the interaction between the centroid motion and the motion around the centroid.

Integrated guidance and control (IGC) is a method to solve the above problems by integrating the guidance loop and control loop as a whole and directly generating control

Chinese Society of Aeronautics and Astronautics (Ed.): CASTC 2023, LNME, pp. 372–381, 2024.
https://doi.org/10.1007/978-981-99-8861-7_38

commands for the actuators based on the motion state information of the aircraft [6]. It fully utilizes the coupling relationship between the loops to directly design an integrated law, which more closely integrates the guidance and control loops and is beneficial to improve the overall performance [7].

Currently, there are various design methods for IGC, such as optimal control [8], sliding mode control [9], backstepping [10, 11], etc. However, there are still some problems in current research. Regarding the IGC design, many studies adopt a separate design approach for each channel, ignoring the coupling relationship of channel motion. This design requires repeated iterations and a large number of simulation tests, which leads to a large design workload and complex process [12]. As far as the existing IGC design scenarios are concerned, the current research is mostly based on missiles, and the flight phase is mostly in the dive phase. There are few studies on HV and its ascent control [13, 14].

A novel three-dimensional IGC algorithm based on dynamic inverse control is proposed in this paper, which solves the strong coupling and high nonlinearity of HV on guidance and control in the wide-range flight state of ascent. Firstly, a nonlinear dynamic inverse attitude loop is designed to meet the requirements of wide-range flight, and the high-precision control of angular rate and airflow angle is realized. Then, the guidance control law is designed by using the dynamic inversion control method, and the trajectory can be tracked accurately. Finally, the nominal trajectory under the designed multi-constraint conditions is used as the target command, the IGC design of the ascent phase is comprehensively simulated and verified, and the height and position can be controlled precisely.

2 Six-Degree-of-Freedom Model of HV

In this section, the six-degree-of-freedom model of the target vehicle is established according to the kinematics and dynamics equations. The motion equations of the six-degree-of-freedom rigid body model are as follows

$$\begin{cases} \dot{V} = \frac{P\cos\alpha\cos\beta - X - mg\sin\theta}{m} \\ \dot{\theta} = \frac{P(\sin\alpha\cos\gamma_V + \cos\alpha\sin\beta\sin\gamma_V) + Y\cos\gamma_V - Z\sin\gamma_V - mg\cos\theta}{mV} \\ \dot{\psi}_V = \frac{P(\sin\alpha\sin\gamma_V - \cos\alpha\sin\beta\cos\gamma_V) + Y\sin\gamma_V + Z\cos\gamma_V}{-mV\cos\theta} \end{cases} \tag{1}$$

$$\begin{cases} \dot{x} = V\cos\theta\cos\psi_V \\ \dot{y} = V\sin\theta \\ \dot{z} = -V\cos\theta\sin\psi_V \end{cases} \tag{2}$$

$$\begin{cases} \dot{\omega}_x = \frac{J_y(M_x - J_{xy}\omega_x\omega_z + (J_y - J_z)\omega_z\omega_y) + J_{xy}(M_y + J_{xy}\omega_y\omega_z - (J_x - J_z)\omega_z\omega_x)}{J_xJ_y - J_{xy}^2} \\ \dot{\omega}_y = \frac{J_x(M_y + J_{xy}\omega_y\omega_z - (J_x - J_z)\omega_z\omega_x) + J_{xy}(M_x - J_{xy}\omega_x\omega_z + (J_y - J_z)\omega_z\omega_y)}{J_xJ_y - J_{xy}^2} \\ \dot{\omega}_z = \frac{1}{J_z}(M_z + J_{xy}\omega_x^2 + (J_x - J_y)\omega_x\omega_y - J_{xy}\omega_y^2) \end{cases} \tag{3}$$

$$\begin{cases} \dot{\alpha} = \omega_z - (\omega_x \cos\alpha - \omega_y \sin\alpha)\tan\beta + \dfrac{1}{mV\cos\beta}(-Y - P\sin\alpha + mg\cos\theta\cos\gamma_v) \\ \dot{\beta} = \omega_x \sin\alpha + \omega_y \cos\alpha + \dfrac{1}{mV}(Z - P\cos\alpha\sin\beta + mg\cos\theta\sin\gamma_v) \\ \dot{\gamma}_v = \dfrac{\cos\alpha}{\cos\beta}\omega_x - \dfrac{\sin\alpha}{\cos\beta}\omega_y + \dfrac{Y + P\sin\alpha}{mV}(\tan\beta + \tan\gamma\sin\mu) \\ +\dfrac{Z}{mV}\tan\gamma\cos\mu\cos\beta \; - \dfrac{g}{V}\cos\gamma\cos\mu\tan\beta - \dfrac{P\cos\alpha}{mV}\tan\gamma\cos\mu\sin\beta \end{cases} \tag{4}$$

$$\begin{cases} m_{now} = m_0 - \int_0^{t_{now}} \dot{m}(t)dt \\ P = P_0(t_{bst}) + S_e(p_0 - p_h) \end{cases}. \tag{5}$$

The model consists of 12 state variables $X = [V, \theta, \psi_v, x, y, z, \omega_x\, \omega_y\, \omega_z\,, \alpha, \beta, \gamma_v]$ and 3 control inputs $u = [\delta_P, \delta_Y, \delta_R]$, Eq. is the dynamic equation of the center of mass motion, where V, θ, and ψ_v are the velocity, angle of track and angle of yaw. Equation (2) is the kinematics equation of the center of mass motion, x, y, z are the displacement of the aircraft in the ground coordinate system. Eq. is the dynamic equation of motion around the center of mass, $\omega_x\, \omega_y\, \omega_z$ are the angular velocities. Eq. is the differential equation of airflow angle, α, β, γ_v are attack angle, sideslip angle and bank angle. $\delta_P, \delta_Y, \delta_R$ are pitch, yaw, roll rudder angle. In addition, X is the total axial force, Y is the total lift force, and Z is the total lateral force. The mass of the assembly decreases due to the loss of fuel. The mass loss rate can be calculated by first-order interpolation to obtain the total mass of the current assembly, In Eq., m_0 is the mass of the HV at the initial full load, m_{now} is the current mass of the HV, $\dot{m}(t)$ is the loss of fuel mass at each time. Actual engine thrust P as in Eq., P_0 is the ground thrust of the combined engine, the thrust data changes with time and the thrust is uncontrollable. The static thrust varies with the static pressure of the gas at the nozzle section, where p_0 is the ground atmospheric static pressure, p_h is the static pressure at the height of the HV, and S_e is the nozzle section area.

The force generated by the rudder of HV includes the aerodynamic force F_R and the force generated by the gas rudder F_G, The torque includes the aerodynamic moment M_R and the moment M_G provided by the gas rudder. The specific expressions are listed as follows

$$\begin{cases} \mathrm{F}_i = F_{\mathrm{R}i} + \mathrm{F}_{\mathrm{G}i} = -C_k Q_s S - \mathrm{F}_k' \\ M_i = \mathrm{M}_{\mathrm{R}i} + \mathrm{M}_{\mathrm{G}i} = (C_j Q_s S) + M_{Gi} \end{cases} i = x, y, z \;\; k = A, N, Z \; j = l, n, m \tag{6}$$

where, C_A, C_N, C_Z are the aerodynamic resultant force coefficients. C_l, C_n, C_m are the aerodynamic resultant moment coefficients. $Q_s = 0.5\rho V^2$ is dynamic pressure, ρ is the current atmospheric density. S is the reference area. L is the reference length. $F_A^{'}, F_N^{'}$, $F_Z^{'}$ are the axial forces, normal force and lateral force of the gas rudder. M_{Gx}, M_{Gy}, M_{Gz} are the resultant moments of the gas rudder in the body shafting.

3 IGC Design Based on Dynamic Inversion

The controlled state variables of the loop around the center of mass are relatively intuitive, which are angle of attack, sideslip angle, velocity inclination angle, and three-channel angular rate. The guidance loop can be realized by controlling the ballistic angle, and the controlled state variables of the center of mass and the center of mass are fully coupled and integrated. Nonlinear dynamic inversion control can decouple between channels without a complex gain schedule. Because the change of the controlled object parameters cannot affect its linear decoupling control structure and gains, nonlinear dynamic inversion is suitable for the control of the aircraft in a wide domain.

Based on the dynamic inversion, the hierarchical design for various subsystems is proposed in this paper. Firstly, the control law is designed by using the dynamic characteristics between the control surface and the angular velocity. Secondly, the relationship between the airflow angle and the angular rate is established. In the third step, connect the airflow angle to the trajectory angle. The fourth step is to map the trajectory angle to the lateral position and altitude. Lastly, the height and lateral position commands to be tracked are connected with the control surface, so that the attitude control commands are obtained by using the relative motion relationship information between the aircraft and the target trajectory, and the aircraft can accurately track the height and lateral position. The following intuitive IGC design model can be obtained. The structure diagram is illustrated in Fig. 1.

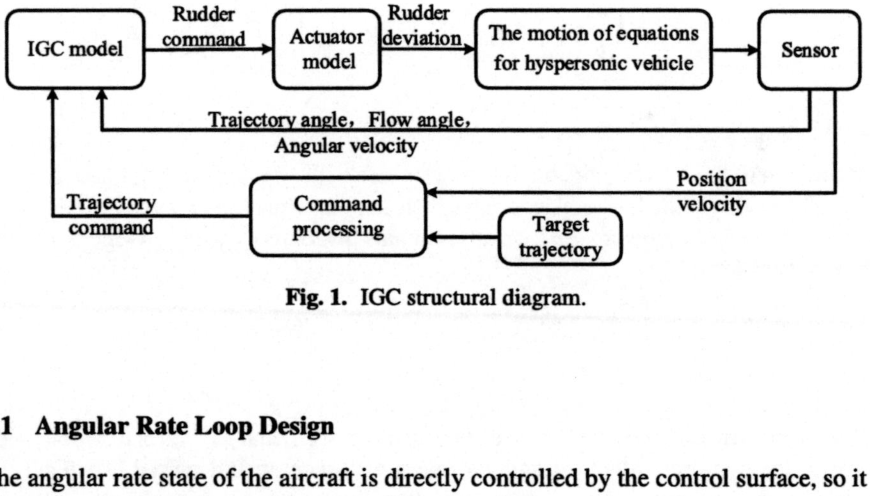

Fig. 1. IGC structural diagram.

3.1 Angular Rate Loop Design

The angular rate state of the aircraft is directly controlled by the control surface, so it is adopted as the inner loop of the control law. The differential equation can be written in the form of the following nonlinear system

$$\begin{cases} \dot{x}_1 = f_f(\overline{x}_f) + g_f(\overline{x}_f)u \\ y_1 = x_1 \end{cases} \tag{7}$$

where, $x_1 = [\omega_x, \omega_y, \omega_z]^T$, $\overline{x}_f = [V, \alpha, \beta, \omega_x, \omega_y, \omega_z]^T$, $u = [\delta_P, \delta_R, \delta_Y]^T$.

According to the differential Eq. (3) of angular rate, $f_f(\overline{x}_f))$ and $g_f(\overline{x}_f))$ can be obtained. As an example, the inversion expressions of the rolling rate ω_x are listed as follows

$$\begin{cases} f_{\omega_x}(\overline{x}_f) = \dfrac{J_y(\hat{M}_x - J_{xy}\omega_x\omega_z + (J_y - J_z)\omega_z\omega_y) + J_{xy}(\hat{M}_y + J_{xy}\omega_y\omega_z - (J_x - J_z)\omega_z\omega_x)}{J_xJ_y - J_{xy}^2} \\ g_{\omega_x i}(\overline{x}_f) = \dfrac{J_yC_{Mxi} + J_{xz}C_{Myi}}{J_xJ_y - J_{xy}^2}QSL, \quad i = \delta_P, \delta_R, \delta_Y \end{cases} \tag{8}$$

where,$\hat{M}_x$, $\hat{M}_y$ represent the rolling and yawing moment except the moment generated by the rudder deflection.

According to the characteristics of the aircraft and the actual engineering experience, this paper constructs the desired control signal: assuming that the error signal is: $\varepsilon = x_c\text{-}x$, x_c is a constant value, $\dot{x}_c$, $\ddot{x}_c$ are 0. What's more, $\ddot{\varepsilon}+2\xi\omega_n\dot{\varepsilon}+\omega_n^2\varepsilon = 0$ is considered. When the error ε approaches 0, $\xi\omega_n > 0$ will be calculated. Then, ω_n^2, $2\xi\omega_n$ are replaced by k_i and k_p respectively, and the instruction model $x/x_c = k_i/(s^2 + k_ps + k_i)$ can be obtained. The desired rudder surface command is obtained by Eq. (9)

$$u = g_f^{-1}(\overline{x}_f)\left(\begin{bmatrix} k_{xi}\displaystyle\int(\omega_{xc} - \omega_x) - k_{\omega_x}\omega_x \\ k_{yi}\displaystyle\int(\omega_{yc} - \omega_y) - k_{\omega_y}\omega_y \\ k_{zi}\displaystyle\int(\omega_{zc} - \omega_z) - k_{\omega_z}\omega_z \end{bmatrix} - \begin{bmatrix} f_{\omega_x}(\overline{x}_f) \\ f_{\omega_y}(\overline{x}_f) \\ f_{\omega_z}(\overline{x}_f) \end{bmatrix}\right). \tag{9}$$

3.2 Airflow Angle Loop Design

In the design of the slower loop, the steering surface deflection effect is ignored, and the steady-state value of the fast loop is adopted instead of its instantaneous value. Therefore, the control law of the slower loop is an approximate inverse control law, which is written in a nonlinear form

$$\begin{cases} \dot{x}_2 = f_m(\overline{x}_m) + g_{m1}(\overline{x}_m)x_1 + g_{m2}(\overline{x}_m)u \\ y_2 = x_2 \end{cases} \tag{10}$$

where,$x_1 = [\omega_x, \omega_y, \omega_z]^T$,$\dot{x}_2 = [\dot{\alpha}, \dot{\beta}, \dot{\gamma}_V]^T$,$\overline{x}_m = [V, \gamma, \alpha, \beta, \gamma_V]^T$.

From the differential Eq. (4) of the airflow angle,$f_m(\overline{x}_m)$ and $g_m(\overline{x}_m)$ can be obtained. Based on the first-order filter, the angular rate expected command is established as

$$\begin{bmatrix} \omega_{xc} \\ \omega_{yc} \\ \omega_{zc} \end{bmatrix} = g_{m1}^{-1}(\overline{x}_m)\left(\begin{bmatrix} k_\alpha(\alpha_c - \alpha) + \dfrac{1}{\tau_\alpha s + 1}\dot{\alpha}_c \\ k_\beta(\beta_c - \beta) + \dfrac{1}{\tau_\beta s + 1}\dot{\beta}_c \\ k_{\gamma_v}(\gamma_{vc} - \gamma_v) + \dfrac{1}{\tau_{\gamma_v} s + 1}\dot{\gamma}_{vc} \end{bmatrix} - \begin{bmatrix} f_\alpha(\overline{x}_m) \\ f_\beta(\overline{x}_m) \\ f_{\gamma_v}(\overline{x}_m) \end{bmatrix}\right). \tag{11}$$

3.3 Trajectory Angle Loop Design

The Eq. (1) can be divided into Eq. (12) by balancing the horizontal and vertical forces. f_θ and f_{ψ_V} are minimal coupling quantities as Eq. (13)

$$\begin{cases} \dot{\theta} = \frac{Y\cos\gamma_V - mg}{mV} + f_\theta \\ \dot{\psi}_V = -\frac{Y\sin\gamma_V}{mV} + f_{\psi_V} \end{cases} \tag{12}$$

$$\begin{cases} f_\theta = \dfrac{mg(1-\cos\theta) + P(\sin\alpha\cos\gamma_V + \cos\alpha\sin\beta\sin\gamma_V) - Z\sin\gamma_V}{mV} \\ f_{\psi_V} = \dfrac{Y\sin\gamma_V}{mV} + \dfrac{P(\sin\alpha\sin\gamma_V - \cos\alpha\sin\beta\cos\gamma_V) + Y\sin\gamma_V + Z\cos\gamma_V}{-mV\cos\theta} \end{cases}. \tag{13}$$

The lift is expressed as

$$Y = QSC_{Y\alpha}\alpha + \Delta Y \tag{14}$$

where, $C_{Y\alpha}$ represents the derivative of the lift coefficient to the angle of attack, and ΔY represents the remaining lift.

Based on Eq. And Eq., the angle of attack command is simplified as Eq.

$$\alpha_c = ((-f_\theta + \dot{\theta}) + \frac{g}{V})\frac{mV}{\cos\gamma_V} - \Delta Y)\frac{1}{QSC_{Y\alpha}}. \tag{15}$$

The command value of the velocity inclination angle is obtained from Eq.

$$\gamma_V = arc\sin(\frac{mV(f_{\psi_V} - \dot{\psi}_V)}{Y}). \tag{16}$$

The designable control law based on Lyapunov stability theory is

$$\begin{cases} \dot{\theta} = \dot{\theta}_c - k_\theta(\theta - \theta_c) \\ \dot{\psi}_V = \dot{\psi}_c - k_{\psi_V}(\psi_V - \psi_{Vc}) \end{cases}. \tag{17}$$

In the ascent phase of HV, it should always maintain a non-sideslip state. Substituting Eq. Into Eq. and Eq., the desired command can be obtained. In summary, the expected command for establishing the aerodynamic angle can be obtained

$$\begin{cases} \alpha_c = ((-f_\theta + \dot{\theta}_c - k_\theta(\theta - \theta_c) + \frac{g}{V})\frac{mV}{\cos\gamma_V} - \Delta Y)\frac{1}{QSC_{Y\alpha}} \\ \beta_c = 0 \\ \gamma_{Vc} = arc\sin(\frac{mV(f_{\psi_V} - \dot{\psi}_{Vc} + k_{\psi_V}(\psi_V - \psi_{Vc}))}{Y}) \end{cases}. \tag{18}$$

3.4 Trajectory Loop Design

The purpose of the controller design in this section is to control the height and the lateral position, then the HV can track their respective expected commands. The uncoupled terms in the differential Eq. of the height and the lateral position are proposed separately, and the following equation can be obtained

$$\begin{cases} \dot{h} = V\sin\theta \\ \dot{z} = -V\sin\psi_V + f_z \end{cases}. \tag{19}$$

where, $f_z = V\sin\psi_V - V\cos\theta\sin\psi_V$.

Similarly, the control law is designed by using Lyapunov stability theory, and shift term of Eq. (19). The command value of the trajectory angle can be obtained

$$\begin{cases} \theta_c = \arcsin(\frac{\dot{h}_c - k_h(h-h_c)}{V}) \\ \psi_{Vc} = \arcsin(\frac{f_z - \dot{z}_c + k_z(z-z_c)}{V}) \end{cases}. \tag{20}$$

According to the motion state information of the aircraft, the control command required by the actuator is directly generated, that is, the relationship between the height and lateral position and the input control surface of the system is obtained, which integrates the guidance and control loop more closely and is conducive to improving the comprehensive performance.

4 Simulation Verification

In order to verify the designed guidance and control integrated controller can effectively deal with the HV in the ascent phase can better follow the designed trajectory command. In this paper, the Gauss pseudospectral method is applied to discretize the nonlinear differential equation and the nonlinear programming is adopted to solve the optimal open-loop reference trajectory. A trajectory command sequence meeting the constraint conditions is designed. By designing the optimal index, that is, the integration of normal and lateral overload in the whole ascent phase is minimized, the trajectory is smooth and the buffeting is minimized during the flight, which ensures the safety of the flight. The height and lateral position command sequences are obtained by interpolation of command values and time. The height varies from 10 km to 110 km, and the lateral position ranges from 0 to 3.1 km. The initial velocity is 3 Mach, and the initial trajectory inclination angle and pitch angle are 10°.

The initial values of aircraft parameters and controller parameters in the simulation are shown in Table 1 and 2, and the simulation results are shown in Fig. 2 , 3, 4, 5, 6 and 7.

Figure 3 is the response curve of the trajectory angle following the expected command. According to Fig. 3, the slight deviation in the front section of the response curve is mainly due to the influence of the velocity change and the reaction time of the controller. In general, the trajectory angle can follow the command curve accurately.

The dynamic response curve of the airflow angles and their command response curves are shown in Fig. 4, and the dynamic response curve of flight angular rate and

Table 1. Parameters of aircraft.

Parameter	Unit	Maximum weight	Empty weight
m	kg	23400	9060
S	m^2	1	1
L	m	11.575	11.575

Table 2. Parameters of controller.

Parameter	Value	Parameter	Value	Parameter	Value	Parameter	Value
k_{xi}	40	k_{ω_x}	13	k_β	0.9	τ_β	0
k_{yi}	40	k_{ω_y}	14	k_{γ_v}	1	τ_{γ_v}	0.1
k_{zi}	40	k_{ω_z}	20	k_θ	0.6	k_h	0.1
k_α	0.9	τ_α	0.5	k_{ψ_V}	0.35	k_z	0.04

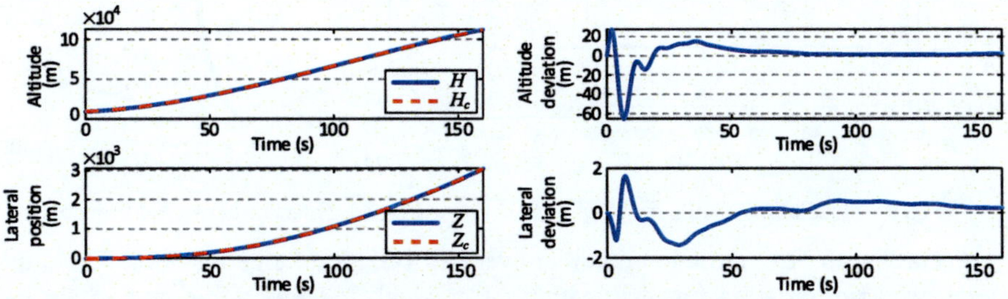

Fig. 2. Simulation results for trajectory response.

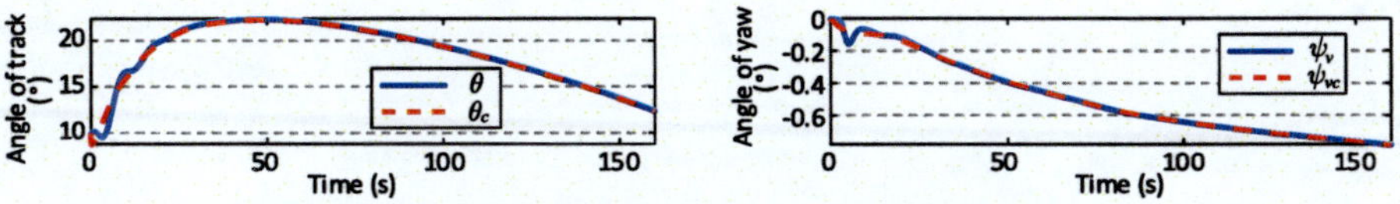

Fig. 3. Simulation results for trajectory angle.

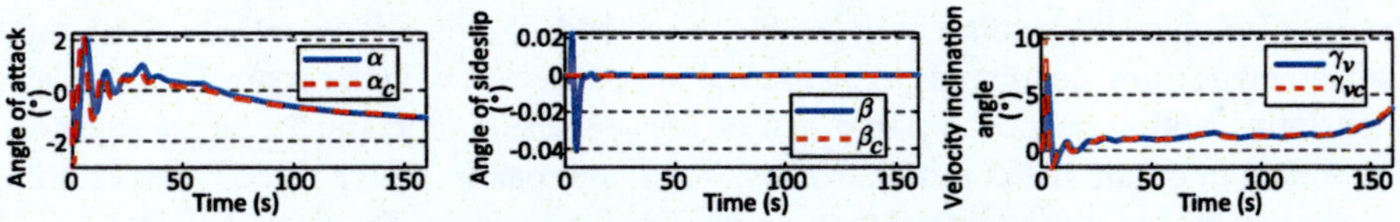

Fig. 4. Simulation results for airflow angle.

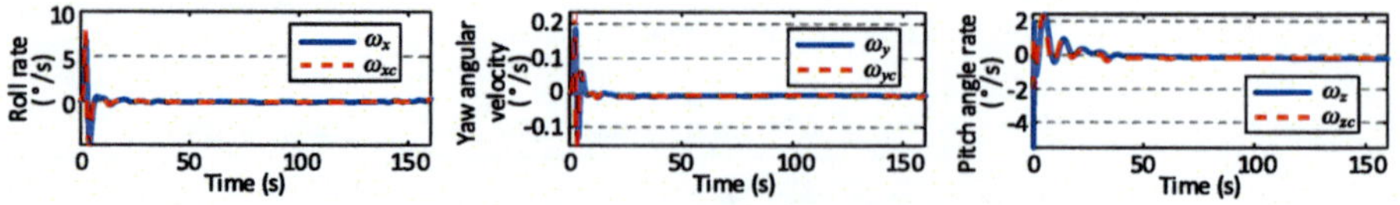

Fig. 5. Simulation results for angular velocity.

their command response curves are in Fig. 5. The sideslip angle remains at 0 almost throughout the ascent phase, and the fluctuation is so small that it can be ignored. The airflow angle and angular rate can better follow the command curve, because of response time, there will be a deviation from the command but does not affect the overall control effect.

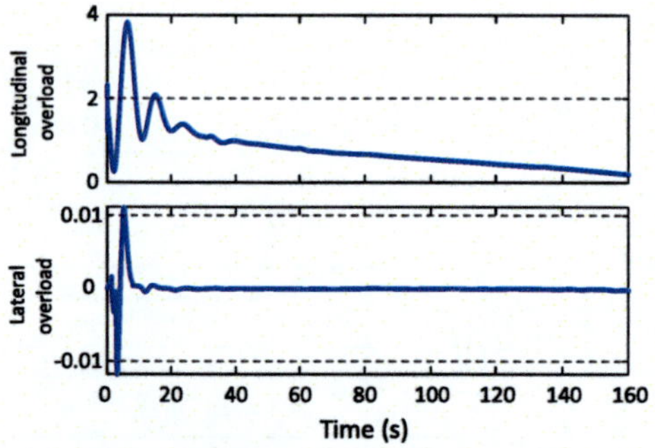

Fig. 6. Simulation results for overload.

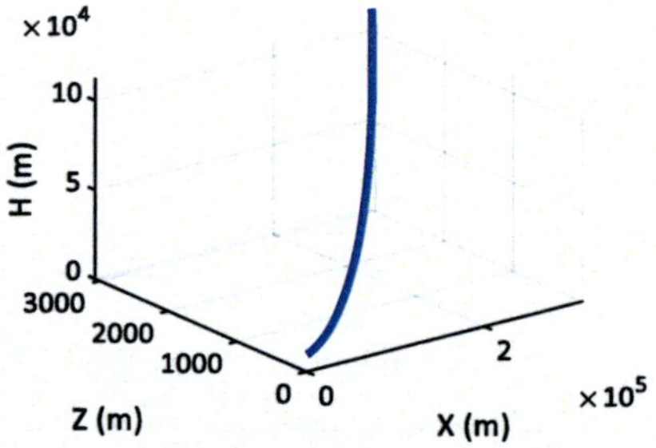

Fig. 7. Three-dimensional trajectory.

According to Fig. 6, the longitudinal overload variation range is within 4, and the lateral overload variation range is within 0.02, which remains stable and within the performance requirements. The flight three-dimensional trajectory, illustrated in Fig. 7, shows that the proposed algorithm ensures the stable control of the ascent phase of the HV.

5 Conclusion

In this paper, the IGC technology in the complicated and varied environment of the ascent phase is studied based on the wide-area flight of the HV. Firstly, a six-degree-of-freedom model of HV is established. Secondly, the IGC algorithm design of the ascent phase is completed by combining the block method of dynamic inversion. The height and lateral position commands to be tracked are connected with the control surface, and the relative motion relationship information between the aircraft and the target trajectory is adopted to obtain the control surface control commands, which directly realizes the tracking of the height and lateral moving position, and ensures the safety and stability of the aircraft. The simulation results show that the integrated controller proposed in this paper can realize the tracking of target height and lateral position in wide-area flight.

References

1. Gstattenbauer, G.J., Franke, M.E., Livingston, J.W.: Cost comparison of expendable hybrid and reusable launch vehicles. In: Proceedings of AIAA/Space Conferences and Exposition, San Jose (2006)
2. Wang, C.Q.: Technological innovation and development prospect of aerospace vehicle. J Astronaut, 2021, 42: 807–819Author, F., Author, S., Author, T.: Book title. 2nd edn. Publisher, Location (1999)
3. Bao, W.M.: Present situation and development tendency of aerospace control techniques. Acta Autom Sin **39**, 697–702 (2013)
4. Siouris, G.M.: Missile guidance and control systems. Springer Science & Business Media (2004)
5. Hou, M., Liang, X., Duan, G.: Adaptive block dynamic surface control for integrated missile guidance and autopilot. Chin. J. Aeronaut. **26**(3), 741–750 (2013)
6. Williams, D., Jack, R., Bernard, F.: Design of an integrated strapdown guidance and control system for a tactical missile. In: Guidance and Control Conference (1983)
7. Chwa, D., Jin Young, C.: Adaptive nonlinear guidance law considering control loop dynamics. IEEE Trans. Aerosp. Electron. Syst. **39**(4), 1134–1143 (2003)
8. Hughes, T., McFarland, M.: Integrated missile guidance law and autopilot design using linear optimal control. In: AIAA Guidance, Navigation, and Control Conference and Exhibit (2000)
9. Xu, H., Maj D.M., Ioannou, P.A.: Adaptive sliding mode control design for a hypersonic flight vehicle. J. Guidance, Contr., Dyn. **27**(5), 829–838 (2004)
10. Guo, C., Xiao-Geng, L.: Integrated guidance and control based on block backstepping sliding mode and dynamic control allocation. Proc. Inst. Mech. Eng., Part G: J. Aerosp. Eng. **229**(9) 1559–1574 (2015)
11. Zong, Q., et al.:Robust adaptive approximate backstepping control of a flexible air-breathing hypersonic vehicle with input constraint and uncertainty. Proceedings of the Institution of Mechanical Engineers, Part I: J. Syst. Contr. Eng. **228**(7), 521–539 (2014)
12. Wang, X., Wang, J.: Partial integrated guidance and control with impact angle constraints. J. Guidance, Contr. Dyn. **38**(5), 925–936 (2015)
13. Wang, J.H., Liu, L.H., Zhao, T., et al.: Integrated guidance and control for hypersonic vehicles in dive phase with multiple constraints. Aerosp. Sci. Technol. **53**, 103–115 (2016)
14. Song, H., Zhang, T.: Fast robust integrated guidance and control design of interceptors. IEEE Trans. Control Syst. Technol. **24**(1), 349–356 (2015)

Civil Aircraft Integrated Standby Navigation System Design Based on BeiDou Satellite Navigation System

Lei Guo(✉), Zhiming Zheng, and Hongyuan Song

Shanghai Aircraft Design and Research Institute, Shanghai 201210, China
guolei@comac.cc

Abstract. The requirements of the integrated standby navigation system based on BeiDou Satellite Navigation System (BDS) are captured and the safety is analyzed. Then, the system interfaces, architecture and three schemes are proposed. In addition, the implementation including hardware and software are introduced. Finally, the test result shows the integrated standby navigation system based on BDS realized in this paper is effective.

Keywords: Standby Navigation System · FHA · FDAL · Civil Aircraft · BDS

1 Introduction

When the aircraft original primary navigation system fails, the integrated standby navigation system based on BDS [1] provides the important navigation reference information for the pilots, which is usually independent of the main display, such as being integrated into the integrated standby instrument (ISI) or other equipment. The external relationships of the system are shown in Fig. 1.

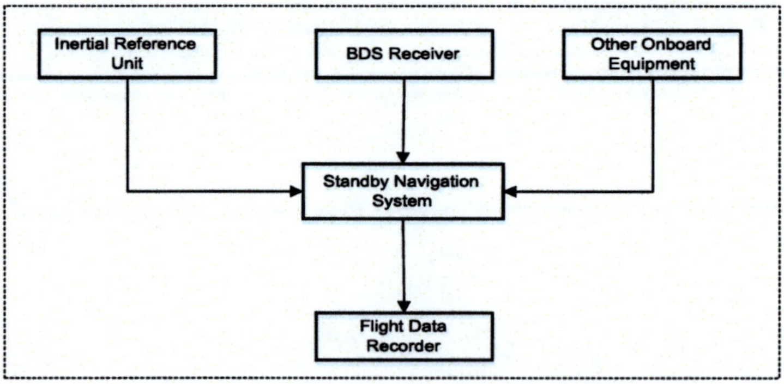

Fig. 1. The external relationships of the integrated standby navigation system based on BDS

Chinese Society of Aeronautics and Astronautics (Ed.): CASTC 2023, LNME, pp. 382–389, 2024.
https://doi.org/10.1007/978-981-99-8861-7_39

The functions of the main cross-linking equipment are as follows: a) BDS receiver: provides the BDS navigation data for the integrated standby navigation based on BDS, such as ground speed, longitude and latitude, and track angle. b) Inertial reference unit (IRU): provides heading, track angle, ground speed, longitude and latitude for the integrated standby navigation system based on BDS. c) Integrated standby navigation display: displays the standby navigation system based on BDS information. d) Flight data recorder (FDR): records the flight data of the system.

2 Requirements and Safety Analysis

2.1 System Requirements

In order to ensure the civil aircraft safety, the integrated standby navigation system based on BDS shall provide but not limited to the following information: a) the current ground speed; b) the current track angle; c) the current latitude and longitude; d) the target latitude and longitude; e) the azimuth angle from the current position to the target waypoint; f) the distance between the current position and the target waypoint; g) the required time from the current position to the target.

2.2 Functional Hazard Assessment

The Functional Hazard Assessment (FHA) plays a significant role in the safety analysis [2]. The failure effect is determined according to the degree of its influence on the aircraft and its crew. It is classified into five categories in the AC 25-1309 [3]: Catastrophic, Hazardous, Major, Minor and No Safety Effect. Each of them corresponds to different probability from 10^{-9} to no requirement. In AC 25-11B [4], the failure effect of completely losing the unrecoverable navigation information display is Major. While the failure effect of the misleading navigation information of both the pilot and the copilot is Major or Hazardous. The loss of the unrecoverable communication is Major, while the loss of the unrecoverable navigation information display and the loss of the unrecoverable communication is Catastrophic.

2.3 FDAL Allocation

The functional development assurance level (FDAL) [5] is assigned depending on the severity classification of failure effect above considering the independence (Table 1).

According to the Table 1, the FDAL of the communication and navigation combined functions is required to be A. If the communication FDAL is C, the navigation FDAL shall be A. Therefore, the FDAL decomposition of the navigation is summarized in Table 2 to meet the FDAL above.

The tree of the navigation FDAL allocation is shown in Fig. 2.

Table 1. Failure effect and FDAL allocation of the Navigation and communication

No	Failure condition	Effect level	FDAL
1	Completely lose the irrecoverable navigation information display	Major	C
2	Misleading navigation information display both on the pilot and copilot	Major or Hazardous	C or B
3	Total loss of communication	Major	C
4	Both the navigation information and communication display are completely lost	Catastrophic	A

Table 2. FDAL decomposition of the navigation

FDAL	Compound mode		Explanation
A	A	-	FDAL of providing the navigation on the display is A
	B	B	FDAL of providing the navigation on the primary display is B and FDAL of providing the navigation on the standby display is B

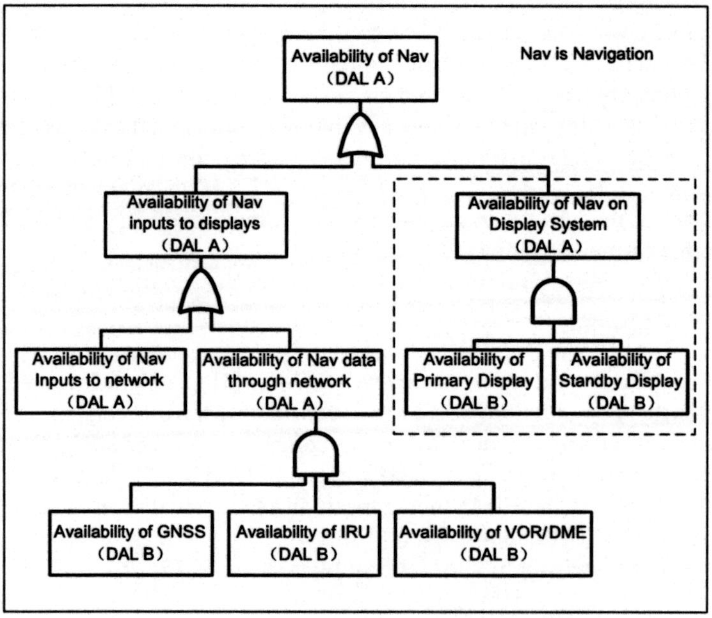

Fig. 2. The tree of the navigation availability

3 System Scheme

3.1 System Architecture

In order to realize backup, the integrated standby navigation device shall receive the navigation data from BDS receiver and IRU through the dependent ARINC 429 [6] input buses. The navigation data includes three types: only processed by BDS receiver, only processed by IRU and the data combined BDS receiver and IRU. The system architecture diagram is shown in Fig. 3.

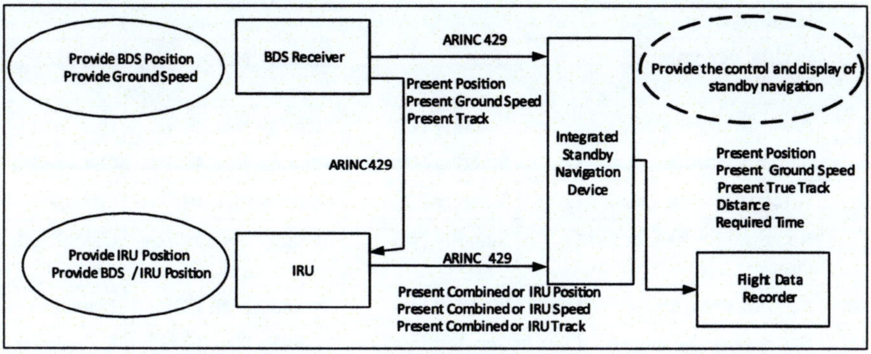

Fig. 3. Architecture diagram of the integrated standby navigation system based on BDS

3.2 System Interfaces

Table 3 is the interfaces of the integrated standby navigation system based on BDS.

3.3 Design Schemes Analysis

Three design schemes of the integrated standby navigation system based on BDS are proposed: a) Scheme 1: based on the tuning control panel (TCP). b) Scheme 2: based on the original ISI (Integrated Standby Instrument) on the aircraft. c) Scheme 3: adopt an independent instrument. Each of the three schemes advantages and disadvantages is shown in Table 4.

If the cockpit installation space is sufficient, the scheme 3 is the optimal. Otherwise, the scheme 2 is the optimal; Scheme 1 is the second choice.

Table 3. The interfaces of the integrated standby navigation system based on BDS

No	Interface Name	I/O	Source	Destination	Comment
1	BDS longitude	I	BDS Receiver	Standby Nav	Obtained
2	BDS latitude	I	BDS Receiver	Standby Nav	Obtained
3	Ground speed	I	BDS Receiver	Standby Nav	Obtained
4	Track angle	I	BDS Receiver	Standby Nav	Obtained
5	UTC time	I	BDS Receiver	Standby Nav	Obtained
6	IRU/BDS combined longitude	I	IRU	Standby Nav	Obtained
7	IRU/BDS combined latitude	I	IRU	Standby Nav	Obtained
8	Aircraft longitude	O	Standby Nav	Standby Nav, FDR	Obtained
9	Aircraft latitude	O	Standby Nav	Standby Nav, FDR	Obtained
10	Aircraft ground speed	O	Standby Nav	Standby Nav, FDR	Obtained
11	Required time to the target	O	Standby Nav	Standby Nav, FDR	Calculated
12	Aircraft track angle	O	Standby Nav	Standby Nav, FDR	Obtained
13	Distance to the target	O	Standby Nav	Standby Nav, FDR	Calculated
14	Track angle to the target	O	Standby Nav	Standby Nav, FDR	Calculated

Table 4. Different design schemes analysis

Advantage/ Disadvantage	Scheme 1	Scheme 2	Scheme 3
Advantage	a) No additional device; b) Numeric keyboard	a) No additional device; b) Readable information; c) Develop on ISI DAL	a) Readable information; b) Convenient to check
Disadvantage	a) Higher DAL of TCP; b) Additional press keys; c) Inconvenient to check	No numeric keyboard	More installation space

4 System Implementation and Test

4.1 Hardware

The integrated standby navigation device is the core component of the calculation and display. It includes the main control unit, the ARINC 429 bus Unit, the LCD unit and the interface Unit, the interface unit, the I/O unit and the power supply unit (Fig. 4). The main control unit can collect and process the position data of the BDS. It can calculate and generate the navigation information with graphic, so as to realize the function of managing the waypoint. The ARINC 429 bus unit can receive the position data in ARINC 743 [7] format sent by the satellite navigation receiver through ARINC 429 bus interface and forward it to the core module. The LCD unit can receive the RGB display output by the display core module and send them to the main control module. The interface unit can realize the debugging port (RS422), the file import interface (USB), the network port (reserved) and so on. The I/O unit can realize the level conversion and protection of I/O interface. The Power supply unit can receive the input DC 28 V and provide the power supply required by each unit at a stable voltage.

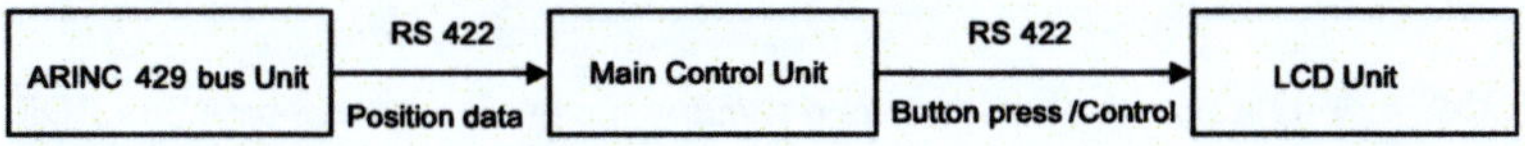

Fig. 4. The Hardware relationship in the integrated standby navigation device

4.2 Software

The development tools of the integrated standby navigation system based on BDS are SCADE and Visual Studio and the software modules are as shown in Table 5.

Table 5. Software modules of the integrated standby navigation system based on BDS

No	Module name	Function
1	Plane_Protocol	Process the ARINC 429 data
2	Air_Route	Calculate and generate the navigation information
3	WPT_List	Display the waypoint and manage them
4	FPL	Display the flight plan and manage it
5	Fix	Display the Fix and manage it
6	Panel	Display the navigation information with graphical form

4.3 Test

As shown in the Fig. 5, connect the BDS signal generated from the BDS simulator to the input interface of the BDS receiver antenna through radio frequency (RF) cable. Then connect the ARINC 429 output interface of the BDS receiver to the tested integrated standby navigation device through matching cable. BDS simulator generates the BDS signal. The BDS receiver receives the BDS signal and sends to the integrated standby navigation system based on BDS. Then the system calculates and generates the navigation information with graphic. The system test includes function test (waypoint management, etc.), interface test (lighting, etc.) and power test.

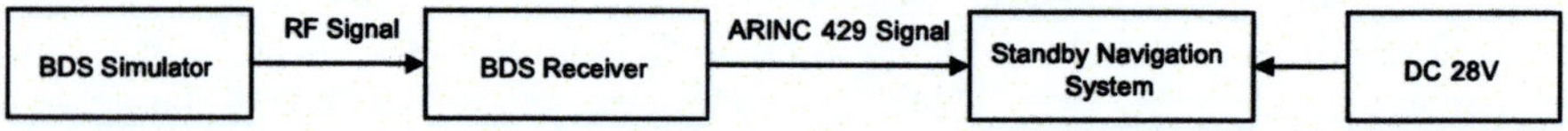

Fig. 5. The test set of the integrated standby navigation system based on BDS

Make a flight plan for the test above is shown in Table 6.

Table 6. Flight plan for the integrated standby navigation system based on BDS

Waypoint ID	Longitude	Latitude
REPOL	119.678333°	37.591667°
DY108	119.039028°	37.547750°
DY206	118.909222°	37.398417°
DY204	118.913694°	37.320528°
DY203	118.800722°	37.316333°
DY202	118.795833°	37.397500°
DY200	118.790000°	37.497139°
DYN	118.788333°	37.530000°

According to the waypoint, the tested fight track is shown in Fig. 6. The result shows the integrated standby navigation system based on BDS designed in this paper is effective.

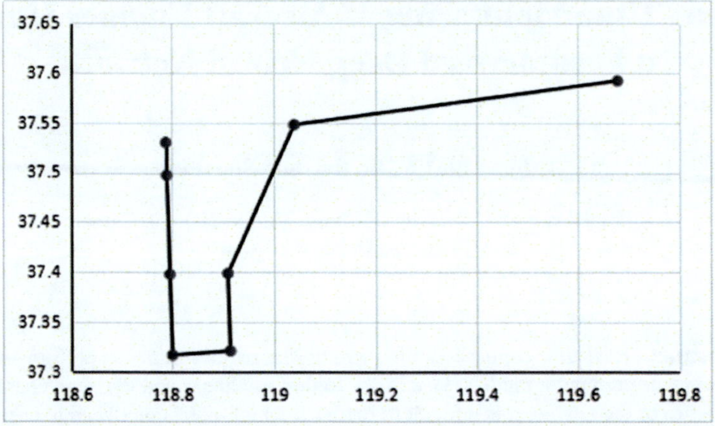

Fig. 6. The simulated fight track of the integrated standby navigation system based on BDS

5 Conclusion

Design of the civil aircraft integrated standby navigation system based on BDS should not only consider the basic elements such as system requirements, functions and performance, but also consider the development cycle, business and other factors. It is a process of comprehensive balance and comprehensive design. As the increasing demanding flight safety, the integrated standby navigation system based on BDS equipped with independent sensors to measure the navigation data may be the future developing trend because of more independence.

References

1. BDS Homepage. http://www.beidou.gov.cn. Accessed 19 Oct 2014
2. Society of Automation Engineering: SAE ARP4761 Guidelines and Methods for Conducting the Safety Assessment Process on Civil Airborne Systems and Equipment, pp. 12–22 (1996)
3. Federal Aviation Administration: AC 25-1309 System Design and Analysis. US Department of Transportation, pp. 99–110 (2016)
4. Federal Aviation Administration: AC 25-11B Electronic Flight Display. US Department of Transportation, pp. 99–110 (2016)
5. Society of Automation Engineering: SAE ARP4754A Guidance for Development of Civil Aircraft and Systems, pp. 37–45 (2010)
6. Airlines Electronic Engineering Committee: Digital Information Transfer System (DITS), pp. 129–181. Aeronautical radio, Inc. (2001)
7. Airlines Electronic Engineering Committee: GNSS Sensor, pp. 9–12. Aeronautical Radio, Inc. (2001)

Real-Time Monitoring of Aircraft Engines Using a Feedforward Deep Neural Network

Jiahuan Liu(✉), Jie Bai, and Shuai Liu

Civil Aviation University of China, Tianjin 300300, China
13759970435@163.com

Abstract. Real-time engine condition monitoring requires efficient performance parameter prediction models. Due to the highly nonlinear and dynamic characteristics of the engine, it is difficult to establish a high-precision dynamic engine model according to the laws of thermodynamics. To solve this problem, a method for modeling engine dynamic systems using feedforward deep neural networks with derived features is proposed, and an engine condition monitoring program is written. The method is verified using the full-flight stage data from the aircraft's fast access recorder, and the results show that the introduction of time-derived features combined with univariate prediction through sliding window data can effectively improve the prediction accuracy of the feedforward deep neural network, even slightly higher than the long short-term memory network of the same structure, and the root mean square error of the prediction of engine performance parameters in the full-flight stage can reach less than 10%. The condition monitoring program enables real-time and offline condition monitoring of the engine through threshold alarms and historical data analysis.

Keywords: Aero engines · Condition monitoring · Neural networks · QAR Data

1 Introduction

As the main source of power for aircraft, engine reliability is critical to flight safety [1, 2]. Aero-engine parts work in harsh environments such as high temperature, high pressure, and high speed for a long time, withstand comprehensive loads such as aerodynamic, centrifugal, and vibration, and have the characteristics of high failure rate and complex failure mode [3–5]. Physical faults can cause deviations from one or more engine health parameters or independent parameters, such as reduced efficiency and varying flow. These changes can further lead to deviations in engine performance parameters such as exhaust temperature, speed, fuel flow, and pressure ratio [6]. Therefore, to achieve real-time engine condition monitoring, it is necessary to accurately predict performance parameters and evaluate the engine status based on deviations from actual parameters.

Since engine benchmark operating data, etc., exist only at the manufacturer, it is difficult to accurately predict engine performance parameters by relying on thermodynamic models. In this case, machine learning methods are gradually being applied in the prediction of engine performance parameters. As early as the 90s of the 20th century,

Chinese Society of Aeronautics and Astronautics (Ed.): CASTC 2023, LNME, pp. 390–400, 2024.
https://doi.org/10.1007/978-981-99-8861-7_40

back-propagation neural networks [7], modular neural networks [8], and multi-neural networks [9] were applied to predict engine performance parameters. With the development of machine learning techniques, more and more deep neural networks are being used to simulate engine performance [10, 11]. Deep neural networks can consider more feature correlations and generally have higher simulation accuracy than shallow neural networks [12]. However, previous studies have often been based on the condition monitoring of engine steady-state operating point performance parameters, and there is little real-time continuous monitoring of engine performance parameters in the whole flight stage [5]. As can be seen from [5], the engine operation is not stable during flight because the flight altitude needs to be ensured by controlling the engine thrust under changing atmospheric conditions (Fig. 1).

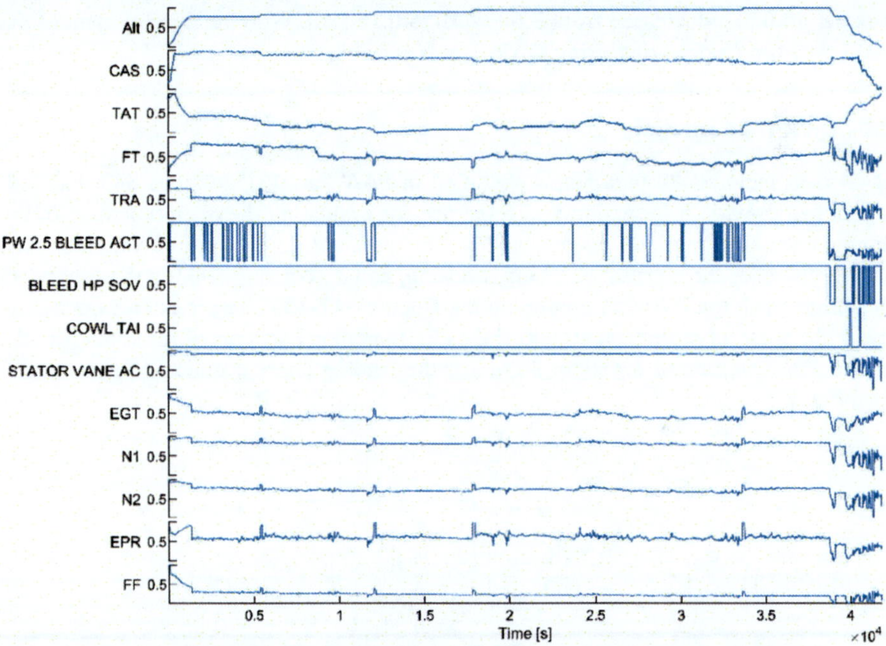

Fig. 1. Some engine-related data for the full flight phase of the sample flight.

Feedforward Neural Network (FNN) is widely used in the industrial field due to their advantages of a simple algorithm, low computing power requirements, and fast computing speed [13]. Compared with recurrent neural networks such as Long Short-Term Memory (LSTM), FNN training requires less computing power and less time to predict each time step, which is more suitable for real-time state monitoring with high time requirements. Previous studies have found that under the same computing power, the LSTM training time for engine dynamic system modeling is nearly 3 h longer than that of FNN. However, because FNN does not have the memory ability like LSTM, it is difficult to directly predict time series data. This paper focuses on how to predict engine performance parameters by improving feedforward neural networks and shows its application in real-time and offline condition monitoring of engines through experiments.

2 Data Analysis and Processing

As early as 1957, the United Kingdom and the United States began to require the installation of flight data recorders (FDR) on aircraft with a take-off weight of more than 12,500 lb, with the development of electronic information technology, civil aircraft gradually installed Aircraft Condition Monitoring System (ACMS), Quick Access Recorder(QAR) as its peripheral equipment began to be used in Flight Operations Quality Assurance (FOQA), which provides support for aircraft maintenance [14, 15]. According to statistics, after the use of the QAR system in American civil aviation, the accident and accident rate per 10,000 h decreased from 3.6 (1950) to 0.15 [16]. With the development of 5G ATG, real-time monitoring using QAR data is possible.

2.1 QAR Data Analysis

QAR data includes 88 parameters specified in FAA Part 121 Sections 343 and 344, which can provide information and reference for evaluating the operating state of the engine and achieving fault diagnosis [16]. The air flight stage of civil aviation aircraft mainly includes three types: climbing, cruising, and descending. Taking the flight data of a flight with the PW4056 engine on the Boeing 747-400 aircraft as an example, in the flight phase of more than 40,000 s, the climbing phase accounted for about 4%, the cruise phase accounted for about 88%, and the landing phase accounted for about 8% (see Fig. 2).

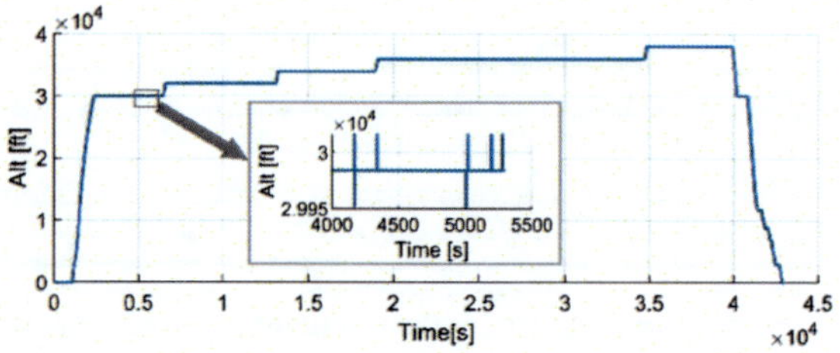

Fig. 2. Example flight altitude QAR data for all flight phases.

2.2 QAR Data Processing

QAR generally records the flight data of the whole flight stage at a frequency of 1 Hz, and there is often noise and numerical loss in the collected data due to sensor errors. For example, as can be seen in Fig. 3, there are noise points in the altitude data of the aircraft between 4000 s and 5500 s during the level flight phase.

QAR data should be preprocessed before training a neural network, including data missing value processing, data outlier discovery and processing, and data standardization. Missing value processing methods include removing missing values, means filling, K-nearest missing value filling, and so on. Outlier identification methods include the 3sigma rule method, scatter plot method, etc. Since the engine QAR data contains both the continuous quantity such as the height data and the discrete quantity (0 or 1) whether the bleed air switch is turned on, it is usually normalized using Min-Max, also known as normalization. However, for real-time condition monitoring, the maximum value in the time series cannot be known in advance, so the maximum value can be selected empirically, and the training and validation data are scaled equally. For example, A single sensor channel of length $n + 1$ input at the beginning of time t is considered a time series.

$$\boldsymbol{x}_{sensor1} = \left\{x_{sensor1}^{(t)}, x_{sensor1}^{(t+1)}, \ldots, x_{sensor1}^{(t+n)}\right\} \tag{1}$$

Scale the time series $\boldsymbol{x}_{sensor1}$.

$$x_1^{(t+i)} = \frac{x_{sensor1}^{(t+i)} - \max(\boldsymbol{x}_{sensor1})}{\max(\boldsymbol{x}_{sensor1}) - \min(\boldsymbol{x}_{sensor1})}(i = 1, 2, \ldots, n) \tag{2}$$

$$\boldsymbol{x}_1 = \left\{x_1^{(t)}, x_1^{(t+1)}, \ldots, x_1^{(t+n)}\right\} \tag{3}$$

In Eq. (2), $\max(\boldsymbol{x}_{sensor1})$ and $\min(\boldsymbol{x}_{sensor1})$ are taken from empirical values, ensuring that no other value exceeds this range. The new sequence after scaling $\boldsymbol{x}_1 \in (0, 1)$ is dimensionless.

The purpose of data normalization is to eliminate the influence of feature dimensions, avoid certain features being ignored, and then affect the results of data analysis, in addition, data standardization can significantly reduce training time. In addition to the Min-Max method, common data standardization methods include Z-Score normalization, Mean normalization, Decimal scaling normalization, vector normalization, and exponential transformation. The preprocessing of training data is of great significance to improve the robustness of the algorithm.

3 Data-Driven Modeling

The engine dynamic system modeling based on FNN can be regarded as a multiple nonlinear regression process, and the FNN is trained using the engine operation data to obtain the performance parameter prediction model. According to the deviation between the predicted and actual parameters, the operating state of the engine can be judged.

3.1 Directly Apply QAR Data to Train FNN

The training process of FNN includes firstly, the training set data being divided into small batches of input FNN. Then, an initial output is calculated by the forward propagation algorithm, and compared with the actual value, a loss value is obtained through the loss function. The loss value is further updated by the backpropagation algorithm for each neuron weight and bias of the FNN, and a new iteration is repeated until the preset requirements are met, such as reaching a predetermined number of iterations or reaching an error target.

Multi-channel sensors collect environmental and control parameters of engine operation to generate time series data X.

$$X = \left\{ \boldsymbol{x}^{(1)}, \boldsymbol{x}^{(2)}, \ldots, \boldsymbol{x}^{(n)} \right\} \tag{4}$$

where each time step $\boldsymbol{x}^{(t)} \in R^m$ of a time series is an m-dimensional vector, m depends on the number of input channels.

$$\boldsymbol{x}^{(t)} = \left\{ x_1^{(t)}, x_2^{(t)}, \ldots, x_m^{(t)} \right\} \tag{5}$$

Performance parameters collected by engine sensors to generate time series data Y.

$$Y = \left\{ \boldsymbol{y}^{(1)}, \boldsymbol{y}^{(2)}, \ldots, \boldsymbol{y}^{(n)} \right\} \tag{6}$$

Each time step $\boldsymbol{y}^{(t)} \in R^5$ is a 5-dimensional vector, including five engine performance parameters: exhaust gas temperature (EGT), low-pressure rotor speed (N1), high-pressure rotor speed (N2), engine pressure ratio (EPR), and fuel flow (FF).

$$\boldsymbol{y}^{(t)} = \left\{ y_1^{(t)}, y_2^{(t)}, \ldots, y_5^{(t)} \right\} \tag{7}$$

The structural design of FNN is an iterative process, usually changing the structure and making the next iteration based on the error of the result of each design. The number of hidden layers and the number of neurons of FNN are hyperparameters, and the best structure within the expected accuracy range can generally be found through iteration. However, after preliminary experiments, it is found that no matter how to adjust the number of layers and neurons of the hidden layer, it cannot effectively reduce the local prediction error, that is, underfitting. And the area with a large error has a certain law. Taking EGT as an example, there are large errors in the prediction results during periods of drastic altitude changes (see Fig. 3).

In the stage where the flight altitude change rate is the largest, the EGT prediction error is extreme, the EGT prediction value is ahead of the actual value, and the EGT prediction value is highly synchronized with the throttle resolver angle (TRA) time. According to the prediction results, it is speculated that this is due to the time-delay characteristics of the engine.

Due to the time lag of the engine, the performance parameter value $\boldsymbol{y}^{(t)}$ of the current time step depends not only on the control parameters and environmental parameters entered at this moment but also on those entered at the previous L moments. Since FNN does not have storage capacity, it is difficult to directly realize the memory function like LSTM.

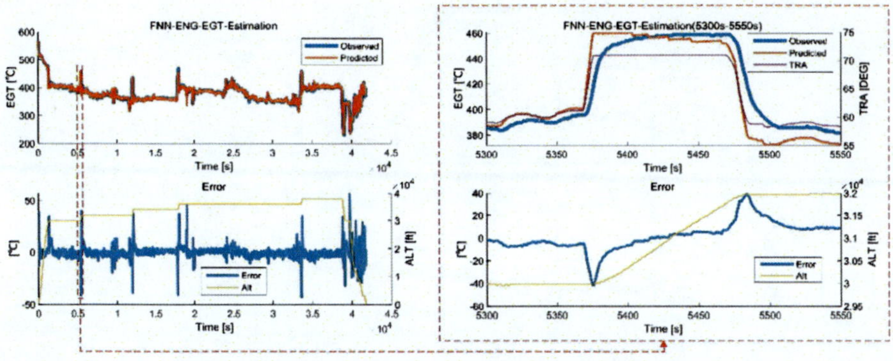

Fig. 3. FNN predicts the full-flight stage engine EGT.

3.2 Construct Data-Derived Features

To enable FNN to consider the temporal dependence of the data, this paper uses the sliding time window method to construct the derived features, adding time-dependent variables as inputs. According to the analysis results of Fig. 3, the TRA value is first used as an example to add the derived feature. The specific method is to set the TRA value to $x_{TRA}^{(t)}$, and the first L time step TRA value to $\left\{x_{TRA}^{(t-L)}, \ldots, x_{TRA}^{(t-2)}, x_{TRA}^{(t-1)}\right\}$, and add the derived feature $x_{TRA.derived}^{(t)}$ to the input $\boldsymbol{x}^{(t)}$ of the time step t.

$$x_{TRA.derived}^{(t)} = \sum_{i=1}^{L-1} k_i x_{TRA}^{(t-i)} \tag{8}$$

where L is the window length and k_i is the weight, and after increasing the TRA-derived feature, the input $\boldsymbol{x}^{(t)}$ of the time step t is expanded to an $m+1$ dimensional vector $\left\{x_1^{(t)}, x_2^{(t)}, \ldots, x_m^{(t)}, x_{TRA.derived}^{(t)}\right\}$. Setting the appropriate weight value k_i can adjust the effect of TRA values at different times within the window on the current forecast value. When $k_i = 1/L$, the derived TRA is equal to the average of the TRA values within the L length window.

$$x_{TRA.derived}^{(t)} = \frac{1}{L-1} \sum_{i=1}^{L-1} x_{TRA}^{(t-i)} \tag{9}$$

In the next study, the TRA-derived feature is equal to the average of the window data, and the influence of window length L on the prediction accuracy is further explored. Taking EGT as an example, it can be seen from Fig. 4 that the FNN prediction accuracy is improved after taking TRA-derived features as input additions, and iterative experiments show that when the window length L = 10, the EGT prediction accuracy is the highest.

In exploring the influence of window length on performance parameters, it is found that because different window lengths cannot be used to optimize all prediction results at the same time, the univariate prediction method is finally adopted. An FNN with derived

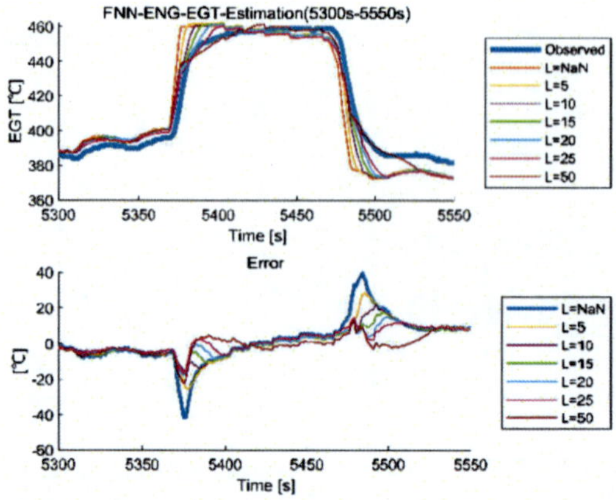

Fig. 4. Effect of window length L on EGT prediction accuracy.

feature inputs is trained for each parameter, and the optimal prediction of each engine parameter is achieved by adjusting it individually. The final window length selected is shown in Table 1.

Table 1. Each performance parameter corresponds to the optimal window length.

Performance parameters	Optimal window length L
EGT	10
N1	3
N2	2
EPR	3
FF	5

In the FNN structure design, the Bayesian optimization method was used to optimize the training parameters, and finally, the FNN structure with 3 hidden layers and 128 neurons per layer was selected. Compared with FNN and single-layer LSTM prediction without derived features, the addition of derived features effectively improves the prediction accuracy of the FNN algorithm on engine performance parameters, in which the root mean square error is less than 0.4%, the average error is less than 0.3%, and the maximum error is reduced to less than 10%, and is slightly better than the single-layer LSTM, and the prediction error is shown in Fig. 5.

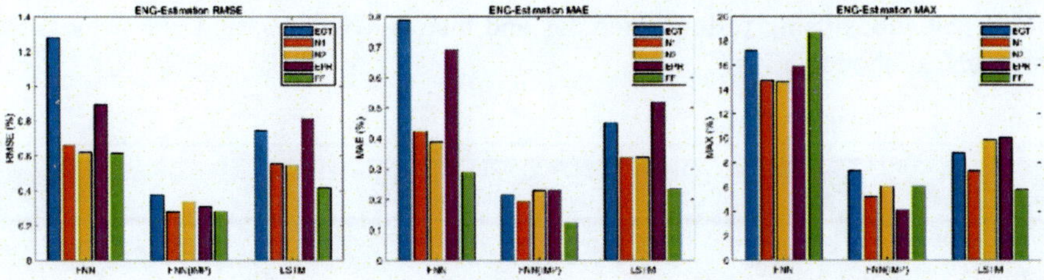

Fig. 5. The error of different neural networks in predicting engine performance parameters.

4 Engine Condition Monitoring

4.1 Engine Condition Monitoring Methods

The process of using trained FNN for engine state prediction mainly includes (see Fig. 6).

Input data. Enter the same environmental parameters and control parameters as the actual engine operation into the FNN model.
Prediction. Predict healthy engine performance parameters through the FNN model.
Calculation deviation. The performance parameters predicted by FNN are used as the baseline, and the deviation is obtained by comparing it with the actual measurement parameters of the engine sensor.
Redline alarm. When the deviation exceeds the pre-set threshold, the alarm is carried out.
Preliminary diagnosis. The expert system is used to realize the preliminary fault diagnosis of the engine.

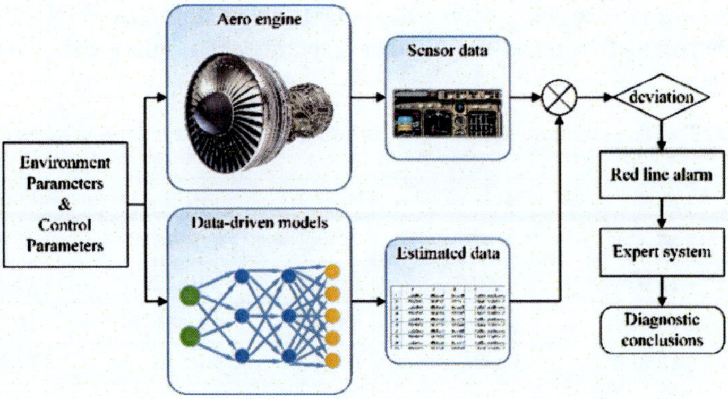

Fig. 6. Real-time engine condition monitoring using FNN models.

4.2 Real-Time Condition Monitoring Experiments

To verify the feasibility of the method, the trained model is used to monitor the engine in real-time. Create a visualization program in the MATLAB Simulink environment, display deviations using rolling windows and history windows, and alarm by preset thresholds, as shown in Fig. 7.

Fig. 7. Real-time engine condition monitoring panel in MATLAB Simulink environment.

The experimental data comes from 4 engines of the same model with different service life. The engine that provides the training set data is defined as ENG1, and the other three engines are ENG2-4. The dataset of 4 engines is the operating parameters of the engine in the full flight phase. The environmental parameters, control parameters, and derived features constructed according to the Chapter 2 method are fed into a data-driven model dominated by FNN, and the output results are compared with the real-time data collected by the sensor. The real-time condition monitoring program can calculate the deviation between the engine performance parameters predicted by FNN and the actual performance parameters in real-time, and the experimental results are shown in Table 2.

Table 2. The maximum deviation of performance parameters for 4 engines (%).

	EGT	N1	N2	EPR	FF
ENG1	7.311	5.191	6.088	4.131	6.094
ENG2	13.438	10.259	10.854	10.427	11.750
ENG3	15.127	13.721	12.413	11.3626	10.937
ENG4	12.880	11.455	10.856	10.151	11.542

It can be seen from Table 2 that the deviation of the maximum performance parameters of different engines is different, mainly due to the different service life of the engine, so the performance degradation is also different. The engine data used in the experiment are all health data, combined with the engine failure data, about 20% of the alarm threshold can be set for condition monitoring, too high or too low threshold will directly affect the fault misjudgment and recognition rate.

4.3 The Impact of Performance Degradation

In addition to threshold alarms, the historical results output by the program can also be analyzed offline engine conditions. Also taking the EGT of the four engines as an example, it can be seen from Fig. 8 that except for ENG1, the other three engines have obvious deviations, and the deviation size distribution in the whole flight stage is relatively stable except for the landing process, so the EGT deviation of each engine is averaged. From the local figure of Fig. 11, the average deviation of ENG1 is 0.417 °C; the average deviation of ENG2 is 4.811 °C; the average deviation of ENG3 is 5.998 °C; and the average deviation of ENG4 is 1.391 °C. This corresponds to the service life of the four engines, from short to long: ENG3, ENG1, ENG4, and ENG2.

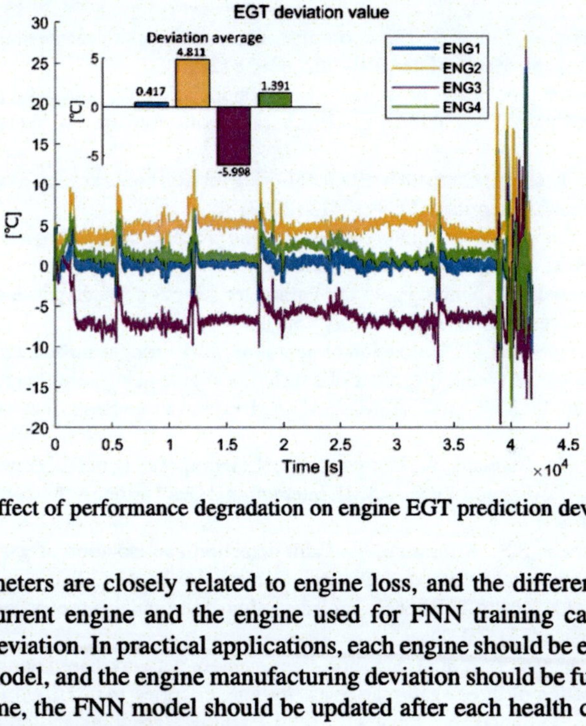

Fig. 8. The effect of performance degradation on engine EGT prediction deviation values.

EGT parameters are closely related to engine loss, and the difference in the state between the current engine and the engine used for FNN training can be identified through EGT deviation. In practical applications, each engine should be established with its own FNN model, and the engine manufacturing deviation should be fully considered. At the same time, the FNN model should be updated after each health check to reduce the impact of performance degradation.

5 Conclusions and Future Work

When designing an FNN model to predict time series data, it is necessary to consider whether the data has a temporal correlation, such as directly treating the data as an independent vector, then the trained neural network will produce a large prediction error. The sliding time window method can construct time-derived features, and the addition of derived features combined with univariate prediction and individual adjustment of window length can make the designed FNN model have LSTM-like memory ability and consider the temporal correlation of more data.

Using the FNN model as a data-driven model, real-time engine condition monitoring can be effectively carried out through performance parameter deviation, but the model should be trained separately for each engine and updated in real-time during the application process. In the future, the performance degradation correlation will be introduced to train the neural network, to obtain a prediction model of engine performance parameters with higher generalization and robustness.

References

1. Chen, Z., Cao, S., Mao, Z.: Remaining useful life estimation of aircraft engines using a modified similarity and supporting vector machine (SVM) approach. Energies **11**(1), 28 (2017)
2. Wen, Z.H., Yin, X., Liu, Y.P.: The uncertainty analysis of health management strategies for aero-engines. Appl. Mech. Mater. **201–202**, 34–38 (2012)
3. Yu, J.: Aircraft engine health prognostics based on logistic regression with penalization regularization and state-space-based degradation framework. Aerosp. Sci. Technol. **68**, 345–361 2017
4. Lu, F., Huang, J., Lv, Y.: Gas path health monitoring for a turbofan engine based on a nonlinear filtering approach. Energies **6**(1), 492–513 (2013)
5. Zhang, B.Y., Zhang, D.Y., et al.: Research status and prospect of fault diagnosis for gas turbine aeroengine. Acta Aeronautica et Astronautica Sinica **43**(8) (2022)
6. Ogaji, S., Sampath, S., Singh, R., et al.: Parameter selection for diagnosing a gas-turbine's performance-deterioration. Appl. Energy **73**(1), 25–46 (2002)
7. Torella, G., Lombardo, G.: Utilization of neural networks for gas turbine engines (1995)
8. Zedda, M., Singh, R.: Fault diagnosis of a turbofan engine using neural networks - a quantitative approach. In: 34th AIAA/ASME/SAE/ASEE Joint Propulsion Conference and Exhibit (1998)
9. Kanelopoulos, K., Stamatis, A., Mathioudakis, K.: Incorporating neural networks into gas turbine performance diagnostics. In: ASME International Gas Turbine & Aeroengine Congress & Exhibition (1997)
10. Tayarani-Bathaie, S.S., Khorasani, K.: Fault detection and isolation of gas turbine engines using a bank of neural networks. J. Process. Control. **36**, 22–41 (2015)
11. Du, X., Chen, J.J., Zhang, H.B., et al.: Fault detection of aero-engine sensor based on inception-CNN. Aerospace **9**(5) (2022)
12. Bai, M., Liu, J., Ma, Y., et al.: Long short-term memory network-based normal pattern group for fault detection of three-shaft marine gas turbine. Energies 14 (2020)
13. Shrestha, A., Mahmood, A.: Review of deep learning algorithms and architectures. IEEE Access **7**, 53040–53065 (2019)
14. Zoller, T.: Advanced Engine Monitoring and Diagnosis Systems: Actual System for the EJ2OO Engine of the EuroFighter 2000 Aircraft and Future Trends (2003)
15. Tsalavoutas, A., Aretakis, N., Mathioudakis, K., et al.: Combining Advanced Data Analysis Methods for the Constitution of an Integrated Gas Turbine Condition Monitoring and Diagnostic System (2000)
16. Gao, Y., Wang, N.: Aero-engine condition monitoring based on Kalman Filter theory. In: 2nd International Conference on Mechatronics and Intelligent Materials, MIM 2012, pp. 176–181 (2012)

Design of the High-Safety Aircraft Squib Drive and Control System

Mingliang Hu(✉), Wen Yan, Jionghui Che, and Fangchun Yu

Xi'an Aeronautics Computing Technique Research Institute, AVIC, Xi'an 710068, China
minglhu@126.com

Abstract. In order to achieve high safety and reliability in the initiation of squibs, this paper designs a high safety aviation squib drive and control system, which mainly includes sensor interfaces, hardware interfaces, software interfaces, and data interfaces. The hardware interface adopts a power output drive method based on inverse time protection to achieve self-recovery and high safety protection of the power drive circuit. The software interface adopts a hierarchical interface fault detection method to identify faults such as squib breakage, terminal output abnormality, achieving accurate isolation and rapid positioning of faults. The data interface uses instruction validity validation to achieve comprehensive processing of instruction data, status data, and BIT data. The experimental results show that the drive and control system has high safety and reliability, which has important practical significance for improving the safety of aircraft systems.

Keywords: Squib · High-safety · Power drive · Inverse time protection · Fault diagnosis

1 Introduction

The aviation weapon system contains a large number of squibs, which are applied to achieve warhead detonation, parachute opening, power startup, fuse safety release or shell separation [1]. A high current DC detonation method is generally applied to the squib. After being energized, the DC current needs to reach 3–5 times the rated current in a short time [2]. The safe and reliable detonation of squibs is crucial to the safety of aircraft [3]. The drive and control of squibs are supposed to meet requirements of safe and reliable detonation when it is needed [4], and should meet requirements for avoiding accidental detonation when the system is abnormal, resulting in incalculable consequences [5]. The traditional squib drive is generally realized by circuit breakers and other methods [6]. The protection is generally carried out by fuses or single overload points, with unique performance indicators, which cannot meet protection requirements of different loads [7]. The traditional fault diagnosis of squib has low fault coverage and isolation rate, which reduces the reliability of the system [8].

This paper designs a high safety aviation squib drive and control system, which mainly includes sensor interface, hardware interface, software interface, and data interface. The power output driving method based on inverse time protection, fault detection

Chinese Society of Aeronautics and Astronautics (Ed.): CASTC 2023, LNME, pp. 401–409, 2024.
https://doi.org/10.1007/978-981-99-8861-7_41

method with hierarchical interfaces, and command validity confirmation are adopted to achieve high safety power driving, fault diagnosis, and isolation. Experimental results show that the inverse time protection curve design is reasonable and can achieve normal overcurrent protection and detonating functions of the squib.

2 Overall System Design

Figure 1 shows the block diagram of the squib drive system, which is mainly composed of sensor interfaces, hardware interfaces, software interfaces, and data interfaces. The sensor interface mainly includes a squib initiating explosive device and cable connection. The hardware interface mainly includes a 28 V/open discrete quantity power output unit, a ground/open discrete quantity power output unit, a 28 V/open discrete quantity input acquisition unit, a ground/open discrete quantity input set unit, and an inverse time limit including circuit and analog quantity acquisition unit. Main functions of the software interface include command parsing, output control, BIT detection, and fault handling. The data interface mainly includes overcurrent state, positive terminal output control command, positive terminal output state, squib load online state, negative terminal output state, and negative terminal output control command. The CPU processor interacts with the aircraft management computer with a data interface, periodically receiving control instructions or sending status information. The hardware interface receives control instructions sent by the software interface, drives the squib to work, and implements BT detection through an output status acquisition circuit. The 28 V/open discrete power output unit monitors the output current in real time to achieve inverse time protection of the power output. The analog mining circuit judges the voltage threshold to achieve online detection of squibs.

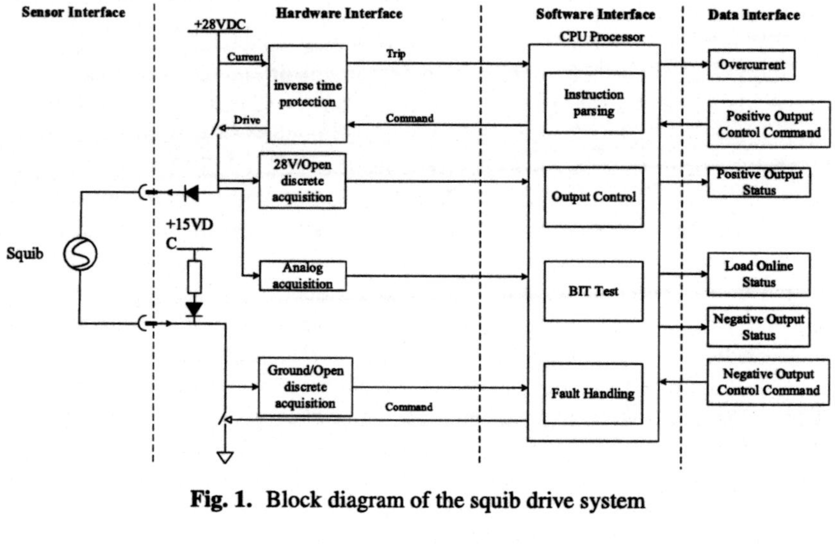

Fig. 1. Block diagram of the squib drive system

3 Design of High-Reliability Hardware

3.1 Design of the Inverse Time Protection Power Output Circuit

Figure 2 illustrates the block diagram of the inverse time protection, which is applied to 28 V/open discrete power output. The inverse time controller collects the load current in real time and accumulates energy based on the inverse current time characteristic to achieve abnormal power output cutoff. It is able to simulate a high order inverse time limit protection curve in a segmented manner, and use a fixed step size as the energy accumulation value for each sampling period through a table lookup. When the energy accumulation value reaches the specified threshold value, it will turn off the output, send a trip signal, and set the hardware to a safe state until the CPU processor releases the trip signal.

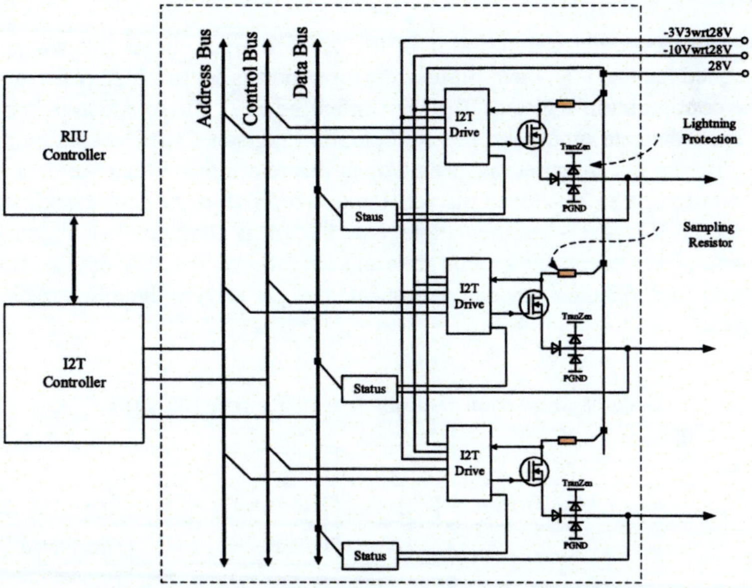

Fig. 2. Block diagram of the inverse time protection

3.2 Design of the Output Acquisition Circuit

Figure 3 shows the block diagram of the output state recovery circuit, using a discrete quantity acquisition chip DEI1282, which has 8 input channels and can be individually configured as "28 V/open discrete quantity input" or "ground/open discrete quantity input" to achieve highly reliable acquisition of the output state driven by the squib power. The DEI1282 discrete quantity acquisition chip has a built-in BIT test path and three-level lightning protection function. It interacts with the CPU processor through the SPI interface, and it achieves acquisition by setting a filter network [9].

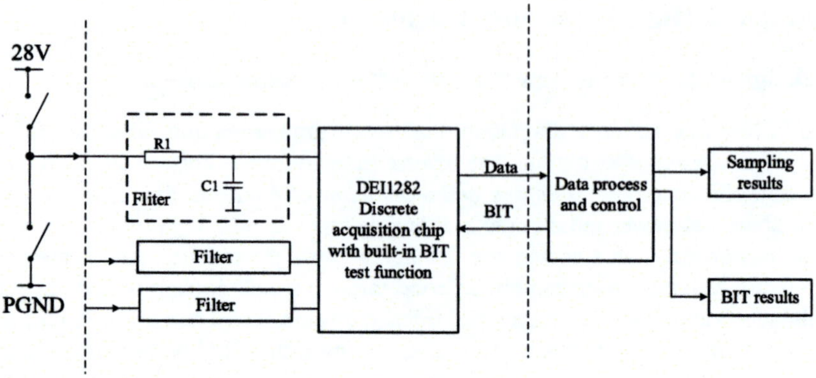

Fig. 3. Block diagram of the output state recovery circuit

Table 1 demonstrates a description of the discrete input circuit BIT status. When BIT is enabled (status is 1), the 8-bit discrete input port is set to a high resistance state. When the configuration register CR[n] is configured as 0, the interface is configured as 28 V/open discrete input, and the configuration register CR[n] is configured as 1, and the interface is configured as ground/open discrete input. When the data register DR[n] is configured as 28 V/open discrete input, 0 represents the open circuit state and 1 means the 28 V state. When the data register DR [n] is configured as a ground/open discrete input, 0 represents the ground state and 1 represents the 28 V open circuit state. BIT achieves fault diagnosis by comparing data register and configuration register data results [10].

Table 1. Description of the discrete input circuit BIT status.

BIT Signals			Fault identification	
BIT Configuration	CR[n]	DR[n]	Interface Fault	External Fault
1	0	1	Failure	Unaffected
		0	Failure-free	Unaffected
	1	1	Failure-free	Unaffected
		0	Failure	Unaffected

3.3 Design of the Output Acquisition Circuit

Figure 4 shows block diagram of the online monitoring of the squib. The positive end output mining circuit of squibs is configured as 28 V/open discrete input, and the negative end output mining circuit of squibs is configured as ground/open discrete input. By judging the acquisition voltage, the online status of squibs is determined. The voltage is collected through a conditioning circuit, a proportional amplification circuit, and an A/D converter. The AD converter utilizes the 12-bit AD conversion chip ADS774 produced by Burr Brown Company of the United States as an analog quantity acquisition chip. The chip has a linear error of +2LSB, a reference voltage of 2.50 V DC, a maximum conversion time of 8.5 us, a full scale error of ±0.25% FS, and a maximum offset error of ±10 mV.

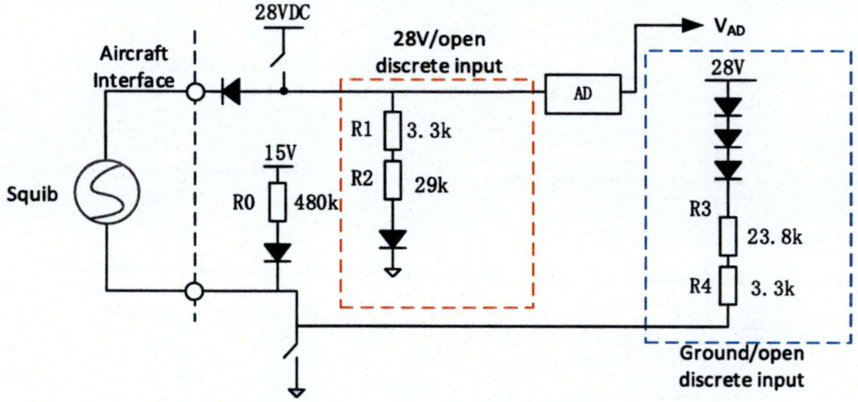

Fig. 4. Block diagram of the online monitoring of the squib

Ignoring the internal resistance of the squib, when the squib is not online, the V_{AD} calculation method is as follows:

$$V_{AD} = \frac{(V15 - 2V_D)(R1 + R2)}{R0 + R1 + R2} + \frac{(V28 - 4V_D)(R1 + R2)}{R1 + R2 + R3 + R4} \quad (1)$$

where, V28 refers to the bus voltage of the bus bar, ranging from 17 to 32 V, V15 represents the internal voltage, ranging from 15 V, while V_D refers to the tube voltage drop of the diode, ranging from 0.7 V, R1 = 3.3kQ, R2 = 29 kΩ, R3 = 23.8kQ2, R4 = 3.3kQ2. The calculated value range of V_{AD} is 8.58 V to 16.74 V.

4 Design of High-Safety Software

Figure 5 illustrates the software operation flow chart. After the system completes initialization, the detonator is set to work in a safe state, and the lower bridge arm power output unit test is conducted. After the lower bridge arm power output unit test passes, the upper bridge arm power output unit test will be conducted. After the upper bridge arm power output unit test passes, online detection of the squib will be started. When the squib is online, it will wait for the bus output command to be valid. When the output command is valid, the squib is turn on monitoring the current in real time. When no current state overcurrent is detected and the squib is not online, the upper and lower bridge arm power output switch tubes are disconnected and the control command is locked. When any item fails the test, the upper and lower bridge arm power output switches is disconnected locking the control command.

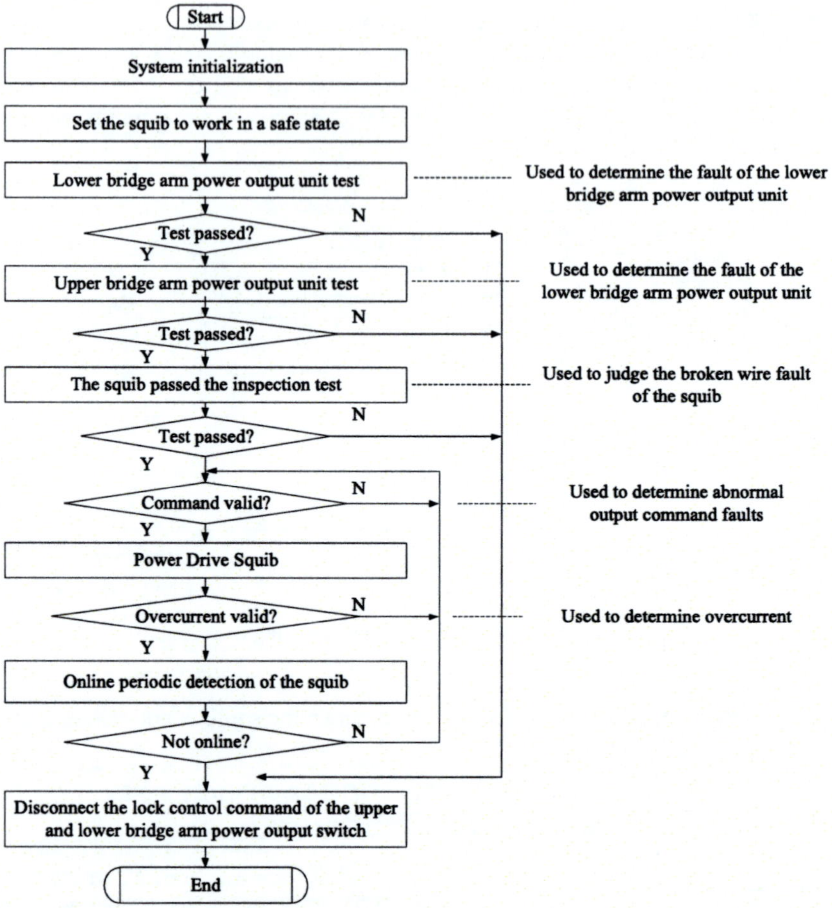

Fig. 5. The flow chart of software operation

5 Experiment

5.1 Inverse Time Protection Curve Test

Figure 6 demonstrates an experimental configuration for inverse time protection testing. The RIU is provided with a DC 28V operating voltage from an external power source. The current of the electronic load is set according to the trip protection time curve, and the bus monitor is used to monitor the RIU status in real time. Figure 7 shows the inverse time limit protection test curve. The abscissa is the protection time, and the ordinate is the rated current. When the current is greater than 8 times the rated current, the shutdown time is 90uS. As the current multiple decreases, the protection time is extended, by which the circuit is able to operate safely and reliably.

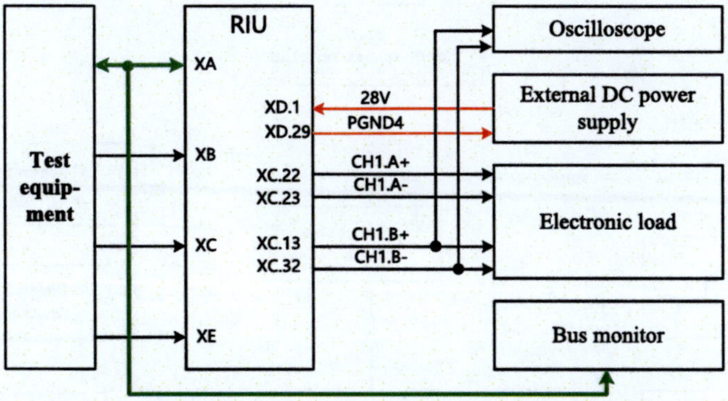

Fig. 6. Experimental configuration of inverse time protection test

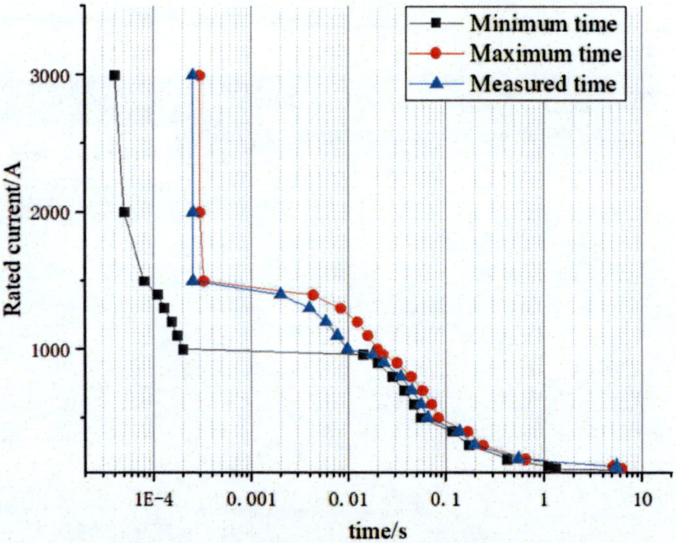

Fig. 7. Inverse time protection test curve

5.2 Explosion Test of the Squib

Figure 8 illustrates the configuration of the squib detonation experiment. Figure 9 shows the physical diagram of the squib detonation experiment. The squib drive and control system are integrated into the Remote Interface Unit (RIU). The RIU injects control commands through a CC simulator, and it receives the squib BIT detection results through a bus monitor. The RIU is connected to the squib by a disconnection box, and the driving current of the squib is observed in real time through an oscilloscope. In the experiment, the equivalent power resistor is applied to replace the squib, and the inverse time protection time of the power drive circuit is tested. When the time is greater than the detonating time of the squib, the actual squib is connected. When injecting a safe state through the

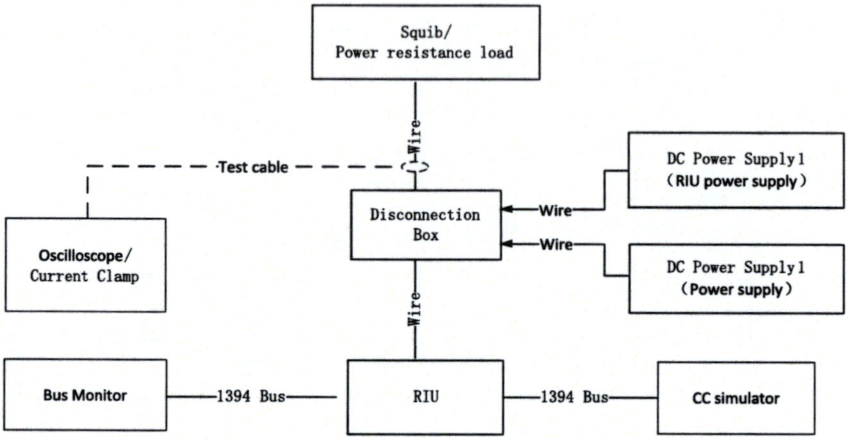

Fig. 8. Experimental Configuration of the squib

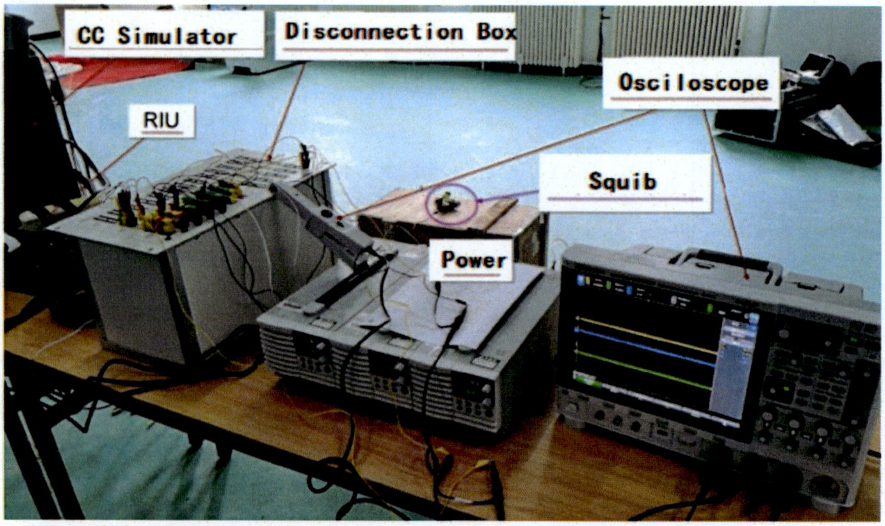

Fig. 9. Wave of phase current based on vibration motor

CC simulator, the BIT detection result on the bus monitor is no output, the power output and read back are both in an open circuit state, the detonator is online, and there is no overcurrent signal. When injecting a command to detonate the detonator through the CC simulator, the BIT detection result on the bus monitor should have an output, with the power output reading back to the "28 V" and "ground" states, respectively. The detonator is online, and there is no overcurrent signal. After the normal detonation of the squib, the BIT detection result is a safe state result. Experimental results are consistent with the above expected results, indicating the effectiveness of the drive and control system design.

6 Conclusion

This paper designs a high safety aviation squib drive and control system, which mainly includes sensor interface, hardware interface, software interface, and data interface. The hardware interface adopts a power output drive method based on inverse time protection to achieve self-recovery, high safety, and high reliability protection of the power drive circuit. The software interface takes a hierarchical interface fault detection method to identify faults such as squib disconnection, positive end output abnormality, and negative end output abnormality, achieving accurate isolation and rapid positioning of faults. The data interface uses instruction validity validation to achieve comprehensive processing of instruction data, status data, and BIT data. The experimental results show that the drive and control system has high safety and reliability, which has important practical significance for improving the safety of aircraft systems.

References

1. Ji, L., Cao, Z., Hong, Q., et al.: An improved inverse-time over-current protection method for a microgrid with optimized acceleration and coordination. Energies **13**(21), 5726 (2020)
2. Choi, C., Lee, W.1.: Fault diagnosis and tolerance on reliability of automotive motor control system: a review. Int. J. Automot. Technol. **23**(4), 1163-1174 (2022)
3. Song, X., Xing, Z., Jia, Y., et al.: Review on the damage and fault diagnosis of wind turbine blades in the germination stage. Energies **15**(20), 7492 (2022)
4. Wang, B., Sun, Y., Wei, G., et al.: Research on test method of ignition temperature of electric explosive device under electromagnetic pulse. Radio Eng. **30**(3), 510–516 (2021)
5. Mahindara, V.R., Rodriguez, D.F.C., Pujiantara, M., et al.: Practical challenges of inverse and definite-time overcurrent protection coordination in modern industrial and commercial power distribution system. IEEE Trans. Ind. Appl. **57**(1), 187–197 (2020)
6. Chang, N., Song, G., Hou, J., et al.: Fault identification method based on unified inverse-time characteristic equation for distribution network. Int. J. Electr. Power Energy Syst. **146**, 108734 (2023)
7. Rassul, F.A., Abdulsada, M.A., Abusief, F.R.: Functional test generation for guided control system. Int. J. Comput. Electr. Eng. **2**(5), 943 (2010)
8. Mahat, P., Chen, Z., Bak-Jensen, B., et al.: A simple adaptive overcurrent protection of distribution systems with distributed generation. IEEE Trans. Smart Grid **2**(3), 428–437 (2011)
9. Zavila, O., Chmelík, R.: Fire and false fire alarm causes in military jet aircraft of Czechoslovakia and the Czech Republic. Adv. Mil. Technol. **12**(2), 229–242 (2017)
10. Al-Ammouri, A., Dmytrychenko, A., Al-Ammori, H., et al.: Development of structures of the aircraft fire alarm system by means of nested modules. East. Eur. J. Enterp. Technol. **2**(9), 14–23 (2019)

Research on Embedded Device Data Standardization Technology Based on OSA-CBM

Yangyang Zhao[1(✉)], Wei Niu[1], Bo Gao[1], Zhenhua Ma[1], TanBao Yan[1], and Qianxiao Cheng[2]

[1] Xi'an Aeronautics Computing Technique Research Institute, AVIC, Xi'an 710068, China
1398655303@qq.com

[2] Germany ZTE Corporation, Xi'an Research Institute, AVIC, Xi'an 710068, China

Abstract. Aiming at the problem that the data communication standards of airborne equipment of different suppliers are obviously different and it is difficult to trace and identify the data transmission between them, it is proposed to standardize the data of airborne equipment based on OSA-CBM standard, create dynamic data models for transmission and static data models for tracing and identification in six functional blocks. In addition, the embedded data processing flow is designed to realize the functions of data acquisition, data processing and state detection in the embedded system. Through instantiation verification of typical embedded equipment on board, evaluate the performance difference between non standardized and standardized embedded equipment data, and verify the interoperability, traceability and identifiability of the data model. It lays a foundation for airborne equipment health management and fault diagnosis, and has engineering application value in airborne embedded equipment condition based maintenance strategy.

Keywords: OSA-CBM · Embedded Device · Data Standardization · Interoperability · Traceability

1 Introduction

At present, the complexity, comprehensiveness and intelligence of airborne equipment are getting higher and higher. The traditional maintenance methods of regular maintenance and post maintenance are difficult to meet the needs of complex equipment maintenance, and the demand for Condition Based Maintenance (CBM) is becoming more and more urgent [1]. The basis of CBM is data, which can assess the current or future health status and provide maintenance decisions by capturing data and monitoring parameters related to the health of airborne equipment. In this process, multiple sensors, processing units and software are required to perform data processing functions, and it is expected that these components will be provided by different suppliers. In order to realize easy integration and interaction between components and realize technology reuse, it is necessary to establish an open CBM system architecture.

Chinese Society of Aeronautics and Astronautics (Ed.): CASTC 2023, LNME, pp. 410–416, 2024.
https://doi.org/10.1007/978-981-99-8861-7_42

Through the use of standardized interfaces and data models to achieve interoperability between CBM components of multiple suppliers, American research institutions have proposed an open standard for CBM application development and integration, and further developed it into an Open System Architecture for Condition Based Maintenance (OSA-CBM) standard. OSA-CBM mainly focuses on PHM functions and related data models. Most OSA-CBM projects mainly come from Boeing, general aviation, ARL and Pennsylvania State University. Literature [2] discussed the benefits of OSA-CBM and reported its use in Boeing research projects. However, due to the proprietary nature of these works, it failed to provide information on how to implement OSA-CBM in engineering. Sreenuch [3, 4] et al. designed the software structure of OSA-CBM module and demonstrated its application through a distributed gearbox CM system. In addition, under the background of IVHM architecture, the distributed data mechanism is designed and analyzed for the communication between OSA-CBM modules. Bayoumi [5] et al. proposed a new scalable CBM architecture for naval fleet maintenance, including data distribution service (DDS) and OSA-CBM, which effectively solved the isolation between shipborne data and shore based analysis tools.

An important component of OSA-CBM is the standardization of data model. In this paper, the data is standardized based on OSA-CBM standard in the embedded environment to ensure that each CBM component uses the same set of data for data processing and data transmission, and each component can correctly identify this standardized data. In addition to the consideration of standardization, the data model also needs to be recognizable and traceable, and needs to be consistent with the data in the reliability and maintenance domain.

2 Data Model Based on OSA-CBM

Open system architecture for condition based maintenance (OSA-CBM) is an open standard, which aims to enhance the interoperability between software components of multiple suppliers and the integration of enterprise information systems. OSA-CBM is divided into data model and interface specification. These specifications are defined using unified modeling language (UML) and are designed to be platform independent and can be mapped to various programming languages and middleware technologies [6].

2.1 A Subsection Sample Data Model

OSA-CBM data model mainly describes the concept and use of metadata (data about data), that is, OSA-CBM data is always identifiable and traceable. In embedded systems, OSA-CBM mainly has two data types, DataEvent and Configuration.

DataEvent are dynamic data related to health management generated by the equipment module, such as sensor measurement data, state information or data processed by algorithms. Configuration provides information about the input source of the device module, an algorithm description for processing the input data, an output list, and various output details.

2.2 Interface Specification

The interface specification serves the data model and mainly describes how information is transmitted between functional blocks. There are four main types of interfaces to adapt to different purposes and technical capabilities: synchronous, asynchronous, service and subscription (see Fig. 1).

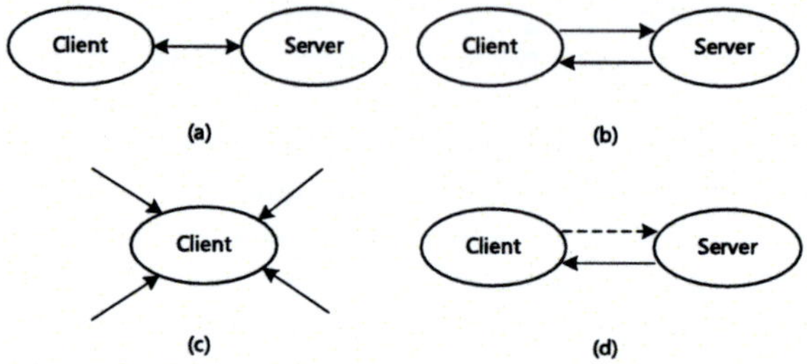

Fig. 1. Schematic diagram of four interface specifications

Figure (a) is a synchronization interface, showing the data returned during the call; Figure (b) is an asynchronous interface that allows any number of higher modules to establish and maintain bidirectional connections when needed; Figure (c) is a service interface, which is used for one-way data input device and can be used for data storage service in the embedded system; Figure (d) is the subscription interface, which is similar to the alarm return or push all communication modes in the asynchronous interface. The upper module subscribes to the data, and the lower module sends the data regularly or in alarm.

3 Standard Method for Embedded Device Data

3.1 Dynamic Data Standardization

The DataEvent class constitutes an important part of the OSA-CBM data model. The DataEvent sub hierarchy is related to the specific abstract state monitoring data processing functions defined in ISO 13374 standard [7] (machine state monitoring and diagnosis - data processing, communication and representation). Under these classes, there are subclasses that describe specific data types. Function definitions can be regarded as knowledge that abstracts data to a higher and higher level of information abstraction. Data Acquisition (DA) data type is the collected sensor data, and its format is consistent. The DA data is then converted into one or more meaningful features in a Data Manipulation (DM) type data format. Compare these functions with the expected values, and the resulting enumeration condition indicators are stored in the State Detection (SD) DataEvent class. The data in Health Assessment (HA), Prognostic Assessment (PA) and Advisory Generation (AG) formats are related to the current health status of the machine, the predicted future faults and the recommended operation steps, respectively. The DataEvent class designed for the embedded system (see Fig. 2).

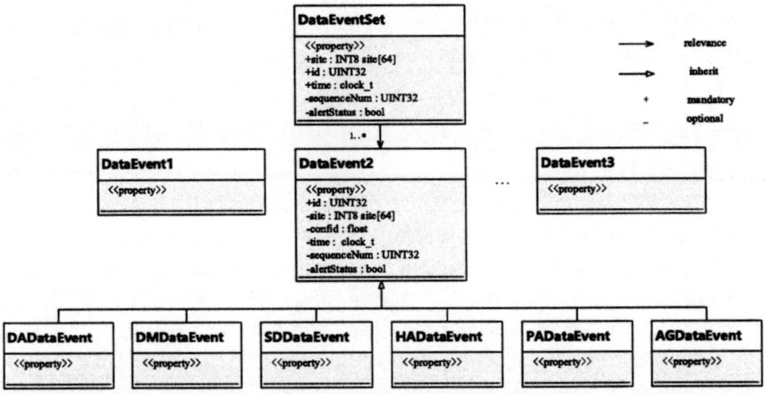

Fig. 2. Schematic diagram of DataEvent composition architecture

The dynamic data DataEvent base class supports unique data identification (ID) and entity object identification (site) to distinguish data from different physical locations; Support data trustworthiness (confid), data sequence number (sequenceNum) and alarm status (alertstatus) to describe data dynamic information; The date/time (time) of referencing a specific time instance is supported.

3.2 Static Data Standardization

An important component of Configuration is Port, which is static data related to PHM events generated by embedded applications. It is associated with DataEvent through its attribute ID and site (see Fig. 3).

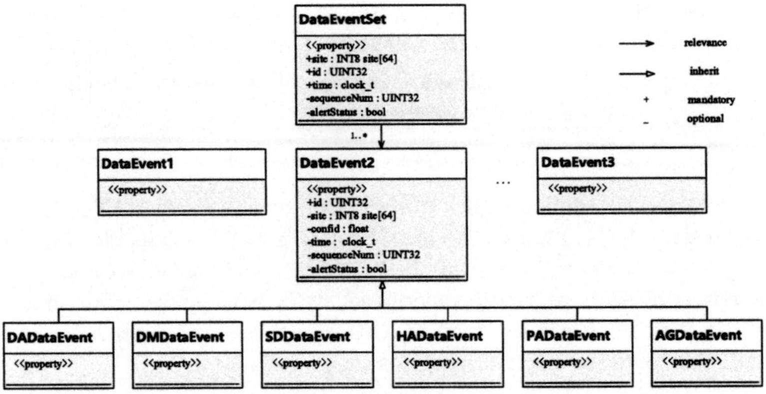

Fig. 3. Schematic diagram of Configuration composition architecture

Port is associated with sensor location information (measloc), equipment asset information (asset), and algorithm description information (algorithm) through foreign keys.

In the initialization stage of the embedded device, the Port and its associated information are designed as static data in the form of XML tables.

4 Data Standardization Verification and Test Analysis of Typical Embedded Devices

In this paper, an embedded device is selected as a typical application case to standardize the data of the device, and carry out data processing and threshold detection. Use the equipment to complete the DA block function, and select two zynq7000 development boards to complete the DM block and SD block functions respectively. Firstly, the data related to equipment health is collected through the DA block. Secondly, the data is preprocessed and sent to the DM block through the bus. The PHM algorithm application is called in the DM block for data processing. Finally, the processed data is sent to the SD block to realize the threshold detection and alarm functions.

The data collected by the DA block includes the low-pressure rotor speed (N), the outlet temperature of the high-pressure compressor (Temp), the static pressure after the fan stator (Ps) and the vibration signal of the turbine rear casing (Vib). The data collection period is 20 ms. Detailed data information is shown in Table 1.

Table 1. DA block generates data information.

Serial number	Identification name	Signal name	Signal source	Acquisition range
1	N	LP rotor speed	Engine N1	[0,5000 Hz]
2	Temp	Outlet temperature of HP compressor	HP compressor outlet	[−140.05 °C, 260.12 °C]
3	Ps	Static pressure behind fan stator	Embedded equipment chassis	[15 kPa, 200 kPa]
4	Ps	Vibration of turbine rear case	Turbine rear case	[0, 200 g]

It is converted into digital signal at the DA block and generates DADataEvent. It calls the interface API conforming to the OSA-CBM interface specification and sends the data packet to the SD block through the RS422 bus. The execution cycle is 20 ms and the baud rate is 230400 bps. On the other hand, static data corresponding to DataEvent is generated, which exists in the form of XML file and is called when the program is initialized. The configuration information is shown in Table 2.

Table 2. DA block generates data configuration information.

ID	Data name	Channel number	Site	Data type	EngUnit	Sensor type	MeasLoc
1	N	0	ZYNQ7000	float	Hz	RotateSpeed	Engine N1
2	Temp	1	ZYNQ7000	float	Degree	Temperature	HP compressor outlet
3	Ps	2	ZYNQ7000	float	Kpa	Pressure	Embedded equipment chassis
4	Ps	3	ZYNQ7000	INT16[8192]	/	Vibration	Turbine rear case

The DM block receives the data sent by the DA block. In order to verify the data standardization function, the PHM application algorithm is designed to process DataEvent, the mean filter algorithm is used to process N, Temp and Ps data, and the Fast Fourier Transform (FFT) is used to process Vib data. Convert the result processed by the algorithm into DMDataEvent, call the interface API conforming to the OSA-CBM interface specification, and send the data packet to the SD block through the extranet bus. The algorithm processing cycle is 40 ms.

In the SD block, the threshold detection algorithm is used to process data. If the threshold setting range is exceeded, an alarm will be triggered, and the alertstatus in the corresponding DataEvent will be set to 1. The ID of DataEvent can be traced to the configuration information, and the configuration information corresponding to the alarm information can be viewed.

In addition, in terms of performance overhead, compare the time required for "non-standardized" and "standardized" data processing tasks. The nonstandard data processing task is to collect data at the DA block, convert it into digital semaphore and send it directly to the DM block. After the DM block calls the PHM algorithm to process the data, the processing result is sent to the SD block, and the threshold detection is performed on the data at the SD block. By counting the running time of programs in DA block, DM block and SD block, (see Fig. 4). It can be found that the nonstandard data processing overhead is 7.541 ms, 15.239 ms and 5.632 ms respectively, while the standardized data processing overhead is 7.623 ms, 15.347 ms and 6.106 ms. The standardized data processing only takes 2.28% of the total overhead to obtain the maximum reusability and configurability. In this case, the performance overhead increased by data standardization is very low.

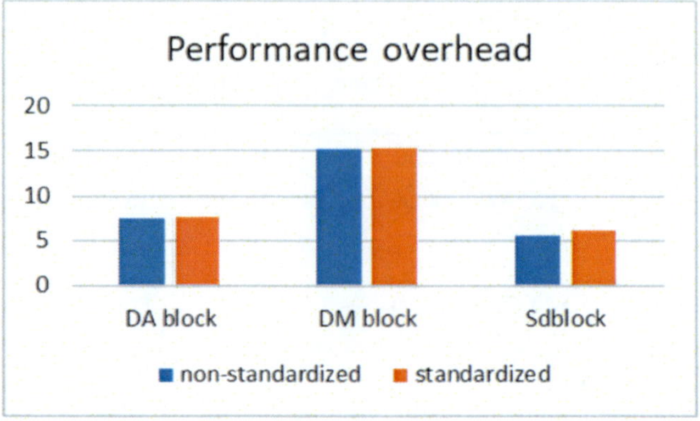

Fig. 4. Schematic diagram of performance overhead comparison

5 Data Model Based on OSA-CBM

In order to meet the data interoperability between airborne equipment of different suppliers and the traceability and identifiability of cross embedded equipment, this paper studies the implementation of embedded data standardization based on OSA-CBM. A hierarchical data model conforming to 13374 standard and OSA-CBM standard is designed to realize the interoperability and identifiability of the data model with dynamic data DataEvent and support the traceability of the data model with static data configuration. And through an example, the functions of data acquisition, data processing and state detection in the embedded system are verified, as well as the higher ability to improve data standardization in the case of very small performance overhead. However, this paper only studies the functions of DA, DM and SD layers in the embedded system in the airborne environment, and will carry out further research on ha, PA and Ag layers on the ground.

References

1. Ma, X., Wang, F., Chen, Y., Li, Y.: Intelligent fault diagnosis method of airborne computer based on health management technology. Comput. Measure. Control **27**(11), 43–47 (2019)
2. Swearingen, K., Keller, K.: Health ready systems. In: Autotestcon 2007, pp. 625–631. IEEE (2007)
3. Sreenuch, T., Tsourdos, A., Jennions, I.K.: Distributed embedded condition monitoring systems based on OSA-CBM standard. Comput. Stand. Interfaces **35**(2), 238–246 (2013)
4. Sreenuch, T., Jennions, I.: Design and evaluation of plug-and-play enabled IVHM architecture. SAE Int. J. Aerosp. **8**(1), 123–136 (2015). https://doi.org/10.4271/2015-01-9001
5. Bayoumi, A., Cao, A., Mccaslin, R., et al.: An Extensible CBM Architecture for Naval Fleet Maintenance Using Open Standards (2015)
6. MIMOSA, OSA-CBM UML Specification 3.3.1 Release, Machine Information Management Open Systems Alliance, Tuscaloosa, AL (2010)
7. ISO, 13374-2 Condition Monitoring and Diagnostics of Machines—Data Processing, Communication and Presentation—Part 2: Data Processing, International Organization for Standardization, Geneva (2007)

A Hierarchical Remote Interface Unit Interconnection Technology Based on IEEE-1394B and Zigbee

Jionghui Che(✉), Mingliang Hu, Wen Yan, Panjang Qu, and Fangchun Yu

Xi'an Aeronautics Computing Technique Research Institute, AVIC, Xi'an 710068, China
66401887@qq.com

Abstract. The data fusion technology developed in this paper is applied to the remote interface units of aircraft, and by taking advantage of the flexibility and reliability of Zigbee networking, it can quickly interact with other nearby remote interface units with sensor data through Zigbee when a remote interface unit sensor collects interface faults. Therefore, the fault remote interface unit does not need to be degraded in the case of interface failure, and can continue to complete its own control task, so as to ensure the completion of this flight mission.

Keywords: Remote interface unit · IEEE-1394B · Zigbee · Interconnection

1 Preface

With the deepening of electromechanical integration and control management, more and more secondary and tertiary controllers or special controllers are included in distributed electromechanical control products. And with the in-depth study of aircraft systems, distributed electromechanical control products have unified the acquisition, control, output and communication interfaces [1]. The remote interface unit came into being.

As an electromechanical terminal processing equipment, the remote interface unit has the advantages of nearby acquisition, light weight, high integration, and interchangeability, so it is widely used by modern aircraft [2].

Modern aircraft electromechanical systems have several remote interface units. Each remote interface unit accepts the control of the electromechanical system core machine, regularly uploads the collected data to the electromechanical system core machine and outputs the control instructions according to the control instructions issued by the electromechanical system core machine [3]. Under this architecture, it is necessary to ensure the reliability of the communication network and ensure that the task is not degraded in the event of an abnormal communication node [4]. Secondly, the remote interface unit is distributed according to the position of the aircraft, and some control objects are far away from the feedback signal, which is difficult to connect to the same remote interface unit [5]; And in the aircraft electromechanical closed-loop control system, the design network speed must be higher than the motor control with the highest requirements for network speed (the control algorithm requires current loop detection to be no more than

Chinese Society of Aeronautics and Astronautics (Ed.): CASTC 2023, LNME, pp. 417–424, 2024.
https://doi.org/10.1007/978-981-99-8861-7_43

50 μs, speed loop requirements for delay of not more than 5 ms, and position loop requirements for delay of not more than 100 ms) [6].

Therefore, in order to realize the mission reliability of the distributed electromechanical system communication network, meet the real-time requirements of the system, reduce the on-board cables to reduce the system weight, this paper proposes a distributed architecture of aircraft electromechanical system based on IEEE-1394B and Zigbee.

2 Architecture Design

Figure 1 shows the block diagram of the electromechanical control system. The system is divided into 3 layers, and the control computer at the highest level is crosslinked to the RIU of the second level via the IEEE-1394B bus. RIU chose the IEEE-1394B bus as the CCDL bus. RIUs use the Zigbee bus to crosslink with smart sensors.

Three homogeneous control computers constitute the data center of the electromechanical system, receive the acquisition results transmitted by the RIU, and generate sensor control instructions through the calculation to the RIU through the IEEE-1394B bus; Seven homogeneous RIUs constitute the sensing unit of the electromechanical system, which is distributed in different areas of the aircraft to realize the collection function of sensor temperature, flow, pressure and other status information; At the same time, these seven remote interface units also serve as actuators for electromechanical systems, driving sensors, motors and actuators to achieve specified functions. GMP is located under the Tier 1 IEEE-1394B bus and is used to complete data logging and cross-link with ground maintenance equipment.

Data sharing between RIUs is completed through CCDL, which helps RIUs complete regional-level closed-loop control. Zigbee wireless units are integrated inside each remote interface unit. Each remote interface unit is networked via the Zigbee bus with nearby smart sensors and other nearby remote interface units.

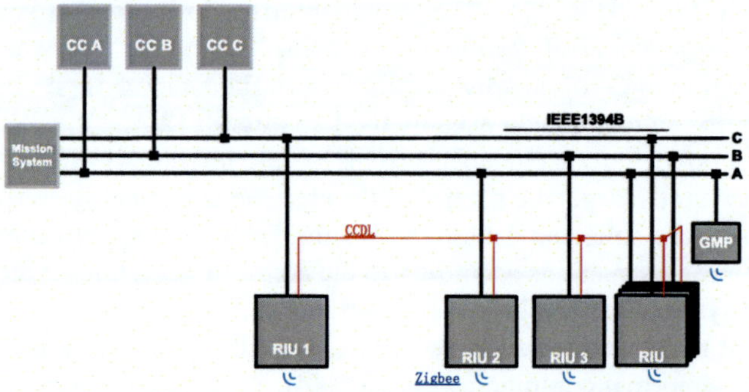

Fig. 1. Block diagram of the electromechanical control system

In this electromechanical control system, the first-level IEEE-1394B bus bus cycle is 12 ms, the second-level CCDL bus cycle is 0.5 ms, and the Zigbee wireless network works at 250 kbps, and the actual test delay is less than 100 μs.

3 Electromechanical System Control System Based on Hierarchical Network

3.1 Level 1 IEEE-1394B Bus High-Speed Data Communication Scheme

The first-level IEEE-1394B bus adopts the loop connection mode, and any network node can ensure normal communication among other nodes if it is disconnected.

The protocol specified in the IEEE-1394B standard is relatively complex, and in the system design, the IEEE-1394B bus is installed in the RIU in the form of a daughter card to realize the remote node (RN) function and complete the bus communication protocol processing and other functions [7].

The IEEE-1394B bus adopts pre-allocated bandwidth technology to improve the determinism of data transmission, pre-allocate fixed sending, receiving and data offsets for each node, plan the number and size of asynchronous stream packets when each node sends offsets, and at the same time provide the ability to change the time offset of remote nodes. Reasonable orchestration of system communication timing with time offset and pre-allocated bandwidth can effectively ensure deterministic time delay in the transmission of critical data.

Taking the transmission of sensor control instructions in the system as an example, assume that RIU takes time $\Delta T1$ from collecting sensor status information to completing asynchronous stream packet assembly and transmission, CC takes time $\Delta T2$ from receiving RIU data to completing control instruction calculation and transmission, and RIU takes $\Delta T3$ from receiving CC instruction to starting to drive sensors. Due to the high total new frequency, little asynchronous stream packet data, and negligible transmission time between bus nodes, then by reasonably designing the time offset of CC and RIU, the response time delay of the system can be controlled within $(\Delta T1 + \Delta T2 + \Delta T3)$, and the system deterministic delay is shown in Fig. 3 (Fig. 2).

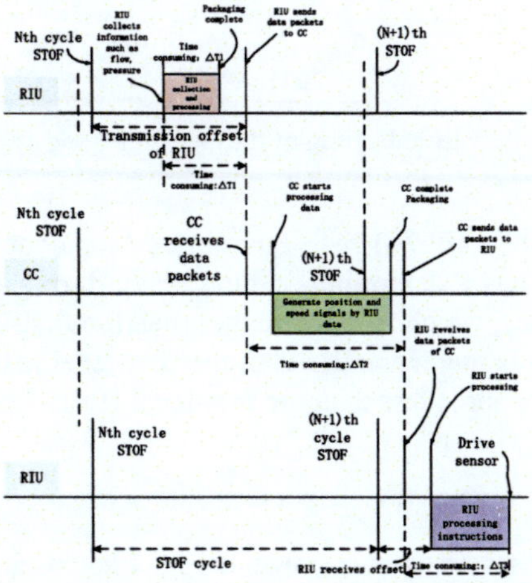

Fig. 2. Deterministic delay description of first-level network system

3.2 IEEE-1394B Bus CCDL Closed-Loop Control Communication Scheme

In order to further improve the reliability of the first-level IEEE-1394B, this paper designs an IEEE-1394B-based CCDL bus. When a RIU L1 bus is dropped, it can transmit data to other RIUs through CCDL and communicate with CCs through the L1 network of other RIUs.

In addition, in order to reduce the closed-loop control pressure of CC, part of the closed-loop control tasks are delegated to the secondary network when the system is designed. As shown in Fig. 3, take the control of a flow regulation flap driven by a motor directly controlled by RIU1, and the data required for its current loop is collected by RIU1. Its position loop is collected by RIU3, and the acquisition results are sent to RIU1 through CCDL (its transmission path is shown in transmission path No. 1 (solid green line) in Fig. 3.

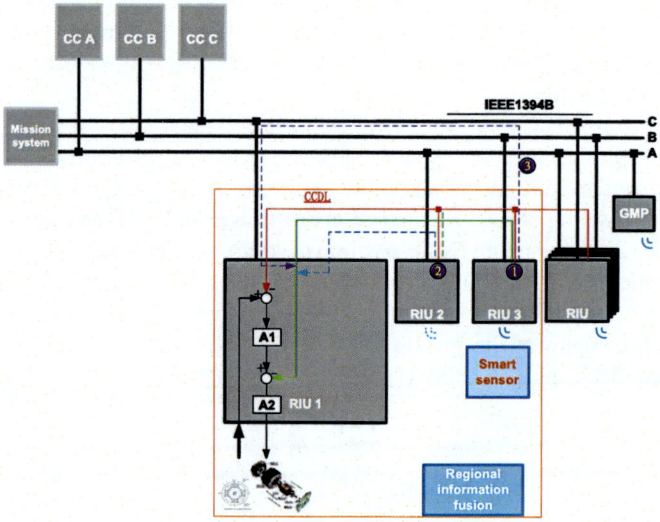

Fig. 3. Block diagram of the date collecting system

Similar to the first-stage IEEE-1394B bus, closed-loop control via CCDL can effectively ensure that data has a deterministic time delay during transmission. Since the CCDL bus period is 5 ms, which is much higher than the first-level network by 12 ms, and the amount of data is much smaller than the first-level network, the deterministic delay on the CCDL network is less than the first-level IEEE-1394B bus.

3.3 RIU Node Zigbee Network Communication Scheme

The Zigbee wireless sensor network is shown in Fig. 1 and uses a mesh structure. In the Zigbee wireless network, GMP acts as a coordinator, responsible for detecting normal mode and ground calibration mode, and forming a network to control the RIU to switch the Zigbee network operating mode. The ground maintenance equipment is connected to

GMP and sends calibration instructions, GMP controls all RIUs into calibration mode, RIU is responsible for sending all wireless sensor data to GMP in ground mode, and GMP sends the collected wireless sensor data to the ground maintenance equipment. In ground calibration mode, it is responsible for collecting all wireless sensing data to assist in the calibration work.

The remote interface unit accesses the network as a routing node and is responsible for collecting data from the terminal wireless sensing node. The smart sensor, as an end node, is responsible for sending the sensing data to the nearest routing node (remote interface unit).

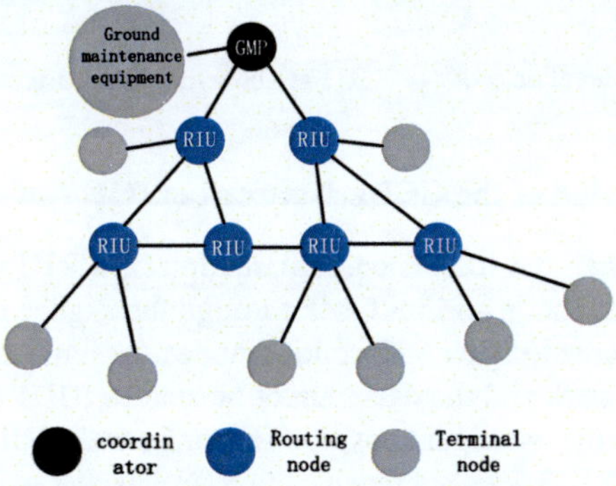

Fig. 4. Structure of Zigbee networking

The tertiary Zigbee network has a transmission speed of 250 kbps, and its smart sensor acts as an end node and is responsible for sending the sensing data to the nearest routing node (remote interface unit).

4 High-Reliability System Bus Network Design

4.1 First-Level IEEE-1394B Bus Network Node Abnormality Bus Reorganization

In order to further improve the reliability of the first-level IEEE-1394B, this paper designs an IEEE-1394B-based CCDL bus. As shown in Fig. 5, when a RIU L1 bus is dropped, data can be transmitted to other RIUs through CCDL and communicated with CCs through the L1 network of other RIUs.

Assume that when RIU1 cannot receive STOF packets from the first-level network for two consecutive cycles, RIU1 sends the first-level network exception information and the external interface collection information to RIU2 through CCDL, and RIU2 sends the information sent by RIU1 to the first-level network as mutual transmission information. After CC receives the RIU1 information sent by RIU2, it can confirm that RIU1 can continue to execute bus instructions through CCDL, RIU2 receives the distribution data and forwards it to RIU1, and RIU1 receives the information forwarded by RIU2 to complete the output task. Therefore, when an exception occurs at a first-level network node, it takes two bus cycles to complete the network recovery.

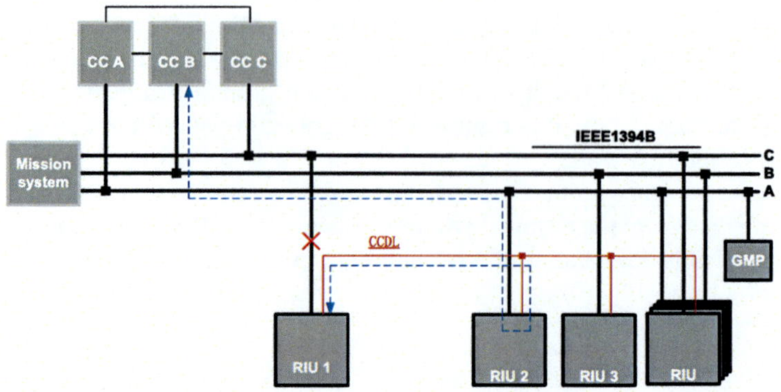

Fig. 5. When the first-level network of RIU1 is abnormal, data is forwarded through RIU2

4.2 Data Reorganization of the CCDL Network of RIU Node Abnormalities

As shown in Fig. 6, when the CCDL communication of a RIU node is abnormal, the communication abnormal node notifies GMP through the Zigbee network to reassign its connected wireless sensors to other RIU routing nodes. Assume that when an exception occurs in the RIU3 network and the data cannot be sent to RIU1 through CCDL, RIU3 notifies GMP to change the intelligent position sensor from the RIU3 branch to the RIU2 branch. Assuming that RIU3 does not receive the STOF packet twice, it takes $2 \times$ STOF, it takes ΔT1 from RIU3 confirming the CCDL connection abnormality to completing the package application and sending ΔT1, from RIU3 issuing the application to GMP receiving the application ΔT2, GMP processing the application and issuing the network reorganization instruction ΔT3, GMP issuing the network reorganization application to RIU2 receiving the instruction ΔT2, RIU2 and the position sensor establishing a connection ΔT4. With reasonable design and arrangement, the system can complete the network reassembly within (2STOF + Δ T1 + 2Δ T2 + Δ T3), and the specific test of its switching process is shown in Fig. 4.

When the network switchover is completed, the new network is crosslinked in the same way as the network before the switchover, and the delay is basically the same, which can ensure that the response time of the system is the same as the original time.

4.3 RIU Node Zigbee Network Abnormality Data Reorganization

In terms of the distribution of RIUs across the entire machine, consider that each terminal node can freely network with at least two RIUs. This eliminates the need for control tasks to be completed due to a single point of network failure. When the wireless network of any RIU is abnormal, the RIU tells GMP through the first-level IEEE-1394B bus, and GMP regains control to connect the disconnected end node to other RIUs near the node.

The switching process is the same as in Sect. 3.2, and the delay of network reorganization is controllable.

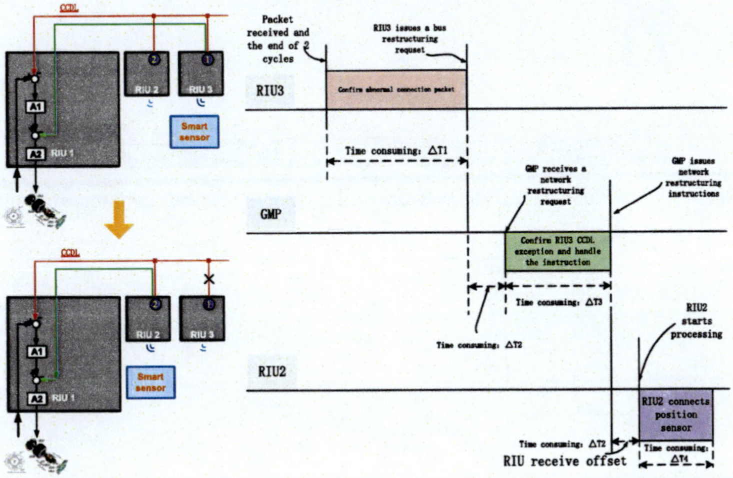

Fig. 6. The network reorganization process and delay description of RIU CCDL network connection exception

5 Test Data Analytics

In the laboratory, the abnormalities of the primary bus node, the CCDL bus node and the Zigbee module of RIU were reconstituted and tested.

Figure 7 shows that the first-level bus node exception is simulated by disabling the first-level IEEE-1394B module of RIU3, and the measured bus can complete the bus reassembly within 36 ms of three bus cycles, and complete the data update in about 48 ms; Fig. 8 shows that by disabling the CCDL module of RIU3, the CCDL bus node exception is simulated, the bus reassembly is completed within 480 ms of the measured bus, and the data update is completed in about 500 ms; Fig. 9 shows that the Zigbee bus node exception is simulated by disabling the Zigbee bus module of RIU3, and the bus reassembly is completed within 660 ms and the data update is completed in about 680 ms. The experimental results show that the three-level bus network designed in this paper has good deterministic and recombination characteristics, which can meet the requirements of high reliability of electromechanical systems.

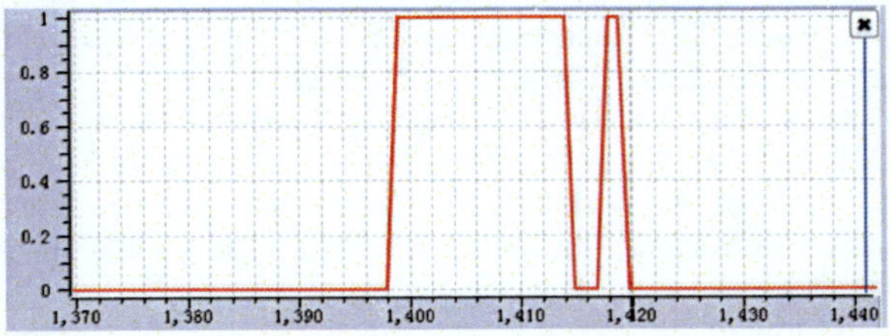

Fig. 7. RIU Level 1 IEEE-1394B network nodes disable post-bus data recovery time

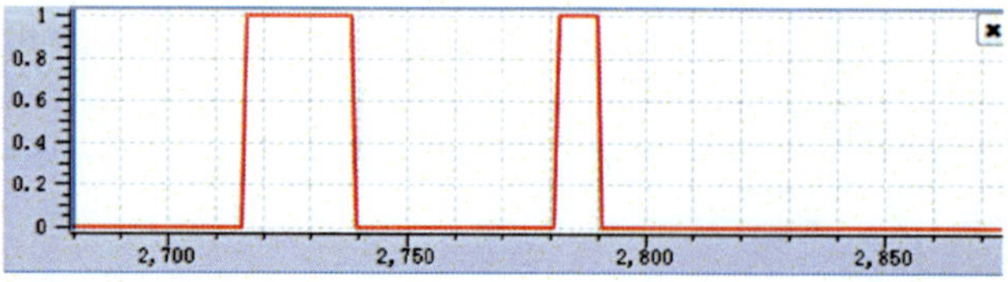

Fig. 8. RIU secondary CCDL network nodes disable post-bus data recovery time

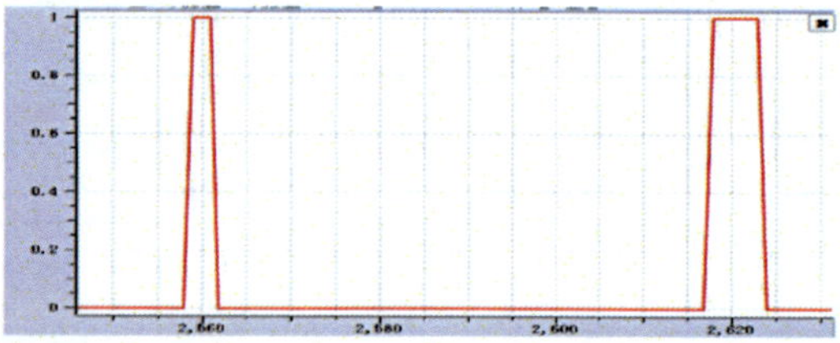

Fig. 9. RIU Level 3 Zigbee network nodes disable post-bus data recovery time

6 Conclusion

The three-level bus network of the electromechanical system developed in this paper uses the high determinism and high real-time of IEEE-1394B bus to complete the information interaction between the main network and the secondary CCDL bus, and uses the Zigbee wireless network to realize the data acquisition of the sensor by RIO, so as to complete the data fusion of the electromechanical system. The bus network has high deterministic and stable bus recombination characteristics to meet the high reliability requirements of electromechanical systems. After system verification, the scheme has the advantages of easy installation, miniaturization, versatility, etc., and achieves a weight reduction of about 150 kg per machine, which has strong practical significance and can be widely used in aviation, aerospace and industry and other fields.

References

1. Lin, W., Zhou, Z., XueWei, W.: The development and application of the data acquisition system. Electr. Measur. Instrum. **41**(8), 4–8 (2004)
2. Olfati-Saber, R., Murray, R.M.: Consensus problems in networks of agents with switching topology and time-delays. IEEE Trans. Autom. Control **49**(9), 1520–1533 (2004)
3. Abdelrahman, M., Rasheed, A.: A methodology for development of configurable remote access measurement system. ISA Trans. **39**(4), 441–458 (2000)
4. Yang, T.: Airborne data acquisition and recording. In: Telemetry Theory and Methods in Flight Test, pp. 151–215. Springer, Singapore (2021). https://doi.org/10.1007/978-981-33-4737-3_4
5. Douglas, R., Charboneau, A., Gertsvolf, M.: Remote time and frequency calibration with holdover traceability from a new treatment of non-white noise in rubidium clocks. Metrologia **58**(5), 055003 (2021)
6. Magdaleno, E., Rodríguez, M., Pérez, F., et al.: A FPGA embedded web server for remote monitoring and control of smart sensors networks. Sensors **14**(1), 416–430 (2013)
7. Li, T.J., Fujimoto, Y.: Control system with high-speed and real-time communication links. IEEE Trans. Ind. Electron. **55**(4), 1548–1557 (2008)

Weight Design of Cross-Day Solar-Powered UAV

Juhong Dang(✉), Yueli Li, Zeying Li, and Yining Gao

First Aircraft Institute, No. 1 Renmin Road, Yanliang District, Xi'an 710089, China
mr.dangjh@163.com

Abstract. In this paper, the development history of solar-powered UAV is briefly reviewed, and the technical characteristics and design difficulties of solar-powered UAV are analyzed. Based on the energy conversion process and the design of a solar-powered UAV, the weight design method and process of cross-day solar-powered UAV are given. In the end, some important problems in the weight design of cross-day solar-powered UAV are emphasized, including the calculation of battery weight, the design and realization of light structure and the choice of flight path, which can be used as reference for the subsequent solar-powered UAV design.

Keywords: Solar Energy · Solar-powered UAV · cross-day UAV · Sustainable Flight · MPPT · Aircraft Design

Solar-powered UAV has the advantages of no pollution, long duration of continuous flight, low signal transmission loss and high precision. In theory, It can fly for months or years. It is called "pseudo-satellite" and can be widely used in military and civilian applications in the case of earth observation, border surveillance, forest fire fighting, power line inspection, communication platform for mobile devices, weather research and forecast, environmental monitoring. It has become a hot spot in the research field of UAV in recent years.

In 1974, "Sunrise-I", a solar powered aircraft, made its maiden flight on the dry lake at Camp Irwin, California. After decades of development, breakthroughs have been made in the field of solar-powered UAV. In 1983, "Pathfinder" realized the partial power flight provided by solar cells. In 1999, "Helios" was completely powered by solar energy. In 2007, "Zhphyr6" first made its day-and-night flight. Since then, "Zhphyr7", "Zhphyr S" and "Zhphyr8" have continuously broken records, and realized the maximum cross-day continuous flight of 64 days. There are many solar aircraft projects in China, such as "Rainbow" series solar-powered UAV, "Qi Mingxing" series solar-powered UAV, etc. These projects are preparing to fly across the day and night in the high altitude.

In this paper, the key problems in the weight design of solar-powered UAV are analyzed and discussed in combination with actual project [1–3].

Chinese Society of Aeronautics and Astronautics (Ed.): CASTC 2023, LNME, pp. 425–433, 2024.
https://doi.org/10.1007/978-981-99-8861-7_44

1 Technical Characteristics and Design Difficulties of Solar-Powered UAV

Solar-Powered UAV, especially cross-day solar-powered UAV, have a higher altitude, lower speed, longer duration, simpler flight profile, etc. These characteristics lead to a bigger lift-drag ratio, larger aspect ratio, smaller wing load. The deep coupling of layout variables and energy parameters brings great difficulty to configuration, aerodynamic, structure, flight control, propulsion and flight path design. Some experienced authors [1–9] summarized and discussed the key technologies and design difficulties of solar-powered UAV, including:

(1) Comprehensive design technology of new aerodynamic layout (involving energy balance, weight and aerodynamic characteristics);
(2) Low Reynolds number aerodynamic design technology for low speed aircraft;
(3) Aeroelastic and gust mitigation techniques for large flexible structures;
(4) Light-weight and high-efficiency structure design and verification technology;
(5) High reliability distributed motor differential control and nonlinear control technology;
(6) Flight trajectory optimization technology.

The research progress of these key technologies has played a key role in the research and development of cross-day solar-powered UAVs. Table 1 shows the main parameters and data of some solar-powered UAVs.

Table 1. Solar-energy UAV main parameters

Aircraft	Country	Span/m	Weight/kg	Configuration	Year	Duration of flight	Maximum flight altitude/km
Helios	American	75.3	720	Flying wing	2003	24 h	29.5
Solong	American	4.75	12.8	Conventional Configuration	2005	48 h	
AtlantikSolar	Swiss	5.65	6.8	Conventional Configuration	2015	81.5 h	
Zephyr-S	England	25	75	Conventional Configuration	2018	26 d	22.5
Zephyr-8	American	25	75	Conventional Configuration	2022	64 d	18

2 The Weight Design of Cross-Day Solar-Powered UAV [4–6]

2.1 Flight Profile Selection of Cross-Day Solar-Powered UAV

The cross-day solar-powered UAV generally uses the gravity energy storage flight profile, on the one hand, it makes the aircraft's battery weight and the aircraft gross weight as small as possible, on the other hand, this method is advantageous to carry more mission system. Figure 1 is a typical flight profile of a cross-day solar-powered UAV. This profile can be divided into take-off, climb, day cruising, descent, night cruising, climb, day cruising... a process that continues to circulate.

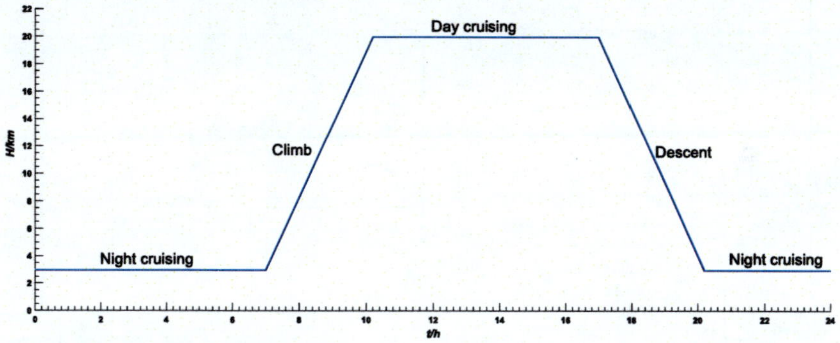

Fig. 1. Typical flight profile of solar-powered UAV across the day and night

2.2 Weight Design of Solar-Powered Cross-Day UAV

In order to realize the cross-day and night flight, the cross-day solar-powered UAV must satisfy three equations at the same time: lifting weight balance (Eq. 1), thrust resistance balance (Eq. 2), and the energy balance (Eq. 3).

$$mg = C_L S_W \left(\frac{1}{2}\rho V^2\right) \tag{1}$$

$$T = C_D S_W \left(\frac{1}{2}\rho V^2\right) \tag{2}$$

$$E_{available} \geq E_{require} \tag{3}$$

NOTH gives the overall design flow of the long-range solar aircraft [2], Guo Zheng gives the overall design method of the solar aircraft [4]. The cross-day solar-powered UAV development process needs to be in accordance with the strong coupling between the various disciplines, and comprehensive multi-disciplinary optimization design is implemented through the development of integrated design platform to improve the design quality. Figure 2 is a flow chart of multidisciplinary optimization design for a typical solar-powered UAV. The aircraft design process is an iterative process. The first stage is conceptual design. According to the aircraft's layout variables, the aircraft gross weight and battery weight are estimated. Based on multidisciplinary design, one or more iterations are completed, and the layout parameters of the aircraft are determined, including the aircraft gross weight and the wing load.

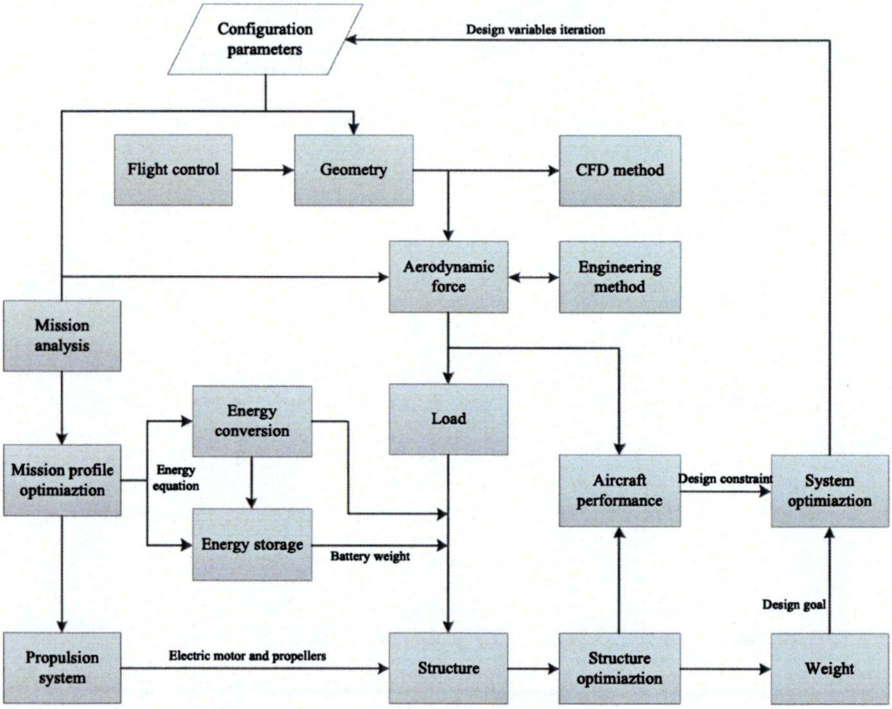

Fig. 2. Multi-disciplinary optimization design chat

Secondly, the aircraft gross weight is decomposed and the weight limit of each sub-system is given. At the same time, according to the overall design requirements and design parameters, the aircraft structure and the system weight are estimated by combining the structure and the system scheme, and the first iteration is completed after comparison. Thirdly, the weight of each sub-system is optimized according to the results of the first round of the design. At the same time, the aircraft's overall performance is calculated and analyzed, and the second and subsequent iterations are completed. Finally, the design scheme of the aircraft is given. According to the basic flow of the weight design of the cross-day solar-powered UAV (Fig. 3), the weight design steps of the cross-day solar-powered UAV are given briefly.

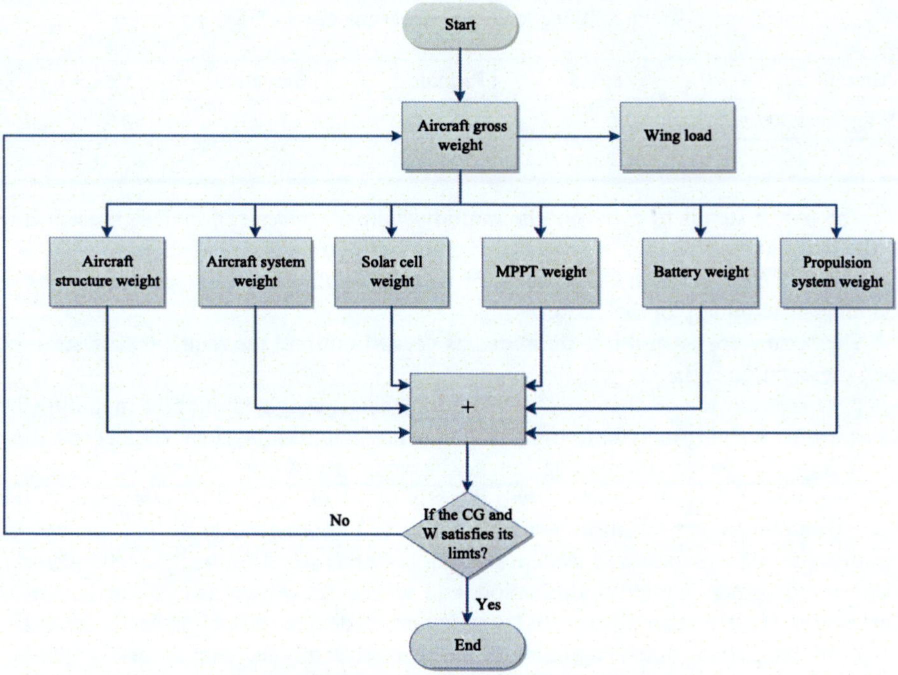

Fig. 3. Weight design chart of cross-day solar-powered UAV

First, determine the wing load, the formula for calculating the upper limit of wing load is given in the thesis [5], and the appropriate wing load can be selected:

$$P_{level} = \frac{P_{tot\_level}}{S_W} = \frac{C_D}{C_L^{3/2}} \left(\frac{mg}{S_W}\right)^{3/2} \sqrt{\frac{2}{\rho}} \tag{4}$$

Wing load (W/S) and thrust-to-weight ratio (T/W) are two important parameters that determine the flight performance of UAV. After the concept scheme of UAV is determined, the wing load and thrust-to-weight ratio of UAV need to be estimated.

In high-altitude and long-endurance flight, the wing load of the UAV should be as small as possible, the smaller the wing load, the larger the wing area, and the larger the solar panel's laying area. Table 2 is the wing load of several solar-powered UAVs. It can be seen that the wing load of the aircraft is below 10 kg/m^2. Combined with the aircraft design, the appropriate wing load is selected to optimize the overall parameters of the aircraft.

The second step is to determine the aircraft gross weight. Since the gross weight of the solar-powered aircraft is almost constant during flight, it can be obtained according to the cruising altitude, combined with the day-night energy balance calculation and the selected wing load.

In the third step, all the sub-system weights are obtained from the weight decomposition relation. At the same time, the structure and the system technicians estimate and compare the sub-system weight according to the aircraft overall parameters.

Table 2. Wing load of several solar-energy UAV

Aircraft	Zephyr	Pathfinder	Centurion	Helios
Wing load (kg/m^2)	0.6	3.4	5.3	4.9

The fourth step is to carry out the multidisciplinary optimization design according to the latest aircraft weight, considering the aircraft performance calculation.

The fifth step, perform a new round of aircraft weight design iteration according to the multi-disciplinary optimization.

The sixth step, do multiple iterations as needed until all the requirements are met, then the iteration ends.

It is necessary to note that weight estimation, that is, weight prediction, including the aircraft gross weight prediction, structure weight prediction and system weight prediction are all important items in weight design. At the beginning of a new aircraft development, the precision of weight prediction will directly affect the aircraft performance. In the multi-disciplinary optimization environment, it is important to carry out the weight optimization of structure and system design, material selection weight optimization, airborne equipment weight optimization, and weight sensitivity analysis of structural parameters, etc. These are all essential steps in aircraft design and weight design iteration.

At the same time, at the beginning of the scheme, combined with multi-disciplinary optimization, we should also carry out the definition of the aircraft characteristic weight.

3 Problems of Concern in Weight Design of Solar-Powered UAV

3.1 Energy Storage Battery Weight [10–12]

Due to the requirement to fly through the night for more than 10 h, cross-day solar-powered UAVs need to carry a large weight of energy storage batteries. Some battery weight even reached 40% of the aircraft gross weight [4], for example, solar power II carried batteries of 633 kg. Zhang Qinling et al. calculated the solar energy UAV energy system, which accounted for about 52% of the aircraft gross weight. Considering the gravity energy storage, the ratio of the battery to the aircraft gross weight is also 21.2% [9]. Therefore, the weight of the battery carried by a solar-powered UAV should not be underestimated, even if gravity energy storage is taken into account.

There are three problems to be solved in battery weight, one is to choose a suitable battery, the other is to accurately calculate battery weight, and the third is to choose a suitable flight strategy (Sect. 3.3).

In recent years, the battery technology witnessed a fast progress. Different types of battery have different energy-to-weight ratio and cycle life. The choice of battery type should consider energy-to-weight ratio, cycle life, as well as cost. This is a comprehensive choice problem. Table 3 shows the performance of different types of battery.

It is a difficult problem to calculate the weight of the cross-day solar-powered UAV battery. In general, the direct calculation method is used, that is, the total energy is calculated according to the integral of the flying stage, and then the weight of the battery

Table 3. Comparison between various types of energy storage systems [6]

Battery types	Sodium-sulfur cell	Lithium sulfur cell	Lithium ion battery	Lithium-polonium cell
Operating voltage/V	2.1	2.5–1.7	3.7	3.7
Energy-to-weight ratio/(Wh/kg)	120–250	350–600	100–250	130–200
Cycle life/Number	>2500	>100	400–1200	>1000
Operating temperature/°C	300–350	−60–60	−20–60	−20–60
Self-discharge rate (%Month)	Non	<15%	8%–31%	5%

is calculated. The problem of direct method is that the constraint conditions are difficult to be used, and it cannot form a robust iterative solution, and it is hard to implement in engineering problems. The calculation method of this paper is: assuming that the aircraft gross weight is all made up of the battery weight, based on the flight profile and the solar radiation model, the time of flight cross-day is discretized, and the energy storage of the battery is obtained by step-by-step integration. If the energy stored under the weight of the storage battery satisfies the constraint conditions, the weight of the storage battery (1 kg) is deducted and calculated iteratively until the constraint conditions are not satisfied. So the battery weight calculated in the last iteration is the minimum battery weight that the aircraft can fly day and night under this test condition. The influence of the surface curvature of the solar panel, the variation of atmospheric attenuation system, the installation angle of the wing, the wind speed, the design lift-drag ratio and the wing load are also considered in the calculation. Figure 4 shows the iterative solution of battery weight.

The key factor of the solar-powered aircraft's cross-day flight is the weight of the battery it carries. The calculation method presented in this paper can calculate the weight of the battery quickly and accurately in the scheme stage, and meet the engineering requirements.

3.2 Ultra-Light Structure Design and Implementation [4]

With large span, low wing load and light weight, all the solar-powered UAVs are made of the most advanced composite materials. These ultra-light structure design and manufacture have become one of the key technologies that affect the flight performance and flight safety of solar-powered UAVs.

SHI Wen presented three problems of ultra-light structure design and manufacture [7]: (1) the characteristics of lightweight and large flexible structure bring multidisciplinary strong coupling problems, however, the optimization design of structure, experimental verification and evaluation methods are only in the preliminary stage; (2) the technology of mutual coupling optimization design between structure and layout, the

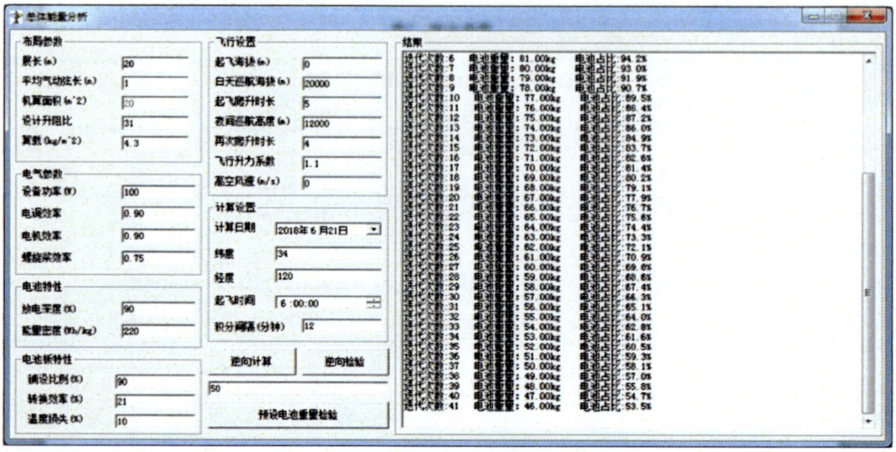

Fig. 4. Iterative solution of battery weight

integrated design technology of structure and function are not mature; (3) the standard specifications of structure design, process, and test are all insufficient and need to be developed urgently. In addition to the design problem, the manufacture of ultra-light structure is also the key problem, because the current solar-powered UAV is basically single piece production, weight control is the main problem of ultra-light structure, and overweight rework often occurs, so the design and manufacture of ultra-light structure is the key problem that the future solar-powered UAV needs to pay attention to.

3.3 Selection of Cross-Day Flight Strategy

The choice of day-and-night flight strategy is not only one of the methods to realize day-and-night flight of solar-powered UAV, but also the concern of weight design, which affects the aircraft weight.

During the cross-day flight, the choice of climbing path during the day, the choice of when to begin to climb, the choice of descending path at night, the choice of descending time are involved in the selection of the whole flight profile, which needs to be optimized many times. Especially in the uncertainty high-altitude wind, how to choose the appropriate flight strategy, whether to frequently correct the flight attitude of the aircraft, increase the differential speed, will affect the work life of the motor, and also increase the consumption of batteries. Frequent corrections of the flight attitude during the flight will cause the engine speed to be too high, which not only affects the engine life, but also increases the energy consumption of the aircraft and adversely effects on the long-range flight of the aircraft.

Therefore, the choice of flight strategy is multi-faceted and needs to be constantly summarized and optimized in actual flight.

4 Conclusion

This paper reviews the history of solar-powered cross-day UAV, summarizes and analyzes its technical characteristics and design difficulties. According to the energy conversion of solar-powered UAV flying process, combined with the design practice of a solar-powered UAV, the weight design method and process of solar-powered cross-day UAV are given. Finally, some key issues in the weight design of cross-day solar-powered UAV are emphasized, which can be used for reference in the future design of solar-powered UAV.

References

1. Ma, D.L., Zhang, L., Yang, M.Q., et al.: Review of key technologies of ultra long endurance solar powered unmanned aerial vehicle. Acta Aeronautica et Astronautica Sinica **41**(3), 623418–623418 (2020). (in Chinese)
2. Noth, A.: Design of solar powered airplanes for continuous flight. Suisse: Ecole Polytechnique Federale de Lausanne (2007)
3. Zhang, D.H., Zhang J., Li, J.F., et al.: Energy balance modeling for solar powered aircrafts. Acta Aeronautica et Astronautica Sinica **37**(S1), S16–S23 (2016). (in Chinese)
4. Guo, Z., Zhu, X.F., Hou, Z.X., et al.: Conceptual Parametric Design Based on Generalized Energy, vol. 12, pp. 40–91. Science Press, Beijing (2021). (in Chinese)
5. Hou, Z.X., Gao, X.Z., Guo, Z., et al.: The Crucial Problem About the High Altitude Long Endurance Solar Powered Unmanned Aerial Vehicle, vol. 12, pp. 17–55. Science Press, Beijing (2021). (in Chinese)
6. Zhang, J., Wang, J.S., Geng, Y.S., et al.: Technology challenges for high altitude long endurance solar powered UAV. Aeronaut. Sci. Technol. **31**(04), 14–20 (2020). (in Chinese)
7. Shi, W., Jia, Y.Q., Li, G.J., et al.: Analysis of key technologies of high altitude ultra long endurance solar powered unmanned aerial vehicle. Aerodyn. Missile J. **2021**(6), 63–66 (2021). (in Chinese)
8. Tao, Y.J.: Development and key technology on near space long voyage solar unmanned aerial vehicle. Aeronaut. Manuf. Technol. **513**(18), 26–30 (2016). (in Chinese)
9. Harasani, W., Khalid, M., Arai, N., et al.: Initialconceptual design and wing aerodynamic analysis of a solar power-based UAV. Aeronaut. J. **118**(1203), 540–554 (2014)
10. Zhang, Q.L., Huang, J., Liu, X.Q.: General design method and analysis of long endurance solar powered UAV. J. Air Force Eng. Univ. Nat. Sci. Edn. **15**(2), 12–15 (2014). (in Chinese)
11. Jin, X., Xiao, W., Wu, H., et al.: Research on influence of flight parameters on performance of photovoltaic modules in solar aircraft. Aeronaut. Sci. Technol. **30**(10), 70–76 (2019). (in Chinese)
12. Zhu, L.H., Sun, G.R., Hu, W.T., et al.: Key technology and development trend of energy system in solar powered unmanned aerial vehicles. Acta Aeronautica et Astronautica Sinica **41**(3), 80–91 (2020). (in Chinese)

Research on Scenario-Based Clustering Model and Analysis Method for Airworthiness Provisions

Yao Yuan(✉)

Airworthiness Engineering Center, Shanghai Aircraft Design and Research Institute, Shanghai 201210, China
yaoyuan2@comac.cc

Abstract. Airworthiness provisions are the requirements that must be followed in the development of civil aircraft. In view of the abstract characteristics of airworthiness provisions, a scenario-based clustering model and analysis method for airworthiness provision is proposed, which is applied to the identification of airworthiness requirements in the development process. Combined with the concepts of knowledge clustering and scenario, the requirements of airworthiness provisions are analyzed based on the elements of the scenario, and a provision clustering spatial model is formed. The verification results of selected CCAR 25.951 show that the proposed method is feasible in identifying the applicability of provisions for specific design features.

Keywords: Airworthiness provision · Scenario · Clustering model · Clustering analysis · Provision correlation

1 Introduction

The development of complex products must comply with a large number of industry standards and specifications [1]. Taking civil transport category aircraft as an example, the airworthiness regulations must be complied in the development process, such as CCAR-25 (Civil Aviation Regulation of CHINA Airworthiness Standard: Transport Category Aircraft) [2]. Airworthiness regulations involve the requirements of aircraft and system functions, performance, structure and other aspects. The provisions have strong characteristic of generality and abstraction, and the interpretation of the provisions need to be combined with the specific accident background and a large number of supporting materials [3]. Based on the above characteristics, there are great difficulties and challenges in determining the applicability and interpreting the requirements of the regulations. Once the airworthiness requirements cannot be captured completely and correctly in the early stage of development, it will probably lead to the iteration of later design schemes, which will greatly increase the development cost and certification period, and affect the commercial operation of aircraft. In the development process of most types of domestic civil aircraft, the judgment of the applicability of airworthiness

Chinese Society of Aeronautics and Astronautics (Ed.): CASTC 2023, LNME, pp. 434–443, 2024.
https://doi.org/10.1007/978-981-99-8861-7_45

provisions depends more on the experience and ability of designers themselves, and there is a lack of systematic technical means to identify and check the integrity of airworthiness provisions.

Airworthiness provisions can be regarded as a series of data sets of aircraft design requirements, and each provision is a relatively independent and interconnected data unit. The method of knowledge clustering to retrieve the most similar or relevant group of airworthiness provisions based on a certain feature or attribute of the design object can be used, so as to initially establish the database of applicable provisions, which are taken as the input to the design of the civil aircraft. At present, the commonly used knowledge clustering methods mainly include [4–8]: partition method, hierarchical method, density-based method, etc. However, the above methods cannot identify the requirements, applicable objects and relevance of complex provisions accurately so this paper takes CCAR-25 airworthiness provisions as an example, proposes a scenario-based clustering model and analysis method of airworthiness provisions, systematically identifies the technical requirements of provisions by drawing on the concept of scenarios, and explicitly calculates the correlations of provisions, which enables the rapid retrieval of other provisions related to specific physical objects/attributes, thus effectively supporting the identification and acquisition of airworthiness requirements in the civil aircraft product design process.

2 Scenario-Based Provision Clustering Model

2.1 Connotation and Expression of the Scenario

"Scenario" is not a new concept, and scenario-based methods have been widely used in software design and other engineering design areas for the expression and clarification of design tasks [9–13].

In engineering design area, due to the great difference between the development of physical products and software systems [14], the concept of "product scenario" has been adjusted and expanded accordingly. Generally speaking, the product scenario is based on the analysis of object behavior, which is used to describe the product behavior in the process of interaction and interaction between the "product to be designed" and environmental elements.

A complete product scenario mainly includes the following elements:

- Time
- Location
- Participant
- The behavior of the participants
- Situation/status

Time and location are the background elements of a scenario. Participants refer to the physical objects involved in the scenario, that is, the subject of a certain behavior, which can be the product to be designed, or the operator, the object of action, such as raw materials, air, sunshine and other environmental elements. Behavior refers to a certain action or action of the scenario participants; A situation or status is a sequential change in the state of an object after an action has been performed/occurred.

2.2 Clustering Spatial Modelling of Provisions Based on Scenario

The airworthiness provision in CCAR-25 is a qualitative or quantitative requirement/constraint for a part of civil aircraft design features/design objects. The "civil aircraft design features" here mainly focus on "structural features" (such as landing gear, fuel tank, etc.), and also include "functional features" (such as cabin fire protection, system lightning protection) and "behavioral features" of aircraft (such as take-off, climb and landing). Functional and behavioral features are ultimately embodied in the structural features of the aircraft, which are the main participants in each "civil aircraft product scenario".

For any CCAR-25 airworthiness provision, it describes the functional, behavioral (or performance) and structural parameter requirements and limitations for the design features of a certain part of the aircraft under certain conditions. This "condition" or "situation" is actually a certain working scenario/usage scenario of the aircraft. Therefore, various application scenarios for a certain part of civil aircraft design features can be separated from these provisions. Through the establishment and expression of these "scenarios", on the one hand, the qualitative or quantitative design requirements contained in the airworthiness provisions can be clarified, and on the other hand, the civil aircraft design features covered by the provisions (i.e., participants in the scenarios) can be identified [15].For example, for CCAR25.631 "Bird Strike Damage", the structural design and verification requirements for the empennage in the bird strike scenario are presented. According to the scenario elements introduced in Sect. 2.1, the scenario model described by natural language is shown in Fig. 1(a), and the corresponding bird strike scenario expression is shown in Fig. 1(b). Of course, for complex provisions, many different scenarios can be stripped out through the analysis of the provisions.

Time	Day and Night
Location	Ground, Sea Level and High altitude...
Participant	
a) Design Object (including features)	System, Structure and Equipment...
b) Environment Element (including features)	Wind, Rain, Thunder, Bird and Flight Crew...
Status after interaction	the constrain for Design Object

(a) Scenario Frame Described by Natural Language

Time	N/A
Location	High altitude
Participant	
a) Design Object (including features)	Nose\Wings\empennage\windshield (angle of inclination is not less than 15 degrees)...
b) Environment Element (including features)	Bird(Weight as 8-pound) Relative velocity(Vc at sea level and 0.85 Vc at 8000 ft)
Status after interaction	continued safe flight and landing of the airplane

(b) Bird Strike Scenario Model

Fig. 1. Scenario Frame Described by Natural Language and bird strike model

3 Analysis of Clustering Model of provisions Based on Scenario

3.1 Correlations of Airworthiness Provisions and Quantification of Correlations

The correlations of airworthiness provisions can be divided into two categories: explicit correlation and implicit correlation. Explicit correlation refers to the reference relationship clearly indicated in the airworthiness provision. For example, for CCAR 25.1193 (cowling and nacelle skin), the cowling design must first meet the CCAR 25.1187 fire zone drainage and ventilation requirements; In addition, in the event of an engine power plant fire, compliance with this CCAR 25.1193(c) is based on compliance with the design requirements of CCAR 25.1191 for firewalls, as shown in Fig. 2.

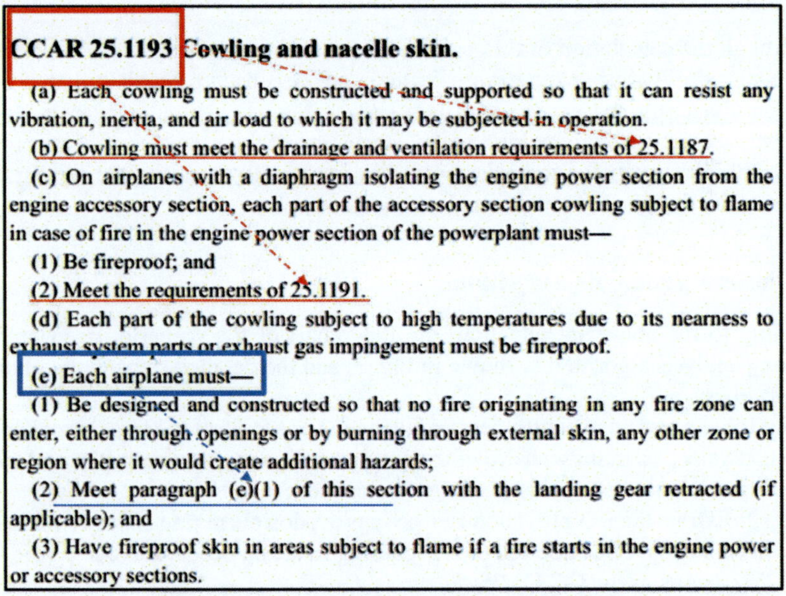
CCAR 25.1193 Cowling and nacelle skin.

(a) Each cowling must be constructed and supported so that it can resist any vibration, inertia, and air load to which it may be subjected in operation.

(b) Cowling must meet the drainage and ventilation requirements of 25.1187.

(c) On airplanes with a diaphragm isolating the engine power section from the engine accessory section, each part of the accessory section cowling subject to flame in case of fire in the engine power section of the powerplant must—

(1) Be fireproof; and

(2) Meet the requirements of 25.1191.

(d) Each part of the cowling subject to high temperatures due to its nearness to exhaust system parts or exhaust gas impingement must be fireproof.

(e) Each airplane must—

(1) Be designed and constructed so that no fire originating in any fire zone can enter, either through openings or by burning through external skin, any other zone or region where it would create additional hazards;

(2) Meet paragraph (e)(1) of this section with the landing gear retracted (if applicable); and

(3) Have fireproof skin in areas subject to flame if a fire starts in the engine power or accessory sections.

Fig. 2. Examples of references to CCAR25.1193

In fact, the correlation of the airworthiness provision is implicit more often, which is, although there is no explicit reference relationship in the text of the provision, there is a certain correlation in the technical content of the provision. For example, CCAR25.1423(a) requires the power supply system to provide continuous power to the in-flight PA system after the engine and APU are shut down or fail; CCAR25.1310 requires that the power supply system must be capable of supplying power to essential loads under all possible operating conditions and for all probable durations. These two provisions both have put forward requirements for the power supply logic and capacity of the power supply system, but there is no explicit reference relationship between these two provisions in the text.

The correlation between provisions may be strong or weak, so the parameter "Conn" is designed to quantify the degree of correlation between provisions. Assume that according to the provision clustering model established in the previous step, the participants of the scenario contained in provision A are described as the design feature set X = {x1, x2, x3 … xn}. The set of design features covered by provision B is Y = {y1, y2, y3, …yn} (the division of design features can be based on the each system unit of civil transport aircraft listed in ATA2200. ATA2200 is a technical information standard provided by the Air Transport Association of America, which contains uniform name and code of design features for various types of civil aircraft.),

If $X \cap Y = \emptyset$, then Provision A is not relevant to Provision B;

If $X \cap Y \neq \emptyset$, then provision A is related to provision B, and the parameter "Conn(A, B)" (the degree of correlation between provision A and B) is calculated as:

$$\text{Conn}(A, B) = \frac{X \cap Y}{X}, Conn(A, B) \in (0, 1] \tag{1}$$

Through the calculation of this parameter, provision clusters based on the overlapping relationship between the design features covered by the provisions in a certain provision clustering space with one item (target item) as the center can be formed. Moreover, according to the ranking of the correlation of the provisions, the priority that the airworthiness provision to be considered in a provision cluster to the current design features in the design process is actually determined.

3.2 Clustering Analysis Algorithm

According to the above introduction, the flow chart of the scenario-based provision clustering analysis algorithm is shown in Fig. 3, and the detailed steps in the algorithm are as follows:

Step 1: Set a certain airworthiness provision in CCAR-25 as the target provision CT;

Step 2: Select a certain scenario of the provision CT as a target scenario, and identify the scene participants as a set X (a design feature set);

Step 3: Extract the i-th provision Pi from a provision clustering space P, wherein the provision cluster space can be selected according to actual needs, and can be the whole CCAR-25 or parts of the CCAR-25;

Step 4: if successful, extract the application scenario set S of the provision established in advance, S = {S1, S2.. Sk}; otherwise, it indicates that all provisions in the current provision clustering space have been traversed, and the program exits;

Step 5: Extract the j-th scenario Sj in S;

Step 6: If successful, then extract its scenario participant set Yj, Yj = {yj1, yj2, … yjn}; otherwise, it means that all scenarios of the current provision Pi have been traversed, set Pi as the traversed provision, and then go to step 3;

Step 7: Traverse Yj, and calculate the parameter Conn between the Yj and the set X of the target scenario according to Formula 1;

Step 8: if Conn is more than 0, identify the current scene Sj as an relevant scenario of the target scenario, otherwise, turn to the step 9;

Step 9: Set he current scenario Sj as a traversed scene, and then turn to step 5;

Step 10: after the above steps is finished, output all the associated scenarios with the association parameter Conn greater than 0.

Extract the source provisions of these scenarios, which are the related provisions of the target provision CT, and then sort them according to the correlation degree of the provisions, and finally form a provision cluster centered on a target provision.

The above process is a clustering analysis process of establishing the association between provisions according to the correlation analysis between the corresponding scenarios of the provisions, and finally forming a provision cluster that restricts the design features of a certain civil aircraft.

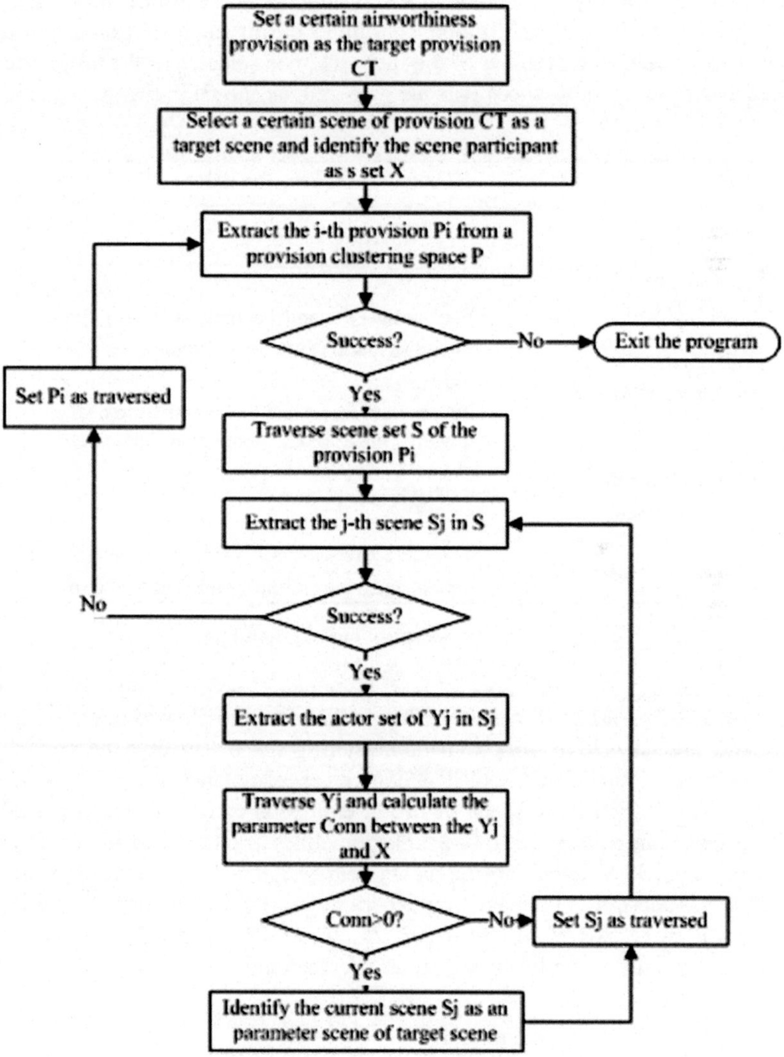

Fig. 3. Provision clustering algorithm

4 Implementation and Application

The method proposed in this paper is realized by using Visual Studio 2008 software development platform to complete the import and storage of provision scenario data and the operation of provision clustering algorithm.

In this paper, CCAR 25.951 Fuel System: General is selected as the target provision to introduce the clustering model and analysis method. In terms of physical structure, the fuel system mainly includes fuel storage system, fuel distribution system, fuel measurement and indication system and other sub systems. The fire protection requirements of the fuel system under various possible conditions during the operation of the aircraft should be considered. Here, the fire of the fuel system is selected as the target scenario, and the relevant provision requirements are gathered, as shown in Fig. 4.

Time	N/A	**Source Provision: CCAR25.951**
Location	Inside the aircraft	
Participant		
a) Design Object (including features)	Fuel storage system(fuel tank and ventilation) Fuel distribution system(fuel supply and transfer) Fuel measurement and indication system (fuel quantity, pressure and temperature indicate)	
b) Environment Element (including features)	TBD(Various elements)	
Status after interaction	The fuel system is ignited by some external action, and a part of the system starts to burn	

Fig. 4. Target Scenario: Fuel System Fire.

Here, CCAR-25 Part C "Structure", Part D "Design and Construction" and Part E "Powerplant" are selected as the provision clustering space. Traverse the provisions in the clustering space, such as the CCAR 25.561 emergency landing condition, in which CCAR25.561(c) describes the scene of fire or explosion caused by the penetration of cabin facilities or large masses into fuel tanks, pipelines or damage to adjacent systems, so the system-level ATA covered by the fuel system design requirements of this provision are 28 (X set); According to the existing experience and scenario analysis, the ATA set (Y set) covered by CCAR 25.561(c) is {03, 21, 24, 25, 26, 28, 30, 36, 38, 44, 52, 71}. According to Formula 1, the following result is calculated.

$$Conn(A, B) = \frac{X \cap Y}{X} = \frac{1}{12}$$

Therefore, CCAR 25.561(c) is a related provision of the target provision CCAR25.951, but the degree of correlation is low.

For another example, CCAR25.581 lightning protection covers the ATA set {14, 21, 23, 24, 26, 27, 28, 29, 30, 32, 34, 35, 46, 50}, which also intersects with the ATA features covered by CCAR 25.951. Therefore, it is also a related provision of the target provision. Then traverse the scenarios contained in other provisions in the provision clustering space to form the associated scenarios of the target scenario, some of which are shown in Fig. 5, so as to finally form the provision clustering of the target provision.

Time	N/A **Source Provision: 25.561(c)**
Location	On the runway
Participant	
a) Design Object (including features)	Fuel storage system(fuel tank and ventilation) Fuel distribution system(fuel supply and transfer)
b) Environment Element (including features)	Cabin facilities or large masses
Status after interaction	During the emergency landing, the fire caused by the penetration or strike of large masses into the fuel system

Time	N/A **Source Provision: 25.581**
Location	High altitude troposphere
Participant	
a) Design Object (including features)	Fuel storage system(fuel tank and ventilation) Fuel distribution system(fuel supply and transfer)
b) Environment Element (including features)	Air and clouds
Status after interaction	The fire of the fuel system caused by the lightning on the airframe

Time	N/A **Source Provision:25.631**
Location	High altitude
Participant	
a) Design Object (including features)	Fuel storage system(fuel tank and ventilation) Fuel distribution system(fuel supply and transfer)
b) Environment Element (including features)	Birds
Status after interaction	After the bird strike, the fuel system is punctured and a fire caused by the leakage of fuel

Time	N/A **Source Provision: 25.863**
Location	Inside the aircraft
Participant	
a) Design Object (including features)	Fuel storage system(fuel tank and ventilation) Fuel distribution system(fuel supply and transfer)
b) Environment Element (including features)	Fuel
Status after interaction	The flammable fluids or vapor caused by leakage of fluid system might be ignited by the fault of electrical circuits or overheating of equipment

Fig. 5. Scenarios related to fire protection of several fuel systems

It can be seen from the above example that CCAR 25.951 is related to CCAR 25.561 and CCAR 25.581. In the case of emergency landing and lightning strike, corresponding protection requirements are made for the scene of fuel system fire. Other relevant provisions for target scenarios in CCAR25, Subparts C, D, and E are shown in Table 1. These provisions, as a cluster of provisions on fuel system fire protection, are airworthiness requirements that must be considered in the design of aircraft fuel systems.

Table 1. Provisions related to the fire protection of the fuel

Serial number	Provision number and title
1	25.561(c) Emergency Landing Condition: General
2	25.581 Lightning Protection
3	25.601 Design and Construction: General
4	25.603 Materials
5	25.611 Accessibility provisions
6	25.631 Bird strike damage
7	25.721 Landing gear: General
8	25.863 Flammable fluid fire protection
9	25.869(a) Fire Protection: System
10	25.899 Electrical bonding and protection against static electricity
11	25.901 Powerplant: Installation
12	25.903 Engines
13	25.943 Negative acceleration
14	25.954 Fuel System Lightning Protection
15	25.963 Fuel tank: General
16	25.965 Fuel tank tests
17	25.967 Fuel tank Installation
18	25.969 Fuel tank Expansion space
19	25.973 Fuel tank filler connection
20	25.975 Fuel tank vents and carburetor vapor vents
21	25.979 Pressure fueling system
22	25.981 Fuel tank ignition prevention

5 Conclusion

Taking the capture of airworthiness requirements in civil aircraft design as an example, this paper studies the clustering model and analysis method of airworthiness provisions based on scenarios. The main contents include:

1. The concept of scenario is introduced to analyze and reconstruct the airworthiness provisions from the elements of the scenario, sort out the technical requirements of the provisions, and form the clustering space model corresponding to the provisions;
2. The concept and calculation method of the degree of correlation are put forward, the degree of correlation of provisions is quantified, the clustering analysis algorithm is developed and implemented, and the relevant provisions of target provisions are identified in the clustering space;
3. The effectiveness of the model and analysis method is verified by the fuel system fire scenario in CCRA25.951.

Through the scenario-based provision clustering modeling, the application scenarios of the provisions and their design requirements can be systematically identified, and through the clustering analysis, the applicable provision set of specific design features can be quickly retrieved and established, so as to improve the integrity and accuracy of the airworthiness requirements capture of civil aircraft.

References

1. Society of Automotive Engineers. SAE ARP 4754A Guidelines for development of civil aircraft and systems. SAE International, Warrendale, PA, USA (2010)
2. Civil Aviation Administration of China.CCAR-25-R4 Airworthiness standards: transport category airplanes. Civil Aviation Authority of China, Beijing (2011)
3. De Florio, F.: Airworthiness: an introduction to aircraft certification. In: Zhang, S., Ke, P., Pan, Q., (eds.) Transl. Beihang UniversityPress, Beijing (2011)
4. Alsabti, K., Ranka, S., Singh, V.: An efficient k-means clustering algorithm. In: Proceedings of the First Workshop on High Performance Data Mining, Ng (1995)
5. Zhang, T., Ramakrishnan, R., Livny, M.: BIRCH: a new data clustering algorithm and its application. Data Min. Knowl. Disc. **1**(2), 141–182 (1997)
6. Karypis, G., Han, E.H., Kumar, V.: Hierarchical clustering using dynamic modeling. Computer **32**(8), 68–75 (1999)
7. Birant, D., Kut, A.: ST-DBSCAN: an algorithm for clustering spatial temporal data. Data Knowl. Eng. **60**(1), 208–221 (2007)
8. Ng, R.T., Han, J.: CLARANS: a method for clustering objects for spatial data mining. IEEE Trans. Knowl. Data Eng. **14**(5), 1003–1016 (2002). https://doi.org/10.1109/TKDE.2002.1033770
9. Potts, C.: ScenIC: a strategy for inquiry-driven requirements determination. In: 4th IEEE International Symposium on Requirements Engineering, pp. 58–65. IEEE Computer Society Press, Los Alamitos, CA (1999)
10. Sutcliffe, A.G.: Scenario-based requirements engineering. In: Proceedings of the 11th IEEE International Conference, Montery Bay, USA, 8–12th, pp. 320–329. IEEE Computer Society Press (2003)
11. Kurakawa, K.: A scenario-driven conceptual design information model and its formation. Res. Eng. Des. **15**, 122–137 (2004)
12. Deng, Y.M., Tor, S.B., Britton, G.A.: A computerized design environment for functional modeling of mechanical products. In: Proceedings of the Fifth ACM Symposium on Solid Modeling Ann Arbor, Michigan, USA, pp. 1–12 (1999)
13. Pahl, G., Beitz, W.: Engineering Design: A Systematic Approach, 2nd edn. Springer, London (1996)
14. Darlington, M.J., Culley, S.J.: Current research in the engineering design requirement. Proc. Inst. Mech. Eng. Part B J. Eng. Manuf. **216**(3), 375–388 (2002). https://doi.org/10.1243/0954405021520049
15. Zelin, L., Zhongyan, Q.: Chen Yong Identification and acquisition for conceptual design requirements of civil aircraft based on conceptual graphs. Comput. Integr. Manuf. Syst. **21**(10), 2548–2557 (2015)

Accurate Reliability Allocation Algorithm of Airborne FC Network Based on Analytic Hierarchy Process

Shikui Wang[1](✉), Dongxu Sun[1,2], and Yuxuan Zhang[1]

[1] AVIC Xi'an Aeronautics Computing Technique Research Institute, Xi'an 710065, Shaanxi Province, China
kuishiwang@126.com

[2] Northwestern Polytechnical University, Xi'an, Shaanxi Province 710072, People's Republic of China

Abstract. As the backbone data network of integrated modular avionics system, the reliability of Fibre Channel network has a great influence on the reliability of the system. At present, network reliability allocation only considers hardware devices, which cannot accurately allocate the reliability index of airborne FC network. This paper analyzes the characteristics of FC airborne network and the influencing factors of node reliability, and propose the model and calculation method of FC network node comprehensive importance based on the analytic hierarchy process. By using AGREE method, an accurate network reliability allocation algorithm is designed, and the reliability allocation of non-redundancy and double redundancy FC network nodes is completed. The verification results show that compared with the traditional reliability allocation method, the reliability allocation method in this paper has the advantages of comprehensive factors, high allocation accuracy and strong practicability.

Keywords: FC Airborne Network · Network Reliability · Comprehensive Importance · Reliability Allocation

1 Introduction

Fibre Channel (FC) network has the advantages of high bandwidth, low latency and flexible topology, and has become the first choice for integrated modular avionics system communication network [1]. Compared with the traditional 1553B bus, the number of nodes, network scale and technical complexity of the FC network of the avionics system have increased by an order of magnitude, and the communication data business of the integrated avionics system undertaken has also increased by dozens to hundreds of times. Therefore, the reliability of the FC network directly affects the reliability of the entire avionics system. The design and evaluation technology of the reliability of the airborne FC network has become a hot research topic [2, 3]. Product reliability allocation is an important work in reliability throughout the stage of comprehensive demonstration, scheme design and engineering development. Traditionally, the reliability allocation of

Chinese Society of Aeronautics and Astronautics (Ed.): CASTC 2023, LNME, pp. 444–454, 2024.
https://doi.org/10.1007/978-981-99-8861-7_46

airborne FC network products only considers the reliability of hardware components, and adopts the reliability allocation method based on series-parallel block diagram, which is not closely combined with the characteristics of airborne FC networks and cannot achieve the accurate reliability allocation of network [4–7].

In order to meet the demand for precise reliability allocation of airborne FC network, this paper first analyzes the FC network composition structure of avionics system and the functional characteristics of network devices, establishes the importance evaluation model of network nodes based on Analytic Hierarchy Process, in which the factors affecting the importance of network nodes such as traffic, software (including logic) and hardware are integrated, and proposes the calculation method of comprehensive importance of network nodes, and then combines AGREE reliability allocation method [8, 9], the accurate allocation of reliability index of FC network is completed. The feasibility and good results of the algorithm are verified through the network reliability assignment of two FC topologies, non-redundancy and double-redundancy.

2 Composition Characteristics of Airborne FC Network

FC network is an open network that can provide a variety of topologies, services and high-level protocol mapping. With high bandwidth and good scalability, the switched FC network can meet the requirements of integrated avionics systems for high-bandwidth and highly reliable interconnection and interoperability between devices. Therefore, switched FC network is preferred as the backbone data network of avionics system, this paper takes it as the object of reliability research.

The FC switch network is mainly composed of FC switches, FC nodes and interconnection links [8]. According to the overall design index, network importance, network reliability index and other requirements of avionics system, FC network can be designed as non-redundancy or dual redundancy. Figure 1 and Fig. 2 show the structure of a non-redundant FC switching network and a dual-redundant FC switching network.

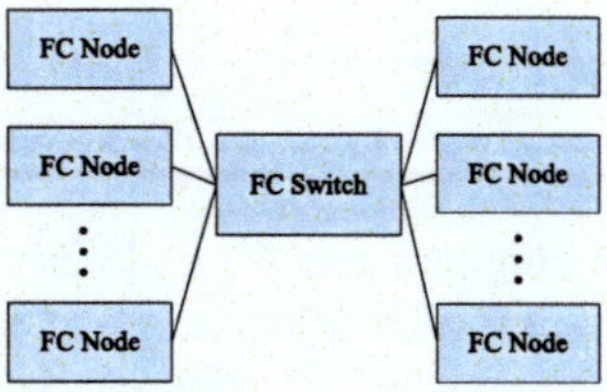

Fig. 1. Non-redundant FC switch network structure diagram

The airborne FC network protocol is divided into FC-0, FC-1, FC-2, FC-3 and FC-4 layers. The FC-0 layer defines the physical interface, transmission medium, etc. The FC-1 layer defines the transmission coding and decoding, etc. The FC-2 layer specifies the framing and flow control protocol, the FC-3 layer provides a set of public services, and the FC-4 layer defines the avionics environment (FC-AE) high-level protocol. To realize the above complex protocol stack, the FC switch and FC node need not only hardware components, but also software and FPGA logic.

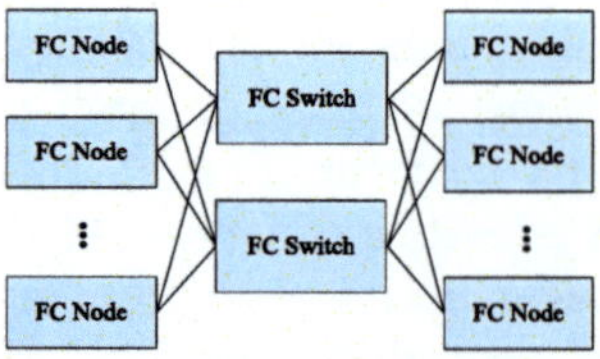

Fig. 2. Structure of dual-redundant FC switch network

In addition to the differences in hardware and software, the types and flows of network communication services undertaken by each device are also different. Therefore, it is necessary to integrate the above factors and accurately calculate the importance of FC network devices, so as to accurately distribute network reliability to each switch and node of the network.

3 Comprehensive Importance Analysis Model of FC Network Nodes Based on Analytic Hierarchy Process

Analytic Hierarchy Process [10, 11] is an effective method to convert non-quantitative factor in system engineering into quantitative analysis, which can reduce the difficulty and workload of analysis. Referring to the analytic hierarchy process (AHP), the flow method of FC network node comprehensive importance evaluation is designed, which can be divided into four steps: constructing hierarchical relationship diagram of influencing factors of FC network node comprehensive importance, constructing node pairwise comparison matrix, calculating absolute importance of influencing factors in criterion layer, and evaluating and calculating FC network node comprehensive importance. The calculation block diagram of FC network node importance evaluation is shown in Fig. 3.

3.1 Construct the Hierarchical Relationship Diagram of FC Network Node Importance

Evaluation of the comprehensive importance of FC nodes is to measure the impact of this node on the whole network and synthesize each factor that affects the importance of FC nodes. The comprehensive importance of FC nodes is calculated by constructing the hierarchical relationship diagram of influencing factors of the comprehensive importance [8]. The factors affecting the comprehensive importance of FC network nodes should be comprehensively analyzed from the composition of FC network nodes, the carrying of communication services, the running software and hardware, etc.

Node importance index is a comprehensive index, which contains different influencing factors. It is analyzed by means of hierarchical relationship diagram of influencing factors. Through the analysis of FC protocol, FC network topology, FC network traffic and FC node information, the importance of nodes should include three factors: traffic, software and hardware.

First, flow factors, referring to FC standard protocol, combined with practical engineering applications, it is summarized into emergency message, event message, flow

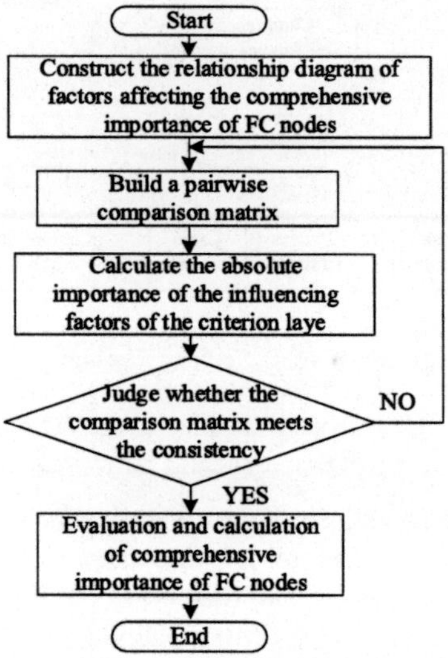

Fig. 3. Flow chart of FC node importance evaluation calculation

message. Emergency message refers to the message with the highest priority. In network scheduling, it is scheduled by the switch according to the priority. Specifically, when multiple messages arrive at a port of the switch at the same time, the switch gives priority to scheduling the first emergency message. Event messages refer to the messages that each node communicates normally, which are used most frequently and in the most quantity in communication. Flow message is a long message, which is longer than the first two, and the maximum length can reach 16Mb. It has the lowest priority and is scheduled last in scheduling. The more messages a node carries and the more high-priority messages, the higher the importance of the node, otherwise, the importance of the node is lower.

Secondly, the influence of software factors on the importance of a node is mainly reflected by the amount of code. The code residing in the node is larger, which shows that the node software realizes more functions, so the node is more important, and vice versa. Specifically, it includes the code amount of application software, the code amount of driving software and the code amount of logic software.

Thirdly, the influence of hardware factors on node importance is measured by the total number of components used in the node and the number of types of components. The more complex the components used in a node and the more the number of components, the higher the importance of the node, otherwise, the importance is lower.

The AHP block diagram for constructing the importance level of FC network nodes is shown in Fig. 4.

3.2 Build Node Pairwise Comparison Matrix

In the hierarchical relationship diagram of influencing factors of FC network node importance constructed above, each element is not in the same scale, for example, the statistics

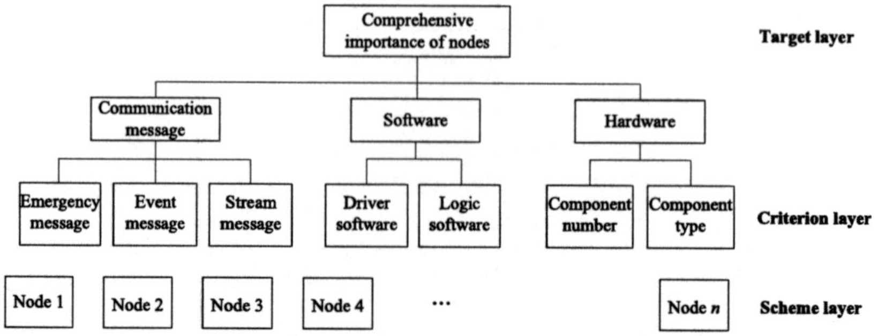

Fig. 4. Hierarchy analysis block diagram of FC node importance

of emergency message and driving software code amount are not in the same unit, so it cannot be calculated directly. Therefore, it is necessary to compare the data of different nodes under the same criterion layer, and then carry out unified processing, so that the elements of the criterion layer can be further calculated. The result of comparing the data of different nodes in the criterion layer is the comparison matrix A of two nodes.

The pairwise comparison matrix A is used to express the importance of each node relative to other nodes with respect to the elements of each criterion layer. In the constructed comparison matrix, the importance of Node *i* relative to Node *j* under the influence factor is expressed by the elements a_{ij} in Matrix *A*.

In the process of node comparison, this paper takes the FC backbone network of avionics system as the reference data of the index system, and then uses the scale of 1–9 provided by the analytic hierarchy process to process and quantify the data linearization, namely $w_1, w_2, \cdots, w_n$. Therefore, according to the definition of pairwise comparison matrix, the elements $a_{ij} = \frac{w_i}{w_j}$ in comparison matrix *A* can be obtained. See formula (1) for pairwise comparison matrix *A*.

$$\mathbf{A} = \begin{bmatrix} \frac{w_1}{w_1} & \frac{w_1}{w_2} & \cdots & \frac{w_1}{w_n} \\ \frac{w_2}{w_1} & \frac{w_2}{w_2} & \cdots & \frac{w_2}{w_n} \\ \cdots & \cdots & \cdots & \cdots \\ \frac{w_n}{w_1} & \frac{w_n}{w_2} & \cdots & \frac{w_n}{w_n} \end{bmatrix} \quad (1)$$

3.3 Determination of Absolute Importance of Criterion Layer

The pairwise comparison matrix *A* reflects the relative importance of network nodes about criteria layer index, but cannot express the absolute importance of each network node about criteria layer index. However, in the process of calculating the comprehensive importance of network nodes, it is necessary to combine the absolute importance of criteria layer index with its relative weight to synthesize the comprehensive importance of network nodes, so the pairwise comparison matrix *A* obtained above cannot be directly applied to the calculation of comprehensive importance of network nodes.

According to the algorithm of calculating the absolute importance of criterion layer provided by AHP, the absolute importance of criterion layer is calculated for each network node by calculating the eigenvector corresponding to the maximum eigenroot of pairwise matrices *A*. See formula (2) for calculation formula.

$$A \cdot W = \lambda_{\max} \cdot W \tag{2}$$

where, $\lambda_{\max}$ is the maximum eigenvalue of the pairwise comparison matrix A, and W is its corresponding maximum eigenvector.

3.4 Evaluation and Calculation of Comprehensive Importance of FC Network Nodes

According to the weight value of criterion layer indicators and the obtained importance of each node on the corresponding criterion layer elements, the comprehensive importance of each FC network node can be obtained and calculated according to formula (3).

$$A = c_1 \cdot \sum_{i=1}^{m_1} b_i \cdot \lambda_i + c_2 \cdot \sum_{j=1}^{m_2} b_j \cdot \gamma_j + c_3 \sum_{z=1}^{m_3} b_z \cdot \mu_z \tag{3}$$

wherein, A is the comprehensive importance of FC network nodes, and λ_i, γ_j, μ_z are the absolute importance of each index in the criterion layer of communication message, resident software, and device hardware. b_i is the weight value of the criterion layer index. c_1, c_2, and c_3 respectively represent the weight value of communication message, software index, and hardware index in the comprehensive importance. m_1 is the number of communication message indicators, m_2 is the number of software indicators, and m_3 is the number of hardware indicators.

To sum up, the evaluation and calculation of the comprehensive importance of FC network nodes is transformed into the comprehensive quantitative analysis of the communication messages, resident software and equipment hardware of the node. Through the comparison matrix, the determination of the absolute importance of the criterion layer, and the establishment of the calculation formula of the importance, the quantitative evaluation and calculation of the comprehensive importance of FC network nodes are solved.

4 Network Reliability Allocation Algorithm Based on Comprehensive Importance

In the design process of avionics system FC network topology, the characteristics of non-redundancy network and double redundancy network are distinguished. If a certain network node fails, it will cause network failure for non-redundant networks; For dual redundancy network, because there is backup, the FC network still works normally through the backup node. Double redundancy network and non-redundancy network have higher reliability.

The comprehensive importance of each node in FC network is obtained by analytic hierarchy process, and AGREE method is selected to allocate the reliability of FC network [9–11]. The AGREE method assigns reliability according to the importance and complexity of each component, which is suitable for systems composed of several series nodes, and the series nodes can be composed of parallel nodes. Therefore, this method is suitable for reliability allocation of airborne FC networks.

Assuming that the lifetime of FC network nodes obeys exponential distribution, the specific allocation method is as follows:

Let the whole network reliability index specified by FC network at t time be R_s, and see formula (4) for calculating the failure rate assigned to the *i-th* component.

$$\lambda_i = \frac{C_i(-\ln R_s)}{A_i \cdot t_i} \cdot k \tag{4}$$

where C_i is the complexity of the node. This paper focuses on the importance of nodes. Therefore, it is assumed that the complexity of each node is defined as *1/n*, where n is the number of nodes. Ai is the comprehensive importance obtained previously, and k is a positive proportional parameter.

Further, according to the definition of exponential distribution, the reliability of each node allocation is shown in formula (5):

$$R_i = e^{-1/(\lambda_i \cdot t)} \tag{5}$$

It can be seen from Formula 4 and Formula 5 that the value of k is inversely proportional to the reliability of node allocation. However, the principle of reliability allocation is $\prod R_i \geq R_s$ and $\prod R_i - R_s \leq \delta$, where δ is the required accuracy of calculation, which is 10^{-12} in this paper. Therefore, it is necessary to solve the optimal solution of k by dichotomy to meet the reliability allocation requirements.

For double redundancy FC network, the nodes that backup each other are regarded as parallel subsystems and calculated. According to the above, the reliability of subsystems is obtained, and then the reliability of backup nodes is calculated according to the reliability calculation method of parallel systems. See Formula (6) for specific calculation.

$$R_b = 1 - \sqrt[m]{1 - R_i} \tag{6}$$

where R_b is the reliability allocated to the backup node, and m is the number of backups of the node.

5 Example Verification and Analysis

In this paper, the reliability allocation problem of ten-node FC network is analyzed and illustrated as an example. The accuracy of the distribution result is set to 10^{-12}. The number of communication messages, the amount of software code and the statistical data of hardware carried by each node are shown in Table 1, Table 2 and Table 3.

5.1 Comparison of Reliability Allocation Based on Traditional Method and Comprehensive Importance Method

For non-redundant FC network data, it is assumed that FC network reliability requirement is 0.99. Traditionally, only hardware factors are considered, and the reliability allocation

Table 1. Node bearer communication message data

Node	Emergency	Event	Flow
1	15	76	12
2	9	146	36
3	20	65	8
4	3	89	10
5	35	213	26
6	12	105	12
7	36	260	35
8	17	142	9
9	10	80	11
10	12	78	7

Table 2. Node resident software code size

Node	Application	Driver	Logic
1	6962	15843	9630
2	5020	19520	9856
3	9721	15630	8896
4	6584	8969	10059
5	15326	21363	15930
6	10056	10562	12563
7	17390	26099	23015
8	11950	15232	9963
9	4540	8610	10563
10	10350	9634	7963

results obtained by AGREE calculation are based on the comprehensive importance of nodes in this paper, and the reliability allocation results calculated by AGREE are shown in Fig. 5.

The results of the calculations in Fig. 5 show that the reliability assigned to Node 5 is the lowest among the 10 nodes, based solely on the total number of hardware. Based on the results of reliability allocation, it can be found that the importance of Node 5 is only lower than that of Node 7 because Node 5 undertakes more communication tasks and the software complexity is high, so the reliability of allocation is higher and more reasonable. It can be seen from Fig. 5 that the high comprehensive importance of Node 5 is in line with the actual network design, while the reliability allocation results calculated by the comprehensive importance of nodes consider the influencing factors of

Table 3. Node hardware statistics

Node	Component Number	Component Type
1	756	23
2	453	16
3	962	29
4	958	35
5	532	29
6	956	18
7	1569	35
8	895	29
9	685	18
10	785	20

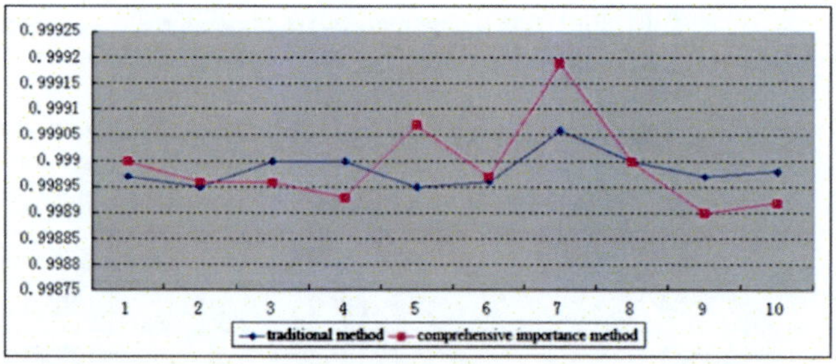

Fig. 5. Comparison of reliability allocation results based on traditional method and comprehensive importance method

nodes more comprehensively, and the reliability allocation results are more reasonable and accurate, which has engineering value.

5.2 Distribution Verification of Different Reliability Requirements

The reliability requirements of FC network are set to 0.999, 0.9995 and 0.9999 respectively. The calculation results of comprehensive importance of each node are shown in Fig. 6. For non-redundant FC network, the reliability allocation results of 10 nodes are shown in Fig. 7. For dual redundancy FC network, the reliability allocation results of 10 nodes are shown in Fig. 8. Comparing Fig. 8 with Fig. 7, it can be seen that under the same FC network reliability requirements, the reliability requirements of each node allocation are reduced, which is consistent with the expectation of double redundancy design.

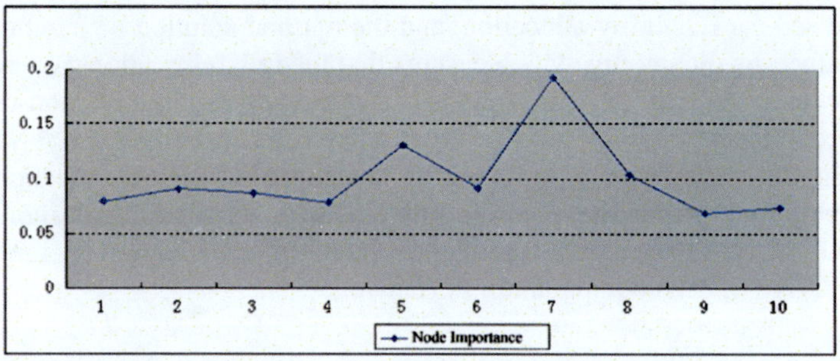

Fig. 6. Calculation results of comprehensive importance of fc network nodes

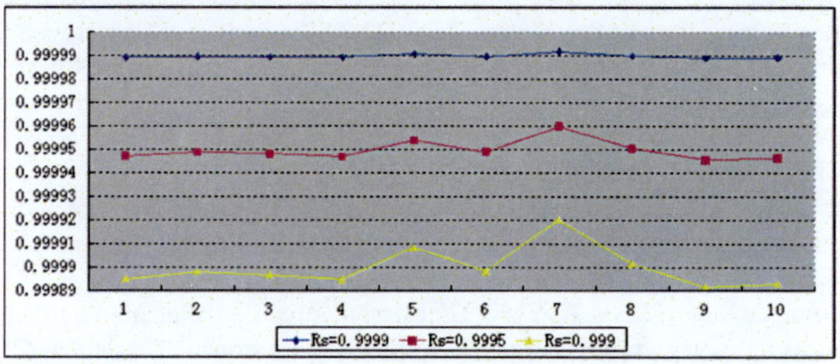

Fig. 7. Reliability distribution results of non-redundant FC network

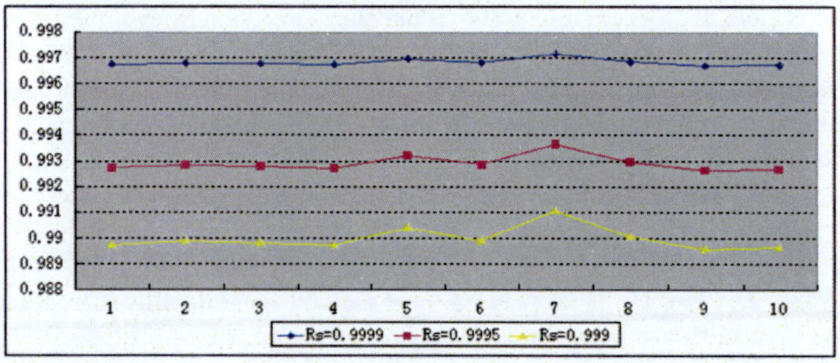

Fig. 8. Reliability distribution results of dual-redundant FC network

6 Conclusion

Based on the analysis of the factors influencing the integrated importance of nodes in airborne FC network, this paper puts forward a strategy to measure the integrated importance of nodes in FC network by three factors: communication message, software and hardware. By using the analytic hierarchy process, the calculation model of the integrated importance of nodes in FC network is established, and the quantitative evaluation results of the integrated importance of nodes in FC network are obtained. The AGREE method is used for reliability allocation, and the optimal solution of reliability allocation is obtained by dichotomy, which ensures that the reliability allocation accuracy of network nodes reaches 10^{-12}. Through the calculation of airborne FC non-redundancy and dual redundancy topology with different reliability requirements, it is verified that the reliability allocation method proposed in this paper has higher allocation accuracy than the traditional method, and is more suitable for the engineering design and application requirements of airborne FC network, which lays a foundation for improving the reliability design of each equipment in FC network.

References

1. Wang, K.S., Zhang, Y., Wang, L., Ning, Y.J.: Design and implementation of FC-AE-ASM data acquisition and forwarding system. In: 2022 7th International Conference on Intelligent Computing and Signal Processing (ICSP), pp. 1374–1377. IEEE, Xi'an, China (2022)
2. Qiao, J., et al.: Design and implementation of FC-AE fault simulation instrument. In: 2020 IEEE International Instrumentation and Measurement Technology Conference (I2MTC), pp 1–6. IEEE, Dubrovnik, Croatia (2020)
3. Liang, F., Zhao, G., Cao, S.: Design of heterogeneous FC-AE-1553 network. In: 2014 IEEE International Conference on Control Science and Systems Engineering, pp. 130–134. IEEE, Yantai, China (2015)
4. Liu, Z., Sun, Y., Xuan, J., Xu, Z., Zhao, G.: Mission reliability allocation method considering multiple factors for repairable systems. J. Beijing Univ. Aeronaut. Astronaut. **45**(4), 834–840 (2019). (in Chinese)
5. Zhang, E., Chen, Q.: Multi-Objective reliability redundancy allocation in an interval environment using particle swarm optimization. Reliab. Eng. Syst. Saf. **145**, 83–92 (2016)
6. Mellal, M.A., Zio, E.: System reliability-redundancy allocation by evolutionary computation. In: 2017 2nd International Conference on System Reliability and Safety (ICSRS). IEEE (2017)
7. Dong, H., Bao, X., Zhao, T., Xin, Y.: Integrated modular avionics fault propagation model establishment and reliability allocation. In: 2018 12th International Conference on Reliability, Maintainability, and Safety (ICRMS). IEEE, Shanghai, China (2018)
8. Li, Q., Li, R.: Reliability allocation of fibre channel network in avionics system. In: 2015 IEEE 12th International Conference on Electronic Measurement & Instruments, pp. 197–201. IEEE, Qingdao, China (2015)
9. Li, R., Wang, J., Liao, H., Huang, N.: A new method for reliability allocation of avionics connected via an airborne network. J. Netw. Comput. Appl. **48**, 14–21 (2015). https://doi.org/10.1016/j.jnca.2014.10.005
10. Coffey, L., Claudio, D.: In defense of group fuzzy AHP: a comparison of group fuzzy AHP and group AHP with confidence intervals. Expert Syst. Appl. **178**, 114970 (2021). https://doi.org/10.1016/j.eswa.2021.114970
11. Liu, Y., Eckert, C.M., Earl, C.: A review of fuzzy AHP methods for decision-making with subjective judgements. Expert Syst. Appl. **161**, 113738 (2020). https://doi.org/10.1016/j.eswa.2020.113738

Helicopter Bionic Landing Gear Design and Verification Technology Based on Multi-link Structure

Jia Ren[1](✉), Jizhen Wang[1], Xiaochuan Liu[1], Feng Gao[2], and Ke Yin[2]

[1] AVIC Aircraft Strength Research Institute, Dianzi No. 2 Road 86, Xi'an 710065, China
renjiabest@163.com

[2] Shanghai Jiao Tong University, Dongchuan Road 800, Shanghai 200240, China

Abstract. For usage scenes and landing stability of the helicopter, a kind of bionic leg landing gear based on multi-link structure is proposed to replace the original landing gear. The landing gear can effectively improve the helicopter's complex terrain take-off and landing adaptability. In this paper, the design concept of the bionic landing gear is given, and the research of design and verification technology is carried out. Firstly, based on the six-leg scheme, the structural design of the bionic leg landing gear is given. Then, the drive/control system and the collaborative control method is proposed based on the landing gear structure. Finally, based on this prototype, the tests in laboratory and in real environment are completed for function verification. These research indicate that lightweight design of multi-link bionic landing gear is achieved. And through its drive/control system, the buffering of the landing gear landing process and the stability control of the fuselage are effectively realized. Compared with the traditional landing gear, this kind of landing gear has the advantages of landing attitude adjustment, and complex terrain adaptation.

Keywords: Bionic Landing gear · Terrain Adaptation · Design and Verification

1 Introduction

The vertical takeoff and landing aircraft (VTOL) is a kind of aircraft which can take off and land vertically without taxiing, which includes helicopter, rotorcraft and the aircraft with tilting wings. Nowadays, the VTOL is used more because of good mobility and lower landing space requirements [1, 2]. In the VTOL using area, the helicopter can be used in remote detection, emergency rescue, fire and disaster relief, and whose structure and control designing are the research hotspots currently [3]. But in field, take-off and landing processes of the helicopter are challenging because of the uncertain ground environment. The helicopter needs to autonomously take off and land on the rugged terrain, slope topography, step floor, and other fields with non-flat surface. And the helicopter needs to autonomously move by walking instead of flying, to avoid frequent take-off and landing in short distance. These new abilities of the helicopter have great significance in practical application and intelligent development, because the environmental adaptability and ground operation capability of the helicopter will be significant improved.

Chinese Society of Aeronautics and Astronautics (Ed.): CASTC 2023, LNME, pp. 455–468, 2024.
https://doi.org/10.1007/978-981-99-8861-7_47

Based on the requirements described above, a special kind of landing gear is necessary and particularly important to replace the original landing gear, for the features of complex terrain landing and taking-off. By watching the take-off and landing processes of birds and insects in the nature, the method of bionics can be used in the landing gear designing. In addition, multi-legged robot technology provides reference for the landing gear designing. So the helicopter with this kind of bionic landing gear can be realized the abilities of complex terrain adaptation and ground walking movement. The increased attendance and survivability of the helicopter have important value in field exploration, rescue, and landing at high sea condition [4].

Similar studies have been carried out abroad. Such as the most famous "Robotic Landing Gear" by DARPA, has been held on a 65 kg unmanned helicopter for laboratory verification test. And it will be used on a 250 kg helicopter, a 3500 kg manned helicopter [5, 6]. Besides, other research findings carried out by Zurich University of Applied Sciences [7], by Edinburgh Napier University [8] and by Solkovo Institute of Technology [9]. In the articles, their physical prototypes designed successively and simplified laboratory verification tests were showed. And the technical maturity of them may reach TRL 4 or lower. However, the research in this area in China has been developed in recent years. A few concept designs and virtual simulation are shown in articles. The design concepts are proposed by Nanjing University of Aeronautics & Astronautics [10] and by Hefei Polytechnic University [11]. Based on these, AVIC Aircraft Strength Research Institute develops the multi-link leggy landing gear designing and complete feasibility study [12, 13].

In this article, a kind of bionic landing gear based on multi-link structure is designed for the helicopter to solve the problems of landing stability and high bearing capacity and mobile convenience. Based on the designed structure, the physical prototype is developed. And then the methods of integration and verification are proposed through the physical prototype used on an unmanned helicopter.

2 Overall Design of Bionic Landing Gear

In this part, the overall design of the bionic landing gear based on multi-link structure is described and departed into two sections including construct collectivity design and normal workflow design.

2.1 Construct Collectivity Design

Based on the application scenarios in military and civilian, the bionic landing gear is defined as a kind of intelligent landing gear which fuses the bionic designing and control algorithm to realize adaptive landing. The bionic landing gear is consisting of mechanical load-carrying legs, modular drive joints, control and driving system, sensing system, and terrain recognition and modeling system. So the helicopter will have the abilities of landing and walking on complex terrain.

Based on the applications and the design principles, the bionic landing gear has six legs and in the matrix arrangement. Because this kind of six-leg arrangement can increase the stability margins in longitudinal and lateral directions, besides, the helicopter can walk in a triangular gait under the leg arrangement [14]. So the overall configuration of the bionic landing gear can be given here as shown in Fig. 1. Each leg is in a structure of two-stage with three joint degrees of freedom. And these joint degrees can make each leg move freely in three-dimensional space and thus the landing gear has the ability to adapt to complex terrain.

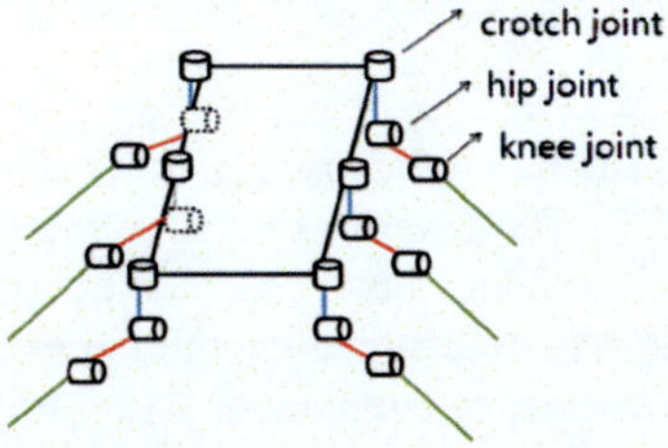

Fig. 1. Overall configuration diagram of the bionic landing gear

Then based on the overall configuration of the bionic landing gear, and considering the application scenarios, the design indexes are given, shown as following.

Design Weight. The design weight of the whole bionic landing gear system is smaller than a quarter of the helicopter maximum takeoff weight.

Landing Terrain Adaptability. The touchdown area of six feet is no more than 20° slope or no more than 200 mm steps.

2.2 Normal Workflow Design

Generally, the landing process of the helicopter can be divided into three stages which are high-altitude high-speed landing, low-altitude low-speed landing, and hover and touch adjustment [15]. The detailed introductions of three stages are shown as following.

Landing Stage of High-Altitude High-Speed. Target landing area is selected through the provide data from the terrain recognition and modeling system. The helicopter lands vertically reaching above the target area and the stage ends when the helicopter at the altitude of 20 m. At this stage, the landing speed of the helicopter is relatively fast.

Landing Stage of Low-Altitude Low-Speed. Under the altitude of 20 m, the vertically landing speed of the helicopter is relatively slow. And this stage ends when a foot of bionic landing gear touches the ground. At this stage, the bionic landing gear should realize the predetermined track achieving.

Stage of Hover and Touch Adjustment. At this stage, the landing gear begins to use impedance control to eliminate the influence of error and interference. All the feet touch the ground in a soft way and the forces on the feet will gradually close to the reference forces. The engine of the helicopter will be powered down when all the feet touch the ground. The impedance control is used to keep the helicopter stability until the landing process is finished.

3 Structural Design of Bionic Landing Gear

Based on the overall design, the kind of bionic landing gear based on multi-link structure is designed. Based on the structural model of one-leg, the structural design of the bionic landing gear is expounded in design of mechanism, drive and control system.

3.1 Kinetic Modeling of One-Leg System

Considering the carrying capacity and applications, the one-leg mechanism of the landing gear should be a hybrid mechanism composed of a five-bar mechanism and a series rotating pair, shown as Fig. 2. The length of each coupler is l_{ab}, l_{bc}, l_{cd}, l_{be}, l_{de}, l_{ef}, and couplers DE and EF are consider as two collinear couplers with one consolidation E. The rotating pair at point A is used to realize the leg overall rotation. The parallel-crank mechanism BCDE is consist of couplers of BC, BE, CD and DE, and the drive bars BC and BE are separately driven by assemblies of motors and reducers. The angle between BC and positive OX is α, and that between BE and positive OX is β, α and β is hypothetically known. The leg coordinate is defined as L and whose origin is at point A.

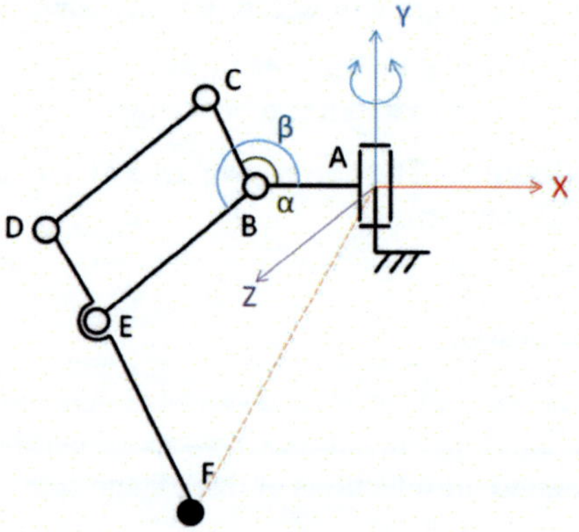

Fig. 2. Mechanism diagram of one-leg

Based on the assumption of rigidity, the kinematics and dynamics models of the one-leg can be constructed by analytic method. So suppose the azimuth (angle between the coupler and positive OX) of couplers, solve process of foot coordinate (x_f, y_f) is shown as following.

In sagittal plane, the spatial position relationship between couplers can be obtained according to vector theory, shown as

$$l_{af}\, e^{i\phi_{af}} = l_{ab} + l_{be}\, e^{i\phi_{be}} + l_{ef}\, e^{i\phi_{ef}} \tag{1}$$

where, let $l_{ab} = d$, $l_{cd} = l_{be} = Lt$, $l_{ef} = Ls$, according to the coordinate decomposition, the coordinate of foot F can be get as follows.

$$\begin{bmatrix} x_f \\ y_f \end{bmatrix} = \begin{bmatrix} Lt \cdot \cos\beta - Ls \cdot \cos\alpha - d \\ Lt \cdot \sin\beta - Ls \cdot \sin\alpha \end{bmatrix} \tag{2}$$

And in leg coordinate, define θ as side slip angle, which is the rotation angle of sagittal plane, the coordinate of foot F can be shown as follows.

$$P_f^L = \begin{bmatrix} x_f^L \\ y_f^L \\ z_f^L \end{bmatrix} = \begin{bmatrix} x_f \cdot \cos\theta \\ y_f \\ -x_f \cdot \sin\theta \end{bmatrix} \tag{3}$$

So, when the coordinate of foot F is known, inverse kinematics solve of the one-leg mechanism can be used to get the drive angles of the joints, and the drive angles are used for motion control. And the side slip angle θ can be got by Eq. (4).

$$\theta = a\tan(-\frac{z_f^L}{x_f^L}) \tag{4}$$

Then, the coordinate of foot F in sagittal plane can be got by Eq. (5).

$$\begin{bmatrix} x_f \\ y_f \end{bmatrix} = \begin{bmatrix} \frac{x_f^L}{\cos\theta} \\ y_f^L \end{bmatrix} \tag{5}$$

And then, through the geometric relationship of the couplers, the drive angles β and α can be got by Eq. (6) and Eq. (7).

$$\beta = a\tan(\frac{y_f}{x_f + d}) - \arccos(\frac{Lt^2 + l_{bf}^2 - Ls^2}{2Lt \cdot l_{bf}}) \tag{6}$$

$$\alpha = \beta - \arccos(\frac{Ls^2 + Lt^2 - l_{bf}^2}{2Lt \cdot Ls}) \tag{7}$$

According to design and test requirements, the length for each coupler can be defined as $l_{ab} = 117$ mm, $l_{bc} = 100$ mm, $l_{cd} = 251$ mm, $l_{df} = 325$ mm. Then, it must to realize the bionic landing gear's movement in the feasible movement space, and ensure the landing and movement abilities on complex terrain ground of the unmanned helicopter. The drive angles of joints should be changed in specific scopes to maintain the leg structure in a parallel-crank mechanism. The constraint condition of the drive angles is shown as follows.

$$\begin{cases} 90^\circ \le \alpha \le 180^\circ \\ 90^\circ \le \beta \le 270^\circ \\ -45^\circ \le \theta \le 45^\circ \\ 0^\circ \le \alpha - \beta + 180^\circ \le 180^\circ \end{cases} \tag{8}$$

And then under this constraint, through scanning every feasible drive angle, the workspace of the foot F is shown as Fig. 3.

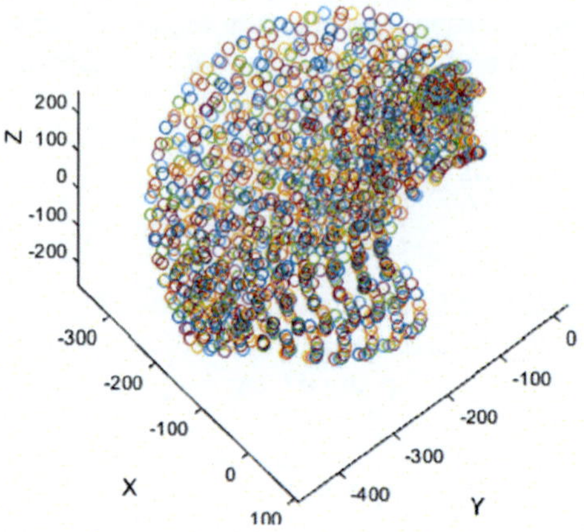

Fig. 3. Workspace axonometric drawing of one-leg foot F

3.2 Mechanical Design of the Landing Gear

Select a 180 kg unmanned helicopter as the verification platform for mechanical design of landing gear and for the performance tests. The mechanical design of the one-leg and the whole landing gear structure should consider of the reducing of lightweight. The mechanical design of one-leg is the basis of structural design because that six legs have the same structure and connect with the helicopter through the connecting plate.

Based on the mechanism diagram of one-leg shown in Fig. 2, the couplers of BE, BC, CD and DF are made of lightweight materials. Cylindrical tubes of BE, DF and CD are made of carbon fibre, and other parts and joints are made of aviation aluminum alloy which processed through Computer Numerical Control (CNC). The frames for installing the drive units are made of aviation aluminum alloy and designed according to the structures. The foot is designed as a rubber hemispherical structure to ensure the uniform contact force and large contact area of foot and ground. All the parts are assembled together by designing appropriate threaded holes. The assembled design drawing of one-leg is shown as Fig. 4.

Then consider the connection between the bionic landing gear and the unmanned helicopter, a connector is designed with many functions as connection conversion, landing gear installation, and control system storage and flight vibration isolation. So shown as Fig. 5, the main body of the connector is made of carbon fiber board, and the part connected with the helicopter is made of aluminum alloy, and the rubber blocks are installed to reduce and isolate vibration. The control system is placed in the empty warehouse between two layers of carbon fiber boards.

To ensure the process stability of landing and walking, six legs of bionic landing gear are distributed at the front, the middle and the rear of the connector. And the design drawing of the connector with six bionic legs installed on the unmanned helicopter is shown as Fig. 6. And the workspaces for three projection planes are shown as Fig. 7. It can be seen that the bionic landing gear can complete the functions of one-leg side away of 100°, standing height of 400 mm, and leg spread of 1400 mm. So the bionic landing

Fig. 4. The assembled design drawing of one-leg

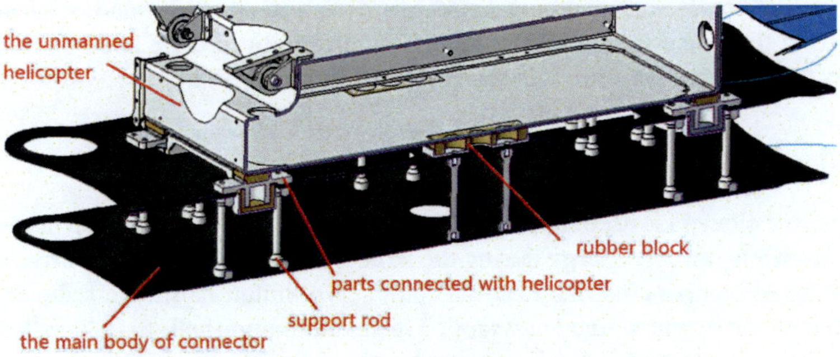

Fig. 5. The design drawing of connector to helicopter

gear can be used to content the functions of walking and adaptive landing, and meet the terrain requirements.

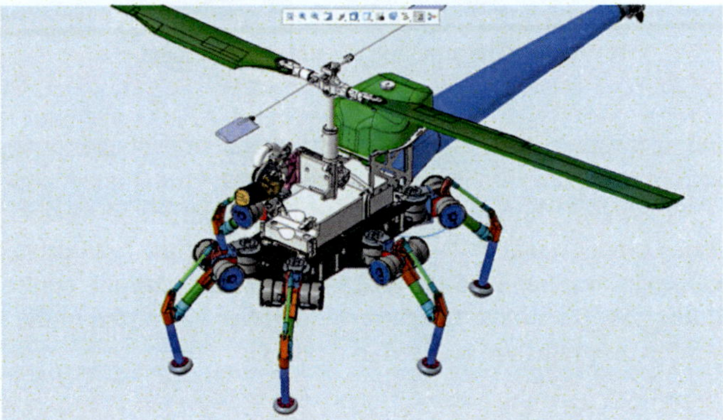

Fig. 6. The design drawing of bionic-leg landing gear system

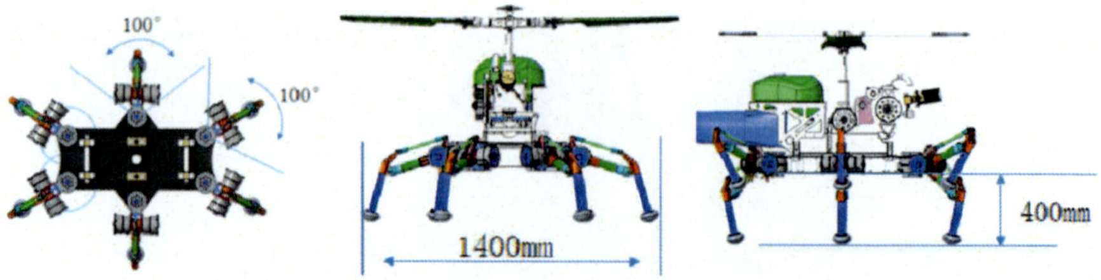

Fig. 7. The workspaces for three projection planes

4 Design of Drive and Control System

The bionic landing gear is consisting of mechanical structures and drive and control system. For the completed mechanical structure designing, the drive and control system is given to fit the function realization. And the lightweight of drive and control system is the key issues to be considered in the system design.

4.1 Design of Joint Drive Element

The joint drive element is used to realize the compliance force control when landing gear actively absorbing impact energy during the helicopter landing. The joint drive element is consisting of components like external rotor servo motor, harmonic reducer, torque sensor, and encoder, and which are wrapped in an aluminum shell, to reduce the weight [16]. The structure of the joint drive element is shown as Fig. 8.

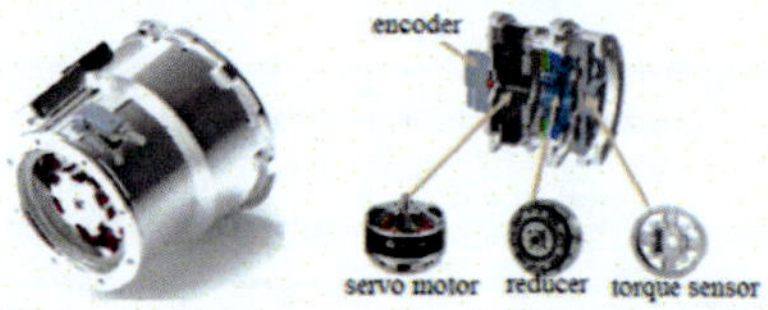

Fig. 8. The structure of joint drive element

The method of torque control is used here to realize impact energy absorbing, and whose purpose is to make the joint drive element have the virtual impedance mechanical characteristics. It achieves through force closed-loop control and to reach the desired torque at the same time. To realize the control algorithm of the joint drive element, the control system program is realized by reference generator, torque controller, current controller and low-pass filter. And the current controller is realized in the servo driver, and other controller programs are realized in the PLC.

4.2 Design of Bionic Landing Gear Control System

The control system of bionic landing gear is realized by a modular IPC computer installed with PLC control software. This modular IPC computer communicates with torque sensor signal acquisition cards and motor servo drivers through EtherCAT bus. In addition, it communicates with IMU through USB serial bus. The torque sensor signal acquisition cards are used to collect the torque signals from torque sensors in the joint driver elements. The network topology of the bionic landing gear control system is shown as Fig. 9. And the modular IPC computer connects to the upper computer, so we can use GUI of the upper computer to realize the functions of dynamic calculation, hardware drive and inter-system communication.

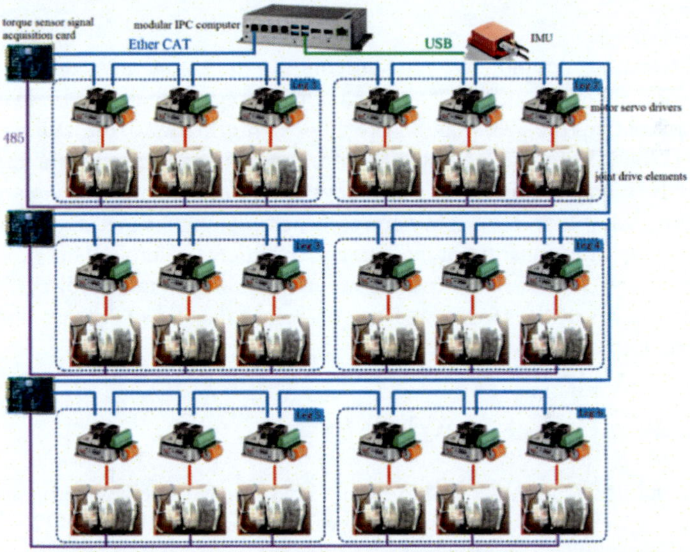

Fig. 9. The network topology of bionic landing gear control system

4.3 System Integration and Collaborative Control

In order to realize the autonomous landing of the unmanned helicopter through bionic landing gear, system integration of flight control system, bionic landing gear control system and terrain recognition system is used to form a unified control system and realize the collaborative control of these three control systems. The system combination diagram is shown as Fig. 10. The flight control system and the terrain recognition system and the bionic landing gear control system use the same power supply and this approach is conductive to data interaction and weight reduction of the whole system. Under the condition that all three systems work reliably, cooperative control and fusion design can be realized by establishing system communication and data link.

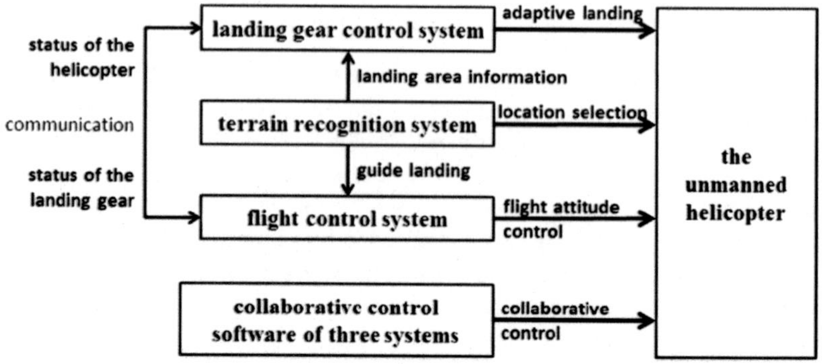

Fig. 10. The system combination diagram

Firstly, terrain recognition and modeling during the autonomous or manual control stage of the unmanned takeoff and landing should be done through the terrain recognition system, which obtains the size, slope and geological information of the landing area based on the point cloud. Then, the landing area information is transferred to the flight control system and the bionic landing gear control system by communication via serial port to guide landing. After that, the flight control system completes the takeoff and landing control and route determination of the unmanned helicopter through the ground station to complete flying. At the same time, the bionic landing gear control system used to control the attitudes of the bionic landing gear, and to achieve adaptive landing of the unmanned helicopter by motion control.

5 Verification Test of the Whole System

The design and verification method of bionic landing gear is given here, and based on this method, the weight of the bionic landing gear physical prototype is 42 kg and 23% of the helicopter maximum takeoff weight. The weight can meet the technical requirement of the weight proposed before. After completing the construction of the bionic landing gear physical prototype, the bearing function and the adaptive landing ability of the prototype are tested through laboratory tests and real environment landing tests. Here, the methods and results of laboratory tests and real environment landing tests are given.

5.1 Laboratory Tests of the Bionic Landing Gear

Laboratory landing tests are necessary methods to verify the design and control effects of the bionic landing gear. Laboratory landing tests are carried out though constructing typical terrain including flat and slope topographies, and though a six-degree motion platform to simulating ship motion. Based on these tests in laboratory, the bearing capacity and adaptive landing function of the bionic landing gear should be reflected.

When testing, the weight of the helicopter is simulated through counterweights of 180 kg. The falling velocity of the landing gear is 1 m/s and the horizontal velocity is 0 m/s. Under the landing condition of flat ground, photo of the bionic landing gear landing process is shown as Fig. 11. The following Fig. 11 shows that (1) is landing readiness, (2) and (3) are aerial descent, (4) is touch ground moment, (5) is the middle position of moving to the lowest position (6), (7) is the middle position of moving to the final stable position (8).

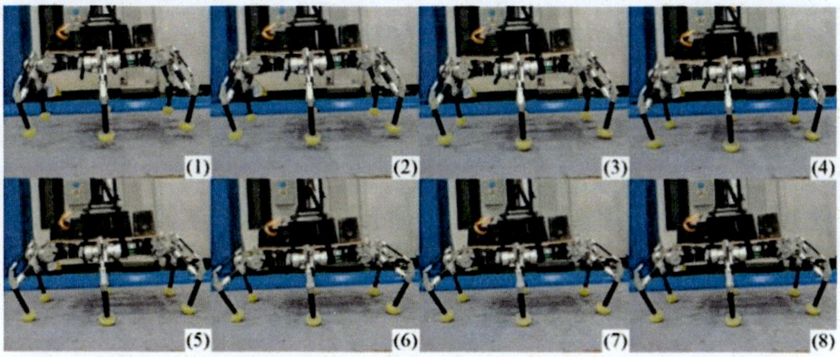

Fig. 11. Landing test of bionic landing gear on flat ground

When the terrain conditions meet the landing requirements, the bionic leg landing gear completes posture preset based on the results of terrain recognition and modeling. Under the slope ground of 20°, photo of the bionic landing gear landing process is shown as Fig. 12. The following Fig. 12 shows that (1) is landing readiness, (2) and (3) are aerial descent, (4) is all legs touch ground moment, (5) is the middle position of moving to the lowest position (6), (7) is the middle position of moving to the final stable position (8).

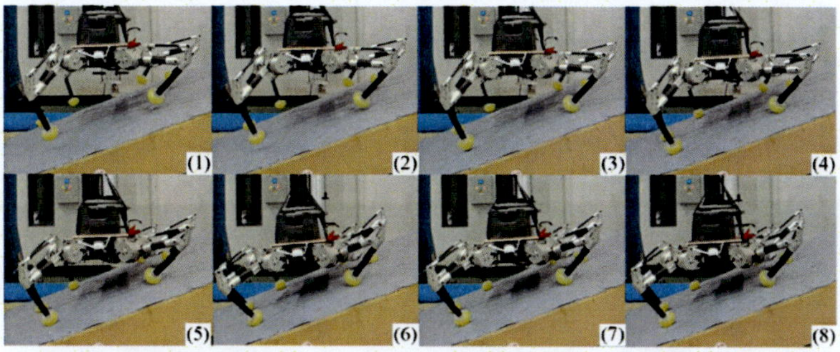

Fig. 12. Landing test of bionic landing gear on slope ground

After landing tests on build terrain, the simulation landing test of the unmanned helicopter with bionic landing gear on shaking ship is realized by simulating the ship motion through six degree platform. The bionic landing gear is always adjust the posture to adapt the motion of the ship before landing until touch the ship surface. And photo of the bionic landing gear landing process is shown as Fig. 13. The following Fig. 13 shows that (1) is pre landing status, (2) and (3) are aerial descent, (4) is all legs touch platform moment, (5) is the lowest position, (6) is the final stable position, (7) and (8) are always adjust the posture of legs to adapt the platform motion.

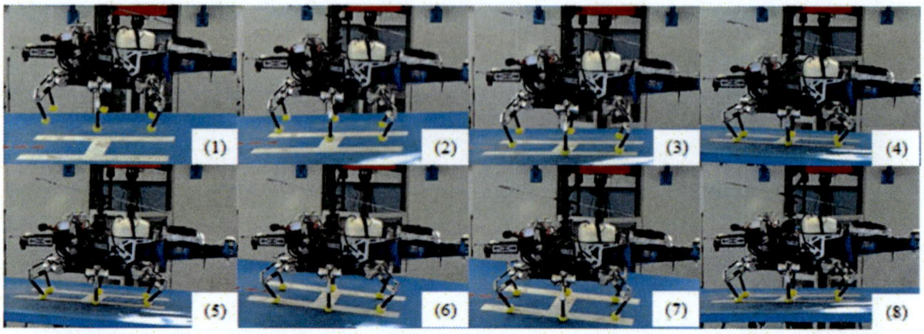

Fig. 13. Landing test of bionic landing gear on simulated dynamic ship surface

5.2 Real Environment Landing Tests of the Bionic Landing Gear

After landing tests in laboratory, the bionic landing gear has been proven to have stability when landing, whose terrain recognition system and control system can help it achieve the functions. The landing tests is to verify the landing cushioning ability of the bionic landing gear in real environments through complex terrain landing. And at the same time, the effects of system integration and collaborative control can be presented and validated in the real environment landing tests.

As the test platform, the unmanned helicopter with bionic landing gear is conducted flight and landing tests many times. It can smooth landing based on terrain recognition and modeling results. The photo of bionic landing gear landing process in real environment is shown as Fig. 14. The following Fig. 14 shows that (1) is flight status, (2) is pre landing status, (3) is all legs touch ground moment, (4) and (5) are adjustment process, (6) is the final stable status.

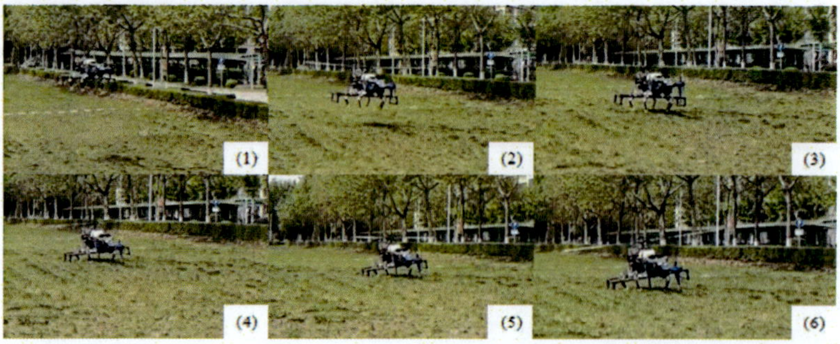

Fig. 14. Landing test of bionic landing gear on outfield real terrain

6 Conclusion

In this article, a new kind of bionic landing gear based on multi-link structure is proposed and the design process and system integration method are discussed here. The bionic landing gear is designed based on the landing stability and high bearing capacity of the unmanned helicopter, and effectively improve the terrain adaptability of the helicopter. Then the physical prototype is developed for the tests in laboratory and real environment. From the article, the following conclusions can be got.

This new kind of landing gear based on multi-link structure has advantages in landing stability and high bearing capacity and mobile convenience. And the bionic landing gear has the functions of autonomous movement, landing attitude adjustment, complex terrain adaption and collapsible structure.

The joint drive element and matched drive and control system constitute the basis of the control system of the bionic landing gear. And based on the flight control system, bionic landing gear control system and terrain recognition system, methods of system integration and collaborative control are proposed to realize the function of the landing gear.

Through the landing tests in laboratory and in real environments by the unmanned helicopter test platform, the rationality of the structural design and the effectiveness of the control have been proven. And the design and verification method of bionic landing gear can be used for all types of vertical takeoff and landing aircraft like the multi-rotor UAV and the helicopter.

References

1. Spanoudakis, P., Tsourveloudis, N.C., Valavanis, K.P.: Design specifications for an unmanned VTOL. In: Proceedings of the 2004 IEEE International Conference on Robotics and Automation, pp. 3616–3621. IEEE Society, New Orleans (2004)
2. Do, K.D., Jiang, Z.P., Pan, J.: Global output feedback tracking control of a VTOL aircraft. In: Proceedings of the 42th IEEE Conference on Decision and Control, pp. 4914–4919. IEEE Society, Hawaii (2003)
3. Xue, P.: Research on Automatic Takeoff and Landing Control Technology for Unmanned Helicopter. Nanjing University of Aeronautics and Astronautics, Nanjing (2012). (in Chinese)
4. Wu, P., Shi, Z., Yan, P.: Research on the unmanned helicopter carrier landing automatically. Ship Sci. Technol. **41**(05), 148–152 (2019). (in Chinese)
5. Kiefer, J., Ward, M., Costello, M.: Rotorcraft hard landing mitigation using robotic landing gear. J. Dyn. Syst. Meas. Contr. **138**(3), 2302–2308 (2016)
6. Manivannan, V., Langley, J.P., Costello, M., et al.: Rotorcraft slope landings with articulated landing gear. In: AIAA Atmospheric Flight Mechanics (AFM) Conference, p. 5160 (2013)
7. Stolz, B., Brödermann, T., Castiello, E., et al.: An adaptive landing gear for extending the operational range of helicopters. In: The Proceedings of 2018 IEEE/RSJ International Conference on Intelligent Robots and Systems (IROS), pp. 1757–1763 (2018)
8. Boix, D.M., Goh, K., Mcwhinnie, J.: Helicopter lands on uneven terrain by means of articulated robotic legs-modelling, simulation and control approach. In: The Proceedings of 2018 IEEE/ASME International Conference on Advanced Intelligent Mechatronics (AIM), pp. 1282–1287 (2018)

9. Sarkisov, Y.S., Yashin, G.A., Tsykunov, E.V., et al.: DroneGear: a novel robotic landing gear with embedded optical torque sensors for safe multicopter landing on an uneven surface. IEEE Robot. Autom. Lett. **3**(3), 1912–1917 (2018)
10. Wang, X., Nan, Y.: Helicopter landing gear design for special terrain based on bionics. Aircr. Des. **34**(04), 46–48 (2014)
11. Sang, Z., Jiang, B., Yang, J., et al.: Design of bionic landing gear for small unmanned helicopter. Mod. Manuf. Technol. Equip. **231**(02), 79–81 (2016)
12. Ren, J., Wang, J., Liu, X.: Design and analysis of terrain-adaptive bionic landing gear system. In: The Proceedings of 2018 IEEE/37th Chinese Control Conference (CCC), pp. 4205–4209. IEEE CCC, Wuhan (2018)
13. Ren, J., Wang, J., Liu, X.: Landing design and control simulation in complex terrain conditions. Aeronaut. Sci. Technol. **31**(09), 84–90 (2020). (in Chinese)
14. He, W., Dong, Y., Sun, C.: Adaptive neural impedance control of a robotic manipulator with input saturation. IEEE Trans. Syst. Man Cybern. Syst. **46**(3), 334–344 (2017)
15. Gao, B., Zeng, G., Liao, Z., et al.: Auto-takeoff and auto-landing control design for small unmanned helicopter. Helicopter Technol. **177**(4), 10–14 (2013). (in Chinese)
16. Gao, F., Yin, K., Sun, Q., et al.: Design and control of legged leaping robot in lunar exploration. Flight Control Detect. **3**(4), 1–7 (2020). (in Chinese)

Tracking Control of Combined Power Aerospace Vehicle in Climbing Phase

Hongyan Zang, Changsheng Gao(✉), and Wuxing Jing

Harbin Institute of Technology, Harbin 150001, Heilongjiang, China
gaocs@hit.edu.cn

Abstract. The trajectory tracking control method of the climbing section of the combined powered aerospace vehicle is studied. Firstly, the configuration of the combined power aerospace vehicle is designed and some parameters are given with reference to the same type of aircraft at home and abroad. On this basis, the longitudinal motion model of the climbing section, the thrust model and aerodynamic model are established. Secondly, the particle swarm optimization algorithm (PSO) and the automatic disturbance rejection control method (ADRC) are combined to design the angle of attack tracking controller of the climbing section. The parameters of the automatic disturbance rejection controller are optimized by using the PSO algorithm. Finally, the simulation of the angle of attack rejection tracking controller is carried out. The simulation results show that the tracking speed of the designed controller is significantly improved, which can effectively reduce the disturbance. Moreover, the controller has a good tracking effect on the input angle of attack command when the aircraft actuator is limited.

Keywords: Combined power aerospace vehicle (CASV) · Tracking control · Automatic disturbance rejection control (ADRC) · Particle swarm optimization algorithm (PSO)

1 Introduction

With the rapid development of aerospace technology, the space launch and transportation systems with fast launch speed, strong mobility and low cost have become a hot demand in various countries [1]. In the 70s of the 20th century, major aerospace technology countries successively proposed the concept of reusable launch vehicle (RLV) [2]. After entering the new century, the breakthrough of aerospace technology gave birth to a new generation of RLV vehicles, Combined power aerospace vehicles (CASV) [3]. CASV are reusable, capable of horizontal/vertical launch, and most of them use single-stage or two-stage orbit mode, which can greatly reduce launch costs [4]. Due to its multi-terrain launch capability, it is less restricted by the launch window than traditional rockets, and can be urgently entered into orbit in wartime. At the same time, the high-speed performance of the CASV enhances its penetration capability, allowing it to conduct rapid reconnaissance of high-threat areas or carry weapons to carry out strike missions [5].

Chinese Society of Aeronautics and Astronautics (Ed.): CASTC 2023, LNME, pp. 469–483, 2024.
https://doi.org/10.1007/978-981-99-8861-7_48

Considering that during the flight of the CASV, the flight airspace span is large, the flight environment is complicated, and it will be subject to unknown disturbances. Combined power engine mode switching will also bring disturbance to the aircraft. Moreover, there may be limitations in the CASV actuator during actual flight [6, 7]. All these situations bring challenges to the accurate tracking and control of CASV. The design of CASV controller should still have good control effect under the conditions of strong uncertainty inside and outside the system [8, 9].

In recent years, some scholars have carried out related research on the control of CASV. National Aeronautics and Space Administration (NASA) has designed the controller for the X-43A aircraft using the variable gain control method and conducted flight tests [10]. For CASV which have high nonlinearity and strong coupling, the variable gain control method has certain limitations, and it is difficult for the controller to ensure the global stability of the system [11]. At the same time, due to the large flight envelope range of CASV, variable gain control is not conducive to engineering realization [12]. Wang et al. [13] linearized the hypersonic vehicle based on the dynamic inverse compensation method, and designed a random robust feedback controller to improve the robustness of the control system due to the possible uncertainty of parameters in the aircraft modeling. Tourne et al. [14] designed the sliding mode controller for hypersonic vehicles in 1998, which is the earliest application of sliding mode control in the aerospace field. Then, Xu et al. [15] accurately linearized the longitudinal motion model of hypersonic aircraft based on the feedback linearization method, and designed the traditional sliding mode controller. Due to the uncertain parameters and unmodeled terms in the system, the robustness of the traditional sliding mode controller will be insufficient, so the adaptive sliding mode observer is used to observe it, which improves the robustness of the system to a certain extent [16]. Montaseri et al. [17] combined the Lyapunov redesign method in each subsystem of the backstep to improve the robustness of the backstep application in strict feedback systems with unknown static parameters, matched or mismatched interference. Zhang et al. [18] designed a neural network-based backstep controller for a class of strict feedback nonlinear systems by using the complex Lyapunov function. Farrell et al. [19] used adaptive backstep method and valuation theory to design the controller for the three-channel flight control system, and used filter estimation to solve the derivative of the virtual control quantity. Fiorentini et al. [20] proposed a robust adaptive dynamic inverse state feedback control method based on dynamic inverse technology and backstepping method, which adopted the strategy of additional ailerons to eliminate the minimum phase behavior of the elastic model of hypersonic aircraft.

Motivated by the above discussions, the objective of this paper is to study the tracking control method of the climbing section of CASV. The rest of this paper is organized as follows. Section 2 describes the establishment of the combined power aerospace vehicle model, including the vehicle configuration, the climbing section motion model, the thrust model and the aerodynamic model. The design method of the automatic disturbance rejection controller based on particle swarm optimization parameters is introduced in Sect. 3. The simulation verification and analysis of the designed controller is given in Sect. 4. Finally, the conclusion section ends this paper in Sect. 5.

2 Combined Power Aerospace Vehicle Model

2.1 Aircraft Configuration

This paper refers to the SR-72 concept aircraft of Lockheed Martin and the reconnaissance/strike hypersonic aircraft concept released by Boeing in 2018, and designs a CASV with a wing-body fusion aerodynamic configuration which have a protruding back, a large sweep delta wing and a single vertical tail aerodynamic layout [21]. The CASV uses independently working left and right lowering ailerons δ_{el}, δ_{er} and rudders δ_r as pneumatic rudders. Lifting ailerons δ_{el}, δ_{er} are equivalent to elevators when deflecting the same angle in the same direction at the same time and they function as ailerons when differential deflection. The thrust system adopts a TBCC cycle engine. The schematic diagram of its basic layout are shown in the Figs. 1 and 2.

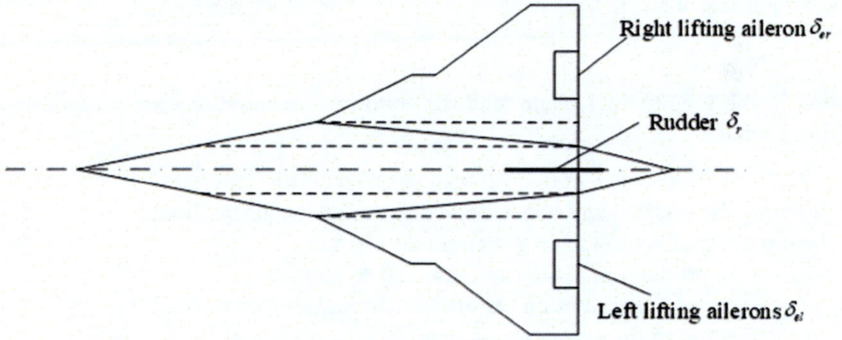

Fig. 1. Top view of the CASV

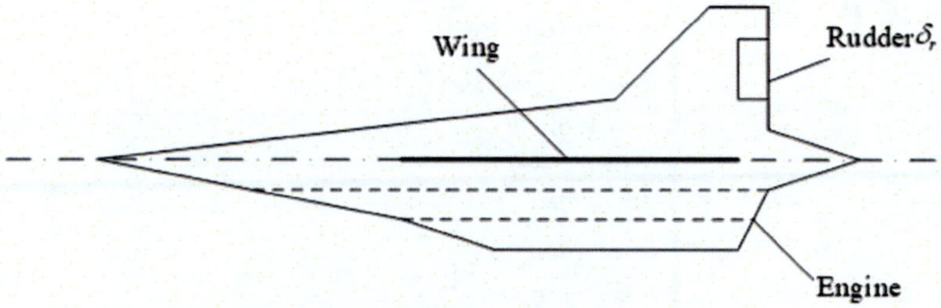

Fig. 2. Side view of the CASV

After design calculation, the overall parameters of this type of aircraft are shown in Table 1.

2.2 The Longitudinal Movement Model of CASV

During the climb section of CASV, the longitudinal motion state changes sharply, and the transverse lateral motion state changes very little. It can be regarded as climbing

Table 1. Overall parameters of CASV

Parameters	Value
Reference area $S\left(\mathrm{m}^2\right)$	92
Average pneumatic chord length $c(m)$	24.4
Wingspan b(m)	12
Length L(m)	32
Total mass of takeoff $m_{tol}(kg)$	30000
Empty weight m_f (kg)	12000
Take-off moment of inertia $I_x(\mathrm{kg} \cdot m^2)$	119605
Take-off moment of inertia $I_y(\mathrm{kg} \cdot m^2)$	769000
Take-off moment of inertia $I_z(\mathrm{kg} \cdot m^2)$	805395

motion in the longitudinal plane, and the following assumptions are made about its motion model:

(1) CASV is regarded as a pure rigid body without elastic vibration.
(2) Ignoring the rotation and non-spherical perturbations of the Earth.
(3) Ignore the shaking of the propellant in the engine.
(4) The inertia products of CASV are assumed to be zero.
(5) The transverse lateral motion parameters are regarded as zero.
(6) The deflection of the elevator mainly affects the pitch motion parameters.

Using the above assumptions, the longitudinal motion equation of CASV is obtained as follows:

$$\begin{cases} \dot{V} = \frac{P\cos\alpha - X}{m} - g\sin\theta \\ \dot{\theta} = \frac{P\sin\alpha + Y}{mV} + \left(\frac{V}{r} - \frac{g}{V}\right)\cos\theta \\ \dot{h} = V\sin\theta \\ \dot{\alpha} = \omega_z - \dot{\theta} \\ \dot{\omega}_z = \frac{M_z}{I_z} \\ \dot{m} = -dm \\ r = h + R \end{cases} \tag{1}$$

where V is the aircraft speed, α is the angle of attack, P is the thrust, X is the drag, Y is the lift, m is the mass of the aircraft, θ is the inclination angle of the speed, h is the height of the aircraft, ω_z is the pitch angular velocity, M_z is the pitch moment, I_z is the moment of moment, dm is the second flow of the engine, r is the geocentric distance. According to the mission requirements of the climb section of the combined power aerospace vehicle, the tracking controller can be designed based on the above motion equations.

2.3 Thrust Model

CASV uses a parallel turboram combined cycle engine (TBCC) as the power system. The engine performance parameters are fitted with reference to the J-58 engine parameters, which are related to the flight altitude and Mach number [22], and its thrust model is such as

$$\begin{cases} P = f(H, Ma) \\ Dm = f(H, Ma) \end{cases} \tag{2}$$

where P is the thrust and Dm is the fuel consumption rate. The combined engine is divided into three working stages, the turbojet stage, the turbojet-ramjet mixing stage and the ramjet stage.

2.4 Aerodynamic Model

During the climb section of CASV, as the speed gradually increases, the influence of aerodynamic force and aerodynamic moment also gradually increases, so the modeling of aerodynamic parameters is also crucial. Since the research of related projects is confidential by various countries, there is currently no detailed aerodynamic data disclosed. This paper refers to the aerodynamic data of the same type of aircraft, and establishes the aerodynamic model as follows [22]:

$$\begin{cases} X = qSC_X \\ Y = qSC_Y \\ M_z = qSbC_{mz} \end{cases} \tag{3}$$

where X is drag forces, Y is lift forces, M_z is pitch moments. In this paper, the aerodynamic parameters of the aircraft at Mach 1.2–10 and 0–20° angle of attack are fitted to the following aerodynamic parameters and the relationship between the rudder surface, angle of attack, side slip angle and Mach number:

$$\begin{cases} C_X = f(\alpha, \beta, Ma, \delta_r, \delta_a) \\ C_Y = f(\alpha, Ma, \delta_e) \\ C_{mz} = f(\alpha, Ma, \delta_e) \end{cases} \tag{4}$$

where $\delta_e, \delta_a, \delta_r$ are the elevator, aileron and rudder deflection angles respectively:

$$\begin{cases} \delta_e = \delta_{er} = \delta_{el} \\ \delta_a = \delta_{er} - \delta_{el} \end{cases} \tag{5}$$

3 Design of Angle of Attack Tracking Controller Based on POS-ADRC

3.1 Angle of Attack Equation of State

During the climbing process, the pitch channel attitude of CASV directly affects its climbing flight state, so this section designs an attitude controller for the pitch channel of the aircraft to verify the ability of the elevator deflection angle to control the attitude of the pitch channel. From Eq. (1), it can be seen that the attitude control of the pitch channel of the aircraft is actually a problem of controlling the angle of attack. In the climbing section, it is believed that the angle of attack of the aircraft α is controlled only by the declination angle of the elevator δ_e, thus transforming the angle of attack control problem into a single-input single-output problem. The differential equation for the angle of attack α obtained by Eq. (1) is:

$$\dot{\alpha} = \omega_z - \frac{P\sin\alpha + L}{mV} - \left(\frac{V}{r} - \frac{g}{V}\right)\cos\theta \tag{6}$$

Order $f_1 = -\frac{P\sin\alpha + L}{mV} - \left(\frac{V}{r} - \frac{g}{V}\right)\cos\theta$, then there is

$$\dot{\alpha} = \omega_z + f_1 \tag{7}$$

Derivation of the above formula:

$$\ddot{\alpha} = \dot{\omega}_z + \dot{f}_1 \tag{8}$$

Substitute $\dot{\omega}_z = \frac{M_z}{I_z}$, $M_z = 0.5\rho V^2 S\left[C_{mz}^0 + C_{mz}^{\alpha}\alpha + C_{mz}^{\alpha^2}\alpha^2 + C_{mz}^{\delta_e}\delta_e\right]$ into the above equation

$$\ddot{\alpha} = \frac{0.5\rho V^2 S\left[C_{mz}^0 + C_{mz}^{\alpha}\alpha + C_{mz}^{\alpha^2}\alpha^2\right]}{I_z} + \dot{f}_1 + \frac{0.5\rho V^2 S C_{mz}^{\delta_e}\delta_e}{I_z} \tag{9}$$

Order $f_2 = \frac{0.5\rho V^2 S\left[C_{mz}^0 + C_{mz}^{\alpha}\alpha + C_{mz}^{\alpha^2}\alpha^2\right]}{I_z} + \dot{f}_1$, $c = \frac{0.5\rho V^2 S C_{mz}^{\delta_e}}{I_z}$, then there is

$$\ddot{\alpha} = f_2 + c\delta_e \tag{10}$$

From above, the equation of state suitable for the design of the ADRC controller with respect to the angle of attack α is obtained:

$$\begin{cases} \dot{\alpha} = \dot{\alpha} \\ \ddot{\alpha} = f_2 + c\delta_e \end{cases} \tag{11}$$

Among them, f_2 is the total disturbance of the system, δ_e is the elevator deflection angle, and c is the control variable coefficient.

3.2 Principle of Active Disturbance Rejection Control

Active Disturbance Rejection Control (ADRC) is developed from traditional PID control technology and has strong robustness and anti-interference performance. ADRC attributes the uncertainties and unmodeled dynamics of the system to the "sum perturbation" of the system. Then it is estimated by the extended state observer (ESO), and real-time dynamic feedback compensation is carried out to achieve linearization of the system. Finally, the nonlinear state error feedback law (NLSEF) is used to suppress the compensation residuals and improve the control performance of the control system [23].

The standard second-order control system equation of state for single-in, single-output is:

$$\begin{cases} \dot{x}_1 = x_2 \\ \dot{x}_2 = f + cu \end{cases} \tag{12}$$

where $x = [x_1, x_2]^T$ is the system state. u is the input control quantity. c is the coefficient of the control variable. The "sum perturbation" of the system f consists of perturbations and unknown disturbances that have been modeled by the system and have specific expressions, which do not need to be accurately calculated, and can be estimated and dynamically compensated based on ADRC. ADRC is mainly composed of three parts: tracking differentiator (TD), extended state observer (ESO) and nonlinear error feedback rate (NLFES).

Tracking Differentiator (TD). TD is one of the core of ADRC, a nonlinear dynamic link proposed in order to reasonably extract differential signals. TD can estimate and track the input signal and obtain the differential signal. At the same time, TD has a strong filtering effect on the noise accompanying the input signal. The discrete formula for TD is:

$$\begin{cases} x_1(k+1) = x_1(k) + h \cdot x_2(k) \\ x_2(k+1) = x_2(k) + h \cdot fhan(x_1(k) - x_c(k), x_2(k), r_{TD}, h) \end{cases} \tag{13}$$

where $fhan(x_1(k) - x_c(k), x_2(k), r_{TD}, h)$ is the fastest control synthesis function, and its calculation steps are as follows:

$$\begin{cases} d = r_{TD}h \\ d_0 = dh \\ c = x_1(k) - x_c(k) + hx_2(k) \\ a_0 = \sqrt{d^2 + 8r_{TD}|c|} \\ a = \begin{cases} x_2(k) + \frac{c}{h}, |c| \le d_0 \\ x_2(k) + \frac{sign(c)(a_0 - d)}{2}, |c| > d \end{cases} \\ fhan = \begin{cases} -\frac{r_{TD}a}{d}, |a| \le d \\ -r_{TD} \cdot sign(a), |a| > d \end{cases} \\ sign(m) = \begin{cases} 1, m > 0 \\ 0, m = 0 \\ -1, m < 0 \end{cases} \end{cases} \tag{14}$$

where $x_c(k)$ is the input signal, $x_1(k)$ is the tracking result of the input signal, $x_2(k)$ is the differential signal of $x_1(k)$. When $x_1(k)$ accurately tracks the input signal, $x_2(k)$ can be approximated as a differential signal of $x_c(k)$. h, r_{TD} are the sampling period and speed factors respectively. And h, r_{TD} can be adjusted to improve the tracking effect when noise is accompanied by the input signal.

Extended State Observer (ESO). ESO is used to observe the quantity of system state z_1, z_2 and the "sum perturbation" of the system z_3 and derive an estimate. ESO is the core of ADRC, which realizes the estimation of system uncertainty dynamics. It is dynamically compensated based on the nonlinear error feedback law to ensure that ADRC has good robustness and strong anti-interference ability. The discrete formula for ESO is:

$$\begin{cases} e(k) = z_1(k) - y(k) \\ z_1(k+1) = z_1(k) + h[z_2(k) - \beta_{11}e(k)] \\ z_2(k+1) = z_2(k) + h\big[z_3(k) - \beta_{12}fal(e(k), \alpha_{11}, \delta) + b_0u(k)\big] \\ z_3(k+1) = z_3(k) + h\big[-\beta_{13}fal(e(k), \alpha_{12}, \delta)\big] \end{cases} \tag{15}$$

where $\beta_{11}, \beta_{12}, \beta_{13}$ are adjustable parameters, which mainly affects the estimation effect, $\alpha_{11}, \alpha_{12}, \delta$ are no-adjustable parameters and often takes a fixed value; b_0 is the compensation factor. The smooth nonlinear function $fal(e, \alpha, \delta)$ selected for ESO can play a good role in estimating variables due to its smooth continuous power structure, and its calculation formula is:

$$fal(e, \alpha, \delta) = \begin{cases} |e|^{\alpha} sign(e), \ |e| > \delta \\ \frac{e}{\delta^{1-e}}, \ |e| \le \delta \end{cases} \tag{16}$$

Nonlinear State Error Feedback Rate (NLSEF). The basic idea of NLSEF is "eliminating error based on error", and its dynamic compensation part is another core of ADRC. Using the initial control instruction u_0, the estimation of the system "sum disturbance" z_3 and the dynamic compensation factor b_0 to dynamically compensate here and then the control quantity u is obtained. Based on the nonlinear unsmooth function, NLSEF performs nonlinear calculation on the generated error signal, which has a significant improvement in error elimination ability and efficiency compared with the traditional linear error feedback law. The discrete formula for NLSEF is:

$$\begin{cases} e_1(k+1) = x_1(k+1) - z_1(k+1) \\ e_2(k+1) = x_1(k+1) - z_2(k+1) \\ u_0(k+1) = -fhan(e_1(k+1), ne_2(k+1), r_N, h_N) \\ u(k+1) = \dfrac{u_0(k+1) - z_3(k+1)}{b_0} \end{cases} \tag{17}$$

where n, r_N, h_N are the controller parameters.

In summary, the advantages of ADRC are very obvious: ADRC determines the system state variables $\boldsymbol{x}$ and control variables $\boldsymbol{u}$, and summarizes the remaining modeled, unmodeled and unknown perturbations in the system into the system "sum perturbation"

f for processing. However, there are eleven parameters that need to be determined in the standard second-order ADRC system, which are r_{TD}, h in TD, $\beta_{11}, \beta_{12}, \beta_{13}, \alpha_{11}, \alpha_{12}, b_0$ in ESO, and n, r_N, h_N in NLSEF.

3.3 Particle Swarm Optimization Algorithm

In this paper, the method based on "parameter tuning and particle swarm optimization (POS)" is designed to optimize the parameter values of ADRC controller, and the basic steps are as follows [24]:

1) Initialize the particle swarm size N, particle dimension q, value range and maximum number $count_{\max}$ of iterations of the parameters to be optimized;

2) Design of fitness function

The fitness function is similar to performance indicators and directly determines the results of PSO parameter optimization. In order to ensure that the system tracking error for the input signal is as small as possible and the tracking speed is fast enough, this article takes:

$$J = \int_0^{t_f} \left[t|e(t)| + c_J u^2(t) \right] dt \tag{18}$$

to minimize J by optimization within the feasible domain.

3) Updates to particle position and velocity

a) Compare the particle's fitness value and individual extremum, if $J[i] < p_{id}$ then order $p_{id} = J[i]$.

b) Compare the fitness value and global extrema of the particle, if $J[i] < p_{gd}$ then order $p_{gd} = J[i]$.

c) The speed and position update formula is as follows:

$$\begin{aligned} v_{id}(t+1) &= \omega v_{id}(t) + c_1\xi_1\left[p_{id}(t) - x_{id}(t)\right] + c_2\xi_2\left[p_{gd}(t) - x_{id}(t)\right] \\ x_{id}(t+1) &= x_{id}(t) + r v_{id}(t+1) \end{aligned} \tag{19}$$

where v_{id}, x_{id} are velocity and position, and the subscript is represented as the d dimensional parameter of the i particle; p_{id} is the historically optimal position of the particle; p_{gd} is the historically optimal position of the group; ξ_1, ξ_2 is a random number evenly distributed in the interval [0, 1]; c_1, c_2 is a learning factor, representing the acceleration trend weight of a single particle and a population, generally taken as $c_1 = c_2 = 2$; r is a constraint factor, and the usual value is 1; ω is the inertia weight, which is used to adjust the global and local search capabilities of the algorithm, and its specific expression is:

$$\omega = \omega_1 - \frac{\omega_1 - \omega_2}{count_{\max}} count \tag{20}$$

where ω_1, ω_2 are the inertia weight of the initial moment and the end moment, and the value is often taken $\omega_1 = 0.9, \omega_2 = 0.4$; $count$, $count_{\max}$ are the number of iterations and the maximum number of iterations respectively.

d) Boundary condition processing is performed to limit the extreme values of particle velocity v_{id} and particle position x_{id}.

4) Determine whether the end conditions are met, and terminate if they are met, otherwise go to step 2.

3.4 Design of ADRC Controller Based on PSO Optimization

Based on the above analysis, the second-order ADRC control system designed to track the angle of attack is shown in Fig. 3:

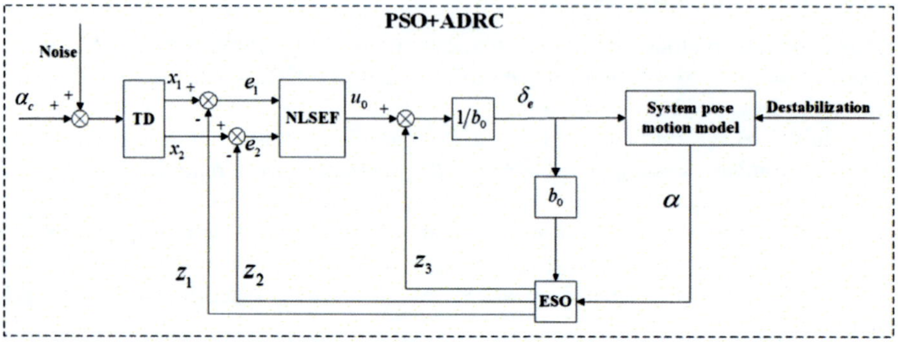

Fig. 3. Angle of attack disturbance rejection control block diagram

It can be seen from the above figure that after the command angle of attack α_c containing random noise is input into the system, the α_c original signal x_1 and differential signal x_2 are extracted by TD tracking. ESO real-time observation estimates the current actual angle of attack z_1 of the aircraft, the rate of change of the angle of attack z_2 and the "sum disturbance" of the system z_3. The state deviation e_1, e_2 is obtained and entered into NLSEF to calculate the initial control quantity u_0. After dynamic compensation, the elevator deflection angle δ_e is generated and acts on the aircraft to generate the real angle of attack α.

From the basic principle of the standard second-order automatic disturbance rejection controller and the angle of attack state equation, it can be seen that there are nine adjustable parameters of the angle of attack tracking controller in the climbing section of CASV, which are r_{TD} in TD, $\beta_{11}, \beta_{12}, \beta_{13}, \alpha_{11}, \alpha_{12}$ in ESO, and n, r_N, h_N in NLSEF.

4 Simulation and Analysis

4.1 Parameter Optimization Simulation

Set the simulation step size h to 0.01 and the sampling period to $h_{TD} = 0.01$.The fitness function of the selected particles is:

$$J = \sum_{t=0}^{t_f} (t \cdot |e(t)|) + \sum_{t=0}^{t_f-1} (u_t(t+1) - u_t(t))^2 \tag{21}$$

where $e = \alpha_c - \alpha$, $t_f = 4s$.

Based on the ADRC controller parameter optimization method, the initial values of the tunable parameters of the tracking controller are as follows:

$$\begin{cases} r_{TD} = 100, \beta_{11} = 100, \beta_{12} = 3000, \beta_{13} = 30000, \\ \alpha_{11} = 1.0, \alpha_{12} = 0.75, n = 1.0, r_N = 0.5, h_N = 0.08 \end{cases} \quad (22)$$

In the PSO algorithm, the number of initial particles is $N = 50$, particle dimension is $D = 9$, and number of iterations is $T = 15$. The initial value of the inertia weight is taken $\omega = 0.9$, and the value range is [0.4, 0.9]. The values of the parameters are:

$$\begin{cases} \text{max}: \\ r_{TD} = 250, \beta_{11} = 500, \beta_{12} = 15000, \beta_{13} = 150000, \\ \alpha_{11} = 5.0, \alpha_{12} = 5.0, n = 5.0, r_N = 5.0, h_N = 0.4 \\ \text{min}: \\ r_{TD} = 5, \beta_{11} = 10, \beta_{12} = 300, \beta_{13} = 3000, \\ \alpha_{11} = 0.1, \alpha_{12} = 0.075, n = 0.1, r_N = 0.1, h_N = 0.008 \end{cases} \quad (23)$$

The input angle of attack signal is set to the step signal of $\alpha_c = 0.25$ rad. The result of controller parameter optimization by PSO algorithm is shown in Fig. 4:

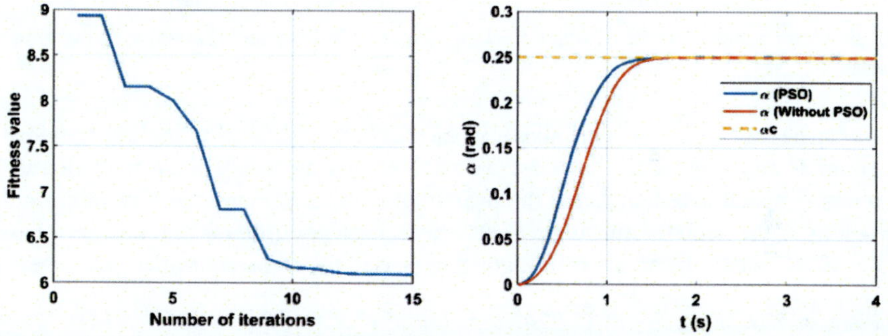

Fig. 4. Schematic diagram of the PSO optimization effect of the tracking controller of angle of attack

The resulting angle of attack tracking controller parameters are:

$$\begin{cases} r_{TD} = 102.5818, \beta_{11} = 117.2432, \beta_{12} = 3517.2974, \beta_{13} = 35172.9743, \\ \alpha_{11} = 1.1724, \alpha_{12} = 0.8793, n = 1.1724, r_N = 0.9329, h_N = 0.0938 \end{cases} \quad (24)$$

As can be seen from the figure above, after a limited number of iterations, the fitness value has decreased significantly. Compared with the angle of attack command tracking of the climbing section of CASV before and after optimization, it can be seen that after optimization based on the PSO algorithm, the tracking speed of the angle of attack tracking controller for the input angle of attack signal is improved, which improves the performance of the ADRC angle of attack tracking controller.

4.2 Consider Angle of Attack Tracking with Limited Actuator Input

This section simulates the angle of attack tracking controller designed in Sect. 3. In fact, during the climbing process of CASV, the elevator as an actuator can rotate in a limited range. So how to ensure that the tracking controller can still accurately and quickly track the change of angle of attack under the limitation of the actuator has become a new problem. In this article, the elevator declination rotation range is set to $[-30°, 30°]$. When the actuator input is limited, the ability of the controller designed in Sect. 3 to track the angle of attack and the elevator declination angle with time are shown in Figs. 5 and 6:

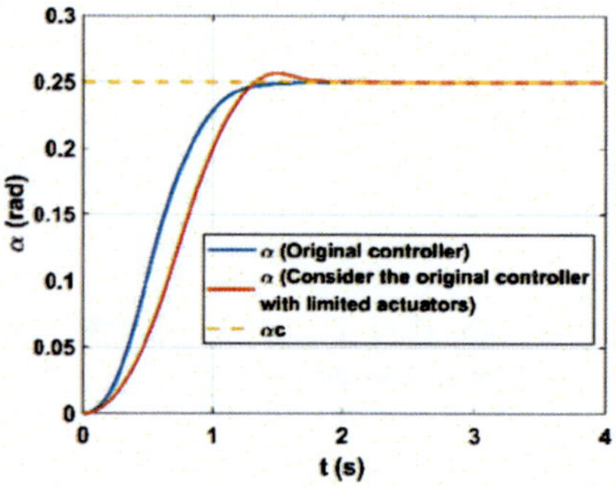

Fig. 5. Comparison of angle of attack tracking results

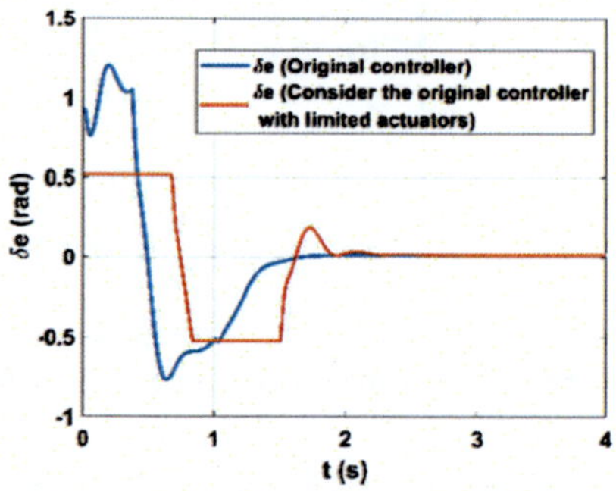

Fig. 6. Comparison of elevator declination curves

As can be seen from the figure above, when the elevator deflection angle is limited, the controller designed in Sect. 3 has a decrease in the tracking ability of the angle of attack, the tracking speed is reduced, and the overshoot amount is increased. Considering the limitations of the actuator, the angle of attack tracking controller is still designed based on the ADRC method, and the PSO algorithm is used to set the controller parameters to

$$\begin{cases} r_{TD} = 102.5818, \beta_{11} = 117.2432, \beta_{12} = 3517.2974, \beta_{13} = 35172.9743, \\ \alpha_{11} = 1.1724, \alpha_{12} = 0.8793, n = 1.1724, r_N = 0.9329, h_N = 0.0938 \end{cases} \tag{25}$$

The controller simulation results are shown in Figs. 7 and 8:

The comparison with the original controller is shown in Figs. 9 and 10:

As can be seen from the figure above, considering the limited actuator, the angle of attack ADRC controller can accurately track the input step signal. Compared with the original controller with limited actuators, the system overshoot is reduced and the tracking speed is improved.

4.3 Climbing Section Angle of Attack Tracking

Under the condition of considering the limitation of the actuator, the optimized controller parameters are used to track the angle of attack of the climbing section of the combined power aerospace vehicle, and the optimal angle of attack curve of the climbing section is shown in Fig. 11, and the tracking control results are shown in Figs. 12 and 13.

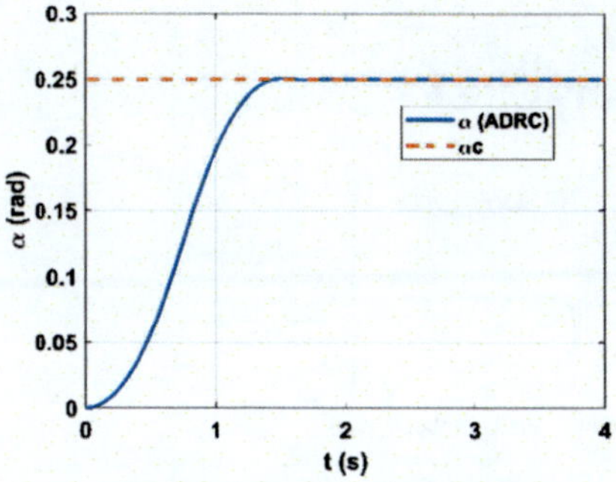

Fig. 7. Angle of attack tracking results

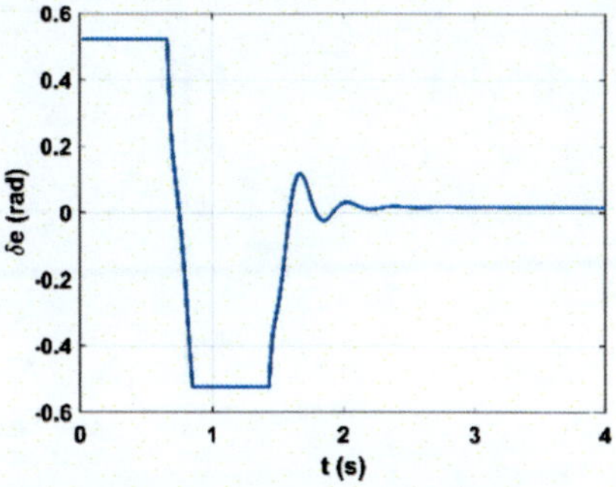

Fig. 8. Elevator declination curve

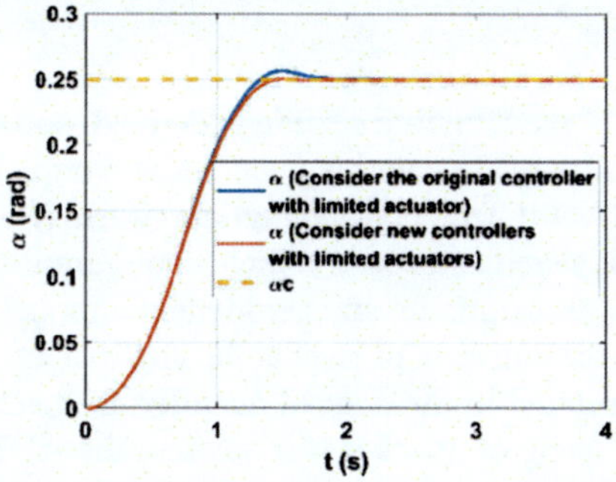

Fig. 9. Comparison of angle of attack tracking results

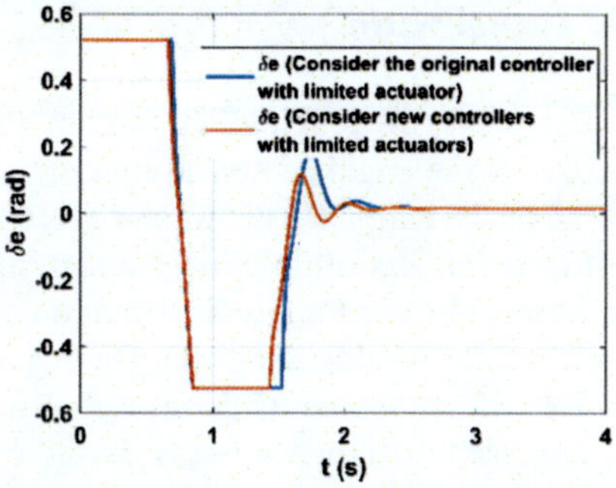

Fig. 10. Comparison of declination curve of elevator

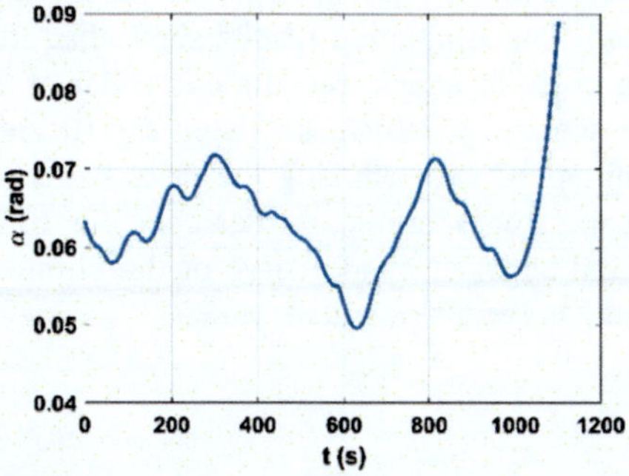

Fig. 11. Optimal angle of attack curve for climbing section

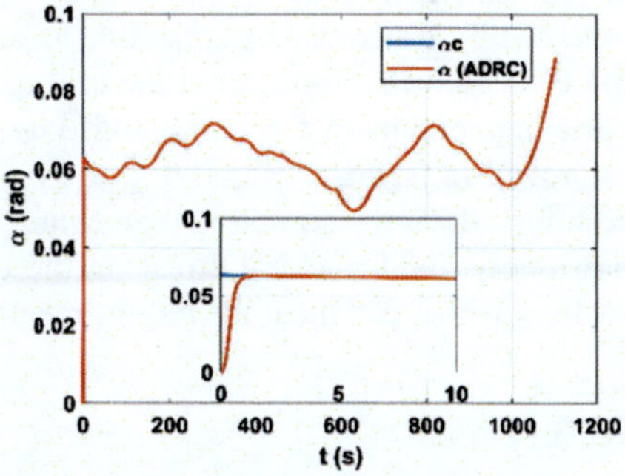

Fig. 12. The climbing section tracks the results of the angle of attack

It can be seen from the above figure that the controller designed in this paper has a very good tracking effect on the optimal angle of attack curve of the climbing section. The deflection amplitude of the elevator in the climbing section is not large, and only jitter in a small range. The controller has certain engineering application value.

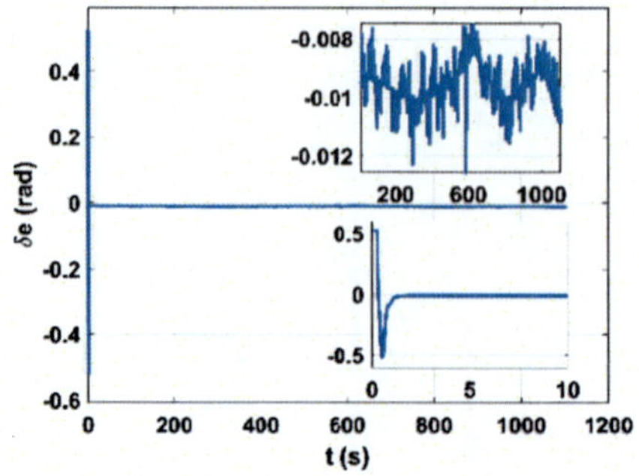

Fig. 13. Climbing section elevator declination change curve

5 Conclusion

In this paper, the longitudinal motion model of the climbing section is derived, and the design of the angle of attack tracking controller of the climbing section is completed based on the automatic disturbance rejection (ADRC) method and particle swarm (PSO) optimization algorithm. Firstly, a combined powered aerospace vehicle configuration is designed, and the longitudinal motion model, thrust model and aerodynamic model of the climbing section are given. On this basis, the equation of state at the angle of attack is derived, and some of the modeled terms related to the control quantity are treated as unknown disturbance terms. Based on the angle of attack state equation, the ADRC angle of attack tracking controller is designed. Aiming at the complex problem of ADRC controller parameter setting, the PSO optimization algorithm is introduced to realize the optimization of controller parameters. Considering the limitation of the actuator of the aircraft, the controller parameters are updated, and the optimal angle of attack curve obtained by trajectory optimization is tracked. The simulation results show that after the PSO algorithm optimizes the tuning, the angle of attack tracking controller of the climbing section of the combined powered aerospace vehicle has been significantly improved in terms of improving the tracking speed and reducing the disturbance. In addition, under the condition that the aircraft actuator is limited, the designed controller has a very good tracking effect on the optimal angle of attack curve of the climbing section, which can meet the requirements and has certain practical value.

References

1. Ju, X., Wu, R.: Geometry and time updaters-based arbitrary-yaw iterative explicit guidance for fixed-thrust boost back of vertical take-off vertical landing reusable launch vehicles final. Aerosp. Sci. Technol. **95**, 105433 (2019)
2. Chavez, F., Schmidt, D.: Analytical aeropropulsive/aeroelastic hypersonic vehicle model with dynamic analysis. J. Guid. Control. Dyn. **17**(06), 1308–1319 (1994)
3. Baris, F., Maj, M., Petros, A.: Flight dynamics and control of airbreathing hypersonic vehicle: review and new directions. In: The 12th AIAA International Space Planes and Hypersonic Systems and Technologies, pp. 1–24, Norfolk, Virginia (2003)

4. Bolender, M., Doman, D.: Nonlinear longitudinal dynamical model of an air-breathing hypersonic vehicle. J. Spacecr. Rockets **44**(02), 374–387 (2007)
5. Mcnamara, J., Friedmann, P.: Aeroelastic and aerothermoelastic analysis of hypersonic vehicles: current status and future trends. In: The 14th AIAA/AHI Space Planes and Hypersonic Systems and Technologies Conference, pp. 1–45, Canberra, Australia (2007)
6. Hao, A., Wu, Q.: Disturbance rejection dynamic inverse control of air-breathing hypersonic vehicles. Acta Astronaut. **151**(10), 348–356 (2018)
7. Rehman, O., Petersen, I., Fidan, B.: Feedback linearization-based robust nonlinear control design for hypersonic flight vehicle. Proc. Inst. Mech. Eng. Part I: J. Syst. Control Eng. **227**(01), 3–11 (2013)
8. Sagliano, M., Mooij, E., Theil, S.: Adaptive disturbance-based high-order sliding-mode control for hypersonic-entry vehicles. J. Guidance Control Dyn. **40**(03), 1–16 (2017)
9. Luo, C., Lei, H., Li, J., et al.: A new adaptive neural control scheme for hypersonic vehicle with actuators multiple constraints. Nonlinear Dyn. **100**(03), 3529–3553 (2020)
10. Bahm, C., Baumann, E., Martin, J., et al.: The X-43A Hyper-X Mach 7 Flight 2 guidance, navigation, and control overview and flight test results. In: AIAA/CIRA 13th International Space Planes and Hypersonics Systems and Technologies Conference, pp. 1–23, Capua, Italy (2005)
11. Jiang, W.A., Dong, C.A., Wang, Q.B.: A systematic method of smooth switching LPV controllers design for a morphing aircraft. Chin. J. Aeronaut. **28**(06), 1640–1649 (2015)
12. Wu, X., Tang, S., Guo, J., et al.: Finite-time reentry attitude control using time-varying sliding mode and disturbance observer. Math. Probl. Eng. **2015**(09), 1–19 (2015)
13. Wang, Q., Stengel, R.F.: Robust nonlinear control of a hypersonic aircraft. J. Guidance Control Dyn. **23**(01), 15–26 (2000)
14. Tournes, C., Landrum, D.B., Shtessel, Y., et al.: Ramjet-powered reusable launch vehicle control by sliding modes. J. Guid. Control. Dyn. **21**(03), 409–415 (1998)
15. Xu, H., Mirmirani, M.D., Ioannou, P.A.: Adaptive sliding mode control design for a hypersonic flight vehicle. J. Guidance Control Dyn. Publ. Am. Inst. Aeronaut. Astronaut. Devoted Technol. Dyn. Control **27**(05), 829–838 (2004)
16. Wang, P., Tang, G., Wu, J.: Sliding mode decoupling control of a generic hypersonic vehicle based on parametric commands. Sci. China (Inf. Sci.) **58**(05), 168–181 (2015)
17. Montaseri, G., Yazdanpanah, M.: Adaptive control of uncertain nonlinear systems using mixed backstepping and Lyapunov redesign techniques. Commun. Nonlinearence Numer. Simul. **17**(08), 3367–3380 (2012)
18. Zhang, T., Ge, S., Hang, C.: Adaptive neural network control for strict-feedback nonlinear systems using backstepping design. Automatica **36**(12), 1835–1846 (2000)
19. Farrell, J., Sharma, M., Polycarpou, M.: Backstepping-based flight control with adaptive function approximation. J. Guid. Control. Dyn. **28**(06), 1089–1102 (2005)
20. Fiorentini, L., Serrani, A., Bolender, M.A., et al.: Nonlinear robust adaptive control of flexible air-breathing hypersonic vehicles. J. Guid. Control. Dyn. **32**(02), 402–417 (2009)
21. Dai, Y., Wang, Y., Xu, X., et al.: An improved method for initial sizing of airbreathing hypersonic aircraft. Aerospace **10**(02), 199–219 (2023)
22. Mixon, B., Chudoba, B.: The Lockheed SR-71 Blackbird - a senior capstone re-engineering experience. In: 45th AIAA Aerospace Sciences Meeting and Exhibit, pp. 1–35, Reno, Nevada (2007)
23. Han, J.: From PID to active disturbance rejection control. IEEE Trans. Industr. Electron. **56**(03), 900–906 (2009)
24. Kennedy, J., Eberhart, R.: Particle swarm optimization. In: Proceedings of ICNN 1995 - International Conference on Neural Networks, pp. 1942–1948, Perth, WA, Australia (1995)

Huber-Based Robust Tracking Method of Hypersonic Cruise Vehicle

Lili Liu, Rongjun Mu(✉), and Naigang Cui

Harbin Institute of Technology, Harbin 150001, Heilongjiang, China
murjun@163.com

Abstract. In order to solve the tracking problem of hypersonic cruise vehicle (HCV) during high-speed flight, a robust tracking method (CHF) for hypersonic cruise vehicle is proposed. Firstly, the hypersonic cruise vehicle tracking model is established, including a vehicle motion model, a tracking model, a ground-based radar measurement model and a flicker noise model. On this basis, for the nonlinear problem of the system, the tracking method based on deterministic sampling is adopted. Aiming at the noise non-Gaussian problem caused by high-speed flight, a robust tracking method based on Huber-based hypersonic cruise vehicle is proposed. Finally, comparative simulations under various conditions are carried out to verify the effectiveness of the robust tracking method based on Huber. The simulation analysis of the tracking and positioning accuracy of hypersonic cruise aircraft was completed from the aspects of angle measurement accuracy and frame rate measurement. The comparison of CKF and CHF algorithms under different intensity flashes shows the superiority of CHF algorithm in hypersonic cruise vehicle tracking, and higher estimation accuracy and system robustness can be obtained.

Keywords: Hypersonic cruise vehicle (HCV) · Flicker noise · Great likelihood estimation · Robust filtering

1 Introduction

Hypersonic cruise vehicles (HCV) can achieve maneuverable penetration flights at more than 5 times the speed of sound in adjacent space [1]. HCV has the advantages of fast response, strong penetration ability, high-speed flight, etc., and has great application prospects, which has significant advantages compared with other aircraft [2]. In aerospace transportation, HCV can greatly shorten the time spent in current aviation flights, while saving fuel and reducing space launch costs [4]. In the military field, with the continuous change of war purposes and forms of war, mission requirements continue to increase, and traditional conventional aircraft systems are no longer suitable for some long-range rapid strike missions [5]. HCV has the characteristics of fast flight speed, long range, rapid response and high hit rate, which can carry out global rapid strikes and is not easy to be intercepted, so it has become a research hotspot for long-range rapid strike missions [6].

Chinese Society of Aeronautics and Astronautics (Ed.): CASTC 2023, LNME, pp. 484–501, 2024.
https://doi.org/10.1007/978-981-99-8861-7_49

At the same time, the rapid development of HCV has brought great challenges to the anti-missile defense system. In order to achieve effective interception of HCV, it is necessary to first complete accurate tracking of the HCV flight trajectory [7]. At present, aiming at the HCV trajectory tracking problem, the research work is mainly carried out in two aspects: tracking model design and estimation algorithm design [8]. Tracking model design often improves tracking models by augmenting the parameters to be estimated into state vectors [9]. Some scholars have established a target tracking model of HCV according to the aerodynamic characteristics and guidance law of the aircraft to achieve accurate tracking, but the target of different models or different guidance laws needs to be re-derived [10]. Ghosh [13] constructs a Wiener process model of the ballistic coefficient of the maneuvering re-entry vehicle, and obtains the pseudo-measurement of the ballistic coefficient through dynamic inversion, so as to estimate it optimally, but this method is difficult to apply to complex models. The estimation algorithm design mainly focuses on the estimation problem of strong nonlinear system [14]. HCV system has strong nonlinearity, and when the traditional Kalman filter method is used for trajectory estimation tracking, it is easy to cause the filter error to diverge and lose the target [15]. Therefore, this problem can be solved by approximating the nonlinear distribution by the sampling method. Cubature Kalman Filter (CKF) is a typical algorithm based on sigma-point filtering, which has the advantages of high accuracy and small calculation amount compared with extended Kalman filter [16]. At the same time, the high-speed flight of HCVs makes them susceptible to interference, and when the noise is polluted, noise non-Gaussian problems occur, which will lead to a sharp decrease in the performance of CKF. Huber [17] proposes a method based on the norm of minimizing error, which can improve the robustness of the system under noise that interferes with the Gaussian distribution by minimizing the variance of the maximum asymptotic estimate. Karlgaard et al. [18] linearized the measurement equation and applied the Huber method to differential filtering. Wang et al. [19] also applied the Huber method to UKF using a linearized method. Chang et al. [20] pointed out that the Huber method to reconstruct the measurement information (measurement noise covariance matrix) can make the filtering algorithm more robust. The above application of the Huber method to the nonlinear filtering algorithm proves that the proposed method can enhance the robustness of the filtering algorithm under non-Gaussian noise. Therefore, this paper applies the Huber method to the CKF framework, and designs a robust tracking algorithm for HCV based on Huber's, which realizes the accurate estimation of the target flight state by a robust filtering tracking algorithm for the high-speed flight of the target.

The robust tracking method of HCV proposed in this paper combines Huber theory and CKF filtering, and uses the fixed-point iterative method to solve the minimization problem based on Huber method, replacing the traditional minimum mean squared error criterion to improve the robustness of the algorithm. By linearizing the measurement equation to construct a linear regression model, the prior information and measurement information are reconstructed by the Huber method, which effectively solves the problem of noise non-Gaussian. Firstly, the dynamics modeling of HCV is carried out, and the model and measurement model of the tracking system are improved. For the nonlinear system of HCV, a nonlinear filtering method based on sigma-point is proposed, and on this basis, a robust filtering method for HCV is proposed for non-Gaussian system noise. Finally, the CHF filtering is analyzed from the aspects of algorithm effectiveness, influencing factor analysis and flicker noise intensity. The simulation verification of the robust nonlinear filtering algorithm of hypersonic cruise vehicle was completed. The simulation results show that the algorithm proposed in this paper has effectiveness and superiority in terms of noise uncertainty caused by system nonlinearity and high-speed flight of hypersonic cruise vehicle, with higher tracking accuracy and better system robustness.

2 Establishment of a Tracking Model for Hypersonic Cruise Vehicle

2.1 Dynamics Modeling of Hypersonic Cruise Vehicle

Figure 1 shows a schematic diagram of the tracking scene of HCV. HCV fly at a certain altitude longitudinally and move at high speed in a certain direction in the horizontal plane. The transverse lateral motion state changes very little, so the transverse lateral motion parameters $\beta, \gamma, \upsilon, \omega_x, \omega_y, \delta_a, \delta_r$ can be treated as zero. Then a longitudinal motion model of HCV can be obtained. Considering the rotation of the Earth and thinking of the Earth as a standard ellipsoid, the aircraft is mainly affected by gravity, engine thrust, aerodynamics, control and the corresponding moment they produce. The longitudinal motion model of the aircraft can be expressed as [21]:

$$\begin{cases} \dot{V} = \dfrac{P\cos\alpha - D}{m} - g\sin\theta \\ \dot{\theta} = \dfrac{P\sin\alpha + L}{mV} + \left(\dfrac{V}{r} - \dfrac{g}{V}\right)\cos\theta \\ \dot{y} = V\sin\theta \\ \dot{\alpha} = \omega_z - \dot{\theta} \\ \dot{\omega}_z = \dfrac{M_z}{I_z} \\ \dot{m} = -dm \\ r = y + R \end{cases} \tag{1}$$

where r is the vector from the center of the earth to the current position of the aircraft (m), V is the speed of the aircraft ($\text{m}\cdot\text{s}^{-1}$), P is the engine thrust vector (N), D is the size of air resistance (N), L is the amount of lift (N), m is the current mass of the aircraft (kg), g is the gravitational acceleration projected in the emission frame ($\text{m}\cdot\text{s}^{-2}$), α is the size of the angle of attack (rad), θ is the aircraft speed inclination (rad), y is the longitudinal plane position component (m), ω_z is pitch angular velocity ($\text{rad}\cdot\text{s}^{-1}$), M_z is the pitch moment ($\text{N}\cdot\text{m}$), I_z is the moment of inertia of the pitch channel ($\text{kg}\cdot\text{m}^2$).

2.2 Hypersonic Cruise Vehicle Tracking Model

Considering a nonlinear discrete system as shown in the following equation:

$$\begin{cases} x_k = f(x_{k-1}) + \omega_{k-1} \\ z_k = h(x_k) + v_k \end{cases} \tag{2}$$

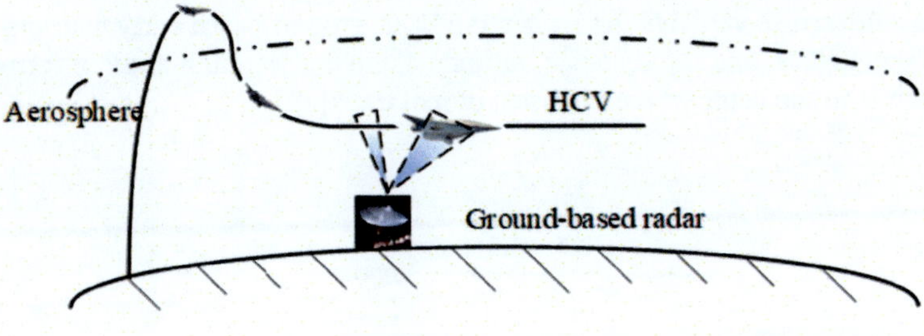

Fig. 1. Schematic diagram of target tracking for a hypersonic cruise vehicle

where $f(\cdot)$ and $h(\cdot)$ are nonlinear equations of state and measurement equations respectively, ω_{k-1} and v_k are known noise sequences with statistical characteristics of noise, which are not correlated with each other, $E(\omega_{k-1}, v_k) = 0$, and their variance arrays are Q_{k-1} and R_k.

Equation of State. Considering the flight characteristics of HCV, its movement can be simplified to uniform linear motion in the horizontal plane Therefore, the equation of state is established using the uniform motion model as follows:

$$\dot{x}(t) = \begin{bmatrix} 0 \cdots 0\, 1 & 0 & 0 \\ \vdots \quad \vdots\, 0 & 1 & 0 \\ 0 \cdots 0\, 0 & 0 & 1 \\ 0 \cdots 0\, 0 & \cdots & 0 \\ \vdots \quad \vdots\, \vdots & & \vdots \\ 0 \cdots 0\, 0 & \cdots & 0 \end{bmatrix} x(t) + \begin{bmatrix} 0\,0\,0 \\ 0\,0\,0 \\ 0\,0\,0 \\ 1\,0\,0 \\ 0\,1\,0 \\ 0\,0\,1 \end{bmatrix} \omega(t) \quad (3)$$

where its discrete form is:

$$x_k = \begin{bmatrix} 1 & 0 & 0 & T & 0 & 0 \\ 0 & 1 & 0 & 0 & T & 0 \\ 0 & 0 & 1 & 0 & 0 & T \\ 0 & 0 & 0 & 1 & 0 & 0 \\ 0 & 0 & 0 & 0 & 1 & 0 \\ 0 & 0 & 0 & 0 & 0 & 1 \end{bmatrix} x_{k-1} + \begin{bmatrix} 0.5T^2 & 0 & 0 \\ 0 & 0.5T^2 & 0 \\ 0 & 0 & 0.5T^2 \\ T & 0 & 0 \\ 0 & T & 0 \\ 0 & 0 & T \end{bmatrix} \omega_{k-1} \quad (4)$$

where $x(t) = \begin{bmatrix} x\ y\ z\ v_x\ v_y\ v_z \end{bmatrix}$ is the system state, $\dot{x}(t)$ is the derivative of $x(t)$, $\dot{x}(t) = \begin{bmatrix} v_x\ v_y\ v_z\ a_x\ a_y\ a_z \end{bmatrix}$, $\omega(t)$ is system noise, x_k is the position velocity of the hypersonic cruise vehicle at k-moment, ω_{k-1} is the system noise at the time of k − 1, T is the sampling interval for the sensor.

Ground-Based Radar Measurement Equations. The long-range and high-speed flight of HCV brings great challenges to tracking and interception. The ultra-long-range, high-precision detection of ground-based radar makes it an effective detection means for hypersonic cruise vehicles. As shown in Fig. 2, ground-based early warning radar is used to detect HCV, and the distance, altitude angle and azimuth angle information of HCV relative to the detector are obtained in real time [22].

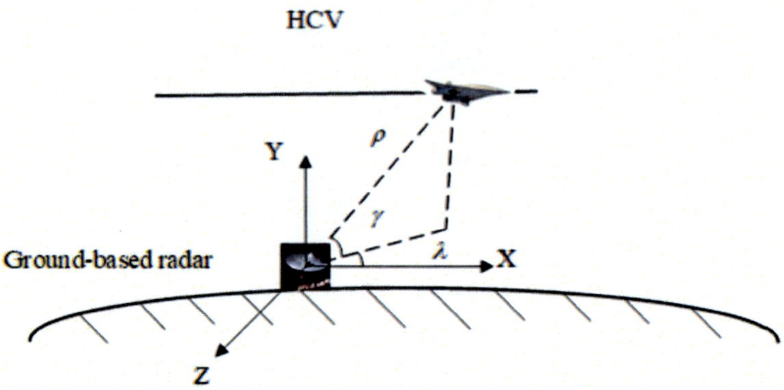

Fig. 2. Schematic diagram of ground-based radar observations

It is assumed that the installation system of the ground-based radar detector coincides with the geocentric solid-connected coordinate system, and the attitude change of the detector is not considered. Then the relative position of the target and the detector can be expressed as in the geocentric solid coordinate system:

$$\begin{bmatrix} x_{ti} \\ y_{ti} \\ z_{ti} \end{bmatrix} = \begin{bmatrix} x \\ y \\ z \end{bmatrix} - \begin{bmatrix} x_i \\ y_i \\ z_i \end{bmatrix} \tag{5}$$

where the target point position status is $\boldsymbol{r} = \begin{bmatrix} x \; y \; z \end{bmatrix}$, and the position of the ground-based radar detector in the geocentric solidly connected coordinate system is respectively $\boldsymbol{R_i} = \begin{bmatrix} x_i \; y_i \; z_i \end{bmatrix}$.

The measurement model of ground-based radar is:

$$\begin{cases} \rho = \sqrt{x_t^2 + y_t^2 + z_t^2} + v_\rho \\ \gamma = \arcsin\left(\frac{y_t}{\sqrt{x_t^2+y_t^2+z_t^2}}\right) + v_\gamma \\ \lambda = \arctan 2(-z_t, x_t) + v_\lambda \end{cases} \tag{6}$$

where ρ is the oblique distance of the target from the radar station; γ is the height angle, λ is the azimuth angle, v_ρ is the oblique measurement noise, v_γ is the height angle measurement noise, v_λ is the azimuth measurement noise.

2.3 Flicker Noise Model

Affected by the curvature of the earth, atmospheric scattering and other factors, the detection of hypersonic cruise vehicles by ground-based radar will be interfered with.

Due to the reflection interference at different positions of the aircraft, the sensor will produce obvious random swing on the target measurement, so that the measurement of the target shows non-Gaussian noise characteristics. Due to the high-speed flight of hypersonic cruise flight, the measurement is not high-profile, which poses a challenge to the tracking of HCV. In this paper, the non-Gaussian measurement noise generated by interference is described by a flicker noise model, and its probability density function is shown in the equation [23]:

$$p(v) = (1 - \varepsilon)p_{v1} + \varepsilon p_{v2} \tag{7}$$

where ε is the flicker probability, $p_{v1} \sim N\left(\mu_1, \delta_1^2\right)$ and $p_{v2} \sim N\left(\mu_2, \delta_2^2\right)$ are both Gaussian noises, the mean of which are μ_1=0 and μ_2=0, The variance satisfies $\delta_1 < \delta_2$, the flicker probability corresponds to the large variance noise, and the flicker intensity is set to $\beta = \delta^2/\delta^1$.

White Gaussian noise with zero mean $p(\xi) \sim N(0, \delta)$ has a probability density function of:

$$p(\xi) = \frac{1}{\sqrt{2\pi}\delta}\exp\left(-\frac{\xi^2}{2\delta^2}\right) \tag{8}$$

Assuming δ_1 is certain, the greater the flicker probability ε and flicker intensity β, the stronger the noise non-Ghost. As shown in the Fig. 3, it can be seen that flicker noise has the characteristics of non-Gaussian and tail length compared with Gaussian noise, also known as "heavy tail noise".

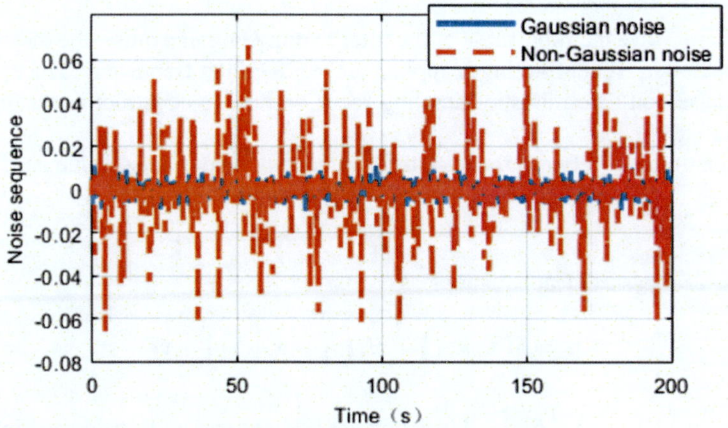

Fig. 3. Gaussian noise and flicker noise

3 Huber-Based Robust Tracking Algorithm

The Kalman filter is based on the minimum mean square deviation estimation under the assumptions of linear system and Gaussian noise, while the tracking of hypersonic cruise vehicles is nonlinear. Nonlinear transformations destroy the properties of the

Gaussian distribution and will make the Kalman filter impossible to apply directly, so the nonlinear function should be approximated. Extended Kalman filtering (EKF) is performed by Taylor expansion of nonlinear equations, which is computationally intensive and inaccurate. Unlike EKF, deterministic sampling-based methods represented by volumetric Kalman filtering (CKF) propagate in a nonlinear system using a set of determined weighted sampling points. On the basis of linearization of nonlinear equations of state and measurement equations, the statistical characteristics of nonlinear systems can be accurately captured, so that high approximation accuracy can be achieved. The nonlinear filtering method based on CKF does not need to solve the Jacobi matrix, and the approximate accuracy is higher [24].

The nonlinear filtering method based on deterministic sampling can only solve the problem of nonlinearity of hypersonic cruise vehicle system, but does not deal with the problem of noise interference caused by high-speed flight. Aiming at the noise non-Gaussian problem caused by high-speed flight, a robust nonlinear filtering method based on Huber is proposed. Generalized maximum likelihood estimation (M estimation) theory is used to solve the problem of symmetric interference in Gaussian distribution attachments. A mixed cost function based on the sum norm l_1 and l_2 (i.e., the Huber kernel function) is used, and the Huber method is used to preprocess the measurement. The tracking of hypersonic cruise vehicles under the CKF filtering framework improves the robustness of the system under non-Gaussian noise while maintaining the estimated efficiency under Gaussian distribution.

3.1 Statistical Linear Regression

In order to improve the robustness of the filter to non-Gaussian noise, the measurement update method in Huber filtering is applied to the CKF filter framework. Huber filtering methods based on deterministic sampling focus on how to construct linear regression problems.

The relationship between the state truth value and the predicted value is:

$$x_k = \widehat{x}_{k|k-1} + \delta_k \tag{9}$$

Linearize measurement equations:

$$z_k = h\left(\widehat{x}_{k|k-1}\right) + H_k\left(x_k - \widehat{x}_{k|k-1}\right) + v_k \tag{10}$$

where $H_k = \left[\left(P_{k|k-1}\right)^{-1} P_{xz}\right]^T$. Measurement updates can be transformed into linear regression problems:

$$\begin{bmatrix} z_k - h\left(\widehat{x}_{k|k-1}\right) + H_k \widehat{x}_{k|k-1} \\ \widehat{x}_{k|k-1} \end{bmatrix} = \begin{bmatrix} H_k \\ I \end{bmatrix} + \begin{bmatrix} v_k \\ -\delta_k \end{bmatrix} \tag{11}$$

Define variables:

$$\begin{cases} S_k = \begin{bmatrix} R_k & 0 \\ 0 & P_{k|k-1} \end{bmatrix} \\ Y_k = S_k^{-1/2} \begin{bmatrix} z_k - h\left(\widehat{x}_{k|k-1}\right) + H_k \widehat{x}_{k|k-1} \\ \widehat{x}_{k|k-1} \end{bmatrix} \\ M_k = S_k^{-1/2} \begin{bmatrix} H_k \\ I \end{bmatrix} \\ \xi_k = S_k^{-1/2} \begin{bmatrix} \nu_k \\ -\delta_k \end{bmatrix} \end{cases} \tag{12}$$

Then the linear regression problem can be transformed into:

$$Y_k = M_k x_k + \xi_k \tag{13}$$

3.2 Generalized Maximum Likelihood Estimation Method

For the results based on statistical linear regression, the generalized maximum likelihood estimation method focuses on how to solve the linear regression problem. The robust volumetric filter measurement update based on Huber filter theory is converted to the minimum value of the index function under the solution.

$$J(x_k) = \sum_{i=1}^{m+n} \rho(\xi_i) \tag{14}$$

where ρ is the indicator function, ξ_i is the i element of ξ.

In generalized maximal likelihood estimation problems, some specific properties of the estimator can be obtained in the form of a selection function ρ. An important property of generalized maximum likelihood estimation is robustness, and Huber gives the function ρ as:

$$\rho(\xi_i) = \begin{cases} \frac{1}{2}\xi_i^2, & |\xi_i| < \alpha \\ \alpha|\xi_i| - \frac{1}{2}\alpha^2, & |\xi_i| \geq \alpha \end{cases} \tag{15}$$

where α is tunable parameter. The function ρ is minimal function of the mixed l_1 and l_2 norm. When $\alpha \to 0$, the function ρ approaches the function l_1 minimum; and when $\alpha \to \infty$, the function ρ approaches the function l_2 minimum. General take $\alpha = 1.345$.

The definition of ξ is as follows:

$$\xi = M_k \widehat{x}_{k|k-1} - Y_k \tag{16}$$

The smallest solution of the indicator function J satisfies the following equation:

$$\sum_{i=1}^{m+n} \phi(\xi_i) \frac{\partial \xi_i}{\partial x} = 0 \tag{17}$$

where $\phi(\xi_i) = \rho'(\xi_i)$. Defining:

$$\begin{cases} \psi(\xi_i) = \dfrac{\phi(\xi_i)}{\xi_i} \\ \Psi(\xi_i) = \mathrm{diag}[\psi(\xi_i)] \end{cases} \tag{18}$$

Then:

$$M_k^T \Psi(M_k x_k - Y_k) = 0 \tag{19}$$

Solving with iteration:

$$x_k^{(j+1)} = \left[M_k^T \Psi^{(j)} M_k\right]^{-1} M_k^T \Psi^{(j)} Y_k \tag{20}$$

where j is the number of iterations, generally $j = 3 \sim 5$. The initial value of iteration is the value after the convergence of the iterative solution, that is, the least squares result is used as the initial value of state estimation.

$$x_k^{(0)} = \left(M_k^T M_k\right)^{-1} M_k^T Y_k \tag{21}$$

3.3 Huber-Based Robust Nonlinear Filtering Method – CHF

The time update method based on deterministic sampling and the measurement update method based on Huber non-Gaussian noise processing are combined to fully solve the nonlinear and non-Gaussian problems of hypersonic cruise vehicles. A robust volumetric filtering algorithm (CHF) for hypersonic cruise vehicles is proposed [25].

1) Time update - consistent with CKF.

Volume points are calculated based on the third-order spherical-radial volume rule:

$$\begin{cases} P_{k-1|k-1} = S_{k-1} S_{k-1}^T \\ \tilde{\chi}_{k-1}^i = \widehat{x}_{k-1} + S_{k-1} \xi_i \end{cases} \tag{22}$$

Recursive volume points and calculate state and covariance predictions:

$$\begin{cases} \tilde{\chi}^{*(i)}_{k|k-1} = f\left[\tilde{\chi}^{i}_{k-1}\right] \\ \widehat{x}_{k|k-1} = \sum \omega_i \tilde{\chi}^{*(i)}_{k|k-1} \\ P_{k|k-1} = \sum \omega_i \left[\tilde{\chi}^{i}_{k-1} - \widehat{x}_{k|k-1}\right]\left[\tilde{\chi}^{i}_{k-1} - \widehat{x}_{k|k-1}\right]^T + Q_{k-1} \end{cases} \tag{23}$$

Calculate the volume points used to measure updates:

$$\begin{cases} P_{k|k-1} = S_{k|k-1} S^T_{k|k-1} \\ \tilde{\chi}^{(i)}_{k|k-1} = \widehat{x}_{k|k-1} + S_{k|k-1}\xi_i \end{cases} \tag{24}$$

Calculate the measured forecast:

$$\begin{cases} \mathbb{Z}^{(i)}_{j,k|k-1} = h_j\left[\tilde{\chi}^{(i)}_{k|k-1}\right] \\ \widehat{z}_{j,k|k-1} = \sum_{i=1}^{2n} \omega_i \mathbb{Z}^{(i)}_{j,k|k-1} \end{cases} \tag{25}$$

2) Measurement Update - Huber-based robust nonlinear filtering.
Calculate the mutual variance:

$$P_{j,xz} = \sum_{i=1}^{2n} \omega_i \left[\tilde{\chi}^{(i)}_{k|k-1} - \widehat{x}_{k|k-1}\right]\left[\tilde{\chi}^{(i)}_{k|k-1} - \widehat{x}_{k|k-1}\right]^T \tag{26}$$

Construct a linear regression model:

$$Y_k = M_k x_k + \xi_k \tag{27}$$

Solve linear regression problems:

$$M_k^T \Psi(M_k x_k - Y_k) = 0 \tag{28}$$

Iteratively solve for state estimates and covariances:

$$x_k^{(j+1)} = \left[M_k^T \Psi^{(j)} M_k\right]^{-1} M_k^T \Psi^{(j)} Y_k \tag{29}$$

where j is the number of iterations. The measurement is updated to:

$$P_k = \left(M_k^T \Psi M_k\right)^{-1} \tag{30}$$

The trace algorithm flow is shown in Fig. 4:

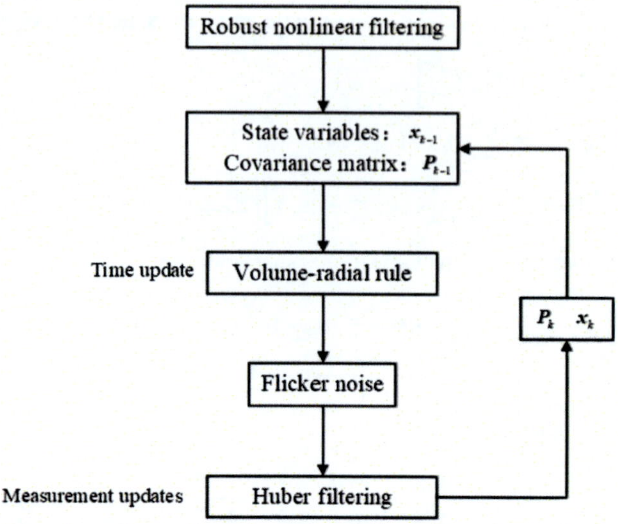

Fig. 4. Flowchart of robust nonlinear filtering algorithm based on Huber

4 Simulation Results and Analysis

4.1 Simulation Scene Settings

The cruising segment of HCV can be regarded as a uniform linear motion, the cruising altitude is $h = 36\,\text{km}$, the speed is $v_x = 1080\,\text{m} \cdot \text{s}^{-1}$, $v_y = 1440\,\text{m} \cdot \text{s}^{-1}$, $v_z = 2\,\text{m} \cdot \text{s}^{-1}$, and the simulation time is 200 s. For the high-speed flight of HCV, the detector is a ground-based radar, which is set on the surface of the earth, corresponding to the projected position of the initial position of the HCV vehicle on the ground $S = \begin{bmatrix} 0 & 0 & R_e \end{bmatrix}$.

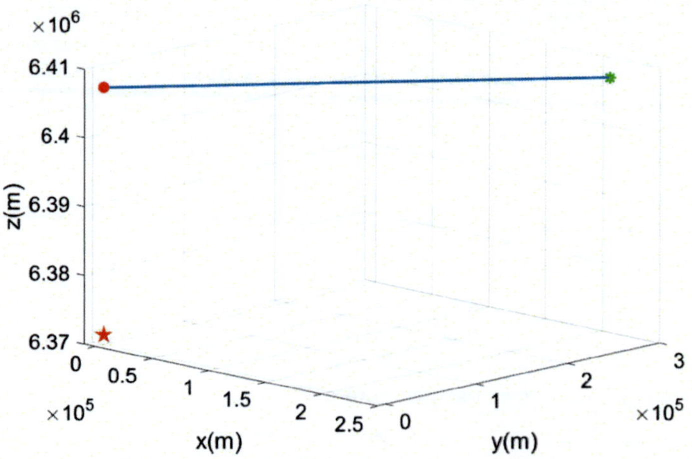

Fig. 5. Three-dimensional trajectory diagram of a hypersonic cruise vehicle

The measurement of the ground-based radar is distance measurement and angle measurement, in which the distance measurement accuracy is 15 m and the angle measurement accuracy is 0.01°. The three-dimensional trajectory map of HCV is shown in Fig. 5.

The filter parameters are set as Table 1:

Table 1. Initial parameters of filter

Parameters	Reference value
Initial location of HCV	$x_0 = \left[0\ 0\ 36e+3\ 1080\ 1440\ 2\right]$
Detector location	$S = \left[0\ 0\ R_e\right]$
Process noise covariance matrix	$Q_k = diag\left(\left[100\ 100\ 100\ 10\ 10\ 10\right].\hat{}2\right)$
Measure the noise covariance matrix	$R_k = diag\left(\left[15\ 0.01\ 0.01\right].\hat{}2\right)$
Dynamics model discrete period, observation update period, total simulation time	$stp = 0.1s$, $stp_z = 0.1s$, $t_s = 200s$
Filter initial covariance matrix	$P_0 =$ $diag\left(\left[1000\ 1000\ 1000\ 100\ 100\ 100\right].\hat{}2\right)$
The initial state vector of the filter	$\widehat{x}_0 \sim N(x_0, P_0)$

The flicker probability and flicker intensity of different intensities of non-Gaussian noise are set as Table 2:

Table 2. High, medium and low flicker intensity noise parameter table

Non-Gaussian flicker noise	Flicker probability ε and flicker intensity β
High-intensity flickering	$\varepsilon = 0.5, \beta = 20$
Medium intensity flashing	$\varepsilon = 0.3, \beta = 10$
Low-intensity flashing	$\varepsilon = 0.1, \beta = 5$

From Fig. 6, the different intensity flicker noise sequences set up for this article are compared.

4.2 Simulation Verification of CHF Algorithm Effectiveness

In order to verify the effectiveness of the robust tracking algorithm of hypersonic cruise vehicle, the position and speed accuracy of the CKF algorithm and the CHF algorithm are compared under Gaussian noise and flicker noise respectively. From Figs. 7 and 10, the CHF accuracy and CKF accuracy are comparable under Gaussian noise. Because of

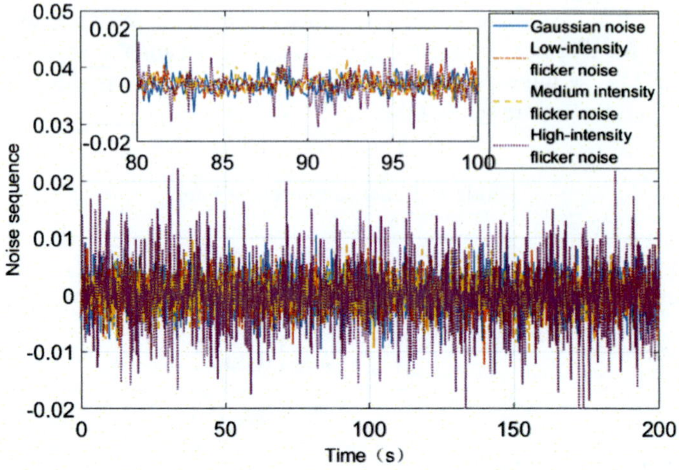

Fig. 6. Flicker noise of varying intensity

the influence of the Huber kernel function, the CHF accuracy is slightly inferior to the CKF tracking accuracy. In non-Gaussian noise, CHF's accuracy positioning and speed measurement accuracy are higher, and the system robustness is stronger. It can be seen that the algorithm of HCV target positioning tracking based on CHF is effective (Figs. 9 and 10).

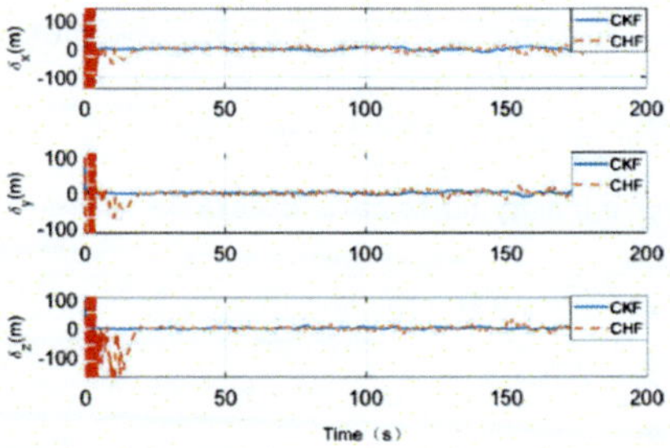

Fig. 7. Position error plot under Gaussian noise

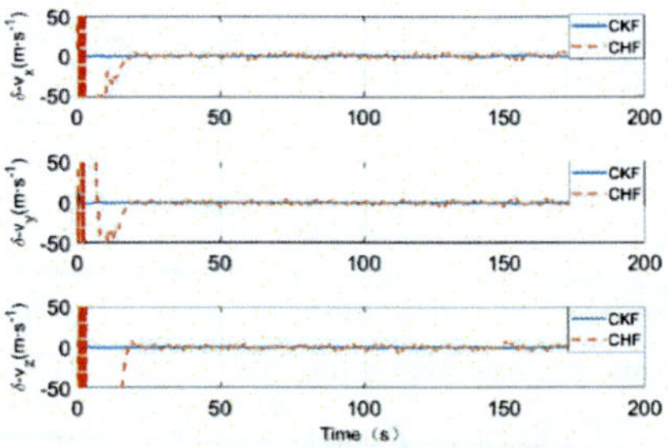

Fig. 8. Speed error plot under Gaussian noise

Single simulation has chance, in order to better illustrate the statistical effect of the algorithm, the Monte Carlo target test is carried out on the above algorithm. The number of target shots is set to 100 times, and the results of its error are statistically averaged. Then the RMSE results of position and speed can be obtained.

As shown in Figs. 11 and 12 the algorithm accuracy of CHF is higher, the positioning accuracy is less than 5m, and the velocity accuracy is less than 1m/s. However, the CKF algorithm cannot effectively deal with non-Gaussian noise. It can be seen that the curve fluctuates more strongly, and the error shows a slightly divergent trend.

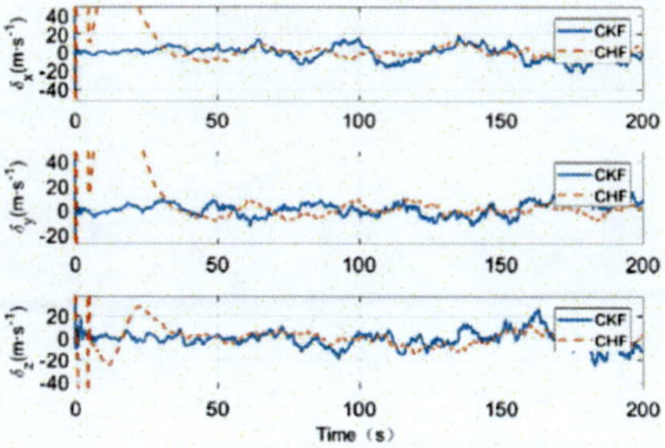

Fig. 9. Position error plot under non-Gaussian noise

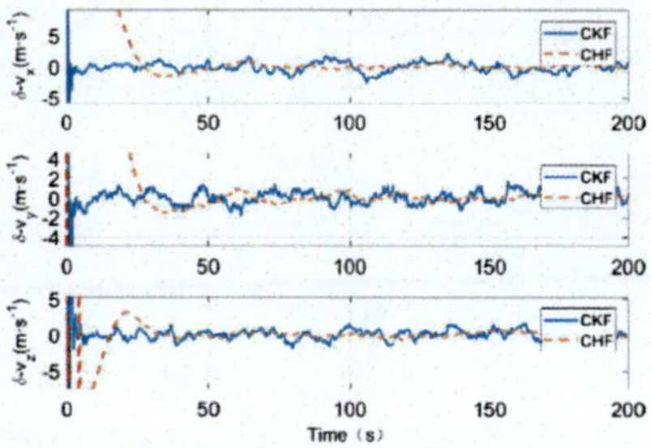

Fig. 10. Speed error plot at non-Gaussian noise

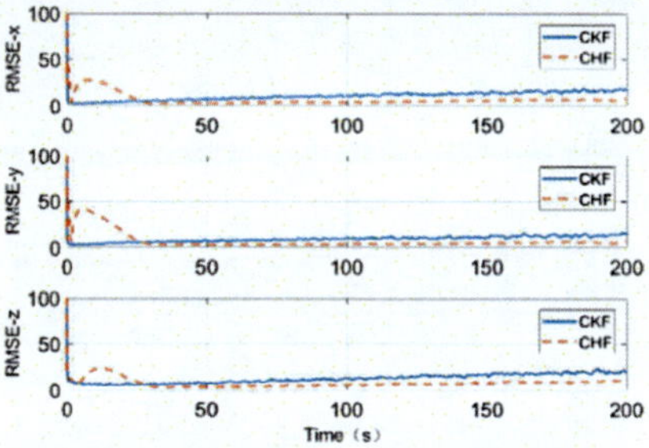

Fig. 11. Positional RMSE curve under non-Gaussian noise

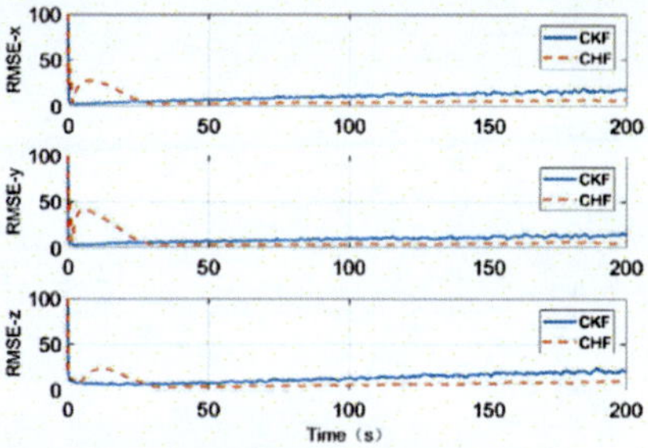

Fig. 12. Speed RMSE curve under non-Gaussian nois

4.3 Analysis of Target Positioning Accuracy Factors of CHF Algorithm

On the basis of verifying the effectiveness of the CHF algorithm, the target positioning accuracy of the CHF algorithm is analyzed for the angular accuracy and frame rate measurement, and the RMSE results of 100 Monte Carlo target statistics shall prevail.

Figures 13 and 14 are the RMSE curve for position and velocity under non-Gaussian noise with angular measurement accuracy $\left[0.001\ 0.005\ 0.01\ 0.05\ 0.1\right]$ degree. It can be seen that the higher the goniometric accuracy, the higher the target tracking accuracy and the better the robustness of the system.

From Figs. 15 and 16, the RMSE curve of the non-Gaussian noise position and velocity when the frame frequency is $\left[10\ 5\ 1\ 0.5\ 0.1\right]$ Hz. It can be seen that the greater the measurement frame rate, the higher the target tracking accuracy and the better the system robustness.

The position and speed RMSE results after filtering are statistically calculated and averaged to obtain RMSE position and speed error results under different influencing factors. As shown in Figs.17 and 18, the RMSE line chart of 100 Monte Carlo target results under different angular accuracy is shown.

As shown in Figs.19 and 20, the RMSE line chart of 100 Monte Carlo target results at different measured frame rates.

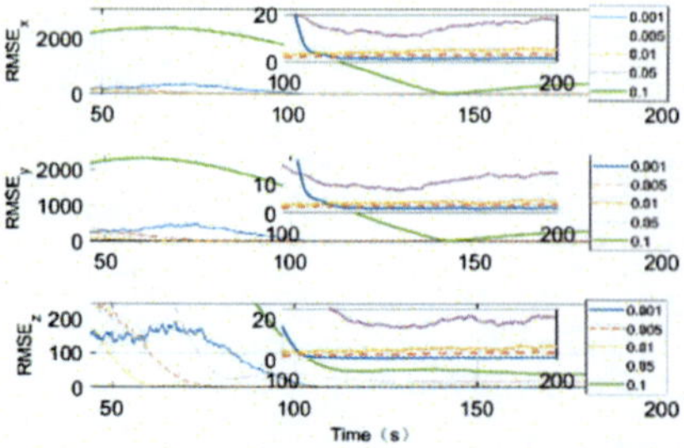

Fig. 13. Position RMSE curves of different angle measurement accuracy under non-Gaussian noise

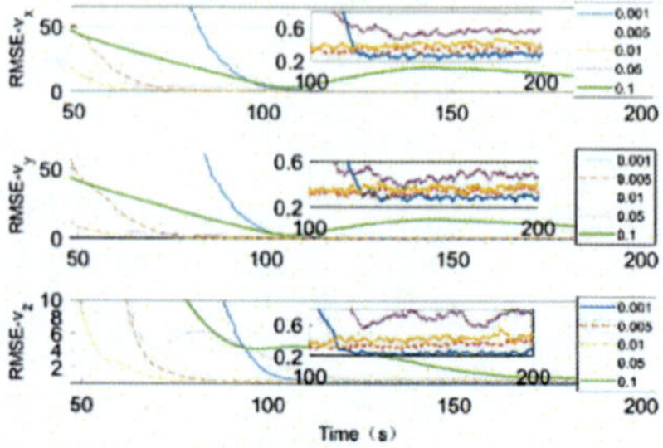

Fig. 14. Velocity RMSE curves with different goniometric accuracy at non-Gaussian noise

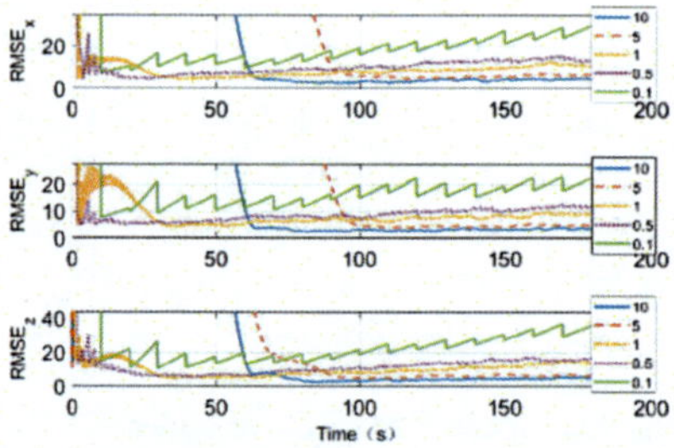

Fig. 15. RMSE curves for measuring the position of frame rates at non-Gaussian noise

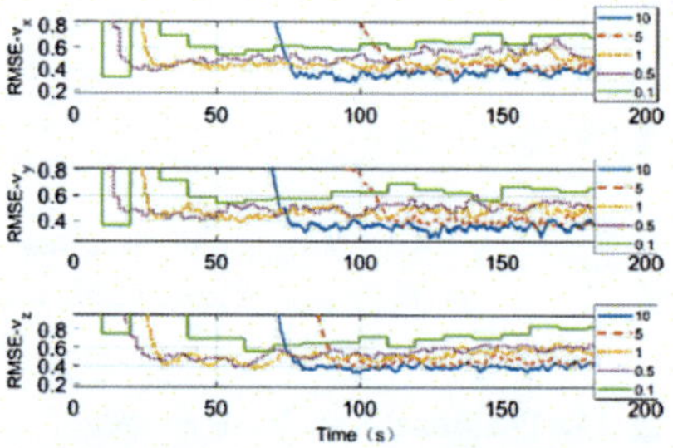

Fig. 16. RMSE curves for measuring the position of frame rates at non-Gaussian noise

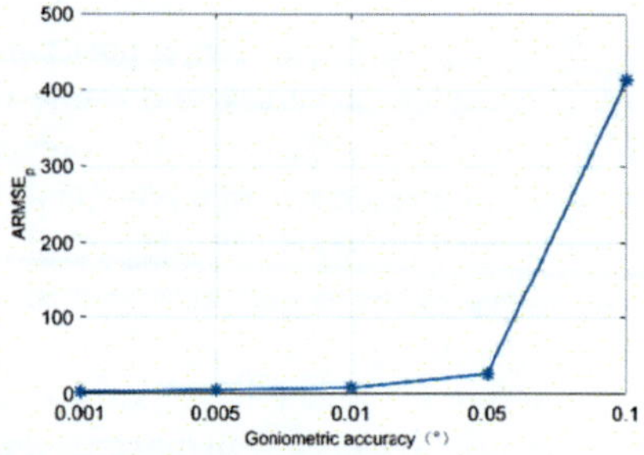

Fig. 17. Position RMSE line chart with different goniometric accuracy

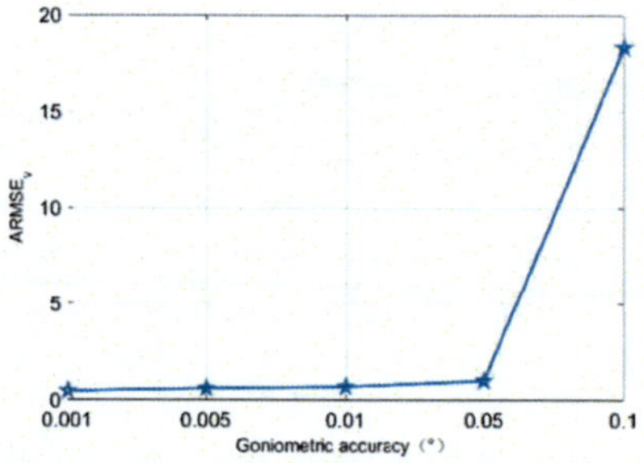

Fig. 18. Velocity RMSE line chart under different angle measurement accuracy

4.4 Comparison of Flicker Noise of Different Intensities

The intensity of non-Gaussian noise flicker is determined by a combination of flicker probability and flicker multiplier, and cannot be quantified. Figures 21 and 22 illustrates the RMSE results of CKF and CHF under the flicker noise of high, medium and low

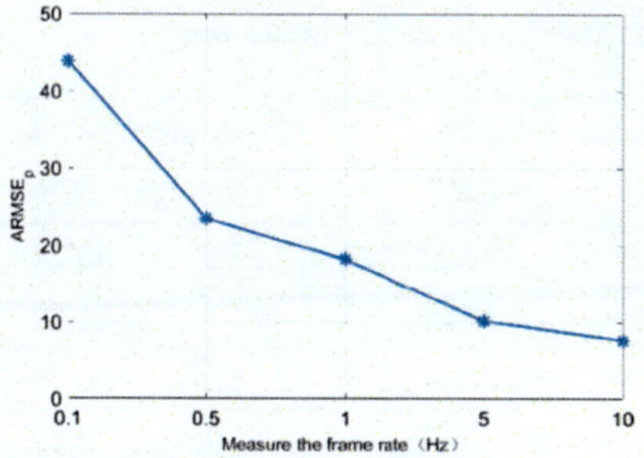

Fig. 19. RMSE line chart of position at different measured frame rates

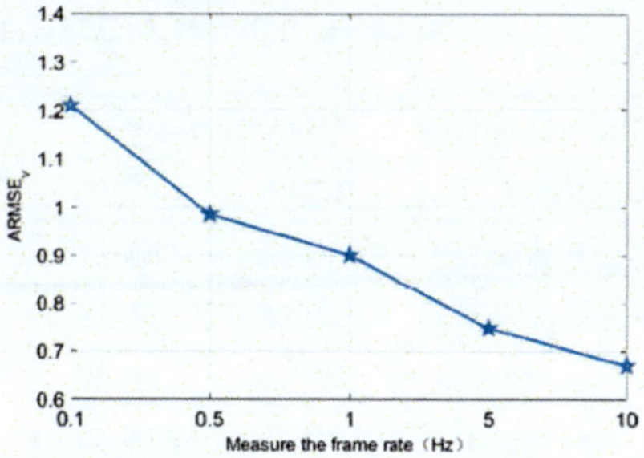

Fig. 20. RMSE line chart of position at different measured frame rates

intensity. It can be seen that the CHF algorithm has higher accuracy than the position and speed estimation accuracy of CKF. The curve of the CKF algorithm has large fluctuations, which is greatly affected by noise non-Gaussian and has poor suppression ability of non-Gaussian noise. While the CHF algorithm shows good system robustness in non-Gaussian noise processing, and has significant advantages for positioning and tracking of hypersonic cruise vehicles. In addition, it can be seen from the influence of flicker noise intensity on the stability of the algorithm that the smaller the flicker intensity, the higher the target tracking accuracy, and the better the system robustness. In general, the accuracy and robustness of the CHF algorithm are better than those of CKF, which shows that the CHF algorithm has strong processing ability for the noisy non-Gaussian problem of hypersonic cruise vehicles.

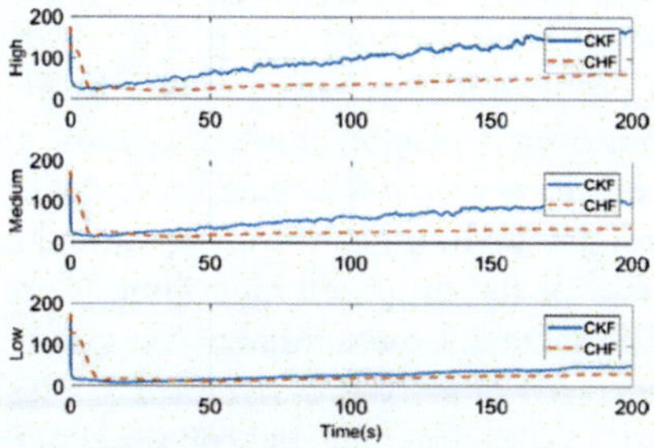

Fig. 21. Comparison of position RMSE curves at different flicker intensities

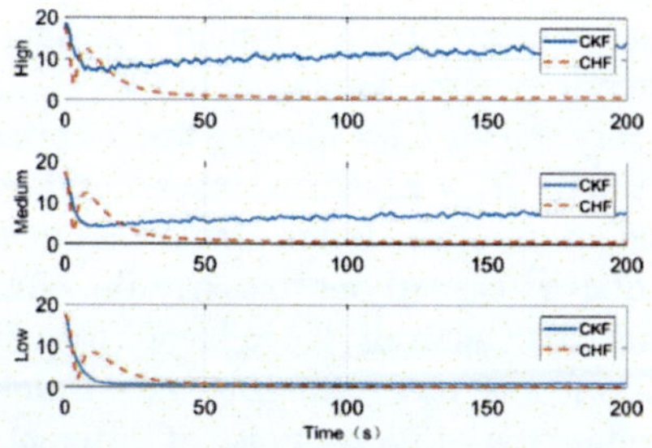

Fig. 22. Comparison of speed RMSE curves at different flicker intensities

The position and velocity RMSE curve results of flicker noise of different intensities were statistically counted, and the RMSE results of 50-200s after stable filtering were averaged. Then the ARMSE statistical results under different flickering intensities can be obtained, as shown in Table 3:

Table 3. Table of ARMSE statistics at different flicker intensities

Flicker intensity	High-intensity		Medium-intensity		Low-intensity	
Algorithm	CHF	CKF	CHF	CKF	CHF	CKF
X-axis position	25.36	65.81	15.80	40.24	11.92	40.24
Y-axis position	20.28	50.88	12.33	31.02	9.87	31.02
Z-axis position	29.05	79.20	21.35	50.06	17.12	50.06
X-axis speed	0.45	6.40	0.30	3.87	0.19	3.87
Y-axis speed	0.39	5.40	0.25	3.26	0.17	3.26
Z-axis speed	0.57	7.50	0.43	4.61	0.36	4.62

Compared with the CKF algorithm, the position estimation accuracy of CHF algorithm is improved by 62.08%, 58.95% and 67.65% compared with the high, medium and low flicker intensities and the speed estimation accuracy was improved by 92.69%, 91.60% and 93.54%.

5 Conclusion

In this paper, aiming at the problem that HCV high-speed flight leads to easy tracking interference, a robust tracking method for hypersonic cruise vehicles is proposed. Firstly, the longitudinal motion model, tracking model, ground-based radar measurement model and flicker noise model of hypersonic cruise vehicle are established respectively. On this basis, a deterministic sampling algorithm is used to solve the nonlinear problem of HCV aircraft system. Aiming at the non-Gaussian problem of measuring noise, the Huber theory and the CKF filtering framework are combined to propose a robust filtering algorithm for HCV based on Huber. Positioning tracking is realized when noise is disturbed and measurement is non-Gaussian. Finally, a comparative simulation under various conditions is carried out to verify the effectiveness of the proposed algorithm. Through simulation analysis, the influence of goniometric accuracy, measurement frame rate and other factors on the tracking and positioning accuracy of HCV is determined. Under the comparison of flicker noise of different intensities, the accuracy and robustness of the CHF algorithm proposed in this paper are proved. The simulation results show that the proposed algorithm has high accuracy and good robustness, and has certain application value.

References

1. Chudoba, B., Haney, E., Gonzalez, L., et al.: Strategic forecasting in uncertain environments: hypersonic cruise vehicle research and development case study. Aeronaut. J. **119**(1211), 45–65 (2015)
2. Wang, K., Zhang, B.: Bilevel trajectory optimization for hypersonic cruise vehicle using bilevel directed search domain. J. Spacecr. Rocket. **58**(02), 1–11 (2021)

3. Feng, X., Lu, Y., Gao, Y., et al.: Adaptive radau pseudo-spectral optimization for descending trajectory of a hypersonic cruise vehicle. Aerosp. Syst. **3**(04), 1–12 (2020)
4. Wu, Z., Liu, Y.: Integrated optimization design using improved pigeon-inspired algorithm for a hypersonic vehicle model. Int. J. Aeronaut. Space Sci. **23**(05), 1033–1042 (2022)
5. Chen, K., Zhou, J., Shen, F.Q., et al.: Hypersonic boost–glide vehicle strapdown inertial navigation system/global positioning system algorithm in a launch-centered earth-fixed frame. Aerosp. Sci. Technol. **98**, 105679 (2020)
6. Liu, J., Luo, Q., Lou, J., et al.: An adaptive infrared tracking method for spacebased surveillance to a hypersonic cruise vehicle. In: 2020 Chinese Automation Congress (CAC), pp. 249–254, Shanghai, China (2020)
7. Zhai, G., Bi, X., Zhao, H., et al.: Non-cooperative maneuvering spacecraft tracking via a variable structure estimator. Aerosp. Sci. Technol. **79**, 352–363 (2018)
8. Hu, Y., Li, J., Zhang, Z., et al.: Parametric modeling of hypersonic ballistic data based on time varying auto-regressive model. Sci. China Technol. Sci. **63**(08), 1396–1405 (2020)
9. Cheng, Y., Yan, X., Tong, S., et al.: An adaptive non-zero mean damping model for trajectory tracking of hypersonic glide vehicles. Aerosp. Sci. Technol. **111**, 106529 (2021)
10. Cheng, Y., Tang, S., Shi, L., et al.: Aerodynamic statistics-based trajectory estimation of hypersonic maneuvering target. IEEE Access **227**(08), 642–656 (2020)
11. Hough, M.: Reentry maneuver estimation using nonlinear markov acceleration models. J. Guid. Control. Dyn. **40**(07), 1–18 (2017)
12. Hough, M.E.: Acceleration characterization for reentry orbit determination with unmodeled maneuvers. J. Guidance, Control Dyn. **41**(7), 1463–1475 (2018). https://doi.org/10.2514/1.G003359
13. Ghosh, S.: Tracking reentry ballistic targets using acceleration and jerk models. IEEE Trans. Aerosp. Electron. Syst. **47**(01), 666–683 (2011)
14. Murata, M., Kawano, I., Inoue, K.: Extended, unscented kalman, and sigma point multiple distribution estimation filters for nonlinear discrete state-space models. IEEE Control Syst. Lett. **4**(04), 982–987 (2020)
15. Chandra, K., Darouach, M.: Two-stage information filters for single and multiple sensors, and their square-root versions. Automatica **98**, 20–27 (2018)
16. Arasaratnam, I., Haykin, S.: Cubature kalman filters. IEEE Trans. Autom. Control **54**(06), 1254–1269 (2009)
17. Huber, P.J.: Robust estimation of a location parameter. Ann. Math. Stat. **35**(01), 73–101 (1964)
18. Karlgaard, C., Schaub, H.: Huber-based divided difference filtering. J. Guid. Control. Dyn. **30**(03), 885–891 (2007)
19. Wang, X., Cui, N., Guo, J.: Huber-based unscented filtering and its application to vision-based relative navigation. Radar Sonar Navig. Let **4**(01), 134–141 (2010)
20. Chang, L., Hu, B., Chang, G., et al.: Huber-based novel robust unscented Kalman filter. IET Sci. Meas. Technol. **6**(06), 502–509 (2012)
21. Wang, W., Hou, Z., Shan, S., et al.: Periodically cruising hypersonic vehicle with active cooling: an optimal-control based design approach. IEEE Access **7**, 65486–65505 (2019)
22. Oh, B.S., Lin, Z.: Extraction of global and local m-DS features from FMCW radar returns for UAV detection. IEEE Trans. Aerosp. Electron. Syst. **57**(02), 1351–1360 (2020)
23. Chen, Y., Yang, Y., Shi, X.: Flicker noise simulation by physical trap model. In: 2019 China Semiconductor Technology International Conference (CSTIC), pp. 1–3. Shanghai, China (2019)
24. Pu, L., Yang, T., Mu, H.: An improved initial alignment method based on adaptive robust CKF algorithm. J. Northwest. Polytech. Univ. **40**(01), 103–109 (2022)
25. Wan, W., Feng, J., Song, B., et al.: Huber-based robust unscented kalman filter distributed drive electric vehicle state observation. Energies **14**(03), 750–756 (2021)

Study and Application of Decompression Loads for Aircraft Interiors

Lei Zhang(✉), Qiudi Xu, Jiafan Jing, Huijun Dou, and Di Sun

AVIC the First Aircraft Institute, Xi'an 710089, China
15191426981@163.com

Abstract. Decompression loads play an important role on aircraft interior design and affect aircraft safety. This paper aims at decompression loads resulting from a sudden release of pressure through an opening in compartment with interiors installed. First of all, a calculation model for interior decompression loads was proposed according to the airworthiness requirements and characteristics of air flow. Based on this model, the magnitude and direction of interior decompression loads for a certain aircraft compartment was computed and concluded. Simultaneously its influence on aircraft safety was also analyzed and more critical case was found. The analysis conclusion could be confirmed by checking recording data from 3U8633, an airline with windshield rupture during its flight. Finally, in order to limit the decompression loads at a low level during aircraft interior design, an appropriate total gap was found, which lay foundation on following detailed interior design, including interior gap size definition and gap style design. Accordingly, the aircraft safety was enhanced.

Keywords: Decompression Loads · Aircraft Interiors · Total Gap · Safety

1 Introduction

A variety of load inputs need to be considered in the interior design of civil aircraft, including actual fuselage overload envelope (flight loads, ground loads), emergency landing loads specified in CCAR 25.561(b) [1], decompression loads specified in CCAR 25.365(e) [1] and abuse loads determined by engineering experience [2]. Among these above, decompression loads closely relates to aircraft safety. If the compartment interior were broken and dropped down due to oversized decompression loads, it may affect passengers' evacuation and pilots' operation. Unfortunately conducting a lab test to verified decompression loads at low cost seems infeasible because the test involves in the entire cabin. Therefore, it would be of great significance for freezing interior design and aircraft airworthiness certification by ensuring interior decompression not cause hazards.

Some study on decompression loads applied on aircraft structure has been conducted [3, 4]. However, there was little research on decompression loads applied on interiors, although the importance of that has been mentioned in some literatures [2]. In this paper, a certain type of aircraft was taken as an example and its interior decompression process was calculated. The further purpose of this study is to guide and improve the interior structure design and improve the safety of aircraft.

Chinese Society of Aeronautics and Astronautics (Ed.): CASTC 2023, LNME, pp. 502–510, 2024.
https://doi.org/10.1007/978-981-99-8861-7_50

2 Definition of Interior Decompression Loads

When a hole appears in the aircraft structure, there will be a significant pressure difference between inside and outside of compartment. In this case, air inside leaks rapidly, while the leakage rate in various compartments, such as cockpit, main cabin and cargo compartment, differs and results in pressure difference between two adjacent cabins. The pressure difference is called decompression loads and is often emphasized during structure design process.

For compartments with interior installed, due to the barrier of interior panels, individual compartment is actually divided into various areas, whose air leakage rate are different and result in pressure difference between inner surface and outer surface of interior panel. The pressure difference could be seen as loads applied on panels, namely interior decompression loads.

3 Calculation Method of Interior Decompression Loads

According to CCAR 25.365(e) [1], any structure, component or part, inside or outside a pressurized compartment, the failure of which could interfere with continued safe flight and landing, must be designed to withstand the effects of a sudden release of pressure through an opening in any compartment at any operating altitude. The opening size Ho must be computed by the following formula:

$$\mathrm{Ho} = \mathrm{P} * \mathrm{As} \tag{1}$$

where,

Ho = Maximum opening in square feet, need not exceed 20 square feet.

$$\mathrm{P} = (\mathrm{As}/6240) + 0.024 \tag{2}$$

As = Maximum cross-sectional area of the pressurized shell normal to the longitudinal axis, in square feet; and.

During the decompressing process, the velocity of air flowing out of the hole / opening is very fast, and the airflow out of the hole can be considered as one-dimensional isentropic flow. In this paper, a nozzle flow model [5] was used for calculating the airflow between inside and outside of the aircraft as well as that between two adjacent compartments:

When $\frac{p}{p_0} > (\frac{2}{\gamma+1})^{\frac{\gamma}{\gamma-1}}$, the airflow rate out of the hole:

$$\dot{m} = A\frac{p_0}{\sqrt{T_0}}\sqrt{\frac{\gamma}{R}}\sqrt{\frac{2}{\gamma-1}\left[\left(\frac{p}{p_0}\right)^{\frac{2}{\gamma}} - \left(\frac{p}{p_0}\right)^{\frac{\gamma+1}{\gamma}}\right]} \tag{3}$$

When $\frac{p}{p_0} \geq (\frac{2}{\gamma+1})^{\frac{\gamma}{\gamma-1}}$, the airflow rate out of the hole:

$$\dot{m} = A\frac{p_0}{\sqrt{T_0}}\sqrt{\frac{\gamma}{R}}\left(\frac{2}{\gamma+1}\right)^{\frac{\gamma+1}{2(\gamma-1)}} \tag{4}$$

where,

$\dot{m}$= mass flow rate out of the hole; A = flow area of the hole or between two adjacent compartments, namely Ho in Eq. (1); p = pressure of compartment with lower pressure; p_0 = pressure of compartment with higher pressure; T_0 = temperature of compartment with higher pressure; R = 287 J / (kg*K); $\gamma = 1.4$.

Based on the mass of air in compartment, the air pressure could be obtained with Ideal Gas Law [6].

$$p = mRT / V \tag{5}$$

where,

m = air mass in compartment; p = air pressure in compartment; V = volume of compartment; R = 287 J / (kg*K); T = temperature of compartment.

Besides, the influence of environment control system and other adjacent pressurized compartments would also be taken into consideration.

4 Analysis of Interior Decompression Loads

4.1 Division of Compartment Area

For compartments with interiors installed, the interior panels divide the cabin into different areas. This paper took the compartment of a certain type of aircraft as an example for calculation and analysis.

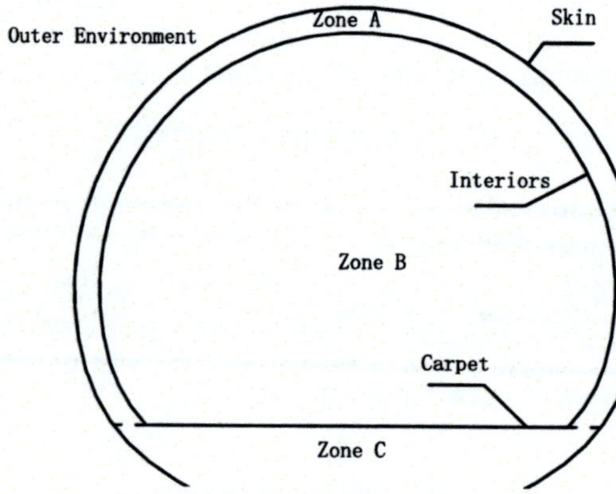

Fig. 1. Compartment zoning of a certain type of aircraft

Figure 1 is a typical cross-section of the compartment, which was divided into 3 pressurized areas: the area from carpet to interior (Zone B in the figure), the area from interior to skin (Zone A in the figure), and the area under floor (Zone C in the figure). Since there was a large flow area between Zone A and Zone C, to simplify the calculation, the both could be considered as one integrated pressurized area (Zone A-C).

In fact, the holes appearing in different areas would lead to different decompression loads, namely different load cases in following Sects. 4.2 and 4.3.

4.2 Hole in Zone A-C

When a hole appeared suddenly in structure skin (Zone A-C) during flight, due to pressure in the Zone A-C much higher than outer environment, the air would flow fast from Zone A-C to outer environment. Accordingly, pressure in Zone A-C would drop down and get lower than that in other adjacent pressurized areas. In this case, the interior panels would bear the decompression loads with direction towards skin, shown in Fig. 2.

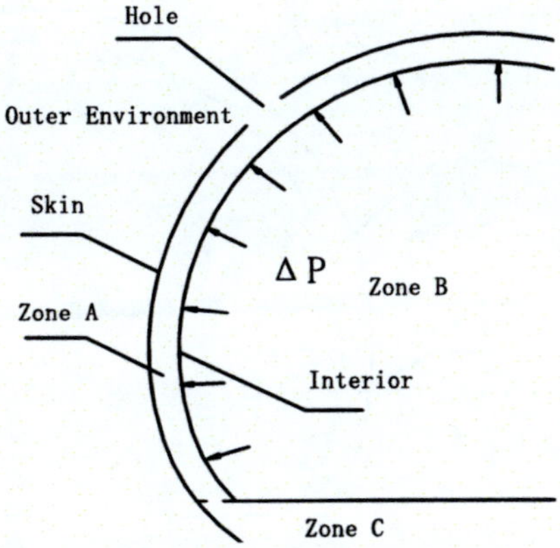

Fig. 2. Decompression loads when a hole appears in Zone A-C

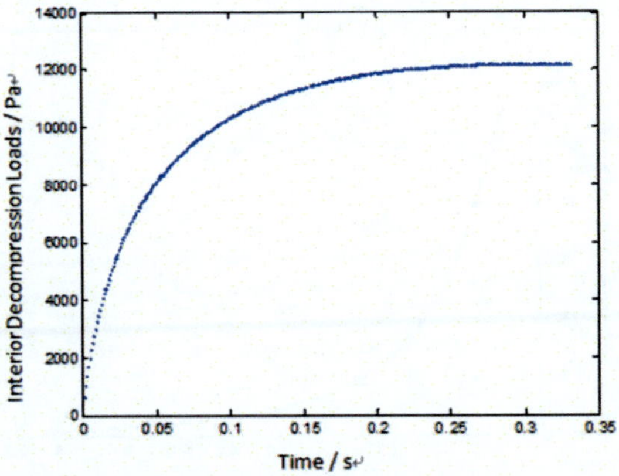

Fig. 3. Curve of interior decompression load change when there was a hole in Zone A-C

The method in Sect. 3 was used to calculate interior decompression loads after the aircraft skin was pierced. As shown in Fig. 3, the interior decompression loads increased rapidly as the air in Zone A-C lost in a short time. Whilst the growth rate of loads had been gradually slowing down over time and eventually stabilized. This is because the volume in Zone B was much greater than that in Zone A, and there was continuous supplementation from the environment control system.

4.3 Hole in Zone B

When a hole appeared suddenly in window glass (Zone B) during flight, due to pressure in Zone B much higher than outer environment, the air would flow fast from Zone B to outer environment. Accordingly, pressure in Zone B would drop down and get lower than that in other adjacent pressurized areas. In this case, the interior panels would bear the decompression loads with direction towards inner of compartment, shown as Fig. 4.

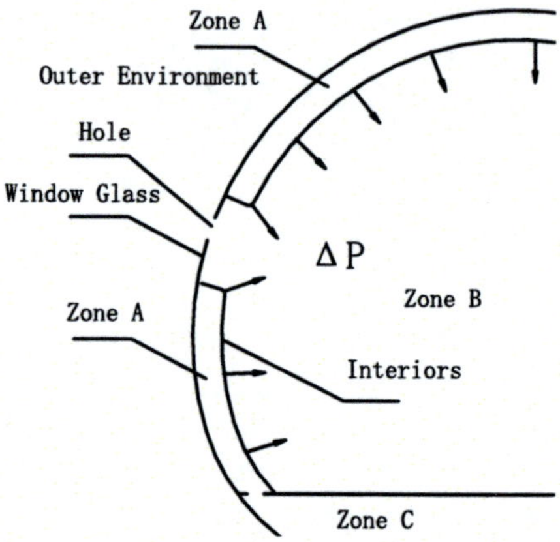

Fig. 4. Decompression loads when a hole appears in Zone B

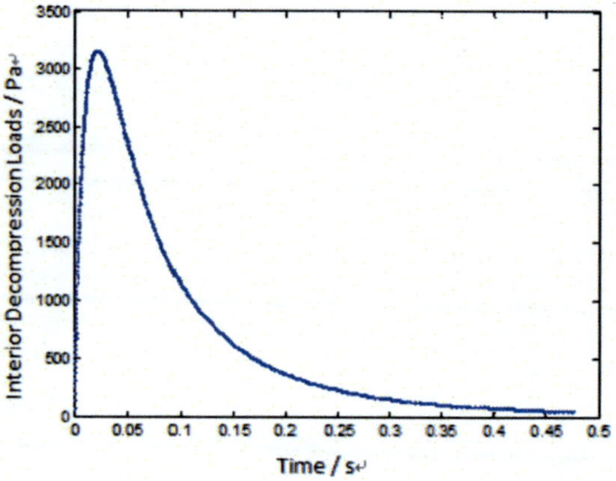

Fig. 5. Curve of interior decompression load change when there was a hole in Zone B

The method in Sect. 3 was used to calculate the interior decompression loads after the window glass was pierced. As shown in Fig. 5, the interior decompression loads in Zone B increased rapidly but then slowed down gradually. This is because pressure in Zone B reached the same level as the outer environment shortly after the window glass broke while the pressure in Zone A didn't change. Therefore the pressure difference between inner face and outer face of interior panel increased fast in a short time. Next, air in Zone A began to flow into Zone B until the pressure in both areas got closed, which means the interior decompression loads dropped down till zero.

It should be noted that the hole produced at cockpit windshield may cause the entire windshield to burst and fall off. In this situation, the actual size of the opening will further exceed that specified in CCAR 25.365(e), but the trend of decompression load change should not be affected. According to the investigation report [7] on windshield rupture incident of Sichuan Airlines Flight 3U8633 on May 14th, 2018, after the windshield glass burst and fell off, the measured cabin altitude increased rapidly closed to standard altitude, and then slowly decreased as the actual aircraft altitude decreased, shown in Fig. 6. Based on these data, it can be seen that the pressure in the Zone B of cockpit decreased rapidly and then decreased slowly. Assuming the pressure in Zone A remained constant (or decreased slowly) in this period, the change process of interior decompression loads also showed a trend of first rapid increasing and then slow decreasing, which is consistent with the calculation results in this paper.

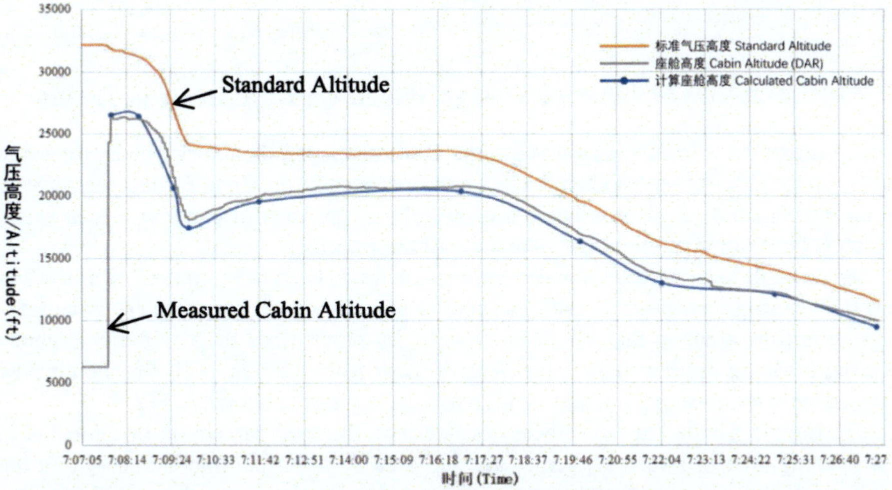

Fig. 6. Change process of cockpit altitude after 3U8633 windshield glass burst and fell off

4.4 Analysis of Safety Impact of Different Load Cases

As shown in Sect. 4.2 and Sect. 4.3, holes in different positions led to different decompression loads, so as to different safety level. The hole in Zone A-C generated a larger interior decompression loads. However, the direction of loads faced aircraft skin, which forced interior panels closer to aircraft structure. There would be little adverse impact on aircraft safety in this case. On the contrast, the interior panels bore lower decompression loads when a hole happened in Zone B but the direction of loads faced inner of cabin. It means the interiors may fell down if the loads were oversized. In this case, the pilot's safety operation and passenger's emergency evacuation might be influenced. Figure 7 shows the ceiling panel of 3U8633 cockpit has fell off after windshield rupture [8]. Therefore, the hole in Zone B is the more critical case to be considered in interiors design.

Fig. 7. 3U8633 ceiling panel of cockpit has fallen off after windshield rupture.

5 Optimization of Interior Design Based on Compression Loads

According to the calculation results in Sect. 3 of this paper, the interior decompression loads, up to 3200Pa, was produced when a hole existed in the Zone B for a certain type of aircraft (see Fig. 5). It was applied uniformly on the interior panel as a load input, which is far more than the bearing capacity of the panel.

As difficult to be verified, the decompression loads need to be controlled at a lower level, for example no more than 400Pa according to engineering experience, by increasing the airflow area, so that it can be covered by other loads such as the emergency landing loads. In order to reduce the peak of loads from 3200Pa to 400Pa, the air flow area, namely the total gap, should be increased to 4 times as shown in Fig. 8.

During the detailed design process of interiors, the total gap would be allocated to the opening area of return air grille, the gap area between the interior panel and the equipment, the gap area between two adjacent panels, etc. In another word, the gap value around interiors and the opening size of the grille can be restricted functionally.

For the interior parts located in various positions of aircraft, the effect of decompression loads is different. So are the gap size and the gap style. Generally, ceiling panel falling down influents safety more than other interior panels and its gap design needs to be considered more critically from decompression factors. In this case, softer silicon rubber seals are often used in gaps between ceiling panels so that the air flow is smooth, while harder plastic trims could be installed between sidewall panels because visual effect in this area plays a more important role, shown in Fig. 9. Instead, air return grille installed in dado could conduct compressing function in sidewall area.

Especially, the interiors in cargo compartment without people are allowed to break under decompression loads as there is no severe safety impact. However, the influence of interiors on the decompressing velocity of cockpit and main cabin should be considered in the process of the whole aircraft decompressing correlation analysis.

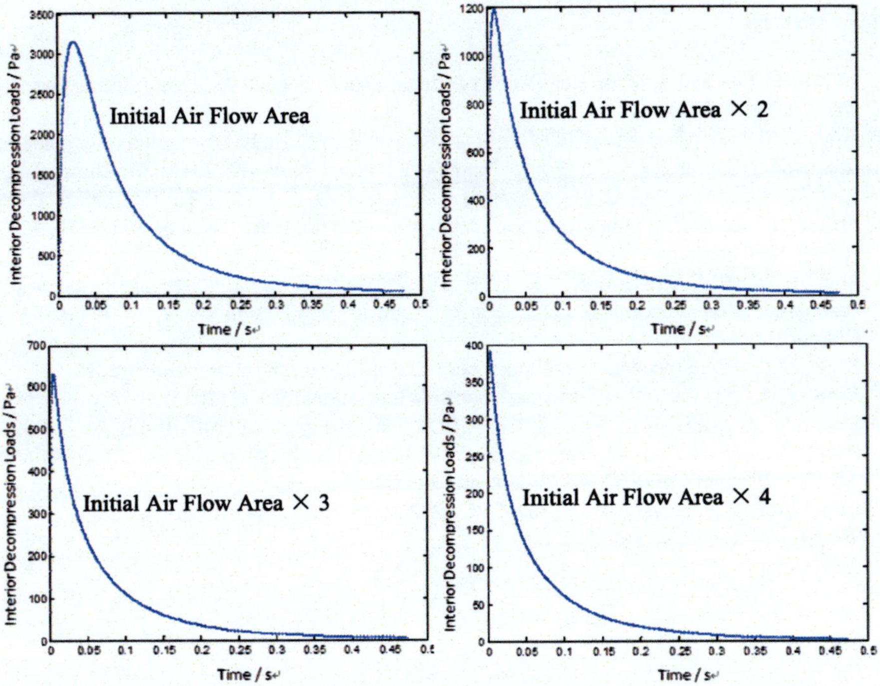

Fig. 8. Curve of interior decompression load change under different air flow area

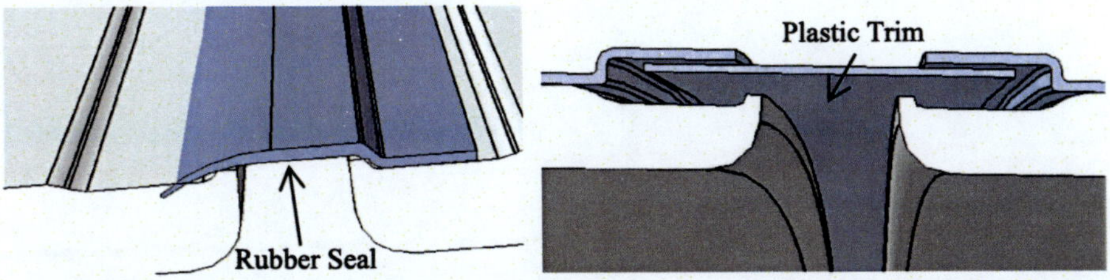

Fig. 9. Rubber seal in gap between ceiling panels and plastic trim between sidewall panels

6 Summary

By taking a certain type of aircraft as an example, a calculation method was found to calculate and analyze the interior decompression load and its changing process in different cases. The structural failure of zone B (represented by the cockpit windshield glass and cabin side window glass) was considered the main working condition in the interior design according to calculation and safety analysis. On this basis, the application of the decompression load in the interior design optimization is studied, including optimize and allocate the total gap, design the gap size and style.

References

1. CCAR-25-R4 Civil Aviation Regulation of China: PART 25. Civil Aviation Administration of China, China (2011)
2. 李枫. 浅谈民用飞机客舱内饰的工程设计. 装备制造技术. Li, F.: Discussion on engineering design of civil aircraft interior. Equip. Manufact. Technol. **8**, 96–99 (2014). (in Chinese)
3. 刘华源, 屠毅. 民用飞机泄压载荷影响因素研究. 科技视界. Liu, H., Tu, Y.: Numerical simulation of decompression in pressurized cabin of civil aircraft. Sci. Technol. Vis. **2016**, 30–31 (2016). (in Chinese)
4. 郁成德. 增压舱突然泄压载荷计算. 民用飞机设计与研究. Yu, C.: Loads calculation on sudden compression of pressurized compartment. Civil Aircraft Design Res. **2**, 43–49 (1997). (in Chinese)
5. Khan, M.K.: Fluid mechanics and machinery. Oxford University Press, India (2015)
6. Rideout M.S.K.: Barron's AP Physics. Barron's Educational Series (2015)
7. SWCAAC-SIR-2018–1. 航空器严重症候调查报告.中国: 中国民用航空局 (2020). SWCAAC-SIR-2018–1 Aircraft Serious Syndrome Investigation Report. Civil Aviation Administration of China, China (2020). (in Chinese)
8. CNR. http://sc.cnr.cn/sc/2014sc/20200602/t20200602_525113645.shtml. Accessed 08 Apr 2023

Fatigue Crack Growth Performance in Refilled Friction Stir Spot Welding and Friction Stir Welding of Aluminum Alloy Joints

Gang Li[(✉)], Haiying Zhang, and Weifeng Zang

National Key Laboratory of Strength and Structural Integrity, Aircraft Strength Research Institute, Xi'an 710065, China

lggang@126.com

Abstract. Friction stir weld (FSW) is a novel solid-state welding technique that has gained significant attention in recent years due to its many advantages over conventional welding techniques. One of the potential applications of FSW is in the aerospace industry, where it could replace riveting as a joining method. In this study, we investigated the fatigue crack growth behavior of four different aerospace materials in three different joining methods, including Refilled friction stir spot weld (FSSW) and FSW, and were compared with those of riveted joints. The experiments were performed using MTS fatigue testing machine. The four aerospace materials investigated in this study were 2A12/2A12, 2A12/7A09, 2024/2024, and 2024/7075. The experiments were conducted in three different joining methods, including FSSW, FSW, and riveting. The fatigue crack growth (FCG) rate was determined for each material and joining method by analyzing the relationship between the crack growth rate and the stress intensity factor. The results showed that the crack growth rate decreased significantly when the crack extended to the welding area and riveted hole in the spot and riveting experiments. However, the crack growth rate did not change significantly in the FSW experiment. In terms of different forms of samples, the crack growth rate was higher in the FSSW samples at lower values of stress intensity factor, but it was equal to or greater than the crack growth rate of the FSW samples as stress intensity factor increased. The crack growth rate of the riveting form was relatively stable and higher than the FSSW and FSW samples.

Keywords: Refilled friction stir spot welding · Friction stir welding · Fatigue crack growth

1 Introduction

The use of high-strength aluminum alloys, such as 7075, 7A09, 2A12, and 2024, is common in aviation structure manufacturing [1]. However, traditional fusion welding techniques can lead to porosity, thermal cracking, and HAZ softening. In 1991, The Welding Institute (TWI) developed a solid-phase joint technology called Friction Stir Welding (FSW), which offers significant advantages over traditional welding methods.

Chinese Society of Aeronautics and Astronautics (Ed.): CASTC 2023, LNME, pp. 511–524, 2024.
https://doi.org/10.1007/978-981-99-8861-7_51

FSW boasts high-quality joints, increased efficiency, low energy consumption, minimal welding deformation, low residual stress, and no pollution [2]. These characteristics make FSW an essential tool for the integrated manufacturing of aviation structures and for reducing structural quality and manufacturing costs. Additionally, Friction Stir Spot Welding (FSSW) has emerged as a new solid-phase joint technology based on FSW [3–5]. Friction Stir Spot Welding (FSSW) is a process that promotes the flow of material to be welded by the stirring head, resulting in a uniform and dense weld core, and producing high-quality joints. FSSW is a simple process compared to traditional resistance spot welding and riveting, while offering excellent bonding strength and stable welding quality. Moreover, FSSW provides a clean working environment and a long life for welding tools. Okada et al. [6]. Conducted a study on the crack growth performance of friction stir welded joints under two welding conditions. The results showed that the fatigue life of friction stir welded joints was longer than that of riveted joints, and the crack growth rate in the welded zone differed from that in the matrix. Takao et al. [7]. Studied the effect of stress ratio on crack growth rate in friction stir welding of 2024-T3 aluminum alloy. Their findings revealed that the maximum crack growth rate did not occur within the maximum residual tensile stress region. Chen et al. [8] explored the fatigue properties of 30 mm thick friction stir welded joints. Their results indicated that the fatigue properties decreased first and then increased from the surface to the bottom, and the fatigue properties varied unevenly along the thickness direction. Qin et al. [9]. Conducted research on the crack growth properties of base metal, argon-arc welding, and friction stir welding. They found that the crack growth rate of argon-arc welding was the highest, while that of friction stir welding was the lowest. This was because argon-arc welding generated larger residual stress and larger grain size than friction stir welding and base metal. Finally, Qiao et al. [10]. Investigated the crack growth behavior of 7075-T6 and LC4CS friction stir welded joints. Their results showed that the anti-crack growth performance of 7075-T6 and LC4CS friction stir welded joints was superior to LC4CS.Research has indicated that Friction Stir Welding (FSW) joints possess better fatigue resistance than traditional welding methods such as MIG, TIG and laser welding. FSW can even achieve the same fatigue performance as the base metal if weld root defects are eliminated [11–13]. However, most studies on FSW fatigue resistance are based on fatigue S-N curves [14, 15], and there is limited analysis on the fatigue crack growth of FSW joints with thin-wall structures, especially on the crack growth characteristics of thin plates with different connection modes such as spot welding (FSSW), seam welding, and riveting of different materials.

This paper aims to conduct crack growth tests on four types of joints in the form of Refilled friction stir spot welding(FSSW), friction Stir Welding (FSW), and riveting of 1 mm thick aluminum alloy sheet 2A12 and 2A12, 2A12 and 7A09, 2024 and 2024, and 2024 and 7075. The study compares and analyzes the fatigue crack growth of different materials and welding forms. The findings can provide an essential experimental basis for evaluating the anti-fatigue crack growth characteristics of friction stir spot welding, friction stir welding, and riveting joints of aluminum alloys.

2 Experimental Procedures

2.1 Specimen Fabrication

The experiment used 1mm thick aluminum alloy plates (2A12-T42, 2024-T3, 7A09-T62, and 7075-T62) with a welding size of 400 mm x 100 mm x 1mm. Before processing the fatigue samples, X-ray nondestructive testing was conducted to ensure that there were no defects in the welds. The Middle tension (M(T)) samples were used for fatigue crack growth rate testing, with their forms and sizes illustrated in Figs. 1, 2 and 3.

Four different combinations of aluminum alloys (2A12 and 2A12, 2A12 and 7A09, 2024 and 2024, and 2024 and 7075) were used for friction stir spot welding, friction stir welding, and riveting. The fatigue prefabricated crack was placed at the center of the sample perpendicular to the loading direction, with an initial crack length of 12 mm. These tests provide valuable insights into the anti-fatigue crack growth characteristics of different welding methods and aluminum alloy combinations, which can serve as an essential experimental basis for future evaluations.

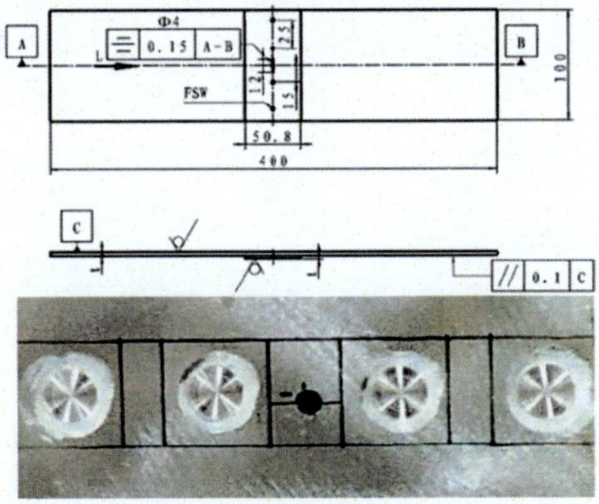

Fig. 1. Refilled FSSW specimen configuration

2.2 Fatigue Crack Growth Test

A fatigue crack growth test was performed in accordance with ASTM E647 standard using an MTS 8801 servo-hydraulic fatigue testing machine. The specimens were subjected to a constant amplitude fatigue load with a stress ratio of $R = 0.1$ until fracture occurred, and the loading frequency was set at 5 Hz. The crack extension was monitored throughout the test using a 50x digital microscope, and the positions of the two crack tips relative to the central line of the specimen (i.e., the half crack lengths) were recorded as a function of load cycles.

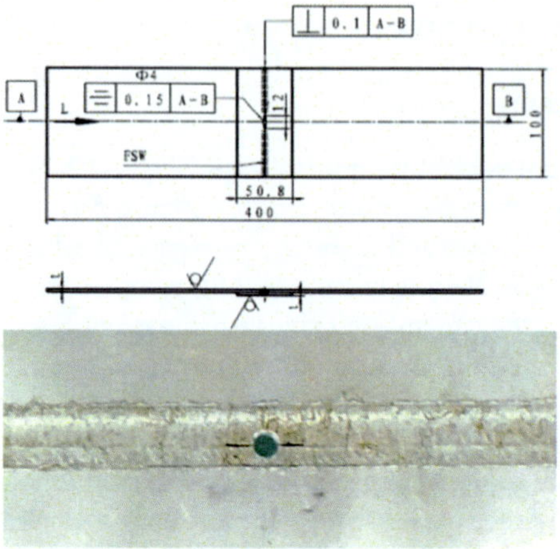

Fig. 2. FSW specimen configuration

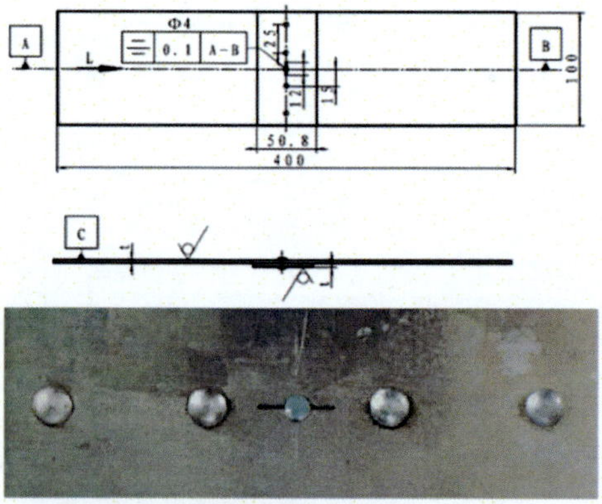

Fig. 3. Riveting specimen configuration

3 Results and Discussion

3.1 a-N Comparison of Crack Growth

To facilitate analysis and comparison of crack growth in the four aluminum alloys (2A12 and 2A12, 2A12 and 7A09, 2024 and 2024, 2024 and 7075), a-N curves were plotted for samples of the same form but different materials on the same coordinate system. The prefabricated crack center point was taken as the x-coordinate point 0, with the left and right crack tips at x = -6mm and x = 6mm, respectively. The a-N curves for the three forms of friction stir spot welding, friction stir welding, and riveting were compared and presented in Figs. 5, 6 and 7.

The results in Fig. 5 indicate that the crack growth rate of 2A12 and 2A12 is similar to that of 2024 and 7075 in the same FSSW structure, while the crack growth rate of 2A12 and 7A09 is similar to that of 2024 and 2024. The propagation rate decreases

Fig. 4. Experimental field

significantly when the crack extends to the spot welding affected area. Figure 6 shows that in the friction stir welding structure, the crack growth rate from large to small is in the order of 2A12 and 7A09, 2024 and 7075, 2A12 and 2A12, and 2024 and 2024, with no significant change in the crack growth rate. As seen from Fig. 7, in the riveted joint, an obvious expansion retardation occurs when the crack extends to the riveted hole, and finally, after a period of fatigue load, it suddenly and rapidly expands until failure.

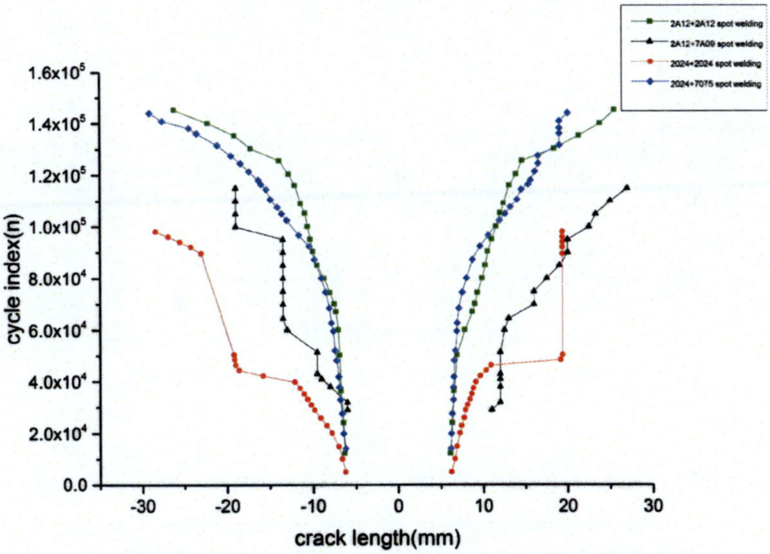

Fig. 5. Measure crack length versus load cycles for refilled FSSW

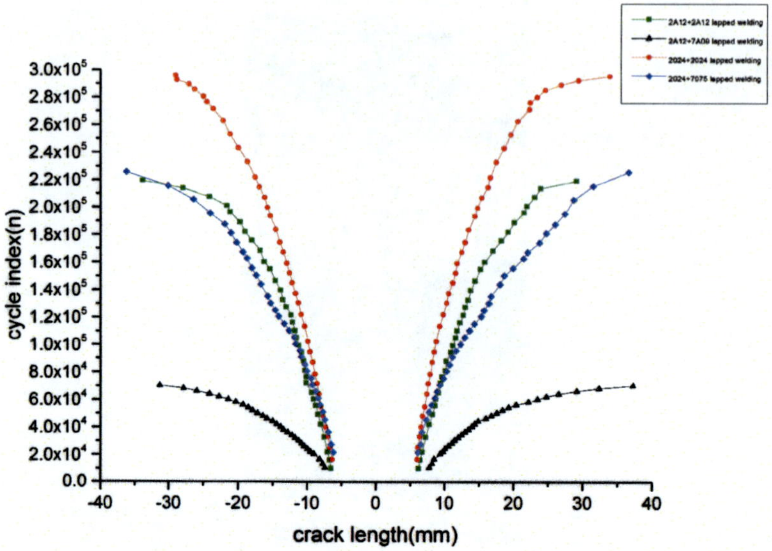

Fig. 6. Measure crack length versus load cycles for FSW

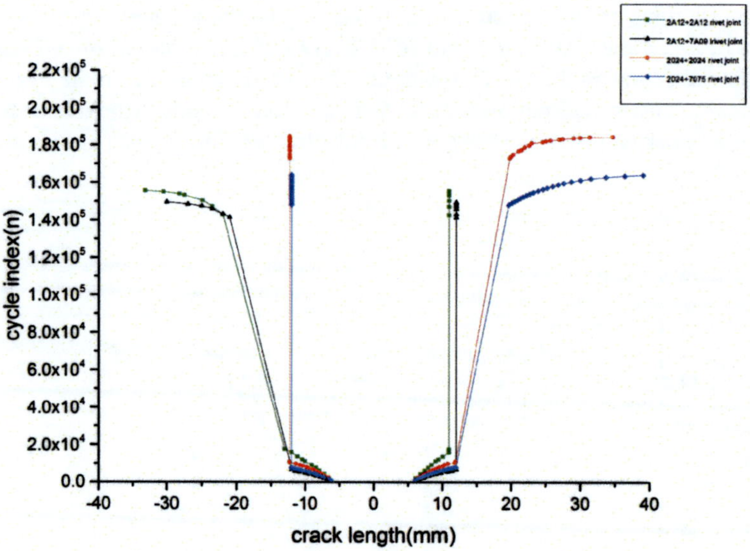

Fig. 7. Measure crack length versus load cycles for riveting

3.2 Comparison of Crack Growth Rate

To better understand how crack growth rates vary between different forms of the same aluminum alloy, da/dN-ΔK curves are plotted for three forms of the same material in the same coordinate system, as illustrated in Figs. 8, 9, 10 and 11.

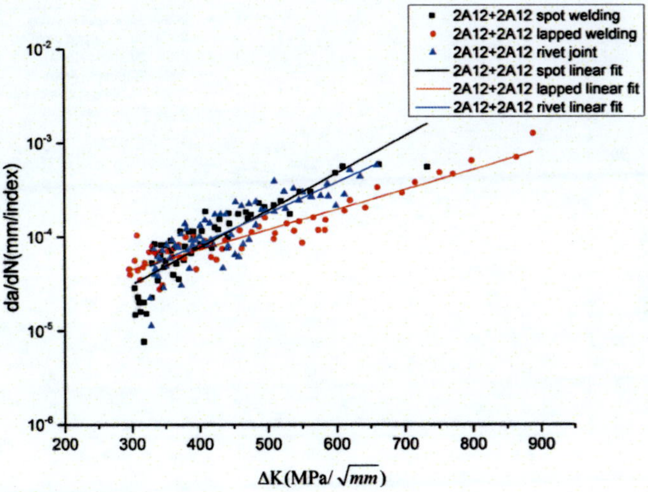

Fig. 8. Comparision of FCG rate between refilled FSSW /FSW and riveted of 2A12-2A12

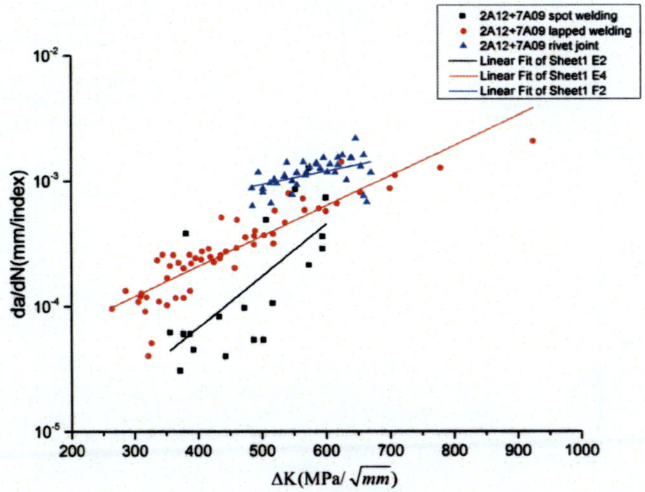

Fig. 9. Comparision of FCG rate between refilled FSSW /FSW and riveted of 2A12-7A09

When ΔK is small, the growth rate of spot welding is slower than that of seam welding and riveting. However, as the ΔK increases, the fatigue crack growth rate of spot welding becomes equal to or exceeds the rate of seam welding, and the expansion rates of the three forms gradually converge. The expansion rate of riveting remains relatively stable and higher than that of spot welding and seam welding.

The slower expansion rate of the FSSW form can be attributed to the formation of fine grains in the core area, which impedes the crack ΔK growth when it enters the refilled stir zone (SZ). However, as the ΔK increases, the rate of FSSW increases, and

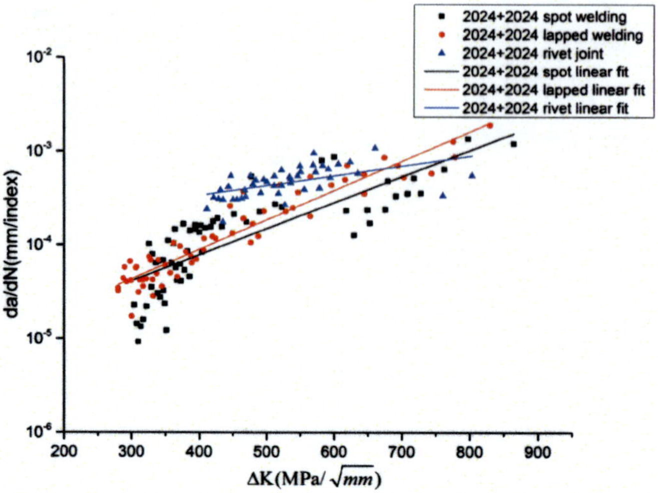

Fig. 10. Comparision of FCG rate between refilled FSSW /FSW and riveted of 2024–2024

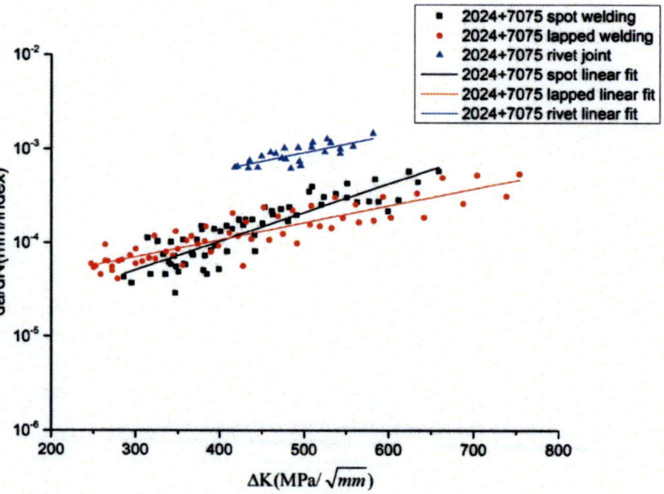

Fig. 11. Comparision of FCG rate between refilled FSSW /FSW and riveted of 2024–7075

the crack propagates through the welding influence area and enters the base metal area, leading to a change in the medium and an increase in the crack growth rate.

3.3 Crack Growth Morphology

As illustrated in Fig. 12, the propagation of cracks in friction stir spot welding occurs in two distinct ways: firstly, by expanding along the periphery of the filled probe hole, and secondly, by penetrating through the weld core. Additionally, Fig. 13 displays the predominant pattern of crack propagation in friction stir welding, which is found to predominantly extend along the Stirred Zone (SZ) and Thermo-Mechanically Affected Zone (TMAZ), perpendicular to the loading direction. Further analysis, depicted in Fig. 14, indicates that in all tested riveted specimens, the crack saw direction at the front edge of the inner rivet hole remained linear. However, as the crack approached the rivet hole, certain specimens in each group exhibited crack branching. It is important to note that the experimental results reveal that rapid fracture occurs when the crack tip reaches the vicinity of the free edge.

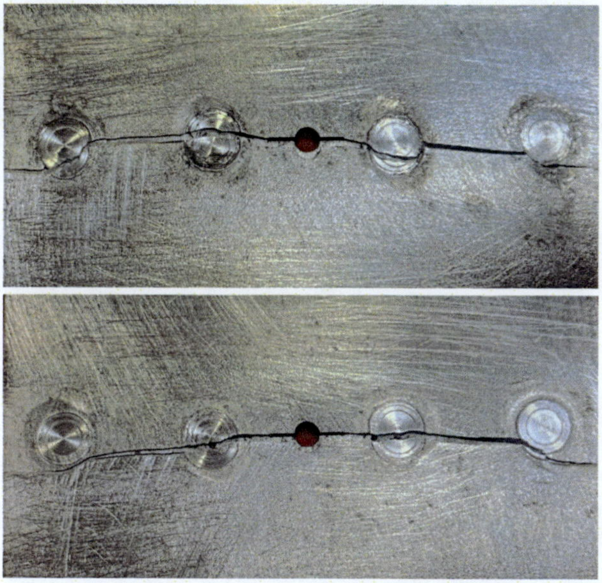

Fig. 12. Crack morphology for refilled FSSW

3.4 Fracture Surface

The scanning electron microscopy (SEM)analysis of the fracture of the crack growth test piece after failure is depicted in Figs. 15, 16 and 17, revealing the typical fracture morphologies of different forms of samples. Figure 15 represents friction stir spot welding, Fig. 16 represents friction stir seam welding, and Fig. 17 represents riveting. Figure 15(a) shows a noticeable difference between the refilled stir zone (SZ) and the thermal-mechanical affect zone (TMAZ). The fatigue striations (fatigue bands), visible in Figs .15(d)–17(d), appear in the fatigue expansion area perpendicular to the direction of crack growth, exhibiting certain spacing, parallelism, and alternating shades. These fatigue bands' continuous direction bending into wavy patterns is a common fatigue fracture feature of ductile materials.

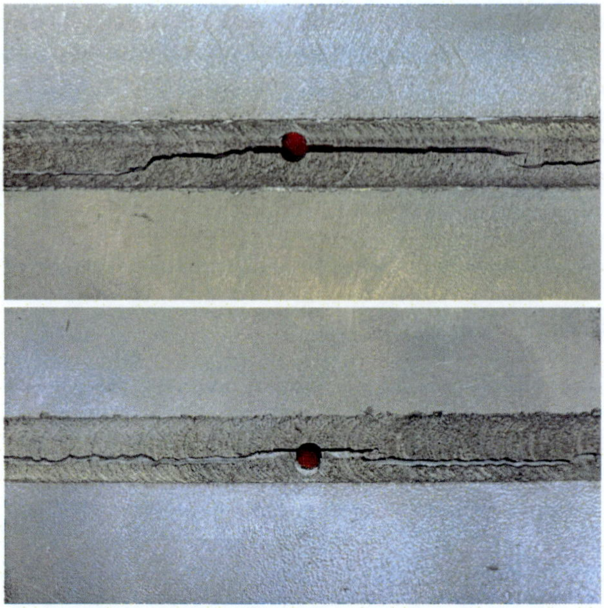

Fig. 13. Crack morphology for FSW

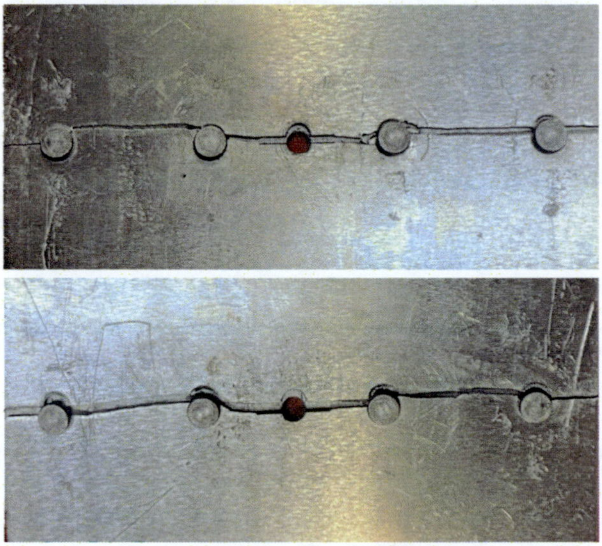

Fig. 14. Crack morphology for riveted

Moreover, Figs. 15(a)–17(a) indicate the presence of several small fault blocks in the crack growth zone, with varying sizes and heights. The fatigue bands on the same small fault block are continuous and parallel, while the bands on adjacent small fault blocks are neither continuous nor parallel. The existence of fatigue steps and the raised tearing edge at the fault block's edge accompany this pattern. The sample transient fracture occurs when the remaining area is inadequate to withstand the cyclic load after the number of cycles increases.

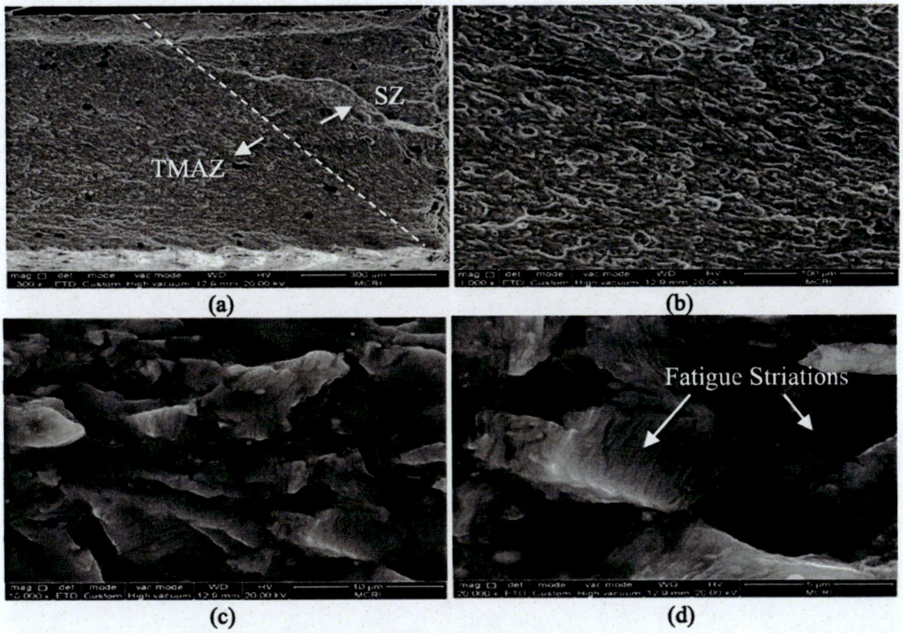

Fig. 15. Fractography of refilled FSSW

Each fatigue strip reflects the loading and unloading of a single cycle in the fatigue test. The spacing of fatigue strips in the crack growth zone represents the distance of crack growth during a cycle of fatigue stress and also the amount of energy released during a cycle of stress. Therefore, under the same test conditions, the size of strip spacing indicates the speed of crack growth, and the wider the spacing, the faster the crack growth. This implies a short life and low fatigue strength. As evident from the variation of fatigue band spacing in the crack growth region, the band spacing initially decreases and then increases with the increase of amplitude, indicating that the crack growth velocity decreases and then increases. This trend is consistent with the crack growth law recorded in the test.

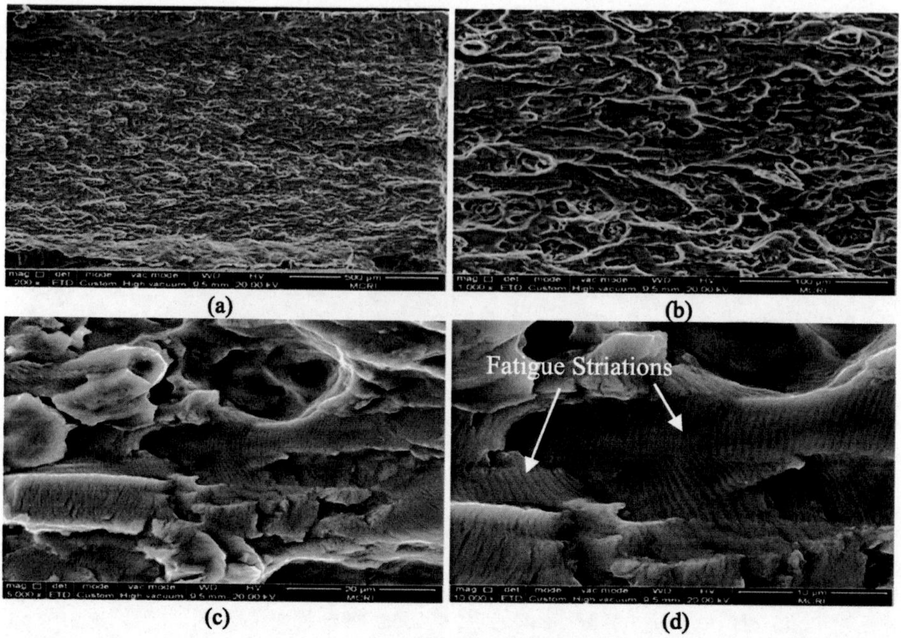

Fig. 16. Fractography of FSW

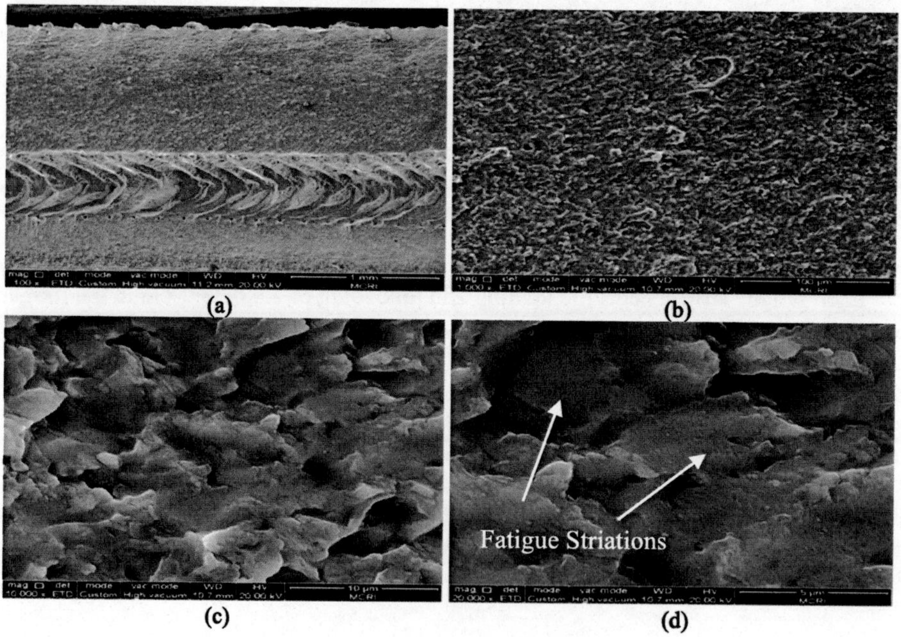

Fig. 17. Fractography of riveted

4 Conclusion

(1) In the refilled FSSW and riveting tests, the rate of crack growth decreased significantly when the crack expanded into the refilled zone or riveted hole. However, in the FSW test, there was no observable change in the crack growth rate.
(2) The crack growth rate in the riveted specimens decreased at a slower rate as the crack expanded into the riveted hole when compared to refilled friction stir spot welding (FSSW) and friction stir welding (FSW). However, when considering the overall fatigue life of both the riveted and refilled FSSW specimens, the fatigue life was almost the same.
(3) The fatigue fracture analysis of the joint revealed compound failure characteristics of fatigue bands and dimples. As the crack extended to the welding area, the spacing of fatigue bands initially decreased and then increased with an increase in the number of cycles, which was consistent with the observed trend in crack growth data with fatigue load.

References

1. Zhou, C.Z., Yang, X.Q., Luan, G.H.: Research progress on the fatigue behavior of friction stir welded joints. Rare Metal Mater. Eng. **35**(7), 1172–1176 (2006). (in Chinese)
2. Nandan, R., Debroy, T., Sandvik, A.: Recent advances in friction stir welding-process, weldment structure and properties. Prog. Mater Sci. **53**, 980–1023 (2018)
3. Mabuwa, S.: Fatigue behaviour of the multi-pass friction stir processed AA8011-H14 and AA6082-T651 dissimilar joints. Eng. Fail. Anal. **118**(1), 104876 (2020)
4. Kumar, R., Kumar, G., Roy, A., et al.: A comparative analysis of friction stir and tungsten inert gas dissimilar AA5082-AA7075 butt welds. Mater. Sci. Energy Technol. **5**, 74–80 (2022)
5. Gao, F., Guo, Y., Yang, S., et al.: Fatigue properties of friction stir welded joint of titanium alloy. Mater. Sci. Eng. A.: Struct. Mater. Propert. Misrostruct. Process **793**, 139819 (2020)
6. Okada, T., Machida, S., Watanabe, N.: Fatigue crack growth behavior of nonthrough cracks in friction stir-welded 2024–T3 aluminum alloy. Trans. Japan Soc. Aeronaut. Space Sci. **62**(1), 20–31 (2019)
7. Okada, T., Machida, S., et al.: Fatigue crack growth of friction-stir-welded aluminum alloy. J. Aircraft **54**(2), 1–10 (2017)
8. Chen, D.G., Liu, J.H., Chen, D.L., et al.: Fatigue properties of welded joint of 77A52 aluminum alloy thick plate by friction stir welding. Ordnance Mater. Sci. Eng. **4**, 90–94 (2016). (in Chinese)
9. Qin, H., Yang, X., Science, S.O., et al.: Performances of fatigue crack growth for aluminum friction stir welds and base materials. J. Aeronaut. Mater. **31**(5), 41–47 (2017)
10. Qiao, Y., Zhang, H., Zhao, L., et al.: Fatigue crack growth properties of AA 5754 aluminum alloy gas tungsten arc welding and friction stir welding joints. J. Mater. Eng. Perform. **29**(4), 2113–2124 (2020).
11. Jordon, J.B., Horstemeyer, M.F., Daniewicz, S.R.: Fatigue characterization and modeling of friction stir spot welds in magnesium. J. Eng. Mater. Technol. Trans. **132**(4), 041008 (2010)
12. Hassanifard, S., Mohammadpour, M., Rashid, H.A.: A novel method for improving fatigue life of friction stir spot welded joints using localized plasticity. Mater. Des. **53**, 962–971 (2014)
13. Rao, H.M., Jordon, J.B., Barkey, M.E., et al.: Influence of structure integrity on fatigue behavior of friction stir spot welded AZ31 Mg alloy. Mater. Sci. Eng., A **564**, 369–380 (2013)

14. Hong, S.J., Kim, S.S., Lee, C.G., et al.: Fatigue crack propagation behavior of friction stir welded Al-Mg-Si alloy. Scripta Mater. **55**, 1007–1010 (2006)
15. Hong, S.J., Kim, S.S., Kim, S.J., et al.: Fatigue crack propagation behavior of friction stir welded 5083–H32 and 6061–T651 aluminum alloys. Mater. Sci. Eng., A **478**, 56–64 (2008)

A Health Assessment Method of Avionics Equipment

Bing Huang[1,2,3](✉), Jing Zhou[4], Liang Cao[1,2,3], Jinglin Wang[1,2,3], Tianmin Shan[1,2,3], Ying Liu[4], Xiaohan Yao[1,2,3], Zexi Luo[1,2,3], Fan Jiang[1,2,3], and Qinyang Li[1,2,3]

[1] Aviation Key Laboratory of Science and Technology on Fault Diagnosis and Health Management, Shanghai 201601, China
huang_bing@bjtu.edu.cn

[2] AVIC United Technology Center for Basic Research on Aircraft System Fault Diagnosis and Health Management, Shanghai 201601, China

[3] AVIC Shanghai Aero Measurement and Controlling Research Institute, Shanghai 201601, China

[4] Shaanxi Aircraft Industry (Group) Company Ltd., Hanzhong 723215, China

Abstract. With the popularization of aircraft digitization, PHM (prognosis and health management) of avionics equipment is becoming increasingly important. Due to the difficulty in collecting performance parameters for avionics equipment, we propose a new method, which achieves component-level health assessment based on fault risk. In response to the issue of assessment weight dilution in the system, considering the impact of serial and parallel structures on system functionality, an improved health assessment method based on reliability logic was proposed to achieve system-level health assessment. And then, based on the above, we integrated a health assessment algorithm for avionics equipment. Through example Analysis, the methodology proposed in this paper can effectively evaluate the health status of the avionics equipment population. The analytical health calculation method proposed in this article has a fast calculation speed, which is easily deployed onboard health assessment systems and enhances the health management capabilities of avionics equipment.

Keywords: Electronic System · Health Assessment · BIT (Built-in Test)

1 Introduction

With the digital development of aircraft, the application of electronic devices in aircraft is increasing. Given the need for flight safety and maintenance economy, the importance of electronic equipment health assessment is increasingly prominent. At present, the health management of avionics equipment relies on the BIT, lacking performance parameters for health assessment. As a result, achieving a health assessment of avionics equipment is difficult. In addition, there is almost no difference between system-level and component-level health assessment methods, and system-level health assessment does not reflect the series-parallel logical structure of the system. For the health assessment of avionics equipment, it is urgent to evaluate the health status of components without performance parameters.

Chinese Society of Aeronautics and Astronautics (Ed.): CASTC 2023, LNME, pp. 525–533, 2024.
https://doi.org/10.1007/978-981-99-8861-7_52

In the absence of performance parameters, literature [1] proposed a general framework for health status assessment, which uses the quantile of life distribution to represent the health of the equipment in the population. On this basis, reference [2] uses the cumulative number of failures to characterize the health level of repairable components. Meanwhile, construct a general path model for the system degradation process, achieving health assessment of bus fleets. Reference [3] defines an exponential relationship between health index and failure probability. Because the aging failure mechanism [4–7] explains this relationship, it has been widely applied in literature, establishing the relationship between failure probability/failure rate and health index, characterizing the health status curve of the equipment population. For system-level health assessment, qualitative evaluation information is inevitable, and most literature usually uses fuzzy theory to quantify qualitative evaluation information [8, 9]. Due to the complexity and multi-level characteristics of the system, there is a problem of weight dilution in the health assessment of the system, resulting in inaccurate system health. From the perspective of system functionality, the health of components directly impacts the system. Reference [9, 10] designs a calculation method for system health based on reliability logic. Through comparative Analysis, based on reliability logic, the health assessment method for series systems has solved the problem of weight dilution. To a certain extent, the above literature solves the problem of health assessment without performance parameters and weight dilution in evaluation. Besides, component-level health assessment does not consider the importance of components, and parallel system health assessment still has the problem of weight dilution. Therefore, this paper proposes a new health assessment method that evaluates the health of components based on fault risk, i.e., the cumulative effect of fault probability and severity level, and improves the system-level health calculation method based on reliability logic, to some extent reducing the impact of weight dilution on the health assessment of series-parallel systems.

2 Methods and Algorithms

In response to the health assessment of avionics equipment, a new health assessment method is conducted a health assessment of avionics equipment, As shown in Fig. 1.

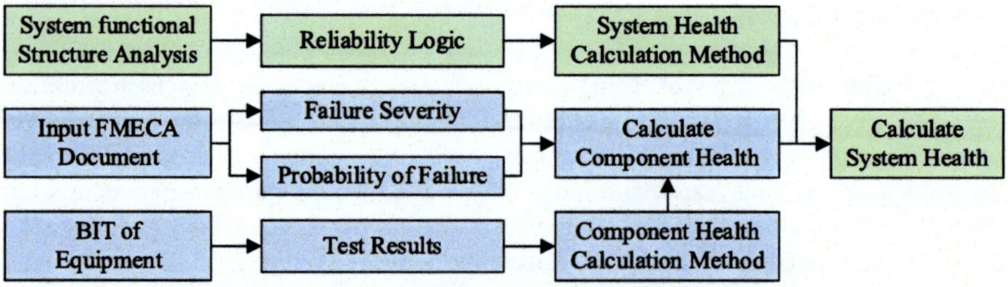

Fig. 1. A flow chart of avionics equipment health assessment

2.1 Component-Level Method

This paper conducts a health assessment of avionics components based on fault risk. We calculate the fuzzy severity and the fault risk, obtaining the health.

Firstly, construct the severity level fuzzy membership degree of the component. After determining the severity level vector $\boldsymbol{V}$, an evaluation group evaluates the severity levels at all levels. We obtain an evaluation of the severity levels of each component at all levels and calculate the number of support personnel x_{ij} for each severity level. Based on this, we obtain the membership degree $b_{ij} = x_{ij}/x_i$, where x_i represents the number of personnel for the evaluated component i; After constructing the fuzzy membership vector $\boldsymbol{B}$ of the component's severity level, based on the severity level vector $\boldsymbol{V}$, the fuzzy severity S is calculated using the center of gravity method and normalized, as shown in formula (1):

$$S = \boldsymbol{V}\boldsymbol{B}^{\mathrm{T}}/maxV \tag{1}$$

Secondly, component-level health is calculated based on fault risk. Based on the failure rate h and initial reliability R_0 of the component, calculate the probability of component failure occurrence OPR, and combine the fuzzy severity level S and its adjustment coefficient $K = 0.5$ to calculate the failure risk RNK of the component, as shown in formula (2):

$$\begin{cases} OPR = 1 - R_0 \exp[-hT] \\ RNK = OPR^{1-KS} \end{cases} \tag{2}$$

Based on the BIT results, the calculation method for component health is selected, where $BIT = 0$ indicates no fault, otherwise fault, as shown in formula (3):

$$HI = \begin{cases} 1 - RNK, \ BIT = 0 \\ 0, \ BIT = 1 \end{cases} \tag{3}$$

2.2 System-Level Method

By improving the health assessment method based on reliability logic, we propose the health assessment of avionics systems. Firstly, we construct the adaptive evaluation weights of the components, and then an improved method based on reliability logic is used to calculate the system-level health. The construction of adaptive component evaluation weights includes a relative importance judgment matrix, component weight calculation, consistency check, and adaptive adjustment of a judgment matrix.

The First Step is to Construct a Relative Importance Judgment Matrix. Based on the fuzzy severity $\boldsymbol{S}$ of each component in the system, calculate the relative importance judgment matrix $\boldsymbol{R}$. Adjust the relative importance matrix according to the range of the nine-scale judgment scale, with an adjustment coefficient of L. The relative importance judgment matrix of the system is shown in formula (4):

$$\boldsymbol{R} \leftarrow \left[\boldsymbol{S}^T\boldsymbol{S}\right]^L \tag{4}$$

The Second Step is to Calculate the Evaluation Weight Vector of the Component. Firstly, sum the row and column directions of the relative importance judgment matrix $\boldsymbol{R}$ to obtain the column sum vector $\boldsymbol{CS_R}$. Then, normalize the column direction of the relative importance judgment matrix $\boldsymbol{R}$ by dividing it by the number of columns and vectors $\boldsymbol{CS_R}$ to obtain the normalized judgment matrix $\boldsymbol{R_C} = \boldsymbol{R}/\boldsymbol{CS_R}$. Then, perform row sum processing on the normalized judgment matrix $\boldsymbol{R_C}$ to obtain the row sum vector $\boldsymbol{RS_R}$. Sum all elements of $\boldsymbol{RS_R}$ to obtain the element sum $TS_{\boldsymbol{R}}$, and then calculate the weight vector $\boldsymbol{W}$ of each component in the system, as shown in formula (5):

$$\boldsymbol{W} = \boldsymbol{RS_R}/\boldsymbol{TS_R} \tag{5}$$

The Third Step is to Conduct a Consistency Check. Firstly, calculate the dot product of the relative importance judgment matrix and the weight vector $\boldsymbol{R}$. Then, divide by numbers with the weight vector $\boldsymbol{W}$ to obtain the vector $\boldsymbol{\lambda}$. Take the average of the vector $\boldsymbol{\lambda}$ to obtain the maximum eigenvalue λ_{max}. Then, based on the number n of components in system, calculate the consistency index CI. Based on the random consistency index RI corresponding to the number n, calculate the consistency ratio CR. Finally, if $CR < 0.1$ passes the consistency check, output the weight vector. On the contrary the judgment matrix is adaptively adjusted, and the relevant calculations are shown in formula (6):

$$\begin{cases} \boldsymbol{\lambda} = \boldsymbol{RW}/\boldsymbol{W} \\ \lambda_{max} = \overline{\boldsymbol{\lambda}} \\ CI = (\lambda_{max} - n)/(n-1) \\ CR = CI/RI \end{cases} \tag{6}$$

The Fourth Step is to Adjust the Relative Importance Judgment Matrix Adaptively. Firstly, initialize the adjustment coefficient $d = 0.5$, adjust the elements in the upper right corner of the relative importance judgment matrix one by one, adjust element $r = r^d$, assign the value r to the elements symmetrical along the main diagonal of element $r = 1/r$, obtain the adjusted relative importance judgment matrix $\boldsymbol{R} \leftarrow \boldsymbol{R_r}$ based on the adjusted elements, recalculate the relative importance judgment matrix and component weights, and obtain the adjusted weight vector $\boldsymbol{W}$. If the consistency check passes, output the weights of the components, otherwise, the judgment matrix adaptive adjustment module will continue to adjust.

The Fifth Step is to Calculate the System Level Health Level. According to the reliability logic, calculate the system health level at each level, as shown in formula (7):

$$HI_s = \begin{cases} \prod_{i=1}^{n} HI_i^{w_i}, \ \textit{Series System} \\ \sum_{i=1}^{n} w_i HI_i^{w_i}, \ \textit{Parallel System} \end{cases} \tag{7}$$

2.3 Integrated Algorithms

Based on the health assessment methods at the component and system levels, the algorithm flow for avionics equipment health assessment as shown in Fig. 2 has been organized. The specific algorithm steps are as follows:

Step S1. Input the component severity level fuzzy membership file, BIT file, component failure rate, and initial reliability file.

Step S2. Input the severity level fuzzy membership file, organize the SEVERITY algorithm according to formula (1), and calculate the component's fuzzy severity.

Step S3. Input the BIT file, component failure rate, and initial reliability file, as well as the fuzzy severity level of the component. According to formulas (2) and (3), organize the HEALTH algorithm, calculating the component health.

Step S4. Input the fuzzy severity, according to formula (4), construct the JUDGEMENT algorithm to calculate the relative importance judgment matrix.

Step S5. Input the relative importance judgment matrix of the components, organize the WEIGHT algorithm according to formula (5), calculating the weights.

Step S6. Input the weight and relative importance judgment matrix of the components, organize the VALIDATION algorithm according to formula (6), and perform consistency check (passed or not).

Step 7. If it does not pass, fine-tune the relative importance judgment matrix by using the ADJUST Algorithm for adaptive adjustment, and repeat steps S5 to S7; If passed, proceed to step S8.

Step S8. Input the health level and evaluation weight of the components, and organize the SYSTEM_HEALTH algorithm according to formula (7), calculating the system-level health.

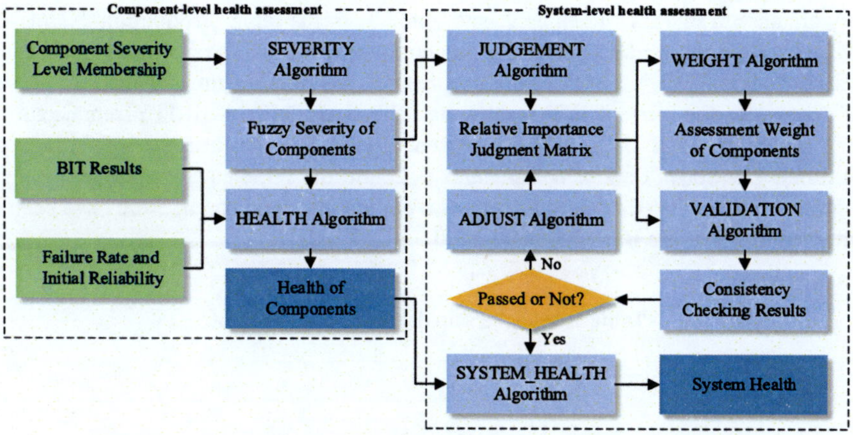

Fig. 2. A health assessment algorithm process for avionics equipment

3 Example Analysis

This paper takes an avionics equipment as an example to Analysis the reliability logic of the system structure and obtain the reliability block diagram shown in Fig. 3. The avionics equipment is a series-parallel hybrid structure, consisting of four identical subsystems. {U2, U3} are level-1 subsystems; {U1, U2, U3} are level-2 subsystems; {U1, U2, U3, U4, U5} are level-3 subsystems.

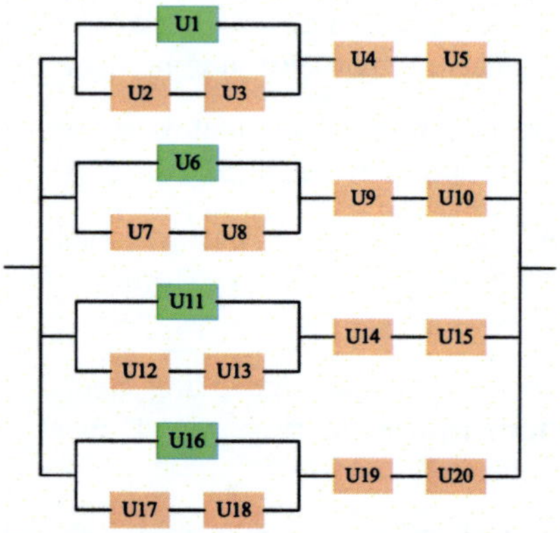

Fig. 3. Reliability block diagram of an avionics equipment

The membership degree of severity level, BIT results, failure rate, and initial reliability are shown in Table 1, Table 2, and Table 3. In the table, Production_NUMBER represents the unique identification of the component, LEVEL represents the severity level (1–5 levels), RUNNING_ TIME represents the running time of the component, BIT represents the current BIT results, and INITIAL_RELIABILITY represents the initial reliability, FAILURE_RATE represents the failure rate (number of failures per unit time). The above data of components 6–20 are the same as components 1–5. The equipment is a new system that has just been put into use, and so far, there have been no fault, meaning that the BIT results are all 0.

Table 1. Membership degree of severity level

PRODUCTION_NUMBER	*V*				
	1	2	3	4	5
1	0.90	0.10	0	0	0
2	0.05	0.80	0.15	0	0
3	0	0.10	0.85	0.05	0
4	0	0	0.05	0.90	0.05
5	0	0	0	0.20	0.80

Table 2. BIT results of components

PRODUCTION_NUMBER	RUNNING_TIME	BIT
1	2.5	0
2	5	0
3	50	0
4	500	0
5	1000	0

First, calculate the component-level health according to algorithm steps S1 to S2; Second, calculate the system-level health according to steps S3 to S8. The calculation results are shown in Fig. 4, where SYSTEM_NUMBER represents the unique identifier of the system/subsystem, and HEALTH identifies the health level of the component/system. From the figure, it can be seen that the health of the system is in a normal state, but there are abnormalities in the level-3 subsystems (SubS31 ~ SubS34) and level-1 subsystems (SubS11 ~ SubS14). Therefore, the health of the components was investigated and it was found that the health of the series components was in an abnormal state. Therefore, it is necessary to implement maintenance and repair measures for the components in abnormal states.

Table 3. Component failure rate and initial reliability

PRODUCTION_NUMBER	INITIAL_RELIABILITY	FAILURE_RATE
1	0.98	0.2
2	0.93	0.1
3	0.95	0.01
4	1	0.001
5	0.95	0.0005

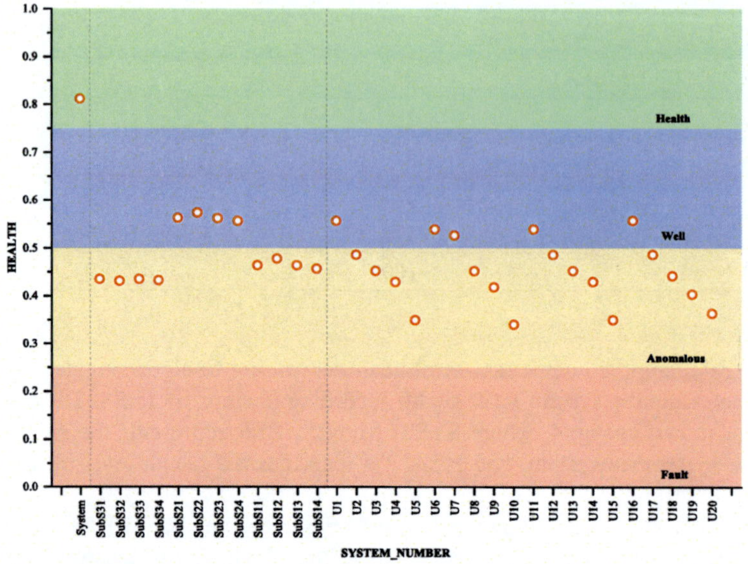

Fig. 4. Distribution diagram of component and system health

4 Conclusion

This paper proposes a new health assessment method for avionics equipment, which solves the problem of component-level health assessment without performance parameters, avoids the impact of weight dilution on system-level health assessment, and the analytical health degree calculation method is very beneficial for the deployment of airborne health assessment systems. In summary, the health assessment method proposed in this article achieves a health assessment of the overall equipment population. Of course, the method proposed in this article still has some limitations. The strong dependence of system-level health assessment on reliability engineers limits the efficiency of application development of this method on other objects. Due to the lack of integration of real-time performance parameters, this method cannot evaluate real-time health status. Therefore, in the future, we will integrate real-time health status and introduce a data-driven methodology that adaptively adjusts the hyperparameter of the health assessment model.

References

1. Jiang, R., Jardine, A.K.S.: Health state evaluation of an item: a general framework and graphical representation. Reliab. Eng. Syst. Saf. **93**(1), 89–99 (2008)
2. Jiang, R., Zhou, Y.: Failure-counting based health evaluation of a bus fleet. In: 2010 Prognostics & System Health Management Conference, pp. 1–4. IEEE, Macau (2010)
3. Hughes, D.: Condition Based Risk Management (CBRM) – enabling asset condition information to be central to corporate decision making. In: 18th International Conference on Electricity Distribution, pp. 1–5. IET, Turin (2005)

4. Li, E.W., Song, B.: Transformer health status evaluation model based on multi-feature factors. In: International Conference on Power System Technology, pp. 1417–1422. IEEE, Chengdu (2014)
5. Shi, C.K., Ning, X., Sun, Z.T.: Quantitative risk assessment of distribution network based on real-time health index of equipment. High Voltage Eng. **44**(2), 534–540 (2018)
6. Yang, C.B., Tao, Q., Zhang, J.: Equipment status evaluation based on comprehensive health index. Power Syst. Protect. Control **47**(10), 104–109 (2019)
7. Song, W.B., Fu, S.B., Wu, K.M.: Assessment method for health status of EMUs by overhaul. China Railway **2022**(2), 47–55 (2022)
8. Zhou, J.Y., E, M.C., Li, Q.: Application of fuzzy FMECA in high speed train bogie wheel failure analysis. Electric Drive for Locomotives **2018**(4), 74–78 (2018)
9. Wang, L.M.: Research on health assessment of electronic equipment of self-propelled guns. Xi'an Technological University, Xi'an (2021)
10. Teng, T., Zhao, Z.H., Cui, X.P.: Study on health assessment method of electromagnetic launching inverter. J. Nav. Univ. Eng. **31**(4), 28–35 (2019)

Modelling Tensile Deformation and Fracture Path of Selective Laser Sintered Nylon12 Lattice Structures

Yining Gao[1(✉)], Jie Niu[2], Wei Sun[3], and Liyong Jia[1]

[1] First Aircraft Institute, No. 1 Renmin Road, Yanliang District, Xi'an 710089, China
gaoyn@avic.com

[2] Wanxiang Qianchao Co., Ltd., No. 1 Wanxiang Road, Xiaoshan District, Hangzhou 311200, China

[3] Faculty of Engineering, University of Nottingham, University Park, Nottingham NG7 2RD, UK

Abstract. Lattice structures fabricated by additive manufacturing such as selective laser sintering (SLS) are promising candidates for lightweight design in aerospace applications due to their superior characteristics such as low density, high specific strength, and specific stiffness, etc. This paper reports an experimental and numerical study on the tensile behaviour of cellular lattices. Nylon 12 lattice specimens, modelled by repeating open-form unit cells consisting of hexagonal prisms, cubic prisms, and triangular prisms are printed via SLS. A finite element (FE) modelling methodology is developed to evaluate the tensile elastic-plastic and fracture behaviour of the lattice specimens. The predicted load-displacement curves and fracture paths are compared to the corresponding experimental results, and the feasibility and capability of the FE model developed is examined.

Keywords: Lattice Structure · Selective Laser Sintering · Fracture Path · Failure Mechanism · Finite Element Modelling

1 Introduction

Selective laser sintering (SLS) and laser powder bed fusion (LPBF) are additive manufacturing (AM) processes that can build solid parts directly from computer-aided design (CAD) models by selectively sintering or partial melting layer by layer of powders using laser technology. Additive manufacturing has the advantages of shorter forming time, higher precision and more design freedom compared to traditional manufacture methods, and is a new approach to realize lightweight design of aircraft structure.

Replacing bulk materials by the designed cellular material or lattice structures and similar constructions is a promising focus of AM. It is inspired by the idea of distributing material where it is exactly needed for a particular utilization. Additively manufactured lattice structures with unit cells offer superior features such as low density, high specific strength and high specific stiffness, more design variables, good energy absorption and good thermal properties [1], and thus have gained widespread attractions in engineering

Chinese Society of Aeronautics and Astronautics (Ed.): CASTC 2023, LNME, pp. 534–546, 2024.
https://doi.org/10.1007/978-981-99-8861-7_53

applications [2]. Many unit cell topologies have been proposed for various purposes [3], like diamond-type base cell for porous biomaterials (Fig. 1(a)), Kelvin cell for energy absorbent structures (Fig. 1(b)), Schoen Gyroid for fluid flow environment in tissue engineering (Fig. 1(c)), and lattice structures with various unit-cell types for honeycombs (Fig. 1(d)). For lightweight and load-bearing benefit, typically unit cells such as triangular, cubic and hexagonal are intensively used (Fig. 1(e)).

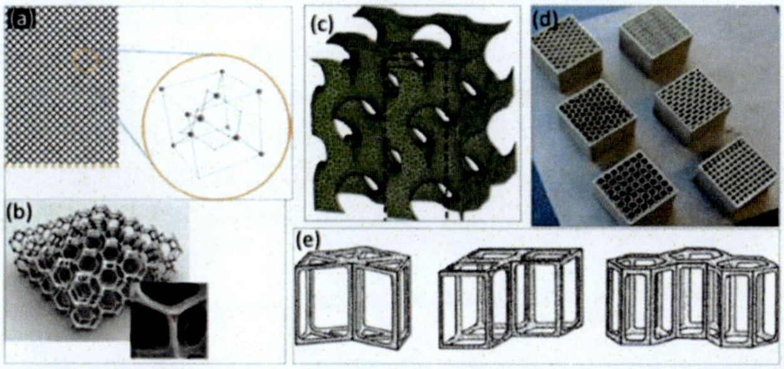

Fig. 1. Classifications of unit cells for different cellular lattice structures: (a) diamond-type base cell [4], (b) Kelvin cells [5], (c) gyroid cells [6], (d) lattice structures with various types of unit cells in structure applications [7], (e) intensely used unit cells for practical engineering [1].

The bulk elastic property of these structures can be represented by effective Young's modulus. However, their elastic-plastic and failure behaviour cannot be determined and could be quite different from 'equivalent' solid materials. Gibson and Ashby [1] laid the foundation for the theoretical study of the deformation and mechanical properties of cellular solids, and extensive research efforts have been made to investigate the mechanical behaviour of AM printed structures, such as to determine effective Young's modulus, yield strength and collapse strength, and to examine the deformation and failure mechanisms of different lattices [8–15]. Niu et al. conducted numerical and experimental study on the mechanical performances of various LPBF printed AlSi10Mg [16] beam-like structures as well as SLS printed Nylon 12 specimens [17] under uniaxial tension, but the main focus is to compare Young's modulus versus different unit-cell types and relative density, and the elastic-plastic and failure behaviour is not investigated.

In this study, a nonlinear FE modelling methodology is developed to investigate the uniaxial tensile elastic-plastic and failure behaviour of SLS fabricated Nylon 12 lattice specimens. Lattice structures are designed by periodic open-form unit cells consisting of hexagonal prisms, cubic prisms, and triangular prisms, respectively. Four types of lattice specimens with varying geometric parameter are SLS printed, and tensile tested. The FE model developed are then used to examine the deformation process, stress distribution and evolution, as well as the failure mode of these lattice structures. The predictive capability of the FE modelling methodology is evaluated by the corresponding experimental results.

2 Material and Methods

2.1 Lattice Specimen Design

From aerospace engineering application perspective, three types of unit cells, namely hexagonal prism, cubic prism, and cubic prism, were chosen to develop the lattice structures in this study. As presented in Fig. 2, the geometric parameters of unit cells were determined by length, l, strut thickness, t, and distance between layers, h. The geometry models of lattices were generated by repeating unit cells in the x and y directions. Then these cellular structures were trimmed to symmetric 20 mm × 20 mm × 50 mm blocks to replace a solid part with three unit-cell topologies for lightweight purpose. Four lattice specimens were constructed using the unit-cell geometric parameters in Table 1 and named Hex-1, Hex-2, Cub-1, and Tri-1, respectively. Their parameters were determined based on the resolution capability of the SLS process. Relative density (RD) in Table 1 is defined as:

$$RD = \left({}^{\rho^*}/_{\rho} \right) \times 100\% \tag{1}$$

where ρ^* is the effective density covered in the general dimensions of the lattice structures, and ρ is the density of the parent material.

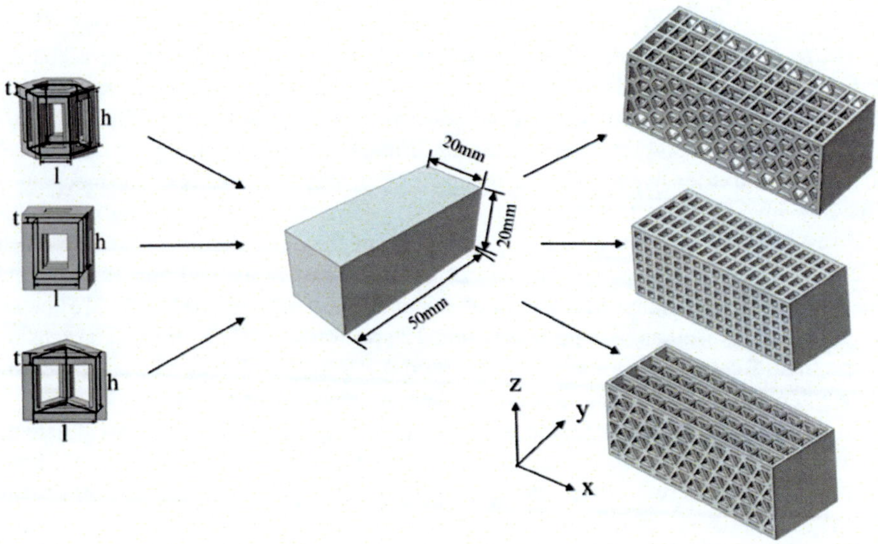

Fig. 2. Schematic illustration of lattice construction by three types of prism unit-cells

2.2 SLS Printed Lattice Specimens

As shown in Fig. 3, Nylon 12 lattice specimens with unit cells of geometric parameters in Table 1 were manufactured using SLS system FORMIGA P110 from Electro Optical

Table 1. Unit-cell geometric parameters

Sample name	Unit-cell dimensions (mm)			Relative Density (%)
	l	t	h	
Hex-1	2.20	1.40	4.65	28
Hex-2	2.80	1.00	4.75	8
Cub-1	3.20	1.40	4.65	25
Tri-1	4.80	0.92	4.77	12

System (EOS). The processing parameters of the sintering process is given in Table 2. For each specimen, two 20 mm × 20 mm cubic solid ends were printed on both sides for gripping and applying load.

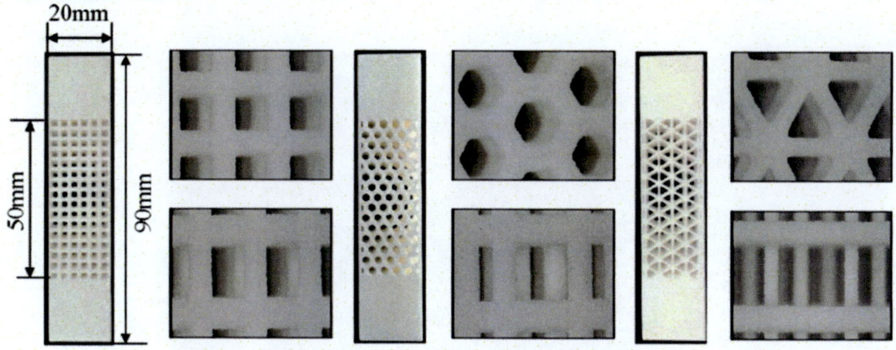

Fig. 3. Lattice specimens and local unit cells of Nylon 12

2.3 Bulk Tensile Lattice Tests

As illustrated in Fig. 4, uniaxial tensile tests were conducted at room temperature using an Instron 8950 universal testing machine at a rate of 2.5 mm/min. The extension speeds were selected according to conventional BS EN ISO 527-1:2012 for plastics. At least two samples of the above four lattice configurations were tested to ensure the effectiveness and repetitiveness of the test results. With both solid ends gripped, the specimens were axially extended until broken.

2.4 Finite Element Modelling

Mesh Generation

The established geometry models of the lattice structures were imported to Abaqus to develop the FE models. As the solid ends attached to the lattice sections on both sides

Table 2. Sintering process parameters

Laser power	16 W
Laser type	CO2, 30 W
Laser spot diameter	0.40 mm
Laser scan spacing	0.15 mm
Layer thickness	0.10 mm
Laser scan speed	5000 mm/s
Laser scan strategy	EOS sorted
Particle size	58 μm
Temperature of process chamber	168 °C
Temperature of removal chamber	157 °C
Environment temperature	23 °C

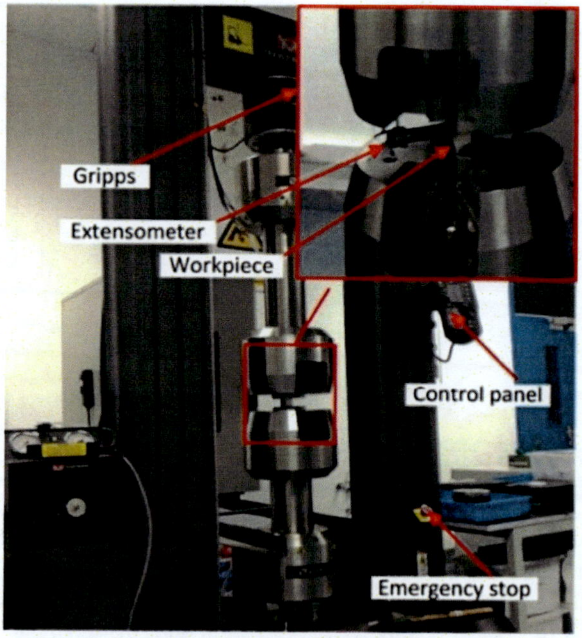

Fig. 4. Tensile test illustration

are much stiffer than the cellular sections, they are assumed to be rigid in the FE model. Continuum three-dimensional (3D) solid elements were employed to mesh the lattice sections. Linear, reduced integration 8-node hexahedral elements were used, combined with 4-node tetrahedral elements at the joints and irregular boundaries to generate the meshes as shown in Fig. 5. The global seeding size was assigned 0.3mm to ensure the consistency of the effect of element size, and element sizes for each specimen were controlled to be similar to avoid excessive distortion. Mesh sensitivity analysis was implemented and the selected meshes were justified as refining the mesh density further until both the load-displacement history and stress distribution converged.

Fig. 5. Mesh generation for the lattice blocks

Material Model

The material inputs were given in true stress-strain relationships for elastic-plastic analysis, where the von Mises yield criterion is used. The damage fracture model in Abaqus is a phenomenological model used to predict the initiation of damage. The model assumes that the equivalent plastic strain at the onset of damage, $\overline{\varepsilon}_D^{pl}$, is dependent on stress triaxiality, η, and strain rate, $\dot{\overline{\varepsilon}}_D^{pl}$ [18]:

$$\overline{\varepsilon}_D^{pl}(\eta, \dot{\overline{\varepsilon}}_D^{pl}) \tag{2}$$

where $\eta = -p/q$ is the stress triaxiality, p is the pressure stress, and q is the von Mises equivalent stress.

Damage occurs when, ω_D, a state variable that increases monotonically with plastic deformation, satisfies the following condition:

$$\omega_D = \int \frac{d\overline{\varepsilon}_D^{pl}}{\overline{\varepsilon}_D^{pl}(\eta, \dot{\overline{\varepsilon}}^{pl})} = 1 \tag{3}$$

Damage evolution defines the post damage-initiation material behaviour and depicts the rate of the material stiffness degradation once the initiation criterion is satisfied. As shown in Fig. 6, damage for an elastic-plastic material with isotropic hardening exhibits by both softening of yield stress and degradation of elasticity. Assume a linear evolution of the damage variable with plastic displacement. Then, once the damage initiation criterion is met, the damage variable increases according to:

$$\dot{d} = L\dot{\overline{\varepsilon}}^{pl} \Big/ \overline{u}_f^{pl} = \dot{\overline{u}}^{pl} \Big/ \overline{u}_f^{pl} \tag{4}$$

where the equivalent plastic displacement at failure, $\overline{u}_f^{pl}$, is computed as:

$$\overline{u}_f^{pl} = 2G_f \big/ \sigma_{y0} \tag{5}$$

where G_f is the fracture energy per unit area, σ_{y0} is the value of the yield stress at the time when the failure criterion is reached. The model ensures that the energy dissipated during the damage evolution process is equal to G_f only if the effective response of the material is perfectly plastic beyond the onset of damage.

The overall scalar stiffness degradation, D, captures the effect of all active damage mechanisms and is computed as follows:

$$D = \max\left\{d_{mult}, \max_{j \in N_{\max}}(d_j)\right\} \tag{6}$$

where d_i is the individual damage variable for each mechanism, d_{mult} is an intermediate variable that combines some of the damage variables in a multiplicative sense described as follows:

$$d_{mult} = 1 - \prod_{k \in N_{mult}} (1 - d_k) \tag{7}$$

In the above expressions, N_{mult} and $N_{\max}$ represent the sets of active mechanisms that contribute to the overall damage in a multiplicative and a maximum sense, respectively, with $N_{act} = N_{mult} \cup N_{\max}$. Fracture occurs when $D = 1$.

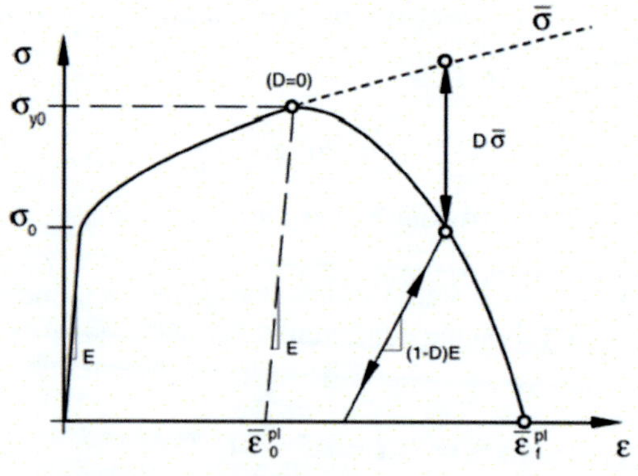

Fig. 6. Schematic representation of elastic-plastic material with progressive damage [18]

Simulation Parameters

The lattice specimens were sintered by PA2200 from EOS GmbH powders. Typical applications of the material are functional parts with high end finish capable of withstanding high mechanical and thermal load. As no tests were conducted to obtain the parameters for elastic-plastic and ductile fracture model, the elastic-plastic material properties as shown in Table 3 were approximated by taking the published nominal material data [19] and determined with a reverse engineering method to better match the predictions with the experimental results. The parameters for ductile fracture modelling were determined so that the damage fracture model of a standard tensile specimen according to ISO 527-2 1A was predicted to rupture in the middle section showing typical cup-and-cone feature.

Abaqus/Explicit was chosen to develop the methodology of simulating the quasi-static deformation process for the lattice specimen under uniaxial tension. To simulate uniaxial tensile test condition, boundary condition and load were applied that the lattice blocks were restrained at one end, while all nodes on the other end were constrained except the displacement along the tensile load direction, meanwhile subjected to a constant uniaxial displacement. The simulation parameters are listed in Table 3.

Table 3. Simulation parameters used for FE analysis

	Parameters	Values
Material properties	Young's Modulus (MPa)	1500
	Poisson's ratio	0.394
	Yield strength (MPa)	41.9
	Density (kg/m3)	930
Step parameters	Step type	Smooth step
	Load rate (mm/s)	2
	Increment order of magnitude	10^{-8}

3 Results and Discussion

3.1 Load-Displacement Relations

Figure 7 is the load-displacement curves obtained from tensile tests. Two specimens of each lattice configuration were printed and tested, and the results demonstrate good repetitiveness in the elastic-plastic stage, but some difference in the extension length before fracture. It shows that the elastic-plastic deformation process follows a smooth and steady progress consisting of three stages: an initial elastic region with a high level of linearity, a yielding and strain-hardening plastic deformation region, and finally, a plateau of near constant stress until fracture. Specimens Hex-1 and Hex-2 of conformal configuration present great difference in the load-displacement curve due to the variation in relative density. The load-bearing capacity of specimen Cub-1 with lower relative density is better than Hex-1 in that Cub-1 is stiffer and reveals superior strain-hardening characteristics. From the above comparisons, it is shown that the mechanical performance of the lattice structures is related to both relative density and unit-cell type. In terms of fracture, it is shown that, in general, the higher the load-bearing capacity, the longer extension the specimen experienced before fracture. Specimen Tri-1 presents multi-stepped fracture that corresponds to the complete rupture of each layer of unit cells observed in the test process.

Comparisons of the experimentally measured and FE predicted load-displacement curves for the four lattice specimens were shown in Fig. 8. The results reveal that the predicted quasi-static tensile behaviour of the lattice structures agree reasonably well with

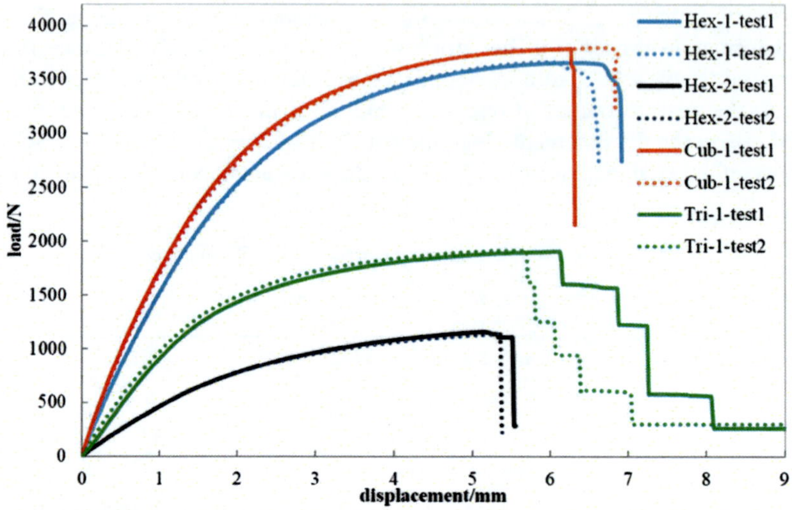

Fig. 7. Load-displacement curves of the lattice specimens from the tensile tests

experimental results. Similar trends are observed, and the predictive curves can reflect the overall features of the elastic-plastic structural responses. The load-bearing capacity of all samples was slightly underestimated compared to the test results, while the stiffness is slightly over predicted. The good agreement between the prediction and experiments validates the effectiveness of the FE modelling methodology. The differences may be partly due to the reproductivity of material property, variation of experimental conditions, residual stresses, processing quality, model simplifications, etc.

3.2 Local Stress Evolution

Figure 9 presents the predicted von Mises stress contours and deformation process of a typical unit cell picked from the central section of the four lattice blocks, respectively, after yielding at displacement equals to 2 mm, 4 mm, and 5 mm. The positions of the maximum von Mises stress in the unit cells are highlighted by the red circle. The von Mises stresses for specimens Hex-1, Cub-1, and Tri-1, primarily homogeneous within the sections, gradually redistribute with increasing tensile load, and become higher in the horizontal struts than those in the other directions, demonstrating that the struts parallel to the tensile load are mainly responsible for bearing load. Plastic deformation initiates near the central portion of the horizontal struts of the unit cells where necking occurs. For specimen Hex-2, however, stress distributes relatively uniform in the horizontal struts while localizes near the joints of adjacent struts. The bevel struts experience plastic bending and are more stretched along the tensile load direction compared to these in Hex-1.

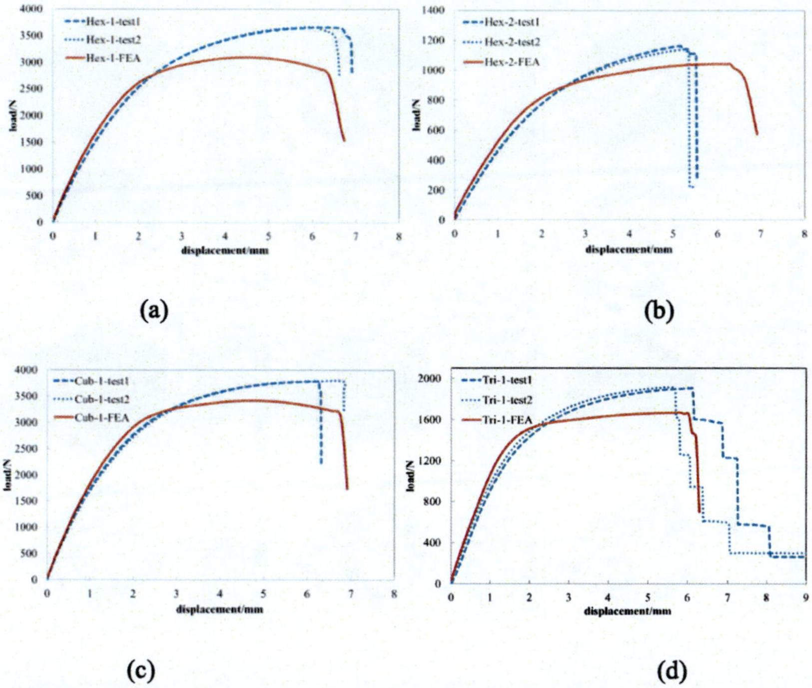

Fig. 8. Comparisons of the predicted and experimental load-displacement curves for (a) Hex-1, (b) Hex-2, (c) Cub-1, and (d) Tri-1

3.3 Fracture Path and Failure Mechanisms

Figure 10 displays the comparison of the physical and FE predicted fracture position. It is shown that all specimens ruptured near the central portion, however, their fracture patterns are different. For specimens Hex-1, Hex-2, and Cub-1, the specimens went through abrupt failure, and the fracture planes were perpendicular to the direction of the tensile load. Specimen Tri-1, however, exhibits multi-stepped progressive fracture, and the fracture path was oriented at 60 degrees to the direction of load due to its geometric nature. Comparison of the observed and predicted rupture features show high conformances in the position of failure and the rupture angle, showing that the phenomenological ductile damage model can give a reasonably accurate prediction of the fracture mode.

Figure 11 compares the predicted and tested local failure of unit cell for the four lattices. The observed fracture paths after tensile test are denoted by the red curves. From unit-cell perspective, failure develops along the centre of horizontal struts in single unit cell where stress localizes for specimens Hex-1, and Cub-1. Unit cell of specimen Hex-2 ruptures near the stress-concentrated joints of adjacent struts, as well as through the horizontal struts during the test. The FE predictions of Hex-2 presents localized stiffness degradation in the joints as well as the horizontal struts, however, the stiffness degradation value within the horizontal strut is predicted higher where damage is assumed to initiate. For specimen Tri-1, the predicted fracture path passes through both the horizontal and

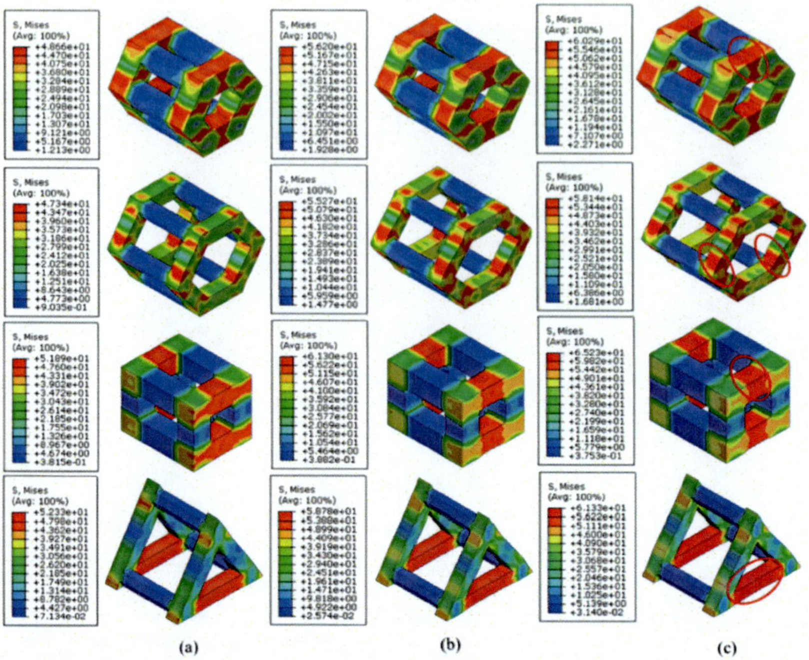

Fig. 9. Illustration of von Mises stress distribution for four samples at various overall deformation (a) $\delta = 2\,\text{mm}$, (b) $\delta = 4\,\text{mm}$, and (c) $\delta = 5\,\text{mm}$

bevel struts in unit-cell scale. In all, the coherence in higher Von Mises stress field and fracture position within unit cells shows that stiffness degradation and failure of the lattice structures under uniaxial tensile test is highly related with stress localization. Softening of material initiates from locations of maximum von Mises stress and develops through unit cells until complete failure, thus determines the overall fracture path.

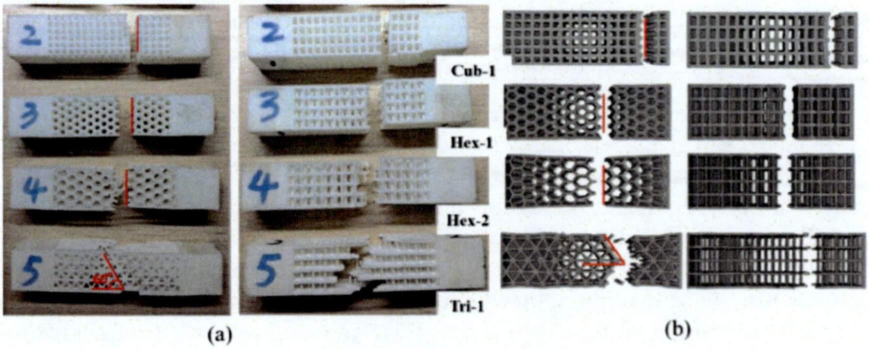

Fig. 10. Comparison of (a) tested, and (b) predicted fracture paths of four lattices

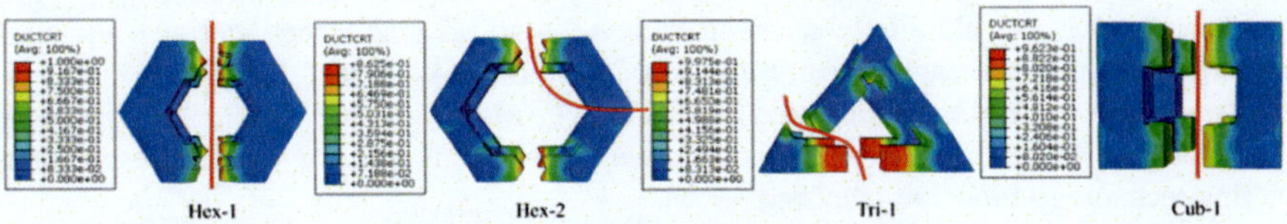

Fig. 11. Local failure within unit cells

4 Conclusions

In this study, Nylon 12 lattice specimens with three types of periodic unit cells are fabricated via SLS to implement uniaxial tensile tests. A FE modelling methodology is developed to study the tensile elastic-plastic behaviour and fracture behaviour of the SLS printed lattice structures, which is evaluated by the experimental results. The main conclusions are:

- The experimental load-displacement curves are examined and compared to the FE predictions. The tensile behaviour of the lattice structures is related to both relative density and unit-cell type, while the extension the specimens experience before fracture increase with load-bearing capacity.
- The von Mises stress evolution and deformation process of typical unit cells are investigated. Stress distribution differs in that the stress localizes near the joints of adjacent struts for Hex-2 instead of yielding from the horizontal struts for Hex-1, Cub-1, and Tri-1.
- All specimen rupture near the central portion, however, their fracture patterns differ in rupture angle and fracture position through unit cells. The coherence in predicted higher von Mises stress field and failure position within unit cells demonstrates that failure and stiffness degradation of the lattice structures under uniaxial tensile test are highly related with the stress localization.
- The predicted bulk elastic-plastic behaviour as well as local failure modes have shown to reasonably agree well with experimental results. The consistency in the tensile load-displacement curves, deformation histories, and fracture path demonstrate the capability of the modelling approach.

References

1. Gibson, L.J., Ashby, M.F.: Cellular Solids: Structure and Properties, 2nd edn. Cambridge University Press, Cambridge (1999)
2. Yan, C., Hao, L., Hussein, A., Young, P., Raymont, D.: Advanced lightweight 316L stainless steel cellular lattice structures fabricated via selective laser melting. Mater. Des. **55**, 533–541 (2014). https://doi.org/10.1016/j.matdes.2013.10.027
3. Pan, C., Han, Y., Lu, J.: Design and optimization of lattice structures: a review. Appl. Sci. **10**(18), 6374 (2020). https://doi.org/10.3390/app10186374
4. Ahmadi, S.M., et al.: Mechanical behavior of regular open-cell porous biomaterials made of diamond lattice unit cells. J. Mech. Behav. Biomed. Mater. **34**, 106–115 (2014)
5. Brennan-Craddock, J., et al.: The design of impact absorbing structures for additive manufacture. J. Phys.: Conf. Ser. **382**, 012042 (2012). IOP Publishing

6. Olivares, A.L., et al.: Finite element study of scaffold architecture design and culture conditions for tissue engineering. Biomaterials **30**(30), 6142–6149 (2009)
7. Association, E.P.M.: Introduction to Additive Manufacturing Technology Brochure Launched (2015). http://www.epma.com/epma-news/introduction-to-additive-manufacturing-technology-brochure-launched
8. Smith, M., Guan, Z., Cantwell, W.J.: Finite element modelling of the compressive response of lattice structures manufactured using the selective laser melting technique. Int. J. Mech. Sci. **67**, 28–41 (2013). https://doi.org/10.1016/j.ijmecsci.2012.12.004
9. Li, P.: Constitutive and failure behaviour in selective laser melted stainless steel for microlattice structures. Mater. Sci. Eng. A **622**, 114–120 (2015). https://doi.org/10.1016/j.msea.2014.11.028
10. Ullah, I., Brandt, M., Feih, S.: Failure and energy absorption characteristics of advanced 3d truss core structures. Mater. Des. **92**, 937–948 (2016). https://doi.org/10.1016/j.matdes.2015.12.058
11. Li, C., et al.: Crushing behavior of multi-layer metal lattice panel fabricated by selective laser melting. Int. J. Mech. Sci. **145**, 389–399 (2018). https://doi.org/10.1016/j.ijmecsci.2018.07.029
12. Li, P.Y., Ma, Y.E., Sun, W.B., Qian, X., Wang, Z.H.: Fracture and failure behavior of additive manufactured Ti6Al4V lattice structures under compressive load. Eng. Fract. Mech. 107537 (2021). https://doi.org/10.1016/j.engfracmech.2021.107537
13. Xca, B., et al.: Compression experiment and numerical evaluation on mechanical responses of the lattice structures with stochastic geometric defects originated from additive-manufacturing. Composites B **194** (2020). https://doi.org/10.1016/j.compositesb.2020.108030
14. Keshavarzan, M., Kaclichodaei, M., Badrossamay, M., Ravari, M.: Investigation on the failure mechanism of triply periodic minimal surface cellular structures fabricated by vat photopolymerization additive manufacturing under compressive loadings. Mech. Mater. **140**, 103150.1–103150.15 (2020). https://doi.org/10.1016/j.mechmat.2019.103150
15. Bhuwal, A.S., Liu, T., Ashcroft, I., Sun, W.: Localization and coalescence of imperfect planar FCC truss lattice metamaterials under multiaxial loadings. Mech. Mater. 103996 (2021). https://doi.org/10.1016/j.mechmat.2021.103996
16. Niu, J., Choo, H.L., Sun, W., Mok, S.H.: Numerical study on load-bearing capabilities of beam-like lattice structures with three different unit cells. Int. J. Mech. Mater. Des. **14**, 443–460 (2017). https://doi.org/10.1007/s10999-017-9384-3
17. Niu, J., Choo, H.L., Sun, W.: Finite element analysis and experimental study of plastic lattice structures manufactured by selective laser sintering. Proc. Inst. Mech. Eng. Part L **231**(1–2), 171–178 (2017). https://doi.org/10.1177/1464420716662296
18. Abaqus. Documentation. Dassault Systemes Simulia Corporation (2016)
19. EOS GmbH. Product Information EOSINT P/PA2200-Pulver, Manufacturer, Supplied Materials Data Sheet (2005). Accessed 15 Feb 2022

An Updating Method of Aero Tyre Model by Shimmy Test

Chongchong Liu[1,2,3](✉), Xiaochuan Liu[1,2], Yong Xu[4], Xiaolei Du[3], Zhengquan Yang[1,2], and Shengli Liu[1,2]

[1] National Key Laboratory of Strength and Structural Integrity, Aircraft Strength Research Institute of China, Xi'an 710065, Shanxi, China
liuchongchong623@163.com

[2] Aviation Key Laboratory of Technology and Science on Structure Impact Dynamics, Aircraft Strength Research Institute of China, Xi'an 710065, Shanxi, China

[3] School of Mechanics, Civil Engineering and Architecture, Northwestern Polytechnical University, Xi'an 710072, China

[4] School of Mathematics and Statistics, Northwestern Polytechnical University, Xi'an 710072, China

Abstract. Speed and oscillation frequency are two key factors that effect tyre lateral dynamic characteristics. Tyre models for previous landing gear shimmy analysis use stationary shape functions for the lateral tyre deformation which eliminate the effects of tyre speed and oscillation frequency. The lateral dynamic characteristics of tyre are studied by test on the dynamometer facility. The variation trend of amplitudes of the tyre self-aligning moment and lateral force with different speeds and oscillation frequencies is observed and fitted based on test results. The tyre model for landing gear shimmy analysis is updated by fitted coefficients. Nonlinear shimmy bifurcation analysis shows the shimmy unstable region enlarges when the tyre speed and frequency effects are considered. The method of updating tyre shimmy model based on test results could be used in landing gear shimmy analysis to improve the analysis accuracy.

Keywords: Tyre model · lateral characteristics · speed effects · dynamometer · bifurcation analysis

1 Introduction

The landing gear shimmy is a state of self-excited oscillations caused by the interaction with elastic tyres [1–4]. It represents a relatively well-studied problem in engineering [5–7]. However, shimmy even often occurs in the landing gear of civil aircraft especially some light airplane or UAVs partly due to the weight reduction design of airplane and landing gear [8, 9]. Lack of accurate shimmy analysis and test verification before test flight may cause landing gear shimmy [10, 11].

Establishing an accurate tyre model is one of mainly challenges to predict landing gear shimmy more accurately due to couple of tyre elasticity and loading in contact with

Chinese Society of Aeronautics and Astronautics (Ed.): CASTC 2023, LNME, pp. 547–564, 2024.
https://doi.org/10.1007/978-981-99-8861-7_54

the ground. Simple physics-based [12–14], empirical-based [15–17] and finite element approaches [18–20] have been utilized to build tyre models. However, the physics and mathematical modelling of transient and speed-dependent tyre behaviours are far from satisfactory especially in tyre lateral dynamic characteristics modelling [21], Calculating the deformation of a rolling tyre is a challenging task since the deformed shape and force wave of the tyre travelling backwards in the contact region which is subject to the effect of partial sticking and sliding governed by friction laws [22–24]. Tyre models for previous landing gear shimmy analysis use stationary shape functions for the lateral tyre deformation calculated at constant yaw angles [1–3, 13–15, 25], the resultant lateral creep force and torque are calculated from quasi-stationary deformations by tyre models for shimmy analysis, that affects the improvement of aircraft shimmy analysis accuracy and the lightweight design of landing gear. Actually the tyre lateral dynamic characteristics vary greatly with the rolling speed and vibration frequency according to test data [26].

Yintao Wei developed a nonlinear finite-element code to model tyre speed-dependent lateral behavior [27]. Takacs rewrited a delay differential equation (DDE) by assuming travelling wave solutions for the deformation of the contact line to express the intrinsic memory effect of the elastic tyre [22]. Nan Xu developed an enhanced brush tire model and the UniTire semi-physical model for tire forces and moments under different traveling velocities [28]. However the methods of the above researches are somewhat large scale computing for application of landing gear engineering shimmy analysis, and may not suitable for the landing gear structural optimization design for shimmy suppressing. Consequently, it is necessary and urgent to build an engineering tyre model considering the speed and frequency effects to improve the landing gear shimmy analysis accuracy.

In this paper, a test method is established to study the tyre lateral dynamic characteristics at different speeds and oscillation frequencies. The variation trend of amplitudes of the tyre self-aligning moment and lateral force with different motion state is observed and fitted based on test results. Tyre model for shimmy analysis is updated by fitted coefficients to consider speed and frequency effects. The updated model is used in the landing gear nonlinear shimmy bifurcation analysis, it is found that the shimmy unstable region is enlarged when considering the effects of speed and oscillation frequency.

2 Tyre Lateral Dynamic Characteristics Test Method

The test is carried on the dynamometer, showed in Fig. 1. Tyre is connected to a landing gear, the landing gear is connected to the 3D force sensors and the sensors are attached to the overhead platform. The overhead platform is supported by side beams and moves on rollers running on guide rails attached to the side beams. The flywheel is driven by a high-power motor and frequency converter, its diameter is 5 m and the maximum speed reaches 125 m/s. The servo hydraulic actuator, the flywheel and the steer system are well controlled by the three sets of controllers.

The vertical load acting on the tyre is simulated by the servo hydraulic actuator, the taxiing speed of the tyre is simulated through the rotating flywheel, the rolling tyre is oscillated by the steer system in yaw at a given frequency when the rolling tyre at the maintained vertical load of interest. The vertical load, lateral load are obtained by the 3D force sensors, torsional torque is obtained by the calibrated strain gauges pasted on the torque arm, yaw angle is obtained by a angular displacement transducer.

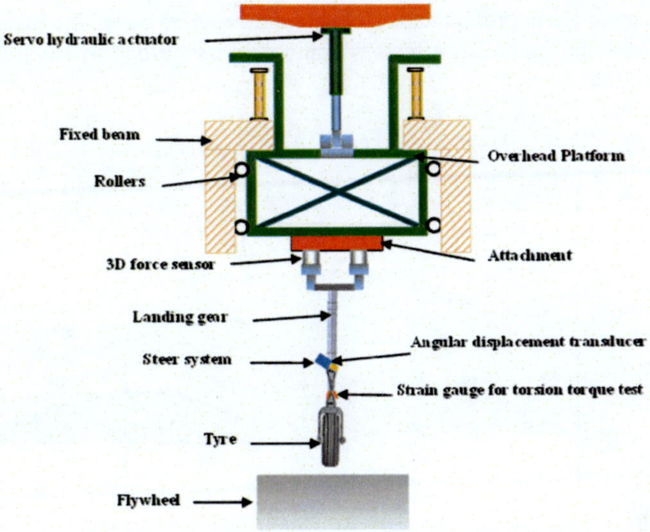

Fig. 1. Schematic diagram of the tyre test system

Since it is difficult to apply loading rates in the shimmy frequency range which is about 10 to 30 Hz, a linear equivalent method is used to simulate the tyre lateral deformations considering speed effects, it is assumed that if the ratio of tyre oscillation frequency to speed are equal, the lateral dynamic characteristics are consistent. Consequently, the flywheel speed needs to be reduced significantly shown in Eq. (1) to match the oscillation wavelength which is recommended in tyre test guide SAE ARP4955A [23].

$$\lambda = \frac{f_e}{v_t} \tag{1}$$

where λ is a certain value to simulate the tyre speed and frequency effects, f_e is the shimmy frequency, v_t is the tyre speed. For example, if the shimmy frequency is 20 Hz and the tyre speed is 100 m/s, it could be simulated by a test frequency of 0.5 Hz and a speed of 2.5 m/s.

The mechanism of this linear equivalent method is not shown in SAE ARP4955A [23]. As shown in Fig. 2, transient lateral deformations of tyre in shimmying and steady-state lateral deformations of tyre at constant yaw angles are quite different in the contact region for brush models [29], the resultant lateral force F and the self-aligning moment M with respect to the contact patch centre O could be calculated by the integral Eq. (2), where the $q(x, t)$ is the lateral deformation of the elastic elements representing the tyre tread, a is the half of the tyre contact patch length, parameters of k and b is the stiffness and damping of tyre linearly distributed spring and damper elements. Consequently, tyre lateral force F and the self-aligning moment M of transient and steady state with the same side slip angle are quite different. The oscillation frequencies and speeds are the main factors affecting the tyre lateral deformations mode, it is assumed that tyre lateral deformations mode and travelling wave are similar with the same value of λ shown in Eq. (1), and the tyre lateral force and self-aligning moment with the same value of λ

and side slip angle are approximated. This method could be called the linear equivalent method of speed and frequency effects which would be verified by the tyre test results of the next section.

$$F = \int_{-a}^{a} (kq(x,t) + b\frac{d}{dt}q(x,t))dx \tag{2.1}$$

$$M = \int_{-a}^{a} (kxq(x,t) + bx\frac{d}{dt}q(x,t))dx \tag{2.2}$$

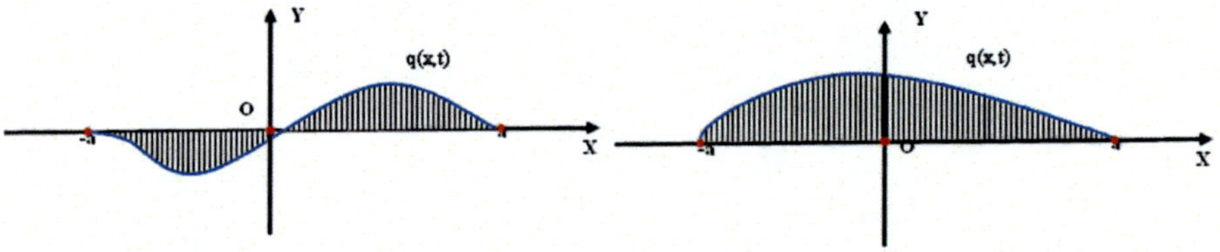

(a) Transient tyre lateral deformations (b) Steady-state tyre lateral deformations

Fig. 2. Tyre lateral deformations in the contact region for brush models

To minimize the tyre test results error, landing gear stiffness should be enough to avoid occurrence of the shimmy and other excess vibrations, the flywheel should be kept at low speed to avoid landing gear vibration in the test. The tyre steering frequency should be lower than 1 Hz, and the natural vibratory frequencies of the landing gear, the platform, the attachment should be much higher than the test yaw frequency, and not be close with each other. Before the test, all bolted connections should be tightened.

The strain gauge is pasted on the torque arm of the landing gear, as shown in Fig. 3. Lateral force F_{la} is at the intersection of upper and lower torque arms, the voltage signal measured by the strain gauge is S_v, Calibration coefficient K_v should be obtained by static test.

$$F_{la} = K_v \cdot S_v \tag{3}$$

The rectangular coordinate system used in the test is shown in Fig. 4. Tyre lateral force F_{lt} was obtained by 3d force sensors attached to the landing gear hinge. Considering the influence of inertia effect, the tyre self-aligning moment M_t could be obtained from Eq. (4).

$$M_t = F_{la}dcos\theta - F_{lt}e - J_\theta\ddot{\theta} \tag{4}$$

where d is the distance between the intersection point of the torsion arm and the axis of the landing gear strut under a certain load, θ is the tyre yaw angle, e is the landing gear caster length, J_θ is the rotational moment of inertia of tyre, wheel and lower landing gear strut around strut axis.

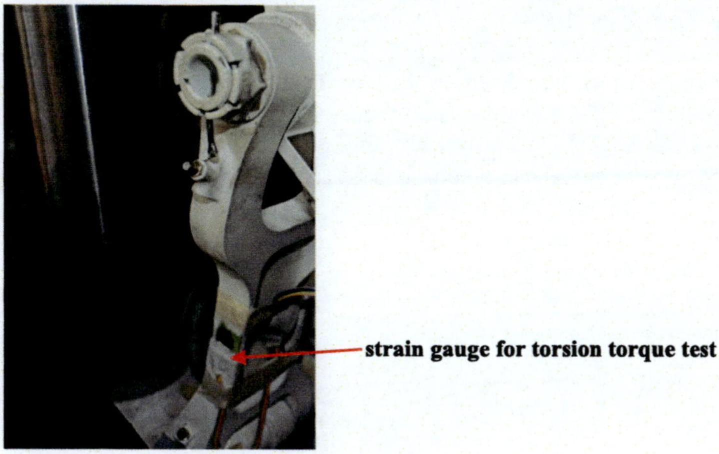

Fig. 3. Torsional torque test

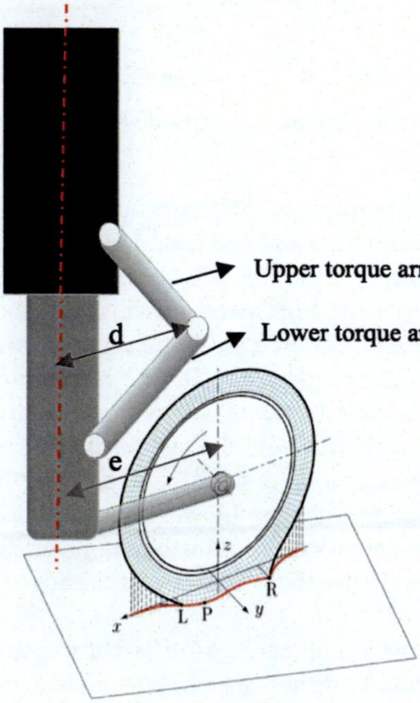

Fig. 4. Coordinate system used in the tyre test

3 Tyre Test Results

A certain vertical load was acted on the tyre, the swivel frequency is 1 Hz, the swive angle amplitude is 2.5°, under this amplitude, the tyre slip angle could be assumed equal to the swive yaw angle [23]. The tyre running speed is 10 m/s, and the dynamic response curves of tyre swive angle, lateral force and the self-aligning moment are shown in Fig. 5.

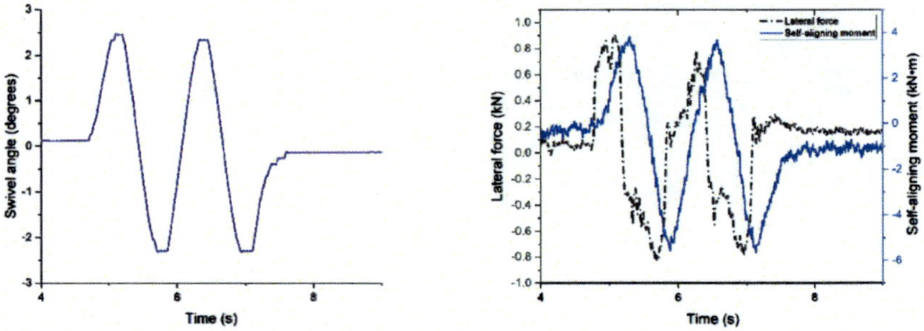

(a) Swivel angle (b) Self-aligning moment and lateral force

Fig. 5. Tyre dynamic response at frequency of 1 Hz, speed of 10 m/s

Tyre lateral loads at different speeds and frequencies were obtained by multiple tests. The amplitudes of self-aligning moment and lateral force varying with speeds at 0.5 Hz and 1 Hz yaw frequencies are shown in Fig. 6. When below a certain speed the amplitudes of self-aligning moment descends with speed growing, and above the certain speed the amplitudes of self-aligning moment increases with speed growing, at 0.5 Hz the critical speed is about 2.8 m/s which corresponds with the research of Wei and Cherry Auto company test results [21]. It could be calculated by Eq. (1) that at 1 Hz the critical speed is about 5.6 m/s which is consistent with the test results at 1 Hz shown in Fig. 6. The amplitudes of tyre lateral force increase with speed growing. When reaching a certain speed, the change range becomes smaller. Because of consistency of the two test results, we assumed that there are some variety rules of the pneumatic tyre lateral dynamic characteristics with the oscillation frequencies and speeds which could be observed from the tyre test results.

In order to better observe the influence of different frequencies and speeds on tyre lateral dynamic characteristics and the speed and frequency effects, the tyre tests at 0.25 Hz and 0.75 Hz frequencies are added, test results are shown in Fig. 7.

In order to verify whether the linear equivalent method of Eq. (1) can simulate the tyre speed and frequency effects, the four experimental test results were linearly processed equivalently according to Eq. (5), and the equivalent vibration frequency is 1 Hz, v_e in Eq. (5) is the equivalent speed, v_0 is the test speed, f_1 is the equivalent frequency, f_2 is the test frequency. As shown in the equivalent results of Fig. 8, it could be seen that the test results of different frequencies processed by the equivalent speed method are consistent better than unprocessed especially at the low speeds, the error of equivalent test results

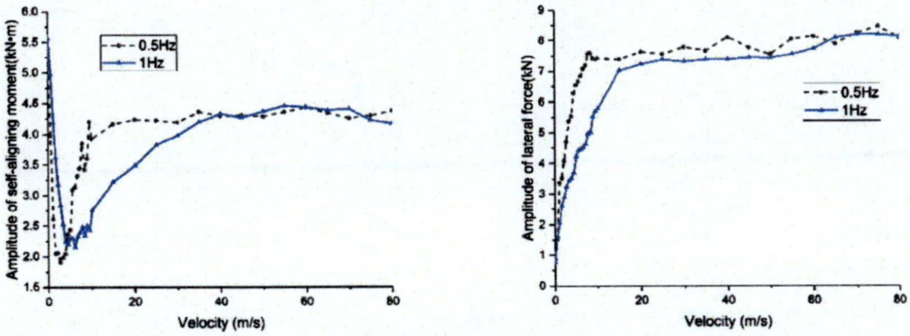

(a) Tyre self-aligning moment varying with speed (b) Tyre lateral force varying with speed

Fig. 6. Lateral load amplitude varying with speed at different frequencies

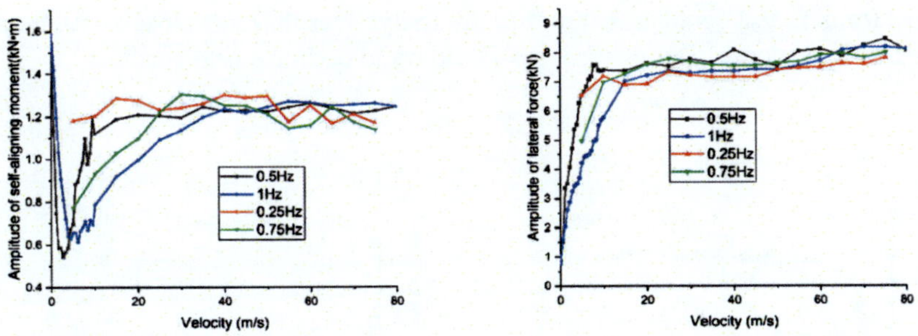

(a) Tyre self-aligning moment varying with speed (b) Tyre lateral force varying with speed

Fig. 7. Lateral load amplitude varying with speed at different frequencies

at different frequencies is engineering acceptable. Consequently, this linear equivalent method could be applied in engineering processing of test data in the research of tyre lateral dynamic characteristics.

$$v_e = \frac{f_1}{f_2} \cdot v_0 \tag{5}$$

Another pneumatic tyre B with different size, produced by different manufacturer is tested by the same method, the yaw frequency is 0.5 Hz, The normalized amplitudes of self-aligning moment and lateral force varying with speed compared with tyre A test results which is described in details above are shown in Fig. 9. The variation trend of amplitude of tyre lateral load varying with speed is approximate, because of the consistency of the variation trend of amplitude of test results of two tyres and Cherry Auto company, we assume that the variation trend of amplitude of lateral load varying with speeds and frequencies of all the pneumatic tyres is approximate, although there may be some errors due to the caster lengths, contact patch length, relaxation length

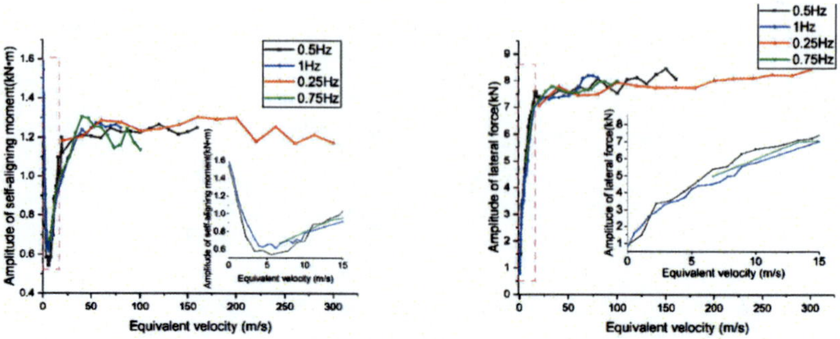

(a) Tyre self-aligning moment varying with speed (b)Tyre lateral force varying with speed

Fig. 8. Processed lateral load amplitude varying with speed

of tyres are different. Updating the quasi-static tyre model through the variation trend obtained from the test could make the shimmy analysis results more close to real landing gear dynamic response. That is what we would try to do in the next sections.

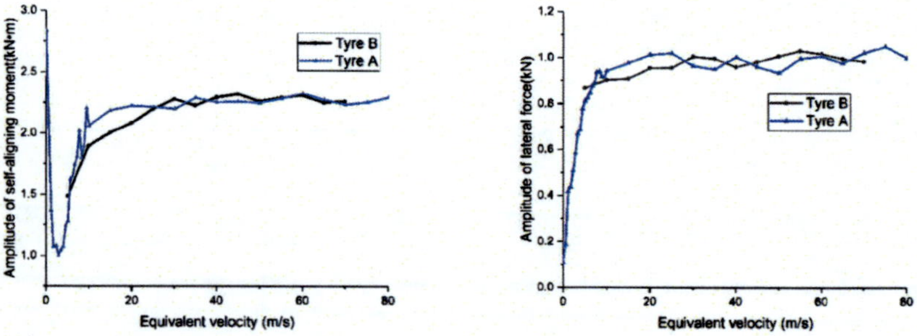

(a) Self-aligning moment varying with speed (b) Lateral force varying with speed

Fig. 9. Lateral load amplitude varying with speed of tyre B compared with tyre A

4 Tyre Model Considering Speed Effects

The landing gear shimmy frequency is generally 10–20 Hz. Here we taking 1 Hz test data for equivalence. Assuming that the shimmy frequency is 15 Hz, the data is equivalently treated by Eq. (5). The tyre self-aligning moment and lateral force amplitudes at different equivalent speeds are shown in Fig. 10. The tyre rolling speeds is generally lower than 150 m/s, and the data within 300 m/s is used for model fitting.

The data should be normalized before fitting. Assuming that the minimum value of the amplitude of the tyre's self-aligning moment in the full speed range is 1, generally

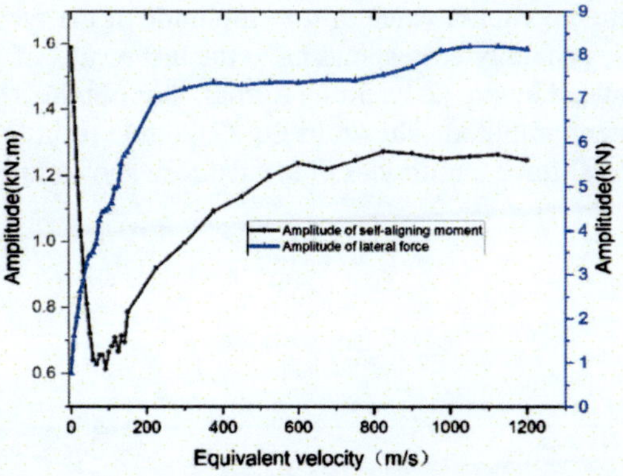

Fig. 10. Amplitude Variation of tyre lateral load with equivalent velocity at 15 Hz

corresponding to the results of the tyre's steady state test on the flat plank tyre test machine or other quasistatic test which could be estimated by Eq. (2.2) [9, 30, 31], Fitting formula is obtained by fitting the variety of the speed dependent amplitude shown in Eq. (6), the curve of fitting coefficient τ_t and speed dependent moment amplitudes test data are shown in Fig. 11.

$$\tau_t = 0.88 + 2.1\mathrm{atan}\left(e^{-\frac{v}{20}}\right) + 0.5\tan\left(\frac{v}{300}\right) \quad (6)$$

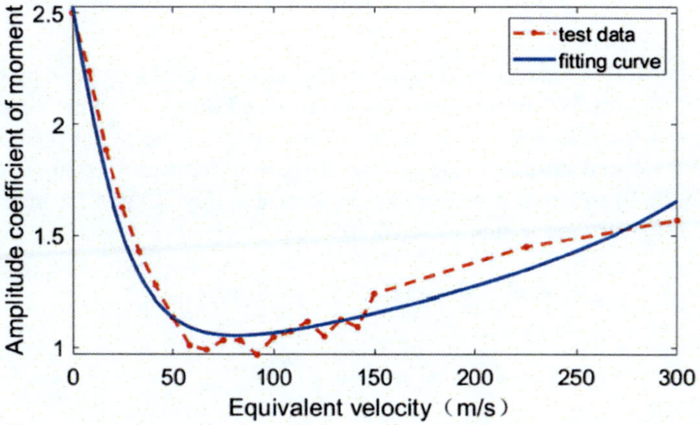

Fig. 11. Fitting formula compared with the test data of speed dependent moment amplitude

Assuming that the maximum value of the amplitude of the tyre lateral force in the full speed range is 1, generally corresponding to the test results of the tyre steady state which could be estimated by Eq. (2.1), fitting formula was obtained by fitting the change of the speed dependent amplitude shown in Eq. (7), curve of fitting coefficient τ_l and speed dependent lateral force amplitudes of test data are shown in Fig. 12.

$$\tau_l = (\frac{v}{300})^{0.4} \tag{7}$$

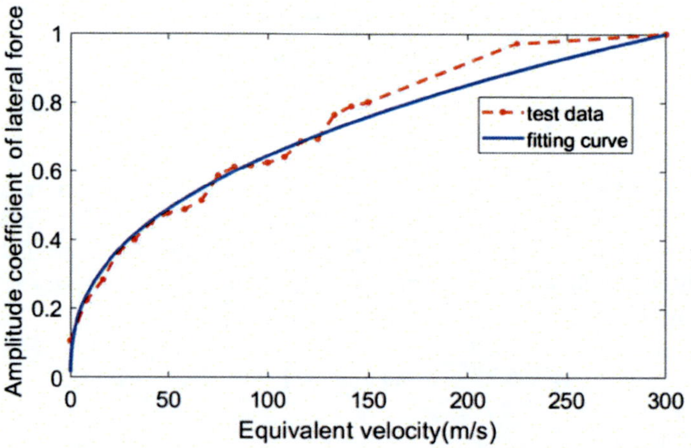

Fig. 12. Fitting formula and test data of speed dependent lateral force amplitude

If the shimmy frequency is not 15 Hz, Eq. (6) and (7) could be updated by Eqs. (8–9), where f is the estimated shimmy frequency. The two fitting coefficient τ_t, τ_l could be used to correct the self-align moment and lateral force of the radial pneumatic tyres to simulate the speed effects. If the estimated shimmy frequency is 20 Hz, the fitting curve by Eq. (8–9) and processed test data are shown in Figs. 13 and 14, in the equation $a_1 = 0.88, a_2 = 0.75, a_3 = 0.5, a_4 = 20, a_5 = 20$.

$$\tau_t = a_1 + 2.1\mathrm{atan}\left(e^{-a_2 \cdot \frac{v}{f}}\right) + a_3\tan(\frac{v}{a_4 \cdot f}) \tag{8}$$

$$\tau_l = (\frac{v}{a_5 \cdot f})^{0.4} \tag{9}$$

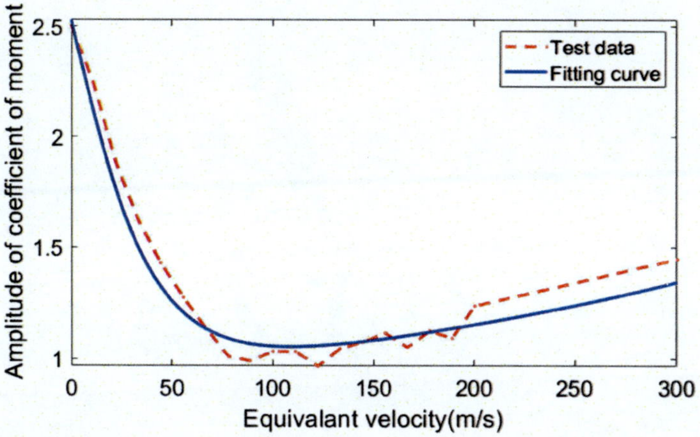

Fig. 13. Fitting formula and test data of speed dependent moment amplitude with estimated frequency is 20 Hz

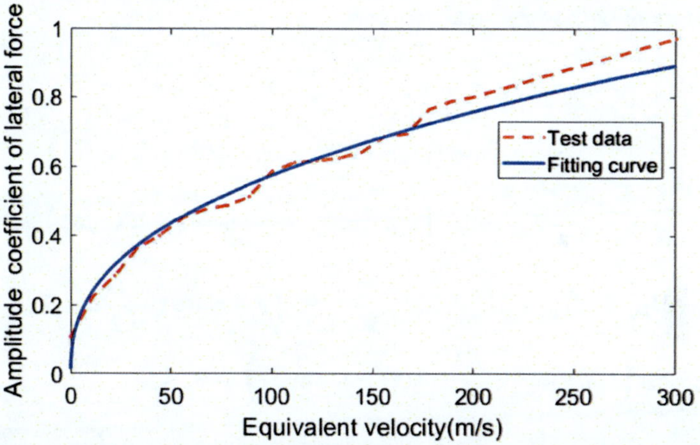

Fig. 14. Fitting formula and test data of speed dependent lateral force amplitude with estimated frequency is 20 Hz

5 Landing Gear Shimmy Analysis by Speed Dependent Tyre Model

To observe the necessity of considering the speeds and frequencies effects on shimmy analysis, the speed dependent tyre model established in this paper is used for landing gear shimmy analysis, the shimmy model and landing gear system parameters was from [25], the tyre model is the magic formula, and the parameters were obtained in quasi-stationary state [15, 22]. The schematic of an aircraft nose landing gear with a dual-wheel and the coordinate system of shimmy analysis is shown in Fig. 15.

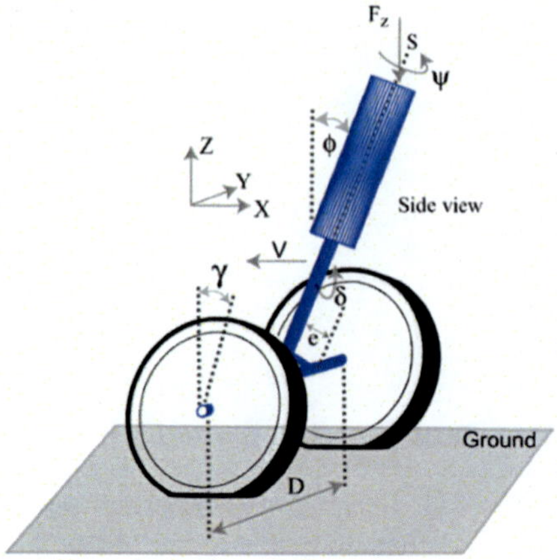

Fig. 15. The schematic of an dual-wheel NLG [25]

The model for the torsional and lateral vibrational modes and their coupling through the elastic tyre of both wheels takes the form.

$$I_\psi \ddot{\psi} + M_{K_\psi} + M_{D_\psi} + M_{K_{\alpha L}} + M_{K_{\alpha R}} + M_{D_{\lambda L}} + M_{D_{\lambda R}} + 2I\dot{\delta}\frac{V}{R} + \left(F_{K_{\lambda L}} + F_{K_{\lambda R}}\right)e_{eff}$$
$$-F_{zL}\sin(\phi)\left(e_{eff}\sin(\theta) + \frac{D}{2}\cos(\theta) + l_g\sin(\delta)\right)$$
$$-F_{zR}\sin(\phi)\left(e_{eff}\sin(\theta) - \frac{D}{2}\cos(\theta) - l_g\sin(\delta)\right) = 0 \quad (10.1)$$

$$I_\delta\ddot{\delta} + M_{K_\delta} + M_{D_\delta} + M_{\lambda_{\delta L}} + M_{\lambda_{\delta R}} - 2I\dot{\psi}\frac{V}{R} - F_{zL}\left(e_{eff}\sin(\theta) + \frac{D}{2}\cos(\theta) + l_g\sin(\delta)\right)$$
$$-F_{zR}\left(e_{eff}\sin(\theta) - \frac{D}{2}\cos(\theta) - l_g\sin(\delta)\right) = 0 \quad (10.2)$$

$$\dot{\lambda_L} + \frac{V}{\sigma}\lambda_L - Vsin(\theta) - l_g\dot{\delta}\cos(\delta) - \left(e_{eff} - h\right)\cos(\theta)\dot{\psi}\cos(\phi) - \frac{D}{2}\dot{\psi}\sin(\theta)\cos(\phi) = 0 \quad (10.3)$$

$$\dot{\lambda_R} + \frac{V}{\sigma}\lambda_R - Vsin(\theta) - l_g\dot{\delta}\cos(\delta) - \left(e_{eff} - h\right)\cos(\theta)\dot{\psi}\cos(\phi) + \frac{D}{2}\dot{\psi}\sin(\theta)\cos(\phi) = 0 \quad (10.4)$$

The effective caster length of the NLG is given by Eq. (11),

$$\mathrm{e}_{eff} = \mathrm{e}\cos\varphi + (\mathrm{R} + \mathrm{e}\sin\varphi)\tan\varphi \quad (11)$$

The overall vertical force F_z on the nose landing gear is asymmetrically divided into two forces F_{z_L} and F_{z_R} on the left and right wheels according to,

$$F_{z_{L/R}} = \frac{F}{2}(1 \mp (\frac{k_y D}{F_z})\sin(\psi\sin(\phi) + \delta)) \quad (12)$$

Here, $\gamma = \psi\ sin(\varphi)$, the restoring or lateral force of each tyre is considered of the form, τ_l is from Eq. (9).

$$F_{K_{\lambda L/R}} = \tau_l \cdot k_\lambda.\mathrm{atan}(7.0\tan(\alpha_{L/R})) \cdot \cos(0.95\mathrm{atan}(7.0\tan(\alpha_{L/R})))F_{zL/R} \tag{13}$$

where $\alpha_{L/R} = \mathrm{atan}(\frac{\lambda_{L/R}}{\sigma})$, σ is the relaxation length of the tyre.

Self-aligning moment of each tyre can be approximated by the Eq. (13), τ_t is from Eq. (8).The tyre model from [31] was corrected to simulate the speed effects by the fitting coefficient of Eqs. (8)–(9) based on test data.

$$M_{K_{\alpha L/R}} = \begin{cases} \tau_t \cdot k_\alpha \frac{\alpha_m}{\pi} \sin\left(\alpha_{\frac{L}{R}} \frac{\pi}{\alpha_m}\right) F_{\frac{zL}{R}} & if \left|\alpha_{\frac{L}{R}}\right| \le \alpha_m \\ 0 & if \left|\alpha_{\frac{L}{R}}\right| > \alpha_m \end{cases} \tag{14}$$

$$M_{D\lambda L} = M_{D\lambda R} = \frac{C_\lambda \psi \cos(\phi)}{V} \tag{15}$$

$$M_{\lambda_{\delta L/R}} = l_g F_{K_{\lambda L/R}} \cos(\theta)\cos(\psi) \tag{16}$$

Detailed definition of parameters and other mechanical models in the equations could be found in Table 1 and [31]. We perform a nonlinear shimmy analysis of the mathematical model represented by Eq. 10. The double parameter bifurcation is carried out to analyze the characteristics of the parameter diagram of the shimmy region [32]. Based on the bifurcation theory [33], suppose the system $\dot{x} = f(x, a), x \in R^d, a \in R$,with smooth f, has for all sufficiently small $|a|$ the equilibrium $x = 0$, with eigenvalues $\lambda_{i,j}(a) = \pm i\rho(a)$,where $\rho(0) > 0$, then the system generates a Hopf bifurcation.

The bifurcation diagram not considering speed effects is shown in Fig. 16, and the shimmy frequency is about 10 Hz as shown in Fig. 17, Substituting the frequency f = 10 Hz into the Eq. (13)–(14), the double parameter bifurcation diagram of the V-F_z plane using tyre speed dependent model is shown in Fig. 18.

The curve composed of the blue torus is the shimmy area of the system of torsional mode. The area above the red curve is the shimmy area of the system with bending mode, and the lower area of the blue curve is the safe region of the system without shimmy. The geometric topological composition of two diagrams is similar, however, the shimmy stability boundary of landing gear is obviously affected. At the same speed, the critical vertical load occurring shimmy descends and shimmy unstable region is enlarged because of tyre motion state effects. This phenomenon could remind landing gear shimmy researchers considering the effect of different speeds and oscillation frequencies on tyre lateral characteristics.

Table 1. System parameters and their values as used in the model

Symbol	Parameter	Value	Unit
Structure parameters			
e	Caster length	0.12	m
φ	Rake angle	0.1571	rad
l_g	Gear height	2.5	m
K_ψ	Torsional stiffness of strut	3.8×10^5	N·m/rad
K_δ	Lateral bending stiffness of strut	6.1×10^6	N·m/rad
C_ψ	Torsional damping of strut	300	N·m·s/rad
C_δ	Lateral bending damping of strut	300	N·m·s/rad
I_ψ	Torsional moment of inertia of strut	100	kg·m^2
I_δ	Lateral moment of inertia of strut	600	kg·m^2
Tyre parameters			
D	Distance between the wheel centres	0.2	m
σ	Relaxation length	0.3	m
h	Contact patch length	0.2	m
R	Radius of nose wheel	0.362	m
α_m	Self-aligning moment limit	0.1745	rad
k_α	Self-aligning coefficient of tyre	1.0	m/rad
k_λ	Restoring coefficient of tyre	0.002	rad^{-1}
c_λ	Damping coefficient of tyre	285	N·m^2/rad
I	Wheel moment of inertia	0.2	kg·m^2
Continuation parameters			
F_z	Vertical force on the gear	30–700	kN
V	Forward velocity	0–180	m/s

The self-aligning moment and lateral force of the other tyre models for shimmy analysis could also be corrected by the updating coefficient of Eq. (8–9), for example Smiley and Moreland tyre model [13, 14] etc. When using the updating coefficient for landing gear shimmy, tyre parameters obtained from the certain test speed and frequency should be considered. Substitute the test speed and frequency into Eqs. (8)–(9) to get a value, and divide the coefficient by this value to get a new updating coefficient.

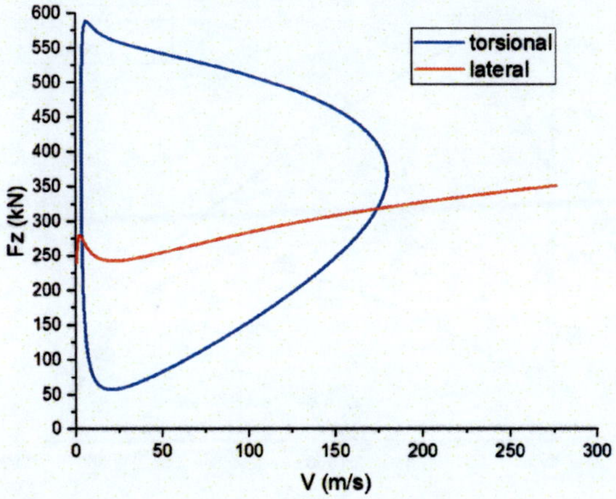

Fig. 16. Double-parameter bifurcation diagrams without considering speed effects

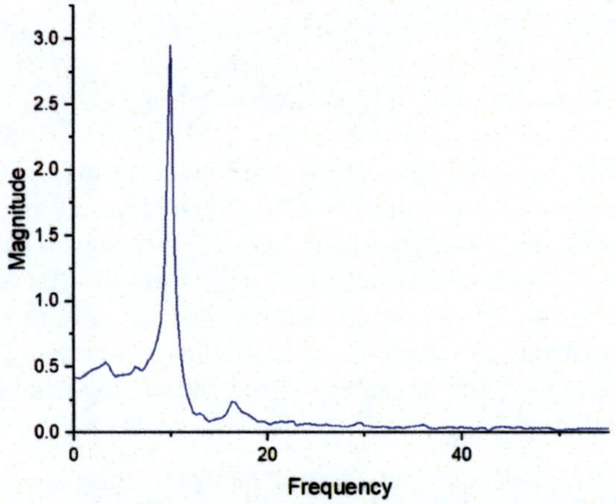

Fig. 17. The frequency spectrum of the landing gear shimmy

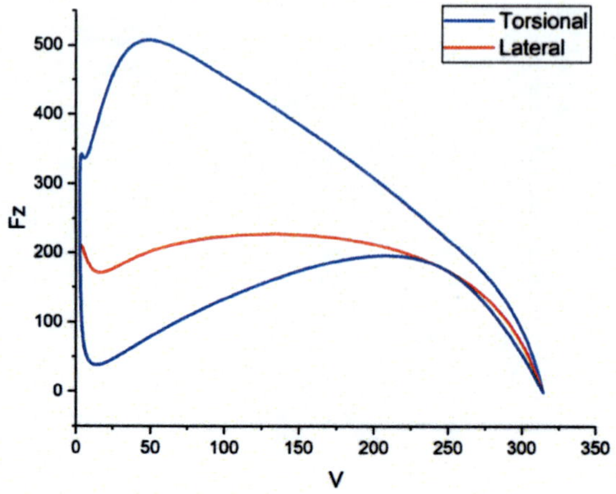

Fig. 18. Double-parameter bifurcation diagrams considering speed effects

6 Conclusions

We present a tyre lateral dynamic characteristics test method and obtain the tyre test data of different speeds and oscillation frequencies. The amplitudes of tyre self-aligning moment and lateral force varies significantly at different speeds and test frequencies. A linear equivalent method recommended in SAE ARP4955A is used to process the test data to simulate the tyre speed and frequency effects. Tyre model for landing gear shimmy analysis is updated by fitted coefficient based on test results. To verify the necessity of considering the tyre speed effects on shimmy analysis, a bifurcation analysis was performed using the updated tyre model. The shimmy unstable region enlarges when the tyre speed and frequency effects are considered, the landing gear become more prone to shimmy. We conclude that:

(1) The tyre lateral dynamic characteristics test method in this paper is practicable since the test results is corresponds with the research of Wei and Cherry Auto company test results. Tyre should be kept at low speed to avoid landing gear vibration in the test.
(2) The linear equivalent method recommended in SAE ARP4955A is verified by the dynamic tyre shimmy properties test results in this paper.
(3) It is suggested that tyre speed and frequency effects should be considered and the method of updating tyre shimmy model in this paper could be used in landing gear shimmy analysis.
(4) When using the updating coefficient in landing gear shimmy analysis, tyre parameters obtained from the test speed and frequency should be considered. And if you predict the updating coefficient using a estimated shimmy frequency and the recalculated shimmy frequency is far from the certain estimated frequency, you can use the new recalculated shimmy frequency to predict the updating coefficient, and resume the shimmy analysis repeatedly till the estimated shimmy frequency is consistent with the recalculated shimmy frequency.

References

1. Sanches, L., Guimarães, T.A., Marques, F.D.: Nonlinear energy sink to enhance the landing gear shimmy performance. Acta Mech. **232**(7), 2605–2622 (2021)
2. Thota, P., Krauskopf, B., Lowenberg, M.: Interaction of torsion and lateral bending in aircraft nose landing gear shimmy. Nonlinear Dyn. **57**(3), 455–467 (2009)
3. Rahmani, M., Behdinan, K.: Interaction of torque link freeplay and Coulomb friction nonlinearities in nose landing gear shimmy scenarios. Int. J. Non-Linear Mech. **119**, 103338 (2020)
4. Feng, F., Nie, H., Zhang, M., et al.: Effect of torsional damping on aircraft nose landing-gear shimmy. J. Aircr. **52**(2), 561–568 (2015)
5. Howcroft, C., Krauskopf, B., Lowenberg, M.H., et al.: Influence of variable side-stay geometry on the shimmy dynamics of an aircraft dual-wheel main landing gear. J. Appl. Dyn. Syst. **12**(3), 1181–1209 (2013)
6. Krüger, W.R., Morandini, M.: Recent developments at the numerical simulation of landing gear dynamics. CEAS Aeronaut. J. **1**(1), 55–68 (2011)
7. Rahmani, M., Behdinan, K.: On the effectiveness of shimmy dampers in stabilizing nose landing gears. Aerosp. Sci. Technol. **91**, 272–286 (2019)
8. Rahmani, M., Behdinan, K.: Investigation on the effect of coulomb friction on nose landing gear shimmy. J. Vibr. Control **25**(2), 255–272 (2019)
9. Somieski, G.: Shimmy analysis of a simple aircraft nose landing gear model using different mathematical methods. Aerosp. Sci. Technol. **1**(8), 545–555 (1997)
10. Krüger, W., Besselink, I., Cowling, D., et al.: Aircraft landing gear dynamics: simulation and control. Veh. Syst. Dyn. **28**(2–3), 119–158 (1997)
11. Tourajizadeh, H., Zare, S.: Robust and optimal control of shimmy vibration in aircraft nose landing gear. Aerosp. Sci. Technol. **50**, 1–14 (2016)
12. Deur, J., Asgari, J., Hrovat, D.: A 3D brush-type dynamic tire friction model. Veh. Syst. Dyn. **42**(3), 133–173 (2004)
13. Moreland, W.J.: The story of shimmy. J. Aeronaut. Sci. **21**(12), 793–808 (1954)
14. Smiley, R.F.: Some considerations of hysteresis effects on tire motion and wheel shimmy (1957)
15. Pacejka, H.B., Bakker, E.: The magic formula tyre model. Veh. Syst. Dyn. **21**(S1), 1–18 (1992)
16. Pacejka, H., Besselink, I.: Magic formula tyre model with transient properties. Veh. Syst. Dyn. **27**(S1), 234–249 (1997)
17. Yuan, Z., Lu, D., Guo, K.: Comparative study on the UniTire model and the magic formula model. Automob. Technol. **36**, 7–11 (2006)
18. Darnell, I., Mousseau, R., Hulbert, G.: Analysis of tire force and moment response during side slip using an efficient finite element model. Tire Sci. Technol. **30**(2), 66–82 (2002)
19. Nackenhorst, U.: The ALE-formulation of bodies in rolling contact: Theoretical foundations and finite element approach. Comput. Methods Appl. Mech. Eng. **193**(39–41), 4299–4322 (2004)
20. Oden, J., Lin, T., Bass, J.: A finite element analysis of the general rolling contact problem for a viscoelastic rubber cylinder. Tire Sci. Technol. **16**(1), 18–43 (1988)
21. Wei, Y., Oertel, C., Liu, Y., et al.: A theoretical model of speed-dependent steering torque for rolling tyres. Veh. Syst. Dyn. **54**(4), 463–473 (2016)
22. Takács, D., Orosz, G., Stépán, G.: Delay effects in shimmy dynamics of wheels with stretched string-like tyres. Eur. J. Mech.-A/Solids **28**(3), 516–525 (2009)
23. Recommended Practice for Measurement of Static and Dynamic Characteristic Properties of Aircraft Tires. SAE ARP 4955A-2012 (2007)

24. Beregi, S.: Delay effects and non-smoothness in the dynamic tyre-road contact and vehicle steering. Budapest University of Technology and Economics (Hungary) (2020)
25. Thota, P., Krauskopf, B., Lowenberg, M.: Multi-parameter bifurcation study of shimmy oscillations in a dual-wheel aircraft nose landing gear. Nonlinear Dyn. **70**(2), 1675–1688 (2012)
26. Daughterty, R.H.: A study of the mechanical properties of modern radial aircraft tires. National Aeronautics and Space Administration, Langley Research Center (2003)
27. Wei, Y., Oertel, C., Shen, X.: Tyre rolling kinematics and prediction of tyre forces and moments: part II–simulation and experiment. Veh. Syst. Dyn. **50**(11), 1689–1706 (2012)
28. Xu, N., Guo, K., Yang, Y.: UniTire model for tire cornering properties under varying traveling velocities. SAE Technical Paper (2016)
29. Takács, D.: Dynamics of towed wheels: nonlinear theory and experiments. Budapest University of Technology and Economics (Hungary) (2010)
30. Guo, K., Liu, Y., Yang, Y.: Development on Tire test technology and prospect of its application in automobile performance research. Automob. Eng. **12**(1), 1–9 (1990)
31. Guo, K., Lu, D.: UniTire: unified tire model for vehicle dynamic simulation. Veh. Syst. Dyn. **45**(S1), 79–99 (2007)
32. Dhooge, A., Govaerts, W., Kuznetsov, Y.A., et al.: New features of the software MatCont for bifurcation analysis of dynamical systems. Math. Comput. Model. Dyn. Syst. **14**(2), 147–175 (2008)
33. Kuznetsov, Y.A.: Two-parameter bifurcations of equilibria in continuous-time dynamical systems. Elem. Appl. Bifurcat. Theory 293–392 (1998)

Design and Analysis of Aero-Engine Acoustic Liner Structure with Adjustable Absorption Frequency

Xin Ju, Jinwu Wu(✉), Tianhang Zhang, and Xili Du

School of Aircraft Engineering, Nanchang Hangkong University, Nanchang 330063, China
wujinwu@nchu.edu.cn

Abstract. In order to solve the problem that the absorption frequency of the traditional honeycomb micro-perforated panel acoustic liner is fixed and can't be adjusted in the face of complex noise environment, Aero-engine acoustic liner with adjustable sound absorption frequency is to be designed, a single cell in the honeycomb micro-perforated panel acoustic liner is designed and analyzed, and a broadband and telescopic micro-perforated panel sound absorption structure is proposed.

Using a structure of micro-perforated plates in series with unequal diameters to expand the range of structural sound absorption frequencies, and achieving the goal of adjustable sound absorption frequencies through a structure designed with telescopic unequal diameters. Simulation results show that it can be tuned in the frequency of 1000–3000 Hz. The acoustic liner design proposed in this paper has a simple structure and strong adaptability in the face of complex noise environment, and has a good engineering application prospect.

Keywords: Honeycomb micro-perforated panel acoustic liner · Broadband sound absorption · Adjustable sound absorption frequency · Telescopic structure

1 Introduction

Due to current strict noise control regulations, the issue of noise reduction in aircraft engines has attracted great attention, and honeycomb micro-perforated plate acoustic liners have been widely used to attenuate engine noise [1, 2]. The traditional honeycomb sandwich structure has a narrow sound absorption frequency, and the maximum sound absorption frequency is at its resonance frequency, which depends on the depth of the honeycomb core. However, the compact structure of the aero-engine does not allow room for an excessively large acoustic liner [3].

Many scholars at home and abroad have designed and studied the single cell of the acoustic liner, that is, the honeycomb core in the honeycomb sandwich structure (as shown in Fig. 1).The main method comprises series-parallel connection of micro-perforated plate sound absorption structures[4, 5] or design the back cavity of the micro-perforated plate[6, 7] to achieve the purpose of broadband, However, in the face of

Chinese Society of Aeronautics and Astronautics (Ed.): CASTC 2023, LNME, pp. 565–570, 2024.
https://doi.org/10.1007/978-981-99-8861-7_55

complex noise environment, the fixed sound absorption band of the above broadband sound absorption structure can't adapt to the changing environment.

In this paper, a telescopic micro-perforated plate structure is proposed, At the same time, it can adjust the sound absorption frequency while maintaining wide frequency, and realize the best absorption of noise by changing the resonant frequency through the telescopic structure.

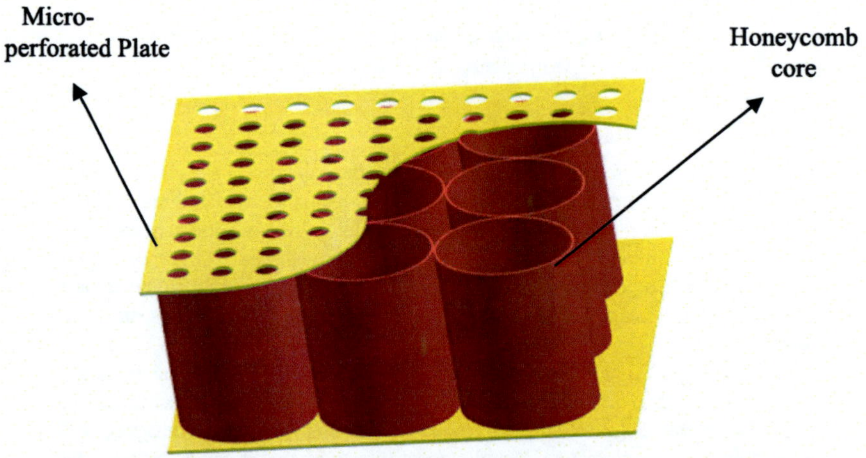

Fig. 1. Simple Model of Honeycomb Micro-perforated Panel Acoustic Liner

2 Structural Design

For the analysis of multi-layer micro-perforated plate structure, the transfer matrix calculation is more accurate, such as the four-layer micro-perforated plate with equal diameter shown in Fig. 2, and the calculation formula is as follows:

$$T_{MPP_i} = \begin{bmatrix} 1 & Z_{MPP_i} \\ 0 & 1 \end{bmatrix} \tag{1}$$

$$Z_{MMP_i} = \rho c(r_i + j\omega m_i) \tag{2}$$

$$T_{D_i} = \begin{bmatrix} cos\left(\frac{\omega D_i}{c}\right) & j\rho c sin\left(\frac{\omega D_i}{c}\right) \\ jsin\left(\frac{\omega D_i}{c}\right)/\rho c & cos\left(\frac{\omega D_i}{c}\right) \end{bmatrix} \tag{3}$$

where T_{MPP_i} is the transfer matrix of MPP_i, Z_{MMP_i} is the acoustic impedance of MPP_i, T_{D_i} is the transfer matrix of the cavity. ρ, c is the density of air and the speed of sound, respectively, D_i is the depth of the cavity. Therefore, the total transfer matrix of a series structure of four equal diameter micro-perforated plates can be written as:

$$T_t = [T_{MPP_1}] \cdot [T_{D_1}] \cdot [T_{MPP_2}] \cdot [T_{D_2}] \cdot [T_{MPP_3}] \cdot [T_{D_3}] \cdot [T_{MPP_4}] \cdot [T_{D_4}] \tag{4}$$

According to the total transfer matrix of the structure, the relative impedance of the structure can be obtained:

$$Z_t = \frac{T_{t_{11}}}{\rho c T_{t_{21}}} \tag{5}$$

Substitute Eq. (5) into the absorption coefficient for normal incidence:

$$\alpha = \frac{4Re(Z_t)}{[1+Re(Z_t)]^2+[Im(Z_t)]^2} \tag{6}$$

For the series connection of four micro-perforated plates of unequal diameter shown in Fig. 3, due to the sudden change of cross section, the calculation equation of the unequal diameter structure is as follows:

$$B_i = \begin{bmatrix} 1 & 0 \\ 0 & s_{i+1}/s_i \end{bmatrix} \tag{7}$$

where, B_i is the ratio matrix, and s_i is the area of MPP_i, then the transfer matrix of the four-layer micro-perforated plate structure with different diameters can be written as:

$$T_t = [T_{MPP_1}] \cdot [T_{D_1}] \cdot [B_1] \cdot [T_{MPP_2}] \cdot [T_{D_2}] \cdot [B_2] \cdot [T_{MPP_3}] \cdot [T_{D_3}] \cdot [B_3] \cdot [T_{MPP_4}] \cdot [T_{D_4}] \tag{8}$$

According to the transfer matrix of the overall structure, the relative impedance of the structure can be obtained by substituting it into Eq. (5). The absorption coefficient under normal incidence can be obtained by substituting the relative impedance of the obtained structure into Eq. (6).

As shown in Fig. 4, by connecting the micro-perforated plates with different diameters in series, we imagine a telescopic micro-perforated plate structure, thus forming a sound absorption structure with adjustable sound absorption frequency. We will use simulation to verify the idea.

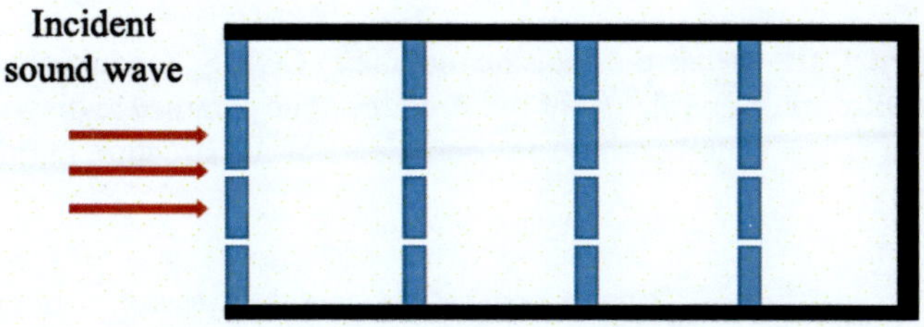

Fig. 2. Four layers of equal diameter micro-perforated plates in series

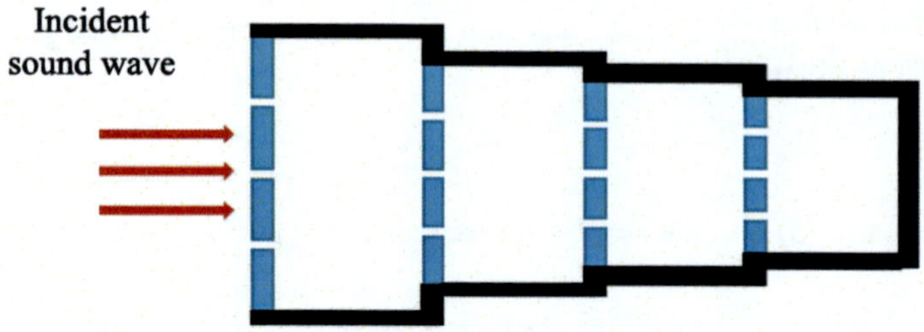

Fig. 3. Unequal diameter micro-perforated plates in series

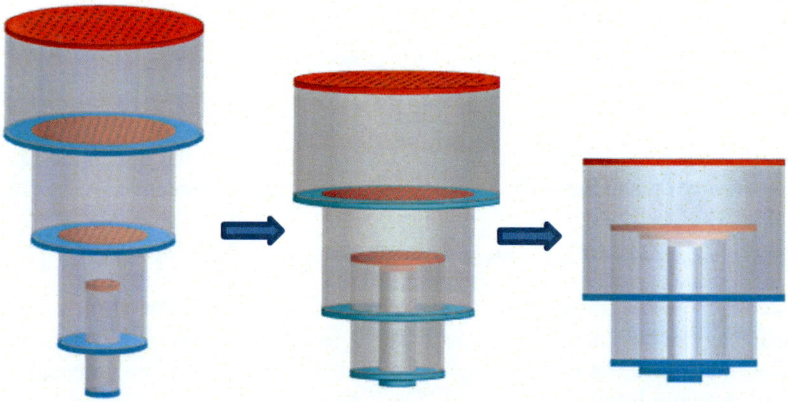

(a) Fourth layer shrinkage (b) Third layer shrinkage (c) Second layer shrinkage

Fig. 4. Model of Sound absorption structure in different stretching states

3 Simulation Analysis

For the above analysis, we use the acoustic module in COMSOL Multiphysics to calculate the sound absorption coefficient of the structure, The parameters of the structure are shown in Table 1. The sound absorption coefficients of the structure in different States are simulated.

Table 1. Parameters of each layer of the structure

Category	First layer	Second layer	Third layer	Forth layer
Aperture(mm)	0.5	0.5	0.5	0.5
Perforation rate(%)	4	4	4	4
Micro-perforated plate diameter(mm)	29	21	13	5
Depth of cavity(mm)	20	20	20	20

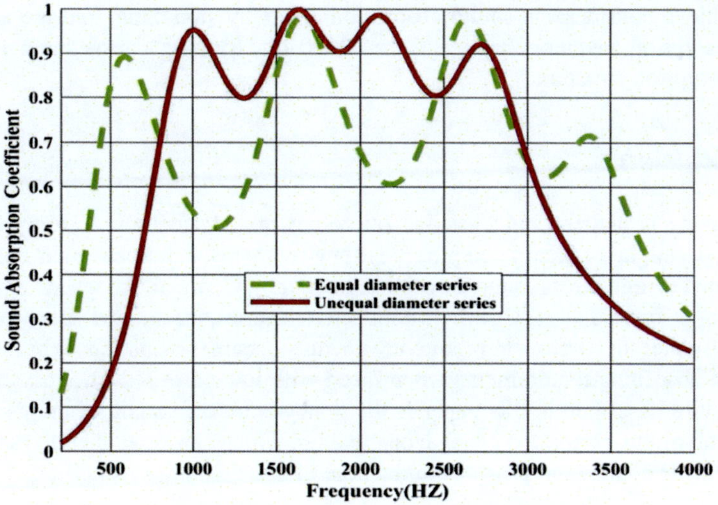

Fig. 5. Comparison of two structures

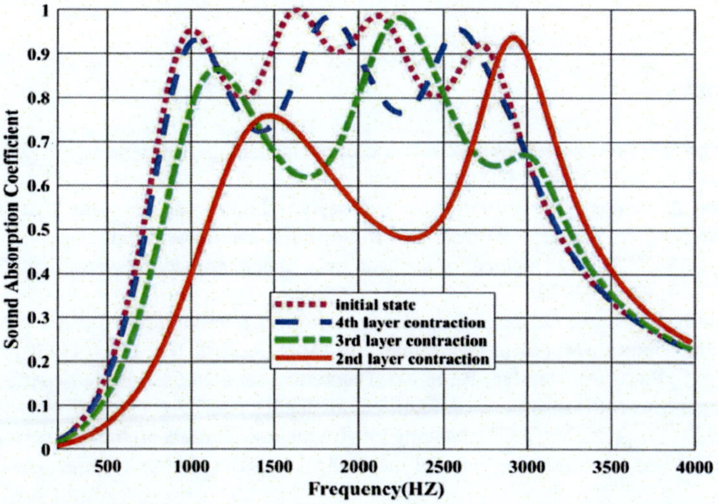

Fig. 6. Sound absorption coefficient under different stretching states of structures

From Fig. 5, it can be seen that the series structure of micro-perforated plates with unequal diameters has better sound absorption effect than the series structure of micro-perforated plates with equal diameters in a certain frequency range, The difference between the peaks and valleys is small and relatively stable, with a wide frequency range of sound absorption performance. When the series structure of micro-perforated plates with different diameters can be telescopic, its sound absorption coefficient under different telescopic states is shown in Fig. 6, In the frequency range of 700–3300 Hz, it can achieve considerable sound absorption effect by stretching, and the adjustable sound absorption frequency range reaches 2000 Hz, forming a broadband adjustable sound absorption structure.

4 Conclusion

In this paper, a broadband and telescopic micro-perforated panel sound absorption structure is proposed, In the face of complex noise environments, the structure can be stretched and adjusted to regulate its sound absorption frequency. Compared to several equal diameter micro-perforated panel absorbers connected in series, the studied telescopic micro-perforated panel sound-absorbing structure not only occupies a small space but also meets the sound-absorbing requirements when faced with low noise reduction requirements. The structure described in this paper is the honeycomb core study of the aero-engine honeycomb micro-perforated panel acoustic liner, which can be applied to the acoustic liner, has wide application prospect in the field of aviation noise reduction.

Acknowledgments. This work was sponsored by the National Natural Science Foundation of China (No. 51965041) and the Jiangxi Postgraduate Innovation Special Fund Project (YC2022-s735).

References

1. Voutsinas, S.G.: Aeroacoustics research in Europe: the CEAS-ASC report on 2005 highlights. J. Sound Vib. **299**, 419–459 (2007)
2. Kempton, A.: Acoustic liners for modern aero-engines. In: Proceedings of the 15th CEAS-ASC Workshop and 1st Scientific Workshop of X-Noise EV. Lausanne (2011)
3. Ma, X.Q., Su, Z.T.: Development of acoustic liner in aero engine: a review. Sci. China Tech. Sci. **63**, 2491–2504 (2020)
4. Bucciarelli, F., Fierro, G.P.M., Meo, M.: A multilayer microperforated panel prototype for broadband sound absorption at low frequencies. Appl. Acoust. **146**, 134–144 (2019)
5. Mosa, A.I., Putra, A., Ramlan, R., et al.: Theoretical model of absorption coefficient of an inhomogeneous MPP absorber with multi-cavity depths. Appl. Acoust. **146**, 409–419 (2019)
6. Zhao, X.D., Wang, X., Yu, Y.J.: Enhancing low-frequency sound absorption of micro-perforated panel absorbers by combining parallel mechanical impedance. Appl. Acoust. **130**, 300–304 (2018)
7. Gai, X.-L., Xing, T., Li, X.-H., et al.: Sound absorption of microperforated panel with l shape division cavity structure. Appl. Acoust. **122**, 41–50 (2017)

Research on the Influence of Multiple Parameters on the Absorption Band Migration of Double Degrees of Freedom Aeronautical Sound Lining

Qi Li and Jinwu Wu(✉)

School of Aircraft Engineering, Nanchang Hangkong University, Nanchang 330063, China
wujinwu@nchu.edu.cn

Abstract. The conventional double-degree-of-freedom aeronautical acoustic lining structure is fixed, as is its absorption band range. In view of the problem that the sound absorption band of traditional double-degree-of-freedom aeronautical acoustic linings cannot be offset adaptively with external noise and the sound absorption band is narrow, a parametric improvement study is conducted on the double-degree-of-freedom acoustic lining structure model, and the influence of multiple parameters on the movement of the sound absorption band is analyzed. The theory of a multi-parameter coordinated adjustment method for the movement of the sound absorption band is proposed, and the correctness and feasibility of the proposed method are verified through numerical calculation and simulation analysis. This paper provides a theoretical reference for the adaptive control design of a double-degree-of-freedom acoustic lining structure with adjustable sound absorption bands.

Keywords: Aeronautical acoustic lining · Sound Absorption Bands · Multi-parameter adjustment

1 Aircraft Engine Nacelle Sound Lining Noise Reduction Research

As the aircraft's heart, the aero-engine generates structural noise, vibrations, and other failure phenomena that pose a serious threat to aircraft safety and may result in fatalities and significant economic losses if left unattended.

Noise control from the propagation of aircraft engine noise has become an important means of noise control. Through the installation of sound lining inside the engine, including nacelle sound lining application parts, the control of noise propagation is improved. Most of the nacelle sound lining is micro-perforated plate structure, which the Chinese Academy of Sciences, Maa Dy [1] proposed in the 1960s and is widely used in noise control engineering. For example, Huo Shiyu [2] carried out a study on the anechoic effect of titanium alloy annular lining under tangential flow and temperature gradient conditions and found that changes in tangential flow and backing plate temperature would shift the noise reduction band. Zheng Wentao [3] used a simulated

Chinese Society of Aeronautics and Astronautics (Ed.): CASTC 2023, LNME, pp. 571–579, 2024.
https://doi.org/10.1007/978-981-99-8861-7_56

annealing optimization algorithm to obtain the geometric parameters of the sound liner with maximum noise reduction effect during the design of a single-degree-of-freedom sound liner and achieved a good noise reduction effect on fan noise. Yang Jiafeng [4] conducted comparative experiments on single-degree-of-freedom and double-degree-of-freedom sound liners in nacelles under tangential flow conditions and showed that double-degree-of-freedom sound liners can effectively broaden the sound absorption band. For the structure of acoustic lining units, Jiang C [5] studied microperforated plates with stepped holes, and AI Mosa [6] proposed non-uniform perforations and double-layer microperforated plate structures with different depths, both of which were used to enhance the sound absorption capacity of microperforated plate acoustic bodies.

However, the traditional sound lining's sound absorption frequency band is largely set, and the adjustable sound lining's sound absorption frequency band hasn't received much attention. The nacelle acoustic lining study solely examined the effects of single parameter adjustments on sound absorption performance; it did not examine the effects of various parameters on the sound absorber's band shift. Aiming at the problem that the sound absorption band of the conventional dual-degree-of-freedom acoustic lining cannot be adaptively offset with external noise and the sound absorption frequency band is narrow, the parameter improvement research of the double-degree-of-freedom acoustic lining structure is carried out. The mechanism of parameter coordination adjustment is analyzed, and the correctness and feasibility of the parameter coordination regulation theory are verified by numerical calculation and simulation analysis, which provides a theoretical reference for the adaptive control design of a double-degree-of-freedom acoustic lining structure with an adjustable sound absorption band.

2 Theoretical Model of the Aeronautical Acoustic Liner

Most of the sound liners used in nacelles are divided into single or multi-layer honeycomb sandwich micro-perforated plate acoustic structures, i.e., single degree of freedom sound liners and multi-degree of freedom sound liners, where the double degree of freedom sound liners have a wider acoustic band and cover a wider range of noise compared to the single degree of freedom sound liners.

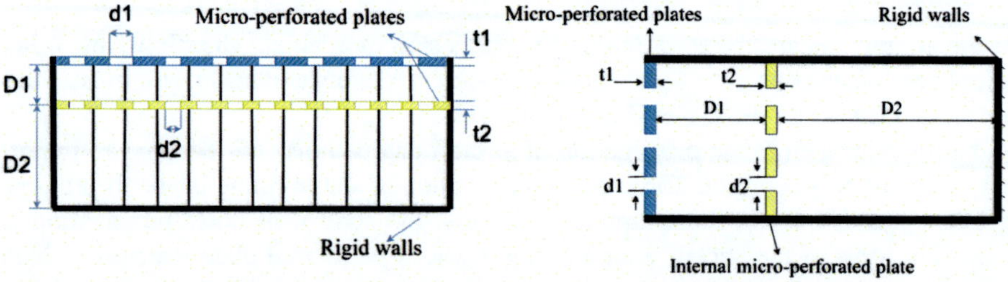

Fig. 1. Structure of the double-degree-of-freedom acoustic liner and structure of the double-degree-of-freedom acoustic liner unit.

Double-degree of freedom sound liners consist of micro-perforated plates, honeycomb cores, micro-perforated plates, honeycomb cavities, and rigid back plates. The overall structure and unit of the double degree of freedom acoustic liner are shown in Fig. 1 above: D is the cavity depth, t is the thickness of the micro-perforated plate, d is the perforated aperture, and p is the perforation rate.

The relative acoustic impedance of the acoustic lining unit was found according to the microperforated plate theory of Maa Dy [1] as:

$$Z = r_1 + j\left(\omega m_1 - \cot\frac{\omega D1}{c}\right) + \frac{\cot^2\frac{\omega D1}{c}}{r_2 + j\left(\omega m_2 - \cot\frac{\omega D1}{c} - \cot\frac{\omega D2}{c}\right)} \tag{1}$$

Through the parallel relationship between multiple sound lining units, the acoustic impedance of the overall sound-absorbing structure of the sound lining is derived as:

$$Z_T = \left(\sum_{i=1}^{n}\frac{a_i}{Z^i}\right)^{-1} \tag{2}$$

where n is the number of lining units and a_i is the ratio of the perforated panel area of the ith unit to the perforated panel area of the overall lining structure.

The sound absorption coefficients of the sound lining structure are as follows:

$$\alpha = \frac{4\mathrm{Re}(Z_T)}{[1 + \mathrm{Re}(Z_T)]^2 + [\mathrm{Im}(Z_T)]^2} \tag{3}$$

Adaptive control of the absorption band of the double-degree of freedom sound lining can be accomplished by coordinating the control parameters: the lower cavity depth D2, the internal micro-perforated plate aperture d2, and the internal micro-perforated plate perforation rate p2. This is evident from the calculation formula (3) of the absorption coefficient of the sound lining structure. The idea of the rate of change of the absorption coefficient [7] is introduced here with the goal of identifying the control law for each parameter in the change. The rate of change of the absorption coefficient is a criterion used to evaluate the change of the absorption coefficient and to reflect the sensitivity of each parameter to the absorption coefficient. It is the ratio of the change in the absorption coefficient to the original absorption coefficient.

3 Simulation Analysis of the Absorption Band Shift

3.1 Analysis of the Effect of Each Parameter on the Absorption Band Shift

For multiple variables, the effect of different parameters on the sound absorption performance of the acoustic body is investigated. Because of the large number of parameters, the effect of a single parameter is investigated separately using the control variable method.

The original parameters are plate thickness t1 = t2 = 0.5 mm, perforation aperture d1 = d2 = 0.5 mm, perforation rate p1 = p2 = 4%, cavity depth D1 = D2 = 30 mm, and simulation frequency range 0–3000 Hz.

In the three parameters of cavity depth, hole diameter, and perforation rate, one of the parameters was controlled to be changed, while the rest of the parameters remained unchanged. The parameter variations are shown in Table 1:

Table 1. Parameter adjustment of the sound lining unit.

Parameter adjustments	Cavity depth D2(mm)	Aperture d2(mm)	Perforation rate p2(%)
Original size	30	0.5	4
30% reduction	21	0.35	2.8
30% increase	39	0.65	5.2

The sound absorption coefficients after varying the cavity depth, aperture diameter, and perforation rate with the remaining parameters held constant are shown below (Figs. 2, 3 and 4):

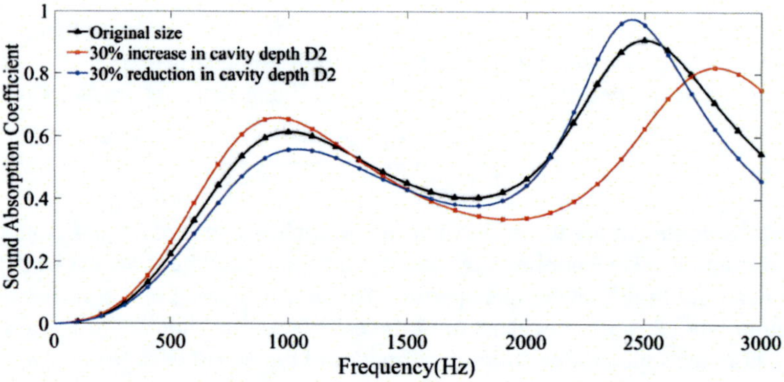

Fig. 2. Variation of sound absorption coefficient when cavity depth D2 is changed.

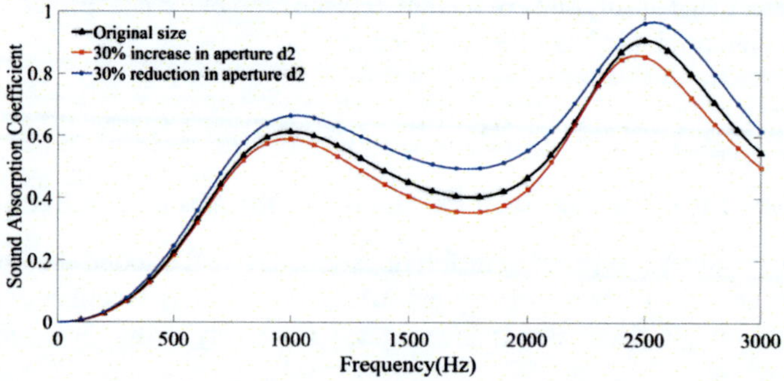

Fig. 3. Variation of sound absorption coefficient when aperture d2 is changed.

The change in the rate of change of the sound absorption coefficient when the single parameters of the sound lining, aperture diameter, cavity depth and perforation rate, are changed respectively, is shown below (Figs. 5 and 6):

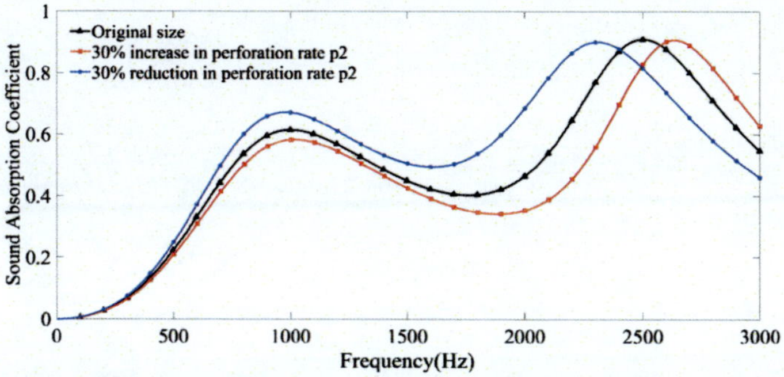

Fig. 4. Variation of sound absorption coefficient when perforation rate p2 is changed.

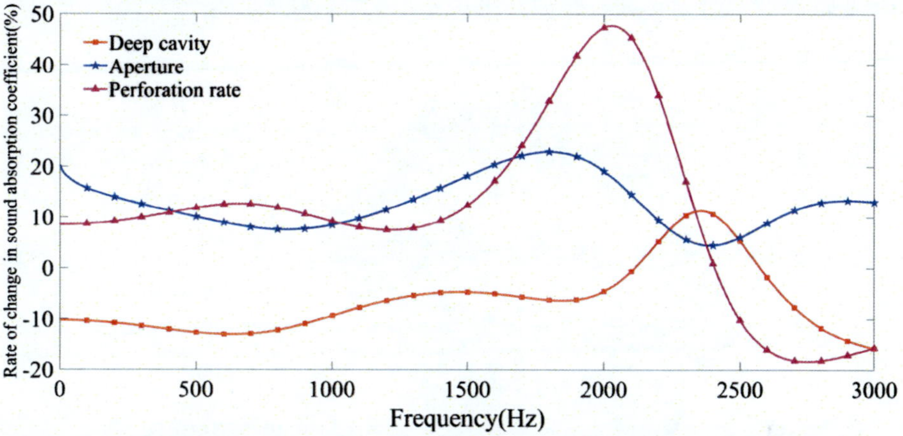

Fig. 5. Graph of the rate of change of the sound absorption coefficient when the cavity depth D2, pore aperture d2 and perforation rate p2 are reduced by 30% respectively.

3.2 Analysis of the Joint Effect of Two Parameters on the Absorption Band Shift

In order to verify the sound absorption band shift, the sound lining model was firstly simulated using the COMSOL multi-physics software in the pressure acoustic (frequency domain) module, and secondly, numerical calculations were carried out using the Matlab software to verify the correctness of the findings.

The acoustic lining model is shown in the figure below, consisting of a honeycomb sandwich structure with several double-layered micro-perforated structures (Fig. 7).

The walls of the structure are set to rigid in the COMSOL simulation and the parameters are synergistically adjusted as in the table below (Table 2):

The COMSOL and Matlab simulations are shown in the following figures respectively (Fig. 8):

As can be seen from Fig. 9 above, the results of the Comsol and Matlab simulations of sound lining absorption band shifts are consistent.

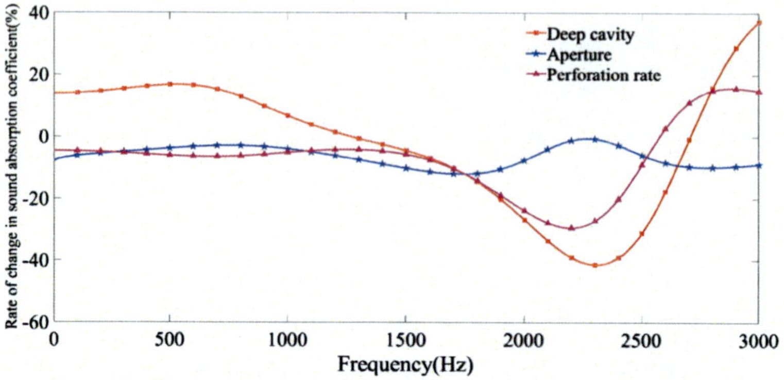

Fig. 6. Graph of the rate of change of the sound absorption coefficient when the cavity depth D2, pore aperture d2 and perforation rate p2 are increased by 30% respectively.

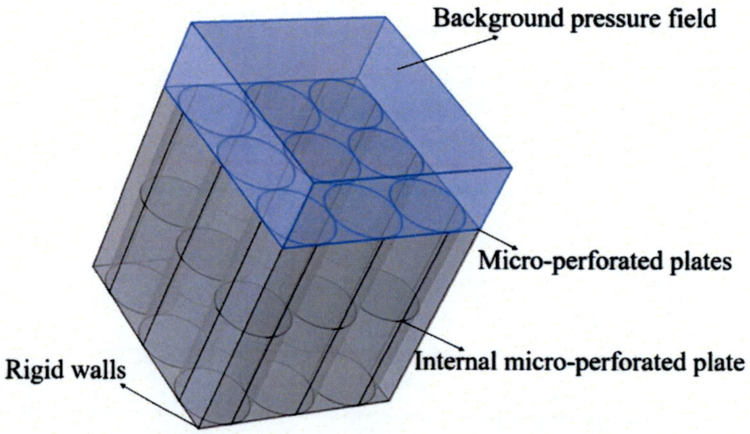

Fig. 7. Diagram of the sound lining model.

Table 2. D2 P2 parameter synergy adjustment.

D2 P2 parameter synergy adjustment	Cavity depth D2(mm)	Perforation rate p2(%)
Original size	30	4
30% increase	39	5.2
50% increase	45	6
30%reduction	21	2.8
50%reduction	15	2

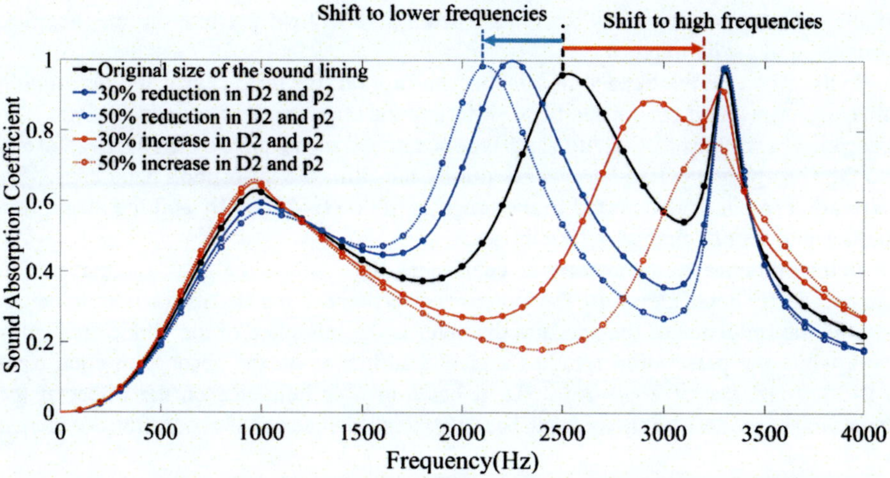

Fig. 8. Matlab simulation: Diagram of the sound absorption band offset of the acoustic lining under the synergistic adjustment of cavity depth D2 and perforation rate p2.

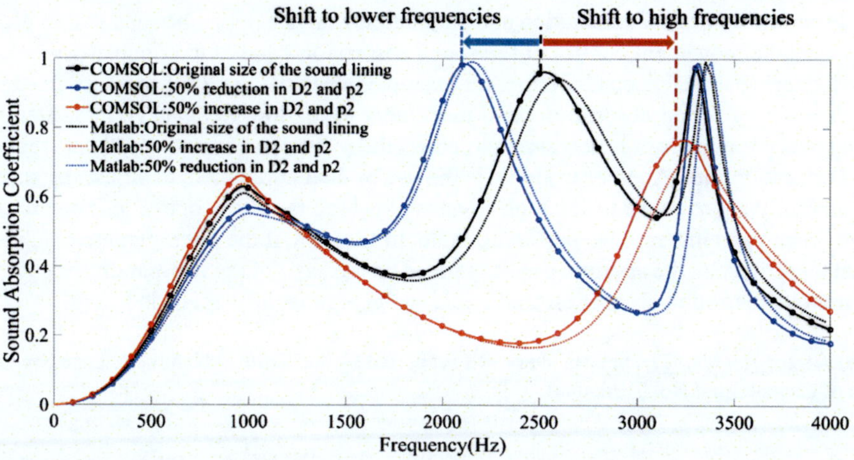

Fig. 9. Comparison of Matlab simulations and Comsol simulations of acoustic lining absorption zone deflection with synergistic adjustment of depth D2 and perforation rate p2.

4 Conclusion

The above simulation calculation shows that: in the three parameters of cavity depth D2, aperture diameter d2, and perforation rate p2, the influence on the sound absorption band shift of the sound lining is from high to low: cavity depth, perforation rate, and aperture diameter; the influence on the sound absorption coefficient is from high to low: cavity depth, perforation rate, and aperture diameter; and the influence on the sound absorption coefficient in some frequency intervals is from high to low: perforation rate, aperture diameter, and cavity depth.

In the effect on the band shift of sound absorption, the deep cavity and perforation rate can cause sound absorption band shift, and the aperture does not cause offset. The absorption band caused by cavity depth is concentrated in the range of 1300 Hz–2500 Hz, and the absorption coefficient changes during the shift; the absorption band caused by perforation rate is concentrated in the range of 2000 Hz–2500 Hz, and the absorption coefficient does not change during the shift.

In the influence on the absorption coefficient, the aperture diameter and perforation rate, and cavity depth can cause the increase or decrease of the absorption coefficient. In general, the influence of the aperture diameter is the influence of the whole frequency band, while the perforation rate has a greater influence on the absorption coefficient at 0–2000 Hz, and at 2000–2500 Hz, it has a greater influence on the offset of the absorption band, and the cavity depth has a greater influence on the absorption coefficient at 0–1300 Hz.

In controlling the shift of the absorption band of the double-degree-of-freedom sound lining, the shift of the absorption band can be controlled by controlling the cavity depth and perforation rate, and the change of the absorption coefficient can be controlled by controlling the aperture diameter and perforation rate.

From the numerical calculation and simulation analysis, it can be seen that in order to realize the movement control of the sound absorption band of the double-degree-of-freedom aeronautical acoustic lining, for example, when the double-degree-of-freedom acoustic lining sound absorption band movement to the low frequency is adjusted by using dual parameters, it is necessary to control the reduction of cavity depth and reduce the perforation rate. Another example is the use of dual parameters to adjust the sound absorption band to move to the high frequency, which requires control to increase the cavity depth and increase the perforation rate. In addition, if the three-parameter adjustment is carried out, the aperture can be reduced on the basis of the two-parameter control to improve the sound absorption coefficient of the aeronautical acoustic lining.

Acknowledgements. The present study was supported by the National Natural Science Foundation of China (Grant No. 51965041).

References

1. Maa, D.Y.: Theory and design of microperforated panel sound-absorbing constructions. Sci. Sinica **18**(1), 55–71 (1975)
2. Huo, S., Yang, J., Deng, Y., et al.: Acoustic characteristics test on aft-fan duct with titanium alloy annular acoustic liner. J. Aerosp. Power **35**(2), 272–279 (2020)
3. Zheng, W., Jiang, Y., Zhao, H., et al.: Design optimization of a single degree of acoustic liner for high bypass ratio fan/booster. Aero-engine **48**(4), 34–39 (2022)
4. Yang, J.F., Xue, D.W., Li, Z.H., et al.: Single and double degree-of-freedom acoustic liners under grazing flow: experiment. Acta Aeronautica et Astronautica Sinica **41**(11), 223860 (2020). (in Chinese)
5. Jiang, C., Li, X., Cheng, W., et al.: Acoustic impedance of microperforated plates with stepwise apertures. Appl. Acoust. **157**, 106998 (2019)

6. Mosa, A.I., Putra, A., Ramlan, R., et al.: Wideband sound absorption of a double-layer microperforated panel with inhomogeneous perforation. Appl. Acoust. **161**, 107167 (2020)
7. Yan, S., Wu, J., Chen, J., et al.: Optimization design and analysis of honeycomb micro-perforated plate broadband sound absorber. Appl. Acoust.Acoust. **186**, 108487 (2022)

A Helmholtz Resonator Array for Low-Frequency Broadband Sound Absorption

Xianze Li and Qibo Mao(✉)

School of Aircraft Engineering, Nanchang Hangkong University, Nanchang 330063, China
qbmao@nchu.edu.cn

Abstract. Helmholtz resonators are widely used to reduce narrow frequency band noise. In this paper, an acoustic absorption structure based on Helmholtz resonator arrays is designed, aimed at achieving continuous broadband and low-frequency sound absorption. The basic unit of the structure is a Helmholtz resonator. The acoustic absorption performance was studied by simulation and experimentation. A 3D printed sound absorber was tested in an impedance tube to verify its sound absorption characteristics. The experimental results are in good agreement with the sound absorption coefficient. The results showed that the coupling effect between the double layer cavity and the neck body enhances the sound absorption performance in the low-frequency band. Experiments show that the designed sound absorption structure has continuous broadband sound absorption in the target range of 300–1100 Hz. This study provides ideas and methods for the structural design of broadband acoustic structures.

Keywords: Sound absorption · Helmholtz resonator · Broadband sound absorption · Lower-frequency range

1 Introduction

The Helmholtz resonator is known for its ability to absorb low-frequency sound waves. It typically consists of a neck and a cavity, and converts sound energy into heat energy by resonance. However, the sound absorption frequency band of a Helmholtz resonator is relatively narrow, making the optimization of its sound absorption performance a subject of ongoing research by scholars.

By changing the structural parameters or shape of the Helmholtz resonator, its resonant frequency can be adjusted. Cai et al. [1] designed spiral and extended necks and applied them to a Helmholtz resonator, then analyzed their acoustic absorption performance. Guo et al. [2] has discussed the absorption characteristics of embedded tube Helmholtz resonators and their combinations in broadband sound absorption, and designed a thin low-frequency absorber consisting of an embedded tube checkerboard-type absorber. Huang et al. [3] studied Helmholtz resonators with embedded tubes to analyze their sound absorption performance. Compared to conventional Helmholtz resonators, the embedded tube Helmholtz resonator structure features compact dimensions

Chinese Society of Aeronautics and Astronautics (Ed.): CASTC 2023, LNME, pp. 580–586, 2024.
https://doi.org/10.1007/978-981-99-8861-7_57

at the same resonance frequency. Liu et al. [4] designed Helmholtz resonators with separators to generate multiple absorption peaks.

It is also a common method to broaden the absorption band by connecting Helmholtz resonators in parallel to form an array. Gai et al. [5] investigated microperforated plates with multiple Helmholtz resonators. The results show that adding Helmholtz resonators leads to an increase in the amount of resonance peaks. It can be used to broaden the absorption bandwidth. Zhu et al. [6] investigated a multilayer overlapping structure consisting of necks and cavities for absorbing low-frequency sound waves in a continuous broadband, focusing on the energy dissipation in the neck, and analysed the sound absorption theory of the overlapping structure from a physical point of view. Ren et al. [7] adopted a hexagonal honeycomb cavity resonator with internal expansion as the element, and proposed an optimization method to account for the influence of impedance coupling effect on the sound absorption coefficient curve, providing inspiration and methodology for the compact and multifunctional design of wideband sound absorption structures. Bi et al. [8] improved the sound absorption performance of Helmholtz resonators by dividing the chambers to achieve high absorption coefficients and wide absorption bands at the same time, and the optimal geometrical parameters were obtained by using a combination of finite element simulations and algorithms. However, simply combining the units may result in a localised degradation of the overall structural acoustic absorption performance, making the design of combined resonant units a complex task.

To sum up, in this paper, a new Helmholtz resonators array is designed. By combining several resonators of different frequencies, continuous broadband low-frequency sound absorption can be achieved.

2 Design Method and Theoretical Modal

The resonant frequency of the Helmholtz resonator is shown in Eq. (1):

$$f = \frac{c}{2\pi}\sqrt{\frac{S}{LV}} \tag{1}$$

where V is the volume of the cavity, S is the cross-sectional area of the neck, and L is the height of the neck.

It is well known that the acoustic absorption in multiple parallel Helmholtz resonators is obtained through the coupling of individual Helmholtz resonators. By increasing the number of resonators, more peaks appear at different frequencies. Therefore, in this paper, a kind of Helmholtz resonator array is studied for continuous broadband low frequency sound absorption effect. As shown in Fig. 1(a). Double neck and double cavity constitute the double-layer Helmholtz resonator.

Then, a Helmholtz resonators array structure is adopted, and the Helmholtz resonator is used as the sound absorption unit. Finally, multiple resonators are connected in parallel, as preseneted in Fig. 1. Obviously, the first Helmholtz resonator is coupled to a second Helmholtz resonator. The second resonator, on the other hand, is influenced by the region of the first resonator.

The acoustic absorption performance of Helmholtz resonator can be adjusted by changing the geometrical parameters of the structure Therefore, the overall acoustic

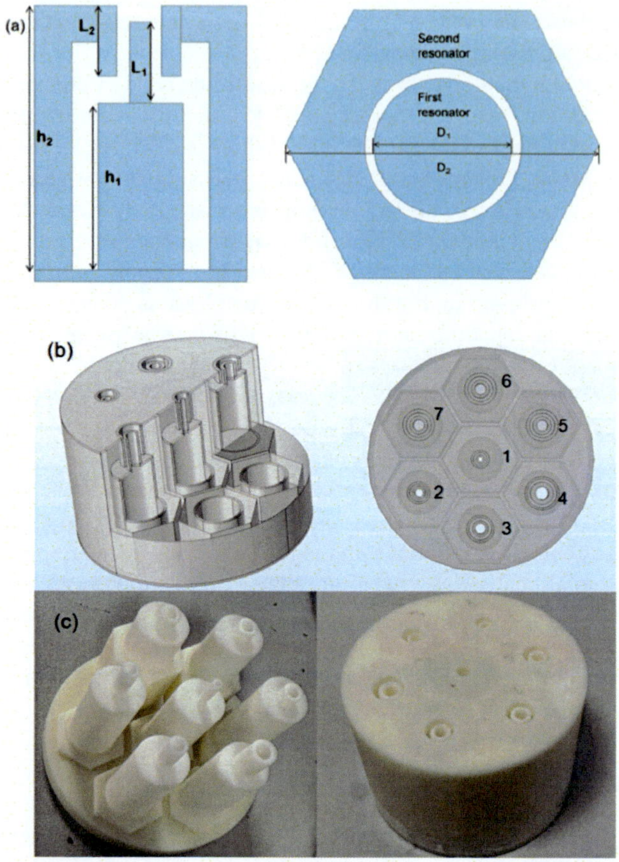

Fig. 1. Geometry of (a) Array of Helmholtz resonators; (b) Three-dimensional view; (c) Fabricated samples by 3D printing.

absorption performance of the array can be changed by adjusting the geometrical parameters of the cavities and necks of the different units. A Helmholtz resonator array consisting of seven units is designed here. Geometric parameters of the absorber and resonance frequency of the each unit are listed in Table 1 and 2, respectively.

3 Simulation and Experimentation

In this section, simulations and experimental verifications was conducted to analyze the sound absorption performance of Helmholtz resonator arrays and its absorption mechanism. The AWA6290T impedance tube was used for measuring the acoustic absorption coefficient. The sample material used in the experiments was photosensitive resin, which was prepared using 3D printing, resulting in a smooth surface with good hardness, meeting the sample production requirements. The AWA6290T dual-channel transfer function sound absorption coefficient measurement system, consisting of a loudspeaker, power

Table 1. Geometric Parameters of the Helmholtz resonator arrays (mm)

Unit	D1	D2	L1	L2	H1	H2
1	2.5	6	15	14	40	57
2	4	8	15	14	40	57
3	5	10	15	14	40	57
4	6	13	15	14	40	57
5	5	11	7	6	40	48
6	5	11	7	6	29	37
7	5	11	7	6	21	29

Table 2. Resonance Frequency of Each Unit (Hz)

Unit	Frequency	Unit	Frequency
1	241	8	568
2	283	9	611
3	316	10	676
4	386	11	738
5	437	12	807
6	489	13	920
7	509	14	1043

amplifier, microphone, and signal analyzer, was employed for the experiments. In order to facilitate measurements using impedance tubes of the same size, the experimental sample was designed as a cylinder with a diameter of 99 mm. In order to satisfy the hard acoustic field boundary conditions, the thickness of the outer wall is not less than 1 mm. As shown in Fig. 2.

The acoustic module of COMSOL Multiphysics is used here to calculate the absorption coefficient of the model. The simulation involved physical modeling, where the small area formed in the neck was considered and the thermal viscosity acoustic model was employed for this region. The other areas, which were large and regular in shape, used the pressure acoustic model. The simulation model of the software is illustrated in Fig. 3, with plane wave radiation as the main entrance and all walls assumed as rigid. Acousto-thermal viscosity coupling was assumed for the entire model.

Figure 4 shows the contours of acoustic pressure and velocity of unit 2 at different resonance frequencies. Two absorption peaks were observed for each unit, with the first resonance frequency of unit 2 being approximately 316 Hz. The second resonance frequency is about 386 Hz. When the excitation frequency corresponds to the unit resonance frequency, the cavity starts to work. It is clearly seen that the resonant cavity has a

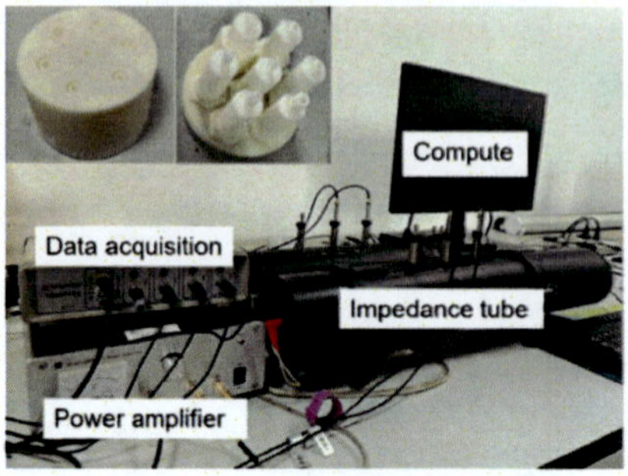

Fig. 2. The impedance tube and testing sample.

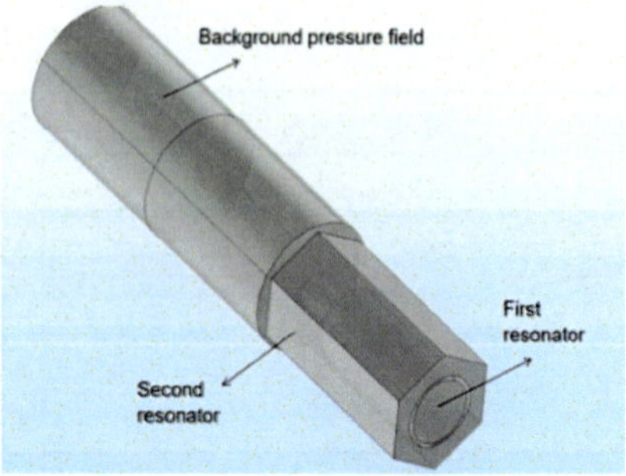

Fig. 3. Unit sound absorber model.

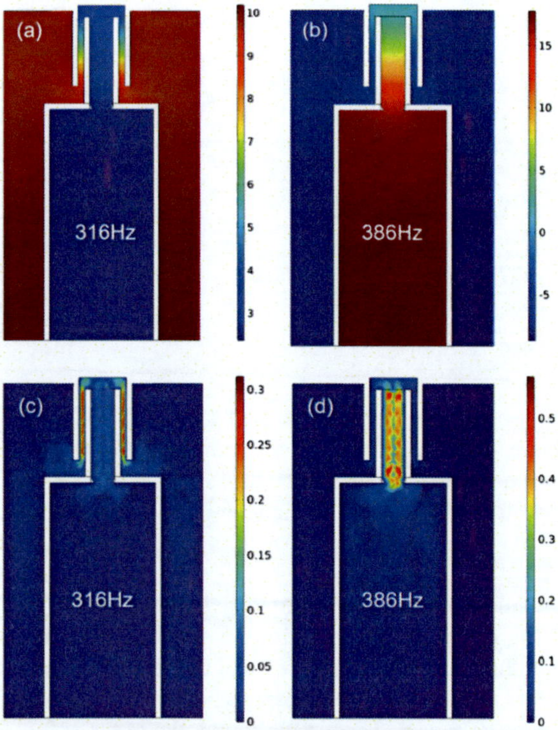

Fig. 4. Sound pressure (Pa) and velocity (m/s) of unit 2: (a) sound pressure at 316 Hz (b) acoustic pressure at 386 Hz (c) velocity at 316 Hz (d) velocity at 386 Hz.

significantly larger sound pressure gradient within its resonant frequency than the other cavities.

This discrepancy in acoustic pressure amplitude results in airflow through the neck with higher velocities, as evident in Fig. 4(c) and (d). In this case, the velocity increases significantly at the neck, resulting in increased friction between the airflow and the resonator wall. As a result, sound energy is rapidly converted to heat.

When the sound absorber is not at the working frequency, the difference in sound velocity and pressure between different necks and cavities becomes insignificant, resulting in poor sound absorption effects.

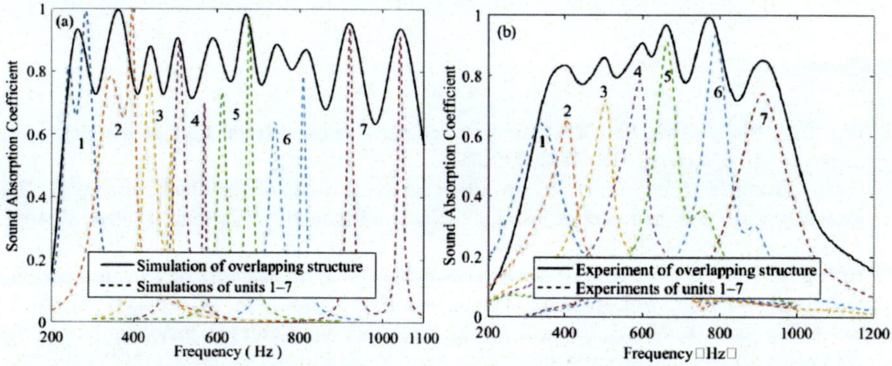

Fig. 5. (a) Simulated Sound Absorption of Units 1–7. (b) Experimental Sound Absorption Coefficient of Absorber and Units 1–7.

The simulation results are shown in Fig. 5(a). There are 14 peak absorption coefficients in the 200–1100 Hz range, corresponding to the resonance frequencies of the individual units. The experimental sound absorption coefficients are shown in Fig. 5(b). Its resonant frequency corresponds to the peak sound absorption frequency corresponding to seven different units. It is noted that the resonance peaks are not clearly reflected in the experimental results, which is due to the larger damping of the actual samples, as well as the existence of some size inaccuracy of the samples. Units 1–7 have different geometrical parameters corresponding to different resonance frequencies.

According to the comparative observation unit 1, 2 and 3 in Table 1, 2 and Fig. 5, when other parameters are the same, increasing D can make the absorption peak generated by HR move towards high frequency. According to the comparative observation in unit 5, 6 and 7, when other parameters are the same, reducing H can make the absorption peak generated by HR move towards high frequency.

4 Conclusions

In conclusion, this study capitalizes on the favorable characteristics of simple structure and effective sound absorption performance offered by Helmholtz resonators. A double-layer Helmholtz resonator with low frequency is designed, incorporating an overlapping structure and parameter adjustments. For the designed arrays, simulations were performed to analyze their sound absorption mechanisms. Additionally, impedance tube measurements are employed to validate the broadband sound absorption performance of the sound absorber. Remarkably, the average absorption coefficient of the Helmholtz resonator array exceeds 0.8 within the frequency range of 300–1000 Hz, thus successfully achieving the objective of broadband efficient noise reduction.

Acknowledgements. This work was sponsored by the National Natural Science Foundation of China (no. 51975266) and the Aeronautical Science Foundation of China (no. 2020Z073056001).

References

1. Cai, C.Z., Mark, C.M., Shi, X.F.: An extended neck versus a spiral neck of the Helmholtz resonator. Appl. Acoust. **115**, 74–80 (2017)
2. Guo, J., Fang, Y., Jiang, Z., et al.: Acoustic characterizations of Helmholtz resonators with extended necks and their checkerboard combination for sound absorption. J. Phys. D Appl. Phys. **53**(50), 505504 (2020)
3. Huang, S., Fang, X., Wang, X., Assouar, B., Cheng, Q., Li, Y.: Acoustic perfect absorbers via Helmholtz resonators with embedded apertures. J. Acoust. Soc. Am. **145**(1), 254–262 (2019)
4. Liu, C.R., Wu, J.H., Ma, F., Chen, X., Yang, Z.: A thin multi-order Helmholtz metamaterial with perfect broadband acoustic absorption. Appl. Phys. Exp. **12**(8), 084002 (2019)
5. Gai, X.L., Xing, T., Li, X.H., Zhang, B., Wang, W.J.: Sound absorption of microperforated panel mounted with Helmholtz resonators. Appl. Acoust. **114**, 260–265 (2016)
6. Zhu, J., Qu, Y., Su, H., Zhang, J., Meng, G.: A multi-layer overlapping structure for continuous broadband acoustic wave absorption at lower-frequencies. Appl. Acoust. **187**, 108496 (2022)
7. Ren, Z., Cheng, Y., Chen, M.: A compact multifunctional metastructure for Low-frequency broadband sound absorption and crash energy dissipation. Mater. Des. **215**, 110462 (2022)
8. Bi, S., et al.: Enhancement of sound absorption performance of Helmholtz resonators by space division and chamber grouping. Appl. Acoust. **207**, 109352 (2023)

Analysis and Prospects of Key Technologies for Hydrogen-Electric Regional Aircraft

Jiaxi Wu and Yuhan Ji(✉)

Aviation Industry Development Research Center of China, Beijing 100029, China
btn5754@163.com

Abstract. As hydrogen fuel can realize zero carbon emissions in full life cycle, developing hydrogen-powered aviation is a well-accepted promising alternative option for the aviation industry to achieve the net zero goal in 2050. Hydrogen-powered aviation can choose two different power solutions: hydrogen fuel cells with electric motor ("hydrogen-electric") or hydrogen fueled turbine engine. Due to high levels of technology maturity and low R&D costs, hydrogen-electric aircraft has developed rapidly, and become the best choice for realizing the rapid application of hydrogen-powered regional aircraft. Worldwidely, there are already two regional aircraft using hydrogen-electric power solutions that have successfully made their maiden flights. The development of hydrogen-electric aircraft cannot be separated from the development of key technologies such as hydrogen storage systems, hydrogen fuel cells, and electric motors. This paper mainly provides an overview and analysis of key technologies research progress of hydrogen-electric aircraft, and finally puts forward expectations for the future development of hydrogen-electric aircraft.

Keywords: Hydrogen-electric · Regional aircraft · Hydrogen fuel cells · Liquid hydrogen tank

1 Introduction

Since the first aircraft was invented 120 years ago, the rapid development of the aviation industry has not only brought convenience to human society but also caused increasing environmental pollution. In 2016, the Paris Agreement set a ambition to limit global warming to no more than 1.5 °C above pre-industrial levels. In order to reduce the negative impact of the aviation industry on the environment, the International Civil Aviation Organization (ICAO) has proposed the Carbon Offsetting and Reduction Scheme for International Aviation ("CORSIA") plan [1], which aims to reduce carbon dioxide emissions to half the level of 2005 by 2050. In response to the carbon footprint reduction targets of the aviation sector, major players in the aviation industry like the United State, the United Kingdom, China, have proposed distinct aviation carbon emissions reduction strategies.

The global aviation industry is working towards achieving the 2050 goal through various means, one of the solutions is to improve existing technology, companies and

Chinese Society of Aeronautics and Astronautics (Ed.): CASTC 2023, LNME, pp. 587–595, 2024.
https://doi.org/10.1007/978-981-99-8861-7_58

institutions such as Airbus, Boeing, and NASA continuously optimizing their aircraft designs and utilizing new composite materials to reduce aircraft fuel consumption and flight efficiency [2, 3]; another solution is to use more electric aircraft (MEA). However, the energy density of lithium batteries limits the payload and range of aircrafts. Therefore, aircrafts powered by lithium batteries are usually less than 10 seats [4]. Moreover, sustainable aviation fuels (SAF) are also a popular substitute for reducing reliance on fossil fuels. Main aero-engine manufacturers such as Rolls-Royce, GE have explored the application of SAF, however, at present, the production costs of SAF is much higher than traditional jet fuel, and its production capacity is severely limited.

Compared with the above three solutions, hydrogen-powered aircraft is a new option for the aviation industry, and maybe the best options for net-zero carbon implementation of regional aircraft at present. There are many advantages for developing hydrogen-powered aircraft. Hydrogen-powered aircraft using green electricity to produce green hydrogen can achieve zero emissions throughout their life cycle. The combustion of hydrogen fuel only produces water, effectively eliminating the carbon dioxide, nitrogen oxides, soot, and other pollutants generated when using fossil fuels. The contrails and induced cirrus cloudiness generated by the traditional aircraft also cause serious greenhouse effects [5, 6], which also can be avoided by adopting hydrogen-powered aircraft. Furthermore, hydrogen fuel can be produced in larger quantities and with lower costs than SAF to meet fuel demand of commercial aviation.

Hydrogen-powered aircraft have two power solutions: one is electric motor powered by hydrogen fuel cells ("hydrogen-electric"), the other is burning hydrogen directly in turbine engine. For the present, the former option has a relatively higher level of technical maturity, and is expected to be put into commercial use in the near future, which may be the most reliable solution for regional aircraft with fewer than 80 seats to turn "green". Restricted by the power density of hydrogen fuel cells and electric motors, aircraft with more than 80 seats are more likely to choose hydrogen fueled turbine engines as power solutions. This paper focuses on the current development status and key technologies of the most popular hydrogen-electric regional aircraft, the following sections will introduce the development status, key technologies of hydrogen-electric regional aircraft, and summarize the findings.

2 Development Status

Currently, there is a global surge in the development of hydrogen-powered aviation. As the first two companies to achieve maiden flights of hybrid turboprop aircraft powered by hydrogen fuel cells, ZeroAvia and Universal Hydrogen have taken the leading place in hydrogen-powered aviation trend. This section provides an overview of the typical products and latest progress of these two companies, and analyzes the global development of hydrogen--electric regional aircraft.

ZeroAvia is developing a hydrogen-electric Dornier 228 testbed aircraft, replacing the turboprop engines on the left wing of the Dornier 228 with ZeroAvia's hydrogen-electric engine. ZeroAvia's hydrogen-electric powertrain comprises electric motors, two hydrogen fuel cell stacks, lithium-ion battery packs, and tanks with compressed gaseous hydrogen. On January 19, 2023, the modified Dornier 228 aircraft successfully completed its maiden flight in the U.K. [7] During the test flight, the hydrogen tanks and fuel cells stacks were installed inside a fuselage. ZeroAvia is developing two sets of hydrogen-electric powertrain simultaneously, one is ZA600 powertrain, designed for 9–19 seat aircraft, expected to be in service by 2025; another is ZA2000 powertrain, designed for 40–80 seat aircraft such as ATR or Dash-8 series aircraft, expected to be in service by 2027 (Fig. 1).

Fig. 1. ZeroAvia's hydrogen-electric Dornier 228 testbed aircraft.

Universal Hydrogen is developing hydrogen-powered commercial aircraft and modular hydrogen capsules distribution and logistics network. On March 3, 2023, Universal Hydrogen successfully completed the maiden flight of their modified Dash-8 hybrid aircraft [8]. The regional turboprop's right-hand engines was replaced with a hydrogen-electric 1-megawatt powertrain, including the Magni650 electric propulsion unit, liquid hydrogen tanks located inside the rear of the fuselage. The test flight reached an altitude of 3500 feet and flew for 15 min, during which the gas engine were almost idle, solely depended hydrogen power. Another aircraft being planned for conversion by Universal Hydrogen is ATR-72, with plans to make it as the first hydrogen-powered regional aircraft in service by 2025. The converted ATR-72 would have a reduced sitting capacity from 68 to 56, and a reduced range of 800 km (50% reduction) (Fig. 2).

Fig. 2. Universal Hydrogen's hydrogen-powered Dash-8 testbed aircraft.

Major aircraft OEMs like Airbus are also advancing various their ambitious hydrogen aviation development plans and have made significant progress in recent years. In 2020, Airbus' ZEROe program proposed to launch the first fully hydrogen-powered commercial aircraft by 2035 [9]. But just one year later, ZeroAvia and Universal Hydrogen announced that the timeline of launching the first fully hydrogen-powered commercial aircraft would be accelerated to 2025. The overall development of hydrogen aviation is accelerating, and assorted industries and core technologies are expected to meet the requirements for commercialization in the near future, especially for hydrogen-electric regional aircraft.

3 Key Technologies

For regional aircraft with seating capacity below 80 and a range less than 1500 km, a propulsion solution consisting of liquid hydrogen tanks, hydrogen fuel cells, and electric motors is fully capable of meeting the requirements. However, for hydrogen-powered mainline aircraft with larger payload and longer range, the power density of hydrogen fuel cells and electric motors are insufficient, and hydrogen fueled turbine engines are a more viable propulsion solution. This paper focuses on hydrogen-electric regional aircraft, and the following sections will primarily discuss the four key technologies that constrain hydrogen-electric regional aircraft development: liquid hydrogen tanks, power management control and delivery module for hydrogen fuel, hydrogen fuel cells and electric motors.

3.1 Liquid Hydrogen Tanks

Despite hydrogen's high calorific value (143 MJ/kg) is three times of jet fuel (43 MJ/kg), the energy volume density of liquid hydrogen is only one-third of jet fuel due to its extremely low density [10]. As a result, the volume of hydrogen fuel that needs to be carried is considerably large. Meanwhile, as liquid hydrogen has a higher density compared to gaseous hydrogen, hydrogen fuel needs be stored in the form of liquid for aircraft, which requiring a redesign of tanks.

Based on the physical characteristics of liquid hydrogen, instead of traditional wing tanks, spherical or cylindrical designs are preferable for liquid hydrogen tanks. However, placing liquid hydrogen tanks in the cabin would increase the size of the fuselage and reduce cabin space, resulting in increased air resistance as well as flight costs. Future potential layout designs for hydrogen tanks are shown in the diagram below, which can be divided into integrated and external types [11]. The integrated layout designs (a) (b) (c) can be retrofitted on existing aircrafts, but they have drawbacks like occupying internal space and reducing payload [12–14]; (d) is designed based on a fuselage-blended layout, which can maximize the use of internal space [15]; external layout designs (e) (f) do not occupy internal space, but may impact the overall aerodynamic layout of the aircraft [16] (Fig. 3).

Hydrogen tanks for aviation are usually divided into integrated fuel tanks and non-integrated fuel tanks. The former directly forms an integral part of the aircraft body structure, while the latter is independent of the aircraft body structure [11]. Integrated fuel tanks can utilize the aircraft body structure to store hydrogen fuel, which is advantageous in aircraft design, as rational-designed fuel tanks can better utilize the space inside the aircraft body. However, on the other hand, integral fuel tanks being a part of the fuselage structure means they must bear the stresses and loads of the fuselage, as well as the weight of the hydrogen fuel itself. Therefore, integral fuel tanks require more complex stress

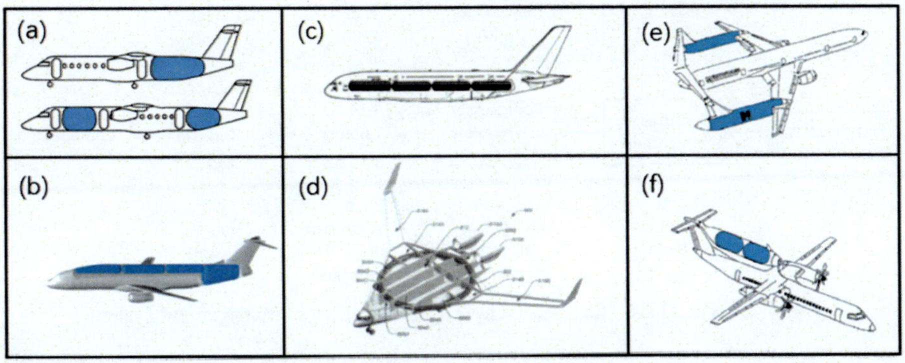

Fig. 3. Layout designs for hydrogen tanks (a, b, c, d are integrated options, d, f are external options).

design compared to non-integral fuel tanks, which only need to deal with the fuel loads. In addition, integral fuel tanks may offer more aerodynamic shape optimization space compared to non-integral fuel tanks, which require significant changes to the aircraft's structural design.

3.2 Power Management Control and Delivery Module for Hydrogen Fuel

The power management control and delivery module for hydrogen fuel comprises a thermal management system for maintaining the low-temperature and high-pressure environment required for liquid hydrogen, a hydrogen delivery system, and hydrogen fuel metering.

Thermal Management System: The thermal management system is a critical component of the power management control and delivery module for hydrogen fuel. The pressure of liquid hydrogen storage needs to be maintained between 0.1 to 0.35 MPa, and the temperature needs to be kept at 20 to 30 K to minimize issues related to hydrogen boil-off and permeation leaks [17]. Maintaining the environmental temperature is primarily achieved through insulation applied to the fuel tank. Insulation can be internal, where the insulation material is in direct contact with the hydrogen fuel; or external, where the insulation material envelops the exterior wall of the fuel tank.

The auto-ignition temperature of liquid hydrogen is approximately 770 K [17, 18]. According to relevant studies, the optimal temperature for the mixture of hydrogen and air entering the combustion chamber is within the range of 400 to 500 K to ensure good thermal performance of the engine. The thermal management system needs to vaporize and heat the hydrogen fuel to the optimal temperature for combustion in the engine, while also providing cooling for the hydraulic oil and air of the engine. A typical architecture of a hydrogen fuel thermal management system for aircraft is illustrated in the diagram below, which includes multiple heat exchangers for heat exchange among hydrogen fuel, hydraulic oil, and air (Fig. 4).

Hydrogen Delivery System: Converting conventional turbojet engines from using jet fuel to hydrogen fuel requires redesigning the engine's delivery system. The core function

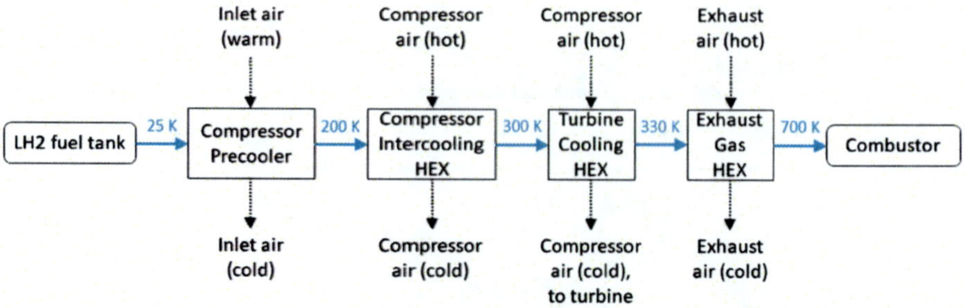

Fig. 4. A typical architecture of a hydrogen fuel thermal management system.

of the hydrogen delivery system is to supply hydrogen fuel to the combustion chamber. One approach is to compress the hydrogen fuel into liquid state using cryogenic pumps to meet the pressure requirements of the combustion chamber, and then evaporate the hydrogen fuel into gaseous state through heat exchangers that recover heat from the engine [17]. The hydrogen delivery system must consider various issues associated with handling low-temperature liquid hydrogen and typically adopts multi-stage compression.

Another issue with the hydrogen delivery system is the use of fuel drain lines. Some studies suggest that excess hydrogen should be discharged and collected back into the fuel tank for storage under low-temperature conditions [17]. Additionally, another approach is to store the excess hydrogen in gaseous form in separate fuel tanks [19].

Hydrogen Fuel Metering: Hydrogen fuel metering is a critical component in the power management control and delivery module for hydrogen fuel, as metering valves provide precise fuel regulation and metering for the engine with rapid response. Metering valves on aircraft must be lightweight and compact, usually utilizing two-stage servovalves [20]. The metering valve can be installed before or after the hydrogen gasification device, where in the former case, liquid hydrogen is gasified in the front-end heat exchanger before passing through the metering valve, and in the latter case, liquid hydrogen is gasified after passing through the metering valve.

Due to hydrogen embrittlement phenomenon, which is the detrimental occurrence of cracks due to hydrogen ingress into metals, the material issue of hydrogen fuel metering valves is also worthy to pay attention. For stainless steel materials, hydrogen embrittlement can cause hydrogen fuel to escape with increasing temperature. Therefore, the selection of materials for hydrogen fuel metering valves must ensure full sealing capability.

3.3 Hydrogen Fuel Cells

The main technical bottleneck that constrains the development of hydrogen fuel cells is power density. In recent years, with advancements in materials and manufacturing technologies, the power density of hydrogen fuel cells has significantly increased, reaching nearly 1 kW/kg at the system level. However, there is still a gap compared to the power density of traditional jet fuel turbofan engines, which is around 3.7 kW/kg [21]. To meet the power requirements of small regional aircraft with 40–80 seats, a power density of hydrogen fuel cell need to reach 1 to 1.5 kW/kg. Most green regional aircraft programs choose proton exchange membrane fuel cells (PEMFC), such as those used by ZeroAvia and Universal Hydrogen. Alternatively, solid oxide fuel cells (SOFC) can also be chosen. ZeroAvia claims to have achieved energy densities of over 2 kW/kg with their PEMFC s in laboratory environments, but this result may not include ancillary systems outside of the fuel cell stack, and there is still a gap to bridge before practical applications can be realized.

One of the challenges faced by hydrogen fuel cells on aircraft is thermal management. The ideal operating temperature for hydrogen fuel cells is between 60 to 80 °C [22], and temperatures that are too high can result in decreased energy conversion efficiency or even system failure. Vehicles that use fuel cells typically employ forced convection heat exchangers as passive heat exchangers, which have a large external radiator area. However, installing such passive heat exchangers on aircraft may adversely affect aerodynamic performance. Some studies have proposed active heat exchanger technologies for aviation applications. In addition, a shared thermal management system with the hydrogen storage system could also be considered [23].

3.4 Electric Motors

The power-to-weight ratio of electric motors directly determines the key performance indicators such as power and efficiency for hydrogen-electric regional aircraft. Currently, there are two main types of high power-to-weight ratio electric motors: permanent magnet synchronous motors (PMSM) and superconducting motors. At present, PMSM is a preferred choice for most aircraft, and examples include Universal Hydrogen's DASH-8 and ATR-72 aircraft are equipped with MagniX's Magni650 series PMSM motors.

Superconducting motors utilize superconductors instead of conventional conductive materials to achieve the conversion of electromagnetic energy to mechanical energy. They have the characteristics of small volume, high efficiency, light weight, low synchronous reactance, and can generate significantly higher torque compared to ordinary motors with the same weight and energy input [24]. Some teams have conducted research on aerospace superconducting motors, such as the high-temperature superconducting motor (operating temperature of 69K) developed by SuperOx, a Russian company, which has been installed on the Yak-40 aircraft and achieved its maiden flight.

Due to the approximate scale independence of motors [25], the power density and efficiency of a single high-power motor and a distributed electric drive system with multiple small-power motors are essentially similar when the total power is the same. Adopting a distributed electric drive system with multiple small-diameter fans driven by small-power motors can effectively increase the bypass ratio, control and fault tolerance of the propulsion system, while the compact size of the electric drive system can be easily integrated into the fuselage, improving the aerodynamic efficiency of the aircraft, especially for electric Vertical Takeoff and Landing (eVTOL) aircraft targeting the Urban Air Mobility (UAM) market. In the future, regional aircraft using hydrogen fuel cells may also adopt distributed layouts with 4–6 or even more motors.

4 Conclusion

This paper discusses the development of hydrogen-powered aircraft in the future, with a focus on regional aircraft using hydrogen fuel cells. The key technologies determining their development include liquid hydrogen tanks, power management control and delivery module for hydrogen fuel, hydrogen fuel cells, and electric motors. This paper provides a detailed discussion on the relevant issues and solutions related to these technologies. Based on the current status of technological and product development, this paper believes that hydrogen-electric regional aircraft are the optimal choice for the current development of hydrogen aviation, as they are relatively mature in terms of technology with lower R&D investment and risks. For aircraft manufacturers, existing platforms can be used for retrofitting experiments once the technology is mature, with the aim of transitioning to commercial applications as soon as possible, while laying the foundation for the subsequent development of larger hydrogen-powered aircraft.

References

1. ICAO Homepage. https://www.icao.int/environmental-protection/CORSIA. Accessed 1 Mar 2023
2. Nesbitt, E., et al.: Boeing-Safran landing gear noise reduction project on 2020 Boeing ecoDemonstrator program. In: 28th AIAA/CEAS Aeroacoustics 2022 Conference, Southampton, pp. 2849–2059 (2022)
3. Warzocha, K., et al.: Transformative use of additive technology in design and manufacture of hydraulic actuator for fly-by-wire system. Appl. Sci. **11**, 4772–4783 (2021)
4. Thapa, N., et al.: All electric aircraft: a reality on its way. Mater. Today Proc. **43**, 175–182 (2021)
5. Shahriar, M., et al.: The current techno-economic, environmental, policy status and perspectives of sustainable aviation fuel (SAF). Fuel **325**, 124905 (2022)
6. Rap, A., et al.: The climate impact of contrails from hydrogen combustion and fuel cell aircraft. Copernicus Meetings, No. EGU23–5520 (2023)
7. ZeroAvia Homepage. https://www.zeroavia.com/do228-first-flight. Accessed 19 Jan 2023
8. Universal Hydrogen Homepage. https://hydrogen.aero/press-release. Accessed 2 Mar 2023
9. Norris, G.: Airbus Reveals Refined ZEROe Blended Wing Body Concept. Aerospace Daily & Defense Report, pp. 279–314 (2022)
10. Sn, A., et al.: Conceptual design and optimization of a general aviation aircraft with fuel cells and hydrogen. Int. J. Hydrogen Energy **46**(64), 32676–32694 (2021)
11. Tamburrano, P., et al.: Fuels systems and components for future airliners fuelled with liquid hydrogen. In: Journal of Physics: Conference Series, vol. 2385, no. 1, p. 012041. IOP Publishing (2022)
12. AIRBUS Homepage. https://www.airbus.com/en/innovation/zero-emission/hydrogen/zeroe. Accessed 15 Sept 2021
13. EUH2020 Homepage. https://www.enableh2.eu/. Accessed 15 Sept 2021
14. Huete, J., et al.: Propulsion system integration for a first-generation hydrogen civil airliner? Aeronaut. J. **125**(1291), 1654–1665 (2021)
15. Suewatanakul, S., et al.: Conceptual design of a hybrid hydrogen fuel cell/battery blended-wing-body unmanned aerial vehicle—an overview. Aerospace **9**(5), 275 (2022)
16. Khandelwal, B., et al.: Hydrogen powered aircraft: the future of air transport. Prog. Aerosp. Sci. **60**, 45–59 (2013)

17. Srinath, A.N., et al.: Thermal management system architecture for hydrogen-powered propulsion technologies: practices, thematic clusters, system architectures, future challenges, and opportunities. Energies **15**(1), 304 (2022)
18. HydrogenTools Homepage. https://h2tools.org/hyarc/hydrogen-properties. Accessed 1 Mar 2023
19. Klug, H.G., et al.: CRYOPLANE: hydrogen fuelled aircraft—status and challenges. Air Space Eur. **3**(3–4), 252–254 (2001)
20. Tamburrano, P., et al.: A review of electro-hydraulic servovalve research and development. Int. J. Fluid Power 1–23 (2018)
21. Wright, S., et al.: Fuel cells and novel thermal management advancements required for aviation. In: 33rd Congress of the International Council of the Aeronautical Sciences ICAS2022. The International Council of the Aeronautical Sciences (2022)
22. Watzenig, D., et al.: Comprehensive Energy Management-Safe Adaptation, Predictive Control and Thermal Management. Springer, Cham (2018). https://doi.org/10.1007/978-3-319-57445-5
23. Nguyen, H.Q., Shabani, B.: Proton exchange membrane fuel cells heat recovery opportunities for combined heating/cooling and power applications. Energy Convers. Manag. **204**, 112328 (2020)
24. Haran, K.S., et al.: High power density superconducting rotating machines—development status and technology roadmap. Supercond. Sci. Technol. **30**(12), 123002 (2017)
25. Deisenroth, D.C., et al.: Thermal management of high-power density electric motors for electrification of aviation and beyond. Energies **12**(19), 3594 (2019)

Active Vibration Control of a Plate Using Piezoelectric Sensors and Loudspeakers

Minghao Chen and Qibo Mao(✉)

School of Aircraft Engineering, Nanchang HangKong University, Nanchang 330063, China
qbmao@nchu.edu.cn

Abstract. The aerospace industry demands high-performance aircraft structures, which has led to the need for low-cost, effective vibration control methods. In this study, the loudspeaker was modified as inertial actuator for vibration control. Furthermore, the piezoelectric ceramic was integrated into the proposed inertial actuator as sensor. The proposed integrated actuator/sensor overcomes the shortcomings of traditional velocity negative feedback control method. Experimental analysis demonstrates the feasibility and effectiveness of the proposed transducer. Significant vibration reduction can be experimentally obtained in wide frequency range. The control circuit is simple and cost-effective. The innovation of this study lies in the proposal and experimental validation of a novel transducer, which has important implications for improving aircraft vibration performance and provides new ideas and methods for vibration control research in the aerospace field.

Keywords: Aircraft structure · Loudspeaker · Inertial actuator · Piezoelectric ceramic sensor

1 Introduction

In recent years, with the rapid development of the aerospace industry, there has been growing interest in the structural vibration problems of various types of aircraft. Due to their small size, light weight, and high driving force, inertial actuators are particularly suitable for reducing the vibration of aircraft, making them a focus of research. Many scholars have studied the application of inertial actuators in aircraft. For examples, Misol [1] applied inertial actuators to the sidewall panels of airplanes to reduce aircraft noise and vibration. Boulandet et al. [2] used inertial actuators and virtual mechanical impedance methods to reduce the vibration of aircraft wall panels.

At the same time, many scholars have conducted extensive research on a new type of inertial actuator with lower natural frequency or improved control performance through phase compensation. Mao et al. [3] designed a tunable absorber using an inertial actuator, which generates a virtual mass by acceleration feedback control to change its inherent frequency and has strong tuning ability. Kras and Gardonio [4, 5] added flywheel elements to the transmission of inertial actuators, reducing their natural frequency and improving the stability of the control system. Zilletti et al. [6] proposed an inertial actuator with inertial elements that effectively reduces the natural frequency of the actuator.

Chinese Society of Aeronautics and Astronautics (Ed.): CASTC 2023, LNME, pp. 596–602, 2024.
https://doi.org/10.1007/978-981-99-8861-7_59

In addition to structural design, there has also been extensive research on the active vibration control of inertial actuators. For example, Debattisti et al. [7] proposed an active vibration control strategy based on multiple independent devices, using wireless networks to share data and compensators to solve the instability problem of inertial actuators. Pan et al. [8] effectively reduced the free and forced vibrations in their study through the combination of inertial actuators and a fuzzy adaptive PID algorithm. It is worth noting that in active control systems, inertial actuators with accelerometers as control units are usually used and control signals are transmitted through an integrator, known as direct velocity feedback [9].

However, in the direct velocity feedback control system, due to the relatively high natural frequency of the inertial actuator, the control effect for the low frequency band is poor. Meanwhile, the control effect of the high frequency band is not significant when the control signal is transmitted by the accelerometer. Misalignment between the inertial actuator and sensor can also reduce the effectiveness of the control.

Therefore, in this study, a low-cost loudspeaker modified as inertial actuator is presented to control vibration of a thin rectangular plate, and a piezoelectric ceramic is used as sensor. First, the physical model of electromagnetic inertia actuator is presented. Then the experimental active control results for plate are given by using proposed inertial actuator. Finally, some substantial conclusions are drawn, which provide valuable reference for optimizing the vibration control of aircraft plate and shell.

2 Mathematical Model

The inertial actuator utilized in this study is designed based on the principles of a loudspeaker. It consists of a permanent magnet with a flux density of B and a coil with a length of l. When a current flows through the coil, the changing electromagnetic field generates a controlling force that acts on the controlled object via a spring supporting element. For simplicity, this paper omits the frequency dependence of complex amplitude. Therefore, the dynamic equation of the mass block and the circuit equation of the inertial actuator can be expressed as

$$M_a\ddot{x}_2 + c(\dot{x}_2 - \dot{x}_1) + k(x_2 - x_1) = BlI \tag{1}$$

$$j\omega L_e I + R_e I = U - Bl(\dot{x}_2 - \dot{x}_1) \tag{2}$$

where c, k, x_1, and x_2 respectively represent the damping coefficient, stiffness coefficient, displacement of the base, and displacement of the mass block.

From Eq. (2), the current can be expressed as

$$I = \frac{U - Bl(\dot{x}_2 - \dot{x}_1)}{j\omega L_e + R_e} \tag{3}$$

Substituting Eq. (3) into Eq. (1), and the expression of acceleration of mass block can be obtained, such as

$$\ddot{x} = -\omega^2 \frac{\left(c + \frac{k}{s} + \frac{Bl}{sL_e + R_e}\right)\dot{x}_1 + \frac{Bl}{sL_e + R_e}U}{-\omega^2 M_a + s\left(c + \frac{(Bl)^2}{sL_e + R_e}\right) + k} \tag{4}$$

The force generated by an inertial actuator, as given by Eq. (4), can be represented as a function of the coil voltage U and the velocity $\dot{x}_1$ applied to the base.

$$f_c = T_a U + Z_a \dot{x}_1 = M_a \ddot{x}_2 \tag{5}$$

$$T_a = \left(s^2 M_a \frac{Bl}{sL_e + R_e}\right) / \left(M_a s^2 + s\left(c + \frac{(Bl)^2}{sL_e + R_e}\right) + k\right) \tag{6}$$

$$Z_a = \left(M_a s^2\left(c + \frac{k}{s} + \frac{(Bl)^2}{sL_e + R_e}\right)\right) / \left(M_a s^2 + s\left(c + \frac{(Bl)^2}{sL_e + R_e}\right) + k\right) \tag{7}$$

where T_a is the driving force of each unit input voltage and Z_a is the mechanical impedance of the inertia actuator (Fig. 1).

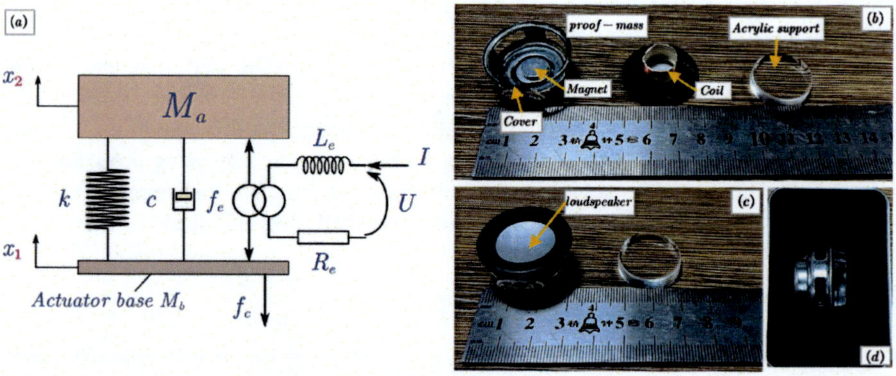

Fig. 1. IA schematic diagram (a) based on feedback; (b) pictures of self-made IA before assembly; (c) partial assembly; (d) after assembly.

Based on the stress condition of a fully clamped rectangular plate under excitation, this paper focuses on the control of the bending vibration of a thin plate. To achieve this, the paper proposes using an inertia actuator to generate a current that is proportional to the local velocity of the plate, which in turn generates a corresponding force for vibration control. The properties of the aluminum plate are summarized in Table 1, including both geometric and physical characteristics.

It is noteworthy that the performance of the control system can be evaluated through the kinetic energy E_k of the structure. Using the Parseval theorem, the kinetic energy of a thin plate with velocity $\dot{w}$, density ρ, and thickness h in the frequency domain can be expressed as

$$E_k = E\left[\frac{1}{2}\int_S \rho h \dot{w}(x, y, t) dS dt\right] = \frac{1}{2\pi}\int_{-\infty}^{\infty} S_K d\omega \tag{8}$$

where: E is the expected value, the area of the plate is A, and the power spectral density S_K of the plate is

$$S_K = E\left[\int_A \frac{1}{2}\rho h |\dot{w}(x, y)|^2 dA\right] \tag{9}$$

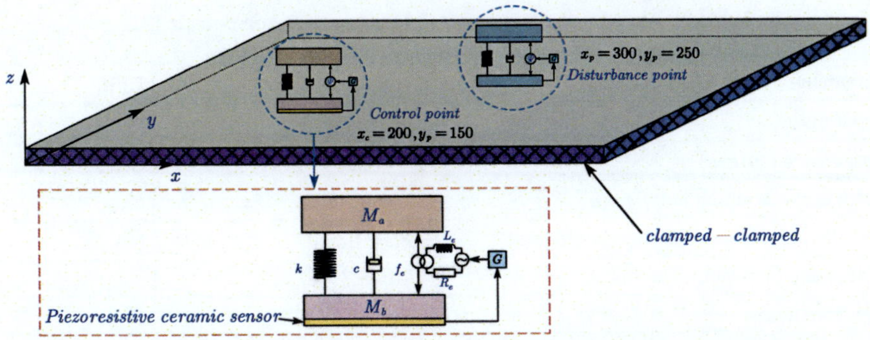

Fig. 2. Implementing feedback control of fully clamped plate via inertial actuator.

Table 1. Geometry and physical parameters of the plate.

Parameter	Numerical value	Unit
Planar dimensions	$l_x \times l_y = 440 \times 340$	mm
Thickness	$h = 2$	mm
Young's modulus	$E = 7 \times 10^{10}$	N/m^2
Density	$\rho = 2700$	kg/m^3

3 Experimental Setup and Results

This study presents an experiment using an aluminum plate as an example to implement a feedback control system with a low-cost inertia actuator designed using a loudspeaker. The inertia actuator modified by a loudspeaker shown in Fig. 2 is designed for convenience of use and effective signal transmission through an integrated design. The physical parameters of the actuator are detailed in Table 2.

The frequency response functions (FRFs) of the inertia actuator in both theory and experiment are shown in Fig. 2. During the experiment, the inertia actuator underwent power amplification, and the experimental results were adjusted to match the theoretical results by adjusting the amplification coefficient, thereby verifying the effectiveness of the inertia actuator modeling process. Notably, the phase of the control force output by the inertia actuator is relatively stable in the frequency band higher than the natural frequency and tends towards 0° (Fig. 3).

Figure 4 illustrates an experimental arrangement where a fully clamped aluminum plate is stimulated by a wide-band disturbance power while feedback control is implemented using a control unit. It is important to note that excitation should not occur at the node line of the low frequency mode. The excitation and control points are also shown in Fig. 4.

Figure 5 reveals that the control position problem results in a 90° phase at the frequency of 876 Hz, which leads to suboptimal control effectiveness. It is important to note that a negative phase in the open-loop channel will render the inertial actuator

Table 2. The physical parameters of the low-cost home-made IA.

Parameter	Numerical value	Unit
Proof mass	$M_a = 0.019$	kg
Suspension stiffness	$k = 3800$	Nm^{-1}
Suspension damping coefficient	$c = 0.2$	Nsm^{-1}
Electromagnetic transduction coefficient	$Bl = 0.018$	NA^{-1}
Modal damping ratio	$\xi = 3.3988$	
Coil resistance	$R_e = 3.5$	Ω
Coil inductance	$L_e = 88.1 \times 10^{-6}$	H

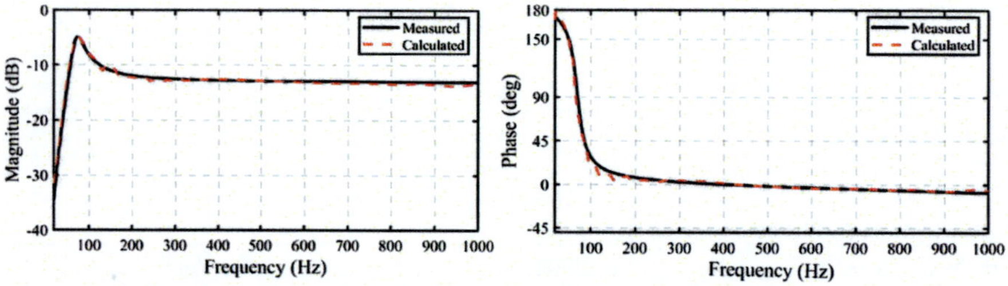

Fig. 3. Scheme of the feedback control system.

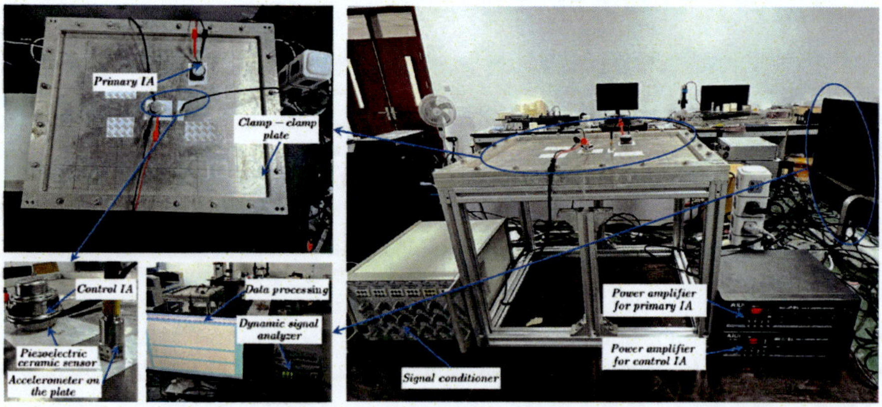

Fig. 4. Experimental setup for feedback control for a clamped-clamped plate by using the inertial actuator of a loudspeaker.

ineffective for active plate control, or even worse than passive control. This is because the inertial actuator's signal output is opposite to the direction of the plate's vibration under negative phase, which is akin to counteracting the vibration while simultaneously pushing the plate in the opposite direction, causing a decrease in control effectiveness or even failure. However, control effectiveness is optimal at approximately 100 Hz and 310 Hz. Meanwhile, Fig. 5 shows the corresponding Nyquist diagram, which is utilized to preliminarily evaluate the control system's stability.

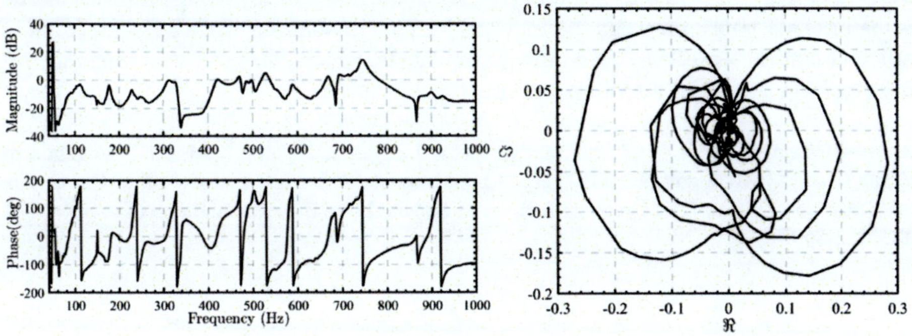

Fig. 5. Bode and Nyquist plot of the measured open-loop FRF.

Figure 6 shows that increasing the feedback gain of the inertial actuator enhances its control effectiveness in the low-frequency range, as illustrated in Fig. 6. At approximately 100 Hz, 313 Hz, and 680 Hz, the vibration energy reduced significantly by 51 dB, 40 dB, and 27 dB, respectively, indicating excellent performance of the control system. To better observe the control effect of the inertial actuator, the time-domain responses before and after control were measured under the excitation of a single frequency signal of 315 Hz. The maximum velocity attenuation is about 65%.

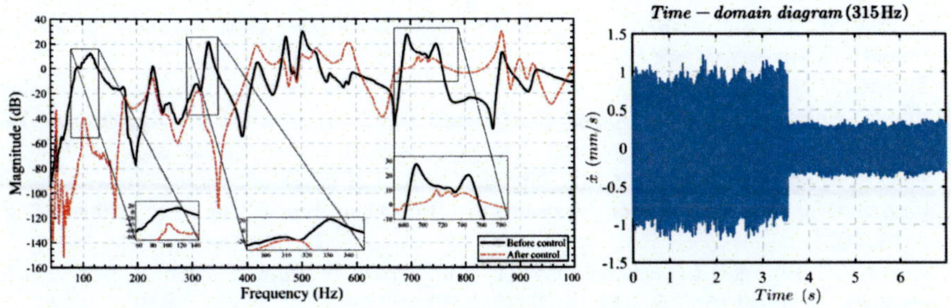

Fig. 6. Experimental control effect of single-channel system under inertial actuator.

4 Conclusions

This paper presents a feedback control scheme that is suitable for the panel structures of aircrafts. It involves the integration of an inertia actuator modified by a loudspeaker and a piezoelectric ceramic sensor. The scheme offers several advantages, including optimized control signals, a simple and direct control loop, and low cost. The experimental results demonstrate that the inertia actuator provides significant control effects in the low-frequency band due to its low inherent frequency, while the piezoelectric ceramic sensor improves the control effect in the high-frequency band. The time-domain analysis reveals a significant decrease in the vibration velocity of the thin plate by using proposed inertia actuator. This paper provides a simple and effective solution that can achieve more efficient control with low-cost control components, which can serve as a reference for the vibration control of aircraft panel structures.

Acknowledgment. This work was sponsored by the National Natural Science Foundation of China (No. 12364058 and No. 51975266), the Aeronautical Science Foundation of China (No. 2020Z073056001), and the Graduate Innovative Special Fund Projects of Jiangxi Province (No. YC2023-S698).

References

1. Misol, M.: Full-scale experiments on the reduction of propeller-induced aircraft interior noise with active trim panels. Appl. Acoust.Acoust. **159**, 107086 (2020)
2. Boulandet, R., Michau, M., Micheau, P., et al.: Aircraft panel with sensorless active sound power reduction capabilities through virtual mechanical impedances. J. Sound Vib.Vib. **361**, 2–19 (2016)
3. Mao, Q., Huang, S.: Design of tuneable vibration absorber by using inertial actuator with proof-mass acceleration feedback. Int. J. Struct. Stab. Dyn.Dyn. **19**(8), 1950087 (2019)
4. Kras, A., Gardonio, P.: Velocity feedback control with a flywheel proof mass actuator. J. Sound Vib.Vib. **402**, 31–50 (2017)
5. Kras, A., Gardonio, P.: Active vibration control unit with a flywheel inertial actuator. J. Sound Vib.Vib. **464**, 114987 (2020)
6. Zilletti, M.: Feedback control unit with an inerter proof-mass electrodynamic actuator. J. Sound Vib.Vib. **369**, 16–28 (2016)
7. Debattisti, N., Bacci, M.L., Cinquemani, S.: Distributed wireless-based control strategy through Selective Negative Derivative Feedback algorithm. Mech. Syst. Sig. Process. **142**, 106742 (2020)
8. Pan, T., Shen, X., Chen, Y.: Active vibration control of thin-walled structure with inertial piezoelectric actuator based on fuzzy adaptive PID algorithm. J. Vib. Eng. Technolog. **12**, 1–16 (2023)
9. Zilletti, M., Elliott, S.J., Gardonio, P.: Self-tuning control systems of decentralised velocity feedback. J. Sound Vib.Vib. **329**(14), 2738–2750 (2010)
10. Chen, M., Mao, Q., Peng. L., et al.: Active vibration control using loudspeaker-based inertial actuator with integrated piezoelectric sensor. Actuators **12**(10), 390 (2023)

Crack Analysis and Improvement of Helicopter Engine Compartment Partition

Zhongwei Yin[1](✉), Shifeng Li[1], Jinhua Cao[2], Changliang Lin[1], and He Zhang[1]

[1] Harbin Aircraft Industry (Group) Co. Ltd., Harbin 150066, China
826455757@qq.com

[2] The Military Representative Office of the Army Aviation Department in Harbin Region, Harbin 150066, China

Abstract. The crack on the engine compartment partition of a helicopter continually occurs, which has affected the flight safety. First, the simplified modeling of engine and local structure detail modeling are studied. Then combined with the flight test data and fracture analysis, the high-cycle failure of the structure under the action of aerodynamic pressure and vibration force is analyzed by using the method of frequency response simulation. The average S-N equation is used to improve the design of the original structure. The fatigue damage resistance of the new structure is obviously improved and has infinite life. A new maintenance platform is also designed for the problem of small engine maintenance space feedback from users. The platform not only increases the maintenance space, ensure the safety of staff, but also reduces the probability of accidental damage to the engine compartment partition.

Keywords: Engine simplified modeling · Crack · High-cycle Fatigue · Helicopter

1 Introduction

The unique working principle and realization method of helicopter make its vibration level significantly higher than that of fixed-wing aircraft [1]. The closer to the rotor system, the worse vibration environment is. Excessive vibration often accelerates the fatigue failure process of the structures on the helicopter. The fatigue design of helicopter structure is more complex than that of fixed-wing aircraft. Improving the high-cycle fatigue resistance is the key issue. The fatigue analysis is usually carried out by the method of S-N characteristic test [2]. The research and analysis of fatigue damage of helicopter structure are endless. For example, Zhang Shaojia [3] analyzed the cause of the main nozzle crack of a helicopter engine. The research of Men Kunfa et al. [4] shows that fatigue load and stress corrosion are the main causes of cracks in a helicopter windshield. Lin Chuxin et al. [5] analyzed that the bolt cracking property of a helicopter fixed tail reducer was also fatigue fracture. Cao Cuiliu et al. [6] gave inspection suggestions for the crack of the main bearing frame of a helicopter.

Chinese Society of Aeronautics and Astronautics (Ed.): CASTC 2023, LNME, pp. 603–612, 2024.
https://doi.org/10.1007/978-981-99-8861-7_60

The engine compartment partition is located on the helicopter engine platform, which can withstand the local strongest flame burning for 5 min [7], and ensure that the flame does not diffuse to other parts during this time. The compartment partition together with the fairing and engine platform forms two closed fire compartments to isolate and protect the engine. The fatigue load borne by the compartment partition mainly includes the rotor aerodynamic force transmitted by the fairing and the vibration force transmitted by the helicopter moving parts. The cracks on compartment partition will lead to debase the protection capability, and the supporting effect on the fairing will also be weakened accordingly, which will have an adverse impact on flight safety.

In this paper, the dynamic modeling of the whole helicopter is established. Based on the experimental data, the cause of the cracks on the engine compartment partition is analyzed, and the targeted improvement design is carried out to improve the anti-fatigue damage ability and ensure the helicopter flight safely.

2 Crack Description

According to the feedback, a helicopter has a crack problem on the engine compartment partition, which has a safety risk and affects the normal use of the helicopter. The engine compartment partition is a three-stage structure, the main material is titanium alloy, and the local reinforcement is aluminum alloy. According to statistics, the occurrence time of cracks are 2342FH, 2410FH, 2738FH and 3429FH. The cracks are concentrated on the upper reinforcement of the engine compartment partition, as shown in Fig. 1.

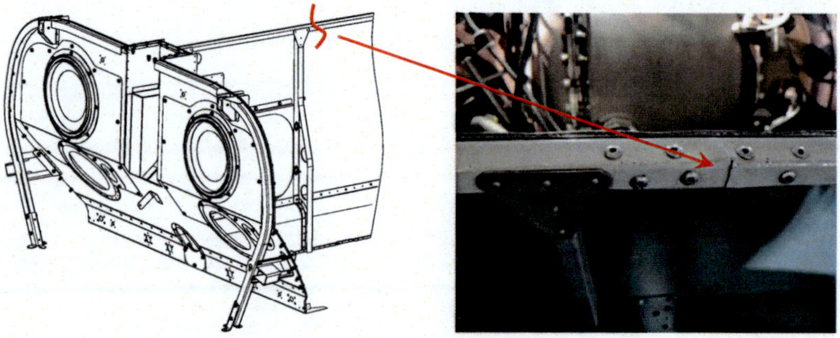

Fig. 1. Crack location.

In order to further determine the crack properties, the fracture analysis is carried out. The analysis result shows that the direction of each sub-crack is consistent, the fracture source area is characterized by a line source composed of multiple sources, and the fatigue band characteristics can be seen at high magnification in the propagation zone, shown Fig. 2. The failure property of the compartment partition is high-cycle fatigue.

The direction of compartment partition crack is consistent, which should be the fatigue failure under the same stress mode. The compartment partition is not the main load-bearing structure and it bears two kinds of load: aerodynamic force and vibration force. The direction of aerodynamic force is random and bears together with the fairing.

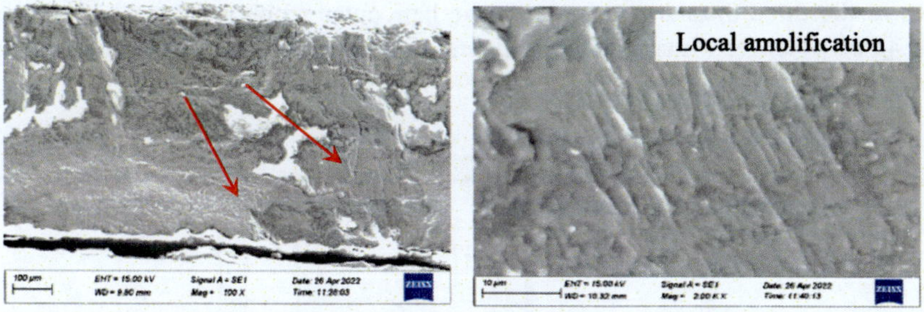

Fig. 2. Fatigue strip characteristics of crack fracture.

The vibration force is mainly derived from the high-cycle alternating load at the hub center, and its frequency is the first-order passing frequency of the rotor system.

3 Vibration Load Measurement

In order to determine the vibration load value on the compartment partition during flight, the vibration level is measured. Three positions are selected for measurement, as shown in Fig. 3. The lateral vibration level during steady-state horizontal flight is measured. The test speed is 120–200 km/h, and the equal interval was 20 km/h. Each speed point remains stable for 60–90 s. The vibration sensor is fixed at the measuring point by using the original bolt hole through a special adapter.

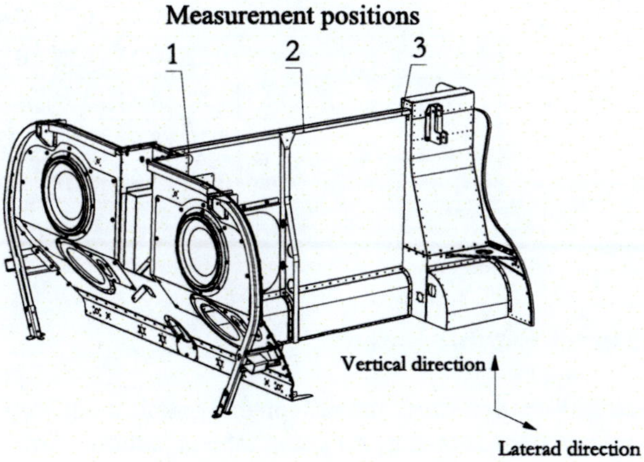

Fig. 3. Vibration measuring positions.

The time history curve of the vibration parameters is obtained by the fast Fourier transform. Figure 4 shows the spectrum curve of the vibration effective value. Then the vibration acceleration values corresponding to the first-order passing frequency of the rotor are selected, shown in Table 1. It can be seen that the lateral vibration acceleration values of the three measuring points on the helicopter with cracks are maintained at about 4g on average, which is about 8 times of the values measured on the helicopter without cracks.

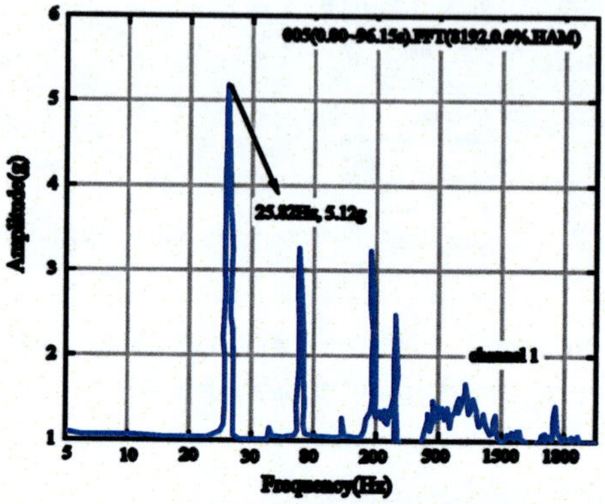

Fig. 4. Amplitude-frequency curve.

Table 1. Measured vibration acceleration value(g).

Test helicopter status	Measurement positions	Flying speed (km/h)					
		100	120	140	160	180	200
Compartment without cracks	1	0.49	0.39	0.30	0.33	0.38	0.47
	2	0.48	0.38	0.22	0.40	0.53	0.77
	3	0.32	0.33	0.28	0.24	0.41	0.67
Compartment with cracks	1	4.51	4.44	3.96	4.64	4.59	5.12
	2	3.19	3.97	3.22	4.56	3.49	3.54
	3	2.19	2.97	3.22	3.56	2.49	2.54

4 Analysis Model Establishment

In order to analyze the vibration stress, the helicopter dynamic modeling is studied. Considering that the vibration load borne by the compartment partition is mainly transmitted by the fairing and the engine, the engine simplified model and the fairing detail model need to be established.

4.1 Engine Simplified Modeling

The two engines on the helicopter are fixed by three points. The front end is connected to the main gear box/engine connecting shaft through the flange, and the back engine bracket is connected to the engine output shaft. Based on this, the shafts are simulated by beam element. One end of the main gear box/engine connecting shaft is connected to the main gear box by MPC(RBE2), and the universal joint is simulated by releasing the Y and Z rotation degrees of freedom. The engine is simplified as a concentrated mass element at its centroid. The engine bracket has a absorber to reduce the vibration transmitted to the airframe. The bracket is simplified as two spring elements, one of them is connected to the engine output shaft and the other one is connected to the airframe. The simplified engine model is shown in Fig. 5.

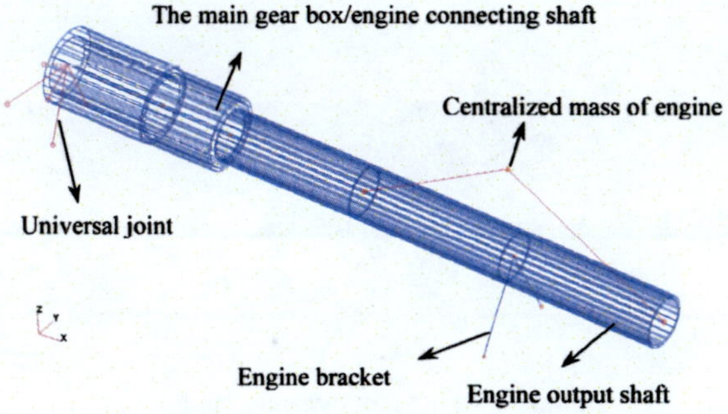

Fig. 5. Simplified engine model.

4.2 Fairing Detail Modeling

Considering that the compartment partition also bears the aerodynamic load transmitted by the fairing, a fairing detail model is established. The model uses shell elements to simulate the structural details. The connection between the compartment partition and the fairing is simulated by MPC(RBE2), and its corresponding rotational degrees of freedom are released. The connection between the detail model and the airframe is simulated by beam elements. The aerodynamic load from CFD calculation is loaded onto the fairing grids by the form of pressure. The detailed model of the fairing is shown in Fig. 6.

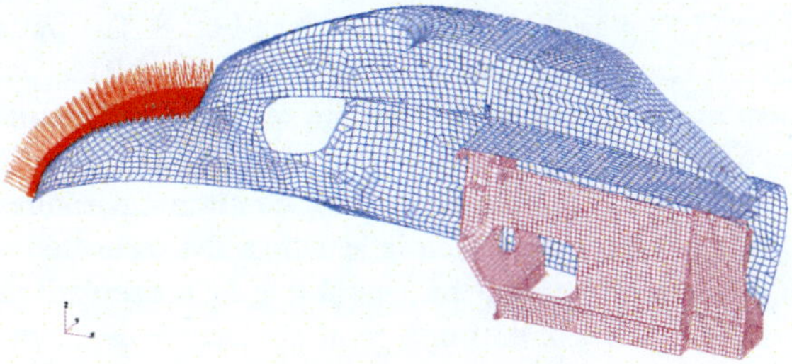

Fig. 6. Fairing detail model.

4.3 Helicopter Dynamic Modeling

On the basis of completing the engine simplified modeling and the fairing detailed modeling, the dynamic model of the whole helicopter is established by using the modular modeling method. In the model, not only the fuselage and tail structure are established into a plate-rod form, but also the main reducer transmission system is modeled in detail. The alternating vibration load is applied to the hub center. The dynamic model is shown in Fig. 7.

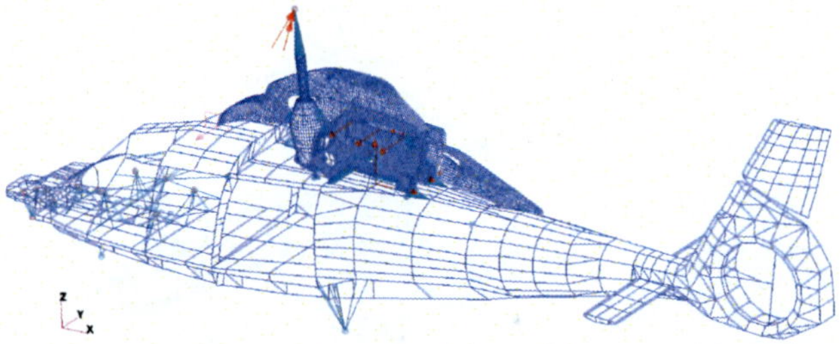

Fig. 7. Helicopter dynamic model.

5 Analysis and Improvement

5.1 Fatigue Analysis

First, the rationality of the model is verified. The detail model is modified to disconnect the grid nodes at the cracks and simulate the crack morphology. At 200 km/h, the measured vibration acceleration of the helicopter without cracks is about 0.6 g on average, and that of the helicopter with cracks is about 4 g. The difference between the two is about 6.67 times. Therefore, the alternative load of the hub center calculated by the method of Reference [8] is expanded by 6.67 times to analyze the vibration acceleration in the presence of cracks. The frequency response analysis of the free-state is carried out. The vibration acceleration values at the grid nodes corresponding to the measuring points are 5.22 g, 3.47 g and 2.71 g, as shown in Fig. 8. The maximum difference between the measured results and the calculated values is 5.21%, indicating that the model establishment and crack simplification are reasonable. This model can be used for vibration stress calculation.

According to the standard requirements, the vibration environment of the engine room is 3.5 g [9] in all directions, which is less than the measured value. Therefore, when analyzing the vibration stress, the lateral 4 g is conservatively loaded on the measuring points. The thickness of the upper reinforcement of the compartment partition is 0.8 mm. The material is 6A02-T62, and its average fatigue limit is 90.4 MPa. The working environment temperature of the engine room is not higher than 150 °C. The influence of temperature is considered when the material property values are chosen.

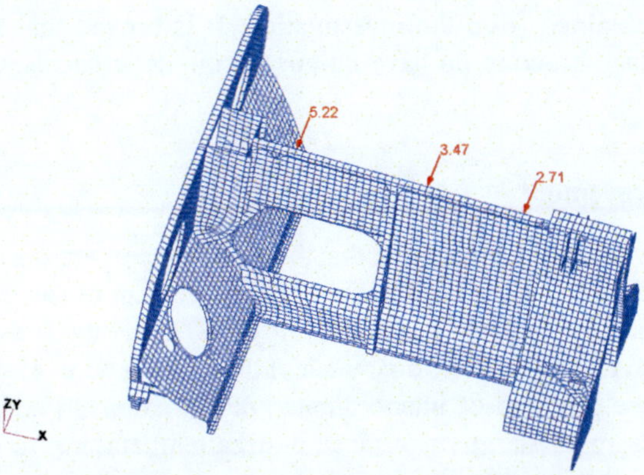

Fig. 8. Vibration acceleration calculation values.

The vibration stress is calculated, and the results are shown in Fig. 9. It can be seen that after ignoring the stress concentration at the connection points, the maximum vibration stress in the crack zone is 93.4 MPa at the first-order passing frequency of the rotor system, which exceeds the average fatigue limit and does not have infinite life.

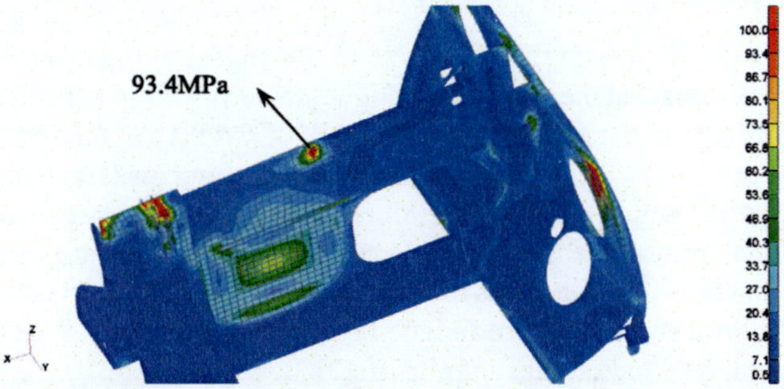

Fig. 9. Vibration stress of compartment partition reinforcement.

The high-cycle fatigue life of compartment partition is calculated by the average S-N equation [2]:

$$\frac{S_a}{S_\infty} = 1 + \frac{A}{N^\alpha} \tag{1}$$

where S_a is equivalent fatigue stress, under constant amplitude load, $S_a = S_{max} = 93.4$ MPa; A and α are shape parameters, $A = 0.892$, $\alpha = 0.5$ for aluminum alloy; $S_\infty = 90.4$ MPa, which is the average fatigue strength.

The fatigue cycle is calculated: $N = 2.57 \times 10^8$, it can be transformed to the flight hours: $2.57 \times 108/23.75/3600 \approx 3000$ *FH*, which is basically consistent with the crack occurrence time obtained from the investigation. It is proved that the analysis accuracy and the original structure do have fatigue design deficiencies, which needs to be improved.

5.2 Design Improvement

Structure Improvement. Considering that the working temperature of the engine room is less than 150 °C, the material of the upper reinforcement of the compartment partition which often occurs cracks is aluminum alloy 6A02, and the compartment partition material is titanium alloy TA2. The mismatch of the heating wires expansion coefficient is easy to cause the problem of uncoordinated deformation. In addition, the original aluminum alloy has poor fatigue resistance at high temperature, so the reinforcement structure material is replaced byTA2.

The Formula (1) is used to design the reinforcement thickness. For the original aluminum alloy reinforcement, the 2342FH is taken as the analysis object to determine the corresponding equivalent fatigue stress. The calculation process is as follows:

$$\begin{aligned} S_a &= (0.892/\sqrt{N} + 1) \times 90.6 \\ &= (0.892/\sqrt{(2342 \times 23.75 \times 3600/10^6} + 1) \times 90.6 \\ &= 96.31\,\text{MPa} \end{aligned} \quad (2)$$

In order to meet the design requirements of the engine compartment partition as a lifeless part, this means that the equivalent fatigue stress $S_a \le 90.6$ MPa. It is necessary to increase the thickness of the reinforcement to $96.31 \times 0.8/90.6 = 0.85$ mm. Considering the metal forming process, the thickness of the upper reinforcement is changed to 1.2 mm.

According to the improved upper reinforcement, the helicopter dynamic model is modified, and the vibration stress is calculated again. The results are shown in Fig. 10. It can be seen that the vibration stress of the modified reinforcement is 80.1 MPa, which is less than one third of the average fatigue limit of titanium alloy TA2 (241.2 MPa at 150 °C) [10], and the structure has infinite life[2].

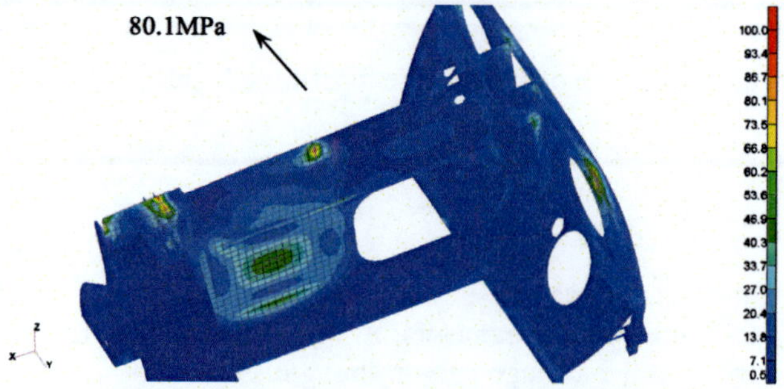

Fig. 10. Vibration stress of modified reinforcement

Maintenance Method Improvement. Due to the small space on the engine platform, it is not easy to maintain the engine. A special engine maintenance platform is designed for engine maintenance, shown in Fig. 11. The platform is portable and can be hung on one side of the fuselage when used. It not only increases the maintenance operation space, prevents the staff from pulling the engine compartment partition in case of emergency, reduces the probability of accidental damage to the compartment partition, but also effectively protects the safety of the staff.

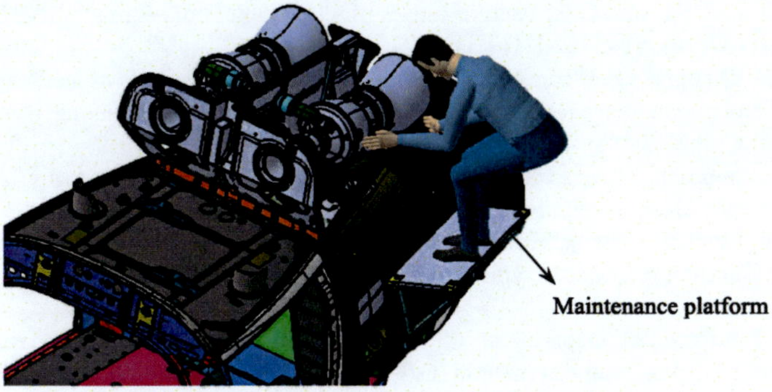

Fig. 11. Maintenance platform using schematic.

6 Conclusion

Based on the fracture analysis and flight test data, this paper analyzes the cause of the cracks on the engine compartment partition of a helicopter and improves its design. The conclusions are drown:

(1) The engine simplified modeling and fairing detail modeling are studied, and the modular idea is used to construct the whole helicopter dynamic model. The model analysis results are in good agreement with the flight data.
(2) Fatigue analysis shows that the cause of engine compartment partition crack is high-cycle fatigue under vibration coupling aerodynamic load.
(3) Considering the working environment temperature and processing requirements, the original structure is improved, and the optimized structure has good fatigue damage resistance. At the same time, the newly designed maintenance platform increases the operating space and effectively reduces the possibility of accidental damage to the compartment partition and staff.

References

1. Han, D.: Helicopter Structural Dynamics. Science Press, Beijing (2020)
2. Jiang, X.T.: Aircraft Design Manual Volume 19 Helicopter Design. Aviation Industry Press, Beijing (2005)
3. Zhao, S.J.: Cracking causes of main nozzle of a helicopter engine. PTCA (Part A Phys. Test) **56**(12), 57–61 (2020)
4. Men, K.F., Zhang, H.X.: Failure analysis of helicopter windshield. Failure Anal. Prevent. **2**(13), 49–53 (2018)
5. Lin, C.X., Lin, S., Sun, G.F.: Facture causes of a helicopter connecting bolt. PTCA (Part A Phys. Test) **57**(10), 51–54 (2021)
6. Cao, C.L., Zhang, G.Q., Wang, H.L., Niu, Y.J., Wang, B.M.: Analysis of load-carrying frame cracking and maintenance requirement. In: 34th Helicopter Annual Meeting, pp. 1–5. CSAA, Shenzhen, China (2018)
7. General Equipmental Department of the Chinese People's Liberation Army: Aero-Engine Fire Protection Design Requirements (GJB3568). Beijing, China (1999)
8. Xia, S.M., Lin, C.L., Zhang, T.L., Cui, Y., Yuan, S.T.: Prediction of hub alternating loads and airframe vibration response in typical flight conditions. Sci. Technol. Eng. **21**(8), 3387–3392 (2021)
9. General Equipmental Department of the Chinese People's Liberation Army: Military equipment environmental test method (GJB150). Beijing, China (1999)
10. Ma, J.M.: China Aviation Materials Handbook Volume 3 Aluminum Alloy, Magnesium Alloy and Titanium Alloy. Standards Press of China, Beijing, China (1989)

Design of Air Curtains for Isolation of the Air Pollutant Transport in Aircraft Cabins

Yun Wei[1](✉) and Tengfei Zhang[2]

[1] Civil Aviation University of China, Tianjin 300300, China
y_wei@cauc.edu.cn
[2] Dalian University of Technology, Dalian 116024, China

Abstract. The outbreak of respiratory infections in recent years has given us a sharp reminder of the importance of controlling the air pollutant transport in aircraft cabins. However, currently no effective strategies have been applied to commercial aircraft cabins. Several studies optimized the air-supply parameters of the environmental control systems in cabins, in order to inhabit the spread of the inside pollutant. But the personnel movements often broke the designed airflow patterns, leading to unsatisfactory effects. This study concentrates on applying air curtains to aircraft cabins, in order to better prevent the spread of the air pollutant in the breathing zones of passengers; the air-supply velocity of the proposed air curtains is optimized by CFD as well, to satisfy the design targets which include mean age of air and air velocity near passengers. Results show that the designed air curtains are able to inhibit the transport of air pollutant in breathing zones by both improving the air distribution and accelerating the fresh air flows in aircraft cabins. In addition, the air-supply velocity of the air curtains is also solved to meet the design targets. Efficient optimization of air-supply parameters for the inlets of air curtains in aircraft cabins requires further exploration.

Keywords: Air curtain · Pollutant control · Cabin environment

1 Introduction

Creating a safe and healthy air environment in aircraft cabins for inside passengers plays a crucial role in public security. However, events on air pollution in commercial aircraft cabins often occurred, especially after the outbreak of COVID-19, H1N1 and SARS in recent years, which shows the urgency of controlling the transmission of air pollutant in aircraft cabins.

Currently, many studies have conducted computational fluid dynamics (CFD) or tracer gas experiments to explore the transmission of air pollutant in aircraft cabins. For example, some researchers conducted tracer gas experiments to observe the longitudinal airflows between the different rows in MD-82 aircraft cabin [1], and to analyze the transport and dispersion of COVID-19 aerosols in Boeing 767 and 777 airplanes that have mannequins seated inside [2]; Yan et al. studied the airflow and droplet transmission around a sneezing or coughing passenger in a full-scale Boeing 767–300 mock-up by

Chinese Society of Aeronautics and Astronautics (Ed.): CASTC 2023, LNME, pp. 613–620, 2024.
https://doi.org/10.1007/978-981-99-8861-7_61

both the tracer gas experiment and CFD simulation [3]. Results from these studies showed that the transmission of air pollutant in a cabin was rampant, the pollutant might spread to all directions from the source.

After recognizing the laws of pollutant transmission in cabins, preventing the pollutant transmission is of primary importance. The proposed strategies can be divided into two categories, optimization of the ventilation system and optimization of other facilities in cabins. For the former strategies, under-floor displacement air distribution systems [4], modified mixing ventilation systems [5], and improved personalized ventilation systems [6, 7] were adopted, to decrease the exposure to pollutants for passengers in the aircraft cabins; some studies even re-designed air supply parameters of the cabin inlets by inverse design methods [8, 9] or other optimization methods [10]. For the latter strategies, Zhang et al. [11] studied the inclination of the seats and the spacing between the adjacent rows to avoid the spread of pollutants in an aircraft cabin. Both the two categories aimed to design better air distributions in cabins. However, personnel movements and behaviors often broke the designed airflow patterns, leading to unsatisfactory effects [12].

Hence, several pioneer studies in recent years introduced air curtains that were often used in buildings into the cabin environmental design, in order to alter the traces of the pollutant transmission in cabins and to isolate the pollutant from breathing zones of the passengers. The air curtain can form strong isolations in air regions by airflows with high speeds, so it may not be easily destroyed by the personnel movements. For example, Kurec et al. [13] installed an air curtain which has an inlet width of 0.025 m and an inlet velocity of 1 m/s in front of one passenger in a cabin model, and compared the effects of the air curtains with two different curtain locations. Results showed that the transport of droplets released from the mouth of the passenger could be effectively inhibited by the proposed air curtain, and the locations of the curtain in the cabin might influence the effects of isolation; Farag et al. [14] also applied an air curtain with an inlet velocity of 10 m/s and an inlet width of 0.05 m to an aircraft cabin for protection of the passengers from being infected by Sarin. Both studies showed the great prospects of the air curtain in aircraft cabins on inhibiting the transmission of the pollutant in breathing zones. However, current studies only conducted preliminary explorations of the air curtain on aircraft cabins. The appropriate air supply velocities of the air curtain have not been designed, which is the aim of our study. Hence, this study first compares the air distribution in a cabin model which has air curtains between rows with that have no air curtains, to show the advantage of the proposed strategy on air curtain, then four air-supply velocities of the air curtains, 2 m/s, 4 m/s, 6 m/s, and 8 m/s, are evaluated, to determine the appropriate one that could both inhibit the pollutant transmission in breathing zones and provide an acceptable thermal environment in the aircraft cabin for emergency cases.

2 Methods

2.1 Case Design

To evaluate the application of air curtains in cabin environments, a half section of a full-scale, three-row cabin is used in this study, as shown in Fig. 1. The width of the cabin is 1.81 m at the floor level, and maximum height is 2.26 m. Nine passengers are seated in the cabin. Surface temperature of these passengers is set as 30.3 °C, and the seats are adiabatic. The translucent red boxes represent the breathing zones of these passengers, each has a volume of $1.22 \times 0.25 \times 0.17$ m^3. The inlet which supplies the conditioned air of 19.5 °C for the ventilation system is on the side wall near the ceiling, and the outlet for the ventilation system is on the side wall near the floor. The total airflow rate is 0.09 m^3/s for the inlet when no sudden air pollution incidents occur, to satisfy the recommended ventilation airflow of 10 L/s per person. For the boundary conditions of walls, temperatures of all the cabin walls are assumed to be 22 °C, except that of the floor, which is assumed to be 23 °C. Boundary conditions of the front and rear walls are set as periodic, and that of the vertical plane in the aisle is set as symmetrical.

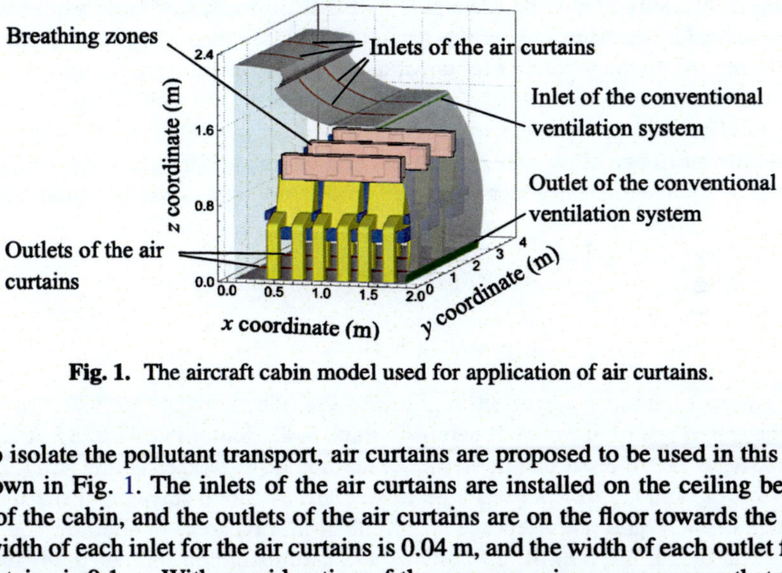

Fig. 1. The aircraft cabin model used for application of air curtains.

To isolate the pollutant transport, air curtains are proposed to be used in this study, as shown in Fig. 1. The inlets of the air curtains are installed on the ceiling between rows of the cabin, and the outlets of the air curtains are on the floor towards the inlets. The width of each inlet for the air curtains is 0.04 m, and the width of each outlet for the air curtains is 0.1 m. With consideration of the energy savings, we assume that the air curtain will only be used for emergency cases, and once the air curtains are opened, the conventional inlet for the ventilation system on the side wall will be closed. Temperature of the supplied air for the air curtains is set as 21 °C, and the turbulence intensity is set as 1.5%. Both values are referred to the study of Kurec et al. [13]. The air-supply velocity can be varied from 2 to 8 m/s, with an equal spacing of 2 m/s. So totally 4 cases with different air-supply velocities of the air curtains were used in this study, to show the advantage of the proposed air curtains and to optimize its air-supply velocity.

2.2 Strategies for CFD Simulations

To obtain the thermal comfort and air quality in the cabin by CFD simulation, the air domain in the cabin was discretized by Gambit first. Totally 2,716,433 tetrahedral meshes were created to satisfy the grid independence, and the meshes were refined near all the solid boundaries, air inlets, air outlets, and surfaces of the passengers. Then the continuity equation, Navier-Stokes equations and the energy equation were solved simultaneously on these meshes by Fluent software, to obtain the airflow field and temperature field in the cabin. The general differential equations used in CFD to solve the above governing equations are expressed as [11]:

$$\frac{\partial(\rho\phi)}{\partial t} + \mathrm{div}(\rho\mathbf{u}\phi) = \mathrm{div}\big(\Gamma_\phi \mathrm{grad}\phi\big) + S_\phi \tag{1}$$

where ρ means the air density, ϕ means the variable in the governing equations such as velocity components and temperature, $\mathbf{u}$ means the velocity vector, Γ_ϕ means the diffusion coefficient, and S_ϕ means the source term. Both Γ_ϕ and S_ϕ are specific to the meaning of ϕ.

Fluent software was used for CFD simulation in this study. The turbulent flows in the cabin were resolved by k-ε RNG model with enhanced wall treatment, and the pressure field and the velocity field were decoupled by SIMPLE method. The pressure variable was discretized by pressure staggering option (PRESTO) scheme, while other variables were discretized by the second-order upwind, in order to capture the turbulences in the cabin with acceptable accuracy [9].

To validate the improvement of air environment in the cabin by the designed air curtains, the mean age of air which assesses the flow time of the air from the inlets to one point in the cabin environment was also calculated, by a scalar transport equation as [15]:

$$\frac{\partial(\rho\tau)}{\partial t} + \mathrm{div}(\rho\mathbf{u}\tau) = \mathrm{div}\big(\Gamma_{\tau,\mathit{eff}}\,\mathrm{grad}\tau\big) + \rho \tag{2}$$

$$\Gamma_{\tau,\mathit{eff}} = 2.88\times 10^{-5}\rho + \frac{\mu_{\mathit{eff}}}{\sigma_\tau} \tag{3}$$

where τ means the mean age of air, μ_{eff} means the effective dynamic viscosity of the air in the cabin, and σ_τ is Schmidt number, which is 0.7 for air in this study. A smaller value of mean age of air at one point means that the air is fresher at this point. In this study, Eqs. (2) and (3) were solved with a fixed airflow field, since the distribution of mean age of air would not affect the airflow field in the cabin. In addition, the mean age of air at the air inlets of the air curtains was assumed to be zero, meaning that the air is purely fresh at these inlets. The passive scalar variable was also discretized by the second-order upwind scheme, which is the same as our previous study [9].

3 Results

3.1 Validation of the Proposed Air Curtains

Figure 2 shows the solved airflow fields in several typical planes for the case without air curtains in the cabin. The supplied air from the inlet on the side wall first flowed along the ceiling because of the thermal plume of the passengers, then flowed into the central region in the cabin and finally flowed to the outlet, as shown in Fig. 2(a). Both the lateral airflow in each row in Fig. 2(b) and the longitudinal airflows between rows in Fig. 2(c) are evident for the cabin without air curtains, it means that once there is a pollutant source in the cabin, the air pollutant released from the source could rapidly mix with the surrounding air, leading to the spread of the pollutant in the cabin.

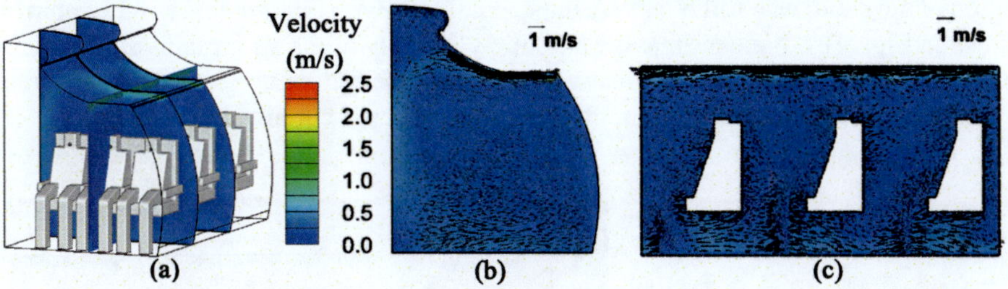

Fig. 2. The solved airflow fields in the typical planes for the case without air curtains in the cabin model: (a) distributions in the typical planes of $x = 0.99$ m, $y = 1.01$ m, $y = 2.06$ m; (b) detailed view in the plane of $y = 1.01$ m; (c) detailed view in the plane of $x = 0.99$ m.

By comparison, Fig. 3 shows the airflow fields in the same planes as Fig. 2, for the case with opened air curtains in the cabin model, and the air-supply velocity of the air curtain was assumed to be 2 m/s. Note that, the inlet of the conventional ventilation system on the side wall is closed for this case. The air spouted vertically from the inlet of the air curtains, and formed air partitions between rows, as shown in Fig. 3(c). In addition, the lateral airflow that flowed in one row was also decreased, as shown in Fig. 3(b), since the air jets with high speeds would entice the surrounding air to flow downwards. Both the lateral airflow in each row and longitudinal airflows between rows are weakened obviously when the air curtains are opened, therefore the air curtain is proved to be a useful tool for control of the pollutant transmission in the cabin.

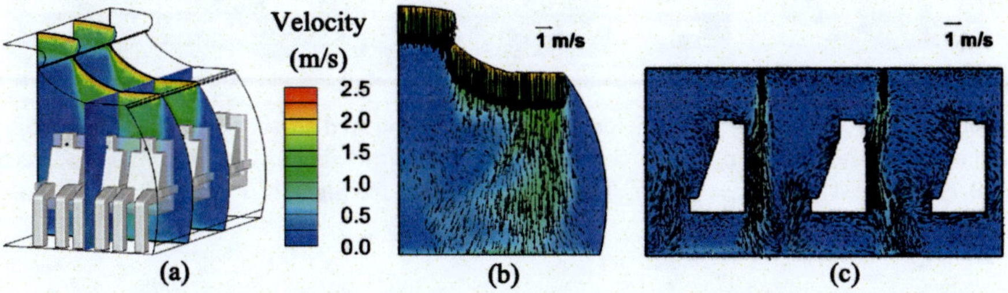

Fig. 3. The solved airflow fields for the case with opened air curtains in the cabin model, and the air-supply velocity of the air curtain was 2 m/s: (a) distributions in the typical planes of $x = 0.99$ m, $y = 1.01$ m, $y = 2.06$ m; (b) detailed view in the plane of $y = 1.01$ m; (c) detailed view in the plane of $x = 0.99$ m.

Figure 4 and Fig. 5 show the mean age of air fields in these typical planes, for cases with and without air curtains. When the air curtains are opened, as shown in Fig. 5, the mean age of air in the cabin is dramatically decreased in comparison with the case shown in Fig. 4. It illustrates that application of the air curtains can accelerate the discharge of the air in the cabin, leading to a healthier air environment for the inside passengers. Hence, besides the improvement of the air distribution, the proposed air curtain strategy could also be helpful to produce a pleasant air quality for the cabin environment. However, for the case with 2 m/s for the curtain inlet velocity, the mean age of air in the central region of the cabin was still relatively large, compared with other regions of the cabin, as shown in Fig. 5(b). It means that the air-supply velocity of the air curtains at the ceiling may be re-designed for further improvement. Then the following section will show the determination of the proper air-supply velocity for the air curtains in the cabin.

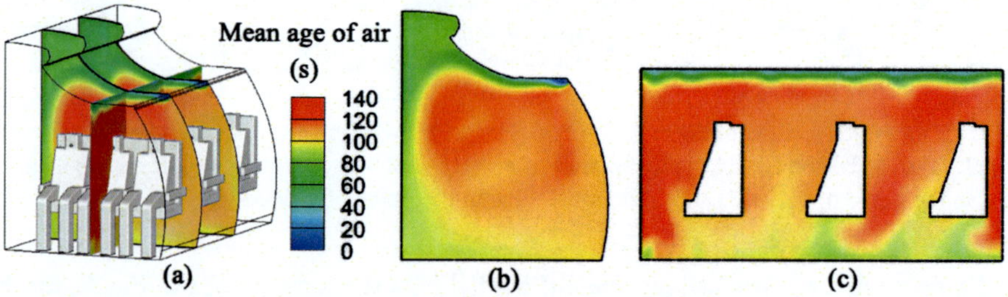

Fig. 4. The solved mean age of air fields for the case without air curtains in the cabin model: (a) distributions in the typical planes of $x = 0.99$ m, $y = 1.01$ m, $y = 2.06$ m; (b) detailed view in the plane of $y = 1.01$ m; (c) detailed view in the plane of $x = 0.99$ m.

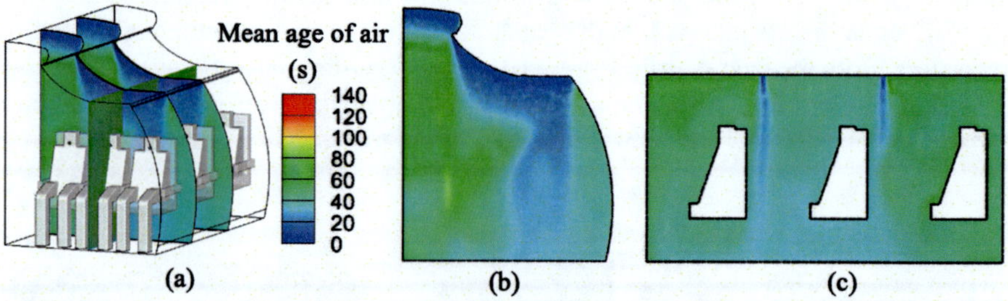

Fig. 5. The solved mean age of air fields for the case with opened air curtains in the cabin model, and the air-supply velocity of the air curtain was 2 m/s: (a) distributions for the typical planes of $x = 0.99$ m, $y = 1.01$ m, $y = 2.06$ m; (b) detailed view in the plane of $y = 1.01$ m; (c) detailed view in the plane of $x = 0.99$ m.

3.2 Determination for the Proper Air-Supply Velocity of the Air Curtains

To determine the proper air-supply velocity of the air curtains that proposed in this study, velocities from 2 m/s to 8 m/s with an equal spacing of 2 m/s were sampled for comparison. Table 1 shows the averaged mean ages of air in the breathing zones as well as the averaged velocities in the surfaces which have a distance of 0.1 m from all the passengers in the cabin, for the four samples, to evaluate the thermal comfort and air quality in the cabin with different air-supply velocities of the air curtains. As the air-supply velocity of the air curtains increased, the averaged mean age of air in the breathing zones dropped dramatically, while the averaged velocity in the surfaces near passengers increased, on the contrary. It means that the increase of the air-supply of the air curtains is beneficial to the improvement of the cabin environment, but may reduce the thermal comfort of passengers by an increasing draft. If we define that the largest averaged velocity near passengers should not be larger than 0.5 m/s, then 4 m/s is the appropriate air-supply velocity of the proposed air curtains. The air curtain with the determined air-supply velocity could then create a good cabin environment which has acceptable thermal comfort and air quality.

Table 1. Averaged mean ages of air in the breathing zones and averaged velocities in the surfaces which have a distance of 0.1 m from all the passengers in the cabin, for cases without air curtains and with different air-supply velocities of the air curtains.

Cases	Averaged mean age of air in the breathing zones	Averaged velocity in the surfaces near passengers
Without air curtains	112 s	0.17 m/s
With 2 m/s for the velocity of curtain inlets	51 s	0.23 m/s
With 4 m/s for the velocity of curtain inlets	26 s	0.47 m/s
With 6 m/s for the velocity of curtain inlets	17 s	0.74 m/s
With 8 m/s for the velocity of curtain inlets	13 s	1.02 m/s

4 Conclusions

This study proposed to apply air curtains to aircraft cabins, in order to inhibit the air pollutant transmission in the cabin. Results show that the proposed strategy is able to reduce both the lateral airflow in each row and longitudinal airflows between rows. The air quality in the cabin could be improved as well, by evaluating the mean age of air in the cabin. With consideration of the air velocities in the surfaces near passengers and the mean age of air in the breathing zones in the cabin, 4 m/s is determined to be the proper air-supply velocity for the air curtains.

Acknowledgement. The work was supported by the Fundamental Research Funds for the Central Universities (Grant No.: 3122020071).

References

1. Li, F., Li, B., Liu, J.: Measuring and containing longitudinal flow: important for airborne pollutants control in an aircraft cabin. Sci. Technol. Built Environ. **21**(8), 1126–1133 (2015)
2. Hanna, S.: Transport and dispersion of tracers simulating COVID-19 aerosols in passenger aircraft. Indoor Air **32**(1), 12974 (2022)
3. Yan, W., Zhang, Y., Sun, Y., et al.: Experimental and CFD study of unsteady airborne pollutant transport within an aircraft cabin mock-up. Build. Environ. **44**(1), 34–43 (2009)
4. Zhang, T., Chen, Q.: Novel air distribution systems for commercial aircraft cabins. Build. Environ. **42**(4), 1675–1684 (2007)
5. Fišer, J., Jícha, M.: Impact of air distribution system on quality of ventilation in small aircraft cabin. Build. Environ. **69**, 171–182 (2013)
6. Gao, N., Niu, J.: Personalized ventilation for commercial aircraft cabins. J. Aircr.Aircr. **45**(2), 508–512 (2008)
7. Mboreha, C.A., Sun, J., Wang, Y., et al.: Airflow and contaminant transport in innovative personalized ventilation systems for aircraft cabins: a numerical study. Sci. Technol. Built Environ. **28**(4), 557–574 (2022)
8. Chen, Q., Zhai, Z., You, X., et al.: Inverse Design Methods for the Built Environment. Routledge, London (2018)
9. Wei, Y., Liu, W., Xue, Y., et al.: Inverse design of aircraft cabin ventilation by integrating three methods. Build. Environ. **150**, 33–43 (2019)
10. Cao, S.J., Ren, C.: Ventilation control strategy using low-dimensional linear ventilation models and artificial neural network. Build. Environ. **144**, 316–333 (2018)
11. Zhang, T., Fan, M., Liu, S.: Impact of seat inclination and misalignment on airborne pollutant transport in a single-aisle aircraft cabin. Appl. Sci. **12**(9), 4538 (2022)
12. Zhang, Y., Li, J., Liu, J.: Experimental study of the impact of passenger behavior on the aircraft cabin environment. Sci. Technol. Built Environ. **27**(4), 427–435 (2020)
13. Kurec, K., Olszański, B., Gumowski, K., et al.: Air curtain as a SARS-CoV-2 spreading mitigation method in a small aircraft cabin. J. Aerosp. Eng. 1–25 (2023)
14. Farag, A.M., Sayed, M.M.A., Abbas, W.: Air curtain CFD model for protection commercial aircraft cabin passengers from Sarin (GB). In: Proceedings of the 2016 IEEE Aerospace Conference, pp. 1–12. IEEE, USA (2016)
15. Sandberg, M.: What is ventilation efficiency? Build. Environ. **16**(2), 123–135 (1981)

Extension of Aircraft Thermal Endurance by Multi-Return Fuel Tank System

Shiyu Yang, Yuanfang Lin, Xianghua Xu(✉), and Xingang Liang

School of Aerospace Engineering, Tsinghua University, Beijing 100084, China
xxh@tsinghua.edu.cn

Abstract. To extend the thermal endurance of the aircraft, a new fuel thermal management system with multi-return flow paths to the fuel tanks (MRFTS) is proposed. Compared with the traditional single return fuel tank system (SRFTS), the biggest difference is the hot fuel can also be returned to the auxiliary tanks in the MRFTS. The flow selector of the MRFTS is equipped with a temperature control strategy, which can keep the system temperature below but approaching the limit values to achieve the acceleration of the waste heat dissipation. And the fuel consuming sequence of the actual fuel tanks has been fully considered in this control strategy. By the numerical calculation of the high-power condition, the thermal endurance of the MRFTS is increased to 3319.8 s while the thermal endurance of the SRFTS is only 389.5 s, which means utilizing the large amount of the heat sink of the auxiliary tanks can significantly improve the aircraft thermal endurance. Furthermore, a mission profile with a long-time aircraft engagement is analyzed as well, which indicates that the MRFTS can control the system temperature not exceeding the limit values by the dynamic selections of the return tanks while the SRFTS cannot meet the thermal management requirements during the engagement stage. Therefore, the MRFTS can extend the aircraft thermal endurance effectively during the entire flight process with the high thermal load.

Keywords: Thermal endurance · Aircraft · Multiple tanks · Fuel · Control strategy

1 Introduction

With the rapid improvement of the speed and maneuverability of the advanced aircraft, thermal load of the airborne equipment and engine is increasing gradually. And the internal thermal load of the aircraft soars on account of the use of high-power weapons, resulting in the severe heat dissipation problems for the aircraft [1]. However, the area of the ram air cooler and the number of the pilot holes are restricted by the stealth requirement of the aircraft. The use of composite materials for the fuselage decreases the heat dissipation ability of the aircraft as well [2]. Therefore, utilizing the internal heat sink to remove the waste heat of the aircraft has become a research hotspot [3]. For the large amount of fuel carried by the aircraft when taking off, of which the property is relatively stable, the thermal management system using fuel as the main heat sink (FTMS) is proposed [4, 5].

Chinese Society of Aeronautics and Astronautics (Ed.): CASTC 2023, LNME, pp. 621–637, 2024.
https://doi.org/10.1007/978-981-99-8861-7_62

Improving the thermal endurance of the aircraft with the limited capacity of fuel heat sink is the core content of the FTMS [6, 7]. The definition of the thermal endurance in this paper is the ability to keep the system temperature not to exceed the limit values under the action of the thermal load, which can be reflected in two aspects. The first aspect is the system can work normally without being over-temperature under a constant thermal load. And the second is the system has the ability to work normally during a full flight mission profile with the variable thermal loads. Two methods were proposed by Doman to extend the aircraft thermal endurance [8]. One is to create a more efficient heat dissipation mechanism by improving the efficiency of components or subsystems [9, 10]. The other is to obtain a better system topology and a temperature control strategy by detailed simulation of the FTMS in the aircraft [11, 12]. In this paper, the thermal endurance of the aircraft is researched from the perspective of the system control, which means the second method is chosen to discuss the related issues.

Doman [13] compared the thermal endurance of the dual tank topology with the single tank topology. The result shows that a small recirculation tank can accelerate the rise of the fuel supply temperature, which is beneficial to the system heat dissipation. The flow distribution in the dual tank system is carried out by a LQI controller. The controller can keep the system temperature below but near the limit values in the entire flight process. Huang et al. [14] proposed a new system structure FLEX for the FTMS. By controlling the valves, the system working mode can be changed under different conditions, realizing the switch between the single and the dual tank topologies. The case shows that the FLEX performs well in extending the thermal endurance. Sigthorsson et al. [15] established transient heat transfer models of the components in the fuel system and designed a flow structure with a control mode of the FTMS. The result shows that the designed system can control each limited temperature within the limit value well and the control mode has a good robustness. Zhao et al. [16] analyzed the performance of the FTMS under different topologies and proved that the dual tank system plays an important role in raising the fuel supply temperature.

Fuel tank system plays an important role in storing the heat sink, which is simplified into one or two fuel tanks in the above studies. However, aircraft carries multiple fuel tanks actually. And the fuel consuming order of the tanks is determined by the center of the aircraft gravity [17]. Therefore, it is very necessary to conduct the analysis of the FTMS with multiple tanks. In the FTMS, the supplied fuel extracted from the feed tank by the booster pump is more than required for combustion, and the excess fuel is returned to the fuel tank after cooled by the ram air [18]. At present, the hot fuel can only be returned to the feed tank with relatively small volume [19], which makes the system temperature too high or even over-temperature under the action of the high thermal load. Fortunately, the fuel capacity in the auxiliary tanks is very large compared with the feed tank. If the heat sink in the auxiliary tanks can be used to resist the high thermal load, the thermal endurance of the system will be significantly improved. Therefore, this paper designs a new fuel tank system with multiple return paths, which extends the aircraft thermal endurance by using the heat sink of the fuel in the auxiliary tanks.

2 Fuel Tank System Scheme

Aircraft carries multiple fuel tanks as shown in Fig. 1. FT is the feed tank, which is used to deliver the fuel for cooling and combustion to the airborne equipment and engine. AT1L and AT1R are the auxiliary tanks located in the left and right wings. AT2 is the auxiliary tank located in the middle of the fuselage. AT3 is the auxiliary tank located in the front of the fuselage. Also, AT4L and AT4R are the fuel tanks located in the left and right horizontal tails respectively. The function of the auxiliary tank is to deliver the fuel to the feed tank in a certain order, which can keep the center of the aircraft gravity. With the fuel from the auxiliary tanks, the feed tank is always full before the available fuel in the auxiliary tanks is exhausted. Therefore, the fuel in the feed tank will not start to be consumed until all the auxiliary tanks are empty.

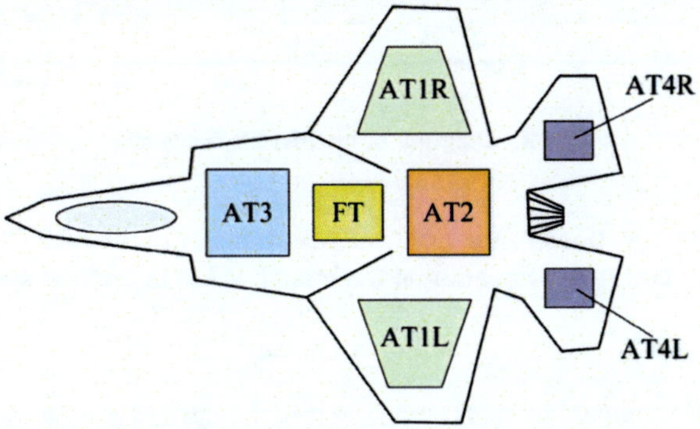

Fig. 1. Diagram of the aircraft fuel tank distribution

The fuel mass flow to the airborne equipment and engine is more than required for combustion. The excess fuel is cooled by the ram air and returned to the fuel tanks. According to the number of the return paths, the fuel tank system is divided into the single return fuel tank system (SRFTS) and the multi-return fuel tank system (MRFTS), both of them will be introduced in detail later.

2.1 Single Return Architecture

The structure of the SRFTS is introduced firstly as shown in Fig. 2. As the AT1L and AT1R in Fig. 1 are symmetric and the fuel consuming conditions of them are synchronized, they can be regarded as a tank AT1. And the same simplification is suitable for the AT4L and AT4R, which are combined into AT4. In this study, the fuel consuming order of the tanks is AT1 → AT2 → AT3 → AT4 → FT.

The supplied fuel from the feed tank is firstly used to cool the electronic cabin system, environmental control system, engine lubricating oil system, etc. (simplified as the Heater in Fig. 2). A portion of the heated fuel enters the combustor and is consumed by the engine. The remaining part is cooled by the ram air and returned to the feed tank.

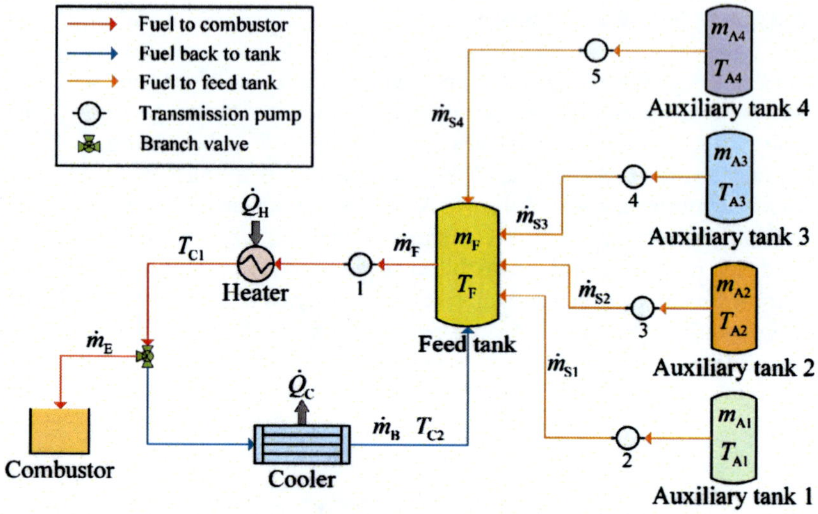

Fig. 2. Diagram of the SRFTS flow path

The governing equations of the flow and heat transfer in the SRFTS are established below. Firstly, the continuity equation at the branch valve in the fuel system is:

$$\dot{m}_F = \dot{m}_E + \dot{m}_B \tag{1}$$

The calculation models of each component in the system can be obtained referring to the modeling method by Doman [8]. Since the fuel is incompressible, the internal energy can be used to represent the total energy when calculating the heat transfer. Also, the fuel tanks are assumed to be adiabatic from the environment and the temperature distribution inside the fuel tank is assumed to be uniform.

For the feed tank, the governing equations of its mass and temperature changes are respectively:

$$\frac{dm_F}{dt} = \sum_{i=1}^{4} \dot{m}_{Si} + \dot{m}_B - \dot{m}_F \tag{2}$$

$$\frac{dU_F}{dt} = \sum_{i=1}^{4} \dot{m}_{Si} u_{Si} + \dot{m}_B u_B - \dot{m}_F u_F \tag{3}$$

where U_F is the internal energy of the feed tank, u_{Si} is the specific internal energy of the auxiliary tank i, u_B is the specific internal energy of the fuel after the cooler and u_F is the specific internal energy of the feed tank. The calculation formulas are:

$$U_F = m_F c_v (T_F - T_r) \tag{4}$$

$$u_{Si} = c_v (T_{Ai} - T_r) \tag{5}$$

$$u_{\mathrm{B}} = c_v(T_{\mathrm{C2}} - T_{\mathrm{r}}) \tag{6}$$

$$u_{\mathrm{F}} = c_v(T_{\mathrm{F}} - T_{\mathrm{r}}) \tag{7}$$

where c_v is the specific heat of the fuel at constant volume, and T_{r} is the reference temperature when calculating the internal energy.

In the SRFTS, as the fuel does not return to the auxiliary tanks, the fuel temperatures inside the auxiliary tanks remain their initial values. The mass change in the auxiliary tank *i* is:

$$\frac{\mathrm{d}m_{\mathrm{A}i}}{\mathrm{d}t} = -\dot{m}_{\mathrm{S}i} \tag{8}$$

If the auxiliary tank *i* is delivering fuel to the feed tank:

$$\dot{m}_{\mathrm{S}i} = \dot{m}_{\mathrm{E}} \tag{9}$$

Otherwise:

$$\dot{m}_{\mathrm{S}i} = 0 \tag{10}$$

The fuel temperature after the heater is:

$$T_{\mathrm{C1}} = T_{\mathrm{F}} + \frac{\dot{Q}_{\mathrm{H}}}{\dot{m}_{\mathrm{F}} c_v} \tag{11}$$

The fuel temperature after the cooler is:

$$T_{\mathrm{C2}} = T_{\mathrm{C1}} - \frac{\dot{Q}_{\mathrm{C}}}{\dot{m}_{\mathrm{B}} c_v} \tag{12}$$

where the cooling capacity can be calculated as follows:

$$\dot{Q}_{\mathrm{C}} = U_{\mathrm{C}} A_{\mathrm{C}} \Delta T_{\mathrm{ln}} \tag{13}$$

where U_{C} is the heat transfer coefficient between the fuel and the ram air, A_{C} is the heat transfer area, ΔT_{ln} is the logarithmic mean temperature difference (LMTD). Since the temperature of the ram air side T_{W} is assumed to remain constant during the cooling process [20], the LMTD can be calculated as follows:

$$\Delta T_{\mathrm{ln}} = \frac{(T_{\mathrm{C1}} - T_{\mathrm{W}}) - (T_{\mathrm{C2}} - T_{\mathrm{W}})}{\ln(\frac{T_{\mathrm{C1}} - T_{\mathrm{W}}}{T_{\mathrm{C2}} - T_{\mathrm{W}}})} \tag{14}$$

In the SRFTS, all the waste heat from the returned fuel can only be absorbed by the heat sink of the feed tank. When the system thermal load is too high, the heat sink in the feed tank cannot meet the system heat dissipation requirement. It is necessary to dissipate the waste heat into the heat sink of other fuels. Therefore, a new topology MRFTS is designed, which will be described in detail later.

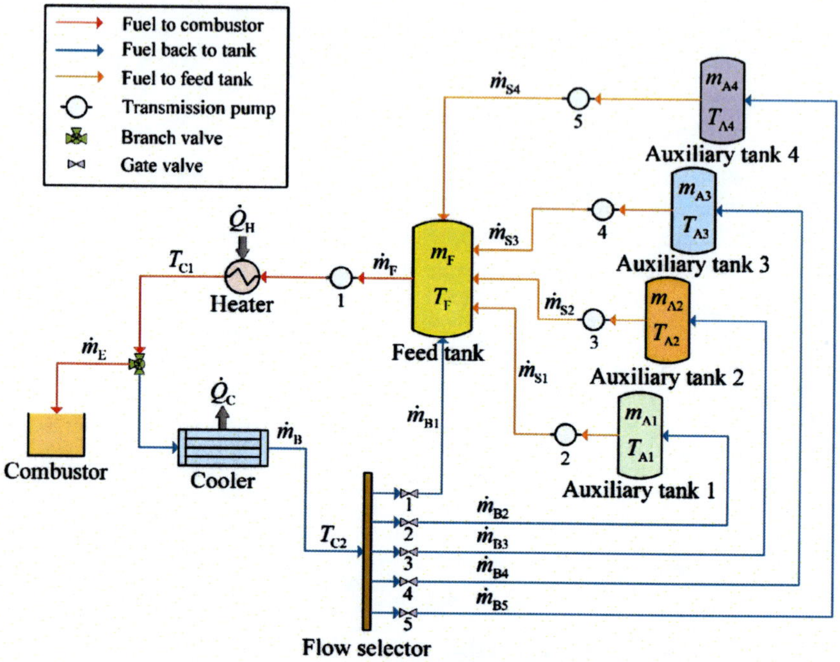

Fig. 3. Diagram of the MRFTS flow path

2.2 Multi-return Architecture

Different from the SRFTS, the fuel cooled by the ram air can also be returned to each auxiliary tank in the MRFTS. The specific structure is shown in Fig. 3.

The returned fuel is controlled by the flow selector, which is an on-off selector. The gate valves determine the opening and closing of each fuel return path, and only one fuel tank is selected as the return tank at a time. The fuel tanks can be used as the return tank include the feed tank and the non-empty auxiliary tanks as shown in Table 1. The auxiliary tanks can be divided into three categories (empty, consuming and full) according to their fuel consuming states. And the tank which is consuming fuel is called the fuel consuming tank (can be the feed tank and any auxiliary tank).

Table 1. Type of the fuel tank and whether can be the return tank

Fuel tank type	Whether the fuel can return to
Feed tank	Yes
Auxiliary tank (empty)	No
Auxiliary tank (consuming)	Yes
Auxiliary tank (full)	Yes

The MRFTS has the same flow path except the return path to the auxiliary tanks comparing with the SRFTS. Therefore, only the auxiliary tanks are analyzed in the discussion of the governing equations of the flow and heat transfer.

The fuel of each tank in the MRFTS is consumed in the same order as the SRFTS. The mass change of the auxiliary tank i can be calculated as follows:

$$\frac{\mathrm{d}m_{\mathrm{A}i}}{\mathrm{d}t} = \dot{m}_{\mathrm{B}(i+1)} - \dot{m}_{\mathrm{S}i} \tag{15}$$

If the auxiliary tank is in the consuming state:

$$\dot{m}_{\mathrm{S}i} = \dot{m}_{\mathrm{B}(i+1)} + \dot{m}_{\mathrm{E}} \tag{16}$$

Otherwise:

$$\dot{m}_{\mathrm{S}i} = \dot{m}_{\mathrm{B}(i+1)} \tag{17}$$

The temperature change of the auxiliary tank i is:

$$\frac{\mathrm{d}U_{\mathrm{A}i}}{\mathrm{d}t} = \dot{m}_{\mathrm{B}(i+1)}u_{\mathrm{B}} - \dot{m}_{\mathrm{S}i}u_{\mathrm{S}i} \tag{18}$$

where $U_{\mathrm{A}i}$ is the internal energy of the auxiliary tank i, and its calculation formula is:

$$U_{\mathrm{A}i} = m_{\mathrm{A}i}c_v(T_{\mathrm{A}i} - T_{\mathrm{r}}) \tag{19}$$

In the MRFTS, the system waste heat carried by the returned fuel can also be stored in the fuel heat sink of the auxiliary tanks, which provides the possibility to improve the thermal endurance of the aircraft.

3 Temperature Control Strategy

In this paper, there are three temperature limits in the fuel system, which are the feed tank limit temperature $\overline{T}_{\mathrm{F}}$, the limit temperature before entering the combustor $\overline{T}_{\mathrm{C1}}$ and the limit temperature after the cooler $\overline{T}_{\mathrm{C2}}$. For the SRFTS, since the working mode is fixed, there is no temperature control strategy. However, as the flow selector is used to determine the return tank in the MRFTS, a temperature control strategy is required for the selector to ensure that the system temperature does not exceed but near the limit values. The system temperature near the limit values can keep the supplied fuel temperature high enough, which can enhance the system heat dissipation rate.

In order to dissipate system waste heat effectively, the feed tank is firstly selected as the return tank. When the system is over-temperature, the next tank will be selected successively according to the fuel consuming sequence of auxiliary tanks. Since the empty auxiliary tank cannot be selected, the fuel return sequence of the flow selector is different under the different fuel consuming states, as shown in Table 2.

Table 2. Fuel return sequence corresponding to the fuel consuming state

Current fuel consuming tank	Fuel return sequence of the flow selector
Auxiliary tank 1 (AT1)	FT/AT1/AT2/AT3/AT4
Auxiliary tank 2 (AT2)	FT/AT2/AT3/AT4
Auxiliary tank 3 (AT3)	FT/AT3/AT4
Auxiliary tank 4 (AT4)	FT/AT4
Feed tank (FT)	FT

There are two conditions that the return tank will not change under the control of the flow selector. One is the current system temperature does not exceed the reference limit values (due to the time delay, an appropriate margin should be retained [15]). The control strategy judges whether the system is over-temperature by the reference limit values, which are lower than the limit values, denoting as $\tilde{T}_{\mathrm{F}}$, $\tilde{T}_{\mathrm{C1}}$ and $\tilde{T}_{\mathrm{C2}}$ respectively. The other condition is that the current system temperature is dropping although it has exceeded the reference limit values. In other conditions, the flow selector will select the return tank which can reduce the system temperature based on the current fuel return sequence. If there is no tank meeting the requirement, the fuel will be returned to the last tank in the fuel return sequence to delay the effect of waste heat in the return fuel on the system temperature.

Considering that the change of the fuel consuming state will lead to the change of the fuel return sequence, while the change of the operating condition will lead to the change of the system thermal load, the reset of the flow selector will be triggered when these conditions happen. The reset here means that the return tank will be selected as the feed tank again, which can make full use of the fuel heat sink and make the dissipation of the waste heat more efficient. As a conclusion, the temperature control strategy of the MRFTS is shown in Fig. 4.

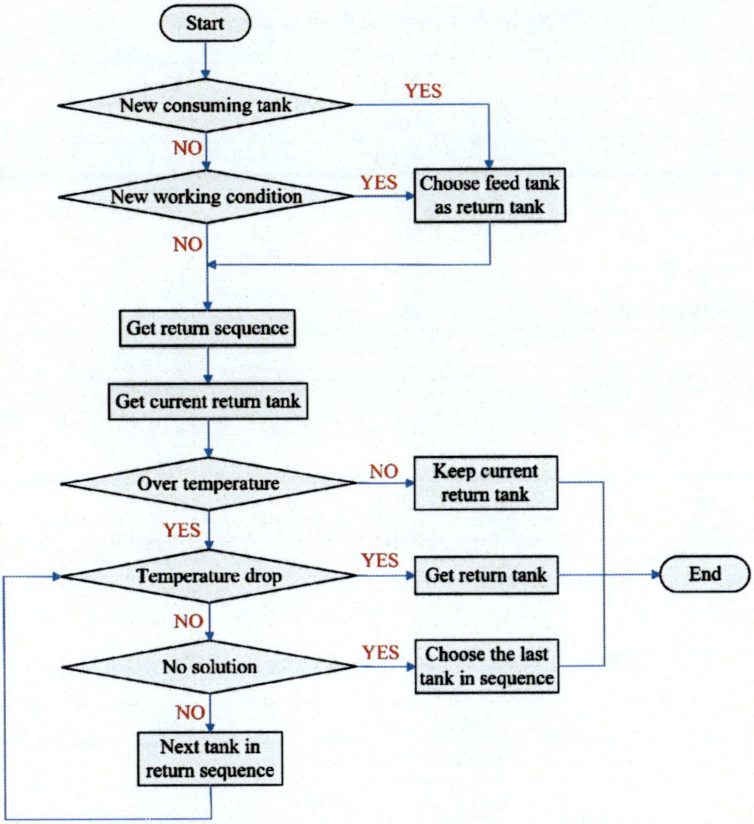

Fig. 4. Temperature control strategy of the MRFTS

4 Case Analysis

In order to reflect the role of the MRFTS in improving the thermal endurance of the aircraft, the abilities of these two structures are compared with each other under the action of high-power weapons. And the flight process analysis is also carried out for the mission profile with a long-time engagement. In this paper, the temperature variation of the system is analyzed by the numerical calculation based on Python.

4.1 Application of High-Power Weapons

In the present-day advanced fighter aircraft, high-power weapons such as direction energy weapons lead to a sharp increase in the heat dissipation demand of the airborne equipment [5], which brings a great challenge to the FTMS. According to the SRFTS in Fig. 2 and the MRFTS in Fig. 3, the thermal endurances of them under the action of high-power weapons are compared. The basic parameters of the fuel tank system are similar to the Doman's [13], as shown in Table 3.

Table 3. Parameters of the fuel tank system

Parameter	Value	Unit
$m_{F,0}$	200	kg
$m_{A1,0}$	1200	kg
$m_{A2,0}$	750	kg
$m_{A3,0}$	500	kg
$m_{A4,0}$	400	kg
Initial fuel temperature	288	K
$\dot{Q}_H$	200	kW
$\dot{m}_F$	1	kg/s
$\dot{m}_E$	0.4	kg/s
$U_C A_C$	500	W/K
T_W	238	K
c_v	2010	J/(kg · K)

Note: the subscript 0 represents the initial state

Table 4. Temperature limit values of the fuel tank system

Parameter	Value	Unit
$\overline{T}_F$	333	K
$\overline{T}_{C1}$	421	K
$\overline{T}_{C2}$	360	K

The temperature limit values [20] of the fuel tank system are shown in Table 4.

Firstly, the performance of the SRFTS under this condition is analyzed. Since there is residual fuel in the tank actually (the booster pump cannot pump all the fuel out of the fuel tank due to the height of the fuel outlet hole), it is assumed that the residual fuel in each tank takes up 2% of the tank capacity in all simulations. Through the simulation of the SRFTS, the variation of the fuel mass in each fuel tank over the time is obtained, as shown in Fig. 5.

Figure 5 shows that the fuel consuming order is in accordance with expectation, and all the system available fuel is exhausted at 7473 s. In this fuel consuming order, the temperature changes of each limit point in the fuel system are shown in Fig. 6.

Figure 6 illustrates that as the temperature of returned fuel is relatively high, the system temperature keeps increasing at the beginning. When the temperature rises to a certain value, the system temperature becomes steady due to the continuous transmission of the cold fuel from the auxiliary tanks to the feed tank. When all the available fuel in the auxiliary tanks is exhausted, the fuel in the feed tank begins to be consumed. In this period, due to the missing of the cold fuel source and the decline of the fuel mass

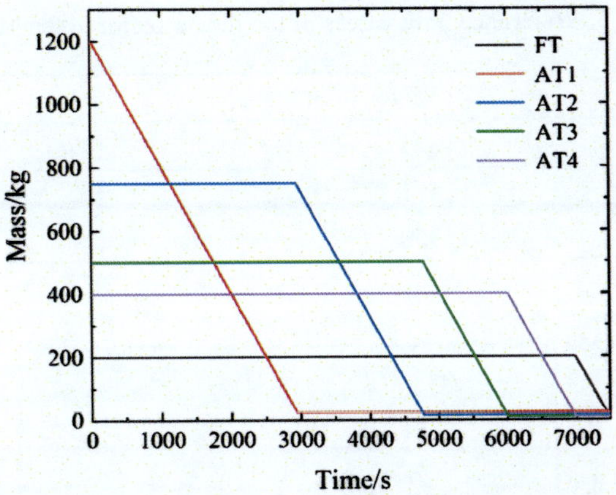

Fig. 5. Variation of the fuel mass in the SRFTS under high-power condition

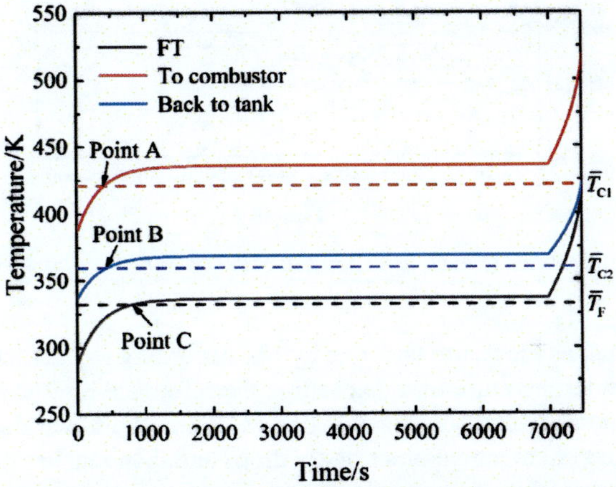

Fig. 6. Temperatures of each limit point in the SRFTS under high-power condition

in the feed tank, the system temperature rises rapidly. Under the action of high-power weapons, the system temperature of the SRFTS exceeds the limit values before reaching the steady state (points A, B and C in Fig. 6) unfortunately. The minimum time when these three limit points exceed the limit values is taken as the thermal endurance of the SRFTS, which is calculated to be 389.5 s.

Before the numerical calculation, the reference limit values of the flow selector in the MRFTS are set, as shown in Table 5.

Based on the control strategy, the temperature changes of each limit point of the MRFTS under the same working condition and system parameters are obtained, as shown in Fig. 7.

Figure 7 indicates that the T_{C2} is the first temperature reaching its reference limit value while the consuming tank is AT1. After the arrival, the flow selector starts to

Table 5. Reference limit values of the flow selector in the MRFTS

Parameter	Value	Unit
$\tilde{T}_F$	323	K
$\tilde{T}_{C1}$	411	K
$\tilde{T}_{C2}$	350	K

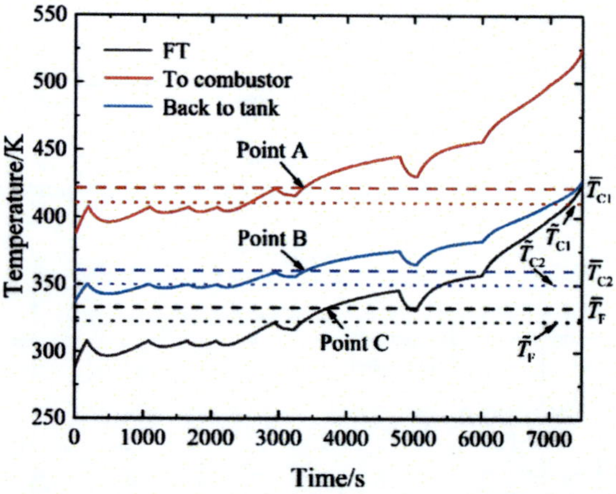

Fig. 7. Temperatures of each limit point in MRFTS under high-power condition

choose the next tank as the return tank one by one according to the fuel return sequence to keep the system temperature from exceeding the reference limit values. The specific selections of the return tanks are shown in Fig. 8. When the flow selector uses AT4 as the return tank, the system temperature firstly drops and then reaches the reference limit value again. However, there is no fuel tank can be used as the return tank to drop the system temperature at this moment. Therefore, the flow selector keeps AT4 as the return tank. The system temperature continues to rise and exceeds the reference limit values. When the available fuel in AT1 is exhausted, the consuming tank becomes AT2 and the flow selector is reset. Since the flow selector keeps AT4 as the return tank in the late stage of the previous fuel consuming state, the return capacity of the first several tanks in the fuel return sequence is recovered by the rise of the system temperature, which means that the system temperature can be reduced by returning the fuel to the first several tanks again. Therefore, in the new fuel return sequence, the flow selector firstly uses AT2 and AT3 as the fuel return tanks to reduce the system temperature slightly, and then chooses AT4 as the return tank. Unfortunately, in the late stage of the current consuming state, the system temperature gradually increases and exceeds the limit values finally (points A, B, and C in Fig. 7). When the fuel consuming tanks are AT3, AT4 and FT, the flow selector adopts the same control strategy as above to select the return tank and the details will not be described here. The time when these three points in Fig. 7 firstly reach the temperature limit values is regarded as the time that the MRFTS can withstand under the high-power condition, which is calculated to be 3319.8 s.

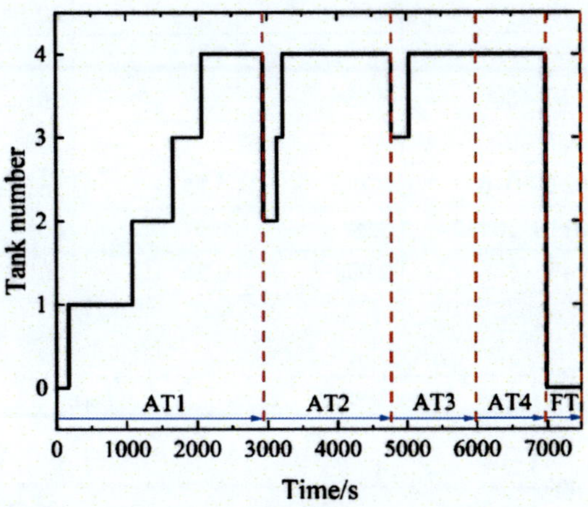

Fig. 8. Selections of the return tanks of the MRFTS under high-power condition (tank number correspondence: 0-FT, 1-AT1, 2-AT2, 3-AT3, 4-AT4)

By comparing the temperature calculation results of the fuel tank systems under two different return structures, it is showed that the thermal endurance can be significantly improved by returning the hot fuel to the auxiliary tanks under the high-power condition. The heat sink of the large amount of fuel in the auxiliary tanks has an important potential in resisting the system high thermal load.

4.2 Mission Profile for Long-Time Engagement

The thermal endurances of these two systems are analyzed below for the mission profile with a long-time engagement. The operation parameters of each stage in the mission profile refer to Doman's paper [13], as shown in Table 6. The difference is that the system thermal load and the time in the engagement stage are improved, which meets the requirements of the present-day advanced fighter aircrafts.

Due to the increase of the engagement time, the aircraft needs to carry more fuel and the fuel mass of the AT1 increases to 1683 kg. Other system parameters are consistent with the Table 3.

In the Table 6, the mass flow and the ram air cooling temperature during taking off and descent are considered to vary linearly with the time. Firstly, the numerical calculation of the SRFTS is carried out in this mission profile, in which the temperature limit values of the system are consistent with the Table 4. The temperature changes of each limit point over the time are shown in Fig. 9.

Figure 9 shows that the system temperature rises suddenly in the engagement stage. The reason is that the thermal load is large while the fuel burning rate is small. In this condition, a large amount of waste heat is stored in the feed tank through the returned

Table 6. Mission profile with a long-time engagement

Flight phase	Time/s	$\dot{m}_E$/(kg/s)	$\dot{m}_F$/(kg/s)	$\dot{Q}_H$/kW	T_W/K
Ground hold	0–1799	0.088	0.7	20	288
Take off	1800–1927	6.3–1.93	7–2.5	40	288–238
Ingress cruise	1928–4800	0.26	0.9	50	238
Engagement	4801–6020	0.4	1	200	238
Egress cruise	6021–10109	0.26	0.85	55	238
Descent	10110–10432	0.29	0.9	40	238–288
Ground hold	10433–14814	0.088	0.7	20	288

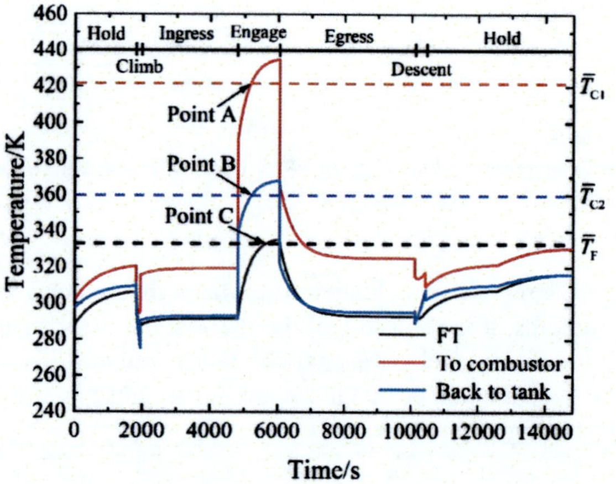

Fig. 9. Temperatures of each limit point in the SRFTS under the mission profile

fuel. Due to the long-time engagement, the system temperature keeps increasing over the time and exceeds the limit values at points A, B and C in Fig. 9 respectively, which indicates that the SRFTS could not meet the long-time and high-power engagement thermal management requirement.

The simulation calculation of the MRFTS is carried out based on the same mission profile, where the reference limit values of the flow selector are consistent with the Table 5. Through the numerical calculation, the temperature changes of each limit point over the time are shown in Fig. 10.

Figure 10 illustrates that the temperature of the returned fuel T_{C2} firstly reaches its reference limit value in the engagement stage. At this time, the flow selector makes dynamic selections of the fuel return tanks according to the temperature control strategy to ensure that the system does not overheat as shown in Fig. 11. Since the thermal load is small in other working conditions except the engagement stage, the flow selector automatically selects the feed tank as the return tank to meet the thermal management

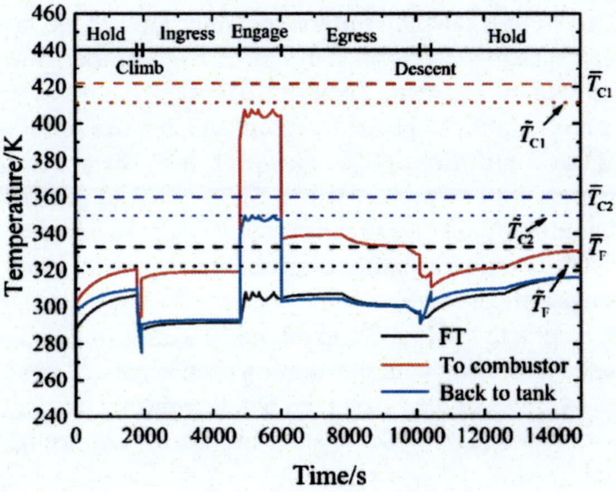

Fig. 10. Temperatures of each limit point in the MRFTS under the mission profile

requirement. Therefore, only the details of return tanks selected by the flow selector in the engagement stage are discussed here.

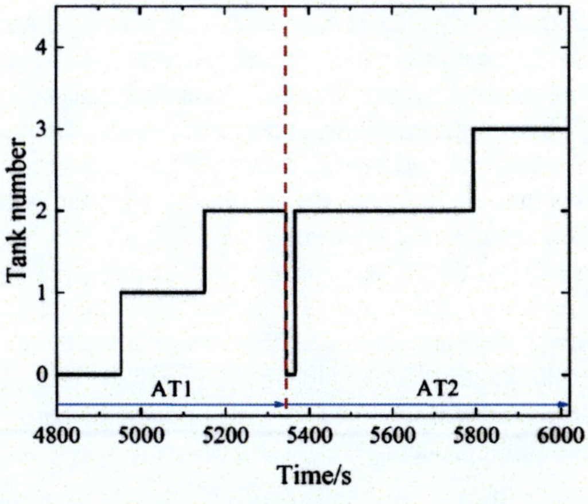

Fig. 11. Selections of the return tanks in the engagement stage of the mission profile (tank number correspondence: 0-FT, 1-AT1, 2-AT2, 3-AT3, 4-AT4)

Figure 11 shows that the consuming tank is the AT1 in the early part of the engagement stage and the initial return tank is the feed tank. When the reference limit value $\tilde{T}_{C2}$ is reached, the flow selector selects AT1 and AT2 as the return tanks in turn to ensure that the system temperature is near but not over the reference limit values. At the 5345 s, the available fuel in the AT1 is exhausted and the fuel in the AT2 begins to be consumed. At the same time, the flow selector is reset due to the change of the fuel return sequence. Because the flow selector is an on-off selector, the system temperature will drop when changing the return tank, which results in an overshoot of the system temperature. In this condition, the fuel return capacities of the previous tanks in the fuel return sequence are improved, which means the system will not be over-temperature when the fuel is returned to the previous tanks again. Therefore, during the time when the consuming tank is the AT2, the return tank can be the feed tank which makes a deep utilization of the heat sink of the fuel in the previous tanks. However, with the action of the high-power weapons, the reference limit value $\tilde{T}_{C2}$ is reached again and the flow selector selects AT2 and AT3 as the return fuel tanks in turn according to the new fuel return sequence. The MRFTS successfully ensures that the system temperature is below but near the limit values during the engagement stage.

The simulation results in the Fig. 9 and the Fig. 10 indicate that the MRFTS has an excellent performance in dealing with the mission profile with a long-time engagement. It can overcome the overheat of the system temperature during the engagement by using the fuel heat sink in the auxiliary tanks to store the waste heat brought by the returned fuel.

5 Conclusions

1. A new fuel tank system with multi-return flow paths (MRFTS) is proposed, which can improve the aircraft thermal endurance by storing the system waste heat brought by the returned fuel to the heat sink in the auxiliary tanks.
2. The flow selector in the MRFTS is equipped with a temperature control strategy, which can enhance the heat dissipation rate of the returned fuel and make full use of the fuel heat sink in the fuel tanks under the actual fuel consuming sequence.
3. Under the action of the high-power weapons, the thermal endurance of the SRFTS is 389.5 s while the thermal endurance of the MRFTS can be extended to 3319.8 s. The result indicates that the fuel heat sink in the auxiliary tanks has a great potential to absorb the thermal load of the system.
4. The calculation results of the mission profile with a long-time engagement indicate that the SRFTS cannot meet the heat dissipation requirement of the system during the engagement stage. Fortunately, the MRFTS can realize the control of the system temperature under the multiple limit values by dynamically selecting the return tanks. The MRFTS ensures the temperatures of each limit point below but near the limit values in the whole flight process.

References

1. Rong, A., Pang, L., Jiang, X., Shi, Y., et al.: Analysis and comparison of potential power and thermal management systems for high-speed aircraft with an optimization method. Energy Built Environ. **2**(1), 13–20 (2021)
2. Huang, H., Spadaccini, L.J., Sobel, D.R.: Fuel-cooled thermal management for advanced aero engines. In: Proceedings of the ASME Turbo Expo 2002: Power for Land, Sea, and Air, pp. 367–376. ASME, USA (2002)
3. Van Heerden, A.S.J., Judt, D.M., Jafari, S., et al.: Aircraft thermal management: practices, technology, system architectures, future challenges, and opportunities. Prog. Aerosp. Sci. **128**, 100767 (2022)

4. Qin, J., Zhang, S.L., Bao, W., et al.: Thermal management method of fuel in advanced aeroengines. Energy **49**, 459–468 (2013)
5. Doman, D.B.: Rapid mission planning for aircraft thermal management. In: AIAA Guidance, Navigation, and Control Conference, Paper No.: 2015-1076. AIAA, USA (2015)
6. German, B.J.: A tank heating model for aircraft fuel thermal systems with recirculation. In: 49th AIAA Aerospace Sciences Meeting including the New Horizons Forum and Aerospace Exposition, Paper No.: 2011-641. AIAA, USA (2011)
7. German, B.J.: Tank heating model for aircraft fuel thermal systems with recirculation. J. Propul. PowerPropul. Power **28**(1), 204–210 (2012)
8. Doman, D.B.: Fuel flow control for extending aircraft thermal endurance part I: underlying principles. In: AIAA Guidance, Navigation, and Control Conference, Paper No.: 2016-1621. AIAA, USA (2016)
9. Oppenheimer, M.W., Sigthorsson, D.O., Doman, D.B.: Extending aircraft thermal endurance by fuel pump sizing. In: AIAA Guidance, Navigation, and Control Conference, Paper No.: 2018-0856. AIAA, USA (2018)
10. Dooley, M., Lui, C., Newman, R.: Efficient propulsion, power, and thermal management integration. In: 49th AIAA/ASME/SAE/ASEE Joint Propulsion Conference, Paper No.: 2013-3681. AIAA, USA (2013)
11. Oppenheimer, M.W., Sigthorsson, D.O., Doman, D.B.: Control of fuel thermal management systems with transport delays. In: AIAA Scitech 2019 Forum, Paper No.: 2019-1917. AIAA, USA (2019)
12. Huang, P.G., Doman, D.B.: Thermal management of single- and dual-tank fuel-flow topologies using an optimal control strategy. J. Thermal Sci. Eng. Appl. **10**(4), 041019 (2018)
13. Doman, D.B.: Fuel flow control for extending aircraft thermal endurance part II: closed loop control. In: AIAA Guidance, Navigation, and Control Conference, Paper No.: 2016-1622. AIAA, USA (2016)
14. Huang, G.P., Doman, D.B., Oppenheimer, M.W., et al.: Topology optimization of a fuel thermal management system. In: AIAA Aviation 2019 Forum, Paper No.: 2019-3471. AIAA, USA (2019)
15. Sigthorsson, D.O., Oppenheimer, M.W., Doman, D.B.: Flight endurance enhancement via thermal management system control subject to multiple limitations. In: AIAA Scitech 2020 Forum, Paper No.: 2020-1825. AIAA, USA (2020)
16. Zhao, Q., Pei, Y., Ge, Y.X., et al.: Analysis of aircraft fuel thermal management system under different architectures. In: Chinese Society of Aeronautics and Astronautics (eds.) Proceedings of the 5th China Aeronautical Science and Technology Conference. LNEE, vol. 821, pp. 841–853. Springer, Singapore (2021). https://doi.org/10.1007/978-981-16-7423-5_84
17. Langton, R., Clark, C., Hewitt, M., et al.: Aircraft Fuel Systems. Wiley, Hoboken (2009)
18. Huang, G.P., Doman, D.B., Rothenberger, M.J., et al.: Dimensional analysis, modeling, and experimental validation of an aircraft fuel thermal management system. J. Thermophys. Heat Transf.Thermophys. Heat Transf. **33**(4), 983–993 (2019)
19. Liu, Y., Lin, G.P., Guo, J.H., et al.: Dynamic prediction of fuel temperature in aircraft fuel tanks based on surrogate. Appl. Thermal Eng. **215**, 118926 (2022)
20. Doman, D.B.: Optimal cruise altitude for aircraft thermal management. J. Guid. Control. Dyn. Guid. Control. Dyn. **38**(11), 2084–2095 (2015)
21. Doman, D.B.: Fuel flow topology and control for extending aircraft thermal endurance. J. Thermophys. Heat Transf.Thermophys. Heat Transf. **32**(1), 35–50 (2018)

Steady State Simulation Program of Aero-Engine Thermal Management System

Shiyu Yang[1], Weilong Gou[2], Yuanfang Lin[1], Xianghua Xu[1](✉), Bo Shi[2], and Xingang Liang[1]

[1] School of Aerospace Engineering, Tsinghua University, Beijing 100084, China
xxh@tsinghua.edu.cn

[2] School of Energy and Power Engineering, Nanjing University of Aeronautics and Astronautics, Nanjing 200016, China

Abstract. In order to study the aero-engine thermal management systematically, a steady state simulation program is developed based on the Python language. Through the analysis of the mass and energy transmission characteristics, the overall system can be divided into four subsystems: the main flow channel subsystem (including the compressor, the combustor, the turbine and etc.), the fuel subsystem, the lubricating oil subsystem and the air subsystem. And there is a strong coupling of the working parameters between these four subsystems. The control equations of the flow and heat transfer of each subsystem are established based on the mass conservation principle, the momentum conservation principle and the energy conservation principle. Further, the efficient solving algorithms of the subsystems are proposed, which can guarantee the accuracy and convergency of the solutions. With the joint simulation of these four subsystems, the steady state operating parameters of the whole aero-engine thermal management system can be obtained. The accuracy of the steady state simulation program is validated by the experimental results of an actual turbofan engine. The calculation process can provide some ideas for the subsequent software development.

Keywords: Thermal management · Aero-engine · Simulation · Program · Algorithm

1 Introduction

With the rapid development of the modern flight technology, the future aircraft demands the high-performance aero-engine [1]. High thrust-weight ratio, low fuel consumption rate and high reliability have become the characteristics of the new generation aero-engines. However, the increase of compressor pressure ratio and turbine inlet temperature makes new thermal management problems for the engine [2, 3]. And there is a strong coupling effect of thermal management in the modern aero-engine subsystems. Therefore, it is very necessary to conduct thermal management of the whole aero-engine and analyze the mass and heat transfer among the subsystems [4].

Aero-engine thermal management is a hot issue at present and many scholars have carried out the related research. However, most of them only studied the component

Chinese Society of Aeronautics and Astronautics (Ed.): CASTC 2023, LNME, pp. 638–651, 2024.
https://doi.org/10.1007/978-981-99-8861-7_63

characteristics to improve the working performance of the components [5–7]. And in the systemic studies, the aero-engine is usually regarded as a thermal load boundary and does not analyze its subsystems in detail [8–10]. Due to the lack of the systemic research on the aero-engine thermal management, it is very import to establish the flow path of each subsystem with detailed components for the aero-engine thermal management system. Since the steady state calculation is an important part for the aero-engine performance check, a thermal management steady state simulation program for aero-engine is developed based on the Python language in this paper.

2 Thermal Management System Architecture

According to the characteristics of mass and energy transmission in aero-engine, the thermal management system of aero-engine is divided into four subsystems: main flow channel subsystem, fuel subsystem, lubricating oil subsystem and air subsystem. The main flow channel system is the core of the engine performance, which includes the air inlet, the compressor, the combustor, the turbine and other important components. The main function of the fuel subsystem is to supply the fuel with certain mass flow rate, temperature and pressure to the combustor precisely. And the supplied fuel also cools the lubricating oil subsystem before entering the combustor. The lubrication oil subsystem is used to cool and lubricate the bearing chambers as well as the gear boxes of the engine, which also takes away the debris caused by the solid surface wear to ensure the safety of the aero-engine. The main effect of the air subsystem is to cool and seal the important internal parts and accessories of the engine, such as the turbine disc cavities and bearing chambers. These four subsystems are coupled with each other in the mass and heat flow analysis, as shown in Fig. 1.

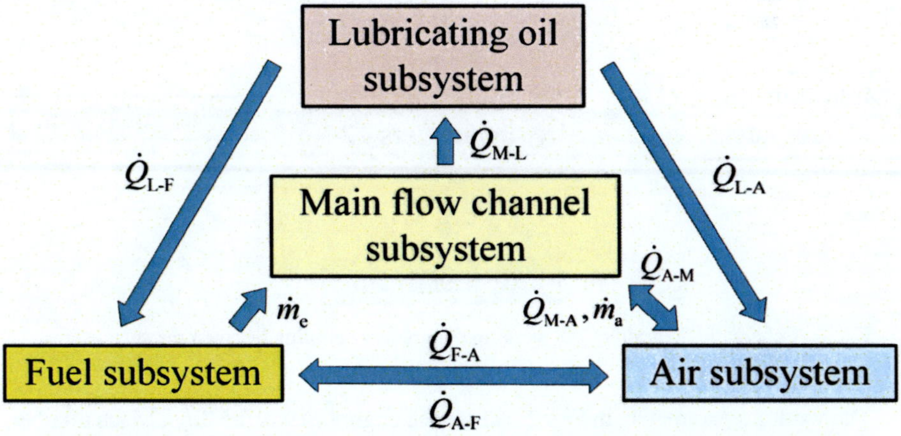

Fig. 1. Schematic diagram of mass and heat flow coupling between the subsystems

The $\dot{Q}_{\text{M-L}}$ in Fig. 1 is the heat that the main flow channel subsystem transfers to the lubricating oil subsystem. When the engine is at different rotate speeds, the heats of each bearing chamber and gear box are different, which are brought into the lubricating oil subsystem through a closed cycle. The heat dissipation of the lubricating oil subsystem consists of two methods. One is to transfer the heat to the fuel subsystem through the fuel-lubricating oil heat exchanger, and the corresponding heat is $\dot{Q}_{\text{L-F}}$. The other method is to transfer the heat to the air subsystem through the air-lubricating oil heat exchanger, and the corresponding heat is $\dot{Q}_{\text{L-A}}$. The fuel subsystem provides the required amount of burning fuel $\dot{m}_{\text{e}}$ to the main flow channel subsystem and also absorbs the heat transferred by the lubricating oil subsystem. If the temperature of the fuel subsystem is too high, the heat can be transferred to the low-temperature air (the external duct air or the air extracted from the compressor low-pressure stage) in the air subsystem through the air-fuel heat exchanger, and the corresponding heat is $\dot{Q}_{\text{F-A}}$. The amount of air extracted from the main flow channel subsystem is $\dot{m}_{\text{a}}$, of which the air used to cool the turbine is usually extracted from the compressor high-pressure stage and the temperature is too high to cool the turbine effectively. To reduce the amount of air used for turbine cooling, the high-pressure air is precooled by the fuel [11] and the heat $\dot{Q}_{\text{A-F}}$ is transferred to the fuel subsystem. In Fig. 1, the cooling capacity of the aero-engine internal parts by the air is denoted as $\dot{Q}_{\text{M-A}}$. And the air will also be used as the heating air for the engine deicing, etc. Under these conditions, the heat $\dot{Q}_{\text{A-M}}$ will be transferred to the main flow channel subsystem. The used air in the air subsystem can be returned to the main flow channel subsystem or expelled from the engine.

The working parameters are also coupled between these four subsystems, as shown in Fig. 2.

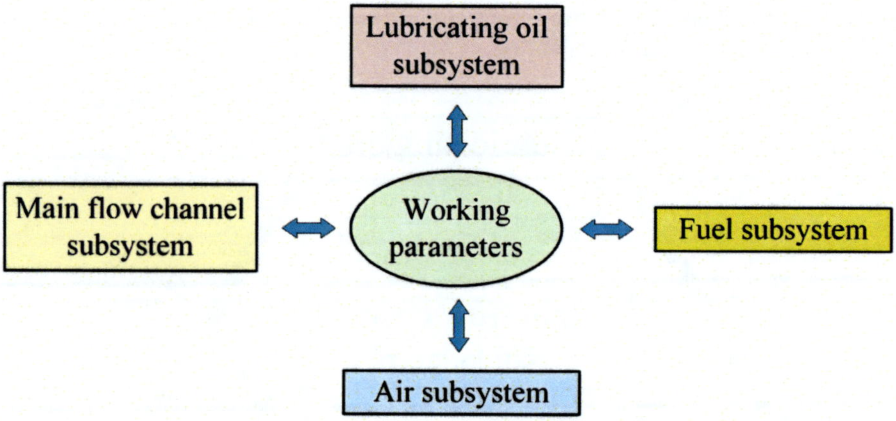

Fig. 2. Schematic diagram of the working parameters coupling between the subsystems

The working parameters in Fig. 2 include the engine speed, the flight Mach number, the flight altitude, the combustor pressure and the burning fuel supply, etc. Through the sharing of the working parameters between the subsystems, the steady state working performance of the aero-engine thermal management system under a certain boundary condition can be obtained.

The architecture of the engine thermal management steady state simulation program is shown in Fig. 3.

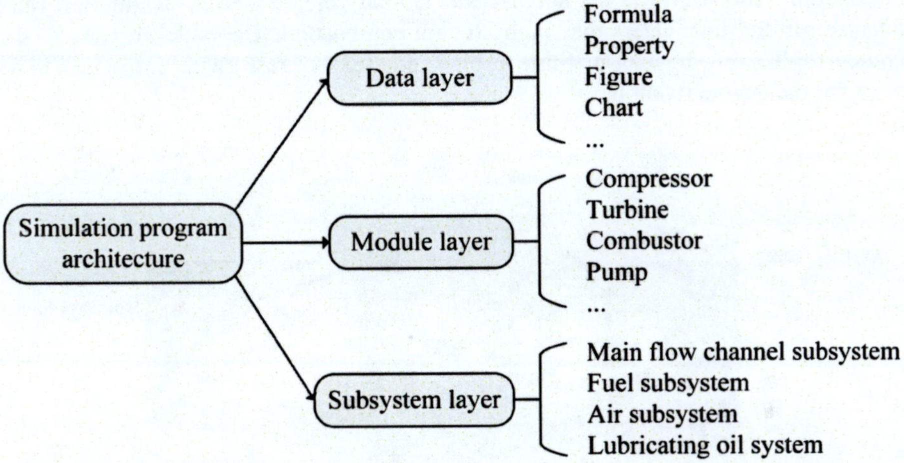

Fig. 3. Architecture of the engine thermal management steady state simulation program

The simulation program in Fig. 3 consists of three layers. The first is the data layer, which includes the characteristics of the components, the thermal properties of the working medium and the related charts, etc. The second is the module layer, which includes the compressor, the turbine, the combustor and the driving pump, etc. The last is the subsystem layer, which includes the frameworks and solving algorithms of the four subsystems in the thermal management system.

The simulation program in this paper is developed based on the Python language. By calling the encapsulated modules to establish the thermal management system and setting the corresponding boundary conditions in the main program, the steady state simulation calculation of the system can be realized.

3 Algorithm and Validation

The components of the aero-engine thermal management system have relatively general models [12–15], and the details are not discussed here. Based on the flow and heat transfer characteristics of each component, the governing equations of each subsystem can be established according to the mass conservation principle, the momentum conservation principle and the energy conservation principle. The establishment of governing equations and their solving algorithms will be emphatically discussed later.

3.1 Main Flow Channel Subsystem

For a two-spool turbofan engine with low bypass ratio, the typical structure of its main flow channel subsystem is shown in Fig. 4. The outside air is sucked into the engine by the fan from the inlet, and the air behind the fan is divided into the external duct air and the internal duct air. The external duct air and the outlet gas of the low-pressure turbine enter the afterburner together, and the mixture is expelled from the nozzle after combustion. The airflow in the internal duct is firstly pressurized by the high-pressure compressor and then enters the combustor for combustion. The outlet mixture of the combustor impacts the high-pressure turbine and the low-pressure turbine in turn to do work for the engine rotation.

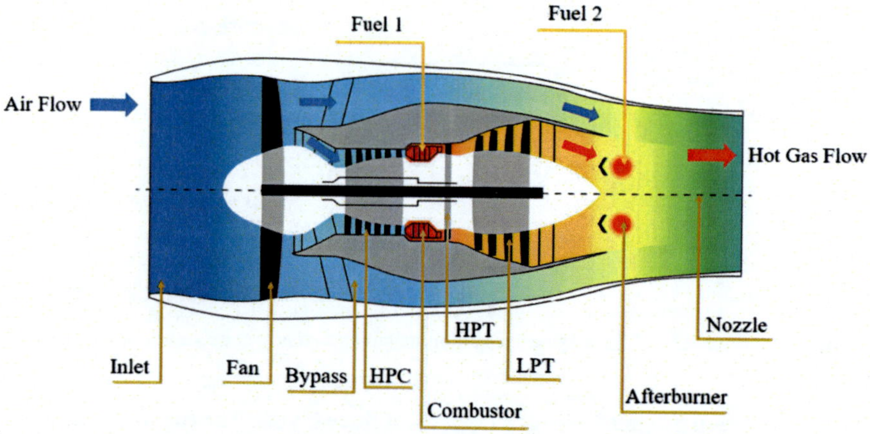

Fig. 4. Typical structure of the main flow channel subsystem

All the components in the main flow channel subsystem need to meet the common working conditions, and the mass conservation equation can be established after the fan:

$$\dot{m}_{\text{fan,outlet}} = \dot{m}_{\text{byp,inlet}} + \dot{m}_{\text{hpc,inlet}} \tag{1}$$

where $\dot{m}_{\text{fan,outlet}}$ is the air mass flow rate at the fan outlet, $\dot{m}_{\text{byp,inlet}}$ is the air mass flow rate at the bypass inlet and $\dot{m}_{\text{hpc,inlet}}$ is the air mass flow rate at the high-pressure compressor inlet.

And the mass conservation equation can be established before the high-pressure turbine:

$$\dot{m}_{\text{hpt,inlet}} = \dot{m}_{\text{hpc,outlet}} + \dot{m}_{\text{com}} \tag{2}$$

where $\dot{m}_{\text{hpt,inlet}}$ is the mixture mass flow rate at the high-pressure turbine inlet, $\dot{m}_{\text{hpc,outlet}}$ is the air mass flow rate at the high-pressure compressor outlet and $\dot{m}_{\text{com}}$ is the fuel mass flow rate to the combustor.

Meanwhile, the mixture mass flow rate at the low-pressure turbine inlet $\dot{m}_{\text{lpt,inlet}}$ equals to that at the high-pressure turbine outlet $\dot{m}_{\text{hpt,outlet}}$ as follows:

$$\dot{m}_{\text{lpt,inlet}} = \dot{m}_{\text{hpt,outlet}} \tag{3}$$

The mass conservation equation can also be established before the nozzle:

$$\dot{m}_{\text{noz,inlet}} = \dot{m}_{\text{lpt,outlet}} + \dot{m}_{\text{byp,outlet}} + \dot{m}_{\text{aft}} \tag{4}$$

where $\dot{m}_{\mathrm{noz,inlet}}$ is the mixture mass flow rate at the nozzle inlet, $\dot{m}_{\mathrm{lpt,outlet}}$ is the mixture mass flow rate at the low-pressure turbine outlet, $\dot{m}_{\mathrm{byp,outlet}}$ is the air mass flow rate at the external duct outlet and $\dot{m}_{\mathrm{aft}}$ is the fuel mass flow rate to the afterburner.

The steady state power balance equation for the high-pressure spool can be established:

$$N_{\mathrm{hpt}}\eta_{\mathrm{h}} = N_{\mathrm{hpc}} + N_{\mathrm{ex}} \tag{5}$$

where N_{hpt} is the power generated by the high-pressure turbine, η_{h} is the mechanical efficiency of the high-pressure spool, N_{hpc} is the power consumed by the high-pressure compressor and N_{ex} is the power extracted from the high-pressure spool by the accessories.

The steady state power balance equation for the low-pressure spool is:

$$N_{\mathrm{lpt}}\eta_{\mathrm{l}} = N_{\mathrm{fan}} \tag{6}$$

where N_{lpt} is the power generated by the low-pressure turbine, η_{l} is the mechanical efficiency of the low-pressure spool and N_{fan} is the power consumed by the fan.

The independent variables of the main flow channel subsystem in Fig. 4 are:

$$\vec{x}_{\mathrm{M}} = [\beta_{\mathrm{fan}}, \beta_{\mathrm{hpc}}, \beta_{\mathrm{hpt}}, \beta_{\mathrm{lpt}}, n_{\mathrm{h}}, n_{\mathrm{l}}]^{\mathrm{T}} \tag{7}$$

where β is the working point parameter and n is the speed of the engine spool.

By constructing the equations from Eq. (1) to Eq. (6), the steady state operating equations of the main flow channel subsystem in a certain condition can be obtained:

$$\begin{cases} s_1(\vec{x}_{\mathrm{M}}) = 0 \\ s_2(\vec{x}_{\mathrm{M}}) = 0 \\ \quad \cdots \\ s_6(\vec{x}_{\mathrm{M}}) = 0 \end{cases} \tag{8}$$

Since Eq. (8) is nonlinear, the Newton-Raphson method is adopted to solve it and the specific calculation process is shown in Fig. 5.

After the solution vector $\vec{x}_{\mathrm{M},i}$ of the current step is obtained, the thermodynamic parameters of each component are updated. Based on this, the Jacobian matrix $\boldsymbol{J}_{\mathrm{M},i}$ and the residual vector $\vec{e}_{\mathrm{M},i}$ are also obtained. Newton-Raphson method is used to update the solution vector, and the next step solution vector $\vec{x}_{\mathrm{M},i+1}$ is obtained. The infinite norm of the difference between two solution vectors is calculated. If the norm is less than the error control value ε, the iteration process stops and the correct solution vector is obtained. Otherwise, update the former solution vector with the newly calculated solution vector. After the solution vector of the main flow channel subsystem is obtained, the flow and heat transfer parameters of the subsystem can be solved completely by updating the working parameters of each component with the independent variables in Eq. (7).

3.2 Fuel Subsystem

A typical flow path of the fuel subsystem of an aero-engine is shown in Fig. 6.

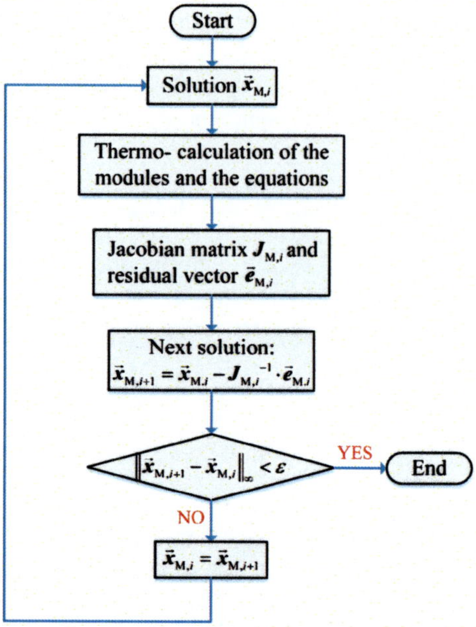

Fig. 5. Schematic diagram of the solving algorithm for the main flow channel subsystem

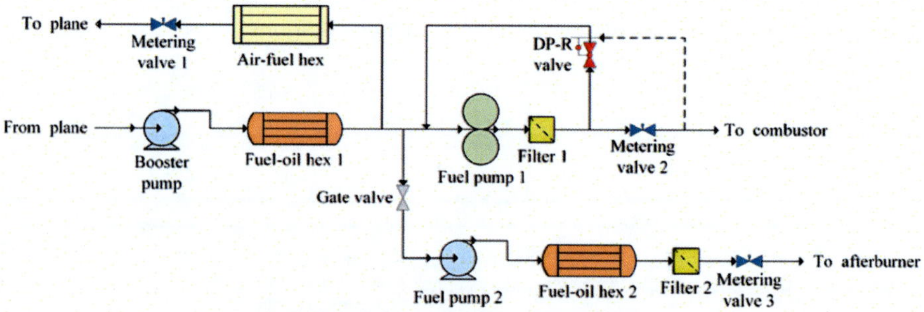

Fig. 6. Diagram of the fuel subsystem flow path

The fuel from the plane is pumped to the engine fuel subsystem by the booster pump and the pressurized fuel firstly enters the fuel-oil heat exchanger 1 to absorb the heat of the lubricating oil subsystem. When the thermal load of the lubricating oil subsystem is too large, the metering valve 1 will be opened to increase the fuel mass flow into the fuel-oil heat exchanger 1, which can improve the cooling capacity of the fuel subsystem. The excess fuel is returned to the plane after cooled by the cooling air. The heated fuel enters the fuel pump 1 for pressurization. When the gate valve is open (in the afterburner state), part of the heated fuel will also enter the fuel pump 2. The fuel entering the fuel pump 1 is firstly filtered through the fuel filter, and then enters the fuel metering device before burning. The fuel entering the fuel pump 2 is firstly used to cool the lubricating oil again through the fuel-oil heat exchanger 2, and then is filtered and metered before the afterburner.

In the flow and heat transfer calculations of the fuel subsystem, each component is simplified as a one-dimensional model, and only the changes at the inlets and outlets of the components are needed to be calculated in the steady state. Fluid network analysis is carried out in the simulation and components connected in series in the same branch are simplified as one component [16]. Based on this, the fluid network will be divided into branches and hubs, as shown in Fig. 7.

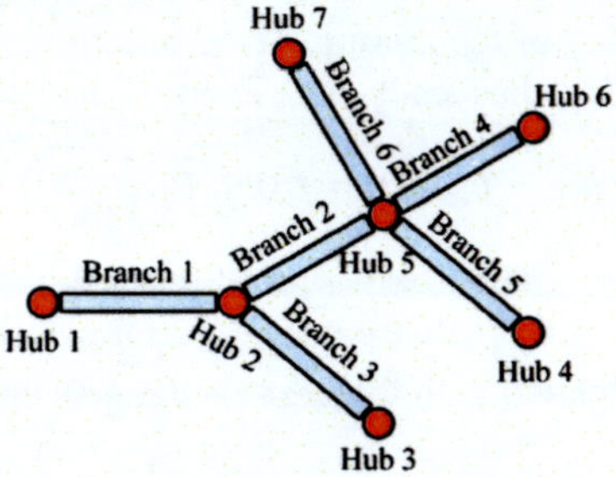

Fig. 7. Diagram of the branches and hubs in fluid network

For the unknown hub *i*, the mass conservation equation can be established:

$$\left(\sum \dot{m}_{\text{in}} - \sum \dot{m}_{\text{out}}\right)_i = 0 \tag{9}$$

where $\dot{m}_{\text{in}}$ is the fuel mass flow rate into unknown hub *i* of a branch and $\dot{m}_{\text{out}}$ is the fuel mass flow rate out of unknown hub *i* of a branch.

For the unknown branch *j*, its momentum conservation equation is:

$$(P_{\text{up}} - P_{\text{down}})_j = g_j(\dot{m}_j, \vec{\boldsymbol{T}}_j, \tilde{\boldsymbol{P}}_j) \tag{10}$$

where P_{up} and P_{down} are the total pressures of the upstream and downstream hubs of the unknown branch *j* respectively, $\dot{m}_j$ is the mass flow rate, $\vec{\boldsymbol{T}}_j$ is the temperature distribution along the flow direction of the unknown branch *j* and $\tilde{\boldsymbol{P}}_j$ is the control pressure set of the branch, which is related to the pressure regulating valve (such as the P-R valve in Fig. 9) and differential pressure regulating valve (such as the DP-R valve in Fig. 6).

Based on Eq. (9) and Eq. (10), the flow control equations of a fluid network with *n* unknown hubs and *m* unknown branches are as follows:

$$\begin{cases} \left(\sum \dot{m}_{\text{in}} - \sum \dot{m}_{\text{out}}\right)_1 = 0 \\ \cdots \\ \left(\sum \dot{m}_{\text{in}} - \sum \dot{m}_{\text{out}}\right)_n = 0 \\ (P_{\text{up}} - P_{\text{down}})_1 = g_1(\dot{m}_1, \vec{\boldsymbol{T}}_1, \tilde{\boldsymbol{P}}_1) \\ \cdots \\ (P_{\text{up}} - P_{\text{down}})_m = g_m(\dot{m}_m, \vec{\boldsymbol{T}}_m, \tilde{\boldsymbol{P}}_m) \end{cases} \tag{11}$$

The corresponding flow solution vector $\vec{x}_{\mathrm{F,flo}}$ is:

$$\vec{x}_{\mathrm{F,flo}} = [\dot{m}_1, \cdots, \dot{m}_m, P_1, \cdots, P_n]^{\mathrm{T}} \tag{12}$$

Then, the steady state heat transfer equations of the fluid network are established. For the unknown hub i, its energy conservation equation is:

$$\bar{h}_i = \left(\frac{\sum \dot{m}_{\mathrm{in}} h_{\mathrm{in}}}{\sum \dot{m}_{\mathrm{in}}}\right)_i \tag{13}$$

where h_{in} is the specific enthalpy of the fuel flowing into the unknown hub i of a branch and $\bar{h}_i$ is the specific enthalpy of the mixture at the hub i.

For the unknown branch j, the steady state energy conservation equation is:

$$(h_{\mathrm{up}} - h_{\mathrm{down}})_j = f_j(\dot{m}_j, \vec{\boldsymbol{P}}_j, \tilde{\boldsymbol{h}}_j, \boldsymbol{E}_j) \tag{14}$$

where h_{up} and h_{down} is the specific enthalpy into and out of the unknown branch j respectively, $\vec{\boldsymbol{P}}_j$ is the total pressure distribution along the flow direction of the branch j, $\tilde{\boldsymbol{h}}_j$ is the specific enthalpy related with the heat exchangers and $\boldsymbol{E}_j$ is the environmental variable set.

Based on Eq. (13) and Eq. (14), the steady state heat control equations of a fluid network with p unknown hubs and q unknown branches are as follows:

$$\begin{cases} \bar{h}_1 = \left(\frac{\sum \dot{m}_{\mathrm{in}} h_{\mathrm{in}}}{\sum \dot{m}_{\mathrm{in}}}\right)_1 \\ \cdots \\ \bar{h}_p = \left(\frac{\sum \dot{m}_{\mathrm{in}} h_{\mathrm{in}}}{\sum \dot{m}_{\mathrm{in}}}\right)_p \\ (h_{\mathrm{up}} - h_{\mathrm{down}})_1 = f_1(\dot{m}_1, \vec{\boldsymbol{P}}_1, \tilde{\boldsymbol{h}}_1, \boldsymbol{E}_1) \\ \cdots \\ (h_{\mathrm{up}} - h_{\mathrm{down}})_q = f_q(\dot{m}_q, \vec{\boldsymbol{P}}_q, \tilde{\boldsymbol{h}}_q, \boldsymbol{E}_q) \end{cases} \tag{15}$$

The corresponding heat solution vector $\vec{x}_{\mathrm{F,hea}}$ is:

$$\vec{x}_{\mathrm{F,hea}} = [\bar{h}_1, \cdots, \bar{h}_p, h_{\mathrm{d},1}, \cdots, h_{\mathrm{d},q}]^{\mathrm{T}} \tag{16}$$

where $h_{\mathrm{d},j}$ is the specific enthalpy out of the unknown branch j.

The flow control equations Eq. (11) can be rewritten as:

$$\begin{cases} F_1(\vec{x}_{\mathrm{F,flo}}, \vec{x}_{\mathrm{F,hea}}) = 0 \\ \cdots \\ F_{m+n}(\vec{x}_{\mathrm{F,flo}}, \vec{x}_{\mathrm{F,hea}}) = 0 \end{cases} \tag{17}$$

And the heat control equations Eq. (15) can be rewritten as:

$$\begin{cases} H_1(\vec{x}_{\mathrm{F,flo}}, \vec{x}_{\mathrm{F,hea}}) = 0 \\ \cdots \\ H_{p+q}(\vec{x}_{\mathrm{F,flo}}, \vec{x}_{\mathrm{F,hea}}) = 0 \end{cases} \tag{18}$$

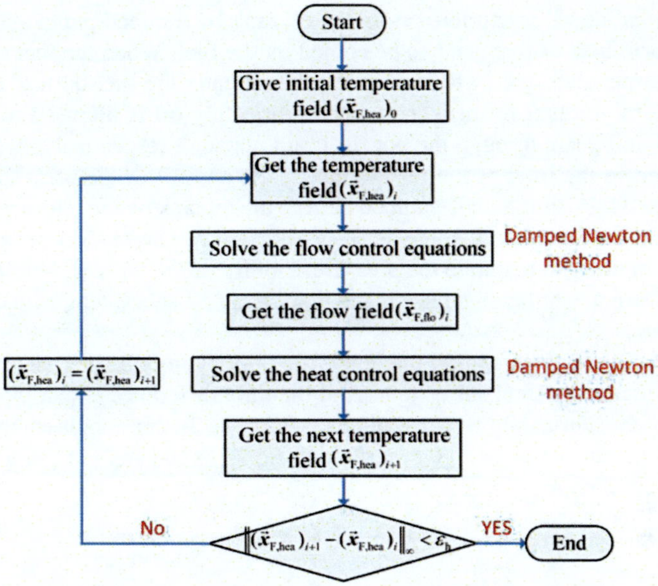

Fig. 8. Decoupling algorithm of the flow and heat transfer

The governing equations Eq. (17) and Eq. (18) are solved by the decoupling of the flow and heat transfer, as shown in Fig. 8.

The function of damped Newton method in Fig. 8 is to ensure the solution convergence by limiting the iteration step size. After the flow and heat transfer parameters of branches and hubs are obtained, the inlet and outlet parameters of the components in series are automatically updated by the branch calculation method.

3.3 Lubricating Oil Subsystem

A typical structure of the aero-engine lubricating oil subsystem is shown in Fig. 9.

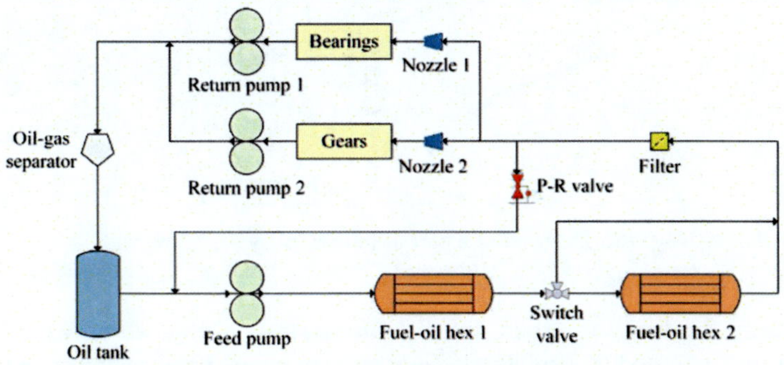

Fig. 9. Diagram of the lubricating oil subsystem flow path

The lubricating oil is extracted from the oil tank by the feed pump and then flows into the fuel-oil heat exchanger 1 to be cooled by the fuel. When the fuel subsystem is in the afterburner state, the switch valve lets the oil enter the fuel-oil heat exchanger 2 to be cooled by the fuel further. The cooled lubricating oil is filtered and then fed to the bearings and gears through the nozzle 1 and nozzle 2 respectively, and the excess oil is returned to the inlet of the feed pump by the control of the pressure regulating valve. The used lubricating oil is pumped back to the oil tank by the return pumps. Since the returned oil is mixed with some air, it is degassed by the oil- gas separator before the oil tank. In order to maintain the pressure stability of the oil tank, bearing chamber, etc., there is also a ventilation device in the lubricating oil subsystem, which will not be discussed here.

The modeling and calculation method of the lubricating oil subsystem is normally the same as that of the fuel subsystem. And the only difference is that the flow in the return path of the lubricating oil subsystem is two-phase. In this condition, the mass flow rate ratio of gas and oil φ will be brought into Eq. (11) and Eq. (15), and the specific enthalpy is rewritten as:

$$h_{\text{dou}} = \frac{\varphi h_{\text{a}} + h_{\text{o}}}{\varphi + 1} \tag{19}$$

where h_{dou} is the specific enthalpy of the two-phase mixture, h_{a} is the specific enthalpy of gas and h_{o} is the specific enthalpy of oil.

After the establishment of the control equations of the lubricating oil subsystem, the decoupling algorithm of the flow and heat transfer is the same as Fig. 8.

3.4 Air Subsystem

A typical flow path of a part of the aero-engine air subsystem is shown in Fig. 10.

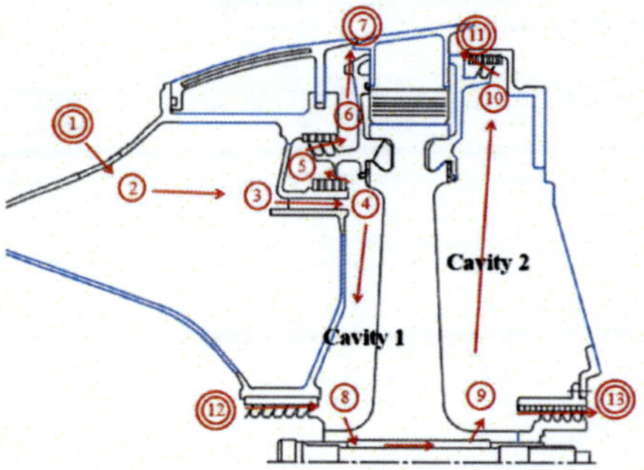

Fig. 10. Diagram of a part of the air subsystem flow path

Similar to the fuel subsystem, the air subsystem is abstracted to a fluid network of components and nodes. The components include pipes, holes, cavities, etc., and the nodes are defined as the inlets and outlets of the components.

The mass and energy conservation equations of the unknown node i can refer to the Eq. (9) and Eq. (13). For the unknown component j, its universal steady state flow formula is:

$$\dot{m}_j = z_j(\theta_j, P_{j,\text{in}}, P_{j,\text{out}}) \tag{20}$$

where $\dot{m}_j$ is the mass flow rate of the component j, θ_j is the term containing thermal property and geometric parameters, $P_{j,\text{in}}$ and $P_{j,\text{out}}$ are the total pressures at the inlet and outlet of component j.

The steady state energy conservation equation of the unknown component j is:

$$(h_{\text{t,out}} - h_{\text{t,in}})_j = [\frac{\omega^2}{2}(r_{\text{out}}^2 - r_{\text{in}}^2) + \frac{K_{\text{av}}A_{\text{w}}}{\dot{m}}(T_{\text{w}} - T_{\text{f}})]_j \tag{21}$$

where $h_{\text{t,out}}$ and $h_{\text{t.in}}$ are the total enthalpies at the outlet and inlet of the component j respectively, ω is the angular rotation velocity, r_{out} and r_{in} are the average rotation radius at the outlet and inlet respectively, K_{av} is the heat transfer coefficient between the fluid and the wall, A_{w} is the heat transfer area, $\dot{m}$ is the air mass flow rate, T_{w} and T_{f} are the temperatures of the wall and fluid respectively.

Since the steady governing equations of the air subsystem are strongly nonlinear, the MC-FN algorithm is developed in this paper to solve the equations by combining the Monte Carlo method and the fluid network method as shown in Fig. 11.

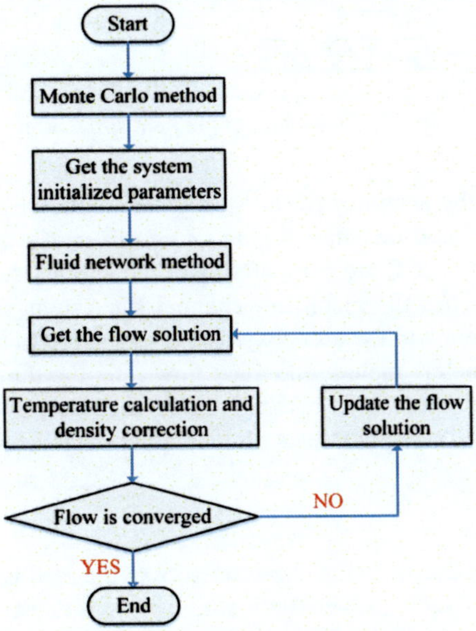

Fig. 11. Flow chart of the MC-FN method

The MC-FN method adopts the layered calculation method, and uses the result of the Monte Carlo method as the initial solution for the fluid network method to improve the convergence of the overall calculation.

3.5 Joint Simulation and Experimental Validation

By associating the calculations of each subsystem with the data transfer process in the Fig. 2, the steady state operating parameters of the whole aero-engine thermal management system can be obtained as shown in Fig. 12. After the initialized system parameters are given, the steady calculations of the four subsystems are carried out. By the data flows, the controller can judge whether the working parameters are converged. And if they are converged, the simulation will stop. Otherwise, the solved parameters in the next step will replace the former parameters and the systematic iteration continues.

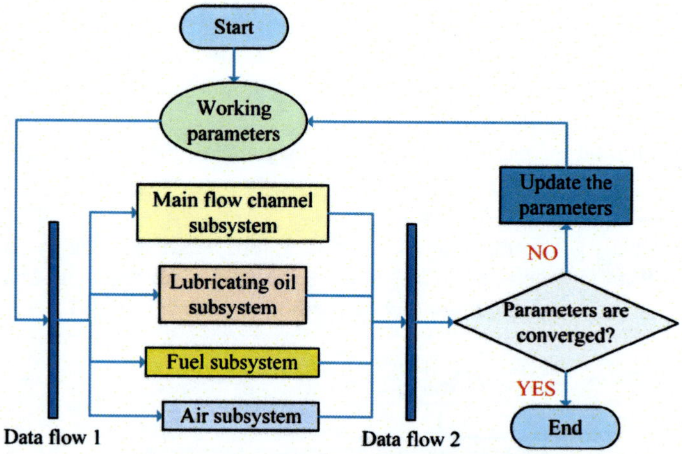

Fig. 12. Flow chart of the joint simulation

In order to verify the accuracy of the steady state simulation program, the experimental results of an actual turbofan engine under different working conditions from the research institution AECC are compared with the simulation results. Five comparison points are selected for the main flow channel subsystem, the fuel subsystem, the lubricating oil subsystem and the air subsystem respectively (a total of twenty points). The comparison parameters include the mass flow rate, the temperature and the pressure of the points. The comparison results indicate that the calculation results of each comparison point meet the requirements of the engineering accuracy.

4 Conclusions

1. The mass and energy transmission characteristics of the aero-engine thermal management system are analyzed and the thermal management system can be divided into four subsystems: the main flow channel subsystem, the fuel subsystem, the lubricating oil system and the air subsystem.
2. The governing equations of the flow and heat transfer for each subsystem are established and the relevant efficient solving algorithms are proposed. And the joint simulation architecture is obtained by the data flows of the operating parameters between these four subsystems.

3. Based on the architecture of the aero-engine thermal management system, the steady state simulation program of the whole system is developed based on the Python language and the calculation accuracy is verified by the experimental results of an actual turbofan engine. This article provides some ideas for the development of the related thermal management software.

References

1. Jiang, Y.G., et al.: Parametric study on the distribution of flow rate and heat sink utilization in cooling channels of advanced aero-engines. Energy **138**, 1056–1068 (2017)
2. Liu, P.H., et al.: Experimental study on the thermal-hydraulic performance of a tube-in-tube helical coil air–fuel heat exchanger for an aero-engine. Energy **267** (2023). Paper No.: 126626
3. Liu, Y.L., et al.: Thermal dynamic and failure research on an air-fuel heat exchanger for aero-engine cooling. Case Stud. Ther. Eng. **42** (2023). Paper No.: 102715
4. Van Heerden, A.S.J., et al.: Aircraft thermal management: practices, technology, system architectures, future challenges, and opportunities. Progr. Aerosp. Sci. **128** (2022). Paper No.: 100767
5. Lin, A.Q., et al.: Comprehensive investigations on fluid flow and heat transfer characteristics of a high-speed rotating turbine disk cavity system of aero-engine. Int. Commun. Heat Mass Transf. **136** (2022). Paper No.: 106170
6. Yang, X., et al.: Study on film cooling for aero-engine turbine endwalls with various small-scale surface structures. Aerosp. Sci. Tech. **123** (2022). Paper No.: 107444
7. Xu, G.Q., et al.: Optimization design with an advanced genetic algorithm for a compact air-air heat exchanger applied in aero engine. Int. J. Heat Mass Transf. **158** (2020). Paper No.: 119952
8. German, B.J.: Tank heating model for aircraft fuel thermal systems with recirculation. J. Propul. Power **28**(1), 204–210 (2012)
9. Huang, P.G., Doman, D.B.: Thermal management of single- and dual-tank fuel-flow topologies using an optimal control strategy. J. Thermal Sci. Eng. Appl. **10**(4) (2018). Paper No.: 041019
10. Huang, G.P., et al.: Control of a switched mode fuel thermal management system. J. Thermophys. Heat Transfer **36**(1), 13–27 (2022)
11. Zhuang, L.H., et al.: Superiority analysis of the cooled cooling air technology for low bypass ratio aero-engine under typical flight mission. Energy Convers. Manage. **259** (2022). Paper No.: 115510
12. Wang, Z.T., et al.: Effect of air properties on a twin-shaft turbofan engine performance during start-up. Appl. Thermal Eng. **218** (2023). Paper No.: 119387
13. Zhang, Y., et al.: Conceptual design and dynamic simulation of thermal management for fighter aircraft fuel system. In: AIAA Modeling and Simulation Technologies Conference, Paper No.: 2010–8087. AIAA, USA (2010)
14. Li, Y.Q., Xuan, Y.M.: Integrated thermal modeling of helicopters. Appl. Therm. Eng. **154**, 458–468 (2019)
15. Yang, X.S., Jian, M.H., Dong, W., et al.: Simulation of the secondary air system of turbofan engines: insights from 1D–3D modeling. Chin. J. Aeronaut. **36**(1), 231–245 (2023)
16. Han, R.B., Xu, X.H., Liang, X.G.: Simplification method of thermal-fluid network with circulation reflux based on matrix operation. Sci. China Technol. Sci. **63**, 1202–1211 (2020)

Research on Optical Testing Method for Laser Damage Effectiveness Evaluation in High-Speed Wind Tunnel

Xiang Liu(✉), Hongyi Gao, Wujun Xie, and Tao Hong

Aerospace Life-Support Industries, Ltd., Xiangyang 441003, China
liux.ali@hotmail.com

Abstract. Under the technical background of testing requirements for laser damage effectiveness evaluation, this project has formed a set of relatively comprehensive optical testing method for evaluating laser damage effectiveness in high-speed wind tunnel from the aspects of optical testing system composition, test layout design, system debugging, parameters setting, data processing, etc., which can comprehensively and effectively meet the testing requirements and complete relevant testing tasks. Starting from preliminary investigation and demonstration, this research delves into the mechanism of laser damage and the macroscopic characteristics of the target, determines the technical route of adopting high-speed photography method, and summarizes the setting rules of photography parameters for different experimental combinations. The research content enriches the technical means and application fields of the current optical testing profession, and accumulates technical experience for undertaking the same type of laser damage verification testing tasks in the future.

Keywords: Optical testing method · Laser damage effectiveness evaluation · High-speed wind tunnel · High-speed photography method

1 Introduction

Laser damage effectiveness evaluation is a very important verification project in the development of laser system, which is of great significance to promote the development of high-energy laser system and the progress of laser damage protection technology in optical system [1–3]. At present, the research on laser damage effectiveness evaluation is mainly based on static power field, and the dynamic damage test verification data is insufficient. It is difficult to accurately realize the transformation and docking between the static damage data of the laser system and the actual dynamic flow field environment damage effect. In order to accurately evaluate the damage performance of the laser system in the actual scene and give scientific and reasonable conclusions, there is still a lack of effective verification methods and sufficient data support. In this context, laser damage verification test under high-speed airflow conditions was carried out in the high-speed wind tunnel of China Aviation Lifesaving Test Base.

Chinese Society of Aeronautics and Astronautics (Ed.): CASTC 2023, LNME, pp. 652–663, 2024.
https://doi.org/10.1007/978-981-99-8861-7_64

As a testing method that can most intuitively demonstrate the damage effect of laser on target materials, optical testing technology is one of the important testing methods for evaluating the effectiveness of laser damage [4, 5]. In view of the rapid and complex characteristics of the laser thermal-mechanical effect destruction process in the laser damage verification experiments, the optical testing mainly uses the high-speed photography method to record the entire process of the experiment at high frequency and accurately interpret the relevant test data. Optical testing technology provides comprehensive image and data support for laser damage effectiveness evaluation, and obtains the spatiotemporal law of laser damage near the irradiation point on the target surface in the form of image data.

2 Mechanism of Laser Damage Effect

The essence of laser damage mechanism is the ionization of electrons after absorbing photons, resulting in the generation of free charge carriers with strong absorption ability for light [6]. The absorbed laser energy is generally converted into thermal energy and causes partial high temperature and high pressure inside the material, thus generating stress on the surface [7, 8]. When the energy density exceeds the damage threshold of the material, the target material will be damaged. The damage effects of laser irradiation mainly include thermal damage, mechanical damage and radiation damage.

- **Thermal Damage Effect.** After laser irradiation of the target, the electrons of the target material collide and convert into thermal energy due to the absorption of light energy. When a certain temperature is reached, the material is melted or even vaporized, resulting in pits or perforations on the material [9].
- **Mechanical Damage Effect.** The absorption of laser pulse energy by material forms a temperature field centered on the laser focal spot, and the material is constrained by external forces or the mutual constraints between internal parts, resulting in force effects such as stretching, compression, and shear.
- **Radiation Damage Effect.** When high-energy laser irradiates the target surface, the vaporized material on the surface will be ionized into plasma cloud to generate ultraviolet and X-ray, and cause damage to the target material [10].

Due to the many factors that affect the laser damage effect, the macroscopic characteristics that directly reflect the target material damage mainly come from the thermal-mechanical effect caused by laser on the material, so it is necessary to accurately record and analyze the process [11]. The evolution process of laser damage mainly involves complex physical and chemical reactions between the high-energy laser beam emitted by the laser and the target, which changes in the structural state and material properties of the target, such as temperature rise, oxidation, expansion, melting, vaporization, breakdown, rupture, and radiation damage [12]. The schematic diagram of the laser damage evolution process is shown in Fig. 1. By understanding the macroscopic characteristics of the target irradiated by laser, an optical testing system based on high-speed photography method can be established, with preset high-speed photography parameters and testing layout, which provides a prior technical reference for the optical testing scheme design of laser damage verification experiment under the condition of high-speed airflow.

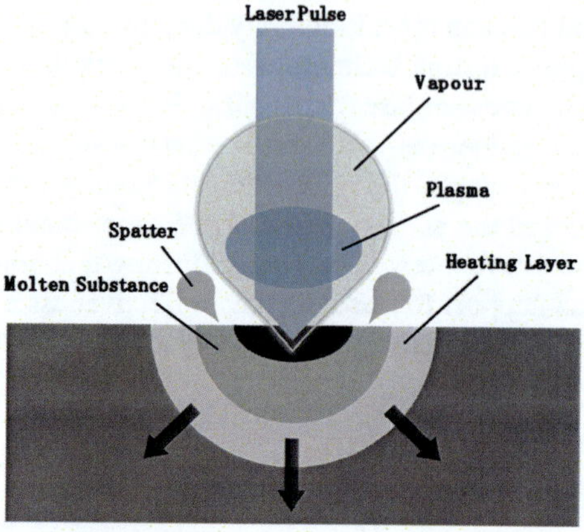

Fig. 1. Schematic diagram of the evolution process of laser damage effect

3 Design of Optical Testing Scheme

3.1 Testing Purpose

On the one hand, by recording the damage morphology of the target irradiated by laser, the macroscopic characteristic rules of different material targets damaged by different power laser under different wind speed conditions are studied. On the other hand, the perforation time of the target irradiated by laser is accurately measured in the microscopic time dimension, and the perforation size is obtained by optical measurement means to evaluate the degree of damage.

3.2 Testing Method

After setting the light output parameters of the laser and the shooting parameters of the high-speed camera, the high-speed wind tunnel begins to blow. When the high-speed airflow velocity reaches the preset value, the console gives the trigger timing signal to the laser and the high-speed camera simultaneously. The laser starts and emits a laser to illuminate the target plate, and the high-speed camera synchronously collects real-time images of the target plate at a certain frame rate under the control of the trigger signal. After the wind tunnel is stopped, the laser is stopped, and the high-speed camera shooting is completed, the test image can be recorded and saved locally, and the damage feature analysis, perforation time interpretation, and size measurement can be performed on the display control terminal.

3.3 Composition and Functions of Optical Testing System

High-Speed Photography System. High-speed photography system is mainly composed of high-speed camera, lens, pan-tilt, tripod and other accessories. The high-speed camera used is the BC2-HD high-speed camera produced by Photron Company. The specific specifications of the high-speed camera are shown in Table 1. Pan-tilt is a component used to connect a high-speed camera to a tripod. The network cable is used to connect the high-speed camera and the control display terminal, and input the shooting information set by the control terminal into the high-speed camera.

Table 1. Main specifications of BC2-HD high-speed camera

Specifications	Parameters
Sensor	CMOS image sensor; Pixel size is 10 μm
Lens interface	F; C; PL; B4
Shutter time	16.7 ms–2.76 μs (Based on frame rate)
Storage	32 GB (5457frames @ maximum resolution)
Digital interface	Gigabit ethernet
Image format	JPEG; AVI; TIFF; BMP; RAW; MRAW; PNG; MOV; FTIF
Working temperature	0 °C–40 °C
Power supply	100 V–240 V AC-1.5A, 50 Hz–60 Hz

Control Display Terminal. Control display terminal is mainly composed of computer and high-speed camera control program. By running the high-speed camera control program on the computer, the functions of setting the relevant shooting parameters, entering the trigger preparation state, downloading and storing the image information, playing and browsing the image can be realized. At the same time, in conjunction with the lens component, the operator can also complete operations such as adjustment of the range and clarity of the target image. Photron FASTCAM Viewer (PFV) is the supporting control program of BC2-HD high-speed camera, which can realize the complete control and real-time browsing of high-speed camera.

Auxiliary Equipment. Auxiliary equipment mainly includes light source system, non-reflective camera, optical lens and scale. Light source system mainly provides supplementary lighting conditions for the test scene in the case of limited light input of high-speed camera. Non-reflective camera is responsible for recording the working state of the craft equipment and target. Scale is the corresponding proportion to achieve the conversion of pixel distance to actual distance. Optical lens is used to protect the lens from damage of target splash during the testing process, while also protecting CMOS from damage caused by high-energy laser irradiation.

3.4 Optical Testing Layout

In the high-speed wind tunnel, the optical testing layout of the laser damage verification experiment is mainly based on the setting state of the laser to deploy a high-speed photography observation system. The layout diagram is shown in Fig. 2. The laser is arranged laterally in an orthogonal manner to the direction of the airflow, and the target material is installed at the intersection of the laser centerline and the wind tunnel outlet centerline. A protective steel plate is placed directly behind the target to avoid the danger caused by laser irradiation to the rear object. The key points of optical testing layout are as follows:

- Two high-speed cameras are horizontally installed on both sides of the laser. The main optical axis forms a small angle of 15° with the laser incident optical path, and the center of the image is aligned with the irradiation point of the target.
- Considering that the stability of the craft equipment installed with the target in the high-speed airflow should be observed while shooting the perforation of the target, two high-speed cameras are equipped with optical lenses in focal lengths of 200 mm and 400 mm respectively.
- The illumination system is placed along the principal optical axis at the rear of the high-speed camera to increase the light intake.
- The signal input terminals of the high-speed camera and laser are connected to the timing controller. During the experiment, the timing controller provides a unified trigger signal to ensure synchronous operation of the high-speed camera and laser.

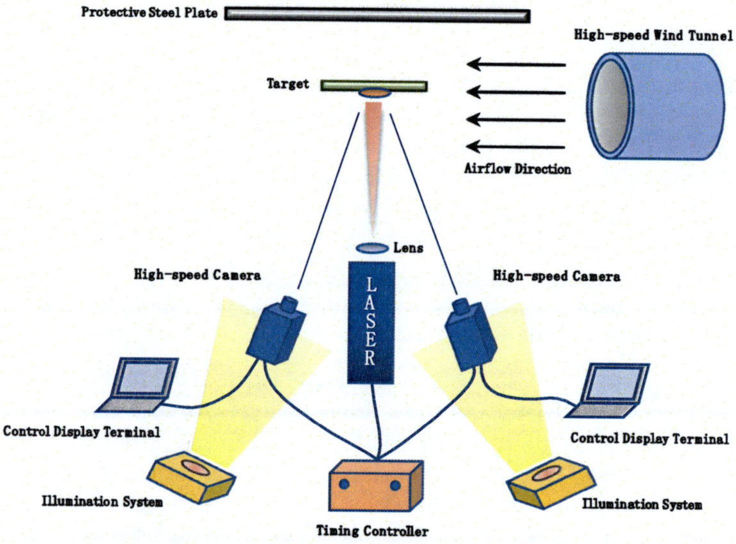

Fig. 2. Diagram of optical testing layout

3.5 Optical Testing Process

According to the purpose of this laser damage verification experiment, the optical testing process throughout the longitudinal experiment is shown in Fig. 3. The specific steps are explained as follows:

1. **Erection of high-speed camera:** According to the optical testing layout, the high-speed cameras are set up to maintain horizontal, and the knobs of the pan-tilt are locked to keep the attitude of the high-speed camera unchanged.
2. **Connection of circuit:** The power line is connected to the 220 V AC power supply, the network cable is connected to the control display terminal, and the trigger line is connected to the timing controller.
3. **Setting of photography parameters:** According to the laser power, target material, blowing time, wind speed and other information of the current test, the photographic parameters such as shooting frequency, shutter time, resolution and trigger time are set up in the control display terminal.
4. **Adjustment of image:** Observe the screen displayed by the control program, adjust the optical lens mechanism to ensure appropriate brightness and sharp edges of the image.
5. **Joint test of timing system:** Entering the state of waiting for trigger, the whole system is tested together with other test equipment to ensure the effectiveness of the timing system.
6. **Real-time synchronous recording:** After receiving the trigger signal, the high-speed cameras synchronously record the damage to the target caused by laser irradiation until the end of the experiment.
7. **Download image:** After the experiment, the test images in the high-speed camera storage are downloaded in time by control display terminal, and the file information is marked according to the naming specification.
8. **Data interpretation processing:** Combined with the image data, the duration of the laser breakdown target, the perforation size, and the light output and stop time are interpreted.

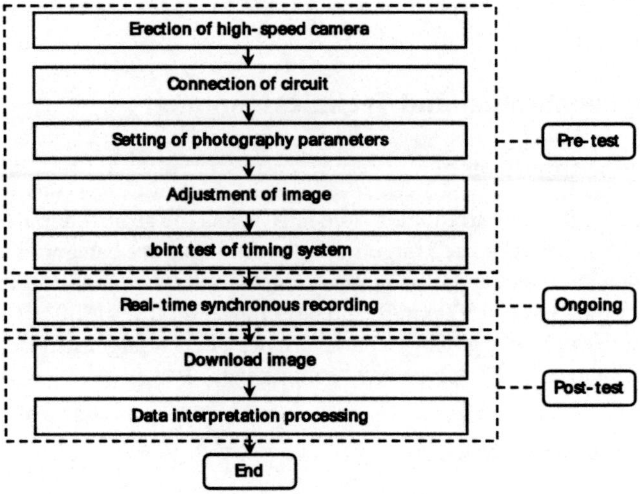

Fig. 3. Optical testing process

3.6 Debugging of Optical Testing System

High-Speed Camera Communication Connection. Configure the TCP/IPv4 option of the computer to set its IP address to be in the same domain as the high-speed camera. Open the PFV control program, the program will automatically search for high-speed camera.

Adjustment of Field of View and Clarity. Based on the laser irradiation area of interest in the experiment, the shooting range is roughly determined by adjusting the lens focal length. The identifiable reference object is placed on the same vertical plane of the target, and the focusing ring of the optical lens is manually focused until the edge of the reference object in the view is clear and sharp.

Determination of Photography Parameters. Under the premise of ensuring that the shooting duration covers the entire process of the experiment, and in response to the specific state of the laser damage verification experiment with the light output time of 10 s and the stable blowing time of 15 s, the high-speed camera shooting frequency is ultimately determined to be 250 fps. To avoid oversaturation of the image, when there is strong light during the shooting process, the aperture and exposure time should be appropriately reduced.

Setting of Trigger Mode. In order to ensure the recording of the entire process images of laser damage verification experiment and leave a certain amount of redundant time to prevent situations that the laser may work in advance, the intermediate trigger mode with 2s reserved before the trigger signal is selected for this experiment.

Calibration and Joint Test. The calibration process refers to placing a scale with a known length at the vertical position of the target installation, and then shooting and saving a still image of the scale. After the trigger mode is set up, timing controller is used to test whether the high-speed camera and the laser are triggered synchronously.

4 Test Implementation and Technical Summary

4.1 Overview of Experiment

In this experiment, the laser irradiation time is 10 s, and the airflow duration with stable wind speed is 15 s. The irradiated targets are divided into three categories: metal target, composite sandwich target and carbon fiber target. Under the combination of different airflow velocity (400 km/h–1300 km/h) and laser power (400 W–1500 W), the evolution process of macroscopic morphology characteristics of each target is observed, and the mechanism of interaction is analyzed and summarized.

4.2 Protection of Image Sensor

There are two kinds of laser interference and damage mechanisms on CMOS, namely soft damage and hard damage [13, 14]. The main phenomena of soft damage include changes in performance such as light saturation, reduced resistivity, reduced breakdown voltage, and loss of responsiveness. Hard damage generally indicates permanent damage to the material or structure of CMOS. In order to protect CMOS from the interference and damage of laser irradiation, optical filters and optical attenuators are used to filter specific wavelengths of laser and reduce the intensity of visible light.

Optical Filter. Optical filter is made by coating an optical substrate, which can pass through the spectrum before a certain wavelength and cut off the spectrum after it. It can effectively eliminate interference light outside the available wavelength, and improve imaging clarity and system sensitivity. Due to the use of near-infrared wavelength laser source in this experiment and the visible light band received by CMOS, the short-wave pass filter of 400 nm–700 nm is selected. On the one hand, the laser beam in the near-infrared band can be cut off to the CMOS through the lens, and on the other hand, clear imaging in the visible band can be achieved. The transmittance wavelength characteristic of optical filter is shown in Fig. 4.

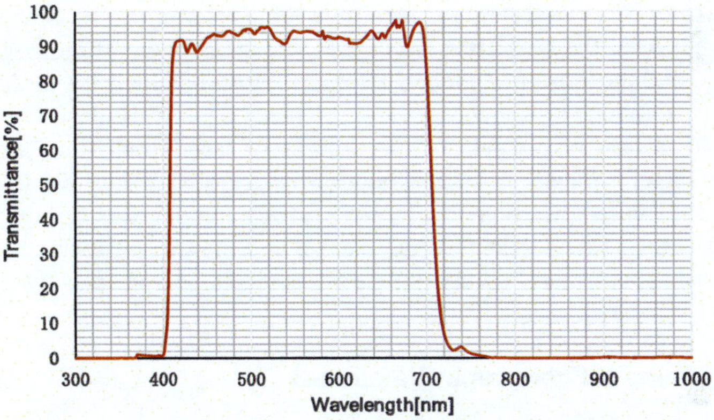

Fig. 4. Transmittance wavelength characteristic of optical filter

Optical Attenuator. The optical attenuator blocks the light intensity by optical density to achieve different transmittance of light. In order to reduce the influence of high intensity visible light on CMOS oversaturation imaging, absorption-type optical attenuator with transmittance of 70% is selected for this experiment. The optical attenuator is made of gray glass, and contains optical absorbing substances that can absorb light in the 400 nm–700 nm wavelength band or a specific wavelength. The transmittance wavelength characteristic of optical attenuator is shown in Fig. 5.

4.3 Morphological Characteristics of Different Targets Under Laser Irradiation

After a series of experimental verification, the damage effects of targets irradiated by laser under high-speed airflow conditions are analyzed, and the typical morphological characteristics of three kinds of targets are summarized.

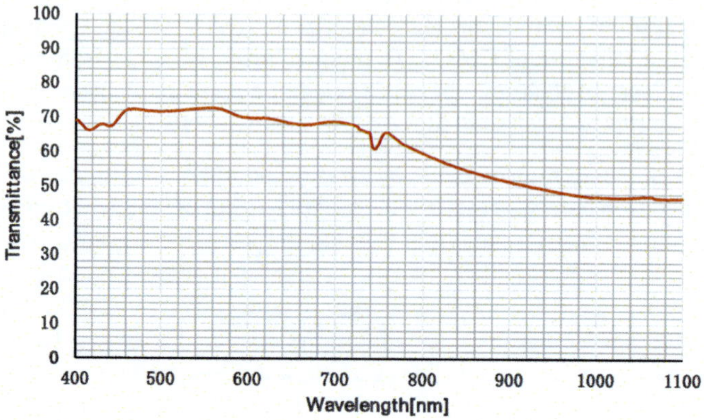

Fig. 5. Transmittance wavelength characteristic of optical attenuator

- **Metal target:** The metal target is mainly in molten form, and no obvious fire light is observed. After the laser stops irradiation, the temperature of the target gradually decreases, and the solidified molten residue is attached along the airflow direction near the perforation.
- **Composite sandwich target:** In the process of laser irradiation, the composite sandwich target is mainly in combustion form and emits obvious fire light.
- **Carbon fiber target:** Carbon fiber target is mainly composed of carbon fiber surface layer and metal matrix. Under the irradiation of high-power density laser, a very bright incandescent spot appears on the surface of the target, and presents the phenomenon of peeling layer by layer.

4.4 Setting Rules of Photography Parameters

On the basis of actual testing practice, based on the principles of filtering out stray light interference and accurately capturing dynamic details to ensure clear imaging during the experimental process, this project summarizes the photography parameter setting rules corresponding to the three targets. The specific content is shown in Table 2. These photography parameters are the best results of continuous exploration and correction during the early experimental process, and are the effective balance between image details and brightness. The summary of photography parameter setting rules can provide technical experience for carrying out the same type of laser verification experiment in the future, and set the photography parameters of high-speed cameras accurately according to the conditions such as target material and laser power.

4.5 Data Processing Method

This laser damage verification experiment uses the proportional conversion method to calculate the perforation size of the target under laser irradiation. The specific perforation size is the diameter of the hole. The origin of the image plane coordinate system *o-xy* is the geometric center of the image, the *x*-axis is parallel to the horizontal sampling direction of the camera, and the *y*-axis is perpendicular to the horizontal sampling direction of

Table 2. Setting rules of photography parameters

Photography parameter	Metal target	Composite sandwich target	Carbon fiber target
Frequency	250 fps	250 fps	250 fps
Resolution	2048 × 1536 pixels	2048 × 1536 pixels	2048 × 1536 pixels
Shutter time	1/3000–1/6000 s	1/4000–1/7000 s	1/367000 s
Aperture value	F5.6–F11	F8–F16	F22

the camera. Due to the horizontal installation of high-speed cameras, the perforation diameter can be obtained by measuring the distance between the upper and lower vertices of the hole. The specific data processing flow is as follows:

- **Interpretation of image coordinates (x, y).** Image interpretation software is used to interpret the image coordinates (x_1, y_1) and (x_2, y_2) of the endpoints on both sides of the reference length in the scale image and the image coordinates (x_3, y_3) and (x_4, y_4) of the upper and lower vertices of the hole in the test image.
- **Calculation of image points distance L.** The specific calculation formulas for the distance between the reference endpoints of the scale L_1 and the distance between the hole vertices L_2 are as follows:

$$\begin{cases} L_1 = \sqrt{(x_1 - x_2)^2 + (y_1 - y_2)^2} \\ L_2 = \sqrt{(x_3 - x_4)^2 + (y_3 - y_4)^2} \end{cases} \quad (1)$$

- **Proportional conversion.** The known reference length of the ruler is S_1. There is an equal proportional relationship between the ratio of the reference length of the scale to the distance of the corresponding image points and the ratio of the hole diameter to the distance of the vertex image points. The actual diameter of the hole S_2 is as follows:

$$S_2 = \frac{L_2}{L_1} \times S_1 \quad (2)$$

5 Prospect for Key Technologies

5.1 Target Image Acquisition Technology Under Flame Occlusion

In the laser damage verification experiment, under the continuous oxygen supply of high-speed airflow, the flame generated on the surface of the target will block the clear imaging of internal erosion. At present, the only way to suppress the interference caused by the flame is to shorten the shutter time and reduce the aperture, and the imaging effect is not good. It can be seen that this photography method based on visible light passive imaging has certain limitations.

In response to the above challenge, research on range-gated laser active imaging technology can be carried out for obtaining target image under flame occlusion. The range-gated laser active imaging technology emits laser signal to the target actively, and controls the gating of CMOS to complete the slice imaging of the target based on the time difference of the reflected optical signal arriving at the receiving system [15]. This active imaging technology can achieve penetration imaging for specific target and greatly improve the perception ability of optoelectronic imaging systems.

5.2 Image Edge Automatic Recognition Technology

Currently, the indicator of perforation size is only represented by calculating a single hole diameter. Strictly speaking, the edge of the hole is not irregular, and this representation is not comprehensive enough.

In view of the above shortcoming, the subsequent research on image edge automatic recognition technology can be carried out to extract the edge of the hole and directly obtain the appearance characteristics and ablation area of the hole. Edge recognition is one of the important technologies in computer vision [16]. Its purpose is to detect and identify a set of pixels with dramatic brightness changes in the image. The correct detection of image edges plays an important role in analyzing the content of the image and realizing the position of object in the image.

6 Conclusion

Based on the laser damage verification test conducted in high-speed wind tunnel, the optical testing method for evaluating laser damage efficiency under high-speed airflow conditions is explored. On the basis of mastering the mechanism and imaging features of laser damage, a set of optical measurement scheme based on high-speed photography method that combines effect observation and quantitative analysis has been designed. Combined with the field distribution of high-speed wind tunnel and the placement state of laser system, the optical testing system composed of high-speed photography system, photographic control system, light source system and other auxiliary equipment is set up in the wind field lateral layout, and the optical testing system is debugging according to the actual test state. The optical testing process is developed, and the damage verification tests for different combinations of airflow velocity and laser power have been completed. The macroscopic damage characteristics of different targets and the corresponding photographic parameter setting rules are summarized, and the key technologies to be improved are forecasted, which provide the testing technology foundation for carrying out the same type laser experiments.

References

1. Han, X.: Research on the evaluation method of damage efficiency of laser-induced optical thin films. Xi'an Technological University, Xi'an (2022)
2. Chen, M.H.: Research on spectrum detection and recognition technology in laser damage. University of Electronic Science and Technology of China, Chengdu (2020)
3. Wu, J.Y., et al.: Experiment and numerical simulation of ablation on aluminum target with high-intensity laser. Trans. Beijing Inst. Technol. **38**(10), 1018–1024 (2018)

4. Lin, X.C.: Thermal and mechanical damage effects of optical lenses irradiated by femtosecond pulsed laser. Shenyang Ligong University, Shenyang (2019)
5. Wu, B.: Experiment and numerical simulation of ablation on aluminum induced by millisecond pulse laser. Beijing Institute of Technology, Beijing (2016)
6. Wei, Y.B.: Evaluation of laser target damage effectiveness. University of Electronic Science and Technology of China, Chengdu (2019).
7. Shen, P.F.: Study on the damage mechanism of multilayer composites under high-energy laser irradiation. Donghua University, Shanghai (2022)
8. Song, N.Q., et al.: Multiple physical modeling for damage mechanism of high energy laser weapon. CIESC J. **67**(S1), 359–365 (2016)
9. Zhao, Y., et al.: Numerical simulation of laser ablation properties of irradiated material. CIESC J. **65**(S1), 426–432 (2014)
10. Hou, J.H., Liu, C., Jing, C.Y.: Analysis of temperature rise of metal targets irradiated by CW laser. Opto-Electr. Eng. **46**(12), 30–37 (2019)
11. Meng, W., et al.: Thermal ablation law of laser irradiation on nylon materials under tangential airflow. Opt. Precis. Eng. **25**(02), 351–357 (2017)
12. Li, C., et al.: Thermal properties of aluminum target irradiated by combined CW/pulsed laser beam. J. Changchun Univ. Sci. Technol. (Natural Science Edition) **41**(02), 1–6 (2018)
13. Wang, X.: Study on laser blindness and damage technology of photoelectric sensors. Xidian University, Xi'an (2018)
14. Liang, Z.Y.: Research on long-range laser blinding CMOS image sensor based on zoom system. Guangdong University of Technology, Guangzhou (2022)
15. Tian, D.K.: 3D image reconstruction based on range-gated laser imaging system. Beijing Institute of Technology, Beijing (2016)
16. Lv, Y.C.: Research on image edge detection and pattern recognition technology. Harbin Institute of Technology, Harbin (2019)

A Systematic Analysis and Exploration of Target Image Acquisition Technology Under Plume Background

Xiang Liu(✉)

Aerospace Life-Support Industries, Ltd., Xiangyang 441003, China
liux.ali@hotmail.com

Abstract. In the integrated ejection test of rocket ejection seat, the working state of the seat at the starting stage is an important examination link. Due to the influence of the seat rocket engine plume, the visible light passive imaging technology based on high-speed photography is difficult to obtain the working details of the target mechanism. Aiming at this technical problem, the feasibility study of using range-gated laser active imaging technology to obtain the target image under the plume background is carried out by analyzing the mechanism of the impact of the seat rocket engine plume on optical imaging. The principle of range-gated laser active imaging technology is described in detail, and key technologies such as laser active illumination, range-gated slice imaging, optical signal filtering, and multi-information image fusion are explored. The implementation approach of image acquisition technology is planned to provide research ideas and directions for subsequent technology realization.

Keywords: Rocket ejection seat · Rocket engine plume · Range-gated technology · Laser active imaging

1 Introduction

As the last safety guarantee for pilots, rocket ejection seat is an aviation life-saving equipment that can eject pilots from the aircraft and make them land safely in an emergency [1]. The ejection exit stage refers to the movement process from the pilot pulling the central pull ring to launch the ejection seat to the human-seat system leaving the cockpit. In a relatively short period of time, the ejection seat has multiple control and power mechanisms to complete the procedures of occupant restraint, propulsive ascent, attitude adjustment, etc. The normal operation of the seat mechanism at the exit stage is the prerequisite to improve the life-saving performance [2].

In the process of integrated ejection test, close-range photography and image recording of the working state of the human-seat system at the exit stage are usually carried out by high-speed cameras mounted on multiple positions, which provides a basis for the performance verification and improvement of the ejection seat. A large amount of flame and smoke are generated by the ejection seat rocket engine during the propulsion

Chinese Society of Aeronautics and Astronautics (Ed.): CASTC 2023, LNME, pp. 664–672, 2024.
https://doi.org/10.1007/978-981-99-8861-7_65

process (see Fig. 1). Due to the oversaturation of the image caused by strong light interference and the attenuation effect of the plume on the visible light, the imaging effect of the high-speed camera is not good, so that the working state of the seat mechanism under the plume background cannot be accurately judged. In view of the test scenario and technical requirements, it is necessary to explore the acquisition technology of clear target image under the plume background, and carry out the feasibility study of using range-gated laser active imaging technology to solve this technical problem.

Fig. 1. Scenario of integrated ejection test

2 Mechanism of Optical Imaging Affected by Plume

2.1 Principle of Light Attenuation Under Plume Background

The propellant of rocket engine produces a large amount of flame and smoke during combustion. The high-speed and high-temperature gas flow ejected from the nozzle is feathery, which is called rocket engine plume [3]. Plume is an artificial aerosol system formed by interaction of solid particles, liquid particles and air, which is mainly divided into primary smog and secondary smog. The primary smog is mainly composed of condensed particles in the plume produced by propellant combustion, and a small part comes from the ablation of nozzle materials and the cracking of coating layer. The secondary smog refers to the liquid mist formed by a large amount of condensable gas contained in the plume under thermodynamic conditions such as ambient temperature and humidity [4–6]. When the beam is incident into the plume, the absorption and scattering effect of the plume greatly attenuates the beam energy [7, 8].

The optical transmittance of the plume is an important parameter to measure the attenuation ability of the plume to the beam energy, which is expressed as the ratio of the intensity of the outgoing light passing through the plume to the intensity of the incident light. The basic principle of plume optical transmittance measurement is Lambert-Beer Law [9], which describes the relationship between the absorption intensity of matter to monochromatic light and the thickness and concentration of matter. The formula is as follows:

$$A = \log_{10} \frac{I_0}{I_t} = \log_{10} \frac{1}{T} = k \times l \times c \tag{1}$$

In the formula: A is the absorbance, I_0 is the incident light intensity, I_t is the outgoing light intensity, T is the medium transmittance, k is the medium absorption coefficient, l is the optical path length, c is the medium concentration.

According to Lambert-Beer Law, when the plume is used as a medium, the optical transmittance is inversely proportional to the optical path length and smog concentration.

2.2 Limitations of Visible Light Passive Imaging

At present, the conventional imaging method is mainly visible light passive imaging, that is, the image sensor (such as CCD, CMOS, etc.) receives the reflected light or self-radiation light of the target to achieve target detection and recognition. Visible light passive imaging is highly dependent on the environment, and its detection ability is greatly limited and interfered by the background radiation environment. The useful return signal may be submerged by the interference light and background light, and it is difficult to detect the target [10–12]. Aiming at the situation of strong flame background and smog environment caused by seat rocket engine plume in the integrated ejection test, it will bring optical observation difficulties in the case of low signal-to-noise ratio, which poses a great challenge to the normal use, efficiency and performance evaluation of optical equipment. The contrast between passive imaging and laser active imaging is shown in Fig. 2.

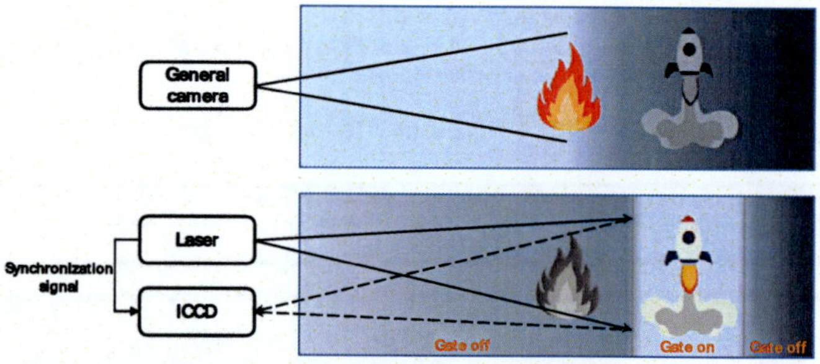

Fig. 2. Comparison between passive imaging and laser active imaging

2.3 Advantages of Range-Gated Laser Active Imaging Technology

The range-gated laser active imaging technology transmits the laser signal to the target actively, and according to the difference in the receiving time of the reflected optical signal, the gating of the image sensor is controlled to complete the slice imaging of the target, so as to achieve the purpose of detecting and identifying the target [13, 14]. The active imaging technology can not only obtain the intensity image containing the physical characteristics of the target, but also obtain the range image containing the position information of the target surface [15]. Range-gated laser active imaging technology has the following unique advantages:

- The laser with high brightness, high directivity and high coherence is used as the illumination source, and the anti-interference ability is strong.
- Target imaging can not only obtain the contour information of the target, but also obtain the three-dimensional distance information of the target, which is conducive to the detection, recognition and tracking of the target.
- When the target is in a smog barrier or low illumination environment, the laser active imaging technology can effectively suppress the backscattering caused by smog particles, and its penetration ability is strong.

Based on the above advantages, the range-gated laser active imaging technology realizes the penetration imaging of specific targets and greatly improves the sensing ability of the photoelectric imaging system.

3 Principle of Range-Gated Laser Active Imaging Technology

The range-gated laser active imaging technology sets the gating time of the gating imager to be consistent with the arrival time of the laser echo pulse, so as to image the target at a specific distance [16–18]. The working principle is shown in Fig. 3.

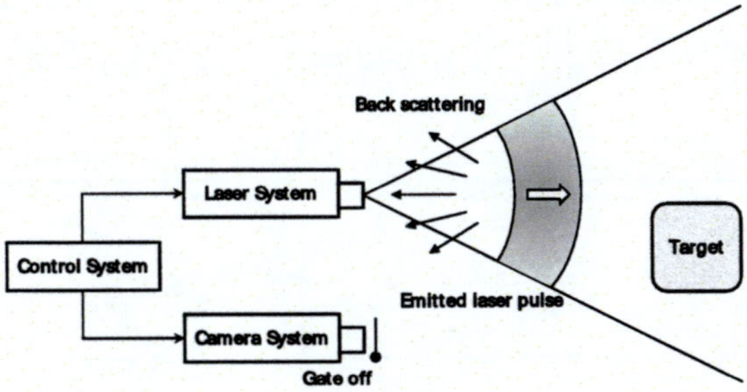

(a) Gate off

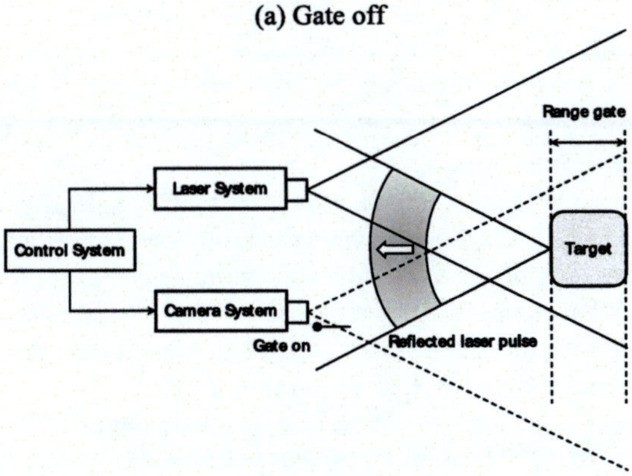

(b) Gate on

Fig. 3. Schematic diagram of range-gated laser active imaging technology

The pulse laser and the gating imager are synchronized by a control circuit. At time of t_0, the laser device emits a laser pulse, and the gating imager is turned off. The gating imager remains closed until the laser pulse is reflected back to its position (at time of t_1), so that the backscattered light generated between time of t_0 and t_1 cannot enter the gated imager to form noise. When the reflected pulse laser reaches the gate imager (t_1), the gate imager receives the returned pulse laser and performs imaging. If the gating time length and laser pulse width are very narrow, the imaging system only receives the reflected light signal near the target, and most of the backscattered light is eliminated, which can greatly improve the signal-to-noise ratio of the imaging system [19, 20]. The acquisition process of laser two-dimensional intensity image is shown in Fig. 4, and the specific steps are as follows:

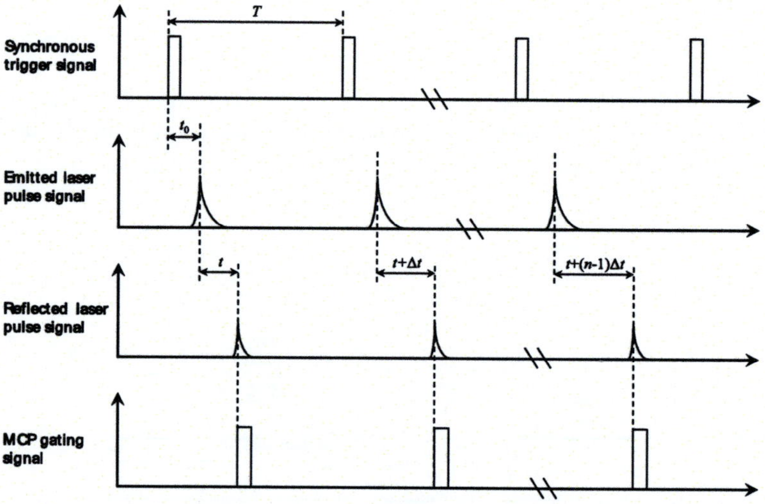

Fig. 4. Distance gating sequence diagram

1. The laser is triggered from the first rising edge of the synchronous trigger signal, and the laser pulse is emitted after t_0.
2. This pulse is loaded onto the microchannel plate (MCP) gate after delaying ($t_0 + t$), so that intensified CCD (ICCD) can detect the laser echo at the distance of $R_1 = ct/2$, and obtain the intensity image of the first slice with the corresponding distance of R_1.
3. When the next pulse comes, the trigger pulse reaches the laser at the same time. At this time, the pulse signal is delayed ($t + \Delta t$) and loaded onto the MCP gate. ICCD will image the target at a distance of $R_2 = c(t + \Delta t)/2$.
4. Repeat this process to get a series of two-dimensional images. This process continues to the set distance of R_n. The number of slices obtained is as follows:

$$n = \frac{2(R_n - R_1)}{c\Delta t} \tag{2}$$

5. Three-dimensional images are synthesized from two-dimensional intensity images. The two-dimensional intensity slices are arranged and numbered in sequence, then the corresponding distance of the two-dimensional image of i is as follows:

$$R_i = \frac{c(t + i\Delta t)}{2n} \tag{3}$$

6. The obtained slice images are synthesized by centroid algorithm, and the target fusion image can be obtained.

According to the above model principle and specific test scenarios, each parameter can be calculated and simulated to provide reference for subsequent system design and technical implementation.

4 Key Technologies and Implementation Approaches

Combined with the reality that the plume of the seat rocket engine interferes with visible light imaging, resulting in poor spatial resolution and loss of picture details, the research method of obtaining clear images of the target under the plume background is explored by using range-gated laser active imaging technology. The following key technologies are summarized to form a complete research route, which provides research ideas and directions for the realization of target image acquisition technology under plume background.

4.1 Laser Active Illumination Technology

The pulse laser active illumination technology suitable for plume background is explored to realize the adaptive adjustment of laser illumination under different plume backgrounds. Combined with the uniform light emission system, the target and region to be observed are illuminated by laser, which provides a suitable light source system for the realization of range-gated imaging [21].

The research contents include the selection of laser light source, laser uniform shaping technology, laser adjustable beam expansion emission system, etc. The aim of this study is to design narrow-linewidth ultrashort pulse laser light source and laser emission system suitable for seat ejection scene. Laser has strong coherence, and the intensity distribution of the beam cross-section is uneven and presents a Gaussian distribution. In the field of laser illumination imaging, the non-uniform distribution of energy will lead to the degradation of image quality. The non-uniformity of the intensity distribution of laser beam limits its application, and it is necessary to transform laser beam into flat-topped beam with uniform intensity distribution to eliminate adverse effects. The diffraction microlens array is used to precisely adjust the wavefront phase of the incident laser to realize the modulation and transformation of the light wave. After passing through each diffraction unit, the laser undergoes diffraction and generates interference at a certain distance to form a specific light intensity distribution. Due to the influence of surface roughness of the object and particle scatterers generated by the plume, the speckle phenomenon of laser imaging will be caused. The application of optical waveguide technology can improve the beam energy distribution and effectively reduce the impact of speckle.

4.2 Range-Gated Slice Imaging Technology

Under the condition of laser active illumination, the range-gated slice imaging technology is studied to overcome the strong light interference and the attenuation effect of the plume, so as to achieve the gated slice imaging of the ejection seat within the range of the ejection path.

The range-gated laser active imaging system realizes the discrimination of the reflected light of the target and the backscattered light of the environment by using the time sequence of reaching the detection surface of the camera. By controlling the switching time of the gate, the target echo is detected at the time when the gate is open, and the backscattered light arriving at other times is rejected, so that only the target of a specific distance can be imaged [22]. According to the environmental conditions of strong scattering, strong background and strong exposure, the theoretical calculation is carried out first. Based on the analysis of receiver indicators, the transfer process between various parameters of the imaging system is derived, and the relationship between each parameter and imaging accuracy is determined. The relative relationship between parameters such as target distance, plume distance, laser pulse width, laser repetition frequency, synchronization timing accuracy, gating time, camera dynamic gain threshold and parameter optimization in fixed scenes are mainly studied.

4.3 Optical Signal Filtering Technology

Combining the optical design of the receiving system with the band-pass interference filter and the neutral density attenuator, the signal-to-noise ratio and imaging quality of the system are further improved by optical filtering and attenuation.

Considering the actual scene of the cockpit and the relative position of the plume and the seat, the influence of most plume backgrounds can be eliminated based on distance gating. However, within the range of the gate distance, there is still interference from plume stray light inevitably. The plume has the optical radiation characteristics such as gas radiation with discrete spectrum and solid radiation with continuous spectrum, and its spectral component is mainly infrared spectrum. Since the wavelength of the active illumination source and the background interference light are different, the band-pass filter can be used to ensure the passage of the specific wavelength of the laser reflected by the target and block the background interference light.

4.4 Multi-information Image Fusion Technology

The multi-information image fusion technology based on polarization modulation is explored. By realizing the illumination imaging of the target with light sources in different polarization directions, the contrast and the detail information of the target image is improved. At the same time, the multi-dimensional information of the target can be obtained by using the slice images and polarization images of different distances, and the multi-information fusion enhanced imaging technology can be developed.

With the development of laser technology, more and more attention has been paid to the polarization characteristics of light in modern target detection and recognition technology. Different substances or the same substance in different states can produce different polarization states and form a polarization spectrum, which is closely related to wavelength and emission system [23]. In practical application, polarization imaging can effectively reduce the impact of adverse factors such as smog, remove image noise, and improve imaging contrast. In the ejection test, the polarization characteristics of the reflected light from the target are different from those of the plume background. In laser active imaging, the parameters of the laser optical emission system and the camera optical receiving system can be controlled to obtain the multi-angle polarization image of the target. The main purpose of studying the fusion algorithm of polarization image and slice image is to take the advantages of both, so that the target image has more dimensional information on the basis of good visual display.

5 Conclusion

The range-gated laser active imaging technology uses laser as the carrier of information, which can be combined with computational imaging and digital image processing technology to present various information of targets in complex environment in the form of images. Based on the above technical analysis and feasibility study, the range-gated laser active imaging technology is introduced into the optical test field of rocket ejection seat as a technical way to obtain the image of seat working characteristics under the background of rocket engine plume. The imaging distance is locked within the range of the seat ejection path by the fast gating scheme, and the echo signal of the remaining distance is shielded to avoid the interference of the rocket engine plume on the imaging. Combined with ultra-narrow-band attenuation filtering and polarization information fusion technology, clear images of the target under the plume background can be obtained to meet the test requirements.

References

1. Jiang, W.: The whole trajectory simulation research of ejection seat and occupant system with Matlab. Nanjing University of Aeronautics and Astronautics, Nanjing (2012)
2. Zhou, H., et al.: Research on ejection attitude of ejection seat based on ADAMS. Aeronaut. Sci. Technol. **32**(02), 66–73 (2021)
3. Chen, X.: Study on the measurement method of light transmittance of engine plume. Xidian University, Xi'an (2020)
4. Liang, M.: Research on true temperature measurement technique of solid propellant rocket engine flame. Harbin Institute of Technology, Harbin (2017)
5. Zhang, B.: The signal detecting and processing in transmissivity dynamic measuring of the plume penetrated by laser. Huazhong University of Science and Technology, Wuhan (2005)
6. Su, R.: Research on the measurement system of smoke plume transmittance. Xidian University, Xi'an (2015)
7. Liu, X.C., Gao, T.C., Liu, Z.T.: Effect of atmospheric aerosols on laser transmission attenuation. J. Atmos. Environ. Opt. **7**(3), 181–190 (2012)
8. Sun, X.L., Zhang, X.J.: Study on attenuation characteristics of pulse laser induced by smoke and dust. Technol. Innov. Appl. **2015**(10), 15–16 (2015)
9. Wu, J.: Research on measurement method of plume light transmittance. Xidian University, Xi'an (2015)

10. Wei, H.P.: Research on laser three dimensional range-gated super-resolution imaging. Changchun University of Technology, Changchun (2019)
11. Li, Z.W.: Research in optoelectronic observation technologies for space objects. Changchun Institute of Optics, Fine Mechanics and Physics, Chinese Academy of Sciences, Changchun (2014)
12. Gu, Q.: Cooled infra-red and low-light-level mixed reconnaissance technique - long distance low-light-level imaging system. Beijing Institute of Technology, Beijing (2015)
13. Ma, Y., Hou, Y.H., Wei, H.P.: The principle and application of laser range-gated three-dimensional imaging technology. Sci. Technol. Innov. **2017**(22), 152–153 (2017)
14. Fan, Y.C., et al.: Research on range gated active laser imaging system for moving target. Laser Infrared **45**(12), 1427–1432 (2015)
15. Wang, N.: Experimental research on dynamic range-gated laser imaging. Harbin Institute of Technology, Harbin (2011)
16. Christnacher, F., et al.: Influence of gating and of the gate shape on the penetration capacity of range-gated active imaging in scattering environments. Opt. Express **23**(26), 32897–32908 (2015)
17. Mariani, P., et al.: Range-gated imaging system for underwater monitoring in ocean environment. Sustainability **11**(1), 1–13 (2018)
18. Tian, D.K.: 3D image reconstruction based on range-gated laser imaging system. Beijing Institute of Technology, Beijing (2016)
19. Ma, Y.: Research on laser range-gated super-resolution 3D imaging technology based on FPGA. Changchun University of Technology, Changchun (2019)
20. Wang, B.N., Gao, Y.Y.: Three dimensional image reconstruction based on range-gated laser imaging system. Laser J. **40**(06), 50–54 (2019)
21. Liu, Y.: Experimental study based on the range-gated laser imaging. Harbin Institute of Technology, Harbin (2010)
22. Fan, Y.C., et al.: Zero signal measurement method based on laser active imaging combined with laser range-gated technology. Infrared Laser Eng. **45**(03), 111–117 (2016)
23. Xie, M.L.: Research on polarization imaging technology of laser active illumination and image fusion algorithm. Xi'an Institute of Optics and Precision Mechanics, Chinese Academy of Sciences, Xi'an (2019)

Feedforward Compensation Enhanced Disturbance Rejection Control for a Spin-Stabilized Guided Projectile Subject to Actuator Rate Limit

Minnan Piao[1], Biao Lu[2(✉)], Yonghui Xie[3], Hailin Zhang[4], Shuyu Zhang[4], and Liangwu Qu[4]

[1] College of Computer Science and Technology, Civil Aviation University of China, Tianjin 300300, China
[2] College of Artificial Intelligence, Nankai University, Tianjin 300350, China
lubiao@mail.nankai.edu.cn
[3] Intelligent Manufacturing Department, CISDI Information Technology Co., Ltd., Chongqing 401147, China
[4] Engineering Department, Eastern Airlines Technic Co., Ltd., Shanghai 200000, China

Abstract. This article addresses the control problem for a spin-stabilized guided projectile which is subject to strong couplings, disturbances, and actuator rate limit. The internal uncertainties and external disturbances are lumped as a state, which is estimated by a reduced-order extended state observer. To enhance the decoupling performance, the dynamics related to the inertia, aerodynamic, and manipulation couplings are compensated by feedforward. Further, to alleviate the adverse effect of the actuator rate limit on the control performance, the describing function method is employed to design the feedforward rudder deflection command. Stability margin tester method is proposed to facilitate the tuning process. Finally, nonlinear 6-DOF simulations are performed to validate the effectiveness of the proposed control method.

Keywords: Spin-stabilized guided projectile · Attitude control · Strong couplings · Disturbance rejection · Actuator rate limit

1 Introduction

Traditional projectiles have advantages of low cost and easiness of use, which can be applied in large-scale to suppress the enemy firepower. However, due to the absence of the guidance and control system, traditional projectiles have large spread, low attack accuracy and poor damage effect, which can no longer meet the requirements of modern warfare. To better resolve the contradiction between expense and precision, the Spin-Stabilized Guided Projectile (SSGP) with simple control devices is developed.

The SSGP usually employs the canard configuration. The tail fins generate the aerodynamic forces that can maintain the spinning around the symmetry axis during the

Chinese Society of Aeronautics and Astronautics (Ed.): CASTC 2023, LNME, pp. 673–684, 2024.
https://doi.org/10.1007/978-981-99-8861-7_66

uncontrolled phase. When the SSGP approaches the apogee, the canards are deployed to control the trajectory. The rapid spinning provides the condition for simultaneously controlling pitch and yaw channels with only one pair of canards. However, the control authority generated by one pair of canards is relatively small, which will limit the maneuverability and disturbance rejection capability [1]. Thus, two pairs of canards are employed in this paper. In some studies, a dual-spin configuration is utilized to avoid the canards spinning with the projectile and simplify the control design. However, development of the dual-spin configuration is expensive. Therefore, we consider the case where the canards spin together with the body and the control problem is more difficult.

Different from non-spinning missiles, the SSGP keeps spinning all the time, which can cause strong and complex cross couplings between pitch and yaw channels. The inertia coupling caused by the gyroscopic effect, the aerodynamic coupling induced by the Magnus effect, and the manipulation coupling related to the rudder delay are the dominant coupling dynamics. Furthermore, the SSGP is also subject to many internal uncertainties and external disturbances. Since the frequencies of rudder deflection commands are equal to the high spinning rate (the average may be 10 Hz), actuator rate limit may cause severe attitude limit cycle oscillations and even instability. Therefore, actuator rate limit must be taken into account during controller design.

With regards to the above problems, researchers have performed many investigations. A unique guidance and control strategy that enables precision and trajectory shaping with reduced sensor and actuator needs was proposed in [2], whereas couplings were not considered in the controller design. In [3, 4], stability boundary analyses for the coining motion were performed for the attitude autopilot and the autopilot with only a rate loop, respectively. Further, more complex pitch/yaw three-loop autopilot was designed for the SSGP, and the sufficient and necessary condition for the coning motion stability was derived [5, 6]. In [7], the seeker disturbance rejection rate parasitical loop effect was considered in the stability analysis for the two-loop autopilot. Since the actuator bandwidth has a significant effect on the spinning stability, the rudder dynamics was added to the analysis for the two-loop autopilot [8]. Reference [9] further considered the actuator backlash for the autopilot with only a rate loop. The control methods employed in the above studies are all linear. To enhance the control performance, nonlinear control methods were developed. In [10], the dynamic inversion method which could achieve fast design was utilized. However, disturbances were not considered. To improve the robustness, the Extended State Observer (ESO) combined with the integral sliding mode control was proposed in [11] to deal with the cross-coupling dynamics, external disturbances, and parametric uncertainties. Note that although many good results were obtained for the SSGP control, actuator rate limit has rarely been investigated. Therefore, in addition to the couplings and disturbances, we also concentrate on the effect and compensation of the actuator rate limit in this paper, and a feedforward compensation enhanced Active Disturbance Rejection Control (ADRC) method is proposed.

The remainder of this paper is organized as follows. In Sect. 2, the complete model is introduced and the problem is formulated. In Sect. 3, the disturbance rejection control with feedforward compensation is presented. The tuning method is described in Sect. 4. Simulations are performed in Sect. 5. In Sect. 6, we give the concluding remarks.

2 Problem Formulation

2.1 Model Description

Kinematic and Dynamic Model. The complete 6-DOF model of a SSGP [12] is

$$\begin{cases} m\dot{V} = P\cos\alpha\cos\beta - X - mg\sin\theta \\ mV\dot{\theta} = P\left(\sin\alpha\cos\gamma_V + \cos\alpha\sin\beta\sin\gamma_V\right) + Y\cos\gamma_V - Z\sin\gamma_V - mg\cos\theta \\ -mV\cos\theta\dot{\psi}_V = P\left(\sin\alpha\sin\gamma_V - \cos\alpha\sin\beta\cos\gamma_V\right) + Y\sin\gamma_V + Z\cos\gamma_V \end{cases} \tag{1}$$

$$\begin{cases} J_x\dot{\omega}_x + \left(J_z - J_y\right)\omega_z\omega_y = M_x \\ J_y\dot{\omega}_y + (J_x - J_z)\omega_x\omega_z + J_z\omega_z\dot{\gamma} = M_y \\ J_z\dot{\omega}_z + \left(J_y - J_x\right)\omega_y\omega_x - J_y\omega_y\dot{\gamma} = M_z \\ \dot{\vartheta} = \omega_z,\ \dot{\psi} = \omega_y / \cos\vartheta,\ \dot{\gamma} = \omega_x - \omega_y\tan\vartheta \end{cases} \tag{2}$$

$$\alpha = \vartheta - \arcsin\left(\sin\theta / \cos\beta\right),\ \sin\beta = \cos\theta\sin(\psi - \psi_V),\ \sin\gamma_V = \tan\beta\tan\theta \tag{3}$$

where m, V, θ, and ψ_V represent the mass, velocity, flight path angle, and course angle, respectively; γ, ψ, and ϑ denote the roll, yaw, and pitch angles, respectively; α, β, and γ_V are the quasi angle of attack, quasi sideslip angle, and quasi bank angle, respectively; ω_i and J_i $(i = x, y, z)$ are the angular rate and moment of inertia of the roll, yaw, and pitch channels, respectively; P, X, Y, Z, and M_i are the thrust, drag, lift, side force, and moment. The pitch and yaw moments can be expressed as

$$\begin{cases} M_y = m_y^{\omega_y}\omega_y + m_y^{\beta}\beta + m_y^{\omega_x}\omega_x + m_y^{\delta_{\text{yaw}}}\delta_{\text{yaw}} \\ M_z = m_z^{\omega_z}\omega_z + m_z^{\alpha}\alpha + m_z^{\omega_x}\omega_x + m_z^{\delta_{\text{pitch}}}\delta_{\text{pitch}} \end{cases} \tag{4}$$

where $m_y^{\omega_y}\omega_y$ and $m_z^{\omega_z}\omega_z$ are the damping moments; $m_y^{\beta}\beta$ and $m_z^{\alpha}\alpha$ are the static stability moments; $m_y^{\omega_x}\omega_x$ and $m_z^{\omega_x}\omega_x$ are the Magnus moments; δ_{yaw} and δ_{pitch} are the mathematical rudder defections; $m_y^{\delta_{\text{yaw}}}\delta_{\text{yaw}}$ and $m_z^{\delta_{\text{pitch}}}\delta_{\text{pitch}}$ are the manipulation moments.

Rudder Deflection Composition/Decomposition. As shown in Fig. 1, F_{y1} and F_{z1} denote the control forces under the body coordinate system generated by the physical rudder deflections δ_y and δ_z, respectively, which can be written as $F_{y1} = k\delta_y$ and $F_{z1} = k\delta_z$ $(k > 0)$. Then, F_{y1} and F_{z1} are transformed into the quasi-body coordinate system as

$$F_{y4} = F_{y1}\cos\gamma - F_{z1}\sin\gamma,\ F_{z4} = F_{y1}\sin\gamma + F_{z1}\cos\gamma \tag{5}$$

According to $F_{y4} = k\delta_{\text{yaw}}$ and $F_{z4} = k\delta_{\text{pitch}}$, we can obtain

$$\delta_{\text{yaw}} = \delta_y\cos\gamma - \delta_z\sin\gamma,\ \delta_{\text{pitch}} = \delta_y\sin\gamma + \delta_z\cos\gamma \tag{6}$$

which can be rearranged into

$$\delta_{\text{yaw}} = \delta\sin(\phi - \gamma),\ \delta_{\text{pitch}} = \delta\cos(\phi - \gamma) \tag{7}$$

where

$$\delta = \sqrt{\delta_{\text{yaw}}^2 + \delta_{\text{pitch}}^2}, \sin(\phi - \gamma) = \delta_{\text{yaw}}/\delta, \cos(\phi - \gamma) = \delta_{\text{pitch}}/\delta \tag{8}$$

Thus, ϕ can be calculated by

$$\phi = \begin{cases} \gamma + \arcsin(\delta_{\text{yaw}}/\delta), & \delta_{\text{pitch}} > 0 \\ \gamma + \pi - \arcsin(\delta_{\text{yaw}}/\delta), & \text{else} \end{cases} \tag{9}$$

wherein $\arcsin(\delta_{\text{yaw}}/\delta) \in [-\pi/2, \pi/2]$. Then, the physical rudder deflections can be derived based on the roll angle and the mathematical rudder deflections as

$$\delta_y = \delta \sin\phi, \delta_z = \delta \cos\phi \tag{10}$$

Subsequently, the transformations in (6) and (10) are called rudder deflection composition and decomposition, respectively.

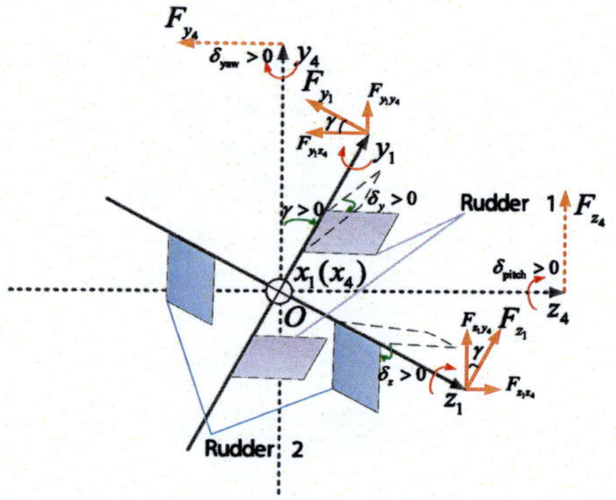

Fig. 1. Physical and mathematical rudder defections

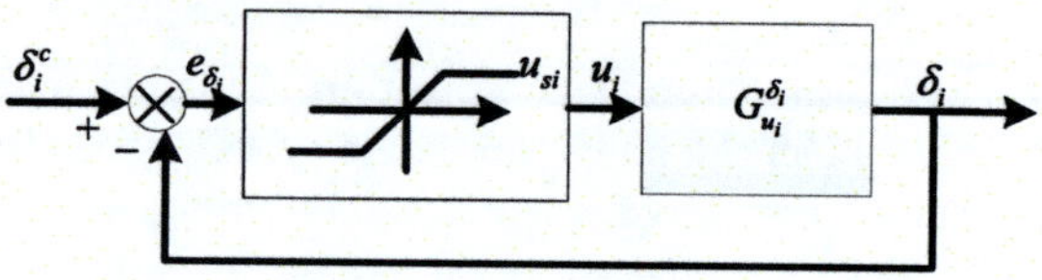

Fig. 2. Nonlinear rudder model with rate limit

Physical Rudder Model. There exists rate limit for a practical rudder system. Based on experimental results, the rudder with rate limit can be approximately modeled as Fig. 2, wherein δ_i^c $(i = y, z)$ is the command of δ_i, $e_{\delta_i} = \delta_i^c - \delta_i$ is the tracking error, u_{si} is the saturation amplitude, and u_i is the input of the following linear part:

$$G_{u_i}^{\delta_i}(s) = b_i/(s(s + a_i)) \tag{11}$$

wherein a_i and b_i are the time constant and the steady gain, respectively.

2.2 Problem Statement

This paper concerns the control phase of the SSGP. The control variables for the pitch and yaw channels are α and β, respectively. The roll channel is not controlled and the projectile keeps spinning freely. α is controlled to achieve the maximal lift-drag ratio, and β is controlled to be zero to reduce the cross coupling. The objective is to realize high-precision and robust control of α and β in spite of the strong couplings, disturbances, and actuator rate limit.

3 Controller Design

3.1 Baseline Controller

The dynamic equations for α and β are

$$\dot{\alpha} = \omega_z + d_z,\ \dot{\beta} = \omega_y + d_y \tag{12}$$

wherein

$$\begin{cases} d_z = -\omega_x \cos\alpha \tan\beta + \omega_y \sin\alpha \tan\beta + (mg\cos\theta\cos\gamma_v - Y - P\sin\alpha)\big/ mV\cos\beta \\ d_y = \omega_x \sin\alpha + \omega_y(\cos\alpha - 1) + (mg\cos\theta\sin\gamma_v + Z - P\cos\alpha\sin\beta)\big/ mV \end{cases} \tag{13}$$

Similarly, the dynamic equations for ω_z and ω_y can be written as

$$\dot{\omega}_z = a_z^{\delta_z}\delta_{\text{pitch}}^c + f_z,\ \dot{\omega}_y = a_y^{\delta_y}\delta_{\text{yaw}}^c + f_y \tag{14}$$

wherein $a_z^{\delta_z} = m_z^{\delta_{\text{pitch}}}\big/ J_z$ and $a_y^{\delta_y} = m_y^{\delta_{\text{yaw}}}\big/ J_y$ are the control gains; δ_{pitch}^c and δ_{yaw}^c are the mathematical rudder deflection commands; f_z and f_y are the lumped disturbances

$$\begin{cases} f_z = \left(-(J_y - J_x)\omega_y\omega_x + J_y\omega_y\dot{\gamma} + m_z^{\omega_z}\omega_z + m_z^{\alpha}\alpha + m_z^{\omega_x}\omega_x + m_z^{\delta_{\text{pitch}}}\left(\delta_{\text{pitch}} - \delta_{\text{pitch}}^c\right)\right)\Big/ J_z \\ f_y = \left(-(J_x - J_z)\omega_x\omega_z - J_z\omega_z\dot{\gamma} + m_y^{\omega_y}\omega_y + m_y^{\beta}\beta + m_y^{\omega_x}\omega_x + m_y^{\delta_{\text{yaw}}}\left(\delta_{\text{yaw}} - \delta_{\text{yaw}}^c\right)\right)\Big/ J_y \end{cases} \tag{15}$$

Then, taking the time derivative of (12) yields

$$\ddot{\alpha} = a_z^{\delta_z}\delta_{\text{pitch}}^c + f_z + \dot{d}_z,\ \ddot{\beta} = a_y^{\delta_y}\delta_{\text{yaw}}^c + f_y + \dot{d}_y \tag{16}$$

ω_z and ω_y can be directly measured in practice, and they are usually used for inner-loop feedback to enhance the damping ratio of the plant. In this way, (16) becomes

$$\ddot{\alpha} = a_z^{\delta_z}\left(\delta_{\text{pitch}}^v - K_\alpha^d\omega_z\right) + f_z + \dot{d}_z,\ \ddot{\beta} = a_y^{\delta_y}\left(\delta_{\text{yaw}}^v - K_\beta^d\omega_y\right) + f_y + \dot{d}_y \tag{17}$$

wherein K_α^d and K_β^d are the feedback gains of angular rates. After damping enhancement, the dynamics from the virtual control signals (δ_{pitch}^v and δ_{yaw}^v) to α and β can be approximated by first-order plants. Therefore, we can have

$$\dot{\alpha} = b_0^z\delta_{\text{pitch}}^v + t_\alpha,\ \dot{\beta} = b_0^y\delta_{\text{yaw}}^v + t_\beta \tag{18}$$

wherein b_0^z and b_0^y denote the equivalent control gains; t_α and t_β are the total disturbances incorporating the modeling uncertainty, coupling effect, and order reduction error. To compensate the disturbances, the Reduced-Order Extended State Observer (ROESO) [13] is designed:

$$\begin{cases} \dot{w}_\alpha = -l_\alpha^2\alpha - l_\alpha w_\alpha - l_\alpha b_0^z \delta_{\text{pitch}}^v \\ z_\alpha = w_\alpha + l_\alpha \alpha \end{cases}, \begin{cases} \dot{w}_\beta = -l_\beta^2\beta - l_\beta w_\beta - l_\beta b_0^y \delta_{\text{yaw}}^v \\ z_\beta = w_\beta + l_\beta \beta \end{cases} \quad (19)$$

wherein w_α and w_β are the intermediate variables; z_α and z_β are the estimates of t_α and t_β; l_α and l_β are the observer gains. Then, the control law can be established as

$$\delta_{\text{pitch}}^v = (K_\alpha(\alpha_c - \alpha) - z_\alpha)/b_0^z,\ \delta_{\text{yaw}}^v = \left(K_\beta(\beta_c - \beta) - z_\beta\right)/b_0^y \quad (20)$$

wherein α_c and β_c are the commands of α and β; K_α and K_β are the control gains. The final control law can be derived as

$$\delta_{\text{pitch}}^c = (K_\alpha(\alpha_c - \alpha) - z_\alpha)\Big/ b_0^z - K_\alpha^d \omega_z,\ \delta_{\text{yaw}}^c = \left(K_\beta(\beta_c - \beta) - z_\beta\right)/b_0^y - K_\beta^d \omega_y \quad (21)$$

3.2 Decoupling Strategy

The main couplings of a SSGP include the inertia coupling caused by the gyroscopic effect, the aerodynamic coupling caused by the Magnus effect, and the manipulation coupling caused by the rudder delay. The inertia and aerodynamic couplings are handled by designing

$$\begin{cases} \delta_{\text{pitch}}^c = (K_\alpha(\alpha_c - \alpha) - z_\alpha)/b_0^z - K_\alpha^d \omega_z + \\ \qquad ((J_y - J_x)\omega_x\omega_y - J_y\omega_y(\omega_x - \omega_y \tan\vartheta) - m_z^{\omega_x}\omega_x)/\left(J_z a_z^{\delta_z}\right) \\ \delta_{\text{yaw}}^c = (K_\beta(\beta_c - \beta) - z_\beta)/b_0^y - K_\beta^d \omega_y + \\ \qquad \left((J_x - J_z)\omega_x\omega_z + J_z\omega_z(\omega_x - \omega_y \tan\vartheta) - m_y^{\omega_x}\omega_x\right)/\left(J_y a_y^{\delta_y}\right) \end{cases} \quad (22)$$

Further, we consider the manipulation coupling. Without consideration of the rate limit, the rudder dynamics can be established as

$$\ddot{\delta}_y = -2\xi\omega_n\dot{\delta}_y - \omega_n^2\delta_y + \omega_n^2\delta_y^c,\ \ddot{\delta}_z = -2\xi\omega_n\dot{\delta}_z - \omega_n^2\delta_z + \omega_n^2\delta_z^c \quad (23)$$

wherein ω_n and ξ are the frequency and the damping ratio. We have the following relations:

$$\begin{cases} \delta_{\text{pitch}} = \delta_y \sin\gamma + \delta_z \cos\gamma,\ \delta_{\text{pitch}}^c = \delta_y^c \sin\gamma + \delta_z^c \cos\gamma \\ \delta_{\text{yaw}} = \delta_y \cos\gamma - \delta_z \sin\gamma,\ \delta_{\text{yaw}}^c = \delta_y^c \cos\gamma - \delta_z^c \sin\gamma \end{cases} \quad (24)$$

Assuming that $\ddot{\gamma} = 0$, then combining (23) and (24) can yield

$$\begin{aligned}\begin{bmatrix}\ddot{\delta}_{pitch}\\ \ddot{\delta}_{yaw}\end{bmatrix} &= \begin{bmatrix}-2\xi\omega_n & 2\dot{\gamma}\\ -2\dot{\gamma} & -2\xi\omega_n\end{bmatrix}\begin{bmatrix}\dot{\delta}_{pitch}\\ \dot{\delta}_{yaw}\end{bmatrix}\\ &+\begin{bmatrix}\dot{\gamma}^2-{\omega_n}^2 & 2\xi\omega_n\dot{\gamma}\\ -2\xi\omega_n\dot{\gamma} & \dot{\gamma}^2-{\omega_n}^2\end{bmatrix}\begin{bmatrix}\delta_{pitch}\\ \delta_{yaw}\end{bmatrix}+{\omega_n}^2\begin{bmatrix}\delta^c_{pitch}\\ \delta^c_{yaw}\end{bmatrix}\end{aligned} \tag{25}$$

The coupling effect and rudder authority get worse with the increase of $\dot{\gamma}$. To compensate the coupling effect caused by the fast spinning, the mathematical rudder commands are designed as

$$\begin{cases}\delta^c_{\text{pitch}} = \delta^c_{\text{pitch}} - \left(2\xi\omega_n\dot{\gamma}\delta_{\text{yaw}} - \dot{\gamma}^2\delta_{\text{pitch}}\right)/\omega_n^2\\ \delta^c_{\text{yaw}} = \delta^c_{\text{yaw}} - \left(-2\xi\omega_n\dot{\gamma}\delta_{\text{pitch}} - \dot{\gamma}^2\delta_{\text{yaw}}\right)/\omega_n^2\end{cases} \tag{26}$$

Note that the coupling terms related to $\dot{\delta}_{\text{pitch}}$ and $\dot{\delta}_{\text{yaw}}$ are not considered since they are difficult to be precisely obtained in practice. After deriving δ^c_{pitch} and δ^c_{yaw}, δ^c_z and δ^c_y can be derived by rudder deflection decomposition.

3.3 Feedforward Compensation of Actuator Rate Limit

To compensate for the rudder rate limit, the describing function method is employed next to modify δ^c_z and δ^c_y. For the nonlinear part of the rudder, assume $e_{\delta_i} = e_i\sin(\omega t + \lambda_i)$ $(i = y, z)$, based on which the output of the saturation unit can be obtained as

$$u_i = \begin{cases}S_i e_i\sin(\omega t + \lambda_i),\ e_i > u_s\\ e_i\sin(\omega t + \lambda_i),\ e_i \le u_s\end{cases} \tag{27}$$

where

$$S_i = 2\left(\arcsin d_i + d_i\sqrt{1-d_i^2}\right)\Big/\pi,\ d_i = u_{si}/e_i \tag{28}$$

It can be seen that the saturation only affects the magnitude. Denote

$$f(e_i) = S_i e_i = 2\left(e_i\arcsin d_i + u_{si}\sqrt{1-d_i^2}\right)\Big/\pi \tag{29}$$

whose derivative with respect to e_i is

$$f'(e_i) = 2\left(\arcsin d_i - d_i\sqrt{1-d_i^2}\right)\Big/\pi \tag{30}$$

We can deduce that $f'(e_i)\big|_{e_i=u_{si}} = 1$ and $\lim\limits_{e_i\to\infty} f'(e_i) = 0$. Further, we can derive

$$f''(e_i) = -2u_{si}\left(2d_i^2\Big/\sqrt{1-d_i^2}\right)\Big/\left(\pi e_i^2\right) < 0 \tag{31}$$

Then, we can obtain that $f'(e_i) > 0$ and $f(e_i)$ is an increasing function. Due to the amplitude saturation of δ_i^c and δ_i, we have $|\delta_i^c| \le 20$ and $|\delta_i| \le 20$, which suggests that $|e_i| \le 40$. Thus, output of the saturation unit (as depicted in Fig. 3) is upper bounded by $f(40)$. For the linear part, the magnitude and phase responses of $G_{u_i}^{\delta_i}$ are

$$\left|G_{u_i}^{\delta_i}(j\omega)\right| = b_i \Big/ \left(\omega\sqrt{a_i^2+\omega^2}\right), \ \angle G_{u_i}^{\delta_i}(j\omega) = -\pi/2 - \arctan \omega/a_i \tag{32}$$

Denote $A_i \sin(\omega t + \varphi_i)$ as the desired rudder deflection. Then, based on the magnitude and phase relations for the closed system, the following equations can be derived:

$$\begin{aligned} &2e_i\left(\arcsin d_i + d_i\sqrt{1-{d_i}^2}\right)b_i \Big/ \left(\pi\omega\sqrt{{a_i}^2+\omega^2}\right) \\ &= A_i, \ \lambda_i - \pi/2 - \arctan \omega/a_i = \varphi_i \end{aligned} \tag{33}$$

wherein e_i and λ_i are unknown variables to be solved. According to (33), the feasible frequency region for a specified A_i can be calculated as shown in Fig. 4. In Fig. 4, the blue line represents that the rudder does not enter the saturated state; the red line denotes that the rudder enters the saturated state, while the desired A_i can be achieved by properly designing δ_i^c; the black line denotes that the rudder enters the saturated state, while the desired A_i cannot be achieved by increasing the magnitude of δ_i^c. Therefore, if the roll angular rate is within the blue and red region, the corresponding e_i can be calculated according to (33). Otherwise, if the roll angular rate is within the back region, we make $e_i = 40$ to minimize the output error. Further, we can directly use the second equation in (33) to calculate λ_i.

After deriving e_i and λ_i, the modified rudder deflection command δ_i^{c*} can just be calculated as follows:

$$\begin{aligned} \delta_i^{c*} &= e_i \sin(\omega t + \lambda_i) + A_i \sin(\omega t + \varphi_i) \\ &= p_{1i} \sin\left(\omega t + \frac{\lambda_i + \varphi_i}{2}\right) + p_{2i} \cos\left(\omega t + \frac{\lambda_i + \varphi_i}{2}\right) \\ &= A_{ri} \sin(\omega t + \varphi_{ri}) \end{aligned} \tag{34}$$

wherein

$$\begin{cases} p_{1i} = (e_i + A_i)\cos\left((x_i - \varphi_i)/2\right), p_{2i} = (e_i - A_i)\sin\left((x_i - \varphi_i)/2\right), \\ A_{ri} = \sqrt{p_{1i}^2 + p_{2i}^2} = \sqrt{e_i^2 + A_i^2 + 2e_iA_i\cos(x_i - \varphi_i)}, \varphi_{ri} \\ \quad = (x_i + \varphi_i)/2 + \chi_i, \\ \cos\chi_i = p_{1i}/A_{ri}, \sin\chi_i = p_{2i}/A_{ri} \end{cases} \tag{35}$$

Until now, the entire controller design is completed. The block diagram of the proposed method is shown in Fig. 5 to facilitate understanding.

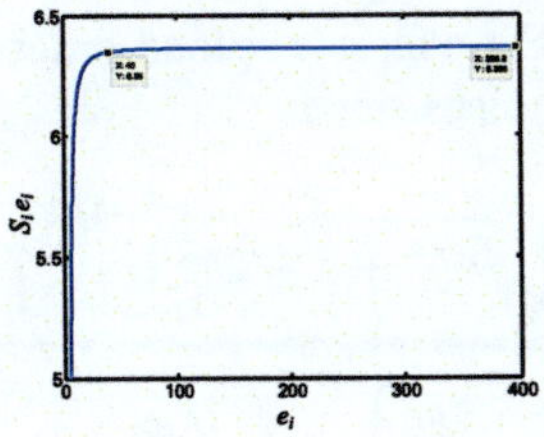

Fig. 3. Input and output of the saturation unit

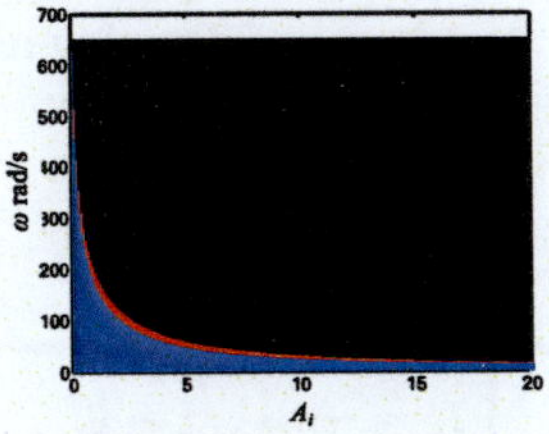

Fig. 4. Ranges of roll angular rates for different A_i under saturation and unsaturation states

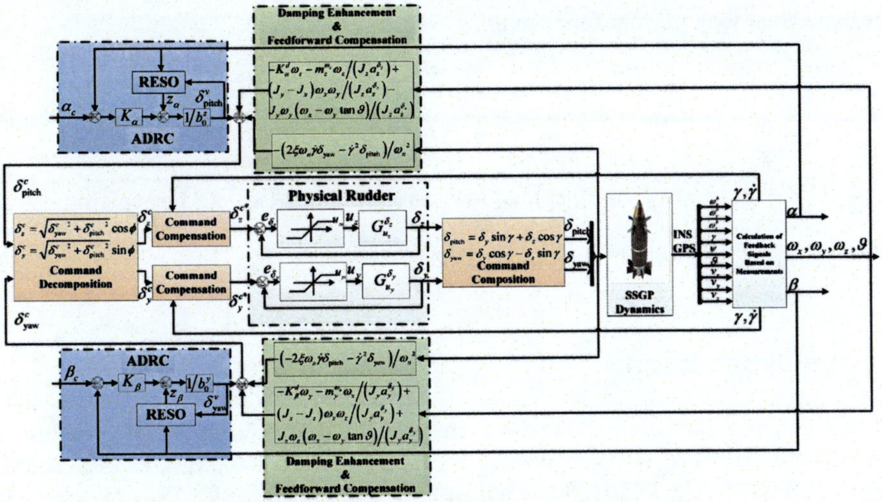

Fig. 5. Block diagram of the proposed control scheme

4 Stability Margin-Based Tuning

Next, parameter tuning is performed by using stability margin. α control is taken as an example to show the tuning process. The loop transfer function for α control is

$$G_{\text{loop}} = (K_\alpha(1 + l_\alpha / s) + l_\alpha)G_{\delta_z}^{\alpha} G_r / b_0^z + K_\alpha^d G_{\delta_z}^{\omega_z} G_r \tag{36}$$

where $G_{\delta_z}^{\alpha}$ denotes the transfer function from δ_z to α; $G_{\delta_z}^{\omega_z}$ denotes the transfer function from δ_z to ω_z; G_r is the transfer function of the rudder. After b_0^z and l_α are determined, let $(B = 1,\ \kappa = 0)$, $(B = 2,\ \kappa = 0)$, $(B = 0.5,\ \kappa = 0)$, and $(B = 1,\ \kappa = 45)$. Then, four boundaries of $\left(K_\alpha, K_\alpha^d\right)$ which satisfy $G_{\text{loop}}(j\omega)Be^{-\kappa j} = -1$ can be obtained. The parameters on these four boundaries represent the critical stability, 6 dB margin, −6 dB margin, and 45° margin, respectively.

To further obtain the $\left(K_\alpha, K_\alpha^d\right)$ that satisfies the transient performance, the closed-loop system is approximated by a unit-feedback one, whose loop transfer function is the traditional second-order plant $G_{\text{loop}} = \omega_l^2 / (s(s + 2\xi_l \omega_l))$, wherein ω_l and ξ_l are the natural frequency and damping ratio, respectively. For a classic second-order plant, the frequency-domain and time-domain performance indices satisfy

$$\begin{cases} \sigma\% = \mathrm{e}^{-\pi\xi_l / \sqrt{1-\xi_l^2}}, \ \kappa_p = \arctan\left(2\xi_l\left(\sqrt{4\xi_l^4+1} - 2\xi_l^2\right)^{-1/2}\right) \\ t_s = -\ln\left(\sqrt{1-\xi_l^2} \times \Delta\%\right)\left(\sqrt{4\xi_l^4+1} - 2\xi_l^2\right)^{1/2} \Big/ (\xi\omega_b) \end{cases} \tag{37}$$

wherein $\sigma\%$ is the overshot percent, and t_s is the settling time defined by the error $\Delta\%$. According to specified $\sigma\%$, $\Delta\%$, and t_s, ω_b and κ_p can be obtained according to (37). Rearrange the original control system into the unit-feedback form, and we can obtain the equivalent loop transfer function as

$$G_{\text{unit}}(s = j\omega) = \frac{K_\alpha \big/ b_0^z (1 + l_\alpha / s) G_{\delta_z}^\alpha G_r}{1 + l_\alpha \big/ b_0^z G_{\delta_z}^\alpha G_r + K_\alpha^d G_{\delta_z}^{\omega_z} G_r} \tag{38}$$

Let $(B = 1,\ \kappa = \kappa_p,\ s = j\omega_b)$, and we can just obtain (K_α, K_α^d) by solving

$$G_{\text{unit}}(j\omega) B e^{-\kappa j} = -1 \tag{39}$$

5 Simulation Results

Complete nonlinear 6-DOF simulations are performed to validate the effectiveness of the proposed method. The step size for model integration is 0.001 s, and the control period is 0.005 s. The INS/GPS is employed to obtain the measurements of velocities, angles, angular rates as shown in Fig. 5. α and β cannot be directly measured and they are calculated based on the measurements and (3). The α for maximal lift-drag ratio is selected as the command, and the command for β is zero. Three cases are considered next.

In the first scenario, there is no rudder rate limit and corresponding feedforward compensation. The control results are shown in Fig. 6. We can see that the control errors of α, β, and γ_v are limited within 0.2°, which can well meet the control requirement.

In the second scenario, the rudder rate limit is added while there is no feedforward compensation. The rudder deflections have large tracking errors due to the rate limit. Therefore, the angles all have large oscillations, and the maximal errors can even reach 3° as shown in Fig. 7, which will cause the loss of flight distance and even the instability.

In the third scenario, the feedforward compensation strategy is supplemented. In addition, measurement noises and a time delay of 0.005 s are added to verify the robustness. Simulation results are presented in Fig. 8 and Fig. 9. According to Fig. 8, all the angle oscillations disappear. The error of β is controlled to be very small, and the error of α is reduced to 1.5°. The rudder deflections are shown in Fig. 9, which demonstrates the effectiveness of the compensation.

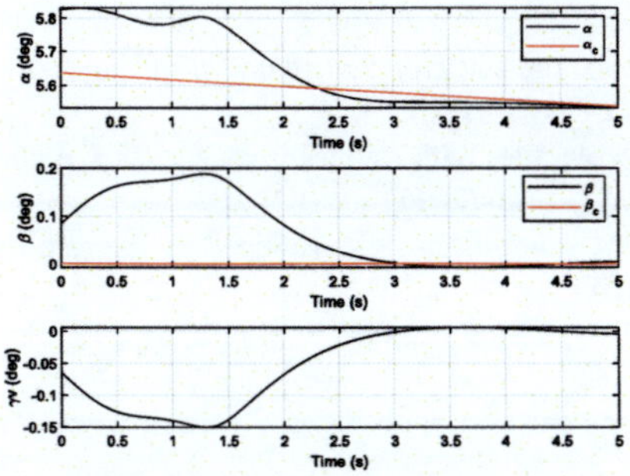

Fig. 6. Angle responses for case 1

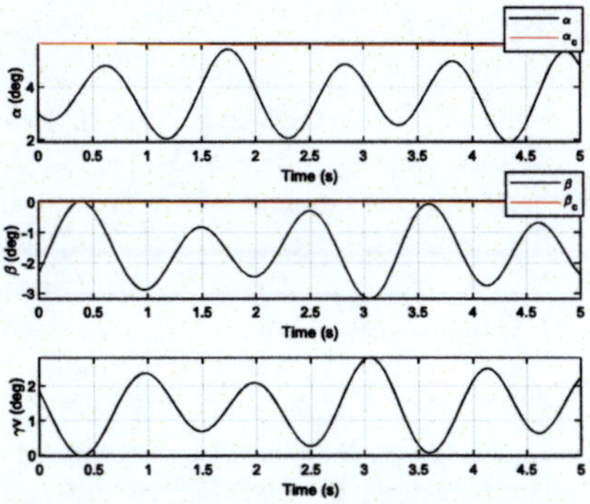

Fig. 7. Angle responses for case 2

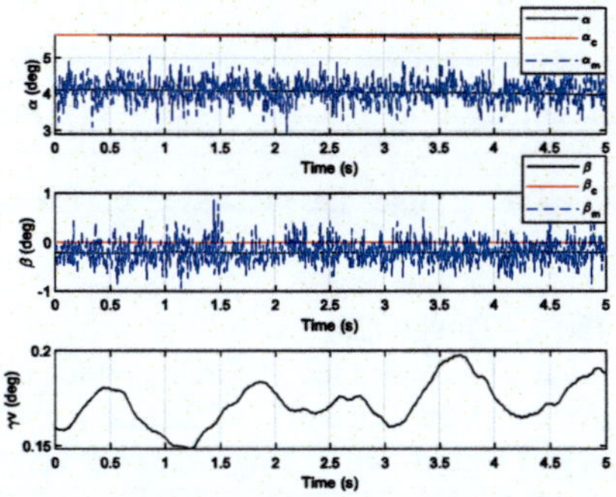

Fig. 8. Angle responses for case 3 (α_m and β_m are calculated based on noise corrupted measurements)

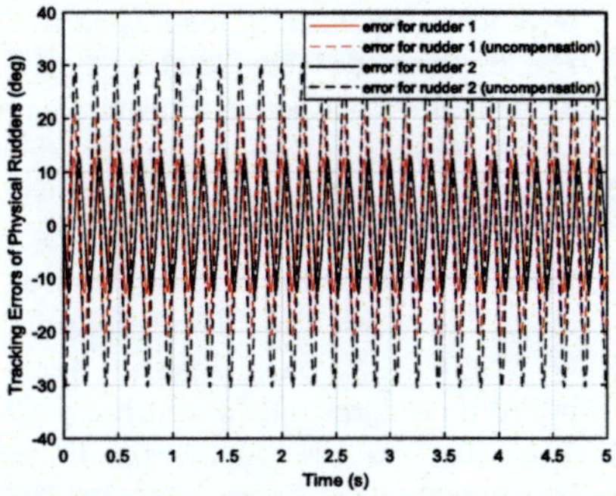

Fig. 9. Tracking errors of physical rudders for case 3

6 Conclusions

In this paper, a feedforward compensation enhanced disturbance rejection control method was proposed for the SSDP. ADRC was first designed as the baseline controller to deal with the internal uncertainties and external disturbances. To improve the decoupling performance, dominant coupling dynamics was also added to the rudder deflection command. Further, the describing function method was employed to design the feedforward command to compensate the rudder rate limit. Stability margin-based method was utilized for tuning. Finally, the comparative simulation results demonstrated the superior performance of our method.

Acknowledgement. The authors acknowledge the financial support from the National Natural Science Foundation (NNSF) of China under Grant 62203450, and the Research Project of Tianjin Municipal Education Commission under Grant 2021KJ043.

References

1. Wang, F., Liang, G., Wang, Z., et al.: Research on dual actuator control law of rolling airframe missile. Aerosp. Shanghai **34**(1), 73–75 (2017). (in Chinese)
2. Fresconi, F.: Guidance and control of a projectile with reduced sensor and actuator requirements. J. Guid. Control. Dyn. **34**(6), 1757–1766 (2011)
3. Yan, X., Yang, S., Zhang, C.: Coning motion of spinning missiles induced by the rate loop. J. Guid. Control. Dyn. **33**(5), 1490–1499 (2010)
4. Yan, X., Yang, S., Xiong, F.: Stability limits of spinning missiles with attitude autopilot. J. Guid. Control. Dyn. **34**(1), 278–283 (2011)
5. Li, K., Yang, S., Zhao, L.: Stability of spinning missiles with an acceleration autopilot. J. Guid. Control. Dyn. **35**(3), 774–786 (2012)
6. Li, K., Yang, S., Zhao, L.: Three-loop autopilot of spinning missiles. Proc. Inst. Mech. Eng. Part G J. Aerosp. Eng. **228**(7), 1195–1201 (2014)
7. Li, W., Wen, Q., Yang, Y.: Stability analysis of spinning missiles induced by seeker disturbance rejection rate parasitical loop. Aerosp. Sci. Technol. **90**, 194–208 (2019)
8. Fan, S., Song, T., Wang, J., et al.: Coupling analysis and dynamic stability boundary of spinning missiles considering actuator dynamics and autopilot. Aerosp. Sci. Technol. **111**(1232), 106481 (2021)
9. Zhou, W., Yang, S., Zhao, L.: Limit cycle of low spinning projectiles induced by the backlash of actuators. Aerosp. Sci. Technol. **69**, 595–601 (2017)
10. Tipan, S., Theodoulis, S., Thai, S., et al.: Nonlinear dynamic inversion flight control design for guided projectiles. J. Guid. Control. Dyn. **43**(5), 975–980 (2020)
11. Qi, C., Wang, X., Yang, J., et al.: Acceleration tracking control for a spinning glide guided projectile with multiple disturbances. Chin. J. Aeronaut. **33**(12), 3405–3422 (2020)
12. Yang, Z.: Dynamics Analysis and Control System Design for Rolling Guided Projectile. Harbin Institute of Technology (2018). (in Chinese)
13. Han, J.: From PID to active disturbance rejection control. IEEE Trans. Industr. Electron. **56**(3), 900–906 (2009)

Elastic Properties Prediction of Composite Materials Using Homogenization Finite Element Method

Yun-qian Zhang(✉), Guo-fan Zhang, Xiao-hua Nie, and Shi-hui Duan

National Key Laboratory of Strength and Structural Integrity, Aircraft Strength Research Institute of China, Xi'an 710065, Shaanxi, China
zhangyunqian1997@163.com

Abstract. Based on the homogenization theory, this paper investigates the macroscopic equivalent properties of unidirectional composite materials using a representative volume element model with a typical micro-configuration. A micromechanical model of the composite material is constructed considering the periodic boundary conditions. Then the equivalent properties are obtained by the homogenization finite element method. The effectiveness and accuracy of the homogenization finite element method in predicting composite material properties are verified by comparing the predicted macroscopic equivalent properties with experimental values from the literature. Additionally, this paper compares the influences of fiber volume content, fiber arrangement, and mesh density on the macroscopic mechanical properties of the composite material. The results demonstrate that the homogenization method can achieve the correlation of macroscopic and microscopic mechanical properties, which is potentially promising for guiding the new materials development.

Keywords: Homogenization · Unidirectional composite material · Equivalent elastic property · Representative volume element · Finite element method

1 Introduction

Fiber-reinforced composites have attracted considerable attention in the aerospace equipment manufacturing and other fields due to their outstanding mechanical properties. The accurate prediction of the mechanical properties of fiber-reinforced composites is a necessary prerequisite for their application in structural design. Analytical methods, such as the Eshelby inclusion model, Mori-Tanaka method, and self-consistent method [1], have limited prediction accuracy, which fail to solve micro-models with complex structural features. In addition to the aforementioned methods, the homogenization method has shown great potential for predicting the equivalent properties of non-uniform materials. For instances, Liu. et al. [2] used homogenization method to predict the macroscopic properties of composites and optimize the design of gradient functional materials. Pan. et al. [3] used homogenization method to study the damaged stiffness of laminated composites.

Chinese Society of Aeronautics and Astronautics (Ed.): CASTC 2023, LNME, pp. 685–695, 2024.
https://doi.org/10.1007/978-981-99-8861-7_67

In this paper, a finite element implementation method based on homogenization theory is developed to predict the equivalent properties of unidirectional fiber-reinforced composite materials. By extracting the micro-mechanical model and combining with periodic boundary conditions, the equivalent stiffness performance of unidirectional fiber-reinforced composite materials is obtained, and compared with experimental values to verify the effectiveness of the proposed method. Several major factors affecting material equivalent properties are further explored to provide guidance for the design and development of new materials.

2 Homogenization Method

The homogenization method establishes a connection between macroscopic and microscopic scales by seeking an effective homogeneous body [4], promising for predicting the equivalent properties and microscopic stress of non-uniform materials. This method has a rigorous mathematical foundation. It assumes that the elastic body of the composite material occupies a space Ω, and its microscopic structure is composed of non-uniform cells arranged periodically, as shown in Fig. 1. The microscopic cell scale Y is a small quantity ε relative to the macroscopic structure scale X, that is, $Y/X = \varepsilon$ $(0 < \varepsilon < 1)$. The volume force f acts inside the elastic body, surface force t acts on boundary Γt, and Γu is the fixed displacement boundary [5].

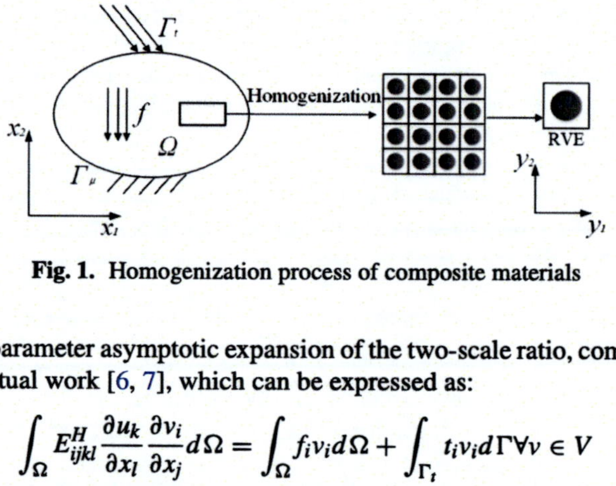

Fig. 1. Homogenization process of composite materials

The small parameter asymptotic expansion of the two-scale ratio, combined with the principle of virtual work [6, 7], which can be expressed as:

$$\int_{\Omega} E_{ijkl}^{H} \frac{\partial u_k}{\partial x_l} \frac{\partial v_i}{\partial x_j} d\Omega = \int_{\Omega} f_i v_i d\Omega + \int_{\Gamma_t} t_i v_i d\Gamma \forall v \in V \tag{1}$$

The formula for calculating the effective elastic modulus of homogenization can be obtained through a series of derivations from the above equation:

$$E_{ijkl}^{H} = \frac{1}{|Y|} \int_{Y} \left(E_{ijkl} - E_{ijmn} \frac{\partial \chi kl_m}{\partial y_n} \right) dY \tag{2}$$

where the generalized displacement χkl_m satisfies:

$$\int_{Y} E_{ijmn} \frac{\partial \chi kl_m}{\partial y_n} \frac{\partial v_i}{\partial y_j} dY = \int_{Y} E_{ijkl} \frac{\partial v_i}{\partial y_j} dY \tag{3}$$

3 Elastic Properties Prediction of Unidirectional Composites

3.1 Micro-mechanical Modeling

The modeling of the microscale mechanics is essential for the prediction of the elastic properties of unidirectional composite materials. A Representative Volume Element (RVE) with stable periodicity is generally referred to an ideal material model at the microscale, connecting the microscale and macroscale properties in multiscale analysis. Establishing a suitable RVE is crucial for ensuring accurate prediction of the macroscopic material properties [8]. The process of extracting the RVE is shown in Fig. 2, and the RVE at the microscale should be small enough compared with the macroscopic scale [9].

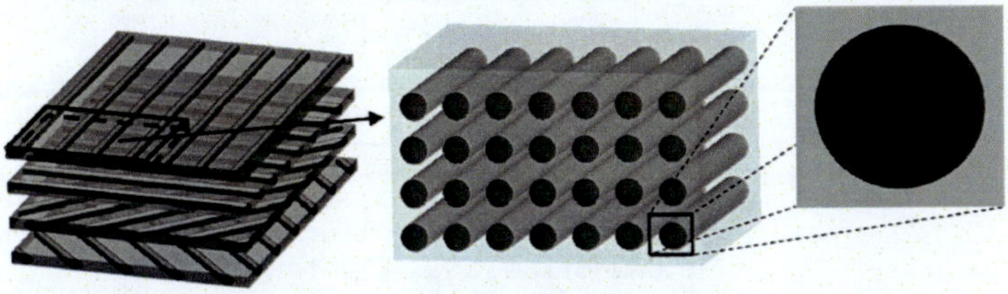

(a) Laminated plate (b) Fiber arrangement (c) RVE

Fig. 2. Composite micro-scale representative element

Taking unidirectional laminate (with its cross section as shown in Fig. 3) as an example, this paper is based on the assumptions of periodicity and symmetry, and considers that the behavior of each RVE under external loading is the same as that of other RVEs, rather than being randomly distributed. Therefore, when appropriate boundary conditions are applied to the RVE, a single RVE can simulate the mechanical behavior of the entire unidirectional laminate. From a microscale perspective, the fiber arrangement can be divided into square and hexagonal arrangements, as shown in Fig. 3(b) and Fig. 3(c), respectively.

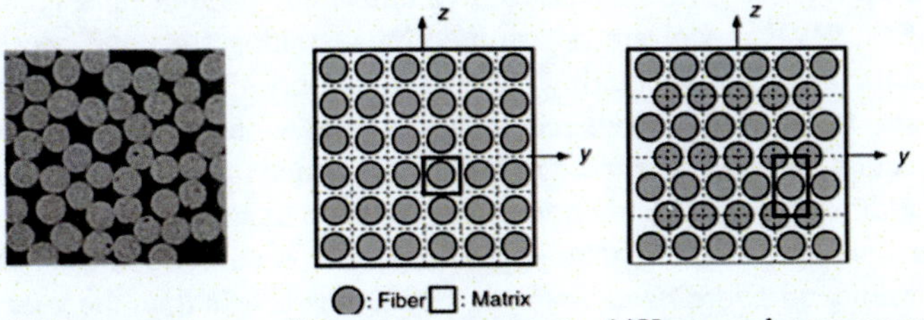

(a)Real fiber arrangement (b)Square arrangement (c)Hexagonal arrangement

Fig. 3. Comparison between real fiber arrangement and ideal fiber arrangement

3.2 Finite Element Implementation of Homogenization Method

The implementation process of the homogenization method using finite element analysis is shown in Fig. 4, which consists of two stages, i.e., pre-processing and post-processing stages. In the pre-processing stage, the geometric dimensions of the RVE are determined, the boundary surfaces are identified, and node sets and corresponding boundary constraint equations are created. In the post-processing stage, stress-strain calculations are performed to predict the macroscopic effective elastic properties of the material.

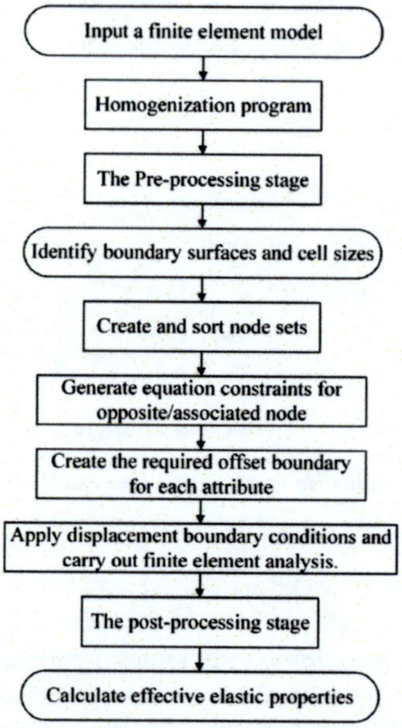

Fig. 4. Process of homogenization finite element method

a. **Pre-processing Stage**

In the pre-processing stage, the maximum and minimum values in the three directions of the RVE model are determined by searching for node coordinates, which serve as the basis for calculating the RVE boundary size and identifying its corners, edges, and faces. Once a node meets the coordinate conditions of a specific node set (corner, edge, or face), as shown in Fig. 5, it is added to an array containing that node set. In addition, the generated node sets also need to be classified in order to identify the corresponding node degrees of freedom. For each node in the associated set, if the difference in coordinates between two nodes is less than the specified mapping accuracy, their corresponding nodes are grouped together.

The calculation of each elastic property requires a combination of different displacement boundary conditions, which are implemented using linear constraint equations to relate the degrees of freedom of nodes. In this process, the boundary set is first converted into a set of single node sets to create matching node labels and positions

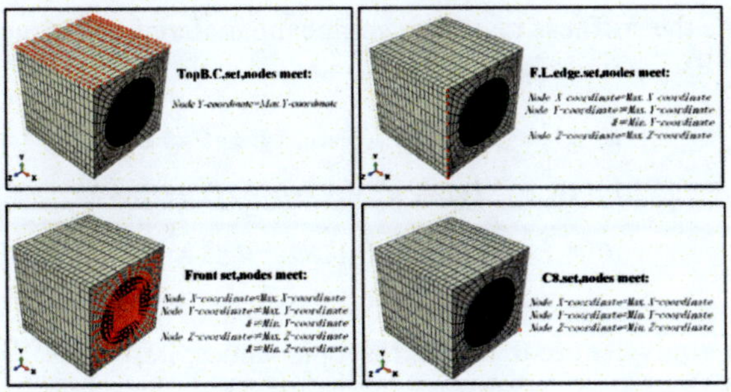

Fig. 5. Node set label

for each corresponding set, so that linear constraint equations can be used to constrain each pair of nodes. To calculate different material properties, displacement is applied to specific reference points (Fig. 6).

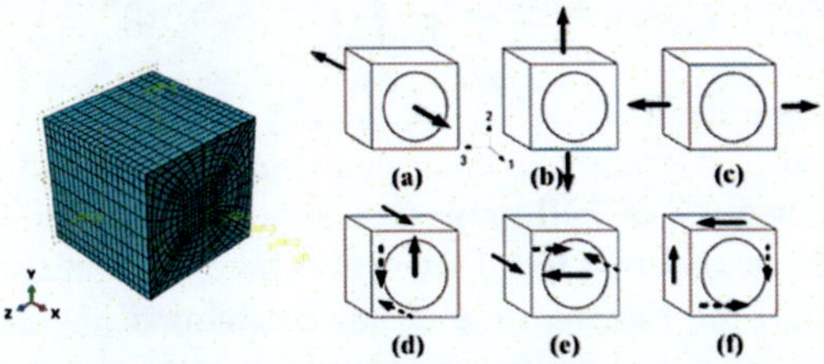

Fig. 6. Selection of reference points and application of macroscopic strain components

b. **Post-processing Stage**

The post-processing stage mainly calculates the macroscopic mechanical properties of composite materials by homogenizing the RVE. According to Hooke's law,

$$\begin{Bmatrix} \overline{\sigma_1} \\ \overline{\sigma_2} \\ \overline{\sigma_3} \\ \overline{\sigma_4} \\ \overline{\sigma_5} \\ \overline{\sigma_6} \end{Bmatrix} = \begin{bmatrix} C_{11} & C_{12} & C_{12} & 0 & 0 & 0 \\ C_{12} & C_{22} & C_{23} & 0 & 0 & 0 \\ C_{12} & C_{23} & C_{22} & 0 & 0 & 0 \\ 0 & 0 & 0 & (C_{22}-C_{23})/2 & 0 & 0 \\ 0 & 0 & 0 & 0 & C_{66} & 0 \\ 0 & 0 & 0 & 0 & 0 & C_{66} \end{bmatrix} \begin{Bmatrix} \overline{\varepsilon_1} \\ \overline{\varepsilon_2} \\ \overline{\varepsilon_3} \\ \overline{\gamma_4} \\ \overline{\gamma_5} \\ \overline{\gamma_6} \end{Bmatrix} \tag{4}$$

when the stiffness matrix of a composite material is given, the effective elastic constants of transversely isotropic materials can be expressed.

To calculate the stiffness matrix of composite materials, the boundary conditions [10, 11] of the RVE should be configured, i.e.

$$\begin{aligned} u_i(a_1, x_2, x_3) - u_i(-a_1, x_2, x_3) &= 2a_1\varepsilon_{i1}^0 \\ u_i(x_1, a_2, x_3) - u_i(x_1, -a_2, x_3) &= 2a_2\varepsilon_{i2}^0 \\ u_i(x_1, x_2, a_3) - u_i(x_1, x_2, -a_3) &= 2a_3\varepsilon_{i3}^0 \end{aligned} \tag{5}$$

The volume-averaged stress and strain fields of composite materials at the microscale are equivalent to their macroscopic values, i.e.

$$\begin{cases} \overline{\sigma_{ij}} = \frac{1}{V}\int_V \sigma_{ij} dV = \sigma_{ij}^0 \\ \overline{\varepsilon_{ij}} = \frac{1}{V}\int_V \varepsilon_{ij} dV = \varepsilon_{ij}^0 \end{cases} \tag{6}$$

When the loading is applied as unit strain, i.e., $\varepsilon_\beta^0 = 1$, the local stress field σ_α and its average value can be calculated by considering the periodic boundary conditions, and the elastic matrix can be determined as follow:

$$C_{\alpha\beta} = \overline{\sigma_\alpha} = \frac{1}{V}\int_V (x_1, x_2, x_3) dV \tag{7}$$

c. **Program Implementation**

In this study, the program was compiled using Python and embedded in commercial software, and the overall implement process is shown in Fig. 7. Since different RVE models is required to be set with different boundary constraints during the determination of the macroscopic elastic modulus of composite materials, the homogenization program should be called for each solution process.

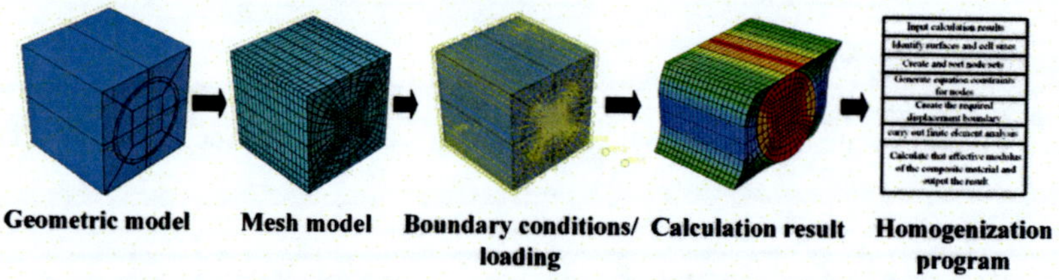

Fig. 7. Homogenization FE method process for predicting macroscopic equivalent property

4 Prediction and Analysis of Equivalent Elastic Properties of UD Composites

4.1 Verification of Homogenization Method

In this paper, a typical unidirectional fiber-reinforced composite square distributed unit cell model is used to verify the homogenization method. The volume fraction of fibers is 0.6. The material properties of the four types [12] of fibers, matrix, and composite are shown in Table 1 and Table 2.

Table 1. Mechanical properties of four fiber types [13]

Property	Type			
	Gevetex E-glass	Silenka E-glass	AS4	T300
Longitudinal elastic modulus/GPa	80	74	225	230
Transverse elastic modulus/GPa	80	74	15	15
In-plane shear modulus/GPa	33.33	30.8	15	15
Transverse shear modulus/GPa	33.33	30.8	7	7
Poisson's ratio	0.2	0.2	0.2	0.2
Transverse poisson's ratio	0.2	0.2	0.07	0.07

Table 2. Mechanical properties of four matrix types [13]

Property	Type			
	LY556 epoxy	MY750 epoxy	3501-6 epoxy	914C epoxy
Elastic modulus/GPa	3.35	3.35	4.2	4.0
Poisson's ratio	0.35	0.35	0.34	0.2

Table 3 compares the predicted elastic constants of the four unidirectional composite materials using the homogenization method with experimental data. The results show that the relative errors are all within 15%, which demonstrates the effectiveness of the proposed homogenization finite element method for predicting the equivalent mechanical properties of unidirectional composite materials.

4.2 Impact of Different Factors on Macroscopic Mechanical Properties

This section mainly studies the impact of fiber volume fraction, fiber arrangement, and the density of finite element mesh on the macroscopic mechanical properties.

a. **Impact of Fiber Volume Fraction**

In practical engineering applications, the macroscopic mechanical properties of composite materials are usually updated by adjusting the fiber volume fraction. Therefore, this study further investigates the changes in the mechanical properties of composite materials under different fiber volume fractions. Typical isotropic unidirectional fiber-reinforced composite materials were selected for analysis The elastic modulus of fiber $E_f = 241$ GPa, Poisson's ratio $v_f = 0.2$. The elastic modulus of matrix $E_m = 3.12$ GPa, Poisson's ratio $v_m = 0.38$. The fiber volume fractions were set as 0.2 to 0.7 with an interval of 0.1.

The macroscopic mechanical properties of the composite material as a function of fiber volume fraction are shown in Fig. 8.

Table 3. Comparison between prediction results and experimental results of four UD composites elastic constants [13]

Laminate type	Method	E_{11}(GPa)	E_{22}(GPa)	G_{12}(GPa)	v_{12}
Gevetex E-glass/LY556	Predicted value	49.27	15.20	4.60	0.25
	Experimental value	53.48	17.7	5.38	0.27
	Relative error	7%	14%	14%	10%
Silenka E-glass/MY750	Predicted value	45.68	14.94	5.05	0.25
	Experimental value	45.6	16.2	5.83	0.27
	Relative error	1%	7%	13%	10%
AS4/3501-6	Predicted value	130	11.23	5.99	0.25
	Experimental value	126	11	6.6	0.28
	Relative error	3%	2%	9%	10%
T300/914C	Predicted value	139	12.3	4.87	0.25
	Experimental value	138	11	5.5	0.28
	Relative error	1%	11%	11%	10%

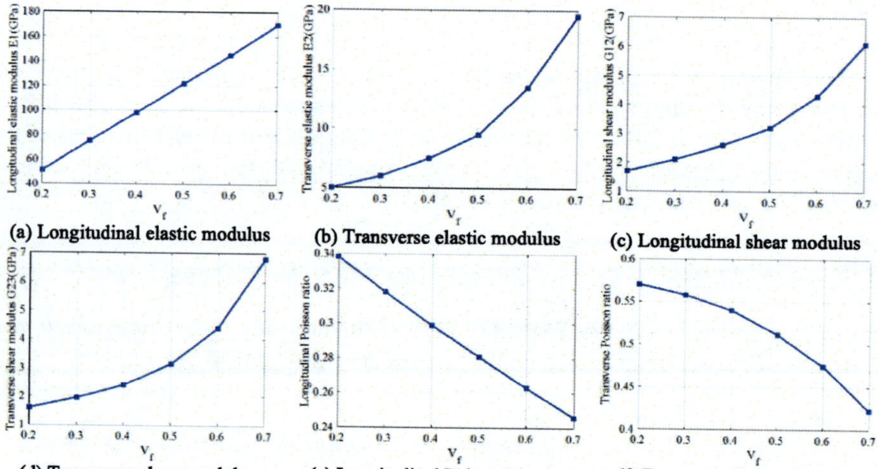

(a) Longitudinal elastic modulus (b) Transverse elastic modulus (c) Longitudinal shear modulus

(d) Transverse shear modulus (e) Longitudinal Poisson's ratio (f) Transverse Poisson's ratio

Fig. 8. The variation of transverse elastic modulus with fiber volume fraction

The following conclusions can be extracted from Figs. 8:

1. The longitudinal elastic modulus is linearly and positively correlated with the fiber volume fraction.
2. The longitudinal and transverse Poisson's ratios are negatively correlated with the fiber volume fraction.

3. The transverse elastic modulus, longitudinal shear modulus, and transverse shear modulus increase exponentially with the increment of fiber volume fraction.

b. **Impact of Fiber Arrangement Mode**

In this section, the impact of different fiber arrangements on the macro-mechanical parameters of RVE model is further studied. The macro-mechanical characteristics of RVE models with square and hexagonal arrangements is compared, as explained below.

RVE models with square and hexagonal arrangements were established (see Fig. 9). The relative error of the longitudinal elastic modulus of the two types of RVE models is shown in Fig. 10. It can be observed that the hexagonal arrangement of RVE models has smaller relative error, and the relative error gradually decreases with the increase of fiber volume fraction ($V_f > 0.6$). For the RVE model with a square arrangement, the relative error first decreases and then increases with the increase of fiber volume fraction. Therefore, it can be concluded that for unidirectional composite materials, when the fiber volume fraction is greater than 0.6, the fiber arrangement tends to be closer to the hexagonal arrangement.

(a) Hexagonal arrangement (b) Square arrangement

Fig. 9. Two RVE arrangements

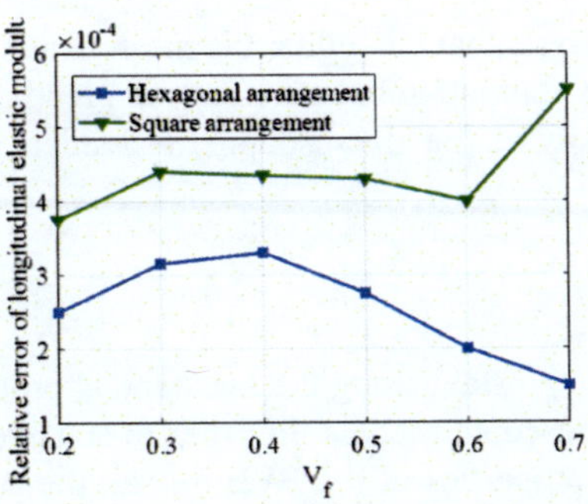

Fig. 10. Relative error of longitudinal elastic modulus of different arrangement modes

Figure 11 presents the variation of transverse elastic modulus with fiber volume fraction for the two fiber arrangements. Both RVE models illustrate an exponential increase in transverse elastic modulus with increasing fiber volume fraction. When the fiber volume fraction is the same, the square fiber arrangement RVE has a

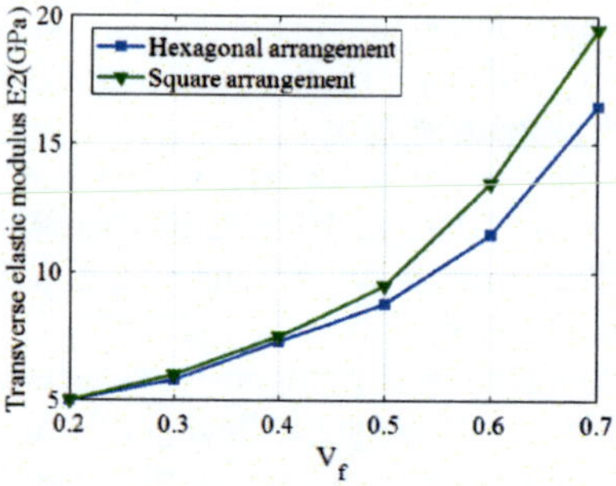

Fig. 11. Comparison of transverse elastic modulus of different arrangement modes

larger transverse elastic modulus, and this difference becomes more pronounced with increasing fiber volume fraction.

c. **Impact of Grid Density**

This section further investigates the effect of mesh density on the accuracy of the homogenization finite element method. Meshes with significantly different densities are used, as shown in Fig. 12.

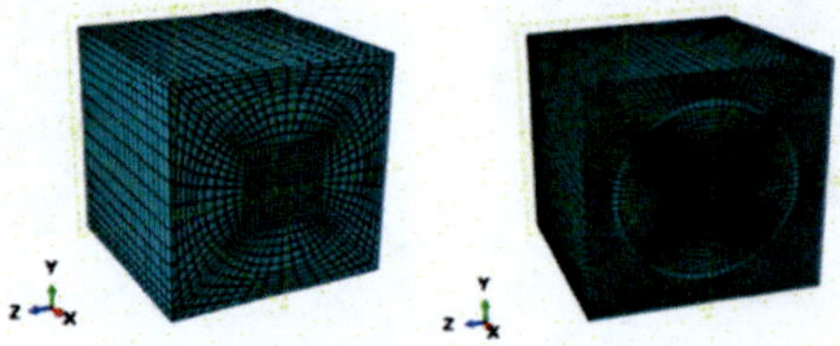

Fig. 12. Comparison of finite element mesh density

It is verified by analysis that the higher the mesh density, the higher the accuracy of the solution, but the computational efficiency is greatly reduced. In addition, the influence of mesh density on the macroscopic mechanical properties of the material can be neglected.

5 Conclusion

The focus of this paper is on establishing the mechanical property correlation between the macroscopic and microscopic models. In order to achieve rapid calculation of the macroscopic mechanical properties of the RVE model, the composite material is idealized and modeled based on its microscopic structure. Homogenization theory is then introduced to load the RVE model, obtain the predicted value of the equivalent stiffness of the unidirectional fiber-reinforced composite material. The predicted values are then compared with the experimental ones in the literature to verify the accuracy and effectiveness of the method used in this paper. The effects of fiber volume fraction, fiber arrangement, and mesh density on the macroscopic mechanical properties of composite materials are discussed, which can be beneficial to developing new materials.

References

1. Aboudi, J.: Mechanics of composite materials-a unified micromechanical approach. NASA STI/RECON Tech. Rep. A **93**, 29778 (1991)
2. Liu, S.: Study on the thermal stress relaxation and constitutive equations. Acta Mater. Compositae Sinica **01**, 152–157 (2005)
3. Pan, Y.: The damaged stiffness analysis of laminated composites. J. Astronaut. **03**, 88–93 (2000)
4. Xie, X., Liao, D.: On the design of compliant mechanisms using homogenization method. China Mech. Eng. **14**(11), 953–955 (2003)
5. Xie, G., Zhao, J., Tian, J.: Prediction macroscopic effective elastic properties of composite laminated plate by homogenization FEM. Fiber Reinf. Plast. Compos. (7), 23 (2014)
6. Swan, C.C., Kosaka, I.: Homogenization-based analysis and design of composites. Comput. Struct. **64**(1–4), 603–621 (1997)
7. Xie, G.: Study of multiscale methods for polymer matrix composites. Ph.D. thesis, Xiangtan University, China (2006)
8. Liu, H.: Macroscopic mechanical property prediction of periodic composites by homogenization finite element method. Ph.D. thesis, Hebei Polytechnic University, China (2018)
9. Zhang, X.: The macro and micro mechanical properties and application of the fiber reinforced composite. Ph.D. thesis, Xidian University, China (2015)
10. Xia, Z., Zhang, Y., Ellyin, F.: A unified periodical boundary conditions for representative volume elements of composites and applications. Int. J. Solids Struct. **40**(8), 1907–1921 (2003)
11. Suquet, P.M.: Elements of homogenization theory for inelastic solid mechanics. In: Sanchez-Palencia, E., Zaoui, A. (eds.) Homogenization Techniques for Composite Media. LNP, vol. 272, pp. 193–278. Springer, Heidelberg (1987). https://doi.org/10.1007/3-540-17616-0_15
12. Shen, G.: Mechanics of composite materials. Fiber Reinf. Plast. **2**, 51–53 (1996)
13. Huang, Z., Zhang, H.: Current status and future trend of researches on the strength of fiber-reinforced composites-a summary of the results from a "failure Olympics." Ad. Mech. **37**(1), 80–98 (2007)